# NATIONAL GEOGRAPHIC
# World Cultures and Geography

Go interactive with **myNGconnect.com**

**Acknowledgments**

Grateful acknowledgment is given to the authors, artists, photographers, museums, publishers, and agents for permission to reprint copyrighted material. Every effort has been made to secure the appropriate permission. If any omissions have been made or if corrections are required, please contact the Publisher.

**Photographic Credits**
**Front Cover:** © Roger Ressmeyer/Corbis
**Back Cover:** © Neale Clark/Robert Harding
World Imagery/Getty Images

Acknowledgments and credits continue on page R124.

Copyright © 2017 National Geographic Learning, a part of Cengage Learning

ALL RIGHTS RESERVED. No part of this work covered by the copyright herein may be reproduced or distributed in any form or by any means, except as permitted by U.S. copyright law, without the prior written permission of the copyright owner.

"National Geographic," "National Geographic Society" and the Yellow Border Design are registered trademarks of the National Geographic Society ® Marcas Registrada

For product information and technology
assistance, contact us at
Customer & Sales Support, 888-915-3276
For permission to use material from this text or product,
submit all requests online at
**www.cengage.com/permissions**
Further permissions questions can be emailed to
**permissionrequest@cengage.com**

**National Geographic Learning | Cengage Learning**
1 N. State Street, Suite 900
Chicago, IL 60602

Cengage Learning is a leading provider of customized learning solutions with office locations around the globe, including Singapore, the United Kingdom, Australia, Mexico, Brazil, and Japan. Locate your local office at www.cengage.com/global.

Visit National Geographic Learning online at NGL.Cengage.com
Visit our corporate website at www.cengage.com

ISBN: 9781305967182

Printed in the USA.

Sheridan, KY, A CJK Group Company

Print Number:09

Print Year: 2022

# PROGRAM CONSULTANTS

### Andrew J. Milson

Andrew Milson is a professor of social science education and geography at the University of Texas at Arlington. He taught middle school history and geography near Dallas, Texas. Andy conducts research on geographic education and the use of geospatial technologies in education. He has published more than 30 articles and is an elected member of the Executive Board of the National Council for Geographic Education. He serves as an associate editor of the *Journal of Geography*.

### Peggy Altoff

Peggy Altoff's experience includes teaching middle school and high school students, supervising teachers, and serving as adjunct university staff. Peggy served as a state social studies specialist in Baltimore and as a K–12 facilitator in Colorado Springs. She was president of the National Council for the Social Studies (NCSS) in 2006–2007 and was on the task force for the new NCSS National Curriculum Standards.

### Mark H. Bockenhauer

Mark Bockenhauer is a professor of geography at St. Norbert College and a former geographer-in-residence at the National Geographic Society. Mark has extensive experience in teacher professional development. He co-wrote *Our Fifty States* and the *World Atlas for Young Explorers, 3rd edition*—both for National Geographic. Mark is coordinator of the Wisconsin Geographic Alliance, and he served as president of the National Council for Geographic Education in 2007.

# PROGRAM CONSULTANTS

### Janet Smith

Jan Smith is an associate professor of geography at Shippensburg University. Jan began her teaching career as a high school teacher in Virginia where she served as a teacher consultant for the Virginia Geographic Alliance for many years. Her primary research interest focuses on how children develop their spatial thinking skills. Jan served as president of the National Council for Geographic Education in 2008, and she is currently the coordinator for the Pennsylvania Geographic Alliance.

### Michael W. Smith

Michael Smith is a professor in the Department of Curriculum, Instruction, and Technology in Education at Temple University. He became a college teacher after 11 years of teaching high school English. His research focuses on how experienced readers read and talk about texts, as well as what motivates adolescents' reading and writing. Michael has written many books and monographs, including the award-winning *"Reading Don't Fix No Chevys": Literacy in the Lives of Young Men*.

### David W. Moore

David Moore is a professor of education at Arizona State University. He taught high school social studies and reading before entering college teaching. He currently teaches teacher preparation courses and conducts research in adolescent literacy. David has published numerous professional articles, book chapters, and books, including *Developing Readers and Writers in the Content Areas* and *Principled Practices for Adolescent Literacy*.

# CONSULTANTS AND REVIEWERS

## Teacher Reviewers

**Kayce Forbes**
Deerpark Middle School
Austin, Texas

**Michael Koren**
Maple Dale School
Fox Point, Wisconsin

**Patricia Lewis**
Humble Middle School
Humble, Texas

**Julie Mitchell**
Lake Forest Middle School
Cleveland, Tennessee

**Linda O'Connor**
Northeast Independent
School District
San Antonio, Texas

**Leah Perry**
Exploris Middle School
Raleigh, North Carolina

**Robert Poirier**
North Andover Middle School
North Andover, Massachusetts

**Heather Rountree**
Bedford Heights Elementary
Bedford, Texas

**Erin Stevens**
Quabbin Regional
Middle/High School
Barre, Massachusetts

**Beth Tipper**
Crofton Middle School
Crofton, Maryland

**Mary Trichel**
Atascocita Middle School
Humble, Texas

**Andrea Wallenbeck**
Exploris Middle School
Raleigh, North Carolina

## Reviewers of Religious Content

The following individuals reviewed the treatment of religious content in selected pages of the text.

**Charles Haynes**
First Amendment Center
Washington, D.C.

**Shabbir Mansuri**
Institute on Religion and
Civic Values
Fountain Valley, California

**Susan Mogull**
Institute for Curriculum Reform
San Francisco, California

**Raka Ray**
Chair, Center for South Asia Studies
University of California
Berkeley, California

## National Geographic Society

The National Geographic Society contributed significantly to *World Cultures and Geography*. Our collaboration with each of the following has been a pleasure and a privilege: National Geographic Maps, National Geographic Education Programs, National Geographic Missions Programs, National Geographic Digital Motion, National Geographic Digital Studio, and National Geographic Weekend. We thank the Society for its guidance and support.

# NATIONAL GEOGRAPHIC EXPLORERS, FELLOWS, AND GRANTEES

**Greg Anderson**
Linguist
*National Geographic Fellow*
Greg Anderson records and preserves many different endangered languages. He is also the co-director of the Enduring Voices Project.

**Thomas Taha Rassam (TH) Culhane**
Urban Planner
*National Geographic Emerging Explorer*
T.H. Culhane works with residents of Cairo to install solar water heaters.

**Katey Walter Anthony**
Aquatic Ecologist and Biogeochemist
*National Geographic Emerging Explorer*
Katey Walter Anthony explores ways to use a greenhouse gas for energy.

**Jenny Daltry**
Herpetologist
*National Geographic Emerging Explorer*
Jenny Daltry saves endangered species of reptiles and inspires local people to protect reptiles and their habitats.

**Ken Banks**
Mobile Technology Innovator
*National Geographic Emerging Explorer*
Ken Banks develops mobile technology to connect groups in remote areas.

**Wade Davis**
Anthropologist and Ethnobotanist
*National Geographic Explorer-in-Residence*
Wade Davis studies plants and people while living within many indigenous cultures around the world.

**Katy Croff Bell**
Archaeological Oceanographer
*National Geographic Emerging Explorer*
Katy Croff Bell uses deep sea technology to explore the depths of the ocean and the Black Sea.

**Sylvia Earle**
Oceanographer
*National Geographic Explorer-in-Residence*
Sylvia Earle's research focuses on exploring and preserving marine ecosystems.

**Christina Conlee**
Archaeologist
*National Geographic Grantee*
Christina Conlee researches geoglyphs and Nasca lines etched into the earth in South America.

**Grace Gobbo**
Ethnobotanist
*National Geographic Emerging Explorer*
Grace Gobbo studies traditional medicine practices in Tanzania and plants native to the region.

**Alexandra Cousteau**
Social Environmental Activist
*National Geographic Emerging Explorer*
Alexandra Cousteau works to educate people to protect water resources and oceans.

**Beverly Goodman**
Geoarchaeologist
*National Geographic Emerging Explorer*
Beverly Goodman uses her skills to uncover past tsunamis and to help prevent disasters in the future.

### David Harrison
**Linguist**
*National Geographic Fellow*

David Harrison studies and archives endangered languages and cultures. He is also the co-director of the Enduring Voices Project.

### Dereck Joubert
**Filmmaker and Conservationist**
*National Geographic Explorer-in-Residence*

Dereck Joubert researches and films big cats and wildlife in Africa, sharing their stories with the world.

### Kristofer Helgen
**Zoologist**
*National Geographic Emerging Explorer*

Kristofer Helgen discovers new species of mammals and researches animals from across the world.

### Albert Lin
**Research Scientist and Engineer**
*National Geographic Emerging Explorer*

Albert Lin uses computer technologies to search for archeological sites without disturbing the land.

### Fredrik Hiebert
**Archaeologist**
*National Geographic Fellow*

Fredrik Hiebert uncovers mysteries of the past and has traced ancient trade routes, including the Silk Road.

### Elizabeth Kapu'uwailani Lindsey
**Filmmaker and Anthropologist**
*National Geographic Fellow*

Elizabeth Lindsey strives to preserve the Polynesian culture by using documentary films and education.

### Zeb Hogan
**Aquatic Ecologist**
*National Geographic Fellow*

Zeb Hogan studies freshwater fish and educates people on how to save these fish species from extinction.

### Sam Meacham
**Cave Diver**
*National Geographic Grantee*

Sam Meacham explores, preserves, and maps aquifers, caves, and other groundwater sources in Mexico.

### Shafqat Hussain
**Conservationist**
*National Geographic Emerging Explorer*

Shafqat Hussain works with herders in Pakistan to protect the endangered snow leopard.

### Kakenya Ntaiya
**Educator and Activist**
*National Geographic Emerging Explorer*

Kakenya Ntaiya established a school for girls in Kenya, and she continues to improve girls' education there.

### Beverly Joubert
**Filmmaker and Conservationist**
*National Geographic Explorer-in-Residence*

Beverly Joubert has spent years filming and protecting big cats and other wildlife in Africa.

### Johan Reinhard
**Anthropologist**
*National Geographic Explorer-in-Residence*

Johan Reinhard conducts field research in South America and investigates the cultural practices of Andes people.

# NATIONAL GEOGRAPHIC EXPLORERS, FELLOWS, AND GRANTEES

### Enric Sala
**Marine Ecologist**
*National Geographic Explorer-in-Residence*

Enric Sala dedicates his life to finding ways to reverse the damage humans have caused to the seas.

### Beth Shapiro
**Molecular Biologist**
*National Geographic Emerging Explorer*

Beth Shapiro studies samples of ancient DNA to find out how species change over time.

### Kira Salak
**Writer/Adventurer**
*National Geographic Emerging Explorer*

Kira Salak is an adventure traveler who writes about her explorations in exotic, often dangerous places.

### Cid Simoes
**Sustainable Agriculturalist**
*National Geographic Emerging Explorer*

Cid Simoes teaches farmers in Brazil how to farm in economically efficient and ecologically sound ways.

### Katsufumi Sato
**Behavioral Ecologist**
*National Geographic Emerging Explorer*

Katsufumi Soto uses technology to study animal behaviors with the goal of conserving their habitats.

### José Urteaga
**Marine Biologist and Conservationist**
*National Geographic Emerging Explorer*

José Urteaga uses creative ideas to help save sea turtles from extinction.

### Paola Segura
**Sustainable Agriculturalist**
*National Geographic Emerging Explorer*

Paola Segura teaches sustainable farming techniques to farmers in Brazil, conserving the rain forest.

### Spencer Wells
**Population Geneticist**
*National Geographic Explorer-in-Residence*

Spencer Wells studies human migration patterns. He is the director of National Geographic's Genographic Project.

# THE ESSENTIALS OF GEOGRAPHY

**National Geographic Atlas** ............... A2
  EXPLORE THE WORLD ............................. 1

## CHAPTER 1
### The Geographer's Toolbox ............... 10

**SECTION 1 • GEOGRAPHIC THINKING**
- 1.1 Thinking Spatially ........................ 12
- 1.2 Themes and Elements .................. 14
- 1.3 World Regions ............................. 16

**SECTION 2 • MAPS**
- 2.1 Elements of a Map ....................... 18
- 2.2 Map Scale .................................... 20
- 2.3 Political and Physical Maps .......... 22
- 2.4 Map Projections ........................... 24
- 2.5 Thematic Maps ............................. 26

  NATIONAL GEOGRAPHIC **PHOTO GALLERY** ............... 28

## CHAPTER 2
### Physical & Human Geography ........... 32

**SECTION 1 • THE EARTH**
- 1.1 Earth's Rotation and Revolution ..... 34
- 1.2 Earth's Complex Structure ............ 36
- 1.3 Earth's Landforms ........................ 38
- 1.4 The Ring of Fire ........................... 40
- 1.5 Waters of the Earth ...................... 42

**SECTION 2 • PHYSICAL GEOGRAPHY**
- 2.1 Climate and Weather .................... 44
- 2.2 World Climate Regions .................. 46
- 2.3 Extreme Weather .......................... 48
- 2.4 Natural Resources ........................ 50
- 2.5 Habitat Preservation .................... 52
- **2.6 EXPLORING THE WORLD'S OCEANS** ........... 54

  NATIONAL GEOGRAPHIC **PHOTO GALLERY** ............... 56

**SECTION 3 • HUMAN GEOGRAPHY**
- 3.1 World Cultures ............................. 58
- 3.2 Religions and Belief Systems ........ 60
- 3.3 Economic Geography .................... 62
- 3.4 Political Geography ...................... 64
  - **GLOBAL ISSUES • HUMAN RIGHTS**
- **3.5 PROTECTING HUMAN RIGHTS** ............. 66

**COMPARE ACROSS REGIONS** ................ 70
Activities ............................................ 72

## TECHTREK

### myNGconnect.com

**Digital Library**
Unit 1 GeoVideo
*Introduce the Essentials of Geography*

**Explorer Video Clip**
Sylvia Earle, Oceanographer
National Geographic Explorer-in-Residence

NATIONAL GEOGRAPHIC **PHOTO GALLERY**

Photos of Earth's physical features and world cultures

**Maps and Graphs**
Interactive Map Tool

**Interactive Whiteboard GeoActivities**
- Draw the Stages of an Earthquake
- Build a Climograph
- Map the Spread of Buddhism

Archaeologists at work

# UNIT 2 NORTH AMERICA

## CHAPTER 3
North America Geography & History ..................................................76

**SECTION 1 • GEOGRAPHY**
1.1 Physical Geography ................................78
1.2 The Great Plains .....................................80
1.3 Western Mountains and Deserts ..............82
1.4 Mexico's Mountains and Plateaus ............84
1.5 EXPLORING THE YUCATÁN......................86

**SECTION 2 • U.S. & CANADIAN HISTORY**
2.1 Exploration and Colonization...................88
2.2 Settling Quebec .....................................90
2.3 Revolution and Independence .................92
2.4 U.S. Constitution ....................................94
2.5 Expansion and Industrialization ..............96
2.6 Civil War and Reconstruction ..................98
2.7 World Conflict .......................................100

**SECTION 3 • HISTORY OF MEXICO**
3.1 The Maya and the Aztecs ......................102
3.2 The Conquistadors ................................104
3.3 Mexican Independence .........................106
3.4 The U.S.-Mexican War ..........................108
**DOCUMENT-BASED QUESTION**
3.5 The Mexican Revolution .......................110

## CHAPTER 4
North America Today ..........................114

**SECTION 1 • UNITED STATES & CANADA**
1.1 North America's Cultural Diversity .........116
1.2 The Media Culture ................................118
1.3 The Changing Economy ........................120
**GLOBAL ISSUES • ENERGY**
1.4 FINDING NEW ENERGY SOURCES ...................122
1.5 Citizens' Rights and Responsibilities .....124

**SECTION 2 • FOCUS ON MEXICO**
2.1 Daily Life in Mexico ..............................126
2.2 The Impact of Globalization ..................128
2.3 Reaching Toward Democracy ................130

COMPARE ACROSS REGIONS .........................134
Activities .......................................................136

## TECHTREK

**myNGconnect.com**

**Digital Library**
Unit 2 GeoVideo
*Introduce North America*

**Explorer Video Clip**
Sam Meacham, Cave Diver
National Geographic Grantee

**NATIONAL GEOGRAPHIC PHOTO GALLERY**
Regional photos, including the Grand Canyon and Great Plains, Mexico City, Vancouver, and New York City

**Maps and Graphs**
Interactive Map Tool

**Interactive Whiteboard GeoActivities**
• Compare Climates
• Illustrate the Rain Shadow Effect
• Create a Sketch Map of Tenochtitlán

Tourists at the Grand Canyon, Arizona

# UNIT 3 Central America & the Caribbean

## CHAPTER 5
### Central America & the Caribbean Geography & History .......................... 140

**SECTION 1 • GEOGRAPHY**
- 1.1 Physical Geography ................................. 142
- 1.2 Earthquakes and Volcanoes .................. 144
- 1.3 Rain Forests of Central America ............. 146
- **1.4 SAVING SEA TURTLES** ............................. 148
- PHOTO GALLERY ........................................... 150

**SECTION 2 • HISTORY**
- 2.1 Trade Across Continents ....................... 152
- **DOCUMENT-BASED QUESTION**
- 2.2 The Columbian Exchange ..................... 154
- 2.3 Paths Toward Independence ................ 156
- 2.4 Comparing Cuba and Puerto Rico .......... 158

## CHAPTER 6
### Central America & the Caribbean Today ...................... 162

**SECTION 1 • CULTURE**
- 1.1 The Impact of Tourism ........................... 164
- 1.2 Caribbean Food and Music .................... 166
- 1.3 The Panama Canal ................................. 168

**SECTION 2 • GOVERNMENT & ECONOMICS**
- 2.1 Comparing Costa Rica and Nicaragua ..... 170
- 2.2 Challenges in Haiti ................................. 172
- **GLOBAL ISSUES • FOOD SUPPLY**
- **2.3 FEEDING CENTRAL AMERICA** ................ 174
- 2.4 Migration and the Caribbean ................. 176
- 2.5 Conserving the Rain Forest .................... 178

COMPARE ACROSS REGIONS .......................... 182
Activities ........................................................ 184

# TECHTREK

## myNGconnect.com

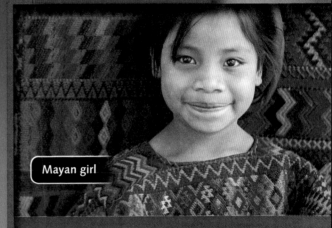

Mayan girl

### Digital Library
**Unit 3 GeoVideo**
*Introduce Central America & the Caribbean*

**Explorer Video Clip**
José Urteaga, Marine Biologist and Conservationist
National Geographic Emerging Explorer

**PHOTO GALLERY**
Regional photos, including the rain forest, the Andes Mountains, and Montserrat

**Music Clips**
Audio clips of music from the region

### Maps and Graphs
Outline Maps

### Interactive Whiteboard GeoActivities
- Research Rain Forest Species
- Build a Time Line of Colonial Rule
- Analyze Push-Pull Factors

# UNIT 4 SOUTH AMERICA

## CHAPTER 7
South America Geography & History .................. 188

**SECTION 1 • GEOGRAPHY**
- 1.1 Physical Geography .................. 190
- 1.2 Life at Different Elevations .................. 192
- 1.3 The Amazon River .................. 194
- 1.4 Cold and Warm Currents .................. 196

  **GLOBAL ISSUES • CLIMATE CHANGE**
- 1.5 RAIN FORESTS AND CLIMATE CHANGE .................. 198

**SECTION 2 • HISTORY**
- 2.1 The Inca .................. 200
- 2.2 EXPLORING NASCA CULTURE .................. 202
- 2.3 People of the Lowlands .................. 204
- 2.4 The Spanish in South America .................. 206
- 2.5 Brazil and the Slave Trade .................. 208

  **DOCUMENT-BASED QUESTION**
- 2.6 Simón Bolívar on Independence .................. 210

## CHAPTER 8
South America Today .................. 214

**SECTION 1 • CULTURE**
- 1.1 Indigenous Cultures .................. 216
- 1.2 Language Diversity .................. 218
- 1.3 Daily Life .................. 220

**SECTION 2 • GOVERNMENT & ECONOMICS**
- 2.1 Comparing Governments .................. 222
- 2.2 The Pampas Economy .................. 224
- 2.3 Chile's Food Production .................. 226
- 2.4 Products of Peru .................. 228

**SECTION 3 • FOCUS ON BRAZIL**
- 3.1 Brazil's Growing Economy .................. 230
- 3.2 São Paulo .................. 232
- 3.3 Impact of the Olympics .................. 234

COMPARE ACROSS REGIONS .................. 238
Activities .................. 240

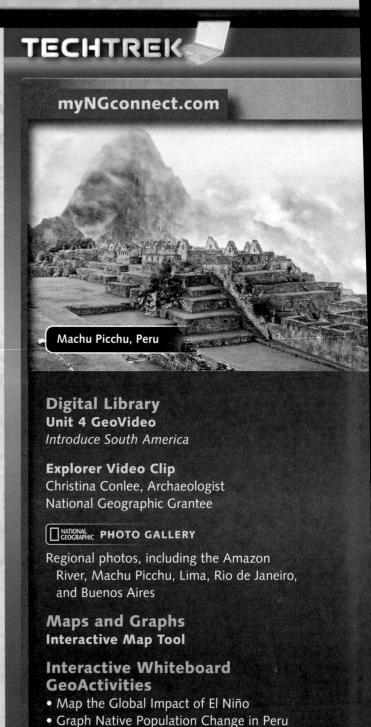

Machu Picchu, Peru

## TECHTREK

**myNGconnect.com**

**Digital Library**
Unit 4 GeoVideo
*Introduce South America*

**Explorer Video Clip**
Christina Conlee, Archaeologist
National Geographic Grantee

**NATIONAL GEOGRAPHIC PHOTO GALLERY**
Regional photos, including the Amazon River, Machu Picchu, Lima, Rio de Janeiro, and Buenos Aires

**Maps and Graphs**
Interactive Map Tool

**Interactive Whiteboard GeoActivities**
- Map the Global Impact of El Niño
- Graph Native Population Change in Peru
- Analyze Sustainable Agriculture

# UNIT 5 EUROPE

## CHAPTER 9
### Europe Geography & History ..........244

**SECTION 1 • GEOGRAPHY**
- 1.1 Physical Geography ...............................246
- 1.2 A Long Coastline ..................................248
- 1.3 Mountains, Rivers, and Plains ...............250
- 1.4 **PROTECTING THE MEDITERRANEAN** ........252
-  **PHOTO GALLERY** .....................................254

**SECTION 2 • EARLY HISTORY**
- 2.1 Roots of Democracy ..............................256
- 2.2 Classical Greece ....................................258
- 2.3 The Republic of Rome ...........................260
- 2.4 The Roman Empire ................................262
- 2.5 Middle Ages and Christianity .................264
- 2.6 Renaissance and Reformation ................266

**SECTION 3 • EMERGING EUROPE**
- 3.1 Exploration and Colonization ................268
- 3.2 The Industrial Revolution ......................270
- 3.3 The French Revolution ..........................272
- **DOCUMENT-BASED QUESTION**
- 3.4 Declarations of Rights ...........................274
- 3.5 Nationalism and World War I ................276
- 3.6 World War II and the Cold War ..............278

## CHAPTER 10
### Europe Today .........................................282

**SECTION 1 • CULTURE**
- 1.1 Languages and Cultures ........................284
- 1.2 Art and Music .......................................286
- 1.3 Europe's Literary Heritage .....................288
- 1.4 Cuisines of Europe ................................290

**SECTION 2 • GOVERNMENT & ECONOMICS**
- 2.1 The European Union ..............................292
- 2.2 The Impact of the Euro .........................294
- 2.3 Democracy in Eastern Europe ................296
- **GLOBAL ISSUES • MIGRATION**
- 2.4 **CHANGING DEMOGRAPHICS** .......................298

- COMPARE ACROSS REGIONS.........................302
- Activities ...................................................304

## TECHTREK

**myNGconnect.com**

**Digital Library**
Unit 5 GeoVideo
*Introduce Europe*

**Explorer Video Clip**
Enric Sala, Marine Ecologist
National Geographic Explorer-in-Residence

**PHOTO GALLERY**
Regional photos, including Florence, Paris, Amsterdam, and Budapest

**Maps and Graphs**
Interactive Map Tool

**Interactive Whiteboard GeoActivities**
- Map Europe's Land Regions
- Compare Greek and Roman Governments
- Analyze Causes and Effects of World War I

Louvre Museum, Paris, France

# UNIT 6 RUSSIA & THE EURASIAN REPUBLICS

## CHAPTER 11
### Russia & the Eurasian Republics Geography & History ...... 308

**SECTION 1 • GEOGRAPHY**
- 1.1 Physical Geography ............... 310
- 1.2 Land of Extreme Climates ....... 312
- 1.3 Natural Resources ................ 314
- **1.4 EXPLORING SIBERIAN LAKES** ..... 316
- 1.5 Central Asian Landscapes ....... 318

**SECTION 2 • HISTORY**
- 2.1 Early History ...................... 320
- 2.2 European or Asian? ............... 322
- 2.3 Defending Against Invaders ..... 324
- 2.4 Serfdom to Industrialization .... 326
- 2.5 The Soviet Union .................. 328

## CHAPTER 12
### Russia & the Eurasian Republics Today ...... 332

**SECTION 1 • CULTURE**
- 1.1 Climate and Culture .............. 334
- 1.2 Trans-Siberian Railroad ......... 336
- 1.3 St. Petersburg Today ............. 338

**SECTION 2 • GOVERNMENT & ECONOMICS**
- 2.1 The Soviet Collapse ............... 340
- 2.2 Russia's Government ............. 342
- 2.3 Unlocking Energy Riches ........ 344
- **GLOBAL ISSUES • POLLUTION**
- **2.4 AFTER CHERNOBYL** ............. 346

COMPARE ACROSS REGIONS ......... 350
Activities ................................. 352

## TECHTREK

**myNGconnect.com**

**Digital Library**
Unit 6 GeoVideo
*Introduce Russia & the Eurasian Republics*

**Explorer Video Clip**
Katey Walter Anthony, Aquatic Ecologist
National Geographic Emerging Explorer

**NATIONAL GEOGRAPHIC PHOTO GALLERY**
Regional photos, including St. Basil's Cathedral in Moscow, the Ural Mountains, Siberian tundra, and St. Petersburg

**Maps and Graphs**
Interactive Map Tool

**Interactive Whiteboard GeoActivities**
- Analyze Central Asian Economies
- Graph Napoleon's March Through Russia
- Explore the Trans-Siberian Railroad

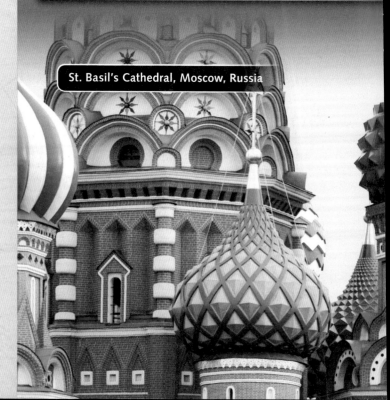
St. Basil's Cathedral, Moscow, Russia

# UNIT 7 Sub-Saharan Africa

## CHAPTER 13
### Sub-Saharan Africa Geography & History ..........356

**SECTION 1 • GEOGRAPHY**
- 1.1 Physical Geography ..............358
- 1.2 East Africa and the Rift Valley ..............360
- 1.3 West Africa's Steppes ..............362
- 1.4 Rain Forests and Resources ..............364
- 1.5 Southern Plateaus and Basins ..............366
- 1.6 EXPLORING AFRICA'S WILDLIFE ..............368

**SECTION 2 • HISTORY**
- 2.1 Bantu Migrations ..............370
- 2.2 Early States and Trade ..............372
- 2.3 Impact of the Slave Trade ..............374
- 2.4 Colonization to Independence ..............376

NATIONAL GEOGRAPHIC PHOTO GALLERY ..............378

## CHAPTER 14
### Sub-Saharan Africa Today ..............382

**SECTION 1 • CULTURE**
- 1.1 Africa's Borders and Cultures ..............384
- 1.2 African Music Goes Global ..............386
- 1.3 Kenya Modernizes ..............388
- 1.4 EXPLORING TRADITIONAL CULTURES ..............390

**SECTION 2 • GOVERNMENT & ECONOMICS**
- 2.1 Prized Mineral Resources ..............392
- 2.2 Nigeria and Oil ..............394
- 2.3 Agriculture and Food Supply ..............396
- **GLOBAL ISSUES • HEALTH**
- 2.4 IMPROVING PUBLIC HEALTH ..............398
- 2.5 Sudan and Somalia ..............400
- 2.6 Ending Apartheid ..............402

COMPARE ACROSS REGIONS ..............406
Activities ..............408

# TECHTREK

## myNGconnect.com

A musician and griot in Senegal

### Digital Library
**Unit 7 GeoVideo**
*Introduce Sub-Saharan Africa*

**Explorer Video Clip**
Dereck and Beverly Joubert, Conservationists
National Geographic Explorers-in-Residence

### NATIONAL GEOGRAPHIC PHOTO GALLERY
Regional photos, including Kenya's savanna, Victoria Falls, Johannesburg, and the Great Rift Valley

### Music Clips
Audio clips of music from the region

### Maps and Graphs
Interactive Map Tool

### Interactive Whiteboard GeoActivities
- Compare Precipitation Across Regions
- Locate a Wildlife Reserve
- Research Vanishing Cultures

# UNIT 8
# SOUTHWEST ASIA & NORTH AFRICA

## CHAPTER 15
Southwest Asia & North Africa
Geography & History ........................ 412

### SECTION 1 • GEOGRAPHY
1.1 Physical Geography ........................ 414
1.2 Tigris and Euphrates Rivers ............ 416
1.3 The Arabian Peninsula .................... 418
1.4 Anatolian and Iranian Plateaus ....... 420

### SECTION 2 • HISTORY
2.1 Mesopotamia .................................. 422
2.2 Birthplace of Three Religions .......... 424
2.3 Diffusion of Religions ..................... 426
2.4 The Ottoman Empire ...................... 428

### SECTION 3 • FOCUS ON EGYPT
3.1 The Nile River Valley ...................... 430
3.2 Egypt's Ancient Civilization ............. 432
3.3 The Great Pyramids ....................... 434

## CHAPTER 16
Southwest Asia
& North Africa Today ........................ 438

### SECTION 1 • CULTURE
1.1 Migration and Trade ....................... 440
1.2 Istanbul: Bridging East and West .... 442
1.3 **EXPLORING ANCIENT ISRAEL** ..... 444
1.4 Dubai: Desert City ......................... 446

### SECTION 2 • GOVERNMENT & ECONOMICS
2.1 Comparing Governments ................ 448
2.2 Oil and Wealth ............................... 450
2.3 Tensions in Southwest Asia ............ 452
2.4 Iraq's Problems and Promise .......... 454
2.5 Afghanistan: Moving Forward ......... 456
  GLOBAL ISSUES • STANDARD OF LIVING
2.6 **BUILDING SCHOOLS** ................... 458

COMPARE ACROSS REGIONS ............... 462
Activities ......................................... 464

## TECHTREK

myNGconnect.com

View from the world's tallest building, Dubai

### Digital Library
**Unit 8 GeoVideo**
*Introduce Southwest Asia & North Africa*

**Explorer Video Clip**
Beverly Goodman, Geoarchaeologist
National Geographic Emerging Explorer

### NATIONAL GEOGRAPHIC PHOTO GALLERY
Regional photos, including the Blue Mosque, Hagia Sophia, and Topkapi Palace

### Maps and Graphs
Outline Maps

### Interactive Whiteboard GeoActivities
• Explore an Ancient Irrigation System
• Compare Major Rivers of the World
• Trace the Benefits of Education

# UNIT 9 South Asia

## CHAPTER 17
### South Asia Geography & History .................. 468

**SECTION 1 • GEOGRAPHY**
- 1.1 Physical Geography .................. 470
- 1.2 **EXPLORING THE HIMALAYAS** .................. 472
- 1.3 Living with Monsoons .................. 474
- 1.4 Resources and Land Use .................. 476
- 1.5 Conservation .................. 478
- **GLOBAL ISSUES • WATER RESOURCES**
- 1.6 **SOUTH ASIA'S WATER CRISIS** .................. 480

**SECTION 2 • HISTORY**
- 2.1 Early Civilizations .................. 482
- 2.2 Historic Empires .................. 484
- 2.3 Religion in South Asia .................. 486
- 2.4 Colonialism to Partition .................. 488
- **DOCUMENT-BASED QUESTION**
- 2.5 Gandhi and the Bhagavad Gita .................. 490

## CHAPTER 18
### South Asia Today .................. 494

**SECTION 1 • CULTURE**
- 1.1 Hinduism Today .................. 496
- 1.2 Changing Traditions .................. 498
- 1.3 Daily Life .................. 500
- 1.4 Popular Culture .................. 502

**SECTION 2 • FOCUS ON INDIA**
- 2.1 The Largest Democracy .................. 504
- 2.2 Economic Growth .................. 506
- 2.3 The Golden Quadrilateral .................. 508

**SECTION 3 • GOVERNMENT & ECONOMICS**
- 3.1 The Impact of Urbanization .................. 510
- 3.2 Pakistan's Changing Government .................. 512
- 3.3 Fighting Poverty in Bangladesh .................. 514

**COMPARE ACROSS REGIONS** .................. 518
**Activities** .................. 520

## TECHTREK

### myNGconnect.com

**Digital Library**
Unit 9 GeoVideo
*Introduce South Asia*

**Explorer Video Clip**
Kira Salak, Writer/Adventurer
National Geographic Emerging Explorer

**NATIONAL GEOGRAPHIC PHOTO GALLERY**
Regional photos, including the Himalayas, the Ganges River, and the Taj Mahal

**Music Clips**
Audio clips of music from the region

**Maps and Graphs**
Interactive Map Tool

**Interactive Whiteboard GeoActivities**
- Draw a Mental Map of South Asia
- Build a Time Line of Colonialism in India
- Graph and Compare Internet Use

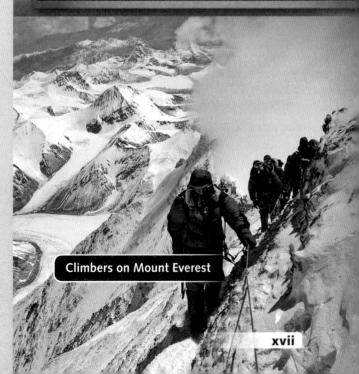

Climbers on Mount Everest

# UNIT 10 East Asia

## CHAPTER 19
### East Asia Geography & History ....... 524

**SECTION 1 • GEOGRAPHY**
- 1.1 Physical Geography ..................... 526
- 1.2 China's Rivers and Plains ..................... 528
- 1.3 The Island Arc of Japan ..................... 530
- 1.4 The Korean Peninsula ..................... 532
- 1.5 Mongolia's Desert Landscape ..................... 534
- 1.6 TRACKING AQUATIC WILDLIFE ..................... 536

**SECTION 2 • CHINA'S HISTORY**
- 2.1 Early Dynasties ..................... 538
- NATIONAL GEOGRAPHIC PHOTO GALLERY ..................... 540
- 2.2 Confucianism ..................... 542
- 2.3 Silk Roads and Trade ..................... 544
- 2.4 Exploration and Isolation ..................... 546
- 2.5 Communist Revolution ..................... 548

**SECTION 3 • HISTORY OF JAPAN & KOREA**
- 3.1 Japanese Samurai ..................... 550
- 3.2 Japan Industrializes ..................... 552
- 3.3 Korea's Early History ..................... 554
- 3.4 The Korean War ..................... 556

## CHAPTER 20
### East Asia Today ..................... 560

**SECTION 1 • CULTURE**
- 1.1 Religious Traditions ..................... 562
- 1.2 China's Inventions ..................... 564
- 1.3 Shanghai's Rapid Growth ..................... 566
- NATIONAL GEOGRAPHIC PHOTO GALLERY ..................... 568
- 1.4 Japanese Anime ..................... 570
- 1.5 Bullet Trains ..................... 572

**SECTION 2 • GOVERNMENT & ECONOMICS**
- 2.1 China's Economy Today ..................... 574
- GLOBAL ISSUES • POPULATION GROWTH
- 2.2 CHINA'S POPULATION POLICY ..................... 576
- 2.3 China's Three Gorges Dam ..................... 578
- 2.4 Republic of China (Taiwan) ..................... 580
- 2.5 Japan's Economic Future ..................... 582
- 2.6 Comparing North and South Korea ..................... 584
- NATIONAL GEOGRAPHIC PHOTO GALLERY ..................... 586

COMPARE ACROSS REGIONS ..................... 590
Activities ..................... 592

## TECHTREK

### myNGconnect.com

**Digital Library**
Unit 10 GeoVideo
*Introduce East Asia*

**Explorer Video Clip**
Albert Lin, Research Scientist and Engineer
National Geographic Emerging Explorer

**NATIONAL GEOGRAPHIC PHOTO GALLERY**
Regional photos, including the Great Wall of China, Mount Fuji, bullet trains, and a Buddhist temple in Korea

**Maps and Graphs**
Interactive Map Tool

**Interactive Whiteboard GeoActivities**
- Follow the Chang Jiang
- Barter on the Silk Roads
- Explore the Forbidden City

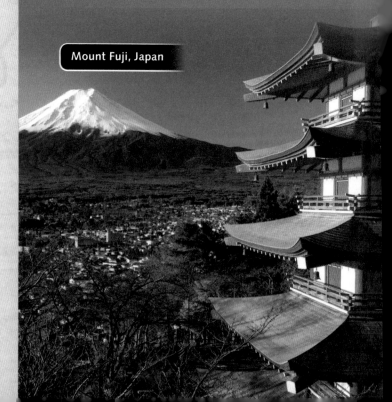

Mount Fuji, Japan

# UNIT 11 Southeast Asia

## CHAPTER 21
### Southeast Asia Geography & History .......... 596

**SECTION 1 • GEOGRAPHY**
- 1.1 Physical Geography .............. 598
- 1.2 Parallel Rivers ..................... 600
- 1.3 The Malay Peninsula ............ 602
- 1.4 Island Nations ..................... 604
- 1.5 **DISCOVERING NEW SPECIES** .... 606

**SECTION 2 • HISTORY**
- 2.1 Ancient Valley Kingdoms ....... 608
- 2.2 Trade and Colonialism ......... 610
- 2.3 Indonesia and the Philippines ... 612
  **DOCUMENT-BASED QUESTION**
- 2.4 The Vietnam War ................. 614

## CHAPTER 22
### Southeast Asia Today .......... 618

**SECTION 1 • CULTURE**
- 1.1 Religious Traditions ............. 620
- 1.2 Thailand Today ................... 622
- 1.3 Regional Languages ............ 624
- 1.4 Saving the Elephant ............ 626

**SECTION 2 • GOVERNMENT & ECONOMICS**
- 2.1 Governing Fragmented Countries ...... 628
- 2.2 Migration Within Indonesia ..... 630
  **GLOBAL ISSUES • GLOBALIZATION**
- 2.3 **SINGAPORE'S GROWTH** ............ 632
- 2.4 Malaysia and New Media ..... 634

COMPARE ACROSS REGIONS.......... 638
Activities .................................. 640

## TECHTREK

myNGconnect.com

A floating market in Thailand

### Digital Library
**Unit 11 GeoVideo**
*Introduce Southeast Asia*

**Explorer Video Clip**
Kristofer Helgen, Zoologist
National Geographic Emerging Explorer

**NATIONAL GEOGRAPHIC PHOTO GALLERY**
Regional photos, including the Mekong and Irrawaddy rivers, Angkor Wat, and Hanoi and Jakarta

### Maps and Graphs
Interactive Map Tool

### Interactive Whiteboard GeoActivities
- Investigate New Species
- Map the Spice Trade
- Analyze Remittances and GDP

# UNIT 12
# AUSTRALIA, THE PACIFIC REALM & ANTARCTICA

## CHAPTER 23
Australia, the Pacific Realm &
Antarctica Geography & History ...... 644

### SECTION 1 • GEOGRAPHY
1.1 Physical Geography ............................... 646
1.2 Indigenous Plants and Animals ............ 648
1.3 Biological Hitchhikers ......................... 650
1.4 The Pacific Islands ............................... 652
   **GLOBAL ISSUES • HABITAT PRESERVATION**
1.5 **SAVING THE REEFS** ....................................... 654

### SECTION 2 • HISTORY
2.1 Indigenous Populations ........................ 656
2.2 Seagoing Societies ................................. 658
2.3 From Convicts to Colonists ................... 660
2.4 **EXPLORING VANISHING LANGUAGES** ........ 662

## CHAPTER 24
Australia, the Pacific Realm
& Antarctica Today ............................. 666

### SECTION 1 • CULTURE
1.1 From Ranch to City ............................... 668
1.2 Immigration to Australia ....................... 670
1.3 Polynesian Cultures ............................... 672
1.4 Human Footprint in Antarctica .............. 674

### SECTION 2 • GOVERNMENT & ECONOMICS
2.1 New Trade Patterns ............................... 676
2.2 Rights for Indigenous People ................. 678
2.3 Adventure Tourism ............................... 680
2.4 New Zealand Today ............................... 682

COMPARE ACROSS REGIONS ...................... 686
Activities ....................................................... 688

### REFERENCE
Skills Handbook ........................................... R2
Economics & Government Handbook ........... R41
World Religions Handbook ........................... R51
English Glossary .......................................... R59
Spanish Glossary .......................................... R78
Index ............................................................. R98
Acknowledgments ........................................ R124

# TECHTREK

**myNGconnect.com**

Opera House, Sydney, Australia

**Digital Library**
Unit 12 GeoVideo
*Introduce Australia, the Pacific Realm
& Antarctica*

**Explorer Video Clip**
David Harrison and Greg Anderson, Linguists
National Geographic Fellows

**NATIONAL GEOGRAPHIC PHOTO GALLERY**
Regional photos, including the Great Barrier
   Reef and glaciers in Antarctica

**Maps and Graphs**
Interactive Map Tool

**Interactive Whiteboard
GeoActivities**
• Research Indigenous Species
• Research and Report on Endangered Languages
• Build a Time Line of Indigenous Rights

# SPECIAL FEATURES

## National Geographic Explorers

Exploring the World's Oceans with Sylvia Earle .........54
Exploring the Yucatán with Sam Meacham .............86
Saving Sea Turtles with José Urteaga ...................148
Exploring Nasca Culture with Christina Conlee ......202
Protecting the Mediterranean with Enric Sala ........252
Exploring Siberian Lakes with Katey Walter Anthony ...316
Exploring Africa's Wildlife
 with Dereck and Beverly Joubert .......................368
Exploring Traditional Cultures
 with Grace Gobbo and Wade Davis....................390
Exploring Ancient Israel with Beverly Goodman .....444
Exploring the Himalayas with Kira Salak ...............472
Tracking Aquatic Wildlife with Katsufumi Sato ........536
Discovering New Species with Kristofer Helgen ......606
Exploring Vanishing Languages
 with David Harrison and Greg Anderson .............662

## Global Issues

Human Rights: Protecting Human Rights ...............66
Energy: Finding New Energy Sources ...................122
Food Supply: Feeding Central America .................174
Climate Change: Rain Forests and Climate Change ..198
Migration: Changing Demographics ....................298
Pollution: After Chernobyl ...............................346
Health: Improving Public Health ........................398
Standard of Living: Building Schools ...................458
Water Resources: South Asia's Water Crisis ...........480
Population Growth: China's Population Policy ........576
Globalization: Singapore's Growth ......................632
Habitat Preservation: Saving the Reefs .................654

## Document-Based Questions

The Mexican Revolution ...................................110
The Columbian Exchange .................................154
Simón Bolívar on Independence .........................210
Declarations of Rights .....................................274
Gandhi and the Bhagavad Gita ..........................490
The Vietnam War ...........................................614

## Compare Across Regions

Climate .......................................................70
Becoming a Citizen ........................................134
Canals as Transportation .................................182
Sports and the Olympics .................................238
World Languages ..........................................302
Natural Resources and Energy ..........................350
Deserts of the World ......................................406
The World's Bread .........................................462
Economic Growth ..........................................518
Mobile Access ..............................................590
Endangered Species ......................................638
Pastures and Crops .......................................686

## Graphs

Belem, Brazil, and Omsk, Russia: Average Monthly
 Temperature and Precipitation ..........................71
Mexico's GDP ..............................................129
Literacy Rates in Mexico (1950–2008) .................131
U.S. Military Naturalizations ............................135
U.S. Naturalizations .......................................135
Workers Leaving the Caribbean, Remittances .........177
Deforestation in Brazil, 2000–2009 .....................198
Mining Exports of Peru (2000–2007) ...................229
Population Growth in São Paulo (1950–2010) ........233
Soy Exports: Argentina and Brazil, United States .....236
Growth of Key Industries in Russia, 1890–1900 .....327
Immigrant Populations in Europe .......................298
Number of Languages Spoken by Continent ..........303
Growth of Key Industries in Russia, 1890–1900 .....327
Energy Consumption Per Country ......................351
Sub-Saharan Africa Wheat Production ................405
Primary School Attendance in Afghanistan ...........458
International Migrants in
 Selected Countries, 2010 ...............................460
Sales of Passenger Vehicles in India and
 the United States (2002 and 2007) ...................509
Population Growth in South Asia, 1971–2011 ........511
Economic Indicators for India and Hungary
 (1990 and today) .........................................519
Religions in East Asia .....................................563
China's GDP by Sector ...................................575
GDP of Selected Countries ..............................575
Chinese Population 2010 .................................576
GDP by Sector, Export Partners, Taiwan ..............581
Japan's Population, 2009 .................................583
Indonesia: Island Populations ...........................631
Singapore's Annual GDP (1960–2010) .................632
Southeast Asia: Internet and Cell Phone Access .....635
Immigration to Australia ..................................671
New Zealand's Export Partners ..........................685
U.S. Inflation Rate, 1976–2008 .........................R44
U.S. National Debt, 1980–2010 .........................R45
U.S. Poverty Rate, 1960–2009 ..........................R45
Comparing Industrial Sectors ...........................R46

## Charts and Tables

World's Longest Rivers ....................................43
Economic Indicators of Selected Countries ............63
Universal Declaration of Human Rights .................67
U.S. and World Endangered Species ....................68
Three Branches of U.S. Government ....................94
Settling Quebec ............................................112
Speakers of Major Languages,
 United States and Canada ..............................117
United States Trade Balance .............................132
By the Numbers: U.S. Naturalization ...................135

Major Earthquakes in Central America
& the Caribbean Since 2000 ............................**145**
Population of Haiti in 1789 ................................**161**
Tourism in Central America and the Caribbean .......**165**
Panama Canal Milestones .................................**169**
Comparing Economic Development Indicators,
  Costa Rica and Nicaragua ..............................**171**
Responding to the Earthquake in Haiti .................**173**
Some Major Crops in Selected
  Central American Countries ...........................**174**
Population and Poverty in Central America ...........**181**
South American Altitude Zones .........................**192**
Amazon River and its Tributaries .......................**194**
Indigenous Populations of South America (2007) .....**217**
Population of Mediterranean Cities .....................**253**
Selected English Words Formed from Greek Roots ...**257**
Government of the Roman Republic ....................**261**
Miles of Railway Track in Selected
  European Countries (1840–1880) ......................**281**
Cost of a Ten-Minute Phone Call
  to the United States (in Euros).........................**301**
Time Industrial Laborers Needed to Work
  to Buy Selected Goods, 1986 ..........................**331**
Fast Facts on St. Petersburg ..............................**338**
Compare Russian and U.S. Governments ...............**343**
Chernobyl Disaster by the Numbers ....................**346**
Natural Gas Reserves (2012) ..............................**349**
Big Cats by the Numbers .................................**369**
The Middle Passage .......................................**375**
Impact of Changes in Land in Africa ...................**380**
Cooperative International Organizations in Africa ....**385**
African Plants and Uses ...................................**391**
Selected Oil and Diamond Exports,
  Sub-Saharan Africa, 2005–2006 .......................**392**
Malaria by the Numbers ..................................**399**
Religion in Southwest Asia and North Africa .........**436**
Population of Dubai .......................................**447**
Oil Reserves/Production ..................................**451**
South Asia Grain Production and Population
  (1950–2012) ..............................................**477**
Pollution by the Numbers, India (2012) ................**480**
Selected Indo-European Languages ....................**483**
The Green Revolution and
  Indian Villages (1955–1986) ............................**492**
India's Governmental Structure .........................**505**
Comparison of Selected Consumer Goods,
  South Asia and the United States.....................**507**
Literacy Rate and GDP in Bangladesh ..................**517**
Rivers in China .............................................**558**
Life in the Two Koreas ....................................**589**
Cell Phone Usage in East and South Asia ..............**591**
Asian Elephants by the Numbers .......................**627**
Major Ethnic Groups in Selected Southeast
  Asian Countries ..........................................**636**
Great Barrier Reef by the Numbers ....................**654**
Comparing Language Use in Australia
  and Papua New Guinea .................................**665**

## Infographics and Models

Earth's Four Seasons: Northern Hemisphere ............**35**
Earth's Structure ............................................**36**
Tectonic Plate Movements .................................**37**

The Continental Margin ....................................**39**
The Hydrologic Cycle .......................................**42**
How a Tornado Forms ......................................**48**
U.S. Constitution ............................................**95**
Wind Turbine ...............................................**122**
Citizens' Rights ............................................**125**
U.S. Naturalization ........................................**135**
Panama Canal Expansion Project .......................**168**
Workers Leaving the Caribbean, Remittances ........**177**
Compare Panama Canal and Suez Canal ..............**183**
Slash-and-Burn Agriculture ..............................**204**
Pampas in Argentina .....................................**225**
Olympic Snapshot
  (Beijing Olympics, Summer 2008) ....................**239**
Manor in the Middle Ages ...............................**265**
Euro Coin of Three Eurozone Countries ...............**295**
Process of Methane Release .............................**317**
Adapting to Climate in Russia
  and Eurasian Republics .................................**335**
Cutaway of European Slave Ship .......................**374**
Selected Deserts of the World ..........................**407**
Diffusion of Religions .....................................**426**
The Great Pyramid of Khufu .............................**435**

**THE GREAT PYRAMID OF KHUFU, 2550 B.C.**

Bread Around the World .................................**463**
The Dynastic Cycle of China .............................**539**
Zheng He's Expeditions ..................................**546**
Spoken Languages in Selected Southeast
  Asian Countries ..........................................**625**
Pacific Island Formation .................................**653**
Grassland Use Around the World ......................**687**
Business Cycle .............................................**R41**
Economic Systems ........................................**R42**
Free Market Economy ....................................**R43**
Gross Domestic Product .................................**R43**
Supply and Demand .....................................**R47**
Responsibilities of U.S. Citizens ........................**R48**
U.S. Executive Branch ....................................**R49**

# NATIONAL GEOGRAPHIC Atlas ............... A2–A39

## Maps

### Unit 1
Western Hemisphere ......................................... 12
World Regions .................................................. 17
Germany's Economic Activity ............................. 18
Latitude, Longitude, and Hemispheres ................. 19
Charlotte, North Carolina .................................. 20
North Carolina ................................................. 21
Eastern United States ........................................ 21
East Asia Political and Physical ........................... 22
Map Projections ............................................... 24
U.S. Alternative Energy ..................................... 26
Thailand's Population Density ............................ 27
Philippines' Earthquakes ................................... 27
North America Political .................................... 30
The Ring of Fire ............................................... 40
Western United States: Climate and Weather ........ 45
World Climate Regions ..................................... 46
Extreme Weather in the United States ................. 49
Selected Natural Resources of the World ............. 51
Earth's "Hope Spots" ........................................ 55
World Religions ............................................... 61
Canada's Biological Resources ........................... 69

### UNIT 2
North America Political .................................... 77
North America Physical .................................... 78
North America Climate ..................................... 79
Agricultural Regions and Great Plains .................. 81
Population Density (Western United States) ......... 83
Mexico's Cities and Resources ........................... 85
European Colonies and Native American
    Groups, c. 1650 .......................................... 89
New France, c. 1740 ........................................ 91
United States in 1783 ....................................... 93
Railroads and Pioneer Trails .............................. 97
Civil War in the United States ............................ 99
Mayan Civilization, c. 900 ............................... 103
Aztec Empire, 1520 ........................................ 103
Mexico in 1840 .............................................. 109
Mexico in 1853 .............................................. 109
Battle of the Alamo, 1836 ............................... 113
Broadband Internet Access by U.S. County ........ 119
Mexico's Resources ........................................ 133

### UNIT 3
Central America and the Caribbean Political ...... 140
Central America and the Caribbean Physical ...... 142
Central America and the Caribbean Climate ...... 143
Plates and Volcanoes ..................................... 145
Central American Rain Forests ........................ 146
Triangular Trade ............................................ 153
European Colonial Claims ............................... 157
Cuba's Economic Resources ............................ 159
Puerto Rico's Economic Resources ................... 159
Central American Land Use ............................ 160
Panama Canal ............................................... 180
Suez Canal .................................................... 183

### UNIT 4
South America Political ................................... 189
South America Physical ................................... 190
South America Climate ................................... 191

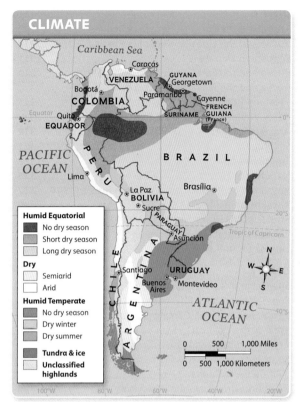

South America Population Density ................... 193
South America's Resources ............................. 193
Amazon River and Basin ................................. 194
Wind Currents ............................................... 197
Ocean Currents ............................................. 197
Inca Empire, c. 1520 ...................................... 200
South America's Lowlands .............................. 205
Explorers in South America ............................. 207
Portuguese Claims in Brazil ............................. 209
Peru Physical ................................................. 212
Indigenous Languages .................................... 218
Argentina's Land Use ..................................... 225
World Mediterranean Climates ........................ 227
Brazil's Resources and Industries ...................... 231
Brazil ............................................................ 237

### UNIT 5
Europe Political ............................................. 245
Europe Physical ............................................. 246
Europe Climate ............................................. 247
Europe's Major Landforms and Rivers ............... 249
Europe Land Use and Natural Resources ........... 251
Mediterranean Europe ................................... 253
Ancient Greek City-States ............................... 257
Alexander's Empire, 330 B.C. .......................... 259
The Hills of Rome .......................................... 260
The Roman Empire at its Height, A.D. 200 ........ 263
Early Colonization of Africa, Asia,
    and the Americas ...................................... 269
Industries in Europe, 1840–1890 ..................... 271
Europe Before World War I, 1914 .................... 277

SPECIAL FEATURES **xxiii**

| | |
|---|---|
| Post-World War II Europe, 1950 | 279 |
| Trade in the Roman Empire | 280 |
| European Union | 293 |
| Christianity in Europe | 300 |

## UNIT 6

| | |
|---|---|
| Russia and the Eurasian Republics Political | 309 |
| Russia and the Eurasian Republics Physical | 310 |
| Population Density | 311 |
| Russia and the Eurasian Republics Climate | 313 |
| Russia and the Eurasian Republics Natural Resources | 315 |
| Katey Walter Anthony's Expeditions | 316 |
| Aral Sea Satellite Maps, 1973 and 2009 | 319 |
| Silk Roads | 321 |
| European and Asian Russia | 322 |
| Napoleon's 1812 Invasion of Russia | 325 |
| Resources and Industries in Western Russia, 1900 | 327 |
| Kazakhstan Political | 330 |
| Routes of the Trans-Siberian Railroad | 337 |
| Oil and Natural Gas Pipelines | 345 |
| Languages in Russia | 348 |

## UNIT 7

| | |
|---|---|
| Sub-Saharan Africa Political | 357 |
| Sub-Saharan Africa Physical | 358 |
| East Africa Political, Climate | 361 |
| West Africa Political | 363 |
| Central Africa's Land Use | 365 |
| Southern Africa's Gross Domestic Product Per Capita, 2009 | 367 |
| Southern Africa's Natural Resources | 367 |
| African Wildlife Reserves | 369 |
| Bantu Migrations | 371 |
| Sub-Saharan African Empires | 373 |
| European Colonies, 1938 | 377 |
| African Nations, 2011 | 377 |
| Africa's Climate | 381 |
| Language Families in Sub-Saharan Africa | 385 |
| Kenya's Parks and Reserves | 389 |
| Sub-Saharan Africa's Resources | 393 |
| Ethnic Groups and Oil | 395 |
| Sudan Political, Somalia Political | 401 |
| Botswana Political | 404 |

## UNIT 8

| | |
|---|---|
| Southwest Asia and North Africa Political | 412 |
| Southwest Asia and North Africa Physical | 414 |
| Southwest Asia and North Africa Climate | 414 |
| Tigris and Euphrates Rivers | 417 |
| The Arabian Peninsula | 419 |
| Anatolian and Iranian Plateaus | 421 |
| Mesopotamia | 422 |
| Birthplace of Three Religions | 425 |
| The Ottoman Empire, 1683 | 429 |
| Nile River Valley | 431 |
| Ancient Egypt | 437 |
| Bosporus Strait | 443 |
| Oil in Southwest Asia and North Africa | 451 |
| Israel Political | 453 |
| Iraq's Religious and Ethnic Divisions | 455 |
| Afghanistan | 456 |
| GDP Per Capita, Arabian Peninsula | 461 |

## UNIT 9

| | |
|---|---|
| South Asia Political | 469 |
| South Asia Physical | 470 |
| South Asia Climate, Population Density | 471 |
| Bhutan: The Snowman Trek | 473 |
| Summer Monsoon, Winter Monsoon | 475 |
| South Asia Land Use and Natural Resources | 477 |
| Population of the Ganges River Basin | 479 |
| Ancient River Valley Civilizations | 483 |
| Empires of South Asia | 485 |
| South Asia's Religions | 487 |
| Conflict Over Kashmir | 489 |
| Aryan Migrations (1500–25 B.C.) | 493 |
| India's Highways and the Golden Quadrilateral | 509 |
| Pakistan Political | 513 |
| Bangladesh Resources and Industries | 515 |
| Languages of South Asia | 516 |

## UNIT 10

| | |
|---|---|
| East Asia Political | 525 |
| East Asia Physical | 526 |
| East Asia Climate, Population Density | 527 |
| China's Rivers and Cities | 529 |
| Earthquake 2011 | 531 |
| Korea's Land Use and Natural Resources | 533 |
| Mongolia Physical | 535 |
| Trade Routes of the Silk Roads | 545 |
| Zheng He's Expeditions | 546 |
| The Long March | 549 |
| Korea's Early Kingdoms, A.D. 500s | 555 |
| The Korean War | 557 |
| Spread of Buddhism | 559 |
| East Asia High-Speed Rail Routes | 571 |
| Three Gorges Dam | 579 |
| Taiwan Physical | 581 |
| Languages of East Asia | 588 |

## UNIT 11

| | |
|---|---|
| Southeast Asia Political | 597 |
| Southeast Asia Physical | 598 |
| Rivers and Population Density | 601 |
| Malay Peninsula Physical, Climate | 603 |
| Tectonic Plates and Volcanoes | 605 |
| Southeast Asia Population Density | 605 |
| Southeast Asia Under Colonial Rule, c. 1895 | 611 |
| Malaysia Physical | 616 |
| Khmer Empire | 617 |
| Philippines Political | 637 |

## UNIT 12

| | |
|---|---|
| Australia, the Pacific Realm, and Antarctica Political | 645 |
| Australia, the Pacific Realm, and Antarctica Physical | 646 |
| Australia's Precipitation | 647 |
| European Rabbits in Australia, 1870–2000 | 651 |
| Pacific Islands | 659 |
| Australia Physical | 664 |
| Australia's Population Density | 669 |
| Antarctica's Human Footprint | 675 |
| APEC Member Countries | 677 |
| Geothermal Energy | 683 |
| New Zealand Physical | 684 |
| Grassland Use Around the World | 687 |

# NATIONAL GEOGRAPHIC
# ATLAS

| | |
|---|---|
| World Physical .................................... A2 | South Asia Physical ............................. A30 |
| World Political .................................... A4 | South Asia Political ............................. A31 |
| Endangered Species ........................... A6 | East Asia Physical ............................... A32 |
| North America Physical ..................... A8 | East Asia Political ............................... A33 |
| North America Political ..................... A9 | Southeast Asia Physical ..................... A34 |
| United States Political ....................... A10 | Southeast Asia Political ..................... A35 |
| Central America & Caribbean Physical .... A12 | Australia, the Pacific Realm & Antarctica Physical ..................... A36 |
| Central America & Caribbean Political .... A14 | |
| South America Physical ..................... A16 | Australia, the Pacific Realm & Antarctica Political ..................... A38 |
| South America Political ..................... A17 | |
| Europe Physical .................................. A18 | |
| Europe Political .................................. A20 | |
| Russia & Eurasian Republics Physical .... A22 | |
| Russia & Eurasian Republics Political .... A24 | |
| Africa Physical ................................... A26 | |
| Africa Political ................................... A27 | |
| Southwest Asia Physical .................... A28 | |
| Southwest Asia Political .................... A29 | |

# World Physical

# World Political

# NATIONAL GEOGRAPHIC
# Endangered Species

**CALIFORNIA CONDOR**
Condors are threatened by pesticides and habitat loss. A captive breeding program has raised numbers in the wild from 22 in 1987 to more than 160 today.

## It's About Habitat

Wild animals need space to live. That means room to roam, room to find food, and room to hide. Quality habitat also means clean water and clean air. However, as the human population grows, the places where people live tend to spread into quality habitat. As the habitat is lost the number of animals decreases. If too much of a species' habitat is lost, the animal is at risk of becoming extinct. The International Union for Conservation of Nature (IUCN) works to save habitat. By identifying the level of endangerment with simple categories, the IUCN helps target the habitats that need to be conserved and the species to be protected.

**CONSERVATION STATUS**
- Critically Endangered
- Endangered
- Vulnerable

- Asian elephant
- Black rhino
- Blue whale
- California condor
- Giant armadillo
- Giant panda
- Hawksbill turtle
- Mountain gorilla
- Polar bear
- Snow leopard
- Tiger
- Whooping crane

**BLUE WHALE**
Commercial whaling is no longer a danger, but climate change threatens the blue whale's food source, krill. Fewer than 5,000 individuals exist in the wild.

**HAWKSBILL TURTLE**
The Hawksbill's habitat is severely threatened, but a thriving trade in turtle products poses a greater danger.

**POLAR BEAR**
Climate change is the polar bear's biggest threat. They hunt on Arctic and sub-Arctic ice flows. As the ice melts, they lose access to their main food source, seals.

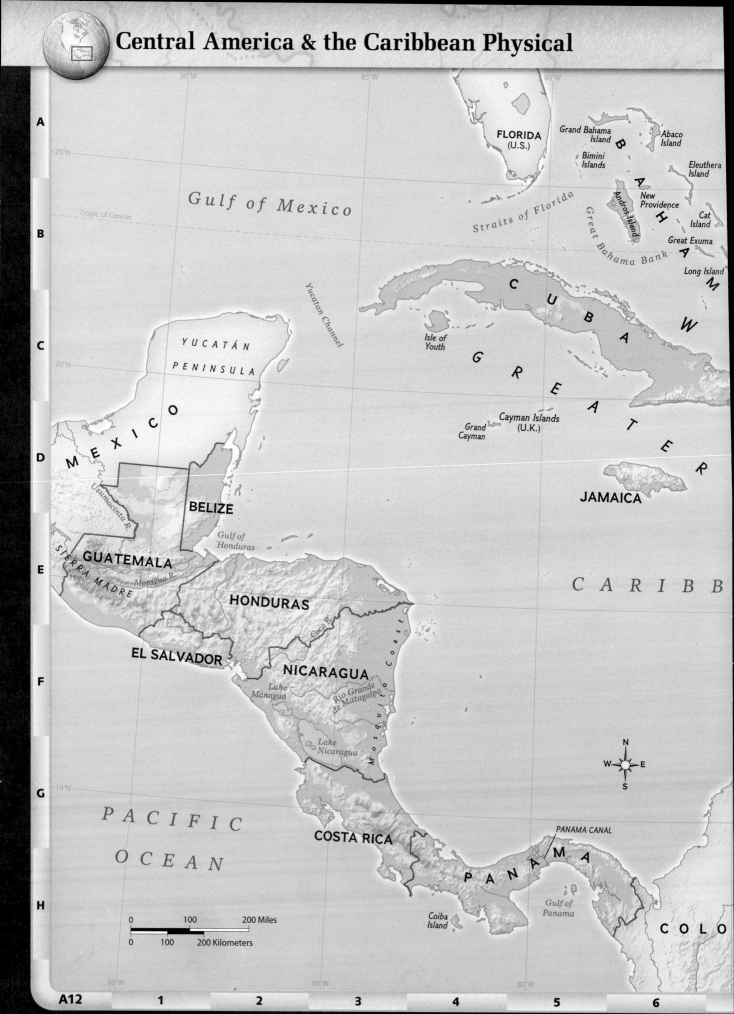

# NATIONAL GEOGRAPHIC

## West Indies

**ATLANTIC OCEAN**

- San Salvador
- Rum Cay
- Crooked Island
- Mayaguana Island
- Acklins Island
- Turks & Caicos Islands (U.K.)
- Caicos Islands
- Turks Islands
- Great Inagua Island
- Windward Passage
- Île de la Tortue
- Île de la Gonâve
- HISPANIOLA
- HAITI
- DOMINICAN REPUBLIC
- Mona Passage
- Isla Mona (Puerto Rico)
- Puerto Rico (U.S.)
- Vieques (Puerto Rico)
- Virgin Islands (U.S.)
- St. Thomas
- St. Croix
- Tortola
- Virgin Islands (U.K.)
- Anegada Passage
- Anguilla (U.K.)
- St. Martin (France)
- St. Maarten (Neth.)
- St.-Barthélemy (France)
- Saba (Neth.)
- St. Eustatius (Neth.)
- St. Kitts
- Nevis
- ST. KITTS AND NEVIS
- Barbuda
- Antigua
- ANTIGUA AND BARBUDA
- Montserrat (U.K.)
- Grande-Terre
- GUADELOUPE (France)
- Basse-Terre
- Marie-Galante
- DOMINICA
- Aves (Bird Island) (Venezuela)
- Martinique (France)
- ST. LUCIA
- BARBADOS
- St. Vincent
- Bequia
- ST. VINCENT AND THE GRENADINES
- Carriacou
- GRENADA
- ARUBA (Neth.)
- CURAÇAO (Neth.)
- BONAIRE (Neth.)
- Tobago
- Trinidad
- TRINIDAD AND TOBAGO

**GREATER ANTILLES**
**LESSER ANTILLES**
**WEST INDIES**
**CARIBBEAN SEA**

- Magdalena R.
- COLOMBIA
- VENEZUELA
- Orinoco R.
- Orinoco River Delta
- GUYANA

Tropic of Cancer

### Elevation

| feet | meters |
|---|---|
| 10,000+ | 3,050+ |
| 5,000 | 1,524 |
| 2,000 | 610 |
| 1,000 | 305 |
| 500 | 152 |
| 0 | 0 |
| Below sea level | |

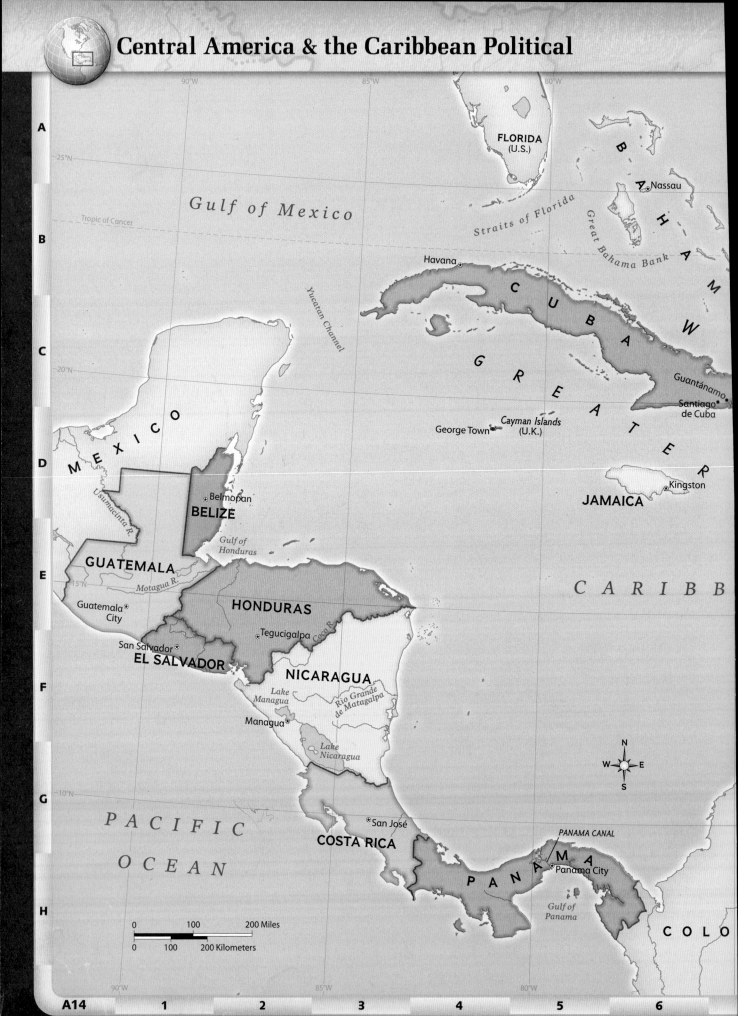

# NATIONAL GEOGRAPHIC

## ATLANTIC OCEAN

Tropic of Cancer

**Turks & Caicos Islands (U.K.)**
Cockburn Town

Windward Passage

**HISPANIOLA**
**DOMINICAN REPUBLIC**
Port-au-Prince
Santo Domingo
**HAITI**

Mona Passage

**WEST INDIES**

Charlotte Amalie
Virgin Islands (U.K.)
Anguilla (U.K.)
The Valley
San Juan
**Puerto Rico (U.S.)**
Road Town
St. Martin (France)
St. Maarten (Neth.)
St.-Barthélemy (France)
Virgin Islands (U.S.)
(Neth.) Saba
(Neth.) St. Eustatius
Basseterre
**ANTIGUA AND BARBUDA**
St. John's
**ST. KITTS AND NEVIS**
Brades
Montserrat (U.K.)
**GUADELOUPE (France)**
Basse-Terre

**GREATER ANTILLES**

**CARIBBEAN SEA**

Aves (Bird Island) (Venezuela)

**DOMINICA**
Roseau

Martinique (France)
Fort-de-France

Castries
**ST. LUCIA**

Kingstown
Bridgetown **BARBADOS**
**ST. VINCENT AND THE GRENADINES**

**LESSER ANTILLES**

**ARUBA (Neth.)**
Oranjestad
Willemstad Kralendijk
**CURAÇAO (Neth.)** **BONAIRE (Neth.)**

St. George's **GRENADA**

**TRINIDAD AND TOBAGO**
Port of Spain

Orinoco River Delta

Magdalena R.
Lake Maracaibo

**VENEZUELA**
Orinoco R.

COLOMBIA

**GUYANA**

# South America Physical

**Elevation**

| feet | meters |
|---|---|
| 10,000+ | 3,050+ |
| 5,000 | 1,524 |
| 2,000 | 610 |
| 1,000 | 305 |
| 500 | 152 |
| 0 | 0 |
| Below sea level | |

- HONDURAS
- EL SALVADOR
- NICARAGUA
- COSTA RICA
- PANAMA
- Caribbean Sea
- Lake Maracaibo
- LLANOS
- Orinoco R.
- VENEZUELA
- GUYANA
- SURINAME
- FRENCH GUIANA (France)
- GUIANA HIGHLANDS
- Magdalena R.
- COLOMBIA
- GALÁPAGOS IS. (ARCHIPIÉLAGO DE COLÓN) (Ecuador)
- ECUADOR
- Negro R.
- Amazon R.
- Equator
- Marañón R.
- ANDES
- PERU
- Ucayali R.
- Purus R.
- AMAZON BASIN
- Selvas
- Madeira R.
- Tapajós R.
- Xingu R.
- Araguaia R.
- Tocantins R.
- BRAZIL
- São Francisco R.
- Campos
- Lake Titicaca
- BOLIVIA
- MATO GROSSO PLATEAU
- BRAZILIAN HIGHLANDS
- Atacama Desert
- Pantanal
- Paraguay R.
- PARAGUAY
- Paraná R.
- Tropic of Capricorn
- Ojos del Salado 22,615 ft (6,893 m)
- Gran Chaco
- Iguazú Falls
- CHILE
- Cerro Aconcagua 22,831 ft (6,959 m)
- ARGENTINA
- PAMPAS
- Entre Ríos
- Uruguay R.
- URUGUAY
- Río de La Plata
- Colorado R.
- PATAGONIA
- ATLANTIC OCEAN
- PACIFIC OCEAN
- Valdés Peninsula
- Gulf of San Jorge
- Laguna del Carbón -344 ft (-105 m)
- Strait of Magellan
- TIERRA DEL FUEGO
- FALKLAND ISLANDS (ISLAS MALVINAS) (U.K.) Administered by the United Kingdom (claimed by Argentina)
- South Georgia (U.K.)
- Cape Horn (Cabo de Hornos)
- Scotia Sea
- ANTARCTICA
- Antarctic Circle

0–1,000 Miles / 0–1,000 Kilometers

# Russia & the Eurasian Republics Physical

# NATIONAL GEOGRAPHIC

## Map Labels

**Oceans and Seas:**
- OCEAN (Arctic)
- Chukchi Sea
- East Siberian Sea
- Laptev Sea
- Bering Sea
- Sea of Okhotsk
- Sea of Japan (East Sea)
- Yellow Sea
- PACIFIC OCEAN

**Islands and Peninsulas:**
- North Pole
- Severnaya Zemlya (North Land)
- New Siberian Islands
- Wrangel Island
- Chukchi Peninsula
- Taymyr Peninsula
- Kamchatka Peninsula
- Sakhalin

**Physical Features:**
- Lake Taymyr
- Putorana Plateau
- CENTRAL SIBERIAN PLATEAU
- SIBERIA
- Verkhoyansk Range
- Cherskiy Range
- Kolyma Range
- Koryak Range
- Central Range
- Dzhugdzhur Range
- Stanovoy Range
- Yablonovyy Range
- Eastern Sayan Mountains
- Western Sayan Mountains
- Sikhote Alin' Range
- Lake Baikal
- Lake Khanka

**Rivers:**
- Kotuy R.
- Olenek R.
- Lena R.
- Yana R.
- Indigirka R.
- Kolyma R.
- Anadyr R.
- Vilyuy R.
- Aldan R.
- Chunya R.
- Angara R.
- Yenisey R.
- Olekma R.
- Vitim R.
- Amur R.
- Zeya R.

**Countries:**
- MONGOLIA
- CHINA
- NORTH KOREA
- SOUTH KOREA
- JAPAN

## Elevation

| feet | meters |
|---|---|
| 10,000+ | 3,050+ |
| 5,000 | 1,524 |
| 2,000 | 610 |
| 1,000 | 305 |
| 500 | 152 |
| 0 | 0 |
| Below sea level | |

Scale: 0 – 200 – 400 Miles / 0 – 200 – 400 Kilometers

# Russia & the Eurasian Republics Political

# Africa Physical

# NATIONAL GEOGRAPHIC

## NORTH PACIFIC OCEAN

HAWAI'I (United States)

Johnston Atoll (U.S.)

Monday / Sunday

Palmyra Atoll (U.S.)

Howland Island (U.S.)
Baker Island (U.S.)

Equator

K I R I B A T I

Phoenix Islands

LINE ISLANDS

VALU

TOKELAU (N.Z.)

Îles Wallis (France)

SAMOA

Îles de Horne (France)

AMERICAN SAMOA (U.S.)

COOK ISLANDS (N.Z.)

Marquesas Islands

FIJI

Niue (N.Z.)

SOCIETY IS.
Tahiti

TUAMOTU ARCHIPELAGO

TONGA

AUSTRAL IS. (TUBUAI IS.)

FRENCH POLYNESIA (France)

PITCAIRN ISLANDS (U.K.)

Raoul Island (N.Z.)

Date Line

Cook Strait

Chatham Islands (N.Z.)

## SOUTH PACIFIC OCEAN

unty Islands (N.Z.)

des Islands (N.Z.)

0  250  500 Miles
0  250  500 Kilometers

A37

NATIONAL GEOGRAPHIC

HAWAI'I (United States)
Honolulu

Johnston Atoll (U.S.)

NORTH PACIFIC OCEAN

Monday / Sunday

Palmyra Atoll (U.S.)

LINE ISLANDS

Howland Island (U.S.)
Baker Island (U.S.)

Equator

KIRIBATI

TUVALU
Funafuti

TOKELAU (N.Z.)

COOK ISLANDS (N.Z.)

Marquesas Islands

Îles Wallis (France)
SAMOA
Apia
Pago Pago
AMERICAN SAMOA (U.S.)

Îles de Horne (France)

TUAMOTU ARCHIPELAGO

Suva
FIJI

Niue (N.Z.)

Papeete
Tahiti

Nuku'alofa

FRENCH POLYNESIA (France)

TONGA

PITCAIRN ISLANDS (U.K.)

Raoul Island (N.Z.)

Date Line

Wellington
Cook Strait
Christchurch

Chatham Islands (N.Z.)

SOUTH PACIFIC OCEAN

Bounty Islands (N.Z.)

Antipodes Islands (N.Z.)

0   250   500 Miles
0   250   500 Kilometers

A39

# Inspiring people to care about the planet
-National Geographic Society Mission

For more than 100 years, National Geographic Society has sparked our curiosity about the world. NGS supports a network of explorers whose work in the field is vital to the planet. Some of them are shown below. Through education and exploration, the Society works to protect the physical environment and preserve the world's cultures.

# Explorers at WORK

**Oceanographer**
NG Emerging Explorer Katy Croff Bell explores underwater archaeology.

**Research Scientist**
NG Emerging Explorer Albert Lin uses technology to study artifacts from Asian civilizations.

**Archaeologist**
NG Emerging Explorer Beverly Goodman researches the ways in which humans affect nature along coastlines.

**Conservationist**
NG Emerging Explorer Shafqat Hussain protects endangered snow leopards.

**Educator**
NG Emerging Explorer Kakenya Ntaiya works to improve girls' education in Kenya.

**Filmmaker & Anthropologist**
NG Fellow Elizabeth Kapu'uwailani Lindsey strives to preserve the Polynesian culture through documentary film.

Across the world, archaeologists, anthropologists, oceanographers, and linguists represent National Geographic Society in their work. Don't let their long titles confuse you—they are all scientists doing exciting work in the field. New information about Earth's physical features is uncovered every day.

**These are some of the explorers at work for National Geographic Society.**

**Oceanographer**
NG Explorer-in-Residence Sylvia Earle works to preserve marine ecosystems.

**Aquatic Ecologist & Biogeochemist**
NG Emerging Explorer Katey Walter Anthony explores ways to use greenhouse gas for energy.

**Technology Innovator**
NG Emerging Explorer Ken Banks develops mobile technology to connect remote groups.

**Anthropologist**
NG Explorer-in-Residence Johan Reinhard studies the cultural practices of Andes people.

# From the classroom to the world

## Tools for EXPLO

National Geographic explorers all depend on tools for exploration. You can use the variety of tools provided with your program at myNGconnect.com to explore the world and its cultures.

my eEdition

**Interactive Map Tool**
A comprehensive online mapmaker at your fingertips

ORATION

Taj Mahal, India

**Digital Library**
GeoVideos, Explorer Video Clips, and hundreds of photographs of the world's physical geography and cultures

**Interactive Whiteboard GeoActivities**
Hands-on activities to learn more about how the world works

**Magazine Maker CD-ROM**
Tool for creating student magazines using program resources or by uploading your own photos

# From the classroom to the world

The knowledge and skills that explorers need are described in the National Geography Standards, shown here. Keep them in mind as you study *World Cultures and Geography*.

**HIGH**

**REPORT**

1. How to use maps and other geographic representations, tools, and technologies to acquire, process, and **REPORT** information from a spatial perspective

2. How to use mental maps to organize information about people, places, and environments in a spatial context

3. How to **ANALYZE** the spatial organization of people, places, and environments on Earth's surface

**ANALYZE**

4. The physical and human characteristics of places

**EXPERIENCE**

5. How people create regions to **INTERPRET** Earth's complexity

6. How culture and **EXPERIENCE** influence people's perceptions of places and regions

**INTERPRET**

7. The physical processes that shape the patterns of Earth's surface

8. The characteristics and spatial distribution of ecosystems on Earth's surface

9. The characteristics, distribution, and migration of human population on Earth's surface

# ER ORDER THINKING

**10.** The characteristics, distribution, and complexity of Earth's cultural mosaics

**11.** The **PATTERNS** and networks of economic interdependence on Earth's surface

**12.** The processes, patterns, and functions of human settlement

**13.** How the forces of cooperation and conflict among people influence the division and control of Earth's surface

**14.** How human actions **MODIFY** the physical environment

**15.** How physical systems affect human systems

**16.** The changes that occur in the meaning, use, distribution, and importance of resources

**17.** How to **APPLY** geography to interpret the past

**18.** How to **APPLY** geography to interpret the present and plan for the future

**PATTERNS**

**MODIFY**

**APPLY**

# NATIONAL GEOGRAPHIC UNIT 1
# THE ESSENTIALS OF GEOGRAPHY

## MEET THE EXPLORER

**NATIONAL GEOGRAPHIC**

Explorer-in-Residence Spencer Wells is director of the Genographic Project. He analyzes DNA samples to trace humankind's ancient migration patterns and learn where we really come from.

## INVESTIGATE GEOGRAPHY

In 2002, NASA released its spectacular "blue marble" image of Earth. It is made from satellite observations gathered over several months, and then carefully combined. The image was inspired by a photo taken by astronauts during a 1972 space mission.

## CONNECT WITH THE CULTURE

Culture is how people of a certain region live, behave, and think, but culture is not limited by geography. A busy urban area like New York City (shown here) attracts people from around the world, representing a vast variety of cultures. Those cultures can live together in a single place.

### THINK LIKE A GEOGRAPHER

Scientists work to uncover the skeleton of an ancient whale in Wadi Al Hitan, Egypt. Fossils, like those shown here, reveal clues about the earth's past.

**NATIONAL GEOGRAPHIC CHAPTER 1**

# THE GEOGRAPHER'S TOOLBOX

## PREVIEW THE CHAPTER

**Essential Question** How do geographers think about the world?

### SECTION 1 • GEOGRAPHIC THINKING

1.1 Thinking Spatially ..................................... 12
1.2 Themes and Elements ............................ 14
1.3 World Regions ........................................... 16

**KEY VOCABULARY**
- spatial thinking
- geographic pattern
- Geographic Information System (GIS)
- absolute location
- Global Positioning System (GPS)
- relative location
- region
- continent
- terrace

**ACADEMIC VOCABULARY**
significant, categorize

**Essential Question** How do people use geography?

### SECTION 2 • MAPS

2.1 Elements of a Map ................................... 18
2.2 Map Scale .................................................. 20
2.3 Political and Physical Maps .................. 22
2.4 Map Projections ....................................... 24
2.5 Thematic Maps ......................................... 26
☐ NATIONAL GEOGRAPHIC Photo Gallery............ 28

**KEY VOCABULARY**
- globe
- map
- latitude
- equator
- longitude
- prime meridian
- hemisphere
- scale
- cartographer
- elevation
- relief
- projection

**ACADEMIC VOCABULARY**
distort, theme

**TERMS & NAMES**
- North Pole
- South Pole
- Northern Hemisphere
- Southern Hemisphere
- Western Hemisphere
- Eastern Hemisphere

# TECHTREK FOR THIS CHAPTER

 Student eEdition
 Maps and Graphs
 Interactive Whiteboard GeoActivities
Digital Library

Global Positioning System (GPS)

Go to myNGconnect.com for information on geographers' tools.

Scientists wear breathing aids while collecting gas samples from a spring in the Cueva (Cave) de Villa Luz in Mexico.

# SECTION 1 GEOGRAPHIC THINKING
## 1.1 Thinking Spatially

**TECHTREK**
myNGconnect.com For an online map and photos of cities on water
Maps and Graphs · Digital Library

### WESTERN HEMISPHERE

**DENVER, COLORADO**

**CHICAGO, ILLINOIS**

⌃ **Critical Viewing** Denver, Colorado, is located in the foothills of the Rocky Mountains. Chicago, Illinois, is located on the shore of Lake Michigan—one of the five Great Lakes. How might these different locations affect the activities of people who live in these cities?

**Main Idea** Geographers study the location of places and the people who live there.

Geography is about more than the names of places on a map. It involves **spatial thinking**, or thinking about the space on Earth's surface, including where places are located and why they are there.

## Ask Geographic Questions

Geographers use spatial thinking to ask questions, such as "Where is a place located? Why is this location **significant**, or important?"

Look at the photographs of Denver, Colorado, and Chicago, Illinois, above. Denver is near the Rocky Mountains, where gold was once discovered. Chicago is on Lake Michigan, which made it an important shipping center for many years.

Now find New York City on the map. It is on a protected bay. Why is New York's location significant? Like Chicago, it is near water, which is good for trade.

## Study Geographic Patterns

By asking and answering many questions, geographers can find patterns. **Geographic patterns** are similarities among places. The location of large cities near water is one example of a geographic pattern.

Many geographers use computer-based **Geographic Information Systems (GIS)**. They create maps and analyze patterns using many layers of data.

**Before You Move On**
Summarize What do geographers study? How do they study it?

### ONGOING ASSESSMENT
# MAP LAB

1. **Location** Where is Rio de Janeiro located on the map? Use directional words in your answer.
2. **Make Inferences** Look at the other cities on the map. Where are most of them located? What pattern do you notice?
3. **Summarize** What are geographic patterns, and how do geographers find them?

SECTION 1.1 **13**

# SECTION 1 GEOGRAPHIC THINKING
## 1.2 Themes and Elements

**Main Idea** Geographers use themes and elements to understand the world.

Geographers ask questions about how people, places, and environments are arranged and connected on Earth's surface. The five themes and six elements will help you categorize, or group, information.

### The Five Themes of Geography

Geographers use five themes to categorize similar geographic information.

1. **Location** provides a way of locating places. **Absolute location** is the exact point where a place is located. Geographers use a satellite system called the **Global Positioning System (GPS)** to find absolute location. **Relative location** is where a place is in relation to other places. The Great Wall is located near Beijing in northern China.

2. **Place** includes the characteristics of a location. A famous place in the western United States is the Grand Canyon. It has steep rock walls that were carved over centuries by the Colorado River.

3. **Human-Environment Interaction** explains how people affect the environment and how the environment affects people. For example, people build dams to change the flow of rivers.

4. **Movement** explains how people, ideas, and animals move from one place to another. The spread of different religions around the world is an example of movement.

5. **Region** involves a group of places that have common characteristics. North America is a region that includes the United States, Mexico, and Canada.

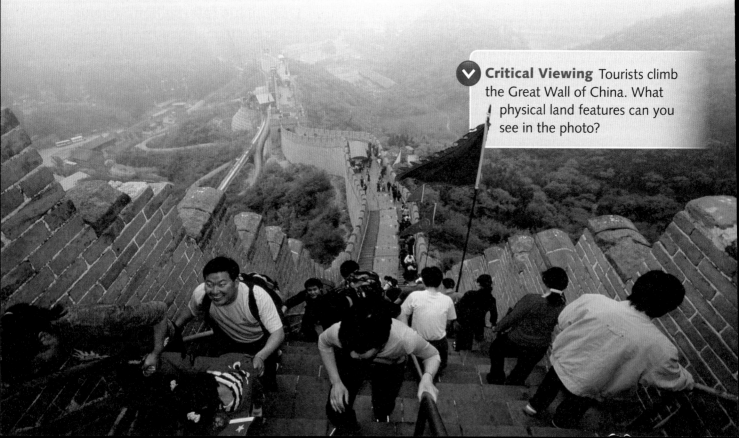

**Critical Viewing** Tourists climb the Great Wall of China. What physical land features can you see in the photo?

The Great Wall runs through the Chinese countryside.

## Six Essential Elements

Some geographers identify essential elements, or key ideas, to study physical processes and human systems.

1. **The World in Spatial Terms** Geographers use tools such as maps to study places on Earth's surface.

2. **Places and Regions** Geographers study the characteristics of places and regions.

3. **Physical Systems** Geographers examine Earth's physical processes, such as earthquakes and volcanoes.

4. **Human Systems** Geographers study how humans live and what systems they create, such as economic systems.

5. **Environment and Society** Geographers explore how humans change the environment and use resources.

6. **The Uses of Geography** Geographers interpret the past, analyze the present, and plan for the future.

### Before You Move On

**Make Inferences** How do geographers use the themes and elements to better understand the world?

### ONGOING ASSESSMENT
# WRITING LAB
*GeoJournal*

1. **Write About Geographic Themes** Write a paragraph in which you use the five themes to describe your community. Explain which theme is the most important in making your community what it is today. Go to **Student Resources** for Guided Writing support.

2. **Categorize** Which theme and which element would you use to categorize information about forms of energy? Explain your answer.

3. **Compare and Contrast** Create a chart like the one below. Use the six essential elements to compare how your town relates to another one that you know about.

| ELEMENT | MY TOWN | OTHER TOWN |
|---|---|---|
| The World in Spatial Terms | North of the highway | South of the highway |
| Places and Regions | | |
| Physical Systems | | |

SECTION 1.2 15

# SECTION 1 GEOGRAPHIC THINKING

## 1.3 World Regions

**TECHTREK**
myNGconnect.com For an online map and photos of world regions

 Maps and Graphs   Digital Library

**Main Idea** Geographers divide the world into regions. Each region is shaped by shared physical and human processes.

In 1413, a Chinese admiral and explorer, Zheng He, sailed from China to Arabia. When he arrived in Arabia, he saw people dressed in ways he had never seen. Yet, like him, these people wanted to trade.

### Regions and Continents

Zheng He saw that regions of Earth have similarities and differences. A **region** is a group of places with common traits. The places within a region are linked by trade, culture, and other human activities. They also share similar physical processes and characteristics, such as climate.

A region often includes an entire continent. A **continent** is a large landmass on Earth's surface. Geographers have identified seven continents: Africa, Asia, Australia, Europe, North America, South America, and Antarctica.

Geographers study the world's regions, but they also take a global perspective when they investigate Earth. They might, for instance, study ocean currents around the globe or how one region affects another. Both ways of looking at the world add to our understanding of it.

### Before You Move On

**Make Inferences** Why do geographers study the world by dividing it into regions?

**Visual Vocabulary** Chinese field workers view the terraced landscape. A **terrace** is a flat surface that is built into a hillside.

## WORLD REGIONS

## Regions in This Book

**A** **North America** contains the United States, Canada, and Mexico, along with the Great Lakes—the largest group of freshwater lakes.

**B** **Central America and the Caribbean** includes the islands of the Caribbean and the countries that connect North and South America.

**C** **South America** includes Brazil, a growing economic power, and the Amazon Rain Forest, the world's largest tropical rain forest.

**D** **Europe** includes approximately 50 countries and has nearly 24,000 miles of coastline.

**E** **Russia and the Eurasian Republics** includes countries that were part of the former Union of Soviet Socialist Republics (U.S.S.R.).

**F** **Southwest Asia and North Africa** spans two continents—Africa and Asia.

**G** **Sub-Saharan Africa** includes Africa south of the Sahara, the world's largest desert.

**H** **South Asia** includes India, one of the world's fastest growing countries.

**I** **East Asia** includes China, the world's most populous country.

**J** **Southeast Asia** includes Indonesia, a country made up of over 17,000 islands.

**K** **Australia, the Pacific Realm, and Antarctica** includes the Pacific island nations north and east of Australia and New Zealand.

### ONGOING ASSESSMENT
## MAP LAB  GeoJournal

1. **Interpret Maps** Which regions span more than one continent?

2. **Pose and Answer Questions** Find your region on the map. Write one geographic question about the region and answer the question.

3. **Region** Identify a physical feature in the region in which you live that makes it different from other regions. How does this feature affect you, or those who live near it?

SECTION 2 MAPS

# 2.1 Elements of a Map

**TECHTREK**
myNGconnect.com For online maps of geographic regions and photos of antique maps

 Maps and Graphs    Digital Library

**Main Idea** Globes and maps are two different tools used to study places on Earth.

**A GERMANY'S ECONOMIC ACTIVITY**

Have you ever needed to figure out how to get to a friend's house? Imagine that the only resource you had was a globe. In order to see enough detail to find your friend's house, the globe would have to be enormous—much too big to carry around in your pocket!

## Globes and Maps

A three-dimensional, or spherical, representation of Earth is called a **globe**. It is useful when you need to see Earth as a whole, but it is not helpful if you need to see a small section of Earth.

Now imagine taking a part of the globe and flattening it out. This two-dimensional, or flat, representation of Earth is called a **map**. Maps and globes are different representations of Earth, but they have similar features.

## Map and Globe Elements

**Ⓐ** A **title** tells the subject of the map or globe.

**Ⓑ Symbols** represent information such as natural resources and economic activities.

**Ⓒ Labels** are the names of places, such as cities, countries, rivers, and mountains.

**Ⓓ Colors** represent different kinds of information. For example, the color blue usually represents water.

**Ⓔ Lines of latitude** are imaginary horizontal lines that measure the distance north or south of the equator.

**Ⓕ Lines of longitude** are imaginary vertical lines that measure the distance east or west of the prime meridian.

**Ⓖ** A **scale** shows how much distance on Earth is represented by distance on the map or globe. For example, a half inch on the map above represents 100 miles on Earth.

**Ⓗ** A **legend**, or key, explains what the symbols and colors on the map or globe represent.

**Ⓘ** A **compass rose** shows the directions north (N), south (S), east (E), and west (W).

**Ⓙ** A **locator globe** shows the specific area of the world that is shown on a map. The locator globe on the map above shows where Germany is located.

## Latitude

Lines of **latitude** are imaginary lines that run east to west, parallel to the equator. The **equator** is the center line of latitude. Distances north and south of the equator are measured in degrees (°). There are 90 degrees north of the equator and 90 degrees south. The equator is 0°. The latitude of Berlin, Germany, is 52° N, meaning that it is 52 degrees north of the equator.

## Longitude

Lines of **longitude** are imaginary lines that run north to south from the **North Pole** to the **South Pole**. They measure distance east or west of the prime meridian. The **prime meridian** runs through Greenwich, England. It is 0°. There are 180 degrees east of the prime meridian and 180 degrees west. The longitude of Berlin, Germany, is 13° E, meaning that it is 13 degrees east of the prime meridian.

Remember that absolute location is the exact point where a place is located. This point includes a place's latitude and longitude. For example, the absolute location of Berlin, Germany, is 52° N, 13° E. You say this aloud as "fifty-two degrees North, thirteen degrees East."

## Hemispheres

A **hemisphere** is half of Earth. The equator divides Earth into the **Northern Hemisphere** and the **Southern Hemisphere**. North America is entirely in the Northern Hemisphere. Most of South America is in the Southern Hemisphere.

The **Western Hemisphere** is west of the prime meridian. The **Eastern Hemisphere** is east of the prime meridian. South America is in the Western Hemisphere. Most of Africa is in the Eastern Hemisphere.

### Before You Move On
**Monitor Comprehension** How are maps and globes different? How is each one used?

---

**ONGOING ASSESSMENT**
# MAP LAB
*GeoJournal*

1. **Interpret Maps** What types of industry are located in Germany? What map elements did you use to find the answers?
2. **Make Inferences** What is the main industry in southern Germany? Why might this industry be located there?
3. **Location** What is the difference between lines of latitude and lines of longitude?

SECTION 2 MAPS

# 2.2 Map Scale

**Main Idea** Maps use different scales for different purposes.

On a walk through a city, such as Charlotte, North Carolina, you might use a highly-detailed map that shows only the downtown area. To drive up the Atlantic coast, however, you would use a map that covers a large area, including several states. These maps have different scales.

## Interpreting a Scale

A map's **scale** shows how much distance on Earth is shown on the map. A large-scale map covers a small area but shows many details. A small-scale map covers a large area but includes few details.

A scale is usually shown in both inches and centimeters. One inch or centimeter on the map represents a much larger distance on Earth, such as a number of miles or kilometers.

To use a map scale, mark off the length of the scale several times on the edge of a sheet of paper. Then hold the paper between two points to see how many times the scale falls between them. Add up the distance.

**LARGE-SCALE MAP** This detailed map shows only the city of Charlotte. On this map's scale, a half inch on the map represents a quarter mile on Earth.

## Purposes of a Scale

The scale of a map should be appropriate for its purpose. For example, a tourist map of Washington, D.C., should be large-scale, showing every street name, monument, and museum.

**Critical Viewing** The Atlantic Ocean rolls in along the dunes on the Outer Banks of North Carolina. Which map labels the location of the Outer Banks?

## NORTH CAROLINA

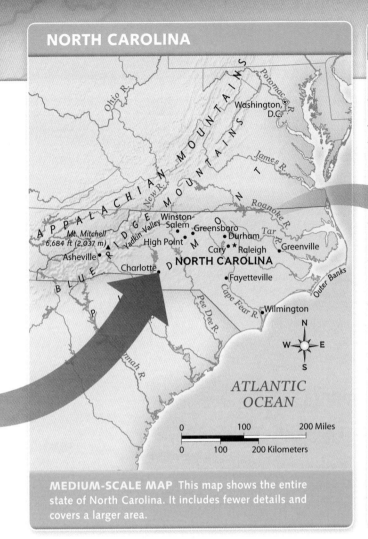

**MEDIUM-SCALE MAP** This map shows the entire state of North Carolina. It includes fewer details and covers a larger area.

## EASTERN UNITED STATES

**SMALL-SCALE MAP** This map shows the Atlantic coast states, from Florida to Maine. It covers a large area and shows even fewer details than a medium-scale map.

Maps of any scale show geographic patterns. The map of Washington, D.C., for instance, would show that many government buildings are in one area.

### Before You Move On
**Summarize** What are the purposes of a small-scale map and a large-scale map?

### ONGOING ASSESSMENT
# MAP LAB

1. **Interpret Maps** On the map of North Carolina, approximately how many inches represent 200 miles? On the map of the eastern United States, how many miles does one inch represent?

2. **Location** How far is Washington, D.C., from the southwestern border of Maine? Which map did you use to find the distance? Why?

3. **Synthesize** What geographic pattern do you see in the location of cities in North Carolina? How might you explain this pattern?

# SECTION 2 MAPS

## 2.3 Political and Physical Maps

**TECHTREK**
myNGconnect.com For online political and physical maps of world regions

Maps and Graphs

EAST ASIA POLITICAL

EAST ASIA PHYSICAL

**Main Idea** Political maps show features that humans have created on Earth's surface. Physical maps show natural features.

The governor of a state needs a map that shows counties and cities. A mountain climber needs a map that shows cliffs, canyons, and ice fields. **Cartographers**, or mapmakers, create different kinds of maps for these different purposes.

## Political Maps

A political map shows features that humans have created, such as countries, states, provinces, and cities. These features are labeled, and lines show boundaries, such as those between countries.

## Physical Maps

A physical map shows natural features of physical geography. It includes landforms, such as mountains, plains, valleys, and deserts. It also includes oceans, lakes, rivers, and other bodies of water.

A physical map can also show elevation and relief. **Elevation** is the height of a physical feature above sea level. **Relief** is the change in elevation from one place to another. Maps show elevation by using color. The physical map at left uses seven colors for seven ranges of elevation.

**Before You Move On**
**Monitor Comprehension** How is a political map different from a physical map?

**Critical Viewing** The Sobaek Mountains cut diagonally across South Korea. Which map best indicates the location of these mountains?

### ONGOING ASSESSMENT
### MAP LAB  GeoJournal

1. **Interpret Maps** What is the most mountainous country in East Asia? How did you find the answer?
2. **Human-Environment Interaction** Based on elevations shown on the map, what economic activity would you expect to find on the North China Plain?
3. **Draw Conclusions** What do the locations of Hong Kong, Shanghai, and Tokyo have in common? What conclusion can you draw about the location of cities around the world?

SECTION 2.3

# SECTION 2 MAPS
## 2.4 Map Projections

**Main Idea** Cartographers use various projections to show Earth's curved surface on a flat map.

The world is a sphere, but maps are flat. As a result, maps **distort**, or change, shapes, areas, distances, and directions found in the real world. To reduce distortion, mapmakers use **projections**, or ways of showing Earth's curved surface on a flat map. Five common map projections are the azimuthal, Mercator, homolosine, Robinson, and Winkel Tripel. Each projection has strengths and weaknesses—each distorts in a different way.

When cartographers make maps, they need to choose a map projection. The type of projection depends on the map's purpose. Which elements are acceptable to distort? Which are not acceptable to distort? For example, if a cartographer is creating a navigation map, it is important that directions are not distorted. It may not matter, however, if some areas or shapes are distorted.

### Before You Move On
**Make Inferences** How do cartographers decide which projection to use?

### AZIMUTHAL PROJECTION

Mapmakers create the **azimuthal projection** by projecting part of the globe onto a flat surface. The projection shows directions accurately but distorts shapes. It is often used for the polar regions.

### MERCATOR PROJECTION

This **Mercator projection** shows much of Earth accurately, but it distorts the shape and area of land near the North and South Poles. This projection shows direction accurately, so it is good for navigation maps.

## HOMOLOSINE PROJECTION

The **homolosine projection** resembles the flattened peel of an orange. It accurately shows the shape and area of landmasses by cutting up the oceans. However, it does not show distances accurately.

## ROBINSON PROJECTION

The **Robinson projection** combines the strengths of other projections. It shows the shape and area of the continents and oceans with reasonable accuracy. However, the North and South Poles are distorted.

## WINKEL TRIPEL PROJECTION

The **Winkel Tripel projection** is a modified version of the Robinson projection. It has less distortion of area and shape. This projection has been the preferred projection of the National Geographic Society since 1998.

### ONGOING ASSESSMENT
# MAP LAB

GeoJournal

1. **Compare and Contrast** Locate Greenland on the Mercator projection and on the Robinson projection. What is similar and different in the two maps? Why?
2. **Location** What does the azimuthal projection show about the relative location of Alaska and Russia?

SECTION 2 MAPS

# 2.5 Thematic Maps

**TECHTREK**
myNGconnect.com For additional examples of thematic maps

Maps and Graphs

**Main Idea** Thematic maps focus on specific topics, such as the population density or economic activity in a region or country.

Suppose you wanted to create a map showing the location of sports fields in your community. You would create a thematic map, which is a map about a specific **theme**, or topic.

## Types of Thematic Maps

Thematic maps are useful for showing a variety of geographic information, including economic activity, natural resources, and population density. Common types of thematic maps are the point symbol map, the dot density map, and the proportional symbol map.

### Before You Move On
Make Inferences Look through this textbook and identify another example of a thematic map. Why did you choose this map?

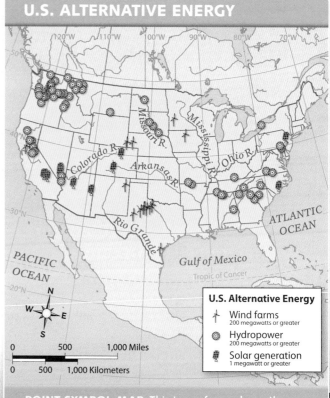

**POINT SYMBOL MAP** This type of map shows the location of activities at different points. For example, this map has symbols that show some sources of wind, water, and solar energy in the United States.

> **Critical Viewing** Solar panels in the Nevada desert absorb light from the sun and turn it into energy. Why might Nevada be a good location for solar fields?

26 CHAPTER 1

## THAILAND'S POPULATION DENSITY

**Population Density**
· One dot represents 50,000 people

**DOT DENSITY MAP** This type of map uses dots to show how something is distributed in a country or region. Each dot represents an amount. For example, the dots on this map show population density in Thailand.

## PHILIPPINES' EARTHQUAKES

Earthquake Magnitude
- 7.3–7.6
- 6.8–7.2
- 6.3–6.7
- 5.5–6.2

**PROPORTIONAL SYMBOL MAP** This type of map uses symbols of different sizes to show the size of an event. For example, the size of the circles on this map shows the severity of earthquakes in the Philippines.

### ONGOING ASSESSMENT
# MAP LAB  GeoJournal

1. **Place** According to the map, which area of the United States has the most wind farms?
2. **Make Inferences** In the Philippines, where are people most at risk for severe earthquakes? What do these places have in common?
3. **Create Sketch Maps** Create a thematic map for your neighborhood or community. Focus on the location of schools, gas stations, and grocery stores. Be sure to include a title and a legend that explains the symbols.

SECTION 2.5

NATIONAL GEOGRAPHIC
# Photo Gallery • The World At Night

For more photos from the National Geographic Photo Gallery, go to the **Digital Library** at myNGconnect.com.

Mayan artifact

Golden Gate Bridge

Neuschwanstein Castle

**Critical Viewing** This view of Earth at night was made by combining images from three satellites over a one-year period. In addition to the lighted areas, fires and natural gas burn-off are shown in orange.

Archaeologist at work

Eastern Hemisphere, 1928

Tokyo, Japan, at night

Lima, Peru

# CHAPTER 1
# Review

## VOCABULARY

For each pair of words, write one sentence that explains the connection between the two words.

1. absolute location; relative location

   *The absolute location of Washington, D.C., is 39° N, 77° W; its relative location is on the Potomac River.*

2. region; continent
3. latitude; longitude
4. relief; elevation
5. distort; projection

## MAIN IDEAS

6. What are Geographic Information Systems? How do geographers use them? (Section 1.1)
7. How are the five themes of geography and the six essential elements similar? How are they different? (Section 1.2)
8. Why is the construction of a highway an example of human-environment interaction? (Section 1.2)
9. What are traits of a region? (Section 1.3)
10. How do latitude and longitude help determine the absolute location of a place? (Section 2.1)
11. Is a map of the world a large-scale map or a small-scale map? Why? (Section 2.2)
12. How do physical maps show elevation and relief? (Section 2.3)
13. How do map projections distort Earth? (Section 2.4)
14. What type of thematic map would a cartographer use to show different types of agriculture in Africa? Explain your answer. (Section 2.5)

## GEOGRAPHIC THINKING

### ANALYZE THE ESSENTIAL QUESTION

How do geographers think about the world?

**Critical Thinking: Describe Geographic Information**

15. How do geographers use spatial thinking to make sense of space on Earth's surface?
16. How would you describe New York City using the five themes of geography?
17. What is an example of a physical process, and how can it affect a region?

### INTERPRET MAPS

**NORTH AMERICA POLITICAL**

18. **Region** According to the map, what is the relative location of Canada's cities? Why do you think they are located there?
19. **Compare and Contrast** How does the relative location of Mexico's cities compare with the relative location of Canada's cities?

## MAPS

### ANALYZE THE ESSENTIAL QUESTION
How do people use geography?

**Critical Thinking: Make Inferences**

20. On a Mercator projection map, Greenland looks larger than South America. However, it is actually much smaller. How could this affect a person's understanding of land areas?

21. A geographer wants to show where economic activities are located. What kind of thematic map should the geographer use? Explain.

22. An explorer is the first to map a region that has hills, canyons, and flat areas. What type of map should the explorer create? Explain.

### INTERPRET PRIMARY SOURCES

In 1953, Sir John Hunt led a group of climbers up Mount Everest, the highest mountain in the world. Read Hunt's description of climbing Everest and answer the questions.

> What makes Everest murderous is . . . its cold, its wind, and its climbing difficulties. . . . At 28,000 feet a given volume of air breathed contains only a third as much oxygen as at sea level. On the ground, even if a man were exercising violently, his lungs would need but 50 liters of air per minute. Near Everest's summit he struggles to suck in as much as 200 liters. Since he inhales his air cold and dry and exhales it warm and moist, the stress on his parched lungs and respiratory passages becomes appalling.
>
> –*National Geographic*, July 1954

23. **Find Details** What are the characteristics of the environment on Mount Everest?

24. **Human-Environment Interaction** How is this description an example of human-environment interaction?

## ACTIVE OPTIONS

**Synthesize** the Essential Questions by completing the activities below.

25. **Pose and Answer Geographic Questions** Choose a place in the world you would like to visit. List questions such as, "Why am I curious about it?" "What are its features?" "Where is it located?" "Why is its location significant?" Research and write your answers. Guess what places your classmates chose. Take turns sharing questions and answers. **Share facts about your place with the class.**

> **Research Tips**
> - Use print or online encyclopedias, or other reliable Web sites. Review the lesson on page R39.
> - Set a period of time for finding answers to help you avoid getting distracted.
> - Be sure to put others' writing into your own words.

**TECHTREK** myNGconnect.com For maps and photographs of the five themes of geography

26. **Create a Digital Presentation** Use software to create a presentation about what geographers do. Show how they organize and categorize what happens in both the natural world and the human world. To illustrate your presentation, select photographs from the **Digital Library** or from other sources. Find maps in the **Online World Atlas**. Use correct spelling and punctuation on your slides.

*Geographers use five themes to study places and people.*

- *Location is an example of one theme. Geographers determine location several ways.*

**NATIONAL GEOGRAPHIC** **CHAPTER 2**

# PHYSICAL & HUMAN GEOGRAPHY

## PREVIEW THE CHAPTER

**Essential Question** How is the earth continually changing?

### SECTION 1 • THE EARTH

1.1 Earth's Rotation and Revolution .............. 34
1.2 Earth's Complex Structure ...................... 36
1.3 Earth's Landforms .................................. 38
1.4 The Ring of Fire ...................................... 40
1.5 Waters of the Earth ................................ 42

**KEY VOCABULARY**
- solstice
- equinox
- tectonic plate
- continental drift
- plain
- plateau
- continental shelf
- butte
- erosion
- earthquake
- tsunami
- volcano
- evaporation
- condensation
- precipitation

**ACADEMIC VOCABULARY**
benefit, essential

**TERMS & NAMES**
- Ring of Fire

**Essential Question** What shapes the earth's varied environments?

### SECTION 2 • PHYSICAL GEOGRAPHY

2.1 Climate and Weather ............................. 44
2.2 World Climate Regions .......................... 46
2.3 Extreme Weather .................................. 48
2.4 Natural Resources ................................. 50
2.5 Habitat Preservation ............................. 52
2.6 Exploring the World's Oceans ................ 54
     NATIONAL GEOGRAPHIC Photo Gallery............ 56

**KEY VOCABULARY**
- climate
- weather
- vegetation
- hurricane
- cyclone
- tornado
- raw material
- nonrenewable resource
- renewable resource
- habitat
- ecosystem
- marine life

**ACADEMIC VOCABULARY**
restore, impact

**Essential Question** How has geography influenced cultures around the world?

### SECTION 3 • HUMAN GEOGRAPHY

3.1 World Cultures ...................................... 58
3.2 Religions and Belief Systems ................. 60
3.3 Economic Geography ............................ 62
3.4 Political Geography ............................... 64
     **GLOBAL ISSUES: HUMAN RIGHTS**
3.5 Protecting Human Rights ...................... 66

**KEY VOCABULARY**
- culture
- civilization
- communal
- gaucho
- culture region
- kimono
- monotheistic religion
- polytheistic religion
- economy
- capital
- entrepreneurship
- free enterprise economy
- gross domestic product (GDP)
- government
- citizen
- democracy
- human rights

**ACADEMIC VOCABULARY**
symbol

**TERMS & NAMES**
- United Nations (UN)
- Universal Declaration of Human Rights

# TECHTREK
**FOR THIS CHAPTER**

Student eEdition

Maps and Graphs

Interactive Whiteboard GeoActivities

Digital Library

Go to **myNGconnect.com** for more on physical and human geography.

A special suit for collecting lava samples

Scientists observe a cloud of ash erupting from Italy's Mount Etna.

# SECTION 1 THE EARTH

## 1.1 Earth's Rotation and Revolution

**Main Idea** Earth's tilt, rotation, and revolution cause weather changes and the four seasons.

In the summer, people who live in the northern parts of Iceland, Norway, Sweden, and Finland have more than 20 hours of daylight. In the winter, these same people have more than 20 hours of darkness. These long days and nights result from Earth's tilt and revolution around the sun.

### Revolution and Rotation

The solar system is formed by the sun, Earth, and seven other planets. Earth, which is the third planet from the sun, revolves around the sun at a speed of about 67,000 miles per hour. It takes one year for Earth to make one revolution.

At the same time, Earth rotates on its axis, an imaginary line that runs from the North Pole to the South Pole through the center. Each rotation takes almost one day.

> **Critical Viewing** Bathers in Iceland celebrate the summer solstice with a midnight swim in a hot spring. Based on the photo, what can you tell about the summer solstice in Iceland?

Earth tilts at an angle of about 23.5°. Because of this tilt, the Northern Hemisphere receives more direct sunlight for half the year, and temperatures are warmer. During these months, the Southern Hemisphere receives less direct sunlight, and temperatures are cooler.

As Earth continues around the sun, the Northern Hemisphere faces the sun less directly and temperatures are cooler. Meanwhile the Southern Hemisphere faces the sun more directly, and temperatures are warmer. This process creates the four seasons in both hemispheres.

### Summer and Winter Solstices

The exact moment at which summer and winter start is called a **solstice**. June 20 or 21 is the summer solstice in the Northern Hemisphere. It is the longest day of the year. Six months later, on December 21 or 22, the Northern Hemisphere has its winter solstice. This is the shortest day of the year.

The Southern Hemisphere is exactly the opposite. June 20 or 21 is its winter solstice and December 21 or 22 is its summer solstice.

## EARTH'S FOUR SEASONS: NORTHERN HEMISPHERE

**SPRING EQUINOX** (March 21)

**WINTER SOLSTICE** (December 21 or 22)

SUN

**SUMMER SOLSTICE** (June 20 or 21)

365 days

24 hours

North Pole
Northern Hemisphere
Southern Hemisphere
South Pole

**AUTUMN EQUINOX** (September 23)

## Spring and Autumn Equinoxes

The beginning of spring and autumn is called an **equinox**. Twice a year, the sun's rays hit the equator directly, and day and night are the same length. In the Northern Hemisphere, the spring equinox occurs around March 21, and the autumn equinox occurs around September 23. The Southern Hemisphere is exactly the opposite.

### Before You Move On
**Monitor Comprehension** How do Earth's tilt, rotation, and revolution cause the seasons?

### ONGOING ASSESSMENT
## VIEWING LAB  GeoJournal

1. **Analyze Models** According to the model above, when do the sun's rays hit the Southern Hemisphere most directly? This is the beginning of which season?

2. **Analyze Visuals** What happens to the sun in Iceland on the day of the summer solstice? Why does this happen?

3. **Compare and Contrast** How are the spring equinox and the autumn equinox alike?

4. **Make Inferences** What happens to the length of days in the Northern Hemisphere after the spring equinox?

# SECTION 1 THE EARTH

## 1.2 Earth's Complex Structure

**TECHTREK**
myNGconnect.com For photos and diagrams that illustrate plate tectonics

Digital Library | Student Resources

**Main Idea** Physical processes within Earth bring about changes on the surface.

If you could dig a tunnel to Earth's center, you would travel through several layers. Each layer would be under tremendous pressure and give off intense heat.

### Earth's Layers

On your journey, you would first pass through the crust. This layer includes landmasses and the ocean floor. It is about 30 miles thick.

Next, you would come to the mantle, which consists of molten, or melted, rocks called magma. The mantle is about 1,800 miles thick and has two parts—the upper mantle and the lower mantle.

Descending even deeper, you would find yourself in the outer core, which is about 1,400 miles thick. This layer is mostly liquid, consisting of molten iron and nickel.

At the very center is the inner core. It is about 700 miles thick. It reaches a temperature of 12,000° F—hotter than the surface of the sun. The inner core is made up of iron, which remains solid because the pressure from all the layers above it is so intense.

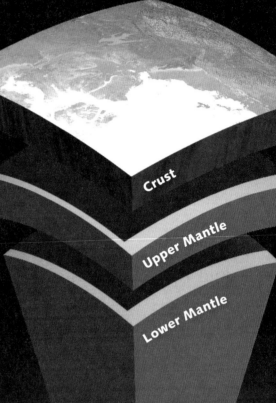

**EARTH'S STRUCTURE**

Crust
Upper Mantle
Lower Mantle
Outer Core
Inner Core

36 CHAPTER 2

## TECTONIC PLATE MOVEMENTS

**1 SUBDUCTION** A dense oceanic plate dives under a less dense continental plate. Volcanoes are often formed by this movement.

**2 DIVERGENCE** As the oceanic plates spread apart, magma pushes up. It creates mountains and an expanding ocean floor.

**3 CONVERGENCE** Two continental plates collide and are pushed upward, causing the formation of steep mountain ranges.

**4 TRANSFORM** Two plates slide past each other, causing a crack, or fault, in the crust. The energy released often causes an earthquake.

## Tectonic Plates

Earth's crust is divided into sections called tectonic plates. The plates float on Earth's mantle. They are constantly shifting and may move up to four inches a year.

The seven continents rest on these tectonic plates. As the plates have shifted over time, the continents have moved into their current positions. This slow movement of the continents is known as continental drift.

Tectonic plates move in four ways, as shown in the diagram above. The enormous force of the movements and collisions creates mountains and causes earthquakes and volcanoes.

**Before You Move On**
Make Inferences How do the movements of tectonic plates change Earth's surface?

### ONGOING ASSESSMENT
## VIEWING LAB  GeoJournal

1. **Analyze Visuals** According to the diagram above, which tectonic plate movement often results in volcanoes? Which movement can cause earthquakes?
2. **Place** The Himalaya Mountains formed by convergence when the Indian plate collided with the Eurasian plate. The Indian plate is still moving almost an inch north every year. How do you predict this will affect the Himalayas?
3. **Summarize** What is the main characteristic of each layer of Earth?

# SECTION 1 THE EARTH

## 1.3 Earth's Landforms

**TECHTREK**

myNGconnect.com For photos and a diagram of landforms

 Digital Library

 Student Resources

> **Main Idea** Landforms are physical features on Earth's surface. They are continually reshaped by physical processes.

The Rocky Mountains rise more than 14,000 feet above sea level. The Grand Canyon is more than 5,000 feet deep. Both are landforms, or physical features on Earth's surface.

### Surface Landforms

Landforms such as the Rocky Mountains in western North America and the Grand Canyon in Arizona provide a variety of physical environments. These environments support millions of plants and animals.

Several common landforms are found on Earth's surface. A mountain is a high, steep elevation. A hill also slopes upward but is less steep and rugged. In contrast, a **plain** is a level area. The Great Plains, for example, are flat landforms stretching from the Mississippi River to the Rocky Mountains. A **plateau** is a plain that sits high above sea level and usually has a cliff on all sides. A valley is a low-lying area that is surrounded by mountains.

### Ocean Landforms

Earth's oceans also have landforms that are underwater. Mountains and valleys rise and fall along the ocean floor. Volcanoes erupt with hot magma, which hardens as it cools to form new crust.

The edge of a continent often extends out under the water. This land is called the **continental shelf**. Most of Earth's marine life lives at this level of the ocean. Beyond the continental shelf, the land develops a steep slope. Beyond the slope, before the ocean floor, the land slopes slightly upward. This landform is called the continental rise. It is formed by rocks and sediment carried by ocean currents. Together, these landforms are known as the continental margin.

> **Visual Vocabulary** A **butte** (BYOOT) is a hill or mountain with steep sides and a flat top. These buttes in Monument Valley, Arizona, are called "the Mittens."

### THE CONTINENTAL MARGIN

## The Changing Earth

Earth is always changing, and the changes affect plant and animal life. For example, a flood can cause severe erosion, which can ruin farmers' fields. **Erosion** is the process by which rocks and soil slowly break apart and are swept away.

Erosion also results from weathering, which is when air, water, wind, or ice slowly wear away rocks and soil. The buttes in Monument Valley, Arizona, were formed in this way over a span of millions of years.

### Before You Move On
**Summarize** How do physical processes reshape Earth's landforms?

### ONGOING ASSESSMENT
## SPEAKING LAB

**Turn and Talk** How does a landform in your community affect your daily life? Make a list of landforms you encounter every day, such as hills or plains. Then turn to a partner and talk about how one of these landforms affects your life. Develop an outline to present to the class. List your topic first. Then list your supporting details.

# SECTION 1 THE EARTH
## 1.4 The Ring of Fire

**Main Idea** Plate boundaries around the Pacific Ocean cause earthquakes and volcanic eruptions.

The **Ring of Fire** is a circle of volcanoes and earthquakes along the rim, or outer edge, of the Pacific Ocean. It exists because a large tectonic plate under the ocean slides against plates in Asia, Australia, South America, and North America. The movements create tremendous pressure, which causes volcanoes and earthquakes.

### Earthquakes

An **earthquake** is a violent shaking of Earth's crust. Many earthquakes occur along faults, which are cracks in Earth's surface. Earthquakes are common in the Ring of Fire, but they also occur in other areas on Earth. One area runs from the land around the Mediterranean Sea through East Asia. Other earthquake zones include the middle of the Arctic Ocean and the Atlantic Ocean.

On May 18, 1980, Mount St. Helens erupted, blasting away one side of the mountain.

Mount St. Helens, Washington

Earthquakes can cause buildings, bridges, and roads to collapse. For example, in 2010, an earthquake in Haiti killed more than 200,000 people. Many people who died were trapped under buildings that collapsed.

Earthquakes beneath the ocean can cause **tsunamis**, which are large, powerful ocean waves that can cause great destruction along the coast.

## Volcanoes

The Ring of Fire contains more than 75 percent of the world's volcanoes. A **volcano** is a mountain that erupts in an explosion of molten rock, gases, and ash. Lava, which is molten rock, flows down the side of the mountain.

Volcanoes can cause severe damage. In 1883, Krakatoa in Indonesia spewed ash and rock fragments over an area of 300,000 square miles. It also triggered a tsunami that killed 36,000 people. Yet volcanoes can also **benefit**, or be useful to, plant and animal life. For example, mineral-rich lava turns into fertile soil.

Scientists have learned how to predict volcanic eruptions, and engineers can design buildings that survive earthquakes. As a result, more people can live safely.

### Before You Move On
**Monitor Comprehension** What is the Ring of Fire and why is it significant?

### ONGOING ASSESSMENT
# WRITING LAB  GeoJournal

1. **Make Inferences** Why do so many buildings collapse during an earthquake?

2. **Human-Environment Interaction** How do earthquakes and volcanoes affect people? How have people tried to solve these problems? Copy and complete the chart.

| DISASTER | PROBLEM | SOLUTION |
|---|---|---|
| earthquake | | |
| volcano | | |

3. **Write an Action Plan** Imagine that you live along the Ring of Fire. With a partner, write an outline of an action plan to help people survive a serious earthquake.

# SECTION 1 THE EARTH
## 1.5 Waters of the Earth

**TECHTREK**
myNGconnect.com For photos of water and a diagram of the hydrologic cycle
Digital Library | Student Resources

**Main Idea** Water is essential for all forms of life on Earth.

The Mississippi River begins as a small stream in northern Minnesota. More than 2,000 miles south, it pours more than 4.7 million gallons of water per second into the Gulf of Mexico. Water flowing in rivers like the Mississippi is **essential**, or necessary, for all forms of life.

### Fresh Water

The Mississippi River contains fresh water. People use fresh water to drink, cook, bathe, and irrigate crops. Early civilizations developed along rivers such as the Nile River in Egypt because of the available fresh water.

Different bodies of fresh water exist for different geographic reasons. A river is a path of water that flows from a higher elevation to a lower elevation. Streams, brooks, and creeks are like rivers, but smaller. A lake is a large body of water that is surrounded by land.

### Salt Water

Salt water contains salt and other minerals. It is a major source of the world's seafood supply and a means of transportation. Oceans are large bodies of salt water. Earth's four oceans are the Atlantic, the Pacific, the Indian, and the Arctic. Continuously moving flows of water, called currents, circulate through the oceans and affect climates on land.

Seas are smaller bodies of salt water. The Red Sea, for example, lies between the Arabian Peninsula and eastern Africa.

### THE HYDROLOGIC CYCLE

The hydrologic cycle is the continual movement of water from Earth's surface into the air and back again.

**2 CONDENSATION**
During **condensation**, cooler temperatures in the atmosphere cause the water vapor to change into droplets that form clouds.

**1 EVAPORATION**
During **evaporation**, the sun heats the ocean, and water vapor rises up into the atmosphere.

### Before You Move On
**Monitor Comprehension** How is water essential for all life on Earth?

42 CHAPTER 2

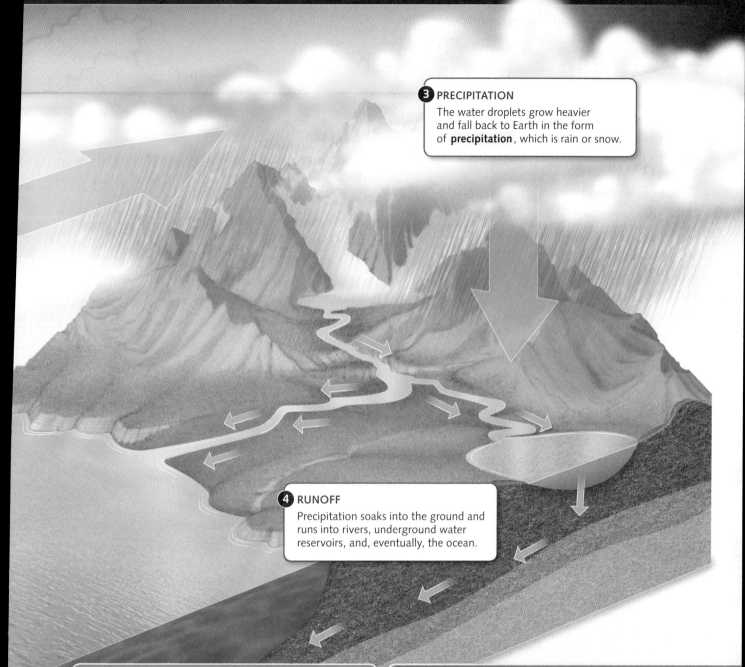

**3 PRECIPITATION**
The water droplets grow heavier and fall back to Earth in the form of **precipitation**, which is rain or snow.

**4 RUNOFF**
Precipitation soaks into the ground and runs into rivers, underground water reservoirs, and, eventually, the ocean.

## WORLD'S LONGEST RIVERS

| River | Location | Length (miles) |
|---|---|---|
| Nile | Africa | 4,241 |
| Amazon | South America | 4,000 |
| Chang Jiang (Yangtze) | Asia | 3,964 |
| Mississippi-Missouri | North America | 3,710 |
| Yenisey-Angara | Asia | 3,440 |

**Source:** *National Geographic Atlas of the World*, 8th ed.

## ONGOING ASSESSMENT
# DATA LAB

1. **Interpret Charts** According to the chart, which continent has two of the longest world rivers, and what are they? How do you think the two rivers have affected that continent?

2. **Interpret Models** How does the hydrologic cycle explain why rivers and lakes do not run out of water?

3. **Location** St. Louis, Missouri, is located just south of where the Missouri River flows into the Mississippi River. Why is this a good location for a major city?

SECTION 1.5

# SECTION 2 PHYSICAL GEOGRAPHY
## 2.1 Climate and Weather

**Main Idea** Climate and weather are different, but they both influence life on Earth.

People who live in Sacramento, California, have mild winters. When they go skiing in the nearby Sierra Nevada Mountains, they wear parkas to protect themselves from the colder temperatures. They have adapted to a different climate.

### Climate Elements

**Climate** is the average condition of the atmosphere over a long period of time. It includes average temperature, average precipitation, and the amount of change from one season to another. For example, Fairbanks, Alaska, has a cold climate. In the winter, the temperature can reach -8°F. Yet the temperature can rise to 90°F in the summer. The city goes through changes from one season to another.

Four factors that affect a region's climate are latitude, elevation, prevailing winds, and ocean currents. Places at high latitudes, such as Fairbanks, experience more change between winter and summer. Places close to the equator have nearly the same temperature throughout the year. Places at higher elevations have generally colder temperatures than places closer to sea level.

Prevailing winds are winds coming from one direction that blow most of the time. In Florida in the summer, the prevailing winds come from the south, making a warm climate even hotter.

Ocean currents also affect climate. The Gulf Stream is a current that carries warm water from the Caribbean Sea toward Europe. Air passing over the water becomes warm and helps create a mild winter climate in England and Ireland.

**Critical Viewing** Skiers head to the top of Clouds Rest in Yosemite National Park, California. What does the photo suggest about the weather at this location?

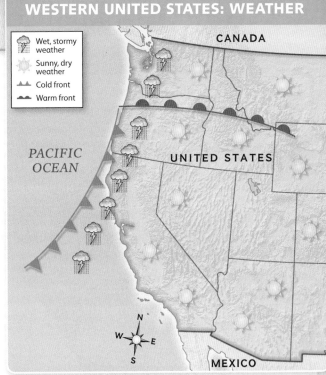

## Weather Conditions

**Weather** is the condition of the atmosphere at a particular time. It includes the temperature, precipitation, and humidity for a particular day or week. Humidity is the amount of water vapor in the air. If a weather forecaster says the humidity is at 95 percent, he or she means that the air is holding a large amount of water vapor.

Weather changes because of air masses. An air mass is a large area of air that has the same temperature and humidity. The boundary between two air masses is called a front. If a forecaster talks about a warm, humid front, he or she usually means that thunderstorms are headed toward the area.

### Before You Move On
**Monitor Comprehension** What is the difference between climate and weather?

### ONGOING ASSESSMENT
## MAP LAB
GeoJournal

1. **Interpret Maps** Compare the weather map and the climate map of the western United States. How do you think ocean currents and mountains affect the climate and the weather?
2. **Make Inferences** How might the climate and weather of the western United States influence everyday life there?
3. **Place** The chart shows average temperatures and rainfall for Los Angeles, California. Ask and answer questions about the data.

| Los Angeles, CA | January | July | November |
|---|---|---|---|
| Temperature (°F) | 57.1 | 69.3 | 61.6 |
| Rainfall (inches) | 2.98 | 0.03 | 1.13 |

Source: National Drought Mitigation Center

SECTION 2.1

# SECTION 2 PHYSICAL GEOGRAPHY
## 2.2 World Climate Regions

**TECHTREK**
myNGconnect.com For an online map and photos of world climate regions

Maps and Graphs  Digital Library

**Main Idea** Geographers identify climate regions to help them understand and categorize life on Earth.

A climate region is a group of places that have similar temperatures, precipitation levels, and changes in weather. Geographers have identified 5 climate regions that are broken down into 12 subcategories. Places that are located in the same subcategory often have similar vegetation, or plant life.

**Before You Move On**
**Make Inferences** How might climate regions help geographers analyze life in a particular place?

## WORLD CLIMATE REGIONS

**A Dry Climates** have little to no rain or snow and both hot and cold temperatures. Plant life includes shrubs and cacti.

Saguaro Cactus, Sonoran Desert, Arizona

**B Humid Temperate Climates** have cool winters, warm summers, and ample rainfall. Plant life includes mixed forests with evergreens and leafy trees.

Mixed Forest, Great Smoky Mountains, North Carolina

**C Humid Equatorial Climates** are found near the equator. They have high temperatures and rainfall all or most of the year. Plant life includes tropical plants and rain forests or grasslands with trees.

Bromeliads, Amazon rain forest, Peru

**D Tundra or Ice Climates** are north of the Arctic Circle and south of the Antarctic Circle. They have long, cold winters and short summers. Plant life includes mosses or no vegetation.

Mosses, Disko Bay, Greenland

**E Humid Cold Climates** have cold winters, warm summers, rain, and snow. Plant life includes evergreen or deciduous (leafy) forests.

Natural Park, Eastern Siberia, Russia

**TECHTREK**
Go to myNGconnect.com to explore this map with the Interactive Map Tool.

**Humid Equatorial**
- No dry season
- Short dry season
- Long dry season

**Dry**
- Semiarid
- Arid

**Humid Temperate**
- No dry season
- Dry summer
- Dry winter

**Humid Cold**
- Dry winter
- No dry season

- Tundra or ice
- Unclassified highlands

## ONGOING ASSESSMENT
### MAP LAB — GeoJournal

1. **Interpret Maps** What is the most common climate in northern Africa? How might this climate affect population?
2. **Compare and Contrast** How does the climate of western Europe differ from that of eastern Europe?
3. **Human-Environment Interaction** What advantages would humid temperate climates have for farming? For logging?

# SECTION 2 PHYSICAL GEOGRAPHY
## 2.3 Extreme Weather

**Main Idea** Extreme weather can cause great destruction, but scientists are lessening its effect.

On August 29, 2005, Hurricane Katrina raced toward New Orleans, Louisiana. The water level in the Gulf of Mexico rose 34 feet and flooded 80 percent of the city. Thousands of people lost their homes and many businesses were destroyed.

### Wild Weather

Katrina is an example of extreme weather, which is weather so powerful it can deeply affect human lives. A **hurricane** such as Katrina is a strong storm with swirling winds and heavy rainfall. Winds rotate fiercely and can reach 200 miles per hour. A hurricane is a type of **cyclone**, which is a storm with rotating winds. In the Eastern Hemisphere, a cyclone is called a typhoon.

A **tornado** is a smaller storm than a cyclone, but it has even more powerful winds that can reach 300 miles per hour. A tornado follows an unpredictable path and can rip buildings from their foundations. Tornadoes occur all over the world, but most of them form in the United States, east of the Rocky Mountains.

Other types of extreme weather are not as dangerous as cyclones or tornadoes, but they still put people at risk. A flood occurs when water covers an area of land that is usually dry. Floods often occur after a cyclone. A blizzard is a heavy snowstorm with strong winds and very cold temperatures. A drought results when the amount of rainfall drops far below the average amount. It is sometimes accompanied by a heat wave, or unusually high temperatures over a period of time.

### HOW A TORNADO FORMS

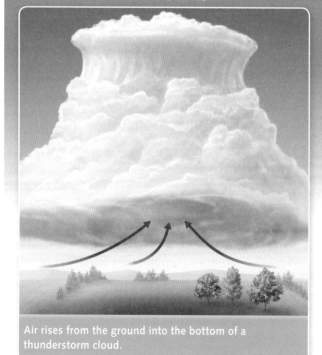
Air rises from the ground into the bottom of a thunderstorm cloud.

The air begins to rotate and extends to the ground in a funnel shape.

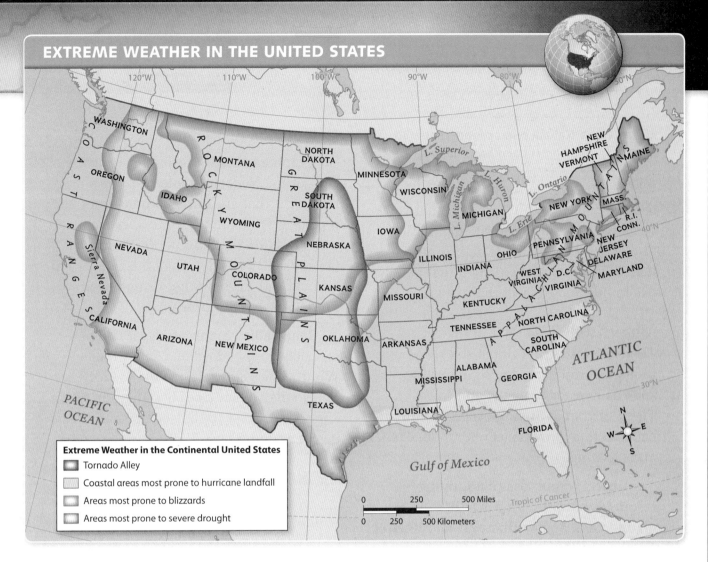

**EXTREME WEATHER IN THE UNITED STATES**

Extreme Weather in the Continental United States
- Tornado Alley
- Coastal areas most prone to hurricane landfall
- Areas most prone to blizzards
- Areas most prone to severe drought

## Scientific Solutions

Scientists are working to lessen the effects of extreme weather on humans. For example, they can often predict the path of a hurricane. They have also worked with engineers to design levees, or walls to hold back floodwaters. Many residents of New Orleans believe that better levees might have limited the damage caused by flooding after Hurricane Katrina.

The ability to predict tornadoes has also improved. Today, the National Weather Service uses radar and satellite imagery, as well as a network of "spotters," to track big storms. Today's instant communications technology makes it possible to broadcast warnings before a storm strikes.

### Before You Move On
**Summarize** How are scientists helping to lessen the impact of extreme weather?

### ONGOING ASSESSMENT
## VIEWING LAB  GeoJournal

1. **Analyze Visuals** What happens in each stage of tornado formation?
2. **Interpret Maps** According to the map, which state is at risk for all four types of extreme weather?
3. **Draw Conclusions** Even though scientists can predict when and where tornadoes may occur, why do these storms still catch people by surprise?
4. **Human-Environment Interaction** How might a severe drought affect people who live in the Great Plains area of the United States?

SECTION 2.3  **49**

## SECTION 2 PHYSICAL GEOGRAPHY
# 2.4 Natural Resources

**Main Idea** Natural resources are central to economic development and basic human needs.

What materials make up a pencil? Wood comes from trees. The material that you write with is a mineral called graphite. The pencil is made from natural resources, which are materials on Earth that people use to live and to meet their needs.

### Earth's Resources

There are two kinds of natural resources. Biological resources are living things, such as livestock, plants, and trees. These resources are important to humans because they provide us with food, shelter, and clothing.

Mineral resources are nonliving resources buried within Earth, such as oil and coal. Some mineral resources are **raw materials**, or materials used to make products. Iron ore, for example, is a raw material used in making steel. The steel, in turn, is used to make skyscrapers and automobiles.

### Categories of Resources

Geographers classify resources in two categories. **Nonrenewable resources** are resources that are limited and cannot be replaced. For example, oil comes from wells that are drilled into Earth's crust. Once a well runs dry, the oil is gone. Coal and natural gas are other examples of nonrenewable resources.

**Renewable resources** never run out, or a new supply develops over time. Wind, water, and solar power are all renewable. So are trees because a new supply can grow to replace those that have been cut down.

> **Critical Viewing** Pumpjacks pump oil at a field in California. Based on the photo, how does this action affect the land?

50 CHAPTER 2

## SELECTED NATURAL RESOURCES OF THE WORLD

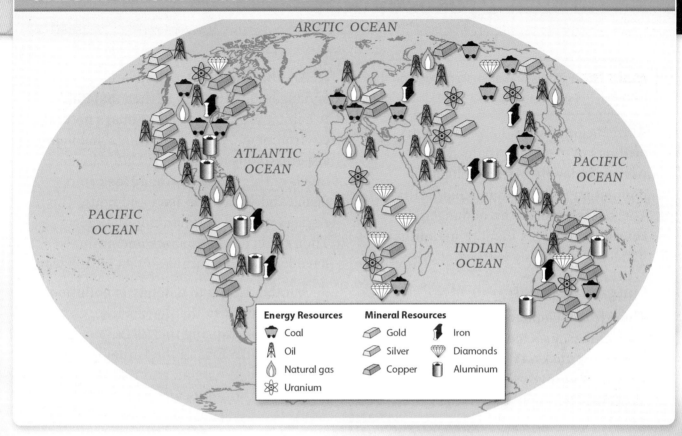

Natural resources are an important part of everyday life, yet countries with a large supply are not always wealthy. Nigeria, for example, is a major supplier of oil, but seven out of every ten Nigerians live in poverty. Japan is one of the wealthiest countries in the world—yet it must import oil from other countries.

### Before You Move On

**Monitor Comprehension** Why are natural resources important?

### ONGOING ASSESSMENT
# MAP LAB

1. **Interpret Maps** Copper is in demand for electric wiring and other uses. What part of the world do you think benefits from the demand for copper? Why?

2. **Location** Where are supplies of oil found in the world? How do these supplies impact people living in these regions?

3. **Describe Geographic Information** What are examples of a biological and a mineral resource? How are these examples different?

SECTION 2.4  51

SECTION 2 PHYSICAL GEOGRAPHY

# 2.5 Habitat Preservation

myNGconnect.com For photos of animal habitats

Digital Library

**Main Idea** Plants and animals depend on their natural habitats to survive.

At the beginning of the 20th century, millions of elephants roamed across Africa. Today the African elephant population is fewer than a half million. These elephants are an endangered species, which is a plant or an animal in danger of becoming extinct.

## Natural Habitats

The African elephant is endangered for several reasons. One is the demand for their ivory tusks. Poachers, or people who hunt animals illegally, slaughtered elephants at a rapid rate in the early 1970s.

Another reason elephants are endangered is the loss of their **habitat**. A habitat is a plant or an animal's natural environment. African elephants' habitats are grasslands and forests. Unfortunately, much of this land is being turned into farms and villages to feed and house Africa's growing human population. Thousands of other plants and animals have lost their natural habitats in this way.

Another threat to habitats is pollution, or human activity that harms the environment. During the 1960s, for example, Lake Erie in the United States was a polluted habitat, and fish nearly disappeared from its waters.

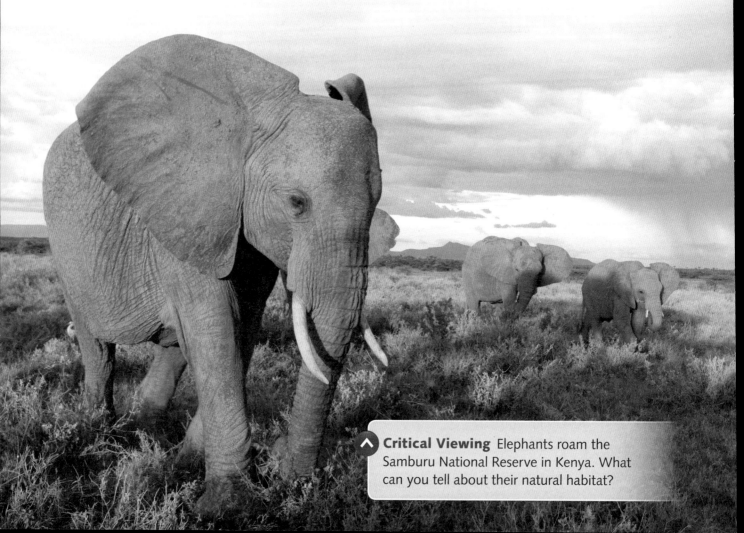

**Critical Viewing** Elephants roam the Samburu National Reserve in Kenya. What can you tell about their natural habitat?

## Habitat Loss and Restoration

The loss of habitats can destroy an entire ecosystem. An **ecosystem** is a community of plants and animals and their habitat. Earth has many different ecosystems that interact with each other. The destruction of one ecosystem affects all the others. For example, many scientists believe the destruction of rain forest habitats has led to global climate change.

People around the world have taken steps to save ecosystems and preserve natural habitats. In 1973, for example, the United States passed the Endangered Species Act, which protects the habitats of endangered species. People have also **restored**, or brought back, habitats such as forests by planting trees.

### Before You Move On
**Monitor Comprehension** How do plants and animals lose their natural habitats?

Tiny spruce trees grow in a cut forest near Olympic National Park in the state of Washington.

### ONGOING ASSESSMENT
## PHOTO LAB  GeoJournal

1. **Analyze Visuals** Based on the photo of the reserve in Kenya, why might an elephant's habitat be desirable to people for farming?

2. **Describe** Based on the photo above, what steps are taken to restore forest habitats?

3. **Make Inferences** In a forest ecosystem, wolves eat deer, and deer carry the parasite that causes Lyme disease in humans. If the wolf population declines, how might it affect the number of cases of Lyme disease?

4. **Human-Environment Interaction** What is your natural habitat? With a partner, think of resources and interactions that are part of your daily life. Write a short paragraph describing your habitat.

## 2.6 SECTION 2 PHYSICAL GEOGRAPHY
# NATIONAL GEOGRAPHIC

**TECHTREK**
myNGconnect.com For photos of oceans and an Explorer Video Clip

 Maps and Graphs   Digital Library

# Exploring the World's Oceans
### with Sylvia Earle

> **Main Idea** The oceans, a natural habitat for thousands of plant and animal species, face many challenges.

### Life in the Ocean

As a child, Sylvia Earle loved oceans. This love continued into her adult life. In over 50 years, Explorer-in-Residence Sylvia Earle has led roughly 70 diving expeditions to explore **marine life**, or the plants and animals of the ocean. During these dives, she has seen the incredible variety of ocean life—more than 30 major divisions of animals.

However, Earle has also seen the different ways people have harmed the oceans. "Taking too much wildlife out of the sea is one way," she claims. "Putting garbage, toxic chemicals, and other wastes in is another." Earle has witnessed a huge drop in the number of fish in the ocean. She has also noted pollution's destructive **impact**, or effect, on the oceans' coral reefs. These "rain forests of the sea" house one-fourth of all marine life.

myNGconnect.com
For more on Sylvia Earle in the field today

## EARTH'S "HOPE SPOTS"

⌃ **Critical Viewing** This satellite map of the world shows the location of 17 "hope spots," places that are important to the overall health of Earth's oceans. What patterns, if any, do you notice about the location of these spots?

### Mission Blue

Many people believe human activity has no effect on the vast oceans. They also do not understand the impact oceans have on all forms of life. "The ocean is the cornerstone of our life support system," says Earle. "Take away life in the ocean and we don't have a planet that works."

Earle is trying to educate the public. In 2009, she launched Mission Blue, a program that seeks to heal and protect Earth's oceans. One of the program's goals is to establish marine protected areas (MPAs) in endangered hot spots, or "hope spots" as she calls them. These spots are ocean habitats that can recover and grow if human impact is limited.

### Voice of the Ocean

Earle's efforts to save Earth's oceans have earned her many awards and honors, including the title "Hero for the Planet." She continues to work tirelessly to protect the world's oceans.

In 2010, the Gulf of Mexico was hit with the largest oil spill in U.S. history. Earle went before Congress to testify about the impact of the spill on the Gulf's natural resources. "I really come to speak for the ocean," she began.

### Before You Move On
**Summarize** According to Sylvia Earle, what challenges do our oceans face?

---

**ONGOING ASSESSMENT**
## READING LAB  GeoJournal

1. **Monitor Comprehension** What does Sylvia Earle hope to accomplish through the Mission Blue program?

2. **Analyze Cause and Effect** How has human activity affected the world's oceans? Use a chart like the one below to list some of the causes and effects of these activities.

# NATIONAL GEOGRAPHIC
# Photo Gallery • Slip Point Tidal Pool

For more photos from the National Geographic Photo Gallery, go to the **Digital Library** at myNGconnect.com.

Grand Canyon

Kathmandu, Nepal

Japanese man with laptop

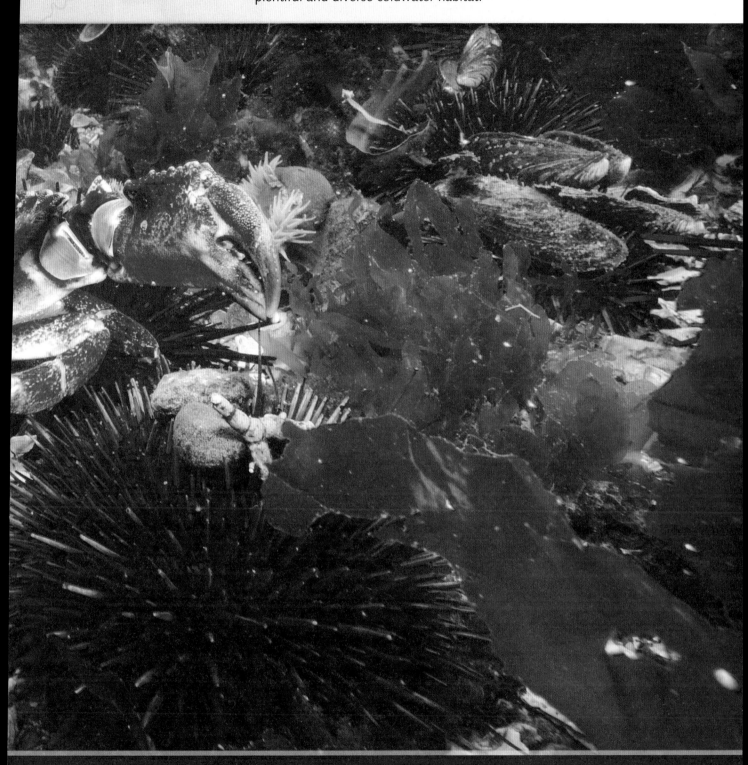

**Critical Viewing** A crab walks over sea urchins in a tidal pool at Slip Point in Clallam Bay, Washington. River outlets and tides create this plentiful and diverse coldwater habitat.

Maori carving

Polar bears

Woman and baby, Peru

Tornado

# SECTION 3 HUMAN GEOGRAPHY
## 3.1 World Cultures

**TECHTREK**
myNGconnect.com For photos of cultures around the world

Digital Library

**Main Idea** The ways people speak, eat, work, play, and worship are all part of culture.

A Japanese tea ceremony is an old tradition. The host greets the guests and prepares the tea. The guests remain silent. Once the tea is poured, though, a lively conversation follows.

### Expressions of Culture

The tea ceremony is an important part of Japan's culture. **Culture** is how people in a region live, behave, and think. Expressions of culture include language, religion, beliefs, and customs. Culture also includes the arts, such as music, dance, literature, theater, and film.

Culture is reflected in symbols that people recognize and respect. A **symbol** is an object that stands for something else. For example, the stars on the American flag are symbols for the 50 states.

### Civilization and Culture

Culture is a main trait of civilizations. A **civilization** is a society that has a highly developed culture and technology. People in a civilization are not born knowing their culture. They learn it by watching and imitating others.

A civilization's culture affects people's lives. It guides how people meet their basic needs of food, clothing, and shelter. It also influences people's values and beliefs.

**Visual Vocabulary** Senegalese children gather around a **communal**, or shared, plate of food. Communal dishes are common in African cuisine.

**Visual Vocabulary** A **gaucho** (GOW cho), or cowboy, herds sheep in South America.

## Culture Regions

Geographers study culture regions, areas that are unified by common cultural traits, or characteristics. For instance, some geographers group Mexico, Central America, the islands of the Caribbean, and South America in a culture region called Latin America. Many of the people in this region speak Spanish or Portuguese and practice the Roman Catholic religion. Many also share a history of Spanish colonization.

**Before You Move On**
**Monitor Comprehension** What are some expressions of culture?

**Visual Vocabulary** Women dressed in **kimonos**, traditional Japanese clothing, bow at a fashion show. In Japan, it is customary to bow when meeting someone.

### ONGOING ASSESSMENT
## PHOTO LAB     GeoJournal

1. **Place** Each photo and caption shows an expression of culture in a different country. How do these expressions compare to various expressions of U.S. culture?

2. **Turn and Talk** What is one activity, such as playing music or dancing, that is important in your culture? Discuss with a partner and prepare a response to share with the class.

# SECTION 3 HUMAN GEOGRAPHY

## 3.2 Religions and Belief Systems

TECHTREK
myNGconnect.com For an online map and photos of world religions
Maps and Graphs | Digital Library

**Main Idea** Religions and belief systems are important parts of cultures around the world.

Religious institutions are basic to all societies. A religion is an organized system of beliefs and practices. Thousands of religions exist in the world. Most teach that one or more gods, or supreme powers, exist. Most religions teach a set of beliefs and a moral or ethical code. Major world religions include Christianity, Hinduism, Islam, Buddhism, Judaism, and Sikhism.

### Elements of Religion

Religion is a powerful influence that helps people answer questions such as, "What is the purpose of life?" At the center of many religions is the belief in a deity. **Monotheistic religions** are those with a belief in one deity. Christianity, Islam, and Judaism are monotheistic religions. **Polytheistic religions**, such as Hinduism, have many deities.

In Judaism and Christianity, the Ten Commandments are a code of conduct telling what to do and not do in one's life.

Scriptures are sacred, or highly respected, texts that communicate the beliefs of a religion. The Bible is the sacred text for Christianity, the Koran is the sacred text for Islam, and the Torah is the sacred text for Judaism.

People celebrate holidays and observe special days to practice and honor their religious beliefs. In today's societies, Christmas, Easter, Yom Kippur, Ramadan, and Diwali are among religious celebrations.

### Origin and Spread of Religions

Several of the major world religions are based on the teachings of an individual. For example, Christianity grew from the teachings of Jesus Christ nearly 2,000 years

### BUDDHISM
**Founder** Siddhartha Gautama, the Buddha
**Followers** 400 million
**Basic Beliefs** People reach enlightenment, or wisdom, by following the Eightfold Path and understanding the Four Noble Truths.

### CHRISTIANITY
**Founder** Jesus of Nazareth **Followers** 2.3 billion
**Basic Beliefs** There is one God, and Jesus is the only Son of God. Jesus was crucified but was resurrected. Followers reach salvation by following the teachings of Jesus.

### SIKHISM
**Founder** Guru Nanak
**Followers** 27 million
**Basic Beliefs** Sikhs believe in one God. One can achieve unity with God through service to humanity, meditation, and honest labor. Vaisakhi is a significant holiday during which Sikhs celebrate their history, community, and values. It is celebrated every April.

### ISLAM
**Founder** The Prophet Muhammad
**Followers** 1.6 billion
**Basic Beliefs** There is one God. Followers must follow the Five Pillars of Islam in order to achieve salvation. The hajj is the annual pilgrimage to Mecca. One of the Five Pillars states that Muslims must make the pilgrimage at least once in their lives.

## WORLD RELIGIONS

ago. Other religions, such as Hinduism, grew from the beliefs of ancient peoples.

Religions have grown and spread around the world. For example, Buddhism began in India, but it spread to Japan, China, Korea, and Southeast Asia through migration and trade. Religions have also spread due to the work of missionaries, people who convert others to follow their religion.

**Before You Move On**
**Make Inferences** How is religion an important part of culture?

### JUDAISM
**Founder** Abraham    **Followers** 15 million
**Basic Beliefs** There is one God. People serve God by living according to his teachings. God handed down the Ten Commandments to guide human behavior.

### HINDUISM
**Founder** Unknown    **Followers** 860 million
**Basic Beliefs** Souls continue to be reborn. The cycle of rebirths ends only when the soul achieves enlightenment, or freedom from earthly desires.

### ONGOING ASSESSMENT
## READING LAB  GeoJournal

1. **Monitor Comprehension** What is the difference between a monotheistic religion and a polytheistic religion?
2. **Compare and Contrast** How are Hinduism and Buddhism similar?
3. **Region** What major religions are found in the area around the eastern Mediterranean Sea? Why do you think they are found in this region?

SECTION 3.2 **61**

# SECTION 3 HUMAN GEOGRAPHY
## 3.3 Economic Geography

**Main Idea** People produce, buy, and sell goods in a variety of ways.

Singapore became a trading colony of the British Empire in 1824. Today, it is an independent country, but trade is still an important part of its economy. An economy is a system in which people produce, sell, and buy things.

### Economic Activity

The production of goods and services is known as economic activity. Geographers divide this activity into different sectors. The primary sector involves taking raw materials from the soil or water. It includes mining, farming, fishing, and forestry. The secondary sector involves using raw materials to manufacture products, such as cars. The tertiary, or third, sector includes services, such as banking and health care.

### Factors of Production

Geographers study where economic activity occurs and how this activity is connected around the world. A country is more likely to have a strong economy if it has all four factors of production—land, labor, capital, and entrepreneurship. Land includes all the natural resources used to produce goods and services. Labor involves the size and education level of the workforce. Capital is a country's wealth and infrastructure. The fourth factor, entrepreneurship, involves the creativity and risk needed to develop new goods and services.

### Economic Systems

Economic systems are ways in which countries organize the production of goods and services. Four main systems are found around the world:

- In a traditional economy, people trade goods and services without money.

- In a free enterprise economy, privately owned businesses create goods that people buy in markets. This is also called a market economy or capitalism.

- In a command economy, the government owns most parts of the economy and decides what will be produced and sold.

- A mixed economy has elements of a free enterprise and command economy.

### Economic Indicators

The strength of a country's economy can be measured by several indicators, or signs. One is gross domestic product (GDP). It is the total value of the goods and services that a country produces. The GDP per capita is the value of products that a country produces per person. Other indicators include income, literacy rate, and life expectancy.

Economies fall in one of two categories. Countries with high GDPs are more developed countries. Most of their economic activity is in the tertiary sector. Countries with low GDPs are less developed countries. Most of their activity is in the primary or secondary sector.

### Before You Move On
**Summarize** What are four ways in which countries organize the production of goods and services?

## ECONOMIC INDICATORS OF SELECTED COUNTRIES*

| Country | Population | GDP (in U.S. dollars) | GDP Per Capita (in U.S. dollars) | Life Expectancy | Literacy Rate (percent) |
|---|---|---|---|---|---|
| Afghanistan | 31 million | 33.5 billion | 1,000 | 50 | 28.0 |
| Brazil | 201.0 million | 2.3 trillion | 12,000 | 73 | 88.6 |
| China | 1.3 billion | 12.3 trillion | 9,100 | 75 | 92.2 |
| Ethiopia | 90.8 million | 103.1 billion | 1,200 | 60 | 42.7 |
| Germany | 81 million | 1 trillion | 39,100 | 80 | 99.0 |
| Haiti | 9 million | 12.9 billion | 1,300 | 63 | 52.9 |
| Mexico | 116.2 million | 7 trillion | 15,300 | 77 | 86.1 |
| Singapore | 5.4 million | 325.1 billion | 60,900 | 85 | 92.5 |
| United States | 316.6 million | 15.6 trillion | 49,800 | 79 | 99.0 |

**Sources:** The World Bank; United Nations

**< Critical Viewing** This shipping terminal in Singapore is busy all day and all night. Based on the photo and the chart, what type of country is Singapore—a more developed or a less developed country?

### ONGOING ASSESSMENT
## DATA LAB  GeoJournal

1. **Interpret Charts** Which country has the largest population? the largest GDP per capita? What can you conclude about the relationship between the two?

2. **Synthesize** What sector is probably the main source of Haiti's economic activity? Explain.

3. **Region** In the region in which you live, what is an example of an economic activity from each of the three sectors?

SECTION 3.3

# SECTION 3 HUMAN GEOGRAPHY
## 3.4 Political Geography

TECHTREK
myNGconnect.com For photos of governments in action

Digital Library

**Main Idea** Countries around the world have different forms of government.

How is trash collection related to the protection of free speech? Both are responsibilities of government. A **government** is an organization that keeps order, sets rules, and provides services for a society. Political geographers study boundaries between different places, where different types of government exist, and how geography affects government.

### Government and Citizens

Governments govern citizens. A **citizen** is a person who lives within the territory of a government and is granted certain rights and responsibilities by that government.

Governments are either limited or unlimited. A limited government does not have complete control over its citizens. The citizens have some individual rights and responsibilities. An unlimited government has complete control over every aspect of its citizens' lives.

**Critical Viewing** Voters in South Africa line up for miles to vote. How important to them is the right to vote? How can you tell?

## Types of Government

In the modern world, five types of government are common. The major differences among them are in the power and rights that citizens have.

- In a **democracy**, citizens elect representatives to govern them. A legislature creates laws, an executive branch carries out laws, and a judicial branch interprets laws. Citizens have many rights, such as those in the Bill of Rights of the U.S. Constitution. The United States was the first modern country to establish a representative democracy.

- In a monarchy, a king, queen, or emperor rules society. The ruler usually inherits, or is born into, the office. Citizens in an absolute monarchy, such as Saudi Arabia, have few or no rights. In a constitutional monarchy, such as the United Kingdom, the queen or king shares power with a government organized by a constitution.

- In a dictatorship, one person, the dictator, rises to power and rules society. The dictator controls all aspects of life, including education and the arts. Citizens have few or no rights. North Korea has been ruled by a dictator for more than a half century.

- In an oligarchy, a group of a few people rules society. The ruling group usually is wealthy or has military power, and citizens have few or no rights. The government of Myanmar (Burma) has been an oligarchy since 1988.

**Critical Viewing** Female soldiers march in a military parade in North Korea to celebrate the country's 60th anniversary. Based on the photo, what qualities are valued by North Korea's government?

- Communism is a type of command economy in which the government, controlled by the Communist Party, owns all the property. Citizens have few or no rights. Cuba has been a communist country since 1959.

### Before You Move On
**Summarize** What are common types of government in the modern world?

## ONGOING ASSESSMENT
## READING LAB  GeoJournal

1. **Monitor Comprehension** Which of the five types of government are limited governments? Which are unlimited governments?

2. **Create Charts** Create a chart in which you compare the five forms of government. Use the following format:

| Type of Government | Source of Power | Leader or Ruler | Citizens' Rights |
|---|---|---|---|
| democracy | | | |

3. **Make Inferences** Look at the map of Russia in the National Geographic Atlas in the front of your textbook. How might Russia's geography affect the government's ability to rule?

SECTION 3.4 **65**

# SECTION 3 GLOBAL ISSUES • Human Rights
## 3.5 Protecting Human Rights

**Main Idea** The Universal Declaration of Human Rights states how all people deserve to be treated.

**KEY VOCABULARY**

**human rights,** n., political, economic, and cultural rights that all people should have

Nesse Godin was 13 years old during World War II when the Nazis occupied her town in Lithuania. Because she and her family were Jewish, they were transferred to a concentration camp. Then Nesse was sent to different labor camps, where she worked digging ditches. In January 1945, she and other prisoners were forced to march in the cold weather with little food. Many prisoners died.

### The Need for a Declaration

Nesse Godin survived the events of the Holocaust, but 6 million other people did not. At the end of World War II, people vowed to never let it happen again.

In 1945, 51 countries from around the world formed the United Nations (UN). The main goals of this organization were to keep peace, to develop friendly relationships among countries, and to protect people's **human rights**. The UN established the Commission on Human Rights to create a "common standard of achievement for all people and all nations." On December 10, 1948, the UN General Assembly approved the commission's **Universal Declaration of Human Rights**.

The need for such a declaration has not diminished since the 1940s. Governments around the world have been, and continue to be, guilty of targeting particular groups of people and denying them their human rights.

The declaration has 30 articles, or sections. Article 1 states that all people should be treated with respect:

All human beings are born free and equal in dignity and rights. They are endowed with reason and conscience and should act towards one another in a spirit of brotherhood.

Twenty-one of the articles explain political rights, such as the right to equality before the law, the right to freedom from torture, and the right to take part in government. Six of the articles address people's economic and cultural rights, such as the right to work, the right to education, and the right to participate in the cultural life of the community.

### The Impact of Human Rights

The declaration has had an impact, or effect, on people and governments. During the 1960s–1980s, countries around the world cooperated to pressure the Republic of South Africa to grant human rights to its non-white population. Many countries refused to trade with South Africa, and the country was barred from participating in the Olympic Games from 1964–1990.

**Inspiring people to care about the planet**
National Geographic Society Mission

## UNIVERSAL DECLARATION OF HUMAN RIGHTS

- Article 26.1
- Article 25.2
- Article 24
- Article 18
- Article 5
- Article 4
- Article 1

No one shall be held in slavery or servitude ...

Motherhood and childhood are entitled to special care and assistance.

## All human beings are born free ...

**Everyone has the right to freedom of thought, conscience, and religion ...**

No one shall be subjected to torture or to cruel, inhuman, or degrading treatment or punishment.

Everyone has the right to education.

Everyone has the right to rest and leisure ...

In 1994, South Africa finally gave in to the pressure and held elections in which all people could vote. The people elected Nelson Mandela, a leader of the African population, as president. This action showed the power of human rights.

### Before You Move On
**Monitor Comprehension** According to the Universal Declaration of Human Rights, how do all people deserve to be treated?

## ONGOING ASSESSMENT
# WRITING LAB  GeoJournal

1. **Form and Support Opinions** Choose one of the rights from the Universal Declaration of Human Rights that are listed above. In a paragraph, explain what this right means and why it is important.

2. **Write Reports** Select a news story in the newspaper or on television. In a short report, explain how the story shows the importance of protecting human rights.

SECTION 3.5 **67**

# CHAPTER 2
# Review

## VOCABULARY

For each pair of vocabulary words, write one sentence that explains the connection between the two words.

1. solstice; equinox

   *Summer begins at the solstice while spring begins at the equinox.*

2. plateau; continental shelf
3. renewable resource; nonrenewable resource
4. habitat; restore
5. free enterprise economy; democracy

## MAIN IDEAS

6. How does Earth's tilt affect the seasons? (Section 1.1)
7. What layers would you pass through on a journey to Earth's center? (Section 1.2)
8. How does erosion change Earth's surface? (Section 1.3)
9. How have scientists tried to lessen the impact of earthquakes and volcanoes? (Section 1.4)
10. In what ways is climate different from weather? (Section 2.1)
11. How is a humid equatorial climate different from a humid temperate climate? (Section 2.2)
12. What is an example of a renewable resource and a nonrenewable resource? (Section 2.4)
13. What are endangered species, and what factors threaten them? (Section 2.5)
14. What are four ways in which culture affects people's lives? (Section 3.1)
15. How is a free enterprise economy different from a command economy? (Section 3.3)
16. What are the common types of government in the modern world? (Section 3.4)

## THE EARTH

### ANALYZE THE ESSENTIAL QUESTION
How is the earth continually changing?

### Critical Thinking: Make Inferences
17. What impact do the changing seasons have on how farmers grow food?
18. Why did early civilizations develop along rivers?
19. How does the hydrologic cycle return water to Earth?

## PHYSICAL GEOGRAPHY

### ANALYZE THE ESSENTIAL QUESTION
What shapes the earth's varied environments?

### Critical Thinking: Draw Conclusions
20. How would plants in humid cold climates and humid equatorial climates differ?
21. What factors might prevent a country rich in natural resources from using them effectively?

### INTERPRET CHARTS

**U.S. AND WORLD ENDANGERED SPECIES**

| Group | United States | Other Countries | Total Number |
|---|---|---|---|
| Mammals | 71 species | 255 species | 326 species |
| Reptiles | 13 species | 66 species | 79 species |
| Fish | 74 species | 11 species | 85 species |
| Birds | 76 species | 184 species | 260 species |

Source: U.S. Fish and Wildlife Service

22. **Analyze Data** What percentage of endangered animals in the United States are mammals? What percentage in other countries are mammals?
23. **Human-Environment Interaction** What steps might wildlife conservationists take to reduce the number of endangered species?

## HUMAN GEOGRAPHY

### ANALYZE THE ESSENTIAL QUESTION

How has geography influenced cultures around the world?

**Critical Thinking: Find Main Ideas and Details**

24. What are some characteristics of American culture?
25. In what ways do different religions spread?
26. What is an example of each sector of economic activity?
27. What are two examples of different ways in which governments govern?

### INTERPRET MAPS

**CANADA'S BIOLOGICAL RESOURCES**

Legend:
- Barley
- Cattle
- Corn
- Fish
- Forest products
- Potatoes
- Swine
- Wheat

28. **Place** Are Canada's resources mostly in the north or the south? Why do you think this is the case?

### ACTIVE OPTIONS

**Synthesize** the Essential Questions by completing the activities below.

29. **Write a Public Service Announcement (PSA)** Write a PSA about what to do during an extreme weather emergency in your community. Focus on one type of weather event that is common in your area. Explain the supplies that a household should have. Provide an escape route or hiding place. Also, explain how people should behave during the emergency. **Share your PSA with the class.**

   **Writing Tips**
   - Do research to find the advice of weather experts.
   - Take notes and organize your information to make it clear to readers.
   - Use a tone that is calm and shows that you are informed.

   Go to **Student Resources** for Guided Writing support.

   **TECHTREK** myNGconnect.com For photographs of elements of culture

30. **Create a Visual Overview** Choose a part of culture that interests you, such as music, dance, language, or sports. Use photos from the **Digital Library** or from other online or print sources to create a visual introduction to that aspect of culture around the world. Show similarities and differences among countries around the world. Copy and complete the following organizer to organize your ideas:

   Element of Culture: _____
   Examples: _____
   Similarities: _____
   Differences: _____

COMPARE ACROSS REGIONS **UNIT 1**

# Climate

**TECHTREK**
myNGconnect.com For climographs about U.S. climates

Maps and Graphs

In this unit, you learned that climate is the average condition of the atmosphere in a region over a long period of time. Climate is determined by the latitude of a location, as well as by ocean currents, air currents, and elevation.

Two important factors of climate are temperature and precipitation. These factors determine the length and timing of the growing season and influence the types of economic activities in a region. For instance, a country with year-round warm temperatures and precipitation probably has a better agricultural industry than a country with low temperatures and a short rainy season.

## Compare
- Brazil
- Russia
- United States

### CLIMATE MAPS AND CLIMOGRAPHS

A climate map provides an overview of the climate in a region. However, sometimes more specific information about a city or country is needed. Geographers use a special tool called a **climograph** (climate + graph) to graphically show a range of temperature and precipitation in a place over a period of time. A climograph includes a bar graph that shows the amount of precipitation for a location. Average monthly temperature is shown by a line connecting 12 points, one for each month of the year.

Climographs are useful in comparing the climates of two different locations. They help geographers and others better understand the effects of climate on human activities in those locations.

### INTERPRET CLIMOGRAPHS

Look at the climographs for the cities of Belem, Brazil, and Omsk, Russia, on the opposite page. The months of the year are listed across the bottom of each climograph. A scale on the left vertical axis measures precipitation, or rainfall, in inches. A scale on the right vertical axis measures temperature in degrees Fahrenheit (°F).

The bars on the climographs show the rainfall for Belem, Brazil, and Omsk, Russia. The lines connecting the dots show the range of temperatures for each location.

Study the data in the climographs to analyze the climate in Belem and Omsk. Then answer the questions at right.

# AVERAGE MONTHLY TEMPERATURE AND PRECIPITATION

**Source:** National Drought Mitigation Center, University of Nebraska–Lincoln

## ONGOING ASSESSMENT
### RESEARCH LAB  GeoJournal

1. **Interpret Graphs** How would you describe the range of temperatures in Belem? What is the average temperature in Omsk in January? In July?

2. **Make Inferences** Most crops need water and warm temperatures in order to grow. Based on this information, what can you tell about the growing season in each city?

**Research and Create a Climograph** Choose a city in the United States and do research to find the average temperature and rainfall for each month of the year. Record the data in a chart. Use the data to create a climograph for that city. Then, with a partner, write three questions to help someone analyze and compare data on your two climographs.

REVIEW & EXTEND UNIT 1

# Active Options

**TECHTREK**
myNGconnect.com
For writing templates

Student Resources

## ACTIVITY 1

**Goal:** Extend your understanding of the environment.

### Compose a Top Ten List

For over 40 years, countries around the world have celebrated Earth Day. The purpose of the day is to appreciate our planet and to raise awareness of the need to protect the environment. With your classmates, create a Top Ten list of actions students in your school can plan to do to show appreciation for Earth on the next Earth Day. See if you can get permission to post copies of your list around the school to share with other students.

A reforestation event in the Philippines

## ACTIVITY 2

**Goal:** Learn about culture through religious architecture.

### Create a Guide to Religious Architecture

Religious institutions have inspired some of the world's monumental architecture. Use the **Magazine Maker CD-ROM** to showcase examples of historic buildings constructed by different religions. Include the name of each structure, when it was built, its location, and interesting details about it. Then analyze why these buildings were built to last over time.

## ACTIVITY 3

**Goal:** Learn about positive uses of technology.

### Write about Technology

Explorer Spencer Wells uses technology to analyze DNA. Look back across the unit for examples of technology used by scientists, especially pages 28–29 and 49. Search NASA on the Internet to find out how satellites collect data about the earth. Then write a paragraph telling how you think satellite imagery could help all societies of the world.

# UNIT 2

# EXPLORE NORTH AMERICA
## WITH NATIONAL GEOGRAPHIC

### MEET THE EXPLORER

**NATIONAL GEOGRAPHIC**

By analyzing ancient DNA, Emerging Explorer Beth Shapiro traces the decline of bison in North America. She has proved the decline was not caused by hunters but by an ice age 20,000 years ago.

### INVESTIGATE GEOGRAPHY

Niagara Falls attracts millions of visitors every year. The waters also generate plenty of low-cost electricity. For nearly 50 years, the United States and Canada have shared Niagara's power production, along with a strong desire to preserve its unique setting.

### STEP INTO HISTORY

The U.S. Capitol in Washington, D.C., houses the legislative branch of government in two chambers of Congress. It symbolizes a federal republic, a federation of states with elected representatives. Mexico has also practiced this form of government since 1824.

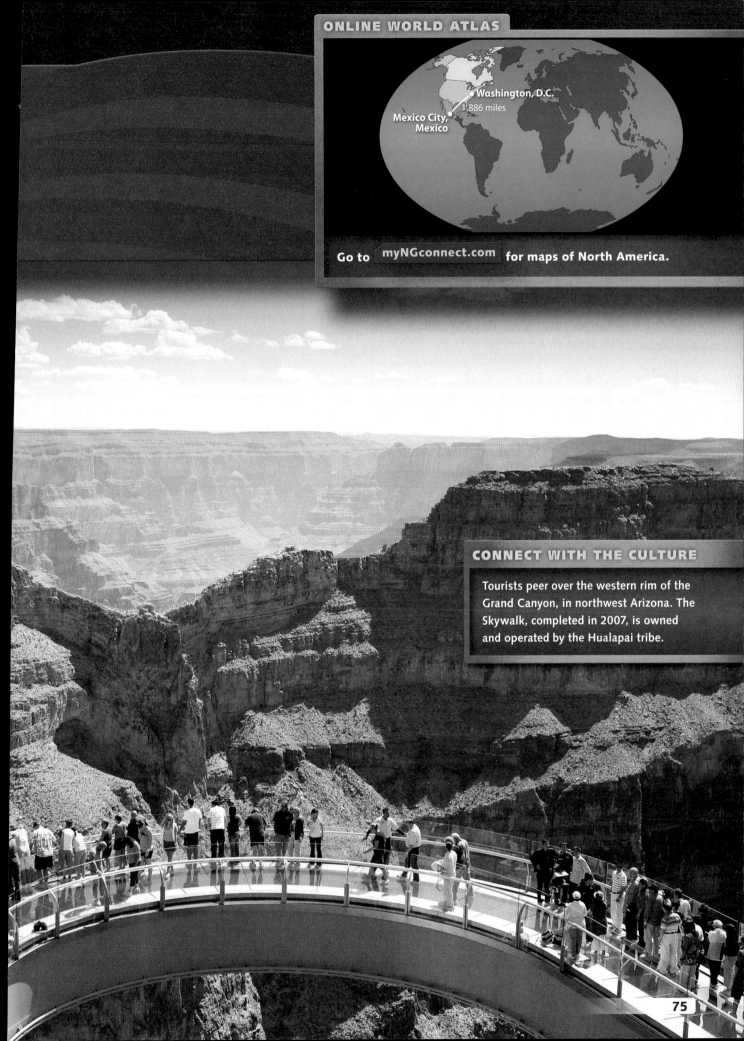

ONLINE WORLD ATLAS

Washington, D.C.
1,886 miles
Mexico City, Mexico

Go to myNGconnect.com for maps of North America.

CONNECT WITH THE CULTURE

Tourists peer over the western rim of the Grand Canyon, in northwest Arizona. The Skywalk, completed in 2007, is owned and operated by the Hualapai tribe.

# CHAPTER 3

# NORTH AMERICA
## GEOGRAPHY & HISTORY

## PREVIEW THE CHAPTER

**Essential Question** What are the significant physical features of North America?

### SECTION 1 • GEOGRAPHY

1.1 Physical Geography .................................. 78
1.2 The Great Plains ........................................ 80
1.3 Western Mountains and Deserts ............... 82
1.4 Mexico's Mountains and Plateaus ............ 84
1.5 Exploring the Yucatán .............................. 86

**KEY VOCABULARY**
- contiguous
- temperate
- glacier
- export
- drought
- commercial agriculture
- cordillera
- rain shadow effect
- dam
- peninsula
- subsistence farming
- aquifer
- cenote
- sustainable

**ACADEMIC VOCABULARY**
modify

**TERMS & NAMES**
- Great Plains
- Great Lakes
- Grand Canyon
- Mexican Plateau

**Essential Question** How did the United States and Canada develop as nations?

### SECTION 2 • U.S. & CANADIAN HISTORY

2.1 Exploration and Colonization .................... 88
2.2 Settling Quebec ........................................ 90
2.3 Revolution and Independence .................. 92
2.4 U.S. Constitution ...................................... 94
2.5 Expansion and Industrialization ............... 96
2.6 Civil War and Reconstruction ................... 98
2.7 World Conflict ......................................... 100

**KEY VOCABULARY**
- colonize
- plantation
- fortify
- missionary
- tax
- revolution
- constitution
- amendment
- pioneer
- industrialization
- transcontinental
- abolition
- secede
- civil war
- alliance
- neutrality
- dictator
- terrorism

**ACADEMIC VOCABULARY**
adapt, protest

**TERMS & NAMES**
- Declaration of Independence
- Bill of Rights
- Louisiana Purchase
- Manifest Destiny
- Trail of Tears
- Emancipation Proclamation
- Gettysburg Address
- Reconstruction
- Pearl Harbor
- Holocaust
- Cold War

**Essential Question** How have various cultures influenced Mexico's history?

### SECTION 3 • HISTORY OF MEXICO

3.1 The Maya and the Aztecs ........................ 102
3.2 The Conquistadors .................................. 104
3.3 Mexican Independence ........................... 106
3.4 The U.S.-Mexican War ............................ 108
  **DOCUMENT-BASED QUESTION**
3.5 The Mexican Revolution .......................... 110

**KEY VOCABULARY**
- civilization
- hieroglyphics
- empire
- tribute
- conquistador
- epidemic
- republic
- exile
- annexation
- reform
- land reform
- tyranny

**ACADEMIC VOCABULARY**
oppose

**TERMS & NAMES**
- Olmec
- Maya
- Aztec
- Hernán Cortés
- Montezuma
- Miguel Hidalgo
- Santa Anna
- Alamo
- Mexican Cession
- Gadsden Purchase

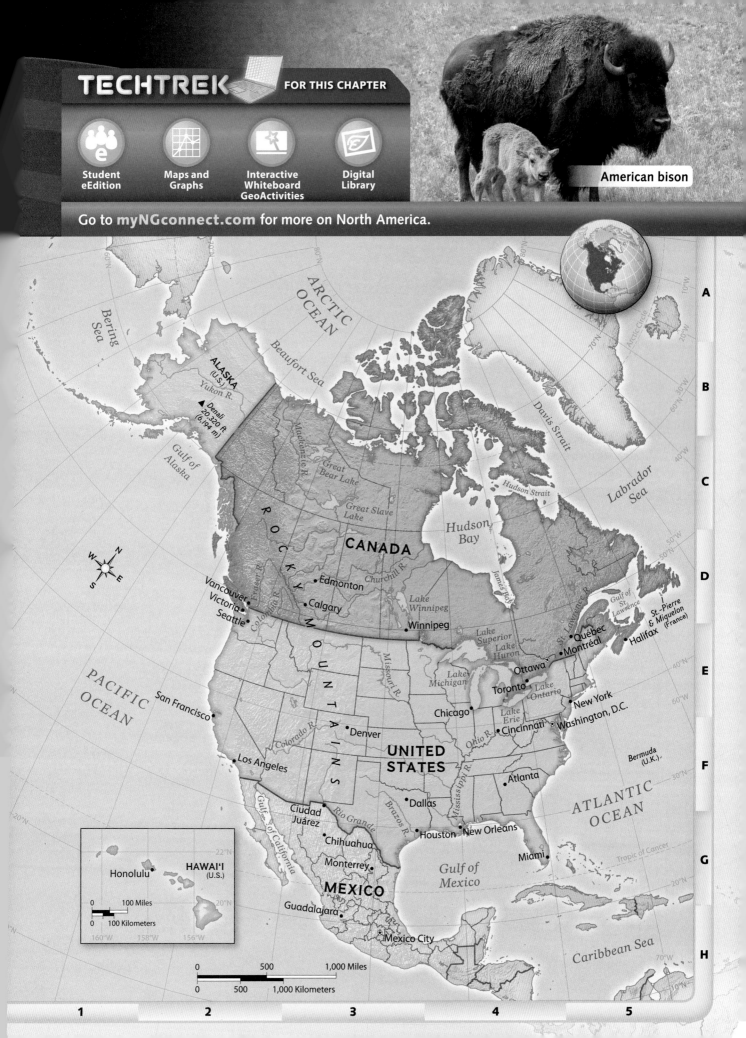

# SECTION 1 GEOGRAPHY

## 1.1 Physical Geography

**TECHTREK**
myNGconnect.com For online maps of North America and Visual Vocabulary

Maps and Graphs | Digital Library

NORTH AMERICA PHYSICAL

Visual Vocabulary: glacier

Visual Vocabulary: Great Plains

**Main Idea** North America has a wide variety of landforms, bodies of water, and climates.

North America stretches from the cold arctic of northern Canada to the warm tropics of Mexico. At the center lie the 48 **contiguous** United States, which means they are all connected in one block. The state of Alaska is in the northwest of the continent. The islands of the state of Hawaii are in the Pacific Ocean.

## Highlands, Plains, and Plateaus

Land elevation in North America generally rises from east to west, though the east has some highlands, or areas of hills and mountains. East of the Rocky Mountains lie the **Great Plains**. Plains are flat areas of land, which make up most of the center of North America. Plains also appear near coasts. Plateaus—flat lands of high elevation—are located between mountains in the western United States and central Mexico.

## Rivers and Lakes

Major cities developed along the region's numerous rivers, such as Cincinnati on the Ohio River, New Orleans on the Mississippi River, and Juárez (HWA rez) on the Rio Grande. The St. Lawrence River provides a waterway from the Atlantic Ocean to the **Great Lakes**. Combined, these five lakes form the largest body of fresh water in the world. Four of the lakes provide a physical boundary between the United States and Canada, just as the Rio Grande is a natural border between the United States and Mexico.

## A Variety of Climates

This region includes **temperate**, or mild, climates as well as extremes of cold and heat. Parts of the far north are covered by **glaciers**, or large masses of ice and packed snow. Much of northern Mexico is warm and dry, and Mexico's tropical south has warm climates and rainfall all year.

### Before You Move On
**Monitor Comprehension** Describe the region's major landforms, bodies of water, and climates.

### ONGOING ASSESSMENT
# MAP LAB
GeoJournal

1. **Interpret Maps** According to the physical map, how might goods be moved from Lake Michigan to the Atlantic Ocean?
2. **Region** According to the climate map, what types of climate occur in both Mexico and Alaska?

SECTION 1.1

## SECTION 1  GEOGRAPHY

# 1.2 The Great Plains

**TECHTREK**
myNGconnect.com For a map and photos of the Great Plains in North America

Maps and Graphs    Digital Library

**Main Idea** The Great Plains of the United States and Canada are a rich agricultural region with valuable energy resources.

The Great Plains region runs through the center of the continent. The crops grown there feed the population of North America with enough left to **export**, or send to other countries for aid or profit.

## Farming on the Great Plains

The Great Plains are well suited to agriculture for two reasons. First, the soil is rich with nutrients, so it produces bountiful crops. Second, the climate on the Great Plains is temperate and the area usually has a plentiful amount of rain.

Some years, rainfall is below normal for a long period of time, causing **drought** that can kill crops. In the 1930s, for example, the region experienced a drought that lasted several years. Plowing of native prairie grasses that helped hold the soil contributed to soil erosion, and the Plains became known as the "Dust Bowl." The persistent winds in the area stirred up great clouds of dry soil, or dust. Today, these winds can be a source of power.

Much of the original prairie land has been replaced by fields of wheat, corn, and other grains. These crops are grown on huge farms where planting and harvesting are done by machine. Such large, highly productive farms are typical of **commercial agriculture**, or the business of producing crops to sell.

Rivers such as the Missouri and the Mississippi transport goods from the Great Plains to lowland areas. Grain from the Canadian plains is moved by rail to the Atlantic Ocean and by ship on the Great Lakes and St. Lawrence River.

## Energy Resources

The Great Plains in the United States and Canada are home to major deposits of oil and natural gas. The main oil fields in the United States are found in the southern part of the Great Plains, from Kansas to Texas. Texas also has many offshore oil fields in the Gulf of Mexico.

> **Critical Viewing** A harvesting machine is used on this wheat farm in Manitoba, Canada. What power sources are in use on this farm?

# AGRICULTURAL REGIONS AND GREAT PLAINS
## United States and Canada

Commercial drilling in the Gulf of Mexico is risky. Oil deposits lie beneath layers of salt that can shift and cause an underwater earthquake. In 2010, human error caused an explosion of a large deepwater drilling structure, resulting in a major oil spill. The spill harmed wildlife and threatened the economy of the region.

Energy resources from the Great Plains are important to the United States, where more energy is used than is produced. High winds in the plains may be utilized as an alternative energy source.

### Before You Move On
**Summarize** What makes the Great Plains an important area for agriculture and energy?

---

### ONGOING ASSESSMENT
## MAP LAB  GeoJournal

1. **Interpret Maps** Which two of the three countries in this region share the Great Plains?

2. **Make Inferences** Look at the climate map in Section 1.1 and the map above. What reason can you suggest for the differences in climate across the Great Plains?

3. **Movement** What methods might be used for transporting goods out of the Great Plains? Use evidence from the map to support your answer.

# SECTION 1 GEOGRAPHY

## 1.3 Western Mountains and Deserts

**TECHTREK**
myNGconnect.com For a map and photos of the western United States and Canada

 Maps and Graphs
 Digital Library

**Main Idea** The resources of the western United States and Canada are rich in some areas and limited in others.

Mountains and high plateaus cover much of the western United States and Canada. These landforms create natural barriers to western coastal regions and offer plentiful resources in some areas.

### Landforms and Climate

The main landform in the western part of the region is the cordillera. A **cordillera** is a system of several mountain ranges that often run parallel to one another. In North America, the cordillera includes the Rocky Mountains and the Sierra Nevadas.

In the United States, the area between the Rocky Mountains and Sierra Nevada Mountains is the Great Basin. A basin is a depression in the surface of the land. The Great Basin is a desert, a dry, often sandy area with little rainfall or plant life.

The Great Basin is marked by smaller mountain ranges and canyons, which are deep, steep-sided valleys formed when rivers cut through soft rock. The best known canyon is the **Grand Canyon** in the southwestern United States. Formed over hundreds of millions of years, it is 277 miles in length and up to 18 miles wide.

The Great Basin is mostly dry. Warm, moist air flows east from the Pacific Ocean toward the mountains of the cordillera. As this warm air rises up the mountains, it cools and releases moisture on the mountains' western slopes. The air that eventually reaches the land east of the mountains is dry. This process is called the **rain shadow effect**.

### Resources and Conservation

Varying climates contribute to a varied supply of resources. The Great Basin and the mountain ranges that surround it contain important mineral deposits.

> **Critical Viewing** A group of horses follows hikers in the Great Basin. What landforms can you identify in the photo?

Areas of southwestern Canada hold reserves of natural gas. This area's heavy rainfall and many lakes allow for the use of water power to provide electricity.

Water for human use is in short supply in the southwestern United States. The population there has grown rapidly in recent years, and the demand for water has increased in this already dry area. A **dam**, which is a barrier that controls the flow of water, can help solve the problem of water shortage. However, dams can also cause problems such as excess soil erosion. The Hoover Dam on the Colorado River makes use of water power to supply electricity, irrigation, and drinking water to parts of Arizona, Nevada, and southern California.

### Before You Move On
**Make Inferences** What does the shortage of certain resources mean for the people who live in the West?

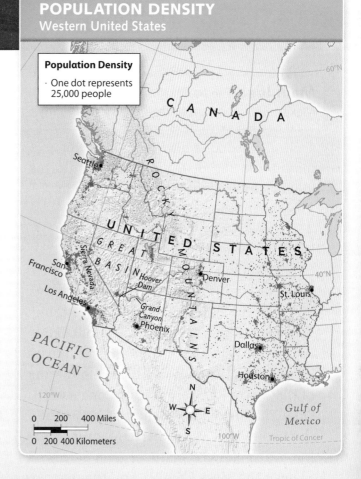

**POPULATION DENSITY**
Western United States

Population Density
· One dot represents 25,000 people

### ONGOING ASSESSMENT
## PHOTO LAB  GeoJournal

1. **Describe Geographic Information** Turn to a classmate and brainstorm words to describe the photo. What information in the lesson supports your description?

2. **Interpret Maps** Where is population density highest and lowest in this part of the region?

3. **Analyze Visuals** What physical features shown in the photo might explain the population distribution in this region?

# SECTION 1 GEOGRAPHY

## 1.4 Mexico's Mountains and Plateaus

TECHTREK
myNGconnect.com For a map and photos of Mexico's resources

Maps and Graphs | Digital Library

**Main Idea** The mountains and plateaus of Mexico are important to the country's economy.

Central Mexico is made up mainly of mountains and plateaus. These landforms yield rich resources that contribute to Mexico's economy.

### Landforms and Climate

Mexico includes two **peninsulas**, narrow strips of land that extend out into a body of water: the Baja (BAH hah) in western Mexico and the Yucatán (yoo kah TAHN) in the southeast. Mexico is shaped like an upside-down triangle with the mountains of the Sierra Madre running along each side. The **Mexican Plateau** lies between the two ranges of the Sierra Madres. Mexicans call the southern area of this plateau the Mesa Central, and the northern area the Mesa del Norte. A mesa—the Spanish word for "table"—is a high, flat plateau. Mexico's highest mountain, volcano Pico de Orizaba, rises at the southern edge of the plateau.

Mexico City, the capital, is on the Mesa Central. The city is home to more than 20 million people, almost 20 percent of the population of Mexico. Some areas of the Mesa Central have been subject to volcanic activity, which results in rich, volcanic soil. This soil helps to produce crops that are important to Mexico's economy, such as sugarcane, corn, and wheat.

Northern Mexico sits in the temperate zone, and the southern half lies in the tropics. On the high Mesa Central, climate is **modified**, or made less extreme, by a higher elevation. Temperatures there are cooler than along lower coastal areas.

### Resources and Agriculture

Mexico's mountains hold resources such as copper, silver, and zinc. However, the richest resource is the oil found in and around the Gulf of Mexico. Over three million barrels are produced each day.

Agriculture is also important to the country. Farmers in the north grow cotton, wheat, and fruit, and also raise cattle. To the south, farms produce sugarcane, coffee, and tropical fruits. Many rural Mexicans live by **subsistence farming**, growing just enough food to feed their families. This type of farming occurs mostly in the southern highlands.

**Before You Move On**
Monitor Comprehension Why is the Mexican Plateau important to the country's economy?

> **Critical Viewing** At a height of more than 18,000 feet, Pico de Orizaba is the highest mountain in Mexico. What words would you use to describe this mountain?

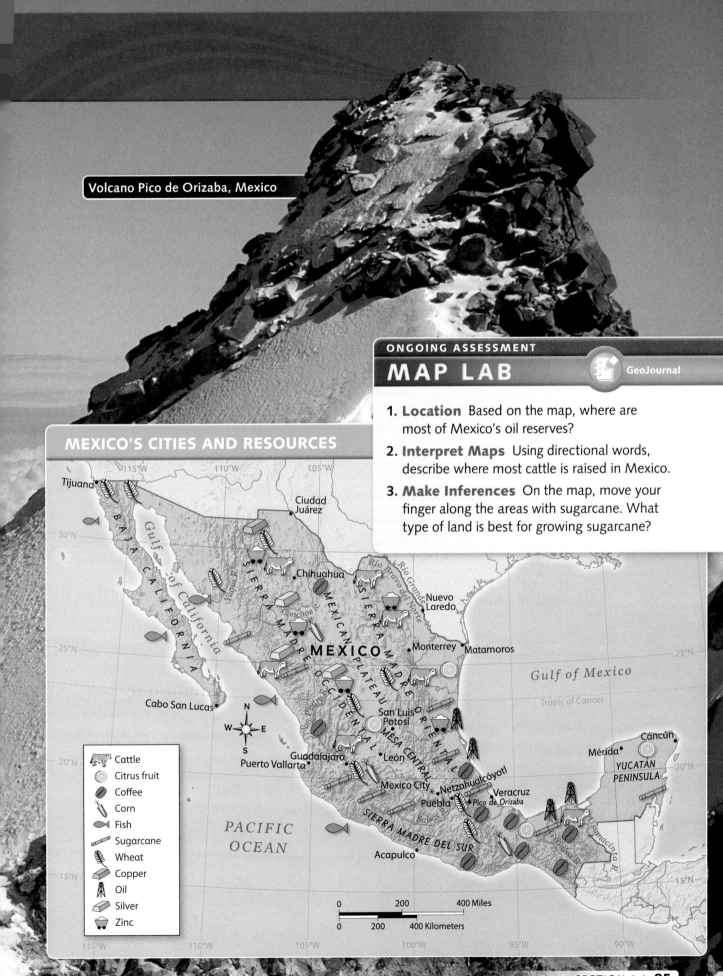

Volcano Pico de Orizaba, Mexico

## MEXICO'S CITIES AND RESOURCES

### ONGOING ASSESSMENT
# MAP LAB
GeoJournal

1. **Location** Based on the map, where are most of Mexico's oil reserves?
2. **Interpret Maps** Using directional words, describe where most cattle is raised in Mexico.
3. **Make Inferences** On the map, move your finger along the areas with sugarcane. What type of land is best for growing sugarcane?

SECTION 1.4 **85**

### SECTION 1 GEOGRAPHIC
# NATIONAL GEOGRAPHIC

**TECHTREK**
myNGconnect.com For photos of the Yucatán and an Explorer Video Clip
Digital Library

# Exploring the Yucatán
### with Sam Meacham

**Main Idea** Exploring untapped resources can help solve the problem of scarce water in dry areas of Mexico.

myNGconnect.com
For more on Sam Meacham in the field today

## Sacred Wells

Mexico's Yucatán Peninsula has both wet and dry seasons. Because of the peninsula's physical geography, very little surface water is available. National Geographic Grantee Sam Meacham is helping to find ways to gain access to this valuable resource.

The peninsula is composed mostly of limestone. Water from rainfall seeps through pores, or openings, in the limestone and flows underground. The weight of this groundwater creates cracks underground. Over time, the cracks expand into caves. These underwater caves function as **aquifers**, or layers of rock beneath the earth that contain water. The Yucatán Peninsula has a dry surface, so the water contained in these aquifers is critical to the water supply.

Sometimes the surface rock collapses to reveal the underground pools. These exposed pools are called **cenotes** (se NO tayz)—Mayan for "sacred wells." Meacham's team explores the cenotes, which provide entry to underground rivers and underwater caves that contain substantial water resources. One cave that Sam's team explored has 112 miles of underwater passages.

## Saving an Important Resource

Meacham's team is helping to reveal not only the importance of this groundwater, but also a potential threat to its safety. The Yucatán Peninsula is a popular tourist destination.

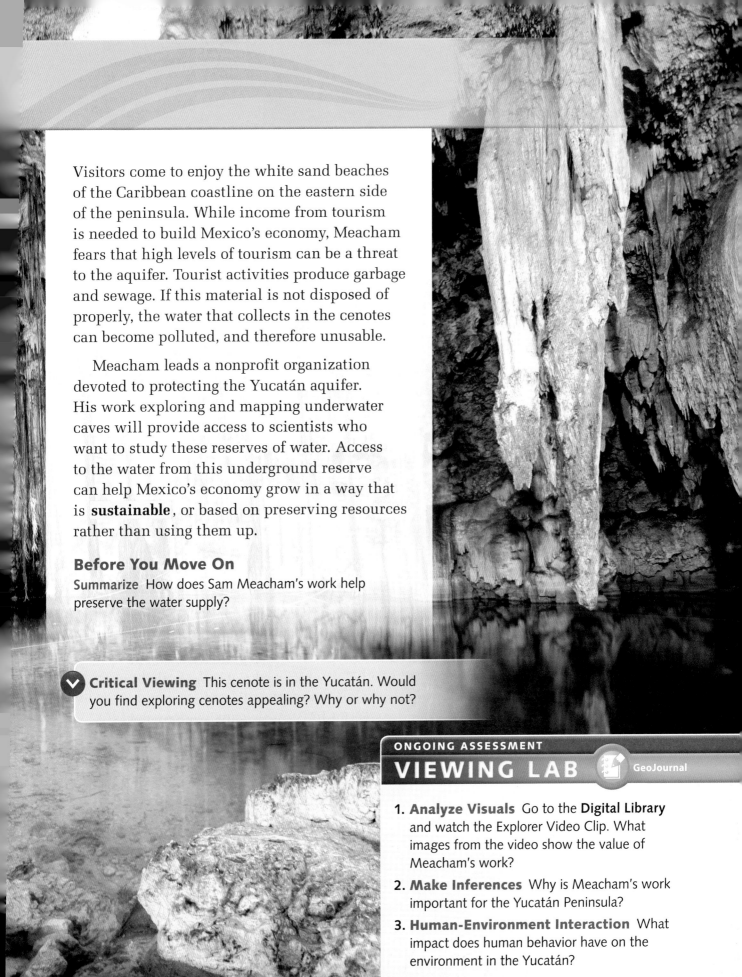

Visitors come to enjoy the white sand beaches of the Caribbean coastline on the eastern side of the peninsula. While income from tourism is needed to build Mexico's economy, Meacham fears that high levels of tourism can be a threat to the aquifer. Tourist activities produce garbage and sewage. If this material is not disposed of properly, the water that collects in the cenotes can become polluted, and therefore unusable.

Meacham leads a nonprofit organization devoted to protecting the Yucatán aquifer. His work exploring and mapping underwater caves will provide access to scientists who want to study these reserves of water. Access to the water from this underground reserve can help Mexico's economy grow in a way that is **sustainable**, or based on preserving resources rather than using them up.

### Before You Move On
**Summarize** How does Sam Meacham's work help preserve the water supply?

> **Critical Viewing** This cenote is in the Yucatán. Would you find exploring cenotes appealing? Why or why not?

## ONGOING ASSESSMENT
## VIEWING LAB  GeoJournal

1. **Analyze Visuals** Go to the **Digital Library** and watch the Explorer Video Clip. What images from the video show the value of Meacham's work?
2. **Make Inferences** Why is Meacham's work important for the Yucatán Peninsula?
3. **Human-Environment Interaction** What impact does human behavior have on the environment in the Yucatán?

# SECTION 2 U.S. & CANADIAN HISTORY
## 2.1 Exploration and Colonization

TECHTREK
myNGconnect.com For a map and images of European colonies

 Maps and Graphs
 Digital Library

**Main Idea** European colonization of North America brought settlers from several countries and permanently changed life on the continent.

In the late 1400s, many Native American groups lived on the continent. Each group **adapted**, or adjusted, to the environment in which it lived. For example, Native Americans in the eastern woodlands hunted deer and farmed the land. In the West, the Lakota hunted the large herds of American bison, a type of buffalo that grazed on the Great Plains.

### Europe Meets Native America

In 1492, Christopher Columbus sailed west from Spain and reached islands in the Caribbean Sea. This voyage was part of a period of European exploration of the Americas. Explorers wanted to find riches and claim new lands for their rulers.

The lives of Native Americans were permanently changed by the arrival of Europeans. Settlers carried diseases such as smallpox, which the native people could not fight. Disease killed a large percentage of certain Native American populations. The spread of settlements also pushed many Native American groups off their traditional lands.

By the late 1500s, European countries had begun to **colonize** the area, or build settlements and develop trade in lands that they controlled. Spain formed the colony of St. Augustine in present-day Florida. In the 1600s, the British settled Jamestown in Virginia and Plymouth in Massachusetts. New Sweden was established in what is now Delaware, and New France arose along the St. Lawrence River in what is now Quebec, Canada. These colonies were mostly small settlements. Most people farmed, traded furs, or did craft work.

The European powers began to compete for land. The British and French fought more than once over their colonies. In 1763, the British would eventually gain control of New France and become the major colonial power north of Mexico.

### Slavery in the Colonies

Over time, the British colonies thrived. In the south, huge **plantations**, or large farms that grow crops for profit, needed more labor than farmers' families could provide. Thousands of Africans were forced into slavery and brought across the Atlantic to the colonies.

Cathedral Basilica of St. Augustine, Florida

1550

1565
Spain forms colony in St. Augustine, Florida.

1600

1607
English establish colony in Jamestown, Virginia.

1608
French establish colony in Quebec, Canada.

88 CHAPTER 3

## EUROPEAN COLONIES AND NATIVE AMERICAN GROUPS, c. 1650

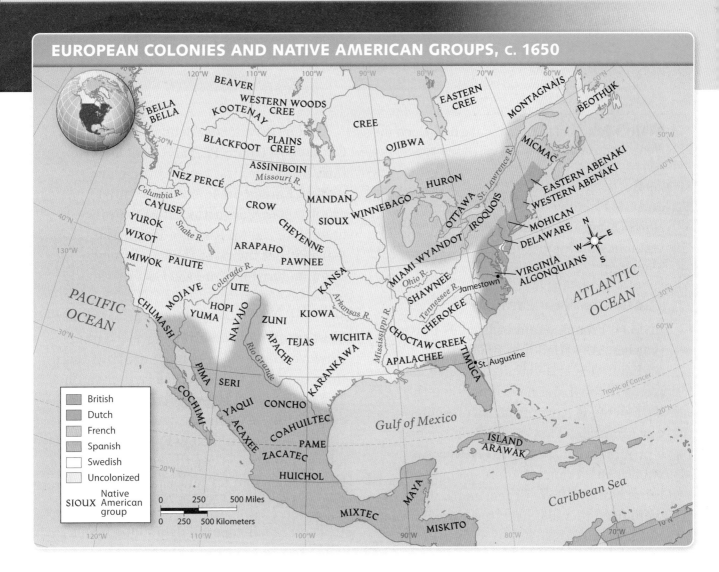

As production of cotton increased on plantations in the South, the population of slaves in these states grew rapidly. The economy of the Southern colonies became more and more dependent on slave labor.

### Before You Move On
**Summarize** How did European colonization change life in North America?

## ONGOING ASSESSMENT
### MAP LAB  GeoJournal

1. **Identify** According to the map, where were most of the European colonies located by 1650?
2. **Interpret Maps** In 1650, which Native American groups might have been most affected by Europeans? Why?
3. **Location** Why did British colonies in the South have the greatest number of slaves?

- **1620** English settlers set up colony in Plymouth, Massachusetts.
- **1650**
- **1664** Dutch colony Beverwyck is renamed Albany, New York.
- **1700**
- **1730s** Swedish colony, present-day Wilmington, Delaware, becomes successful market town.

British settlers arrive in Plymouth, Massachusetts.

SECTION 2.1 **89**

# SECTION 2 U.S. & CANADIAN HISTORY

## 2.2 Settling Quebec

**Main Idea** Conflict between the British and the French shaped the development of Canada.

As European colonies grew, the British and the French competed for land in North America. Their conflict had a major influence on the development of Canada as a nation.

### Founding New France

In 1534, French explorer Jacques Cartier (kar tee AY) sailed into the Gulf of St. Lawrence and claimed the area as New France. Later he sailed down the St. Lawrence River as far as present-day Montreal. This exploration opened a profitable fur trade with Native Americans. French traders exchanged European goods for beaver pelts, which were used to make hats that were popular in Europe.

In the early 1600s, Samuel de Champlain (sham PLAYN) built Quebec (keh BEK), the first major settlement of New France. In 1672, Louis Jolliet and Jacques Marquette left New France to explore the Mississippi River. They learned that the river flowed through Spanish territory into the Gulf of Mexico.

### Life in New France

Most people in New France were farmers. However, farming was more difficult there than in the British colonies farther south. The soil was not as productive, and the colder climate made the growing season shorter. As a result, New France never had a large population. By the early 1700s, New France had far fewer people than the British colonies in North America.

The French generally got along better with Native Americans than did the British. Certain French groups, such as *voyageurs* and missionaries, had positive contact with some Native Americans in New France.

> **Critical Viewing** Walls were built around Quebec City to **fortify**, or strengthen, the city. Why do you think these were needed?

French explorer Samuel de Champlain

*Voyageurs* were adventurous men who traveled the area to trade with Native Americans for furs. **Missionaries** were sent by the Roman Catholic Church to convince Native Americans to accept Christianity. This contact was often cooperative and usually peaceful.

## The British Take Control

The British and French were rivals for power in the region. War erupted in 1754 over whether a specific area of the upper Ohio River belonged to Britain or France. A bigger issue at the time, though, was which culture—British or French—would gain a stronger hold in North America. With greater military and financial resources, the British conquered Quebec and in 1760 gained control of the rest of New France.

With tension growing in the American colonies, Britain needed loyalty from its French subjects in Quebec. In 1774, the British passed the Quebec Act, which strengthened elements of French culture.

The Quebec Act established a French system of law and allowed the mostly Catholic population freedom to practice their religion. After the American Revolution, some people loyal to Britain migrated to Quebec from the United States. As a result, Quebec today has both French and English influences.

### Before You Move On
**Make Inferences** Why were there elements of French culture in Quebec after British conquest?

### ONGOING ASSESSMENT
## MAP LAB

1. **Interpret Maps** Look at the map in Section 2.1. Where was New France in relation to the British colonies?
2. **Compare** Based on maps in 2.1 and 2.2, how did the two countries' colonies compare in size?
3. **Draw Conclusions** Why were the British able to defeat the French?

SECTION 2.2 **91**

# SECTION 2 U.S. & CANADIAN HISTORY
## 2.3 Revolution and Independence

**Main Idea** The American colonists fought for and won independence from Britain.

As the British colonies in America grew, conflicts arose between the colonists and the government of Britain about how to govern the new land.

### Trouble in the Colonies

In the 1760s, the British government passed several laws taxing the colonists. A **tax** is a fee demanded by the government to pay for public services. These taxes angered the colonists, who had no representation in British government. The colonists **protested**, or objected to, British control. In 1773, in an event that came to be known as the Boston Tea Party, angry colonists dumped a shipment of tea from Britain into Boston Harbor because they felt the tea was unfairly taxed.

In response, Britain enacted measures meant to punish the colonies. These actions led some colonists to become more determined to govern themselves. Many were ready to overthrow the government and replace it with one of their own making. In other words, some colonists were ready for a **revolution**.

### The American Revolution

On April 19, 1775, violence erupted in Lexington and Concord, Massachusetts, when British soldiers were sent to destroy military supplies of colonial rebel groups. One of these rebels was Paul Revere. He alerted the Minutemen, colonists who were standing ready to fight at only a minute's warning, that the British were coming. The trouble between Patriots—colonists who wanted American independence—and Britain was now an armed conflict.

Over the next year, fighting continued. In 1776, colonial leaders signed the **Declaration of Independence**, written by Thomas Jefferson and claiming rights for America as an independent country.

Benjamin Franklin, a colonial political leader, traveled to France to help convince the French government to support the Patriot cause against the British. France decided to loan money to the Patriots and to send troops and ships. Marquis de Lafayette, a French nobleman, became an officer in the Patriot army and helped General George Washington.

Liberty Bell rung for the first time on July 8, 1776

Tax stamp issued by the British government for use in the colonies

**1760**

**1765** British Parliament passes Stamp Act, creating a new tax; colonists protest.

**1767** Parliament enacts new taxes, resulting in more protests.

**1770**

**1773** Colonists protest tea tax in the "Boston Tea Party."

General Washington led the Patriots to final victory at Yorktown, Virginia, where British troops surrendered in 1781. The two sides signed the Treaty of Paris in 1783, which was a formal agreement that resolved the issues between Britain and the United States. The Treaty of Paris recognized American independence. It also set the borders of the United States from the Atlantic Ocean to the Mississippi River and from Canada to the northern border of Florida. The American Revolution had created the United States of America.

## Lasting Effects

The Revolution had worldwide impact. Assistance from the French had helped Americans win the war. However, France suffered economic trouble because of the aid it had provided, and this contributed to its own revolution. The American example also helped inspire revolutions in Haiti and Central America. Finally, the Revolution also affected Canada. Many who were still loyal to Britain moved to Canada, increasing British presence there.

### Before You Move On
**Summarize** What factors contributed to the Americans' victory over the British?

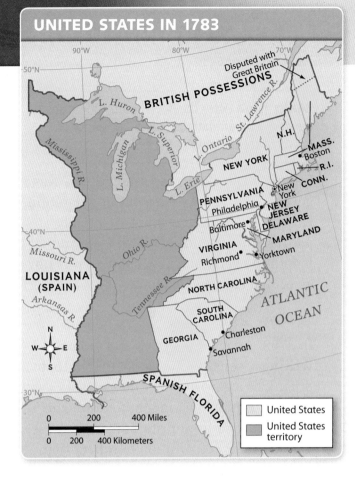

### UNITED STATES IN 1783

### ONGOING ASSESSMENT
## READING LAB  GeoJournal

1. **Make Inferences** Why would an action such as the Boston Tea Party be an effective way to show disagreement?
2. **Summarize** In what ways were colonists ready for a revolution?
3. **Analyze Cause and Effect** In what ways did the Revolution influence other nations?

Replica of General Washington's Patriot Army uniform

- 1775 Fighting breaks out in Lexington and Concord, Massachusetts.
- 1776 Jefferson writes Declaration of Independence.
- 1781 Americans defeat British at Yorktown.
- 1783 With Treaty of Paris, Britain recognizes American independence.

# SECTION 2 U.S. & CANADIAN HISTORY
## 2.4 U.S. Constitution

**Main Idea** After the Revolution, Americans created a government based on a balance of power and the rights of the people.

From 1781 to 1789, the United States was governed by the Articles of Confederation. However, this plan did not effectively address the balance between state and federal powers. The new country was not as united as it needed to be.

### A New Plan

In 1789, American leaders met to begin writing a **constitution**, a document that organizes a government and states its powers. Having just fought for and won independence from Britain, Americans wanted their constitution to reflect that spirit of independence. The document needed to grant certain freedoms to individuals and limit government control. Differing concerns of northern and southern states also had to be addressed.

Officially approved in 1789, the Constitution of the United States has become a model for other countries. It establishes a government based on five principles, shown in the graphic (right). The federal government consists of three branches, shown in the diagram below.

### Bill of Rights

One reason the U.S. Constitution has lasted for so many years is that it can be changed, which makes it a living document. It allows for **amendments**, which are formal changes to a law.

The first ten amendments to the U.S. Constitution make up the **Bill of Rights**. The rights and freedoms it guarantees are summarized below.

1. Freedom of religion, speech, press, assembly, and petition
2. Right to bear arms
3. Freedom from quartering of troops
4. Freedom against unreasonable search and seizure
5. Freedom from self-incrimination (testifying against yourself) and right to due process
6. Right to a speedy trial and to confront accusers
7. Right to a trial by jury
8. Freedom from cruel and unusual punishment
9. Protection of rights not named in the Constitution
10. Powers not given to the federal government are given to states and people

### Before You Move On

**Make Inferences** Why did Americans want to ensure that the U.S. Constitution limited the powers held by their own government?

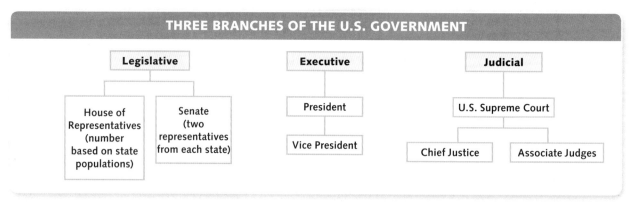

The U.S. Constitution was written in Independence Hall in Pennsylvania.

**THE U.S. GOVERNMENT**
The Constitution is based on five principles, or basic ideas.

**SEPARATION OF POWERS**
Government power is divided among three separate branches.

**CHECKS AND BALANCES**
Each branch partly needs another branch to do its work.

**FEDERALISM**
Specific powers are granted to the central government and state governments.

**LIMITED GOVERNMENT**
Individual rights and the powers of state governments limit the U.S. government.

**DEMOCRACY**
Government is based on the rights and the equality of every citizen.

## ONGOING ASSESSMENT
### SPEAKING LAB  GeoJournal

**Pose and Answer Questions** With a partner, state questions to ask the Constitution writers. Use information from the lesson for answers. If you cannot answer a question, check other sources.

SECTION 2.4 **95**

# SECTION 2 U.S. & CANADIAN HISTORY

## 2.5 Expansion and Industrialization

**Main Idea** During the 1800s, the United States expanded its territory and industries.

The treaty that settled the Revolution expanded U.S. territory to include land east of the Mississippi River. In 1803, President Thomas Jefferson doubled the size of the country by making the **Louisiana Purchase**.

### Settling the West

Some Americans expected that the United States would expand its territory all the way to the Pacific Ocean. Many believed it had the right to do so. This idea came to be known as **Manifest Destiny**.

In 1804, Army officers Meriwether Lewis and William Clark were hired to lead an exploration of the newly purchased land. The expedition set out from St. Louis and traveled to the Pacific Ocean and back. Its arrival at the Columbia River in Oregon ▶ was important because it helped establish a U.S. claim as far west as the Pacific Ocean.

Starting in the 1840s, thousands of Americans became **pioneers**, or settlers of new land. Their westward journey was marked by rough lands, deep rivers, the threat of disease, and even possible attack by Native Americans. The trails they used, such as the Sante Fe and Oregon Trails, still show deep ruts from wagon wheels.

Settlers' thirst for land also led to the forced removal of Native Americans. The Indian Removal Act of 1830 relocated tribes to the West. Some tribes in the southeast had highly developed farming and governments, and no interest in land to the West. If tribes refused to leave, U.S. troops forced them. The Cherokee, who tried to negotiate to keep their land, were eventually forced into a strenuous 116-day journey to Oklahoma. This route became known as the **Trail of Tears**.

> **Critical Viewing** Camps like this one were set up for workers along the construction of the transcontinental railroad. What would be the advantages and disadvantages of being a worker on this project?

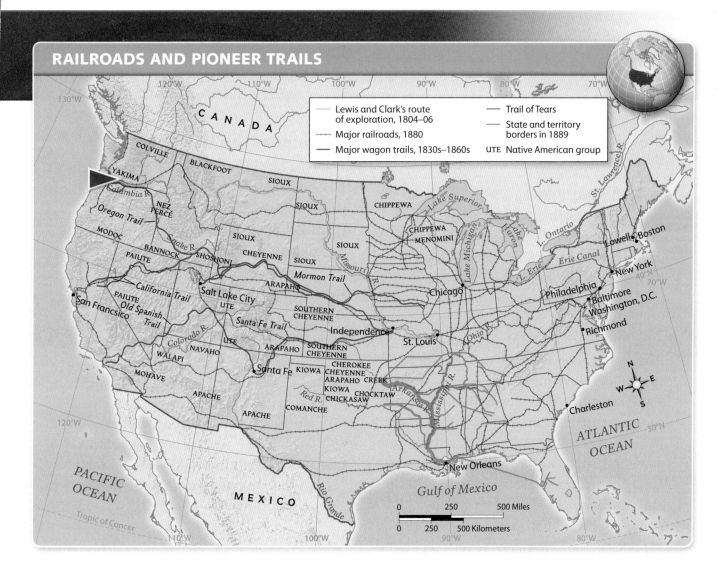

## Industrialization

After independence, industry grew quickly in the east, especially textiles. By 1813, for example, the production of cloth was completely mechanized, or made by machines rather than by hand. The town of Lowell, Massachusetts, with its early textile mills, was the first U.S. town to be planned around an industry.

**Industrialization**—the shift to large-scale production—expanded throughout the 19th century. The construction of the Erie Canal in 1825 provided a water route from the Atlantic to the Great Lakes. As a result, New York City became a major American port. Railroads allowed goods and people to travel even greater distances, especially the **transcontinental** railroad. Completed in 1869, this railroad crossed the entire continent. Movement to the West was faster and easier for Americans than it had ever been before.

### Before You Move On
**Monitor Comprehension** In what ways did the United States expand its territory and industries during the 1800s?

---

**ONGOING ASSESSMENT**

## READING LAB   GeoJournal

1. **Summarize** What is Manifest Destiny? What effect did it have on the United States?
2. **Location** How did the Indian Removal Act of 1830 affect tribes in the southeastern states?
3. **Make Inferences** How did the construction of canals and railroads benefit manufacturers?

SECTION 2.5 **97**

# SECTION 2  U.S. & CANADIAN HISTORY

## 2.6 Civil War and Reconstruction

**TECHTREK**
myNGconnect.com For a map and images of the Civil War
Maps and Graphs
Digital Library

**Main Idea** Differences between Northern and Southern states led to the Civil War.

As the United States grew in the early 1800s, a powerful division arose within the country. Southerners wanted slavery to be allowed in new territories in the West. However, **abolition**, or the movement for ending slavery, was growing in the North.

### Causes of the Civil War

Abraham Lincoln was elected president in 1860. Southerners, who depended on slave labor to run their plantations, feared the new president would try to end slavery in the South. As a result, 11 Southern states **seceded**, or formally withdrew, from the Union in 1860 and 1861. These states formed the Confederate States of America. Richmond, Virginia, became the capital and Jefferson Davis was president.

Lincoln declared the states in rebellion and vowed to reunite the Union. Soon after, Confederate troops fired upon Fort Sumter, a federal fort located in South Carolina. This event was the beginning of the Civil War in the United States. A **civil war** is war between opposing groups of citizens in the same country.

### Conduct of the War

The war lasted four years, from 1861 to 1865. The Confederates, led by General Robert E. Lee, won many early battles. However, the Union had a larger military, a stronger economy, and more resources.

In the midst of the war, on January 1, 1863, the **Emancipation Proclamation**, which freed all slaves in Confederate territory, became effective. The same year, Lincoln delivered the **Gettysburg Address** to honor those who died there in a key battle. These two events helped to highlight a moral purpose to the war—freedom and equality. Eventually, the resources and economy of the North proved too strong, and the Confederacy surrendered in 1865.

### Reconstruction After the War

Racial tensions, desperate poverty, and hostility toward the U.S. government lingered in the South after the war. These problems presented challenges to **Reconstruction**, the effort to rebuild and reunite the states as one nation.

Although former slaves had been granted new rights, many white Southerners did not allow them new freedoms.

President Abraham Lincoln

Emancipation Proclamation

**1860** Lincoln elected president; Southern states begin to secede from Union.

**1862** President Lincoln issues Emancipation Proclamation.

**1860**

**1861** Civil War begins at Fort Sumter.

**1863** Union victory at Gettysburg halts northern advance of the Confederates.

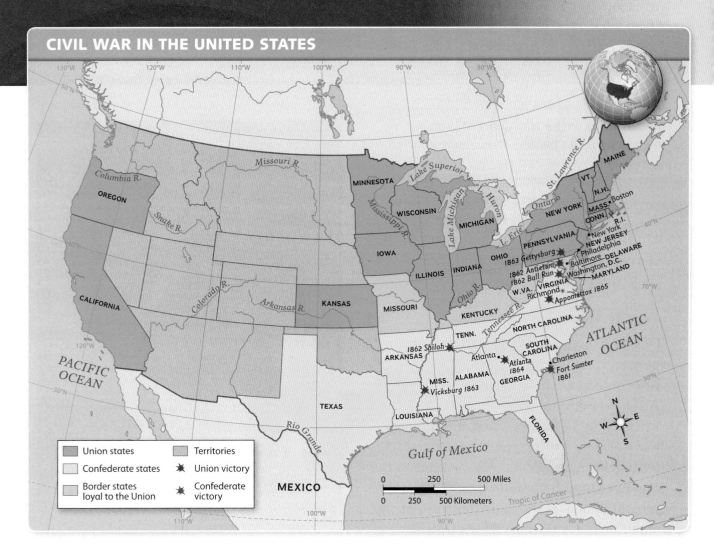

## CIVIL WAR IN THE UNITED STATES

The government had to send federal troops to protect former slaves. By 1877, Reconstruction had abruptly ended. Almost a century passed before the federal government would again protect the civil rights of African Americans.

### Before You Move On
**Summarize** What differences between Northern and Southern states led to the Civil War?

### ONGOING ASSESSMENT
## MAP LAB

1. **Interpret Maps** How did the number of Union and Confederate states compare?
2. **Location** Based on the map, where were most battles fought?
3. **Make Inferences** Which side was most likely to suffer damage from the war? Why?

- **1864** Union captures Atlanta; Lincoln is re-elected president.
- **1865**
- **April 11, 1865** Confederacy surrenders. **April 14, 1865** President Lincoln is shot.
- **1870**
- **1877** Reconstruction ends.

A schoolroom where freed black slaves learned to read

SECTION 2.6

# SECTION 2 U.S. & CANADIAN HISTORY

## 2.7 World Conflict

**Main Idea** In the 20th and 21st centuries, the United States became increasingly involved in world affairs and conflicts.

As the United States grew in population and prosperity, so did its economic power. This new strength made it difficult for the country to remain uninvolved in world issues and conflicts.

### World War I

Due to growing conflicts between major powers in Europe, war broke out there in 1914. Russia, France, and Britain—the Allies—fought the Central Powers, which were led by Germany. The United States had economic **alliances**, or partnerships, with France and Britain. Nevertheless, the country stood firm in its **neutrality**, or refusal to take sides, and worked to bring about peace in Europe. When German submarines sank the Lusitania, a ship carrying U.S. passengers, the United States could no longer remain neutral. In 1917, America joined the Allies and helped them win the war. The terms of the peace treaty signed after the war were thought by many Germans to be unfair.

### World War II and the Cold War

After a period of prosperity following World War I, the U.S. stock market crashed in 1929. The crash set in motion a series of events that led to a worldwide economic downturn called the Great Depression. In 1933, Adolf Hitler rose to power with promises to restore Germany's political and economic strength. He became a **dictator**, a ruler with complete control, and he planned to conquer all of Europe.

**Critical Viewing** A woman works in an engine shop in 1942. Why would women be doing this kind of work at this time?

Hitler's invasion of Poland in 1939 started World War II. The United States again stayed neutral, until the Japanese bombed **Pearl Harbor**, a U.S. naval base in Hawaii, on December 7, 1941. The next day, the United States joined Britain, France, and the Soviet Union against the Axis Powers, led by Germany, Italy, and Japan.

After years of fighting, Germany surrendered in May, 1945, but Japan fought on. In August that year, U.S. planes dropped atomic bombs on Japan. Bombs hit Hiroshima on August 6, and Nagasaki on August 9. Both cities were completely destroyed, and as many as 140,000 civilians were killed. Japan surrendered on September 2, 1945.

The war had a great human cost. The **Holocaust**—Hitler's organized murder of Jews and other groups—killed an estimated six million people. Overall, an estimated 50 million people were killed, primarily in Europe and the Soviet Union.

Even though the United States and the Soviet Union had been allies, the two countries had political differences. They engaged in a **Cold War**, a long period of political tension without fighting. The United States promoted global democracy and freedom, while the Soviet Union promoted communism, in which the state controls all parts of the economy. The Cold War ended in 1991 when the government of the Soviet Union broke apart.

## Terrorism and Modern Conflict

As the 21st century approached, a new type of warfare was becoming more common: **terrorism**. Terrorists use violence to achieve political results.

### ONGOING ASSESSMENT
### WRITING LAB  GeoJournal

1. **Find Main Idea and Details** What role did the United States play in World Wars I and II?
2. **Summarize** What ideas did the United States promote during the Cold War?
3. **Write a Paragraph** Write a short paragraph that describes U.S. involvement in world affairs in the 1900s and beyond. Go to **Student Resources** for Guided Writing support.

On September 11, 2001, terrorists attacked the United States. They hijacked four passenger planes and crashed them into civilian and military targets, killing over 3,000 people. The United States invaded Afghanistan where the terrorist group originated. The campaign to defeat terrorism led to a U.S. operation in 2011 that resulted in the death of Osama bin Laden, believed to have planned the September 11 attack.

### Before You Move On
Summarize In what ways did the United States become increasingly involved in world conflict in the 20th and 21st centuries?

Rescue workers provide medical aid in New York City on September 11, 2001.

# SECTION 3 HISTORY OF MEXICO

## 3.1 The Maya and the Aztecs

**TECHTREK**
myNGconnect.com For maps of the Maya and Aztecs and photos of artifacts

 Maps and Graphs   Digital Library

**Main Idea** The Mayan and Aztec civilizations made important cultural contributions to Mexico.

Present-day Mexico was settled several thousand years earlier than the United States. About 11,000 years ago, Native American groups lived in the Valley of Mexico, the area around modern Mexico City. They survived by hunting and gathering plants to eat. Around 7,000 years ago, the settlers began growing maize, or corn, which was a native grass. High yields from these crops allowed the population to grow.

### The Maya

By about 1000 B.C., an organized society called the **Olmec** lived along the southern coast of the Gulf of Mexico. Their culture had a strong influence on later cultures in Mexico, such as the Maya. The **Maya** lived on the present-day Yucatán Peninsula in Mexico and in northern Central America.

Around 100 B.C., the Maya began to develop into a **civilization**, a society with highly developed culture, politics, and technology. Evidence the Maya left behind provides information about their culture, such as their system of writing.

The Maya used **hieroglyphics**, a system of writing that consisted mostly of pictures and symbols—or hieroglyphs—as characters. The recorded history they left behind in hieroglyphic paintings reveals a highly developed written language. The Maya also studied the sun, moon, stars, and planets, which allowed them to develop an accurate calendar. They used the calendar to mark dates that were important in their religion.

After about A.D. 900, Mayan civilization apparently declined. Historians do not fully understand why this happened. Possible theories include violent conflict between cities, overpopulation, or the overuse of land available for farming.

**Critical Viewing** Tenochtitlán was built on islands within a lake. What might be an advantage of building the Aztec capital city on islands rather than on mainland?

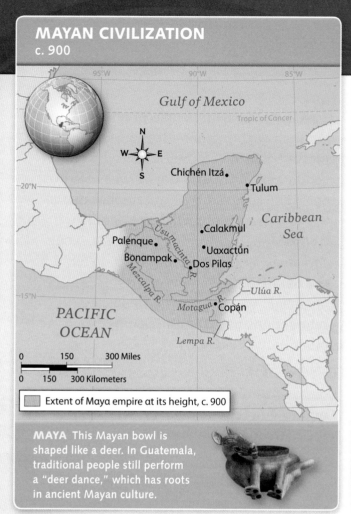

**MAYAN CIVILIZATION** c. 900

**MAYA** This Mayan bowl is shaped like a deer. In Guatemala, traditional people still perform a "deer dance," which has roots in ancient Mayan culture.

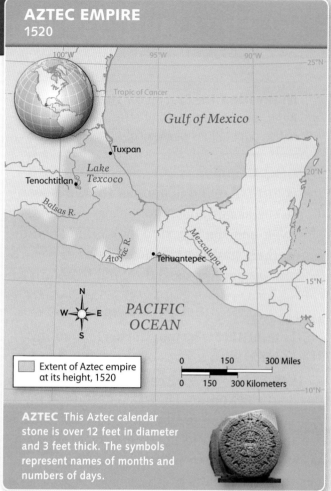

**AZTEC EMPIRE** 1520

**AZTEC** This Aztec calendar stone is over 12 feet in diameter and 3 feet thick. The symbols represent names of months and numbers of days.

## The Aztecs

After the Mayan decline, other native groups gained power in Mexico. The dominant group was the **Aztecs**, who settled in the area of modern Mexico City around A.D. 1325. The Aztec people built a city called Tenochtitlán (teh nohch teet LAHN) on islands in Lake Texcoco. They constructed human-made islands called *chinampas*, or "floating gardens," on the lake to give them more area to grow food.

The Aztecs built a broad **empire**—an extensive group of peoples governed by one ruler—through military conquest of neighboring lands. They gained further power by enslaving conquered people and using their labor to build more cities. Rulers also collected **tribute**, or fees, in the form of money, crops, or other goods.

The Aztec Empire continued to thrive until Spanish explorers arrived in the 1500s, which changed life in the region.

### Before You Move On

**Monitor Comprehension** What cultural contributions did the Maya and the Aztecs make?

### ONGOING ASSESSMENT
## MAP LAB  GeoJournal

1. **Interpret Maps** Use directional words to describe the location of the Maya and Aztecs.
2. **Make Inferences** What might account for so little overlap in the land areas of each group?
3. **Analyze Visuals** What does the circular shape of the Aztec calendar suggest about the Aztec idea of a year?
4. **Summarize** How was the Aztec Empire able to gain so much power?

# SECTION 3 HISTORY OF MEXICO

## 3.2 The Conquistadors

**Main Idea** The Spanish conquest of the Aztec Empire had a lasting effect on the native populations of Mexico.

After Columbus's voyages, Spain seized control of many islands in the Caribbean in the early 1500s. From there, Spanish explorers continued to look for new lands. In their travels, they heard tales of wealth among the Aztecs and set out to obtain it.

### The Conquest of Mexico

Late in 1518, **Hernán Cortés** landed in Mexico with a force of about 500 soldiers. Cortés was a **conquistador**, a Spanish soldier-explorer. He was determined to conquer more of the Americas. Cortés quickly realized that the Aztecs were widely hated by other native groups—and that the rumors of Aztec wealth were true.

The Aztec ruler, **Montezuma**, welcomed Cortés and his men to Mexico's capital, Tenochtitlán. Within three years, however, Cortés had killed Montezuma and captured the city. He ordered the city burned and built a new capital—Mexico City—on its ashes. By 1525, Spanish control reached as far south as Central America.

### Spanish Advantage

Cortés was a bold, skillful leader. The Spanish had guns, swords, and armor, all superior to the Aztecs' weapons. The Spanish also used horses and dogs in battle. The Aztecs were slow to act at first because they believed Cortés was a god. When their leader, Montezuma, was killed by Cortès in 1520, the Aztec armies fell apart in confusion.

Native Mexican people provided the Spanish with further advantage. Many native tribes hated the Aztecs who had conquered them, and Cortés was able to convince thousands of warriors in native tribes to fight along with the Spanish troops. A native woman called Malinche served as an interpreter and guide for the Spanish. Her knowledge of native languages was important for successful interactions between the Spanish and other native tribes.

One of the biggest reasons for the Aztec defeat was unexpected. The Spanish unknowingly brought over diseases that the natives had never encountered and could not resist.

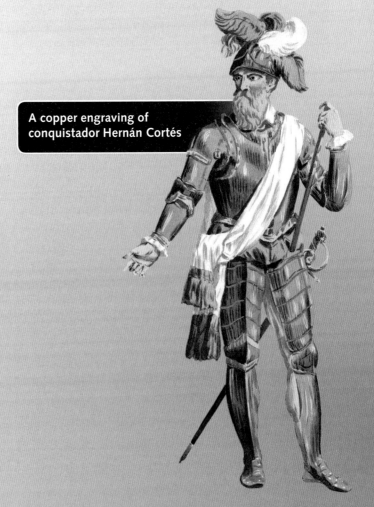

A copper engraving of conquistador Hernán Cortés

Illnesses like measles and smallpox became epidemics. An **epidemic** is the outbreak of a disease that spreads widely. These diseases infected and killed huge numbers of the native population. The deaths weakened the Aztecs' ability to resist the Spanish conquest.

## Results of the Conquest

From their base in Mexico, the Spanish armies conquered all of Central America. They went on to conquer most of South America, building their own vast empire. European diseases continued to wipe out native populations for many years.

Natives who survived were enslaved by the Spanish and forced to work on farms and in silver mines. Using slave labor, the Spanish mined vast wealth from the large amounts of silver in the mountains of Northern Mexico.

Spain's success and sudden wealth had another effect. It drove Spain's rivals in Europe to send explorers to the Americas in search of wealth. These European powers also created new empires there.

### Before You Move On
**Monitor Comprehension** What did the Spanish conquest mean for the Aztecs and other native populations of Mexico?

An engraving of Aztec ruler Montezuma

## ONGOING ASSESSMENT
# VIEWING LAB  GeoJournal

1. **Compare and Contrast** Look at the engravings of Hernán Cortés and Montezuma. What similarities and differences do you notice? Copy the Venn Diagram and complete it with details from the images to compare and contrast the two leaders.

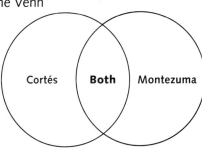

2. **Analyze Visuals** Based on what you see in the engravings, what features of Cortés and Montezuma suggest their position as leaders?

3. **Summarize** What factors contributed to the Spanish army's ability to conquer the Aztecs?

4. **Analyze Visuals** Based on what you see in the illustrations, what would you expect to have happened when these two fighting men met in battle? Why?

# SECTION 3 HISTORY OF MEXICO

## 3.3 Mexican Independence

TECHTREK
myNGconnect.com For images of Mexican independence
Digital Library

**Main Idea** Conflict between social classes in colonial Mexico led to a movement for Mexican independence from Spain.

Because of the conquistadors, Mexico became a major part of Spain's colonial empire. The empire included Central America and Spanish-held areas in the Caribbean and present-day United States.

### Colonial Mexico

In the early 1800s, the society of colonial Mexico was divided into social classes. At the top were people born in Spain, called *peninsulares* (pen in soo LAH rayz). They held the highest offices in the government and the church. Next were *criollos* (kree OH yohz), people of Spanish background born in Mexico. Next came mestizos (mes TEET sohz), people of mixed Spanish and native heritage. Native Americans and slaves were the lowest class of society.

### Independence

In 1810, **Miguel Hidalgo**, a Catholic priest, led a revolt against the *peninsulares*. He led an army of almost 100,000 men, including many mestizos. His army killed many *peninsulares* and *criollos* and captured several of Mexico's cities.

Government troops resisted Father Hidalgo's army, and in 1811, the troops defeated the priest's army and seized him. On July 30, 1811, they executed him, but his rebellion inspired others in Mexico to continue the fight.

José Morelos, a priest and revolutionary, took over leadership of the Mexican independence movement. In November 1813, he and his followers declared independence from Spain. They drafted a constitution that made Mexico a **republic**, a government of elected officials. The constitution also guaranteed freedom, equality, and security to Mexico's citizens. However, Spain refused to accept this constitution, and Spanish troops refused to surrender. For nearly two years they pursued Father Morales's army throughout southern Mexico. He was captured by Spanish troops and executed in 1815.

### The Struggle Continues

Small groups of rebels continued to fight. In 1820, one leader united them—Colonel Augustín de Iturbide (ee toor BEE day). Early in 1821, he issued a plan that included Three Guarantees. First, Mexico would be independent of Spanish control.

Father Miguel Hidalgo

○ 1800s
Colonial Mexico class system solidified

**1800**

1810 ○
Miguel Hidalgo leads a revolt against the *peninsulares*.

**1810**

1811 ○
Hidalgo is executed.

Second, *peninsulares* and *criollos* would be equal to each other under the law. Third, Roman Catholicism would be the only religion in the country.

Iturbide's army rapidly defeated Spanish troops in Mexico. In 1821, Spain signed a treaty granting independence to Mexico. In 1822, with a tremendous thirst for power, Iturbide crowned himself the emperor of Mexico. His reign was a disaster. Even though he had great success leading the rebels to defeat Spain, as emperor he could not unify the country. One of his generals, Antonio López de **Santa Anna**, rebelled against him. In 1823, Iturbide gave up his crown and was **exiled**, or forced to leave the country.

Santa Anna's new government wrote the Constitution of 1824. That year, the United States, Britain, and other countries recognized Mexico as an independent country. However, Spain continued to fight for control of Mexico. Santa Anna's troops finally defeated Spanish forces in 1830. That same year, Mexico's people elected Santa Anna as their president.

### Before You Move On
**Summarize** How did the colonial structure in Mexico contribute to demands for independence?

**Critical Viewing** A crowd surrounds Colonel Iturbide before he crowns himself emperor. What details in the illustration show that the Mexican people were in support of Iturbide?

## ONGOING ASSESSMENT
# READING LAB  GeoJournal

1. **Make Inferences** Do you think the Catholic Church was on the side of Spain or the native people? Explain your answer.
2. **Categorize** Identify the classes of society in Mexico from top to bottom.
3. **Draw Conclusions** How was General Iturbide able to both unite and divide the country?

**1820**

1821 — Spain grants Mexico's independence.

1822 — Iturbide proclaims himself emperor of Mexico.

1823 — Santa Anna and others force Iturbide out of power.

1824 — Mexico adopts a new constitution as a republic.

**1830**

Mexican general and political leader Antonio López de Santa Anna

# SECTION 3 HISTORY OF MEXICO

## 3.4 The U.S.-Mexican War

TECHTREK
myNGconnect.com For maps and images from the U.S.-Mexican War

Maps and Graphs
Digital Library

**Main Idea** As a result of the U.S.-Mexican War, Mexico lost some territory to the United States.

When Santa Anna became president of Mexico, the country included land from present-day Texas to California and north to Utah. This territory soon faced a crisis.

### The Texas Revolution

To encourage new settlers in the area of Texas, Mexico offered land grants to Americans to set up colonies there, and they had begun arriving in 1821. The American population grew so quickly that in 1830 Mexico passed a law blocking further settlement. This move started a growing hostility between the Mexican government and the settlers in Texas.

In 1835, Texans began to rebel against Mexican forces. In response, President Santa Anna led a large army to Texas to end the rebellion. In 1836, in a battle that lasted 13 days, Mexican forces killed almost all of the approximately 200 Texan rebels defending a fort called the **Alamo**. Despite this defeat, Texans continued their fight for independence. The next month, under the leadership of General Sam Houston, the Texan army defeated Santa Anna's troops at the Battle of San Jacinto.

### War with the United States

After winning independence from Mexico, Texans created the Republic of Texas. For nine years, the country struggled with debt, disputes with Mexico, and violence between settlers and Native Americans. In 1845, Texas joined the United States through **annexation**, or adding territory.

Texas and the United States claimed that the southern border was the Rio Grande, but Mexico said the border was the Nueces River, farther north. In early 1846, American troops were sent to occupy the disputed area between the two rivers. When Mexican and U.S. troops fought near the Rio Grande in May 1846, the United States declared war on Mexico.

Santa Anna's troops battled the U.S. army but were defeated after two years. As a result, in 1848 Mexico gave up the area from Texas to California. This land became known as the **Mexican Cession**. In 1853, Santa Anna sold further Mexican land to the United States in what came to be known as the **Gadsden Purchase**. In Mexico, many people **opposed**, or objected to, this decision.

The Alamo

General Sam Houston

**1820**

**1830** Mexico blocks further American settlement of Texas, limits rights of Texans.

**1836** Texans revolt and win independence, but Mexico refuses to recognize it.

**1840**

**1845** United States annexes Texas.

**1846** War between Mexico and United States begins.

MEXICO IN 1840

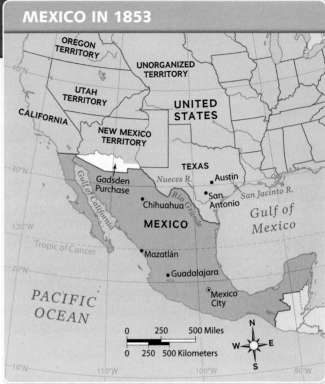

MEXICO IN 1853

## La Reforma—The Reform

Due to growing opposition to Santa Anna and his policies, he was forced out of leadership in 1854. New leaders wanted to change Mexican society. In 1857, President Benito Juárez proposed major **reforms**, or changes, to promote social equality. Religious and military leaders refused to accept these reforms, and the country entered a long period of civil unrest and violence. Conflicts between those who wanted reform in Mexico and those who did not continued throughout the 1860s.

### Before You Move On
**Make Inferences** Why would the Republic of Texas have agreed to become part of the United States after winning independence?

### ONGOING ASSESSMENT
## MAP LAB
GeoJournal

1. **Location** Find the Nueces River and the Rio Grande on the map. Describe the distance between the two rivers.
2. **Compare and Contrast** Based on the two maps, what territories did the United States gain after war with Mexico?
3. **Analyze Cause and Effect** How did the dispute over the border between Texas and Mexico lead to war?

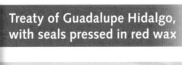

Treaty of Guadalupe Hidalgo, with seals pressed in red wax

○ 1848
War between Mexico and United States ends; United States gains Mexican Cession.

○ 1858
Reformer Benito Juárez becomes president of Mexico.

1860

○ 1853
In Gadsden Purchase, Mexico sells parts of modern Arizona and New Mexico to United States.

President Benito Juarez image on Mexican banknote

SECTION 3.4 **109**

SECTION 3 DOCUMENT-BASED QUESTION

## 3.5 The Mexican Revolution

The U.S.-Mexican War and the resulting civil conflict had weakened Mexico's economy and government. After Juárez's death, General Porfirio Díaz became dictator. Díaz's policies mainly benefited the wealthy.

The Mexican Revolution was partly a fight for land reform, in which large estates would be broken up and the land would be given to the poor. Emiliano Zapata and Pancho Villa led the struggle, which lasted from 1910 to 1920 and killed more than one million Mexicans.

General Porfirio Diaz

### DOCUMENT 1
### Plan of Ayala

Zapata issued the Plan of Ayala in 1911 to state goals for the revolution.

> [W]e who undersign ... declare solemnly .. ... to end the **tyranny** [harsh government] which oppresses us. ...
>
> [T]he pueblos [villages] or citizens who have the titles [to property being held by the government] ... will immediately enter into possession of that real estate [property] ....
>
> [T]he immense majority of Mexican [settlements] and citizens are owners of no more than the land they walk on [and] suffer ... the horrors of poverty without being able to improve their social condition in any way.

**CONSTRUCTED RESPONSE**

1. What does the plan say about conditions in Mexico?

### DOCUMENT 2
### Constitution of 1917

One important goal of the constitution was to bring about changes in land ownership.

> Article 27. Ownership of the lands and waters within the boundaries of the national territory is vested [included] originally in the Nation, which has ... the right to transmit [give] title thereof to private persons ....
>
> [N]ecessary measures shall be taken to divide up large landed estates; to develop small landed holdings [smaller farms] in operation; ...
>
> to encourage agriculture ... and to prevent the destruction of natural resources. ... The rights of small landed holdings in operation [will be] respected at all times.

**CONSTRUCTED RESPONSE**

2. How might the division of large estates into small holdings change Mexico?

# [T]HE IMMENSE MAJORITY OF MEXICAN . . . CITIZENS ARE OWNERS OF NO MORE THAN THE LAND THEY WALK ON.

— PLAN OF AYALA, 1911

Pancho Villa

## DOCUMENT 3

### Photo of Pancho Villa

Pancho Villa (VEE yuh), the son of a poor farm laborer, became a folk hero in Mexico. He gained popularity by stealing from the rich in order to give to the poor. Villa's knowledge of the physical geography of northern Mexico helped him avoid being captured.

**CONSTRUCTED RESPONSE**

3. What does this photo of Villa suggest about him as a leader?

## ONGOING ASSESSMENT
## WRITING LAB — GeoJournal

**DBQ Practice** Think about the Plan of Ayala, the Constitution of 1917, and the photograph of Pancho Villa. What do these show about the conditions that led to the Mexican Revolution?

**Step 1.** Review the description of the revolution and the statements from the Plan of Ayala and the Constitution. Reread the biographical information about Villa and look at the photograph.

**Step 2.** On your own paper, jot down notes about the main ideas expressed in each document.

> Document 1: Plan of Ayala
> Main Idea(s) _____
> Document 2: Constitution of 1917
> Main Idea(s) _____
> Document 3: Photo of Pancho Villa
> Main Idea(s) _____

**Step 3.** Construct a topic sentence that answers this question: What do the statements in the Plan of Ayala and the Mexican Constitution and the photograph of Pancho Villa show about the conditions that led to the Mexican Revolution?

**Step 4.** Write a detailed paragraph explaining what the documents tell you about the conditions that led to the Mexican Revolution.

SECTION 3.5

# CHAPTER 3
# Review

## VOCABULARY

For each pair of vocabulary words, write one sentence that explains the connection between the two words.

1. temperate; glacier

   *Many areas of North America have a temperate climate, but others are so cold they are covered by glaciers.*

2. export; commercial agriculture
3. colonize; plantation
4. abolition; secede
5. civilization; hieroglyphics
6. empire; tribute

## MAIN IDEAS

7. How does the geography of eastern and western North America differ? (Section 1.1)
8. What are important resources of the Great Plains? (Section 1.2)
9. What resources are available from Mexico's mountains and plateaus? (Section 1.4)
10. Why were European countries interested in settling North America? (Section 2.1)
11. Why has the U.S. Constitution been able to last so long? (Section 2.4)
12. What were the causes and results of the Civil War? (Section 2.5)
13. How did the United States change in the late 1800s? (Section 2.6)
14. What events led to U.S. involvement in three world conflicts? (Section 2.7)
15. What were the cultural achievements of the Maya? (Section 3.1)
16. Why were the Spanish able to conquer the Aztec Empire? (Section 3.2)

## GEOGRAPHY

### ANALYZE THE ESSENTIAL QUESTION
What are the significant physical features of North America?

### Critical Thinking: Compare and Contrast

17. In what ways is the physical geography of Mexico different from that of the United States and Canada? In what ways is it similar?
18. What energy resources are found in all three countries of North America?
19. How are the western United States and the Yucatán similar in terms of resources?

## U.S. & CANADIAN HISTORY

### ANALYZE THE ESSENTIAL QUESTION
How did the United States and Canada develop as nations?

### Critical Thinking: Make Inferences

20. How did Reconstruction attempt to address issues that resulted from the Civil War?
21. Which groups suffered as a result of U.S. expansion west? Which groups gained?
22. Why did the population of the United States grow faster than that of Canada?

### INTERPRET CHARTS

| SETTLING QUEBEC | |
| --- | --- |
| 1534 | Jacques Cartier claims St. Lawrence River Valley for France. |
| 1608 | Samuel de Champlain founds colony of Quebec for France. |
| 1671 | France claims the Great Lakes and the Mississippi River. |
| 1759 | British capture Quebec during Seven Years' War. |

23. **Interpret Charts** What land and water claims did France make in the 1500s and 1600s?
24. **Make Inferences** What can you infer about the French explorations in the Americas?

## HISTORY OF MEXICO

### ANALYZE THE ESSENTIAL QUESTION

How have various cultures influenced Mexico's history?

**Critical Thinking: Summarize**

25. Why did the Spanish place the capital of colonial Mexico where they did?
26. What are the social divisions that led to the Mexican Revolution?
27. How did this social class structure lead to Mexican independence from Spain?

**INTERPRET MAPS**

BATTLE OF THE ALAMO, 1836

28. **Interpret Maps** Which areas of the Alamo appear to have been avoided by Mexican forces?
29. **Make Inferences** Which part of the Alamo did Mexican forces not use as entry? Why might this be?

## ACTIVE OPTIONS

**Synthesize** the Essential Questions by completing the activities below.

30. **Write a Comparison Speech** Write a speech that compares the groups that were part of settling the United States, Canada, and Mexico. Your speech should point out similarities and differences based on the groups that settled those three countries. Use the tips below to help you prepare the speech. **Offer to deliver your speech to other classes who are studying North America.**

> **Writing Tips**
> - Identify the groups that settled each country and think about their similarities and differences.
> - Decide whether you think the groups are more similar than they are different.
> - Write that conclusion as your main idea.
> - Support your main idea with details.

31. **Conduct Internet Research** Choose a city in one of the three countries. Research online to gather information about the history of that city.

    Find out when and how the city was settled, how it has grown over time, and what groups contributed to its history. Then choose one of the following options to show what you found:

    - Construct a time line that shows the sequence of events that make up the city's history. Illustrate your time line.
    - Draw a diagram of the city, including its historical landmarks, such as statues or other historical markers. Also label the locations of historical events. The labels should explain why that landmark or location is important. Add your time line or diagram to a class exhibit on North America.

REVIEW **113**

# CHAPTER 4

# NORTH AMERICA TODAY

## PREVIEW THE CHAPTER

**Essential Question** What issues do the United States and Canada face today?

### SECTION 1 • FOCUS ON THE UNITED STATES & CANADA

1.1 North America's Cultural Diversity ......... 116

1.2 The Media Culture ................................. 118

1.3 The Changing Economy ........................... 120

**GLOBAL ISSUES: ENERGY**
1.4 Finding New Energy Sources ................... 122

1.5 Citizens' Rights and Responsibilities ...... 124

**KEY VOCABULARY**
- diversity
- indigenous
- immigrant
- tolerance
- mass media
- smartphone
- global
- mobile
- manufacturing
- fiber optic
- recession
- petroleum
- ethanol
- hybrid
- wind turbine
- naturalization
- due process

**ACADEMIC VOCABULARY**
dynamic

**TERMS & NAMES**
- Internet

**Essential Question** How has globalization affected Mexico?

### SECTION 2 • FOCUS ON MEXICO

2.1 Daily Life in Mexico ............................... 126

2.2 The Impact of Globalization ................... 128

2.3 Reaching Toward Democracy .................. 130

**KEY VOCABULARY**
- descendant
- ancestry
- mural
- artifact
- globalization
- nationalize
- economic sector
- multi-party democracy

**ACADEMIC VOCABULARY**
reform

**TERMS & NAMES**
- North American Free Trade Agreement (NAFTA)
- Institutional Revolutionary Party (PRI)

# TECHTREK FOR THIS CHAPTER

**Student eEdition**
**Maps and Graphs**
**Interactive Whiteboard GeoActivities**
**Digital Library**

Go to myNGconnect.com for more on North America.

New York City's Empire State Building stands out against the Hudson River in the background. Manhattan, a densely populated area of the city, is a major economic center in the global economy.

# SECTION 1 UNITED STATES & CANADA

## 1.1 North America's Cultural Diversity

**TECHTREK**
myNGconnect.com For a chart of languages in North America
Student Resources

**Main Idea** The opportunities and freedoms offered in the United States and Canada have attracted a diverse population.

**Diversity**, or variety, is a major feature of U.S. and Canadian culture. Both countries have a diverse mix of races, languages, religions, and nationalities.

### Diversity in the United States

Early cultures included **indigenous**, or native, tribes, European settlers, and Africans who had been forced onto the continent as slaves. After independence, immigrants came from many other parts of Europe seeking freedom and opportunity. An **immigrant** is a person who takes up permanent residence in another country.

Many historians describe immigration as a series of waves, starting in the 1600s with European colonists. From 1820 to 1870, a second wave brought more than seven million people, many from Ireland and Germany, fleeing poor conditions. Immigrants also came from China, seeking opportunity in the American West. Starting in 1880, a third wave, mostly from parts of southern and eastern Europe, brought 23.5 million more. Changes to U.S. immigration laws in the 1960s brought a fourth wave, many from Asia and the Caribbean. Today, most immigrants come from Mexico, China, the Philippines, and India.

Immigrants bring not only their culture, but also their skills and ambitions, and can play a major role in economic growth. In fact, many immigrants come here for that reason. In the United States they have the freedom to start a business.

**Critical Viewing** Street signs in Chicago's Chinatown are posted in two languages. In what way would signs like this be helpful to immigrants?

American culture has come to be defined by its diversity. **Tolerance**—acceptance of different views—is an important American value.

### Diversity in Canada

Native groups such as the Inuit have been in Canada for thousands of years and are still there today. Historically, Canada has encouraged immigration in order to attract workers to help build its economy.

In the 1600s and 1700s, British and French colonies in Canada had attracted large numbers of immigrants from France, Ireland, and Britain. After World War II, many Europeans who had lost their homes moved to Canada. Immigrants from Asia and Latin America began to arrive in the late 1900s. However, most people in Canada today still have European roots.

**Before You Move On**
**Summarize** Why are the cultures of the United States and Canada so diverse?

### SPEAKERS OF MAJOR LANGUAGES, UNITED STATES AND CANADA

| Language | United States | Canada |
|---|---|---|
| Arabic | 845,396 | 261,640 |
| Chinese | 2,600,150 | 1,102,065 |
| English | 228,699,523 | 17,882,775 |
| French | 1,305,503 | 6,817,655 |
| German | 1,109,216 | 450,570 |
| Italian | 753,992 | 455,040 |
| Polish | 593,598 | 211,175 |
| Portuguese | 731,282 | 219,275 |
| Spanish | 35,468,501 | 345,345 |
| Tagalog (Filipino) | 1,513,734 | 235,615 |

Sources: 2009 American Community Survey; Statistics Canada, Census of the Population, 2006

People gather in Times Square in New York for a New Year's celebration.

## ONGOING ASSESSMENT
## LANGUAGE LAB  GeoJournal

1. **Compare and Contrast** After English, what is the next major language spoken in the United States and in Canada?

2. **Place** What do the top two languages spoken in Canada suggest about its history?

3. **Make Predictions** Which languages do you predict will have a higher number of speakers in the United States in future years? Why?

SECTION 1.1  **117**

# SECTION 1 UNITED STATES & CANADA

## 1.2 The Media Culture

**TECHTREK**

myNGconnect.com For photos and a map on media culture

 Maps and Graphs  Digital Library

**Main Idea** The movement of ideas through various media has shaped communication and cultures in the United States and Canada.

**Mass media** refers to communication that reaches large audiences. Traditionally this meant newspapers, radio, and television. Today, mass media also includes the rapid and nearly continuous flow of information through personal electronic devices.

### Print Media to Electronic Media

During the 1800s, newspapers provided information that often influenced public opinion. By the 1920s, radio provided news and entertainment. It was the most important source of news during World War II. President Franklin Roosevelt used radio for his "fireside chats," in which he shared information about the war with the American public. By this time it was clear that mass media was a powerful tool.

In the 1950s, many U.S. households acquired their first television sets, and this medium surpassed radio in its popularity as a source for news and entertainment. In 1963 the entire nation viewed film footage of the assassination of President Kennedy. In the same decade television became an important medium for witnessing the Vietnam War and the protests against it.

In the 1990s, new media technologies exploded in popularity. Cell phones connected people around the world, even if they were on the move. The **Internet**, a communications network, gave people the power to access huge amounts of information from a variety of sources. It also allowed individuals to distribute their own media, such as blogs and videos, to a mass audience. With so much information available electronically, people rely less on print publications.

**Critical Viewing** Mobile media today combines communication and visual technology. Which details in the photo best illustrate American culture?

*American Gothic*
Grant Wood
Art Institute of Chicago

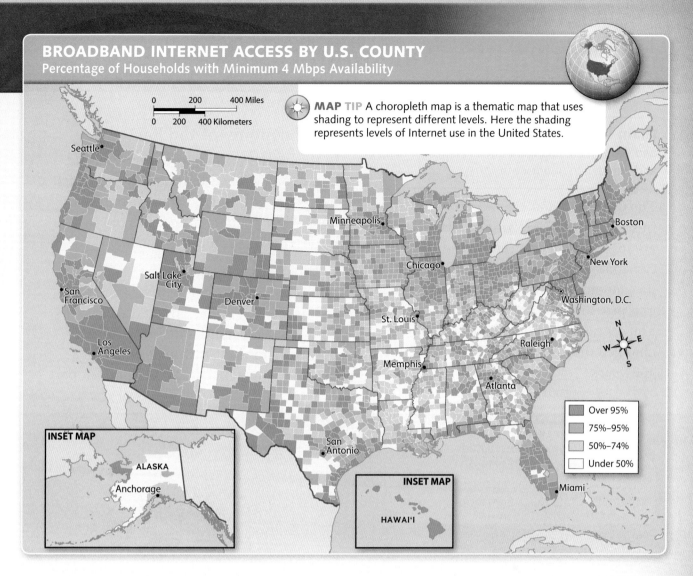

## Global and Mobile Media Today

Today, texting, instant messaging, and social networking allow fast and easy communication between people around the world. Much of this is done with **smartphones**, which combine communication and software applications into one handheld device. Media technology encourages a culture that is **global**, or worldwide. Access to information is available from anywhere in the world. Media culture forms around shared ideas or interests, rather than physical location. Most technology is now **mobile**, which means it moves along with the people who use it. For many people, new media has entirely replaced traditional methods of communication.

### Before You Move On
**Make Inferences** Why do you think mass media has such a strong influence on culture in the United States and Canada?

### ONGOING ASSESSMENT
# MAP LAB

1. **Region** Based on what you know about population density in the United States, how would you describe the pattern of Internet access across the country?

2. **Compare and Contrast** Find where you live on the map. How does the level of high speed Internet access in your county compare to the rest of the country? Why might this be?

3. **Draw Conclusions** What effect do you think widespread availability of the Internet has had on print publications? Why?

SECTION 1.2

# SECTION 1 UNITED STATES & CANADA

## 1.3 The Changing Economy

**TECHTREK**
myNGconnect.com For photos of manufacturing and current events

Digital Library

**Main Idea** The economies of the United States and Canada are changing due to new technology and worldwide trends.

The Canadian and U.S. economies differ in some ways. For example, businesses are less regulated in the United States. Canada's government plays a larger role in its economy by owning some businesses and providing health care for all citizens. The economies of both countries have changed dramatically over the past 150 years.

### The Economy of the Past

In the early 1800s, most Americans and Canadians worked in industries in which people make a living directly from the land. Most people farmed, worked in mines, fished, or logged timber.

During the mid-1800s, the Industrial Revolution, which began in Europe, swept the United States and Canada. **Manufacturing**, or using machines to make raw materials into a usable product, became an important part

**Critical Viewing** Automated robot arms weld trucks in a Michigan assembly plant. What details in the photo illustrate this automation?

of the economy. In vast factories, workers manufactured products such as iron and steel. The manufacturing industry led the economy from the 1870s until the 1950s.

During the 1950s, the service industry expanded. Workers in this industry provide a service rather than make a product. Jobs in areas such as health care, education, entertainment, banking, and retail are all part of the service industry.

## The Information Economy

One important similarity between the U.S. and Canadian economies is that they are **dynamic**; that is, they change quickly. Part of the reason for rapid change is the availability of new computer and information technologies.

In 1974, a tiny company called MITS built the first personal computer, the Altair. It had limited memory but worked well. The only problem was that you had to build it yourself from a kit! Today, more than 8 out of every 10 U.S. and Canadian households have computers, and **fiber optics** uses digital technology to transmit voice, text, and visual messages. These types of advancements have shifted the economy in a new direction.

Today's technologies allow people to store, organize, and retrieve enormous amounts of financial information with a few mouse clicks. New technologies also can bring about more jobs in the communication and information industry. Faster communication across greater distances helps support a global economy, unlimited by physical distance.

**Visual Vocabulary** **Fiber optics** is a method of sending light through glass fibers. Fiber optic cables can transmit digital code quickly across great distances.

## Global Recession

In 2007, the economy faced a **recession**, or a slowdown in economic growth. Lending practices from earlier in the decade led banks to grant large loans to consumers and businesses—loans they could not pay back. These unpaid debts led to a global financial crisis, including the collapse of some major worldwide banks. A recession is officially over after six consecutive months of economic growth. The U.S. economy reached this mark in early 2009.

### Before You Move On
**Make Inferences** In what ways might new technologies bring about more jobs?

### ONGOING ASSESSMENT
## PHOTO LAB

1. **Analyze Visuals** What do you expect to see on an assembly line that is missing from the photo? How might this technology affect jobs?
2. **Synthesize** What similarities and differences do you notice in the photos?
3. **Make Inferences** In what ways can new technologies affect a country's economy?

# SECTION 1 GLOBAL ISSUES • Energy

## 1.4 Finding New Energy Sources

**Main Idea** The growing demand for nonrenewable fossil fuels for energy has led to exploration into renewable, alternative energy sources.

**KEY VOCABULARY**

**petroleum**, n., the raw material used to create oil products

**wind turbine**, n., an engine powered by wind to generate electricity

**ethanol**, n., a fuel obtained from plants that can be used alone or blended with gasoline

**hybrid**, n., a vehicle that can run using either an electric motor or a gas-powered engine

In April 2010, the *Deepwater Horizon* oil drilling rig in the Gulf of Mexico exploded. The explosion opened a hole in a well drilled in the seafloor. Oil gushed from that hole every day for nearly three months, leaking nearly 5 million barrels of oil. Before this accident, the worst U.S. oil disaster at sea was the grounding of the *Exxon Valdez* tanker in 1989. That disaster lost around 260,000 barrels of oil—a large amount, but far smaller than the amount of oil leaked after the explosion of the *Deepwater Horizon*.

### Oil Supply and Demand

The United States runs on energy, and oil provides about a third of that energy. **Petroleum** is a nonrenewable resource, meaning that it will eventually run out. However, no one knows exactly how much petroleum might be left in reserves that lie deep beneath the earth's surface.

Experts try to predict when the United States will use up all of its own oil supply, and they don't all agree. However, they do agree that the demand for oil is growing, rapidly shrinking the supply of this nonrenewable resource. Some experts say that as early as 2020, there won't be enough oil in the United States for everyone who wants it.

Today, the United States depends on other countries for more than two-thirds of its oil. As demand rises and supply falls, the price of oil will rise. Rising oil prices can slow economic growth and drive countries into competition for oil.

**Before You Move On**
**Summarize** What are some of the effects of an increasing demand for oil in the United States?

WIND TURBINE

## Alternative Solutions

Energy sources other than oil include nuclear power, **ethanol**, solar power, and wind power. However, each has issues that make it less than ideal. Nuclear power plants take a long time to build and are very expensive. They also create radioactive waste. Making ethanol from grains means some agricultural production is used for fuel instead of food. Solar panels and wind turbines cannot always produce electricity on demand.

Auto manufacturers are now making more electric and **hybrid** vehicles. Hybrids use fuel cells to make energy. Scientists hope to use common gases like hydrogen and oxygen in these cells. If so, the supply of energy will be cheap and almost unlimited.

**COMPARE ACROSS REGIONS**

## Innovation in Other Countries

Countries around the world are using a variety of energy innovations. For example, Brazil has been producing ethanol since the 1970s and is a leading producer of this fuel. Britain has banned conventional light bulbs in favor of bulbs with lower wattage, which use less energy. Nuclear power supplies more than 76 percent of France's electricity. Denmark's coal plants produce both electricity and hot water, making them more efficient. Some nations, such as Iceland and New Zealand, use geothermal energy, heat produced within the earth. These alternatives are some energy-efficient solutions and illustrate new ways of looking at energy production.

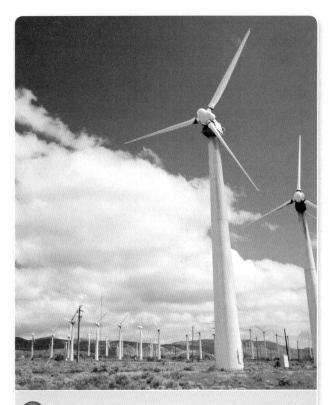

**Visual Vocabulary** A **wind turbine** is an engine that is powered by wind in order to generate electricity. A group of wind turbines placed together is a wind farm.

**Before You Move On**
Make Inferences In what ways might the use of alternative energies in other countries affect energy use in the United States?

### ONGOING ASSESSMENT
## READING LAB  GeoJournal

1. **Make Inferences** Which alternative energy sources depend on physical geography, and in what ways?
2. **Summarize** What issues prevent certain alternative energy sources from being used?
3. **Form and Support Opinions** Which alternative energy source do you think is most promising? Write a paragraph stating your choice. Support your opinion with evidence from the lesson.

SECTION 1.4

# SECTION 1 UNITED STATES & CANADA

## 1.5 Citizens' Rights and Responsibilities

**TECHTREK**
myNGconnect.com For photos and a chart on citizens' rights
Digital Library
Student Resources

**Main Idea** Citizens in the United States and Canada enjoy many rights and also have certain responsibilities.

The governments of the United States and Canada share many similarities. Leaders are elected by voters. Government power is shared by the national government and smaller units—states in the United States and provinces in Canada. Laws are made by a legislative body. Perhaps most importantly, all citizens in these countries have rights and responsibilities.

### Government Structures

Both governments consist of three branches—the executive, the legislative, and the judicial—each with specific powers. The legislative, or law-making, branches are the U.S. Congress and the Canadian Parliament. In the United States, the head of the executive branch is the president, elected by the people. In Canada, by contrast, the chief executive is the prime minister, who is chosen by the party with the most seats in Parliament.

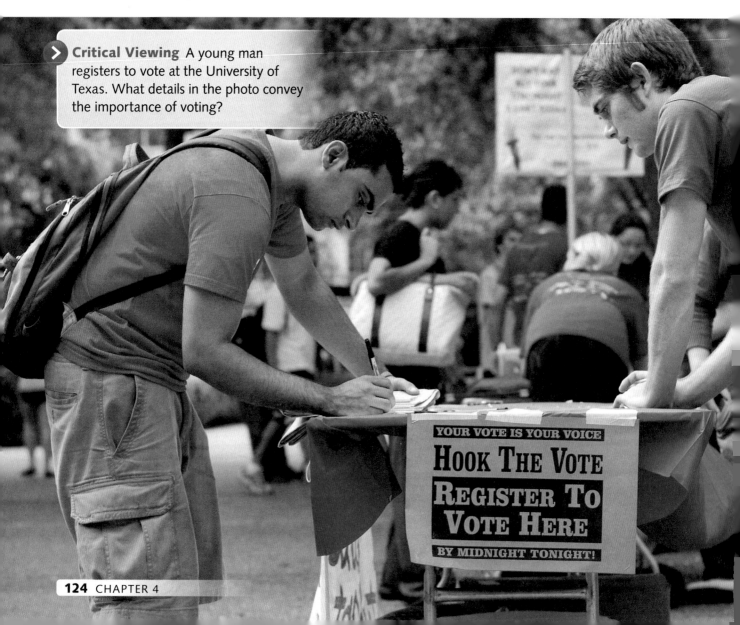

> **Critical Viewing** A young man registers to vote at the University of Texas. What details in the photo convey the importance of voting?

## Rights and Responsibilities

In both countries, all citizens enjoy basic rights, such as the right to citizenship, equality, and fair treatment. People become citizens by being born in the country or by **naturalization**, the process that someone born in another country follows to become a citizen.

The Bill of Rights, which is the first ten amendments to the U.S. Constitution, and Canada's Charter of Rights guarantee basic freedoms. Among these are freedom of speech and religion, and certain legal rights. For example, citizens have the right to **due process**, or specific rules that authorities must follow. Canada's charter also identifies two official languages, English and French. Citizens in Canada have the right to be educated in either one.

The citizens of the United States and Canada enjoy a great deal of freedom. They can live where they wish and speak their own languages. They can express their opinions freely and practice any religion they choose. They can gather together and speak out against their government. Many people who live in these two countries immigrated from countries that did not allow these basic freedoms.

In both countries, citizens with rights also have responsibilities. The right to vote brings with it the responsibility to cast a vote fairly, through formal elections. Legal rights bring about the responsibility to obey laws and to serve on a jury, which is a group of fellow citizens who listen to both sides of a legal matter and decide whether laws were broken.

**CITIZENS' RIGHTS**

**UNITED STATES**
- Protection of rights not guaranteed by the Constitution
- Right to bear arms

**UNITED STATES & CANADA**
- Right to a speedy trial
- Right to trial by jury
- Right to vote
- Freedom of speech, religion, press, assembly
- Protection against unreasonable search
- Right to have a lawyer

**CANADA**
- Right to be educated in a minority language (French)
- Right to be protected against discrimination on basis of race or gender

Some responsibilities are not identified in any document, but are just as important as those required by law. For example, citizens are expected to treat each other fairly. The United States and Canada are home to a wide variety of races, religions, and customs. Citizens in diverse societies like these have the responsibility to accept one another's differences.

### Before You Move On
**Make Inferences** Why is responsibility an important part of having rights and freedoms?

### ONGOING ASSESSMENT
## SPEAKING LAB  GeoJournal

1. **Make Generalizations** Based on the graphic above, which document, the Bill of Rights or Canada's Charter of Rights, provides stronger protection for rights? How?
2. **Express Ideas Through Speech** Write and deliver a speech that explains your ideas about which citizens' right and responsibility you think is most important in a society, and why.

# SECTION 2 FOCUS ON MEXICO
## 2.1 Daily Life in Mexico

myNGconnect.com For photos of Mexico's daily life

Digital Library

**Main Idea** Life in Mexico today reflects a blend of traditional and modern elements and Native American and Spanish cultures.

Much of Mexican life is modern, but many Mexican people hold a deep respect for the past. Mexican culture blends contributions from the Native American and Spanish cultures that shaped its history.

### A Blending of Cultures

Native people of Mexico had created successful empires before the Spanish conquest in 1521. **Descendants**, or future generations, of these indigenous people still make up a large part of Mexico's population. The largest population group is the mestizos, who are Mexicans of mixed **ancestry**, or heritage. This ancestry can include native, European (particularly Spanish), and African family roots. Mixed ancestry is a source of pride among mestizos in Mexico.

The blend of cultures is seen in the many languages and religions of Mexico. While nine out of ten Mexicans speak only Spanish, the country's official language, the rest speak one of more than 50 Native American languages. Similarly, more than 80 percent of Mexico's people follow the traditional Roman Catholic religion. However, many Native Americans practice traditional religions, and many Mexican customs show the blending of cultures. For example, a celebration called the Day of the Dead, which honors one's ancestors, mixes native and Catholic traditions.

Many Mexican artists have blended native and European culture. Diego Rivera and David Alfaro Siqueiros (sih KAY rohs) used European style to paint **murals**, or large paintings on walls, that highlight Native American culture in Mexico.

### Modern and Traditional Life

Today, life in Mexico is a balance of the traditional and modern. Once largely rural, Mexico is now an urban nation. More than three-quarters of its people live in cities. The capital, Mexico City, is one of the world's largest cities, and is home to one-fifth of Mexico's population.

Mexico embraces past and present. The country has preserved sites that include Aztec and Mayan art and **artifacts**, or tools and other ornaments that show something about how a culture lived. The National Palace in Mexico City, built to house colonial viceroys, is still used as a government building. Mexico also has bold, modern architecture, such as the Lighthouse of Commerce in Monterrey.

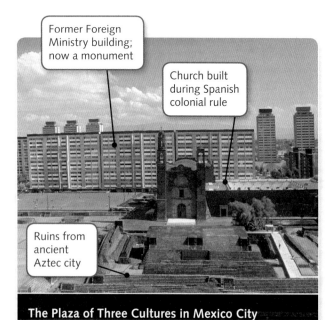

The Plaza of Three Cultures in Mexico City represents three elements of Mexican culture: Aztec, Spanish, and mestizo.

- Former Foreign Ministry building; now a monument
- Church built during Spanish colonial rule
- Ruins from ancient Aztec city

Mexican cooking continues to include staple foods of its indigenous people: corn, beans, and squash. Corn is ground to make *tortillas*, a flat bread served with many meals. Traditional foods now must fit into a more modern lifestyle. Families used to share a large midday meal, but shorter lunch hours and long commutes for modern workers make it difficult to keep up this tradition.

Some Mexican people follow traditional practices as a way of life. This is especially true among rural people. For example, *vaqueros*, or cowboys, still work on some ranches in northern Mexico, using their traditional skills.

**Before You Move On**
**Summarize** In what ways is Mexican culture a blend of influences?

### ONGOING ASSESSMENT
# PHOTO LAB

1. **Synthesize** What details in the photos reveal the importance of tradition in Mexico?
2. **Place** In what ways does the Plaza of Three Cultures represent Mexican culture?
3. **Make Inferences** Why might the mixed cultural heritage of the *mestizos* be a source of pride in Mexico?

> **Critical Viewing** A quinceañera (keen sin NYAR ah) is a traditional celebration of a girl's 15th birthday. It is both a religious and a social event. What details in the photo remind you of other ceremonies?

# SECTION 2 FOCUS ON MEXICO
## 2.2 The Impact of Globalization

**Main Idea** Mexico has a developing economy that faces challenges as a result of global trends.

As parts of Mexico are modernizing, the country's economy is growing. However, **globalization**, the development of a world economy based on free trade and the use of labor from other countries, has presented economic challenges.

### A Drive for Growth

For much of the 20th century, some major industries in Mexico were **nationalized**, or placed under government control. All income from a nationalized industry goes to the government. The profitable oil industry, for example, was nationalized in 1932. For periods of time during the 1900s, the government also controlled Mexico's banking, transportation, and telecommunications systems.

In the 1980s, in an effort to improve the country's economy, Mexico released some control of these industries to private and foreign investors. Another factor of Mexico's economic growth was the adoption of the **North American Free Trade Agreement (NAFTA)** in 1994, which removed many barriers to trade. Since then, Mexican trade with the rest of North America—mostly with the United States—has increased nearly 300 percent. However, NAFTA's critics say the agreement unfairly favors commercial agriculture over small-scale farmers.

### Three Economic Sectors

Economic activity in Mexico takes place in three **economic sectors**, or subdivisions of the economy: agriculture, manufacturing, and service. The agricultural sector includes large-scale production for export.

**Critical Viewing** This worker cleans molds that are used to make shoe soles in a factory in Leon, Mexico. What skills would be required of factory workers?

Major exports include tropical fruits, coffee, sugarcane, and cotton. Silver is also an important mineral resource for Mexico's export income in this sector. Export income also comes from oil sold on world markets.

Manufacturing in Mexico includes automobiles, food processing, and metal products. Heavy foreign investment followed the adoption of NAFTA, which increased manufacturing activity. This investment led to more *maquiladoras*, foreign-owned factories where parts made elsewhere are put together into finished goods and then exported worldwide.

The dollar amount provided to the GDP (Gross Domestic Product) by service industries increased by a factor of more than 25 between 1970 and 2009. This sector includes services such as banking and transportation. Mexico's warm climate and cultural treasures attract tourists from around the globe, making tourism another big part of the service economy.

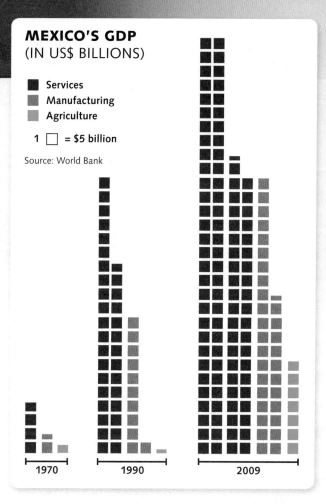

**MEXICO'S GDP (IN US$ BILLIONS)**

- Services
- Manufacturing
- Agriculture

1 ☐ = $5 billion

Source: World Bank

1970  1990  2009

## Economic Challenges

Mexico still faces challenges. Economic growth hasn't happened for most poor people. Many Mexicans have difficulty finding work and migrate to the United States. Some Mexican-owned businesses have transferred jobs to countries where workers are paid less.

### Before You Move On
**Make Inferences** In what ways has globalization affected Mexico's economy?

---

**ONGOING ASSESSMENT**
# DATA LAB  GeoJournal

1. **Analyze Data** Based on the graph, which economic sector has grown the most?
2. **Synthesize** How might NAFTA account for the growth in agriculture between 1990 and 2009?
3. **Region** Why is tourism important to Mexico?

# SECTION 3 FOCUS ON MEXICO
## 2.3 Reaching Toward Democracy

TECHTREK
myNGconnect.com For a graph and photos on democracy in Mexico

Maps and Graphs | Digital Library

**Main Idea** Mexico has made progress toward democracy but still faces serious obstacles.

Mexico faces not only economic challenges but also political ones. The country's political history is a factor in today's struggles toward democracy.

### Stability Without Democracy

The biggest limit on democracy came from the dominance of the **Institutional Revolutionary Party (PRI)**, which controlled Mexico's government from 1929 to 2000. The PRI formed as a result of major political instability that included the assassination of Mexico's newly elected president in 1928. The party's goal was to enact the reforms that were fought for during the Mexican Revolution.

### Multi-Party Democracy

Economic and political problems during the 1990s weakened the PRI's hold on power in the Mexican states. In 1994, the Zapatista Liberation Army staged a violent rebellion in response to new economic policies that were harmful to poor, indigenous people. This rebel group seized control of several Mexican states.

In 1996, the government enacted election **reforms**, or changes aimed at correcting problems. These reforms made voting more fair and made it easier for parties other than the PRI to run candidates. Elections between candidates from the PRI and other parties were closer each year. In 2000, the election of Vicente Fox of the National Action Party (PAN) broke 71 years of PRI rule.

Fox was appealing to many Mexican people. He promised to improve the economy, clean up government corruption, and resolve disputes with rebellious political groups, particularly the Zapatistas. The results of Fox's presidency were complex. He had raised the expectations of many Mexican people who were suffering economically. However, the PRI still controlled Congress. Therefore, Fox's reform efforts moved slowly and many Mexicans who originally supported Fox were disappointed.

> **Critical Viewing** Felipe Calderón greets supporters after being declared winner of the 2006 presidential election. How would you describe the emotions of the crowd?

Perhaps the most important result of Fox's presidency was that it brought about **multi-party democracy**, a political system in which elections include candidates from more than one party. Presidents in Mexico serve only one six-year term, so the 2006 election included two new candidates: Felipe Calderón of the PAN and López Obrador from the leftist party of the Democratic Revolution, which had become another important political voice in Mexican politics. Calderón was declared winner, but with less than a one percent lead over Obrador. This election further proved that multi-party democracy was in place in Mexico.

## Challenges Ahead

One of the obstacles to Mexico's democracy comes from the illegal drug trade, which is controlled by several drug cartels. The activities of the cartels brought about heavy violence in the early 2000s, especially along the U.S.-Mexico border. In 2009, President Calderón sent troops to the area to control the violence, but the cartels still challenge the authority of Mexico's government. Democracy is also challenged by the Zapatistas' ongoing control in several Mexican states.

A positive trend in Mexico is its increasing literacy rate. By the start of the 21st century, nearly 90 percent of Mexico's population was literate, or able to read and write. A high literacy rate is one factor that helps a society be more productive and contribute to the country's economy. It is also an important element of a democracy, because people need to be informed to take part in decision-making.

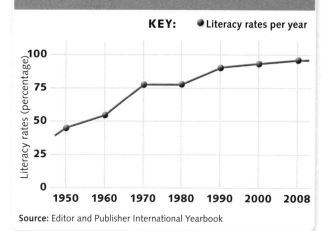

LITERACY RATES IN MEXICO (1950–2008)

KEY: ● Literacy rates per year

Source: Editor and Publisher International Yearbook

Many children in rural Mexico speak a regional dialect and must learn Spanish in school.

### Before You Move On
**Monitor Comprehension** How has the Mexican government become more democratic?

## ONGOING ASSESSMENT
# READING LAB  GeoJournal

1. **Identify Problems and Solutions** What problems does Mexico's government now face? How might it solve them? Explain your answer.

2. **Interpret Graphs** During which 10-year period did Mexico's literacy rate increase the most? When did it increase the least?

3. **Draw Conclusions** How might Mexico's growing literacy rate help achieve democracy?

# CHAPTER 4
# Review

## VOCABULARY

On your paper, write the vocabulary word that completes the following sentences.

1. North America's _____ shows in its mix of races, languages, and religions.
2. With new media able to reach people all over the world, culture has become _____.
3. A _____ occurs when the economy shows a sharp decline in business activity.
4. A vehicle that runs on a combination of electric and gas power is called a _____.
5. Some Mexican artists create _____ that show the country's native cultures.
6. When an industry is _____, it is placed under government control.

## MAIN IDEAS

7. How did the United States and Canada become so diverse? (Section 1.1)
8. How has mobile technology affected cultures in the United States and Canada? (Section 1.2)
9. How did the information age transform the U.S. and Canadian economies? (Section 1.3)
10. Why are researchers looking at alternative sources of energy? (Section 1.4)
11. How do immigrants from other lands gain the rights and responsibilities of American and Canadian citizenship? (Section 1.5)
12. What are two examples of the continuing influence of Spanish culture on Mexico? (Section 2.1)
13. What economic success has Mexico seen in recent decades? (Section 2.2)
14. Why is it important for a country struggling toward democracy to have more than one political party? (Section 2.3)

## FOCUS ON UNITED STATES & CANADA

### ANALYZE THE ESSENTIAL QUESTION

What issues do the United States and Canada face today?

**Focus Skill: Identify Problems and Solutions**

15. What do you think is the single biggest challenge facing the United States? Why do you think so?
16. What do you think is the single biggest challenge facing Canada? Why do you think so?
17. Choose one of the challenges you identified in the previous two questions. What do you think the country should do about that challenge? Why would those steps help?

### INTERPRET TABLES

| UNITED STATES TRADE BALANCE ($ BILLIONS) | | | |
|---|---|---|---|
| | 2006 | 2007 | 2008 |
| Total | −760.4 | −701.4 | −696.0 |
| With main trading partners | | | |
| Canada | −71.8 | −68.2 | −78.3 |
| China | −234.1 | −258.5 | −268.0 |
| Germany | −89.1 | −94.2 | −43.0 |
| Japan | −89.7 | −84.3 | −74.1 |
| Mexico | −64.5 | −74.8 | −64.7 |

Source: 2010 Statistical Abstract of the United States, Table 1264 and Table 1271

18. **Analyze Data** A negative trade balance means that a country imports more than it exports. What trend do you see in the overall trade balance from 2006 to 2008?
19. **Synthesize** If the United States eliminated its trade deficit with Canada and Mexico, would that have much effect on the overall trade balance? Why or why not?

## FOCUS ON MEXICO

### ANALYZE THE ESSENTIAL QUESTION

How has globalization affected Mexico?

**Focus Skill: Summarize**

20. How has the balance of rural and urban populations shifted in Mexico?
21. What factors have contributed to economic growth in Mexico?
22. What challenges come along with economic growth in Mexico?
23. What qualities will help Mexico become a successful democracy and take part in the global economy?

### INTERPRET MAPS

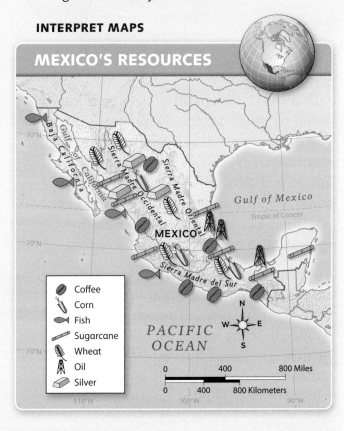

24. **Interpret Maps** Which natural resource is found only in the mountains? Which resource besides fish and oil is found on coastal plains?
25. **Synthesize** Which resources on the map are important to Mexico's success in the global economy? Explain why.

## ACTIVE OPTIONS

**Synthesize** the Essential Questions by completing the activities below.

26. **Write a Speech** Imagine that you are president of the United States preparing to visit either Mexico or Canada. In a speech to that country's people, you want to emphasize the cooperation and good relations between your two countries. Choose one of those countries and then write and deliver a two-minute speech that a president might deliver. Use the following tips to help you write.
**After all the speeches are delivered, discuss the ideas that were presented and evaluate which are best.**

> **Writing Tips**
> - Identify a clear theme.
> - Use details in history and the current situation to support the idea that the two countries cooperate well.
> - State your goals for relations in the future.

27. **Create a Multimedia Presentation** Make a chart like the one below. Use it to plan a multimedia presentation on one of the three countries in this chapter. Use online sources to gather facts and photos.

| Country | |
|---|---|
| Population | |
| Culture | |
| Economy | |
| Government | |
| Challenges | |

REVIEW **133**

COMPARE ACROSS REGIONS  UNIT 2

# Becoming a Citizen

TECHTREK
myNGconnect.com For graphs of U.S. naturalizations

Maps and Graphs

Naturalization is the process by which someone who was not born in a country can become a citizen of that country. Every country spells out in detail who can or cannot become a citizen and what steps must be taken as part of the naturalization process.

Requirements for naturalization in most countries include specific guidelines in several areas: residency (the amount of time spent living in the country); language proficiency (the ability to speak the country's official language); cultural and historical knowledge of the country (usually a written test); and character, or the kind of person you are (usually determined by a background check).

**Compare**
- China
- Cuba
- Ethiopia
- India
- Iran
- Mexico
- Ukraine
- United States

### NATURALIZATION AND CITIZENSHIP

Most countries require a person who wishes to become a citizen to first gain permanent residency status before even applying for naturalization. A permanent resident is allowed to live and work in a country, but legally is still a citizen of the home country.

Permission for a non-citizen to immigrate and live as a permanent resident in a country can be very difficult to obtain. Eligibility may depend on the need a country has to fill certain categories of jobs, or the country's political situation. Permission for permanent residency also may be limited by a variety of requirements. Background checks for any illegal activity or criminal conviction are typical and can disqualify a candidate for both the right to immigrate and the right to apply for naturalization.

Naturalization requirements may vary based on other factors. For example, the years of permanent residency a country requires may be fewer for a person who has served or is willing to serve in that country's military. Permanent residency requirements are usually lower for the spouse, or married partner, of a citizen.

Most countries require a person applying for naturalization to be a competent speaker of the country's official language. Many countries offer and even require applicants to complete a series of citizenship classes. People applying for naturalization usually have to pass a test on the country's history or government, in addition to proving their language proficiency. The final step in obtaining citizenship is often a naturalization ceremony, in which new citizens take an oath, or make a formal promise to live by their new country's laws and customs.

# U.S. NATURALIZATION*

## BY THE NUMBERS

**899,162**
Number of applications filed for U.S. naturalization, 2012

**65,874**
Number of denied applications for U.S. naturalization, 2012

**$680**
U.S. naturalization application filing fee

**40**
Median age at naturalization, 2010–2012

**7**
Median years as permanent resident, 2012

**22%**
Percentage of naturalized citizens with a Bachelor's degree, 2010

**$51,926**
Median household income of naturalized citizens, 2011

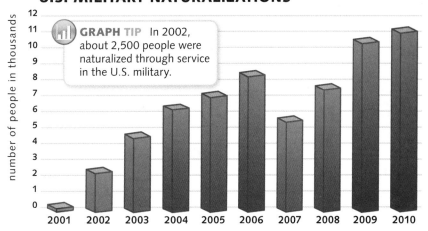

**U.S. MILITARY NATURALIZATIONS**

**GRAPH TIP** In 2002, about 2,500 people were naturalized through service in the U.S. military.

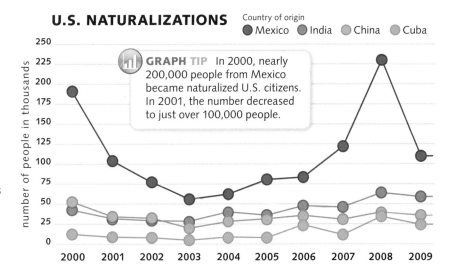

**U.S. NATURALIZATIONS**
Country of origin: Mexico, India, China, Cuba

**GRAPH TIP** In 2000, nearly 200,000 people from Mexico became naturalized U.S. citizens. In 2001, the number decreased to just over 100,000 people.

*Sources: U.S. Department of Homeland Security; Congressional Research Service

## ONGOING ASSESSMENT
### RESEARCH LAB  GeoJournal

1. **Draw Conclusions** Why would a country require knowledge about its history, government, and culture to gain citizenship?
2. **Make Predictions** What would you predict about the number of U.S. military naturalizations in 2011? Use evidence from the U.S. Military Naturalizations graph to explain your answer.
3. **Interpret Graphs** What is similar about the number of U.S. naturalizations from all four countries from 2000 to 2001? What about from 2007 to 2008?

**Research and Create Graphs** Do research to find out the number of people from Ethiopia, Iran, and Ukraine who became U.S. naturalized citizens each year from 2000 to 2009. Create a line graph like the one above to show your data. Then write a short paragraph explaining the changes in the numbers throughout the time period. For example, in which year(s) did the number go down for all three countries? In which year(s) did it increase? What inferences can you draw from the data?

COMPARE ACROSS REGIONS

REVIEW & EXTEND UNIT 2

# Active Options

## ACTIVITY 1

**Goal:** Evaluate important physical structures.

### Write Seven Wonders List

Since ancient times, people have made "Seven Wonders" lists of magnificent structures around the world, such as the Great Pyramids in Egypt or the Statue of Zeus in Greece.

Create such a list for the continent of North America. Review the photos in the **Digital Library** and use the research links to help you decide which structures belong on the list. Your list can include both natural structures and structures built by humans. For each item on your list, write the reasons why the structure deserves to be named as one of the Seven Wonders of North America.

Mount Rushmore in South Dakota

## ACTIVITY 2

**Goal:** Research a national park.

### Conduct a Walk in the Park

The U.S. National Park Service protects about 400 parks, and Parks Canada protects over 40 national parks. Choose a national park in North America from the list below or from another source. Do research and use the **Magazine Maker CD-ROM** to create a "show and tell" route through the park to show what makes it so important and worth preserving.

**National Parks**
- Denali, AK
- Everglades, FL
- Glacier, MT
- Yosemite, CA
- Zion, UT
- Yellowstone, WY, MN, ID
- Glacier, British Columbia, Canada
- Prince Edward Island, Canada

## ACTIVITY 3

**Goal:** Learn about National Geographic Explorers.

### Explore the Explorers

With a group, find out more about National Geographic Explorers Beth Shapiro or Sam Meacham. Research the explorer's background and how he or she became interested in the work he or she does today. Find out more details about the explorer's current project and possible future plans. Then create a poster that highlights the explorer and his or her work.

# UNIT 3

## Explore Central America & the Caribbean with NATIONAL GEOGRAPHIC

### MEET THE EXPLORER

**NATIONAL GEOGRAPHIC**

Emerging Explorer Ken Banks created a communication tool for farmers in El Salvador. They can discuss crop pricing and remain competitive using an inexpensive cell phone.

### INVESTIGATE GEOGRAPHY

Central America was once entirely covered by rain forests. Farmers have cleared many of these hot, tropical areas for cattle ranching and sugar plantations. The remaining rain forests still contain many plants and tropical birds, including this toucan.

### STEP INTO HISTORY

Toussaint L'Ouverture led a slave revolt in Santo Domingo, later called Haiti, and laid the groundwork for its independence in 1804. The former slave was a brilliant general, defeating the powerful French army.

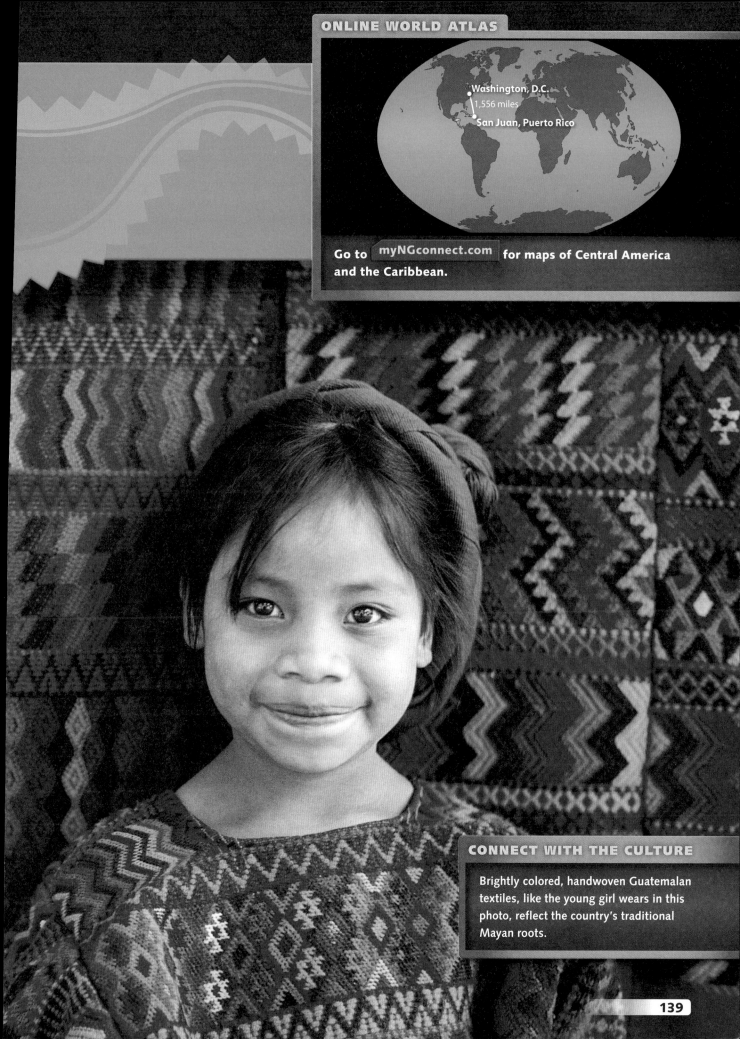

## ONLINE WORLD ATLAS

Washington, D.C.
1,556 miles
San Juan, Puerto Rico

Go to myNGconnect.com for maps of Central America and the Caribbean.

## CONNECT WITH THE CULTURE

Brightly colored, handwoven Guatemalan textiles, like the young girl wears in this photo, reflect the country's traditional Mayan roots.

# CHAPTER 5

# Central America & the Caribbean
# GEOGRAPHY & HISTORY

## PREVIEW THE CHAPTER

**Essential Question** How has physical geography been a positive or negative influence on the economy of the region?

### SECTION 1 • GEOGRAPHY

| | |
|---|---|
| 1.1 Physical Geography | 142 |
| 1.2 Earthquakes and Volcanoes | 144 |
| 1.3 Rain Forests of Central America | 146 |
| 1.4 Saving Sea Turtles | 148 |
| NATIONAL GEOGRAPHIC Photo Gallery | 150 |

**KEY VOCABULARY**
- isthmus
- coastal plain
- rain forest
- archipelago
- tectonic plate
- seismic
- ecosystem
- deforestation
- fertile
- tourism
- canopy
- extinction
- poacher

**ACADEMIC VOCABULARY**
critical

**TERMS & NAMES**
- Caribbean Sea

lemur frog, Costa Rica

Go to myNGconnect.com for more on Central America and the Caribbean.

**Essential Question** How have economic resources influenced the history of the region?

**KEY VOCABULARY**
- cash crop
- scarcity
- triangular trade
- multitude
- staple
- viceroy
- province
- harbor
- dictator
- commonwealth

**ACADEMIC VOCABULARY**
exploit

**TERMS & NAMES**
- Columbian Exchange
- Toussaint L'Ouverture

### SECTION 2 • HISTORY

2.1 Trade Across Continents .................. 152

  DOCUMENT-BASED QUESTION
2.2 The Columbian Exchange .................. 154

2.3 Paths Toward Independence .................. 156

2.4 Comparing Cuba and Puerto Rico .................. 158

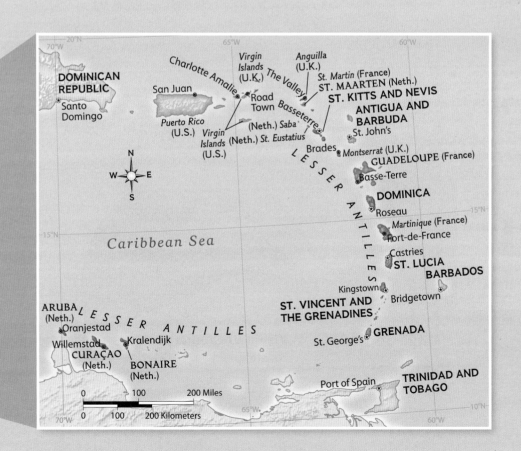

# SECTION 1 GEOGRAPHY
## 1.1 Physical Geography

**TECHTREK**
myNGconnect.com For online maps of the region and Visual Vocabulary
- Maps and Graphs
- Digital Library

### CENTRAL AMERICA AND THE CARIBBEAN PHYSICAL

142 CHAPTER 5

## CLIMATE

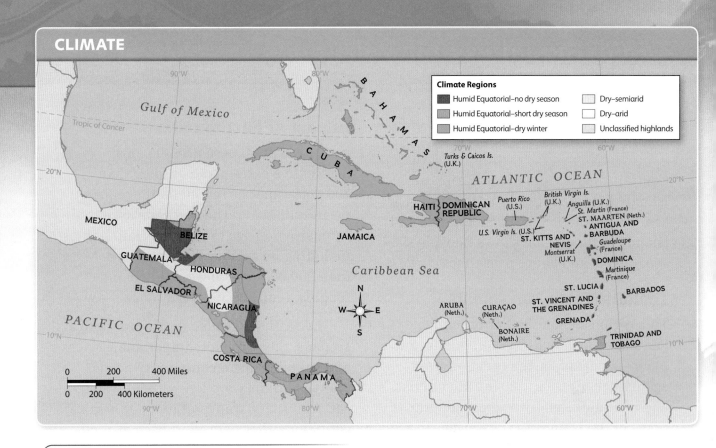

**Main Idea** Mountains, coastal plains, and rain forests are the main landforms that support the region's economy.

Central America and the Caribbean islands are located between the continents of North America and South America. The physical geography of the region supports the agriculture and tourism that are so valuable to the region's economy.

### Central America

Central America is an **isthmus**, a narrow strip connecting two large land areas. A range of volcanic mountains spreads across the seven countries of the region. The climate in the mountains is cool and the rich volcanic soil is ideal for growing coffee beans, an important export. The **coastal plains**, which are the lowlands next to the seacoast, and the tropical **rain forests**—heavily wooded forests that may receive more than 100 inches of rain per year—provide resources that boost the economy.

### The Caribbean Islands

The Caribbean islands curve in an **archipelago**, or chain of islands, between the Atlantic Ocean and the **Caribbean Sea**. Sugarcane, grown on coastal plains, is the leading crop. The climate is mild in winter and hot in summer, and attracts visitors year-round.

**Before You Move On**
Summarize In what ways do the main landforms of the region support the economy?

### ONGOING ASSESSMENT
## MAP LAB

1. **Location** Based on the physical map, where are the mountainous regions of Central America located?

2. **Interpret Maps** What is the climate like on the Central American and Caribbean coasts?

3. **Turn and Talk** Discuss with a classmate the differences in climate in the mountains and on the coastal plains. Where would you like to live?

SECTION 1.1 **143**

# SECTION 1 GEOGRAPHY
## 1.2 Earthquakes and Volcanoes

**Main Idea** Earthquakes and volcanoes affect daily life and economic activities in the region.

Central America and many Caribbean islands were underwater millions of years ago. The formation of the mountains and islands in this region shows how physical geography can change over time.

### Tectonic Plate Movements

This region sits on the Caribbean Plate. For tens of millions of years, this **tectonic plate**—a part of Earth's crust—moved against other plates, causing land to rise and mountains and volcanoes to form. Some of the islands in the Caribbean are really just the tips of ancient volcanoes created by tectonic plate movement.

The location of the islands on these plates is significant. Today, plates slowly continue to move, crashing into or sliding over or under each other. This movement causes earthquake, or **seismic**, activity and volcanic eruptions. An eruption in Montserrat in 1996 caused two-thirds of the population to flee their homes.

### Impact of Earthquakes

In 2010, Haiti experienced its most destructive earthquake in more than 200 years, measuring 7.0 on the moment magnitude scale. The earthquake caused extensive damage to many of Haiti's cities, including its capital, Port-au-Prince. More than 200,000 people were killed and roughly a million were left homeless.

The earthquake also destroyed **critical** systems, the extremely important services and supplies that a community needs. Many Haitians were left without water, gas, electricity, transportation, or medical care. Donations poured in from around the world in an international effort to help Haiti recover and rebuild.

### Before You Move On
**Summarize** What causes earthquakes and volcanoes in this region, and what are some of the effects?

> **Critical Viewing** This courthouse was buried in a landslide caused by a volcano on Montserrat. What does this photo suggest about volcanoes?

## PLATES AND VOLCANOES

| | MAJOR EARTHQUAKES IN THE REGION SINCE 2000* | |
|---|---|---|
| ① El Salvador | January 2001 | 7.7 |
| ② Honduras | May 2009 | 7.3 |
| ③ Nicaragua | October 2004 | 7.0 |
| ④ Haiti | January 2010 | 7.0 |
| ⑤ Guatemala | June 2007 | 6.7 |
| ⑥ Nicaragua | July 2005 | 6.6 |
| ⑦ El Salvador | February 2001 | 6.6 |
| ⑧ Dominican Republic | September 2003 | 6.4 |
| ⑨ Costa Rica | November 2004 | 6.4 |
| ⑩ Costa Rica | January 2009 | 6.1 |

Source: U.S. Geological Survey http://www.usgs.gov/
*based on moment magnitude scale

### ONGOING ASSESSMENT
## MAP LAB  GeoJournal

1. **Interpret Maps** What is significant about Central America's location near where the Caribbean Plate and the Cocos Plate meet?

2. **Location** According to the map, where do the most and the fewest volcanic eruptions happen in Central America? Explain.

3. **Make Inferences** Trace the boundary of the Caribbean Plate. Why are the land masses clustered around the edges of the plate?

# SECTION 1 GEOGRAPHY

## 1.3 Rain Forests of Central America

**TECHTREK**

myNGconnect.com For an online map and photos of the rain forest

Maps and Graphs · Digital Library

**Main Idea** Central American rain forests are an important economic resource for the region.

Central America's rain forests cover a large portion of the region. Countries in the region are working to save rain forests and grow economically at the same time.

### The Importance of Rain Forests

Central American rain forests have tall trees with broad leaves, and grow in tropical areas with heavy rainfall. A rain forest is an individual **ecosystem**, a place where plants and animals rely on the environment to survive. Many rare species make their homes in rain forests. The quetzal (ket SAHL), for example, is the national symbol of Guatemala. It makes a nest in rain forest trees.

### Rain Forest Destruction

In the 1500s, rain forests covered most of Central America. They thrived for hundreds of years until the 1900s. From 1990 to 2005, Nicaragua, Honduras, Guatemala, and El Salvador lost 14 to 30 percent of their rain forest land.

There are several reasons for forest loss, or **deforestation**. Many forests are cleared for commercial farmland and for grazing cattle to supply the beef industry. Trees provide valuable timber, or wood for building. Rain forests are usually located in rural areas, where people use the land to survive. Trees are cut down for firewood and for small farms. However, rain forest soil is not very **fertile**, or able to produce plentiful crops. Rural farmers need to clear more land after nutrients in the soil are used up. The forests are being cleared faster than they can regrow.

### The Future

Many countries encourage rain forest tourism. **Tourism**—the travel industry—brings income to the region and provides jobs so rural people do not need to depend on rain forest land for farming. Certain farming methods can also prevent further damage to rain forests. Shade coffee, for example, grows under the protection of trees. This crop can grow on the same soil as other crops, such as beans. This practice preserves soil quality.

### Before You Move On
**Make Inferences** How do Central America's rain forests contribute to the economy?

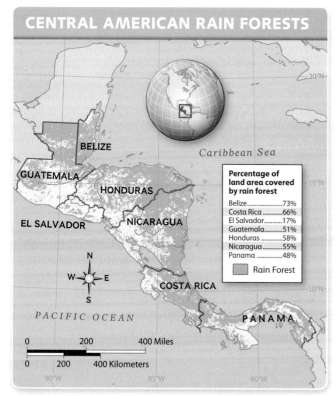

CENTRAL AMERICAN RAIN FORESTS

Percentage of land area covered by rain forest

Belize............73%
Costa Rica.....66%
El Salvador....17%
Guatemala.....51%
Honduras......58%
Nicaragua.....55%
Panama........48%

**Visual Vocabulary Canopy** refers to the tops of trees that create a roof over the rain forest. Many visitors take rain forest canopy tours.

**Critical Viewing** A tourist ziplines through the rain forest in Honduras. A cable is strung at an angle between two points, and the rider uses a pulley to "zip" from the top of the line to the bottom. What would you see by exploring the canopy of the rain forest that would be different from exploring at ground level?

ONGOING ASSESSMENT
# DATA LAB
GeoJournal

1. **Analyze Data** Based on the percentages of land covered by rain forest, in which country would you expect rain forest tourism to be most important? Explain your response.
2. **Create Graphs** Using data shown on the map in this lesson, make a bar graph showing the percentage of forested land in Central American countries. Which method of showing the data do you find the most clear, and why?

SECTION 1.3 **147**

# Saving Sea Turtles

with José Urteaga

**Main Idea** People can save their region's valuable resources by revisiting traditional ideas and finding new economic solutions.

## Identifying the Problem

In the early 1980s, sea turtles in Nicaragua began to disappear. Their decline in population was unusual, considering that they had outlived dinosaurs. For more than 100 million years, sea turtles migrated to Nicaragua's beaches to lay eggs, sometimes hatching 600,000 babies at a time. Emerging Explorer José Urteaga (uhr tay AH gah) is a young marine biologist who decided to find out why sea turtles were disappearing so fast that they were close to **extinction**, or dying out completely.

Urteaga discovered that **poachers**, people who hunt or fish illegally, were stealing sea turtle eggs from beach nests. Poor people in Nicaragua live on less than $1 a day, but a poacher can earn as much as $5 selling just 12 eggs a day, because of the great demand for the eggs. Sea turtle eggs are highly valued in Nicaraguan culture. Turtle meat and turtle eggs are traditional foods of the country. People also use sea turtle shells to make jewelry.

myNGconnect.com
For more on José Urteaga in the field today

### Finding a Solution

Urteaga knew that to save the turtle eggs, he had to help poor people earn money in other ways. He also had to encourage people to see their culture in a new way. Starting in 2002, he put together a team that worked with poachers, celebrities, young people, and conservation groups.

First, he convinced poachers to sell their turtle eggs to him so he could hatch them. Then he taught poachers how to make a living by farming, raising bees, guiding tourists, and making crafts. Urteaga even hired former poachers to patrol the beaches and protect the nests. He started hatcheries, places designed for hatching eggs, to make sure new generations of turtles would survive.

To inspire cultural change, Urteaga focused his message on young people. He launched a huge media campaign to reach this audience.

The campaign included sold-out rock concerts to raise awareness as well as celebrities stating, "I don't eat turtle eggs." His school programs teach children how important it is to protect the species. Children also have the opportunity to work with the hatcheries. In 2008, young people celebrated the first release of hatched baby turtles in their town's hatchery by wearing t-shirts showing a tiny turtle breaking out of its egg. Urteaga's goal is to end the demand for turtle eggs and save sea turtles from extinction by educating a new generation.

By 2010, thanks to Urteaga and his team, almost 90 percent of sea turtle nests became protected. Urteaga is committed to continue saving sea turtles by "motivat[ing] people through their brains and their hearts."

### Before You Move On

**Monitor Comprehension** What actions are helping to change how people view Nicaragua's sea turtles?

**Critical Viewing** Baby sea turtles head to the sea. What difficulties might they meet along the way?

---

**ONGOING ASSESSMENT**

## VIEWING LAB  GeoJournal

1. **Pose and Answer Questions** Write three questions about the efforts to save the sea turtles. Go to the **Digital Library** to watch the Explorer Video Clip. Then work with a partner to answer each other's questions.

2. **Analyze Visuals** What did you find to be the most powerful image in the video?

3. **Human-Environment Interaction** Why would it be important for Urteaga to get young people involved in his cause?

# NATIONAL GEOGRAPHIC
## Photo Gallery • Orange Cup Coral

For more photos from the National Geographic Photo Gallery, go to the **Digital Library** at myNGconnect.com.

Waterfall in Costa Rica

Satiny parrot snake

Vegetable market, Guatemala

**Critical Viewing** A cluster of orange cup coral clings to a support piling on Bonaire Island, West Indies. This type of coral catches its food with sticky tentacles, preying primarily on shrimp and small fish.

Bucket orchid

Musician with steel drum

Coiba Island, Panama

Red-legged honeycreeper

# SECTION 2 HISTORY
## 2.1 Trade Across Continents

**TECHTREK**
myNGconnect.com For a trade map on global trade

Maps and Graphs

**Main Idea** European exploration of the Americas led to new trade routes among the world's continents.

Shortly after Columbus landed in the Caribbean in 1492, the Spanish began to make use of the region's rich resources. Their actions were the first steps toward global trade.

### Farming for Profit

Spanish settlers in the Caribbean grew **cash crops**, or crops for profit. Land on the islands was well suited for growing certain raw materials that were in short supply in Europe. For example, sugarcane grew well on the islands, and could be sold in Europe where there was a **scarcity**, or shortage. Spanish settlers **exploited**, or took advantage of, the labor of the native people for the heavy work on their farms.

The native people were also weakened by sickness. They could not fight the diseases, like smallpox and malaria, that settlers brought with them from Europe. As large numbers of native people died from disease, Spanish settlers had to look elsewhere for the workers they needed to farm their cash crops. So, after minerals and raw materials arrived in Spain, manufactured goods went to Africa to pay for slaves, beginning more than 300 years of slavery in the Americas.

### New Trade Changes the World

Beginning in the 1500s, goods were exchanged in what became known as **triangular trade**—trade among three continents: the Americas, Europe, and Africa. The new trade established economic patterns that continue today.

**Visual Vocabulary** A **cash crop** is a crop grown for profit. Tomatoes, native to the Americas, were an important cash crop for Caribbean farmers to sell to Europe.

152 CHAPTER 5

## TRIANGULAR TRADE

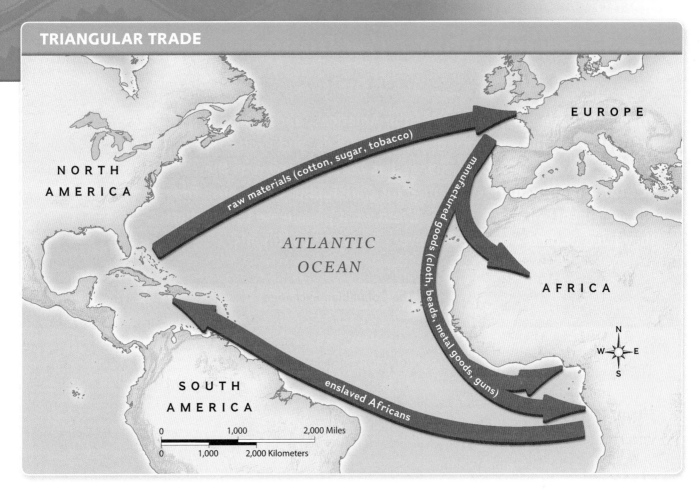

Triangular trade soon brought about competition among European countries for the resources of the region. Spain was gaining great wealth from its settlements in the Americas, causing other European countries to join the race for profit. Portugal, France, and Britain began competing with Spain for control over trade and colonization of the Americas.

### Before You Move On
**Monitor Comprehension** In what ways did European exploration lead to new trade routes?

### ONGOING ASSESSMENT
## READING LAB  GeoJournal

1. **Summarize** What effects did triangular trade have on Europeans, native groups, and Africans?

2. **Make Inferences** What role did native people play in the Caribbean economy?

3. **Movement** Trace the pattern of the triangular trade from the Caribbean to Europe to Africa. Describe the pattern of the trade using directional words. Use the scale to estimate the distance from Africa to the Caribbean.

# SECTION 2 DOCUMENT-BASED QUESTION
## 2.2 The Columbian Exchange

Triangular trade between Europe, the Americas, and Africa brought about the exchange of plants, animals, and even diseases. This worldwide exchange has become known as the **"Columbian Exchange,"** named for Christopher Columbus. The basic elements of everyday life—from what people ate to whether they lived or died—changed forever on several continents.

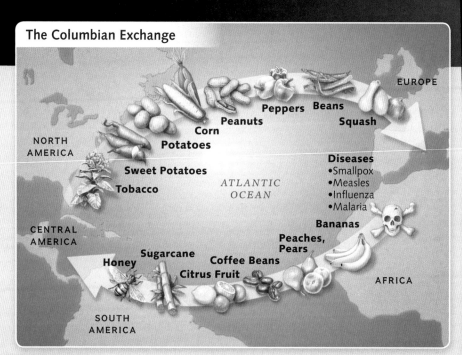

The Columbian Exchange

### DOCUMENT 1
The diagram at right illustrates crops and diseases exchanged among continents.

**CONSTRUCTED RESPONSE**

1. Based on the diagram, which crops from Europe became important to the economy of the Caribbean? In which direction did disease travel in the **Columbian Exchange**?

### DOCUMENT 2

#### Columbus's Journal
Columbus kept a journal detailing his arrival in the Caribbean in 1492. This excerpt describes the first trades made with Native Americans.

**CONSTRUCTED RESPONSE**

2. Explain what can be inferred from Columbus's statement "I was very attentive to them, and strove to learn if they had any gold."

> Saturday, October 13, 1492
>
> At daybreak great **multitudes** [numbers] of men came to the shore... loaded with balls of cotton, parrots, javelins, and other things... These they exchanged for whatever we chose to give them. I was very attentive to them, and strove [tried] to learn if they had any gold... [They] readily bartered [traded] for any article we saw fit to give them... such as broken platters and fragments of glass.

154 CHAPTER 5

> THEY CAME LOADED WITH BALLS OF COTTON, PARROTS, JAVELINS, AND OTHER THINGS. . . . THESE THEY EXCHANGED FOR WHATEVER WE CHOSE TO GIVE THEM.
> — CHRISTOPHER COLUMBUS

Columbus and his crew arrive in the Caribbean.

## DOCUMENT 3

### The Historian's Viewpoint

Alfred Crosby is the historian who invented the phrase "Columbian Exchange." Below is an excerpt from an essay he wrote on the global impact of the Columbian Exchange.

> What is the significance of the Columbian Exchange [?]... What is the **staple** [main food] of the Bantu of southern Africa? Maize [corn], an American food. What is the staple of Kansas and Argentina? Wheat, an Old World [European] food...
>
> How many of the six billion of us are dependent for our nourishment on crops and meat animals that didn't cross the great oceans until after 1492?

**CONSTRUCTED RESPONSE**

3. What point is Alfred Crosby trying to make about the main foods of southern Africa, Kansas, and Argentina?

## ONGOING ASSESSMENT
## WRITING LAB  GeoJournal

**DBQ Practice** Look at the illustration, and reread the diary entry and the passage from Crosby's essay. What do these documents tell you about the Columbian Exchange?

**Step 1.** Review what you know about how trade developed between continents.

**Step 2.** Take notes about the main ideas in Documents 1, 2, and 3.

> Document 1: Diagram of the Columbian Exchange
>
> Main Idea(s) _____
>
> Document 2: Quotation from Columbus
>
> Main Idea(s) _____
>
> Document 3: Excerpt from Alfred Crosby essay
>
> Main Idea(s) _____

**Step 3.** Construct a topic sentence that answers this question: How did the Columbian Exchange change the world's food supply?

**Step 4.** Write a detailed paragraph that explains the effect of the Columbian Exchange on the food supply of Europe and the Americas. Go to **Student Resources** for Guided Writing support.

SECTION 2.2

SECTION 2 HISTORY

# 2.3 Paths Toward Independence

TECHTREK
myNGconnect.com For an online map of European colonies in the region

Maps and Graphs

**Main Idea** Ideas of freedom led to fights for independence in 19th-century Central America and the Caribbean.

The Columbian Exchange led to increased competition for global trade among European countries. By the 1780s, Europe had colonized nearly all of Central America and the Caribbean. Forced labor and other harsh practices used by European settlers stirred thoughts of independence among native groups, who wanted to regain control of their lands.

## Haiti Leads the Way

In 1791, Haiti (then Saint-Domingue) had become the leading producer of sugarcane in the Caribbean. A small number of wealthy French colonists there used about a half million slaves. Their labor provided the backbone of this economy. As demand for sugar grew, more slaves were brought to Haiti from Africa. By 1791, slaves far outnumbered white planters.

Two years earlier, a revolution, or an overthrow of the government, had taken place in France. By August of 1791, the conflict spread to Haiti. As European planters, free people of color, and the English and French armies fought, Haitian slaves began a rebellion.

In 1794, the French government abolished slavery, but it kept control of the island. **Toussaint L'Ouverture** (too SAN loh ver CHOOR), a former slave, began a movement for independence. L'Ouverture's army struggled with the French for power until 1803, when he died in prison. Soon after his death, his army won victory over the French military. Haiti declared itself independent on January 1, 1804.

## Calls for Freedom Spread

Spanish colonies in the Americas were ruled by Spanish **viceroys**—governors who represented the Spanish king and queen. Viceroys controlled the land's resources, such as gold, silver, and crops, and native labor. Most of the region fell under official control of the Spanish viceroy in Mexico City.

In 1821, Mexico seized control of much of present-day Central America. In 1823, it broke from Mexican control and became the United Provinces of Central America. **Provinces** are smaller parts of a larger nation. However, internal conflict renewed calls for independence. Over the next 20 years, each province—Guatemala, Honduras, El Salvador, Nicaragua, and Costa Rica—would declare independence.

INDEPENDENCE IN CENTRAL AMERICA AND THE CARIBBEAN

Toussaint L'Ouverture, leader of slave rebellion in Haiti

1820s–40s
Guatemala, Honduras, El Salvador, Nicaragua, Costa Rica, Dominican Republic

1800

1850

1804 Haiti

1823 Central America becomes United Provinces of Central America.

Juan Pablo Duarte, father of independence in the Dominican Republic

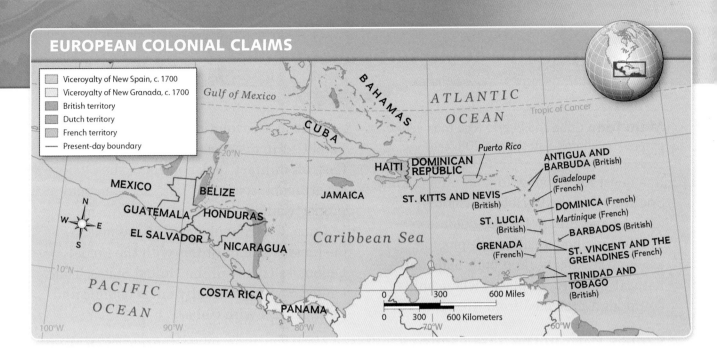

## Caribbean Independence

As you've learned, Haiti was the first Caribbean island to become independent. The United States and Europe wanted to keep control of the islands' many resources. As a result, most of the islands would not be independent until the 1900s.

In fact, several islands became independent as recently as the second half of the 20th century, as the time line shows. Today, some islands remain connected to the European countries that settled them. For example, Bermuda continues to be a British territory, and Aruba is still part of the Netherlands.

### Before You Move On
**Make Inferences** How might the Spanish desire to control resources have brought about the fights for independence?

## ONGOING ASSESSMENT
## MAP LAB   GeoJournal

1. **Location** With your finger, trace the outline of the Viceroyalty of New Spain shown on the map. What present-day countries are included in this area?

2. **Interpret Maps** Find Panama on the map. Which viceroyalty did it belong to? Why do you think Panama could be considered more like South America than Central America?

3. **Make Inferences** What fact about Toussaint L'Ouverture might explain his commitment to gaining independence for slaves in Haiti?

4. **Interpret Time Lines** Between which years was there the biggest gap in Central American and Caribbean countries declaring independence?

- 1902 Cuba
- 1903 Panama
- The flag of Panama
- 1960s Trinidad & Tobago, Jamaica, Barbados
- 1970s–1980s Grenada, Dominica, Bahamas, St. Lucia, St. Vincent & the Grenadines, Antigua & Barbuda, Belize, St. Kitts & Nevis

1900 — 1950 — PRESENT

# SECTION 2 HISTORY

## 2.4 Comparing Cuba and Puerto Rico

TECHTREK
myNGconnect.com For resource maps and photos of Cuba and Puerto Rico

Maps and Graphs   Digital Library

**Main Idea** Cuba and Puerto Rico took very different economic paths in the 20th century.

In a quest for silver, gold, and other riches, Spain settled the Caribbean islands of Cuba and Puerto Rico as colonies. The islands had ideal conditions for growing sugarcane, and both had good natural **harbors**, or places where ships could land protected from the open sea.

### Cuba's Path to the 21st Century

Spain ruled Cuba as a colony from 1511 to 1898. The Spanish built the city of Havana, where European ships carrying cargo such as silver and corn stopped before crossing the Atlantic. Spanish colonists also built sugarcane plantations.

However, native Cubans wanted to control their own resources and political destiny. During the 1800s, they staged several failed rebellions. They finally won independence from Spain in 1898, after the United States defeated Spain in the Spanish-American War. The U.S. military continued to occupy Cuba and control much of the country's economy. Over the next 50 years, Cuba's government was controlled by a series of leaders. Many of them were corrupt, or dishonest, and led with complete control as **dictators**.

In 1959, a revolutionary leader named Fidel Castro overthrew Cuba's dictator. Castro's military took control of the government, seized all land and personal property, and established communism in Cuba. Castro took over U.S.-owned businesses and built ties with the Soviet Union, an enemy of the United States. The United States eventually cut economic and political ties with Cuba.

Until the early 1990s, the government in Cuba controlled all economic activity. In 1993, to improve Cuba's economy, the government started to allow citizens to open their own businesses.

### Commonwealth of Puerto Rico

Like Cuba, Puerto Rico was a Spanish colony from the 1500s until 1898. During this time the Spanish mined gold and built huge sugarcane plantations. Spanish control made most native Puerto Ricans poor, so they tried to rebel against Spain.

During the Spanish-American War, the United States sent troops to Puerto Rico. The island's location was important to U.S. military and economic interests. After Spain surrendered to the United States in 1898, the Treaty of Paris made the island a U.S. territory. In 1917, Puerto Ricans were granted U.S. citizenship.

**CUBA**

1850

**PUERTO RICO**

**1898** Treaty of Paris grants Cuban independence; U.S. continues to occupy.

**1898** Treaty of Paris grants control of Puerto Rico from Spain to U.S.

1900

**1901** U.S sets up naval base at Guantánamo Bay, Cuba.

**1917** Puerto Ricans are made U.S. citizens.

## CUBA'S ECONOMIC RESOURCES

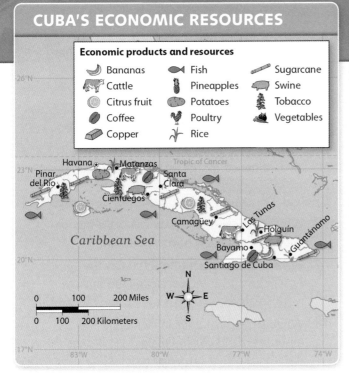

## PUERTO RICO'S ECONOMIC RESOURCES

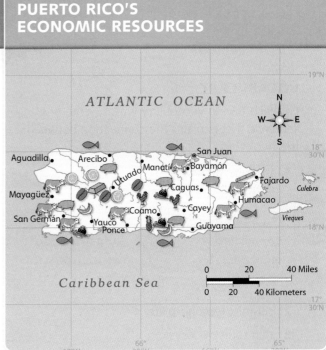

For many years, Puerto Ricans worked for more freedom from the United States. In 1952, Puerto Rico became a U.S. **commonwealth**, a nation that governs itself but is part of a larger country.

Unlike Cubans, Puerto Ricans have political freedom. Like the United States, Puerto Rico benefits from a free-enterprise economy. The government has done what it can to help Puerto Ricans start new businesses, and shift the economy away from farming toward manufacturing. Thousands of Puerto Ricans work in factories that make high-tech products. Many also work in the tourist industry.

### Before You Move On

**Summarize** How did differences in the governments of Cuba and Puerto Rico affect economic opportunities in the two countries?

### ONGOING ASSESSMENT
## MAP LAB
*GeoJournal*

1. **Interpret Maps** Which resource is most of the land used for in Cuba? in Puerto Rico?
2. **Region** Look at the climate map in Section 1.1 and the maps above. Why would the economic resources be similar in both countries?
3. **Make Inferences** Why would the United States have wanted control over Cuba and Puerto Rico after the Spanish-American War?
4. **Interpret Time Lines** Which events show each country's connection to the United States?

# CHAPTER 5
# Review

## VOCABULARY

For each pair of vocabulary words, write one sentence that explains the connection between the two words.

1. isthmus; archipelago

   *An isthmus is one strip of land connected to two larger land areas, but an archipelago is a group of islands.*

2. seismic; tectonic plate
3. rain forest; ecosystem
4. deforestation; fertile
5. cash crops; scarcity
6. province; viceroy
7. harbor; commonwealth

## MAIN IDEAS

8. What feature of Central America makes the soil fertile? (Section 1.1)
9. How do earthquakes and volcanoes affect Central America and the Caribbean? (Section 1.2)
10. What are some causes of rain forest deforestation? (Section 1.3)
11. In what way will creating new jobs help save Nicaragua's sea turtles? (Section 1.4)
12. How did Columbus's landing in the Caribbean lead to international trade? (Section 2.1)
13. How did the Columbian Exchange affect the Europeans and the native peoples of the Caribbean? (Section 2.2)
14. What actions by the Spanish led Central American and Caribbean countries to seek independence? (Section 2.3)
15. How did the economies of Cuba and Puerto Rico develop after independence from Spain? (Section 2.4)

## GEOGRAPHY

### ANALYZE THE ESSENTIAL QUESTION

How has physical geography been a positive or negative influence on the economy of the region?

**Critical Thinking: Compare and Contrast**

16. What physical features do Central America and the Caribbean have in common?
17. What is similar about how the Central American and the Caribbean land masses were created?
18. What is the difference in climate in the mountains and coastal plains? How does this difference affect farming?

### INTERPRET MAPS

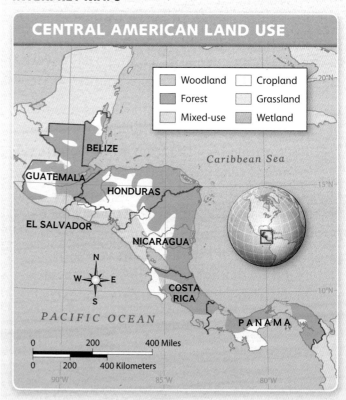

19. **Interpret Maps** How is most land used in Central America?
20. **Draw Conclusions** Locate the wetland areas on the map. What conclusion can you draw about where wetlands are located in Central America?

## HISTORY

### ANALYZE THE ESSENTIAL QUESTION
How have economic resources influenced the history of the region?

**Critical Thinking: Analyze Cause and Effect**

21. How did volcanoes contribute to the success of farming in the region?
22. What did the Spanish hope to gain through control of the region?
23. What effect did outside control of Caribbean islands have on their independence?
24. What political effect did Fidel Castro and his military have on Cuba?
25. Why has manufacturing become a greater part of Puerto Rico's economy than agriculture?

### INTERPRET CHARTS

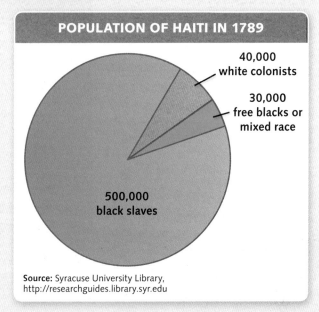

**POPULATION OF HAITI IN 1789**
- 40,000 white colonists
- 30,000 free blacks or mixed race
- 500,000 black slaves

Source: Syracuse University Library, http://researchguides.library.syr.edu

26. **Analyze Data** How many more black slaves were there than white colonists in Haiti in 1789?
27. **Make Inferences** How does the racial makeup of the population in Haiti in 1789 help to explain why a rebellion started?

## ACTIVE OPTIONS

**Synthesize** the Essential Questions by completing the activities below.

28. **Create a Poster** Persuade people to help raise funds for a Central American cause. Choose either rain forest preservation or sea turtle protection. Define the group you want to persuade—for example, your school, your friends, or the government. **Display your poster to the class. Discuss what features of the poster are persuasive.**

    **Poster Tips**
    - Think about the group you want to convince. Ask: Who might be able to help the most? How could this group help the cause?
    - Take notes about what information might convince this group to join your campaign.
    - Create an original slogan for your campaign that will attract attention.

29. **Create a Chart** Make a chart showing comparisons among three countries in the region. Be sure one of your countries is in Central America and one is in the Caribbean. Use online sources to gather data on economic resources and current leaders. In your chart, show the following:

|  | (Central American country) | (Caribbean country) | (Your choice) |
|---|---|---|---|
| Population |  |  |  |
| Square Miles of Land |  |  |  |
| Types of Land |  |  |  |
| Main Economic Resources |  |  |  |
| Current Leaders |  |  |  |

REVIEW 161

# CHAPTER 6
# Central America & the Caribbean TODAY

## PREVIEW THE CHAPTER

**Essential Question** How do trade and globalization affect the cultures of the region today?

### SECTION 1 • CULTURE

1.1 The Impact of Tourism ............................. 164
1.2 Caribbean Food and Music ..................... 166
1.3 The Panama Canal .................................. 168

**KEY VOCABULARY**
- tourism
- intersection
- fuse
- canal
- terrain
- lock

**ACADEMIC VOCABULARY**
distinct, eliminate

**TERMS & NAMES**
- Taino
- Calypso
- Panama Canal Zone

**Essential Question** How is the region trying to improve the standard of living?

### SECTION 2 • GOVERNMENT & ECONOMICS

2.1 Comparing Costa Rica and Nicaragua ..... 170
2.2 Challenges in Haiti .................................. 172
   **GLOBAL ISSUES: FOOD SUPPLY**
2.3 Feeding Central America ......................... 174
2.4 Migration and the Caribbean .................. 176
2.5 Conserving the Rain Forest .................... 178

**KEY VOCABULARY**
- infrastructure
- reserve
- policy
- marketing
- standard of living
- food security
- global warming
- malnutrition
- surplus
- migrate
- remittance
- habitat
- ecotourism

**ACADEMIC VOCABULARY**
displace

**TERMS & NAMES**
- Human Development Index
- Port-au-Prince

Agriculture and tourism drive the economy in Soufriere, a small town on the island of St. Lucia in the Caribbean Sea.

# SECTION 1 CULTURE

## 1.1 The Impact of Tourism

**Main Idea** Tourists who visit Central America and the Caribbean have a major effect on income and resources of the region.

**Tourism**, or the business of travel, is an important source of income for Central America and the Caribbean. However, with more than 20 million tourists each year, the environment can be damaged.

### Diversity Attracts People

For centuries, this region has been an **intersection** of cultures, a meeting point for traders and settlers from many countries. Rich resources attracted European groups, including the English, French, and Dutch. The slave trade brought African culture to the region. The blend of cultures is part of the region's appeal.

Since the 1980s, global air travel and advertising have made tourism an important industry. Visitors usually stay in one of the many island resorts, or on cruise ships that sail around the islands. This "overnight tourism," though, can have long-term negative effects.

### Efforts to Improve

Overnight tourists use a great deal of electricity and consume vast amounts of water and food. This causes shortages for the local people. Large resorts and cruise ships release pollution into the air and water, which threatens marine life.

Organizations such as the United Nations seek to increase environmental protections. However, because the tourism industry is such a large part of the region's economy, local governments are sometimes resistant to new limitations. Many travelers and businesses are making efforts to offset the damage.

Cruise lines are beginning to use recyclable materials and conserve fuel in an effort to continue business without further harming the environment. Some travelers even contribute to programs that plant trees in the region.

### Before You Move On
**Monitor Comprehension** In what ways does tourism affect the region's income and resources?

> **Critical Viewing** These pyramids in Guatemala are temples where ancient Mayans worshipped. What aspects of these ruins might make them appealing to tourists?

### TOURISM IN CENTRAL AMERICA AND THE CARIBBEAN

| Country | International Tourist Arrivals* (number of people) | Receipts from Tourism ** (US $billions) | Percentage of Total Receipts from Tourism ** (percent) |
|---|---|---|---|
| Bahamas | 5,003,967 | $2.2 | 64.6 |
| Barbados | 1,272,772 | $1.2 | 56.6 |
| Belize | 1,082,268 | $0.3 | 35.0 |
| Dominican Republic | 4,239,686 | $4.0 | 34.0 |
| El Salvador | 966,416 | $0.8 | 20.9 |
| Guatemala | 1,181,526 | $1.1 | 12.1 |
| Honduras | 1,056,642 | $0.6 | 8.5 |
| Nicaragua | 734,971 | $0.3 | 9.4 |
| Panama | 1,004,207 | $1.2 | 12.6 |
| St. Lucia | 802,240 | $0.3 | 66.0 |

* **Source:** Association of Caribbean States
** **Source:** World Bank Online

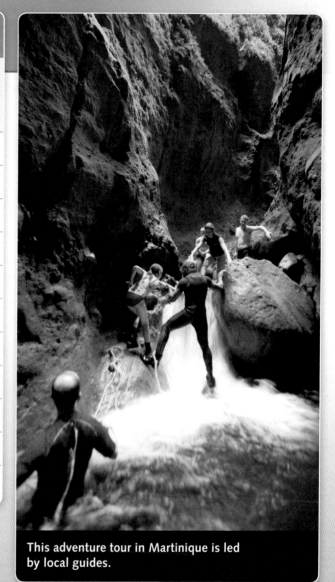

This adventure tour in Martinique is led by local guides.

### ONGOING ASSESSMENT
## DATA LAB  GeoJournal

1. **Interpret Charts** Which country in the region received the most money from tourism? Which ones received the least?

2. **Analyze Data** Which country depends most on money received from tourism?

3. **Identify Solutions** What can travelers and the tourism industry do to cause less harm to the environment in this region?

SECTION 1.1  **165**

# SECTION 1 CULTURE

## 1.2 Caribbean Food and Music

**Main Idea** Caribbean food and music blend influences from indigenous cultures and other world cultures.

Worldwide trade and global communication have spread Caribbean culture traits to the world. At the same time, other cultures continue to influence the food and music in the region.

### Caribbean Food

Since the Columbian Exchange, new influences continued to add to the region's diet. Foods from the native **Taino** (TY noh) and from Europe, Africa, and Asia have **fused**, or blended, into a rich cooking tradition.

This pepper shrimp is a very spicy dish popular in Jamaica.

The basic foods in the Greater Antilles include rice, beans, yams, peppers, plantains (similar to bananas), and avocados. Poultry and fish are also local foods. Caribbean cooks use blends of spices to flavor their foods. For example, Jamaican jerk is a **distinct**, or easily recognized, blend of strong spices used for grilling meat. Freed African slaves living in the Jamaican mountains developed the spice blend in order to preserve meat. Today, many Jamaican families follow the tradition of sharing a large lunch on Sundays. Foods served might include jerk chicken, fish, fried plantains, and a popular dish of rice and black-eyed peas.

In the past, islanders ate healthy diets based on fruits and vegetables, mixed with meat or fish. When open trade was firmly established in the 1990s, fast food restaurants arrived in the region, adding modern foods to the islanders' diet.

### Caribbean Music

Native cultures used wind instruments and drums in their music. European colonists brought stringed instruments to the islands. Island cultures blended European and African instruments and rhythms to make their own musical styles.

**Calypso** began on the island of Trinidad as a type of folk music. It uses simple rhythms and local language to tell stories. *Soca*, developed in the 1970s, is a mixture of calypso and East Indian music. Popular Afro-Cuban styles made it to New York in the 1940s, where they combined with jazz to create *salsa*. Some other well-known Caribbean styles are Dominican Republic *merengue* (meh RENG gay) and Jamaican *ska* and *reggae* (REG ay).

### Before You Move On
**Monitor Comprehension** What foods and styles of music have blended to become part of culture in the Caribbean Islands?

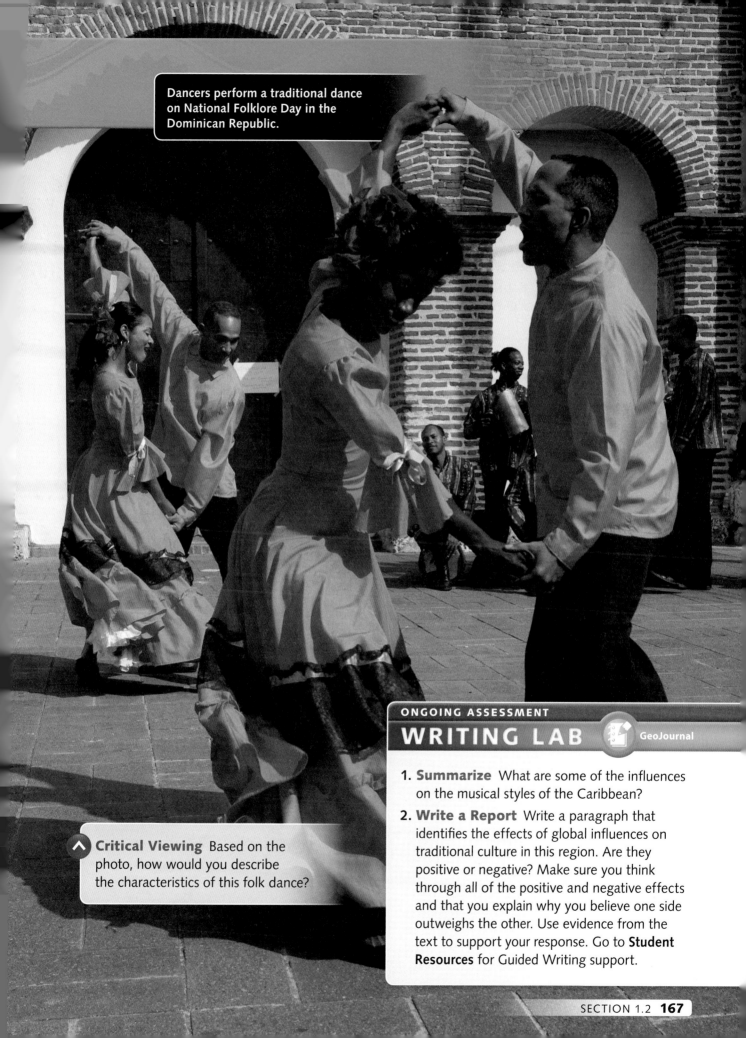

Dancers perform a traditional dance on National Folklore Day in the Dominican Republic.

**Critical Viewing** Based on the photo, how would you describe the characteristics of this folk dance?

### ONGOING ASSESSMENT
# WRITING LAB  GeoJournal

1. **Summarize** What are some of the influences on the musical styles of the Caribbean?
2. **Write a Report** Write a paragraph that identifies the effects of global influences on traditional culture in this region. Are they positive or negative? Make sure you think through all of the positive and negative effects and that you explain why you believe one side outweighs the other. Use evidence from the text to support your response. Go to **Student Resources** for Guided Writing support.

SECTION 1.2 **167**

# SECTION 1 CULTURE

## 1.3 The Panama Canal

**TECHTREK**
myNGconnect.com For photos and an illustration of the Panama Canal
 Digital Library
 Student Resources

**Main Idea** The Panama Canal provides a water route connecting the Atlantic and Pacific oceans.

As early as the 1500s, Spanish explorers wanted to create an artificial water route, or **canal**, through the Central American isthmus. A canal would significantly reduce the time it took for ships to travel from Europe to the Pacific.

### Connecting Two Oceans

Trade routes from the Atlantic to the Pacific required a difficult journey across land or a long water passage around the tip of South America. In 1855, the United States completed the first railroad across Panama. Goods and people could now travel across the isthmus by rail. However, the idea for a canal remained.

### Building the Canal

Panama declared independence from Colombia in 1903. Then the government signed a treaty that gave the United States control of the **Panama Canal Zone**, the area where the canal would be built.

Many obstacles had to be overcome in building the canal. To try to keep builders healthy, doctors worked to **eliminate**, or get rid of, mosquitoes that caused serious diseases. The **terrain**, or physical land features, of Panama also posed challenges. As many as 40,000 workers at a time made plans, cleared brush, drained swamps, and drilled through rocks. The canal took 100 steam engines and 10 years to finish, from 1904 to 1914—and more than 20,000 workers died in the process.

> **Critical Viewing** This illustration shows a portion of the original canal (right) and the planned expansion (left). In what ways will the new canal improve the current canal?

**PANAMA CANAL MILESTONES**

**1855** The United States completes a railroad across Panama.

**1881** A French company attempts to build a canal across Panama.

**1889** The French company's plan for a canal collapses.

**1903** Panama declares independence from Colombia; canal rights are granted to the United States.

**1914** The Panama Canal is opened to water traffic.

**1977** Treaty with the United States grants ownership of the Canal Zone to Panama.

**1999** Panama is granted complete control of the canal.

**2016** Expansion of the canal is scheduled for completion.

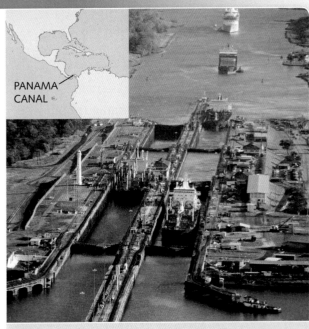

**Visual Vocabulary** Locks are devices that help to equalize the water levels of the waterways being connected. These locks in the Panama Canal are the first step down to the Pacific Ocean.

## Canal Connects the World

The Panama Canal's slogan, "The Land Divided, the World United," reflects the importance of this water passage. Before its completion, a ship traveling from New York to San Francisco had to travel 14,000 miles. The canal's 51-mile length shortened the travel distance by almost half, to approximately 6,000 miles.

As technologies advanced, larger ships were built. In 2006, Panama voted to expand the canal so it could accommodate these ships and the increased trade that caused traffic jams. The project is scheduled for completion in 2016.

**Before You Move On**
**Make Inferences** What made building a water route connecting the Atlantic to the Pacific important?

The "Panamax," the largest ship that can fit through the pre-expansion canal, is 106 feet wide.

**PANAMA CANAL EXPANSION PROJECT**

### ONGOING ASSESSMENT
## VIEWING LAB  *GeoJournal*

1. **Location** Look at the locator map above that shows the position of the Panama Canal. Why do you think builders picked that spot instead of a different one?
2. **Analyze Visuals** Based on the illustration, how will the expansion of the Panama Canal save money?
3. **Summarize** Based on the milestones, how long did the United States control the canal?

SECTION 1.3  **169**

# SECTION 2 GOVERNMENT & ECONOMICS

## 2.1 Comparing Costa Rica and Nicaragua

**Main Idea** Costa Rica and Nicaragua both work to build a more stable economy, despite different political conditions.

During the 20th century, Costa Rica and Nicaragua (nik uh RAHG wah) had very different histories. As a result, Costa Rica became a strong and stable democracy, while Nicaragua became the poorest country in Central America.

### Different Political Paths

Costa Rica has enjoyed peace for over 60 years. In fact, since 1949 the country has never had an army. Costa Rica's capital city of San José is home to several global human rights organizations. Without political conflict to interfere, Costa Rica has had the opportunity to create a stable economy. Tourism is the country's greatest source of income. This industry provides jobs to more than 50 percent of the working population.

Unlike Costa Rica, Nicaragua has had an unstable government since its independence from Spain in 1821. This instability included dictatorships and civil wars that lasted through the 1990s. In addition, the country was hit by a major hurricane in 1998. It destroyed Nicaragua's **infrastructure**, the basic systems a society needs, such as roads, bridges, and electricity. Thousands were left without homes, jobs, or medical care.

Unstable conditions make the fight against poverty more difficult. Nicaragua entered the 21st century facing challenges. It needed to build its economy and develop social programs to help poor people, who make up about a fourth of the population.

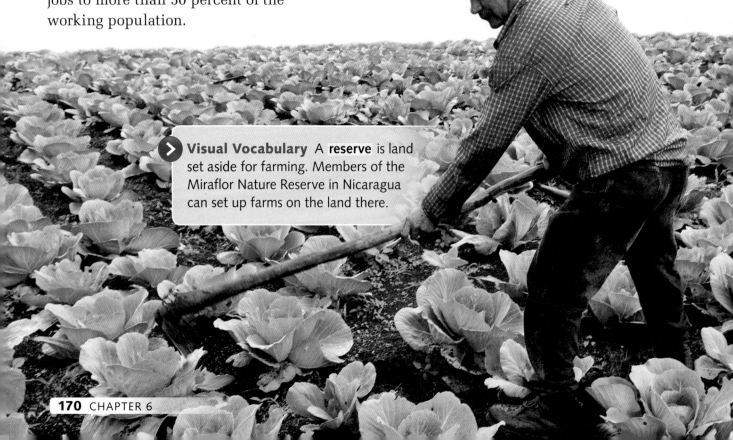

> **Visual Vocabulary** A **reserve** is land set aside for farming. Members of the Miraflor Nature Reserve in Nicaragua can set up farms on the land there.

## Economic Challenges

Costa Rica's economy has grown over the past 20 years. However, the poverty rate has remained between 15 and 20 percent. Changes in government **policy**—official guidelines and procedures—might account for this lack of improvement.

In the 1980s, after years of heavy spending and borrowing, the country was running out of money. Over the next decades, various policies were enacted to address this economic crisis. The government restricted its spending and raised taxes. Funding was reduced for social programs, many of which were created to help poor people. In 2007, in an effort to bring about economic growth, Costa Rica joined other Central American countries in a free trade agreement with the United States. The agreement went into effect in Costa Rica in 2009.

### COMPARING ECONOMIC DEVELOPMENT INDICATORS

|  | Costa Rica | Nicaragua |
|---|---|---|
| Life Expectancy at Birth | 79 years | 74 years |
| Adult Literacy Rate aged 15 and above | 94.9% | 67.5% |
| Per Capita Income (U.S. Dollars) | $7,640 | $1,510 |
| Population Below the Poverty Line | 24.8% | 42% |

Source: CIA World Factbook; World Bank, 2011, 2012 (est.)

Nicaragua joined the same free trade agreement. In 2012, the Nicaraguan economy grew by 4%, just behind Costa Rica's. However, rural poverty in Nicaragua is still a problem. Rural businesses have used **marketing** to promote their products, and new roads have helped for transporting goods. Yet many Nicaraguans continue to migrate to Costa Rica for seasonal work, such as for the coffee and banana harvests.

### Before You Move On
**Summarize** In what ways have both countries built a more stable economy?

### ONGOING ASSESSMENT
## DATA LAB  GeoJournal

1. **Analyze Data** How does literacy rate relate to the percentage of people living below poverty? Answer by completing this sentence: As the literacy rate decreases, the poverty rate _____.

2. **Place** Based on the chart, which country seems to have a better standard of living, or quality of life? How can you tell?

3. **Make Inferences** Why might small farmers want to set up their own reserve in Nicaragua, rather than depend on government funding?

# SECTION 2 GOVERNMENT & ECONOMICS
## 2.2 Challenges in Haiti

**TECHTREK**
myNGconnect.com For photos of the international aid effort in Haiti

Digital Library

**Main Idea** Haiti faces many great challenges in its efforts to build a strong economy and decrease poverty.

In the 1700s, the French colony of Saint-Domingue (san do MANG yuh), which is today's Haiti, was the richest in the Caribbean. Today, Haiti is the poorest country in the Western Hemisphere with 80 percent of its people living in poverty.

### Poverty's Historical Roots

Haiti gained independence from France in 1804. Because most of Haiti's citizens had been slaves, they did not have any money or means of income. European nations feared slave revolts in their colonies in the Caribbean, and so did not support Haiti financially when it became an independent nation. In the 20th century, political conflict and outbreaks of disease kept tourists away, making it difficult for Haiti to grow economically.

> **Critical Viewing** Members of a Chinese emergency rescue team work in Port-au-Prince two days after the 2010 earthquake. Based on the photo, how might past building practices have contributed to the extent of the devastation?

## 21st Century Haiti

The **Human Development Index** (HDI) is used by geographers to compare quality of life in different countries. HDI combines measures of the health, education, and **standard of living**—the level of goods, services, and material comforts—of people in a country. People in countries with a low HDI, such as Haiti, are often less healthy, less educated, and poorer than people in countries with a high HDI, such as the United States.

Politics in 21st century Haiti have been marked by instability and corruption. Due to violence between political groups, U.S. forces were sent in 2004 to Haiti's capital city, **Port-au-Prince**, to maintain security. Over the next few years, further efforts toward peace in Haiti were unsuccessful.

Adding to Haiti's challenges, a massive earthquake struck in January of 2010. Port-au-Prince and the surrounding area were nearly destroyed. Over 1.5 million people were **displaced**, or forced from their homes. Damage to the airport and to seaports made it difficult to receive immediate help from other countries.

As Haiti tried to recover, organizations from around the world began donating money and supplies such as food and medicine. Some countries also sent emergency rescue workers. Many nations and global organizations excused Haiti from paying back billions of dollars in loans, so they could rebuild at home.

### Before You Move On
**Summarize** What factors contribute to the difficulties Haiti faces in overcoming poverty?

## RESPONDING TO THE EARTHQUAKE

**1.5 million**
Number of Haitians displaced from their homes and living in temporary housing after the earthquake

**28,000**
Number of displaced Haitians that had moved into new homes six months after the earthquake

**1,340**
Number of tent cities and camps still being used six months after the earthquake

Source: 2010 United Press International

The day after the earthquake, Haitians set up this tent city amid the rubble. In this aerial photo, the colorful squares in the center are the roofs of tents used as temporary shelters.

### ONGOING ASSESSMENT
## DATA LAB

1. **Analyze Data** How many displaced Haitians were in new homes six months after the earthquake? How many were still homeless?
2. **Make Inferences** What factors might explain this difference in numbers?
3. **Draw Conclusions** What do the photos show about how Haitians and the international community responded after the earthquake?

# SECTION 2 GLOBAL ISSUES • Food Supply

## 2.3 Feeding Central America

TECHTREK
myNGconnect.com For photos on food security in Central America

Digital Library
Global Issues

**Main Idea** Food supply in Central America is affected by natural disasters and human activities.

**KEY VOCABULARY**

**food security,** n., easy access to enough food

**global warming,** n., an increase in temperature around the world

**malnutrition,** n., a lack of healthy food in the diet, which leads to physical harm

**surplus,** n., amount beyond what is needed

**Food security**, or easy access to enough food, is an important issue around the world. Understanding the causes of shortages can help local governments and global organizations take the right steps toward improving food security in Central America.

## Impact on Food Supply

Natural disasters have a major impact on the food supply. Many countries struck by Hurricane Mitch in 1998—El Salvador, Guatemala, Honduras, and Nicaragua—barely had time to recover before major flooding hit Central America in 2008. Disasters can destroy a country's major crops, such as bananas in Honduras. The 2010 earthquake in Haiti made it difficult for Haitians to gain access to nutritious meals and safe water.

In some countries, such as Guatemala and Nicaragua, the rainy season doesn't always provide enough water for the crops that make up the food supply. Even one dry season can greatly reduce food production.

Human activity can also threaten the food supply. Many countries have not managed their natural resources well. Lack of water for irrigation, declining soil quality from growing the same crop year after year, and deforestation all have led to food shortages.

Today, climate change adds a new threat to the food supply. Climate scientists predict that **global warming** will cause more extreme weather patterns, leading to flooding in some areas and drought in others. Many Central Americans survive mainly on corn and beans, but drought has dried up these crops in many areas.

### Before You Move On
**Summarize** How do nature and human activities affect the food supply in Central America?

**SOME MAJOR CROPS IN SELECTED CENTRAL AMERICAN COUNTRIES**

EL SALVADOR
GUATEMALA
HAITI
HONDURAS
NICARAGUA

- eggs
- beans
- maize
- bananas
- avocados
- mangoes
- plantains
- oranges
- rice
- nuts

Fresh mangoes from Haiti

## Childhood Hunger

Childhood **malnutrition**, or the lack of healthy food in the diet, is one of the most serious effects of food shortages in Central America. In some countries, many pregnant women do not get the nutrition they need. As a result, some babies are very small at birth. Poor nutrition continues throughout childhood and can slow down a child's healthy growth.

In Guatemala, 23 percent of children under five are underweight and almost half are small for their age because they don't get proper nutrients in their diet. Many poor families spend their whole day trying to grow or buy enough food to get to the next day. There is no time to tend to health issues or for children to go to school. In these conditions, cycles of malnutrition are difficult to break.

## Solutions For the Future

Most experts believe that the best way to improve food security in Central America is to increase each country's own crop production. Growth in agriculture means larger food supplies and lower food prices. It can also mean higher incomes for farmers both large and small.

Land quality is a determining factor in a country's ability to increase crop production. Honduras and Guatemala have large areas of good quality land for farming. In most years, much of Guatemala and the coastal areas of Nicaragua and Honduras typically experience enough rainfall for their crops to thrive.

Areas with low amounts of quality soil or inadequate rainfall, such as El Salvador and southern Honduras, benefit from programs that provide fertilizer and irrigation methods to help rural farmers.

Education is another way to improve food security in Central America. The United Nations recommends rural people be educated in productive farming methods. Programs have been designed to teach rural farmers in Central America how to keep soil healthy, and how to grow crops and sell the **surplus**, or extra.

In emergency situations such as the earthquake in Haiti, a country can benefit from the aid of foreign countries. However, when a poor country can improve its food supply by learning how to increase its own crop production, it can achieve long-term food security without outside aid.

### Before You Move On
**Make Inferences** How does education provide a long-term solution to the problem of food security in Central America?

---

**ONGOING ASSESSMENT**

## READING LAB  GeoJournal

1. **Location** What geographic features and conditions of Central America contribute to food shortages in the region?
2. **Summarize** Why is it hard for a poor family to break out of the cycle of malnutrition?
3. **Turn and Talk** What kinds of programs would help improve food security in one country in the region? Turn to a classmate and use information from the lesson to develop some specific ideas.

# SECTION 2 GOVERNMENT & ECONOMICS

## 2.4 Migration and the Caribbean

**Main Idea** Many Caribbean people migrate to other countries to find economic opportunites and help support their familes back home.

People **migrate**, or move from one place to another, because of push-pull factors. Push factors make people move away from an area because of difficulties like war or drought. Pull factors draw people to a place because it offers more security and better job opportunities.

### Migration Within the Caribbean

Today, it is hard to make a living on many Caribbean islands. The collapse of major businesses, including the sugar industry, has pushed workers out of rural areas into cities to find work. As a result of this **internal migration**, or migration within a country or region, two-thirds of the population now lives in cities such as Santo Domingo in the Dominican Republic or San Juan, Puerto Rico.

Many of these cities have become overcrowded. Urban unemployment is high in the Caribbean. Migrants seeking better jobs and a higher standard of living have been forced to travel to other islands within the Caribbean region.

By the 1990s, the tourism industry had expanded dramatically across the region. The demand for workers pulled many people to islands with large or growing tourist industries, such as Aruba, the Bahamas, and the Virgin Islands.

### Migration Out of the Caribbean

At the same time, push-pull factors played a part in workers leaving to go the United States, Canada, Europe, and other places. For example, political conflict in Cuba and Haiti pushed people to migrate to the United States.

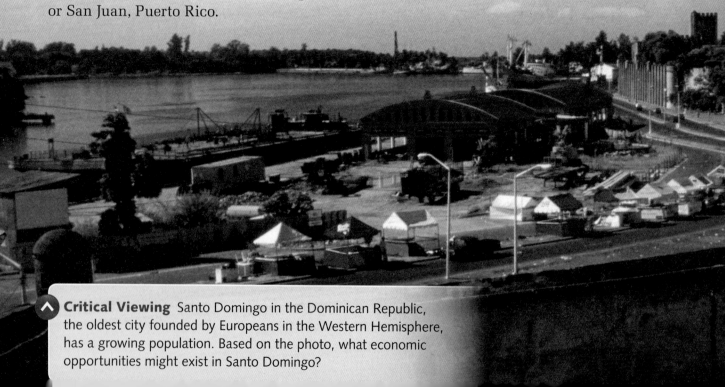

**Critical Viewing** Santo Domingo in the Dominican Republic, the oldest city founded by Europeans in the Western Hemisphere, has a growing population. Based on the photo, what economic opportunities might exist in Santo Domingo?

## Remittances

Most migrants who find work in another country send money back to their families in the form of **remittances**, money sent to a person in another place. Remittances have become a significant part of the economy of some Caribbean countries. For example, Jamaica receives more than $79 million each year in official aid, or money from other governments or organizations. However, the island receives 27 times that figure (see chart at right) in remittances. These remittances are sent from relatives in other countries, such as the United States, Canada, and France.

### Before You Move On
**Summarize** How are people who migrate out of the Caribbean able to help support their families back home?

**WORKERS LEAVING THE REGION (2009/2010)**

SELECTED DESTINATIONS: UNITED STATES & CANADA, EUROPE, MEXICO — CARIBBEAN

→ workers
→ $ remittances

**SELECTED REMITTANCES** — remittances / official aid

Honduras: $564 million
El Salvador: $233 million
Dominican Republic: $152 million
Jamaica: $79 million

Source: World Bank Online

---

### ONGOING ASSESSMENT
## DATA LAB  GeoJournal

1. **Analyze Visuals** Based on the graphic, what leaves the Caribbean, and what comes back in?
2. **Analyze Data** What brings more money into countries in this region, official aid or remittances? How does the graphic show this?
3. **Draw Conclusions** How do remittances help to ease the strain of poverty in the region?

# SECTION 2 GOVERNMENT & ECONOMICS

## 2.5 Conserving the Rain Forest

**TECHTREK**
myNGconnect.com For photos of ecotourism in Central America and the Caribbean
Digital Library

**Main Idea** Ecotourism provides a new opportunity for this region to protect rain forest habitats and fight poverty.

The rain forests and other natural resources of Central America and the Caribbean are a draw for large numbers of travelers every year. Tourism, however, can damage or even permanently destroy these valuable resources.

### Rain Forest Habitats

A rain forest is an important animal **habitat**, a natural home or environment for certain species. Plants in the rain forest provide nourishment for the animal population and replenish the air with oxygen. Some animal and plant species in the rain forest are not known to exist anywhere else.

The features of the rain forest that attract visitors are the very same features that are threatened by excessive tourism. If a species loses its habitat, or the habitat is altered in any way, that species may face extinction. Conserving the resources of the rain forest is an important step towards protecting species that live there.

quetzal

three-toed sloth

**Visual Vocabulary** A **habitat** is a place where certain species have what they need to live. The rain forest canopy is an important animal habitat.

## Rain Forest Ecotourism

**Ecotourism** is a way of visiting natural areas that conserves the resources of the region. The purpose of ecotourism is to allow a visitor to experience an environment in its most natural form. A rain forest ecotourist might explore from the ground on a guided hike or a birdwatching tour. Another ecotourist might explore from above by taking a tour of the rain forest canopy, home to a great variety of animal species.

This method of travel not only protects plants and animals of the rain forest, but also improves the lives of the local people. Ecotourism helps stop deforestation and soil destruction by providing alternative jobs to poor farmers who live off rain forest land. Local people can work in hotels, or as tour guides or rangers.

Ecotourism helps to preserve the environment by allowing a country to profit from and sustain its resources in their natural form. Housing for ecotourists is designed to have very little effect on the environment. For example, the construction of the Lapa Rios Resort in Costa Rica caused the loss of only one tree.

However, even ecotourists can have a negative impact. If too many ecotourists visit a single location, they can interfere with the habitat there. International conservation organizations have become involved in ecotourism. They have begun to point travelers to destinations that are working to preserve the natural resources of the region.

### Before You Move On
**Summarize** How does ecotourism help protect against deforestation?

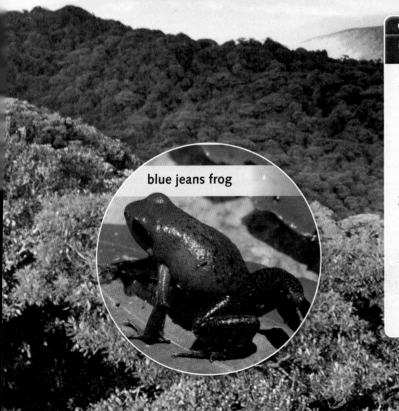
blue jeans frog

## ONGOING ASSESSMENT
### SPEAKING LAB — GeoJournal

1. **Turn and Talk** You are the owner of a resort in a Central American country. With a partner, discuss what you would include on your website to convince people that your resort preserves the environment and helps local people. Share your ideas with the class.

2. **Compare and Contrast** In what way is the focus of ecotourists different from that of other tourists? Provide an example.

3. **Analyze Visuals** From what you observe in the photos of the canopy and some of its animal inhabitants, what do you think draws ecotourists to the rain forest?

# CHAPTER 6
# Review

## VOCABULARY

For each pair of vocabulary words, write one sentence that explains the connection between the two words.

1. tourism; intersection

> The Caribbean was once an intersection of cultures, which makes it a popular area for tourism today.

2. canal; terrain
3. lock; canal
4. infrastructure; policy
5. food security; global warming
6. malnutrition; food security
7. migrate; remittance
8. ecotourism; habitat

## MAIN IDEAS

9. What technological advances helped bring tourists to Central America and the Caribbean? (Section 1.1)
10. What native foods are part of the diet of people in the Caribbean? (Section 1.2)
11. Why is Panama rebuilding its canal? (Section 1.3)
12. What has made it possible for Costa Rica to establish a more stable government than Nicaragua? (Section 2.1)
13. What were the most important needs of the people in Haiti after the earthquake of 2010? (Section 2.2)
14. What is one way to improve food security in Central America? (Section 2.3)
15. Why are remittances so important to many countries in the region? (Section 2.4)
16. In what ways does ecotourism protect the resources of the rain forest? (Section 2.5)

## CULTURE

### ANALYZE THE ESSENTIAL QUESTION

How do trade and globalization affect the cultures of the region today?

**Critical Thinking: Analyze Cause and Effect**

17. What characteristics of Caribbean culture attract tourists today?
18. How might global communication continue to affect cultures in the region?
19. How did Caribbean music become a blend of several different styles?
20. How will an increase in ships passing through the Panama Canal help the region?

### INTERPRET MAPS

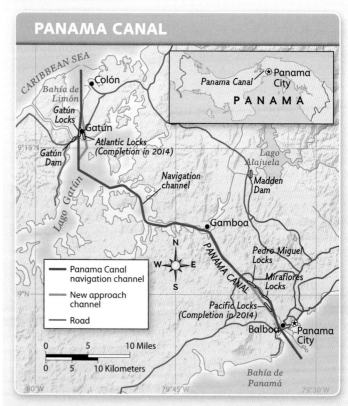

21. **Interpret Maps** What is the length of the Panama Canal in miles and in kilometers?
22. **Analyze Visuals** Will the new locks replace or add to the old ones? How can you tell?

## GOVERNMENT & ECONOMICS

### ANALYZE THE ESSENTIAL QUESTION

How is the region trying to improve the standard of living?

**Critical Thinking: Compare and Contrast**

23. What challenges does Nicaragua face in fighting poverty that Costa Rica does not?
24. Why was it helpful for Haiti to be excused from paying back some of its foreign debt?
25. Why is improving land quality and education a better way to ensure food security than providing foreign aid?

**Critical Thinking: Make Inferences**

26. Assume that the price of sugar and bananas has fallen and the production costs are high due to global competition. How might this affect migration in the region?
27. What economic opportunities does ecotourism provide for poor people?

### INTERPRET CHARTS

**POPULATION AND POVERTY IN CENTRAL AMERICA**

| Country | Population in Millions | Percent Living Below Poverty Line |
|---|---|---|
| Benchmark: United States | 316.6 | 15.1 |
| Guatemala | 14.3 | 54.0 |
| Costa Rica | 4.6 | 24.8 |
| Honduras | 8.4 | 60.0 |
| Dominican Republic | 10.2 | 34.4 |

Sources: CIA World Factbook

28. **Analyze Data** Which Central American country has the lowest percentage of people living in poverty?
29. **Make Inferences** From which country or countries do you think more people might migrate? Why do you think so?

## ACTIVE OPTIONS

**Synthesize** the Essential Questions by completing the activities below.

30. **Write and Design a Web Page** Pick a destination in Central America or the Caribbean, such as a rain forest or historic monument. Design a web page to attract people to visit your destination as ecotourists. Make sure to include how your tour preserves the environment and helps stop poverty. **Display your web page in your class.**

> **Writing and Designing Tips**
> - Describe the destination and include eye-catching visuals.
> - Explain how lodging, food, and transportation help the environment.
> - Show how the tour helps give jobs to local people.

TECHTREK myNGconnect.com For photos of tourism in the region today

31. **Create a Chart** With a partner, talk about the difference between regular tourism and ecotourism. Make a Comparison-Contrast Chart using photos from the **Digital Library** or other online sources to illustrate each type of activity. Explain why the photo represents one type of tourism and not the other. Use the example below as a model for your chart.

| PHOTO | REGULAR TOURISM | ECOTOURISM | WHY? |
|---|---|---|---|
| Lots of people sun-tanning at a beach | X | | People aren't concerned with how beach is affected by so many people. |

REVIEW **181**

COMPARE ACROSS REGIONS  UNIT 3

# Canals as Transportation

myNGconnect.com For a graphic comparing canals

Student Resources

Waterways can often be effective routes for moving people and goods from one place to another. However, natural waterways such as rivers and lakes don't always connect to each other, so water traffic comes to a stop. From ancient times to today, people have built canals to connect natural waterways for transportation and to gain access to water.

Sometimes the bodies of water being joined by a canal have different water levels. In that case, the canal will require the use of locks, which are devices that help to equalize the water levels of the waterways being connected. The Dutch (Netherlands) are believed to be the first to make use of locks, as early as the 1300s.

**Compare**

- China
- Egypt
- France
- Netherlands
- Panama
- Scotland
- Sweden
- United States

### ECONOMIC IMPACT OF CANALS

Early Middle Eastern civilizations are believed to have built canals for irrigation and drinking water, and the Romans built them to transport their military throughout Europe. Later, canals played an important role in Europe and the United States during the Industrial Revolution. These waterways provided cheaper, faster ways to move goods to new markets and to bring raw materials to factories, which helped the economy to grow.

Eventually, railroads took over much of the work of the canals. In fact, it is not uncommon to find a railroad running alongside certain canals. However, some cities today still use canals for transportation, and some of those canals have become popular tourist attractions.

### IMPORTANT CANALS

In the United States, the Erie Canal provides a 363-mile water route from the Atlantic Ocean to the Great Lakes.

Like the Erie Canal, the Panama Canal (Central America) and the Suez Canal (Egypt) are important shipping routes. Use the visuals on the opposite page to compare these two major canals.

Other canals around the world serve the same important function:

- Caledonian Canal (Scotland)
- Canal du Midi (France)
- Göta Canal (Sweden)
- Grand Canal (China)
- Amsterdam-Rhine Canal (Netherlands)

# COMPARE TWO IMPORTANT CANALS

## PANAMA CANAL

**1904** year building began

**1914** year the canal opened

 Connects

PACIFIC → ATLANTIC

**Approximate Distance SAVED**
7,900 miles

San Francisco → New York

Locks

52 miles

## SUEZ CANAL

**1859** year building began

**1869** year the canal opened

 Connects

MEDITERRANEAN SEA → RED SEA

**Approximate Distance SAVED**
5,530 miles

London → Persian Gulf

101 miles

Source: www.britannica.com

## ONGOING ASSESSMENT
## RESEARCH LAB  GeoJournal

1. **Summarize** How have canals been used from ancient times to today?
2. **Compare** Based on the visual, which canal saves more miles between major cities?

**Research and Create Charts** Choose two of the canals from the list on the opposite page. Do research and create a chart to compare data about the two canals. Based on the information you find, you may want to change the categories in your chart from those shown on the visual above

COMPARE ACROSS REGIONS  **183**

# REVIEW & EXTEND UNIT 3
# Active Options

## ACTIVITY 1
**Goal:** Research and prepare a speech.
### TALK ABOUT BATS
Countries of Central America and the Caribbean are home to many species of bats. Over 70 species live on the small island of Barro Colorado in Panama. Research bats in the region. As you outline your talk, incorporate a main idea, and ideas that support it, to explain the unique features of bats that help them function in their habitats. (See lessons on pages R3 and R10.)

big-eared bat

## ACTIVITY 2
**Goal:** Learn about a new technology.
### WRITE A FEATURE ARTICLE
National Geographic Emerging Explorer Ken Banks created software to allow rural groups to communicate even if they don't have Internet access. Rural farmers can get up-to-date information on crop prices—without access to phone or wireless connections—and stay competitive. Use the Internet to find out more about this mobile technology. Then use the **Magazine Maker CD-ROM** to write a feature article that explains the ways this technology can be used to help societies.

## ACTIVITY 3
**Goal:** Extend your knowledge of the geography of the Caribbean.
### CREATE A SKETCH MAP
The Caribbean is famous as the setting for pirate activity through the ages. Create a sketch map of the region that shows places where pirate activity took place. Write captions that describe some famous pirates and explain their routes around the islands.

NATIONAL GEOGRAPHIC
World Cultures and Geography

GEO

# UNIT 4

# Explore South America
## with NATIONAL GEOGRAPHIC

### MEET THE EXPLORERS

**NATIONAL GEOGRAPHIC**

Husband-and-wife team and Emerging Explorers Cid Simoes and Paola Segura work to teach farmers in Brazil to conserve their land while growing sustainable, profitable crops.

### CONNECT WITH THE CULTURE

Since 1940, this stadium, the Estádio do Pacaembu in São Paulo, Brazil, has hosted important soccer matches, including several FIFA World Cup matches in 1950. Soccer helps to unify Brazil. It has won five World Cup championships, more than any other country.

### INVESTIGATE GEOGRAPHY

The Amazon River and its tributaries border eight countries and supply one-fifth of the world's river flow. The rain forests found along its winding course are the largest on Earth and home to millions of plants and animals—many found nowhere else.

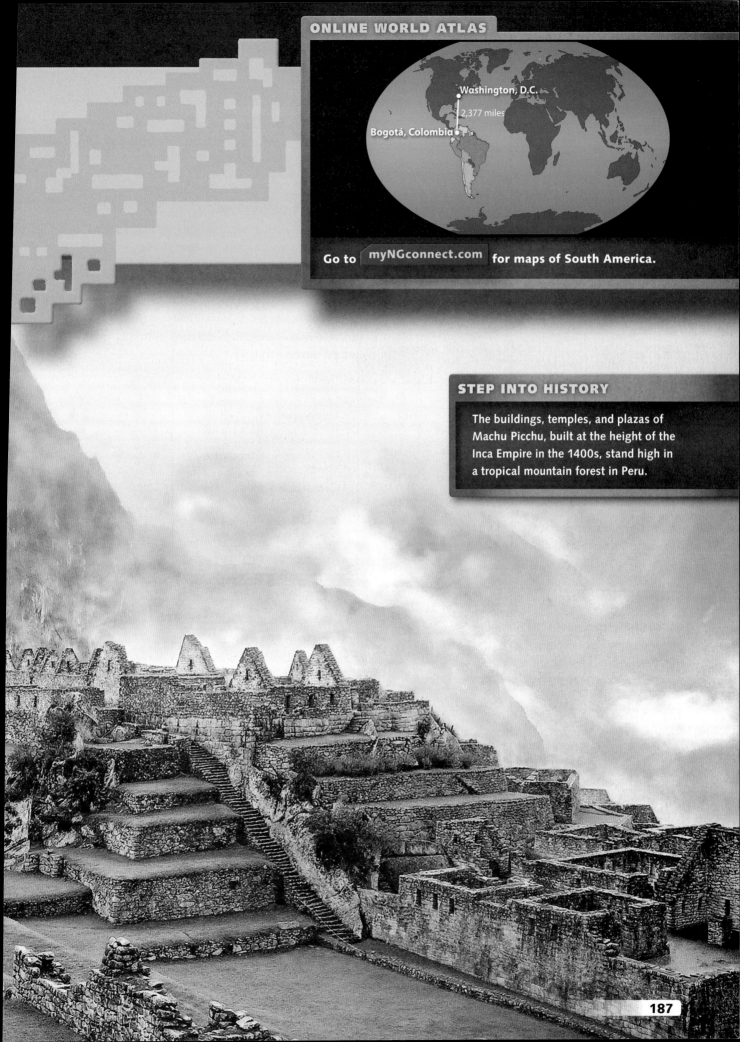

**ONLINE WORLD ATLAS**

Washington, D.C.
2,377 miles
Bogotá, Colombia

Go to myNGconnect.com for maps of South America.

**STEP INTO HISTORY**

The buildings, temples, and plazas of Machu Picchu, built at the height of the Inca Empire in the 1400s, stand high in a tropical mountain forest in Peru.

187

CHAPTER 7

# SOUTH AMERICA
## GEOGRAPHY & HISTORY

## PREVIEW THE CHAPTER

**Essential Question** How does elevation influence climate in South America?

### SECTION 1 • GEOGRAPHY

1.1 Physical Geography .................. 190

1.2 Life at Different Elevations ............ 192

1.3 The Amazon River .................. 194

1.4 Cold and Warm Currents ............ 196

**GLOBAL ISSUES: CLIMATE CHANGE**
1.5 Rain Forests and Climate Change ...... 198

**KEY VOCABULARY**
- vegetation
- grasslands
- adapt
- subsistence farming
- tributary
- biodiversity
- current
- rain shadow
- transpiration
- greenhouse gas

**ACADEMIC VOCABULARY**
saturate, acknowledge

**TERMS & NAMES**
- Andes Mountains
- Amazon River Basin
- Llanos
- Angel Falls
- Pampas
- Atacama Desert
- El Niño

**Essential Question** How did mountains, plateaus, and rivers shape the region's history?

### SECTION 2 • HISTORY

2.1 The Inca .................. 200

2.2 Exploring Nasca Culture ............ 202

2.3 People of the Lowlands ............ 204

2.4 The Spanish in South America ........ 206

2.5 Brazil and the Slave Trade .......... 208

**DOCUMENT-BASED QUESTION**
2.6 Simón Bolívar on Independence ...... 210

**KEY VOCABULARY**
- descendant
- kinship
- terraced
- suspension bridge
- artifact
- geoglyph
- excavate
- lowland
- nomad
- hunter-gatherer
- slash-and-burn
- treaty
- convert
- monopoly
- rebellion
- exile
- liberate

**ACADEMIC VOCABULARY**
utilize, transform, transition

**TERMS & NAMES**
- Machu Picchu
- Guaraní
- Tupinambá
- Yanomami
- Treaty of Tordesillas
- Francisco Pizarro
- Atahualpa

jaguar

# SECTION 1 GEOGRAPHY
## 1.1 Physical Geography

SOUTH AMERICA PHYSICAL

190 CHAPTER 7

**Main Idea** South America contains diverse physical features.

South America's physical features vary widely. The continent contains the dramatic Andes Mountains, the massive Amazon River Basin, and wide open grasslands and plains.

## High Mountains, Large Basin

The **Andes Mountains** are a string of mountain ranges that stretch about 5,500 miles along the western side of the continent. Many of the mountains in the Andes rise higher than 20,000 feet above sea level. The climate in the Andes is generally cool and dry. Because of low temperatures and high elevation, few types of **vegetation**, or plants, grow here.

The **Amazon River Basin** is the largest river basin on Earth. It covers nearly 2,700,000 square miles in the north-central part of South America—nearly the width of the continent. The river drains this basin, flowing from the Andes to the Atlantic Ocean. The climate is warm and wet. Many different species of plants and animals thrive in the Amazon Basin.

## Northern Grasslands, Southern Plains

The northern part of South America has a warm climate, and it contains both low and high elevations. Cattle ranching dominates the **Llanos**, or **grasslands**—wide open areas used for grazing and crops. The Guiana Highlands in the north boast unusual plants and animals. The highest waterfall in the world—**Angel Falls**—is located in Venezuela.

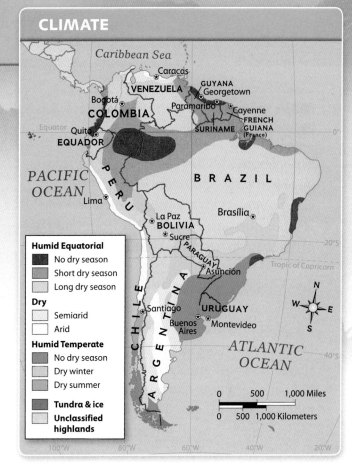

In much of the southern part of the continent, the climate is mild and the elevation is low. The rich soil of the grassy plain in Argentina called the **Pampas** is ideal for growing alfalfa, corn, and wheat.

### Before You Move On

**Monitor Comprehension** What are the main physical features in South America?

## ONGOING ASSESSMENT
## MAP LAB — GeoJournal

1. **Location** According to the physical map, what is the elevation of Cerro (or Mount) Aconcagua, and where is it located?
2. **Interpret Maps** Look at the two maps. What is the difference in climate between the Andes and the Amazon Basin? Which climate supports more diverse vegetation?
3. **Draw Conclusions** Based on climate and elevation, which area of South America is best for growing crops and why?

SECTION 1.1 **191**

# SECTION 1 GEOGRAPHY
## 1.2 Life at Different Elevations

**TECHTREK**
myNGconnect.com For online maps and photos of different altitude zones

 Maps and Graphs   Digital Library

**Main Idea** Elevation and climate affect where people live and how they use the land.

The people of South America must **adapt**, or modify, their economic activities to fit the different elevations and climates of the region. Life can vary greatly from one part of South America to another.

### Elevation and Climate

People, animals, and plants adapt to a range of climates across the continent. More people live in the mild climates of the plains and grasslands than live in the extreme climates of high mountains and tropical rain forests. The low-elevation plains, such as the Llanos, the Pampas, and the Coastal Plain, are moderate, both in temperature and rainfall.

### SOUTH AMERICAN ALTITUDE ZONES

| Altitude Zone | Climate | Elevation Range | Crops |
|---|---|---|---|
| Tierra Caliente | hot; adequate to extreme rainfall | 0 to 2,500 feet | bananas, peppers, sugarcane, cacao |
| Tierra Templada | warm; adequate rainfall | 2,500 to 6,000 feet | corn, beans, wheat, coffee, vegetables |
| Tierra Fría | cool; some rainfall | 6,000 to 12,000 feet | wheat, barley, potatoes |
| Tierra Helada | cold; little rainfall | 12,000 to 15,000 feet | no substantial crops |

**Source:** H.J. deBlij, *The World Today: Concepts and Regions in Geography*. Hoboken, NJ: John Wiley & Sons, 2009.

**Critical Viewing** At an elevation of 11,800 feet, La Paz, Bolivia, is the world's highest capital city. What details in this photo show how people have adapted to living in a high-elevation urban environment?

## POPULATION DENSITY

## RESOURCES

They are located in the *tierra templada* (tee EHR ah tem PLAH dah) elevation zone. The cool, dry elevations of the Andes are called *tierra fría* (FREE ah), or cold land.

Further up in the Andes, a very high and cold elevation range is called *tierra helada* (he LAH dah), or frozen land. At the other extreme, the hot and humid Amazon River Basin is *tierra caliente* (kay YEN tay), or hot land.

## Using the Land

Elevation and climate determine how people use the land. Rain is scarce at high elevations in the mountains and highlands of the continent. Some farmers in the Andes herd animals and grow only enough food for their families. This is called **subsistence farming**. However, others have become part of the global economy, selling wool to European and other manufacturers.

On the plains, higher rainfall provides opportunities for ranching and profitable, large-scale farming. Crops produced at these lower elevations in the region include tropical fruits, sugarcane, coffee, corn, wheat, and soybeans.

### Before You Move On
**Make Inferences** Why do people settle in areas with moderate climates and low elevations?

---

### ONGOING ASSESSMENT
## DATA LAB

1. **Interpret Charts** According to the chart, in what altitude zone do potatoes grow? Which altitude zone experiences the most rainfall?

2. **Draw Conclusions** Based on the chart, why do no substantial crops grow in the *tierra helada* zone?

3. **Human-Environment Interaction** According to the population density map, where do most people in South America live? Use the resources map to help you explain why those areas have the most population.

SECTION 1.2 **193**

# SECTION 1 GEOGRAPHY

## 1.3 The Amazon River

**TECHTREK**
myNGconnect.com For an online map and photos of river basin life

 Maps and Graphs   Digital Library

**Main Idea** The Amazon River supports life in its vast rain forest.

The Amazon River begins high in the Andes Mountains of Peru. It flows east for 4,000 miles across the continent and empties into the Atlantic. This massive river system has more than 1,000 **tributaries**, or small rivers that drain into a larger river. Seven of these tributaries are more than 1,000 miles long.

### Life Along the River

Though the Amazon River is the world's second longest river, it is the largest river in the world by volume. The Amazon creates the largest river basin—2,700,000 square miles—and includes the world's largest tropical rain forest. The Amazon rain forest is home to thousands of plant and animal species and millions of insect species. The variety of species in an ecosystem is called its **biodiversity**.

Annual flooding of the river from melting snow and rain in the Andes usually occurs between June and October. The floodwaters deposit rich nutrients, or substances that support life, in the lowland forest soils. These nutrients support the biodiversity of the rain forest.

Today, a growing population puts pressure on rain forest lands through mining, logging, farming, and land development. Conservationists suggest that setting limits on development will help protect the Amazon rain forest.

### Before You Move On
**Summarize** In what ways does the Amazon River support life in its rain forest?

**AMAZON RIVER AND ITS TRIBUTARIES**

| Source | Andes Mountains |
|---|---|
| Length | 4,000 miles |
| Basin | 2.7 million square miles |
| Directional Flow | East-northeasterly |
| Number of Tributaries | More than 1,000 |
| Three Notable Tributaries | 1. Madeira (2,082 miles long)  2. Tocantins (1,677 miles long)  3. Negro (1,400 miles long) |

Source: Encyclopædia Britannica

**Visual Vocabulary** A **tributary** is a small river that drains into a larger river. The Purus River is a tributary that runs through continuous forest on its way to the Amazon.

- ~ Amazon River drainage area
- ▢ Areas prone to flooding
- ▢ Rain forest
- ~ Shoreline or river
- ~ Area where the country border and a river coincide
- ⊛ Capital city
- ⊙ Administrative capital city
- • Other city

## ONGOING ASSESSMENT
# MAP LAB
*GeoJournal*

1. **Location** Locate the Purus River and Marañón rivers on the map. Use the map legend to describe the land that surrounds each river.
2. **Interpret Maps** What are three cities located on the Amazon River? Which tributary flows into Manaus?
3. **Monitor Comprehension** How is the biodiversity of the Amazon River and rain forest threatened by a growing population?

# SECTION 1 GEOGRAPHY

## 1.4 Cold and Warm Currents

**TECHTREK**
myNGconnect.com For current maps and photos of Peru, Chile, and Brazil

Maps and Graphs | Digital Library

**Main Idea** Wind currents and ocean currents influence climate across South America in powerful and unpredictable ways.

Like elevation, wind currents and ocean currents influence climate too. **Currents** are the continuous movement of air or water in the same direction. As you read about these currents, follow their patterns on the maps at right.

### Currents and Climate

Cold wind and ocean currents flow from the high latitudes near the South Pole toward the equator, making the west coast of South America generally cool and dry. Warm wind and ocean currents flow in the other direction, from the equator toward the South Pole, and create a warm and humid climate on the east coast.

The Peru Current brings cold waters to the Pacific coast in the west. It flows along the southern coast of Chile and northward along the coast of Peru. The Peru Current carries nutrient-rich waters from deep in the Pacific Ocean, so fish thrive off the coasts of Chile, Peru, and Ecuador.

On the eastern side of the continent, the Brazil Current brings warm waters from the Atlantic. The coasts and inland areas of Brazil and Argentina receive warm, humid wind currents and, in some areas, plenty of rainfall. This rainfall nourishes crops and vegetation. The plains of the Pampas benefit from the Brazil Current.

These moist wind currents do not reach the **Atacama Desert**, located on the western side of the Andes. This desert lies in a **rain shadow**, a dry region on one side of a mountain range. In South America, the Andes prevent moist Atlantic winds from reaching west of the mountains. Instead, moisture condenses into rain on the eastern side of the Andes. So, even though it lies along one of the world's largest bodies of water, the Atacama Desert is one of the driest places in the world. On average, only a half inch of rain falls each year.

▲ **Critical Viewing** Vicunas live in the Atacama Desert in Chile. From what you can see in the photo, how would you describe their habitat?

▲ **Critical Viewing** This lush sugarcane plantation in Argentina contrasts with the dry Atacama Desert. What details in the photo tell you that this plantation does not lie in a rain shadow?

## El Niños on the Pacific Coast

El Niños influence climate on the western coast of South America. An **El Niño** (ehl NEEN yoh) happens when the usual wind and ocean currents reverse. This reversal brings warm air currents and water currents that produce high rainfall in coastal areas. El Niños occur in Peru because of its location in the current system and on the Pacific coast.

El Niños do not occur every year, but they do happen somewhat regularly, at least once every 12 years. They are difficult to predict and their results can be devastating. Heavy rains saturate, or soak, coastal areas. These rains cause severe flooding and wipe out habitats. They can also cause damage to crops, canals, bridges, and roads. Scientists try to predict when an El Niño is likely to arrive. This helps prepare local populations to deal with the impacts of heavy rainfall.

### Before You Move On
**Monitor Comprehension** How do wind and ocean currents influence climate in South America?

### ONGOING ASSESSMENT
## MAP LAB

1. **Interpret Maps** Locate the Atacama Desert on the wind currents map. Do wind currents from the Pacific flow over the desert? How might this contribute to its lack of rain?

2. **Describe Geographic Information** Scientists use complex maps like these to predict weather patterns and climate changes. In what ways is the information on these maps similar, and in what ways is it different?

# SECTION 1 GLOBAL ISSUES • Climate Change

## 1.5 Rain Forests and Climate Change

**TECHTREK**
myNGconnect.com For a graph on deforestation and photos of the rain forest

 Maps and Graphs     Digital Library

**Main Idea** The health of the Amazon rain forest can influence climate change across the globe.

Climate change and global warming are topics that are often in the news. In order to understand how these topics are connected to rain forests, it helps to understand the scientific processes that keep rain forests healthy.

Rain forests are complex ecosystems. As you have read, thousands of different animals, birds, and insects live in the Amazon rain forest. Many unusual plants grow there, too. Some of these plants are used to fight deadly diseases like malaria and cancer.

### KEY VOCABULARY

**transpiration**, n., the process by which plants release water vapor into the air

**greenhouse gas**, n., a gas that absorbs heat energy and reflects it back to Earth

### ACADEMIC VOCABULARY

**acknowledge**, v., to recognize

### How Rain Forests Work

Through a process called **transpiration**, plants and trees release water vapor into the air. As the water vapor rises, it cools, forming thick clouds that then produce rain. The warm air and moist soil support the growth of vegetation. In fact, trees grow so high and thick that sometimes sunlight does not reach the forest floor.

The plants in the Amazon rain forest perform an important function. They absorb greenhouse gases from Earth's atmosphere. A **greenhouse gas** is a gas that traps heat energy, warming the earth. Burning fossil fuels such as coal and oil produces greenhouse gases like carbon dioxide.

Too much carbon dioxide in the air causes the atmosphere to warm because it reflects heat energy back to Earth. Rain forest plants and trees naturally absorb this carbon dioxide from the air. In this way, rain forests help clean Earth's air.

### Before You Move On

**Summarize** Why are rain forests important?

**DEFORESTATION IN BRAZIL, 2000–2009**

Source: National Institute for Space Research, Brazil, 2010

**198** CHAPTER 7

Inspiring people to care about the planet
National Geographic Society Mission

## Changes in the Rain Forest

Since the mid-20th century, large sections of the Amazon rain forest have been cut down. In *National Geographic*, journalist Scott Wallace reported that 20 percent of the rain forest has been lost in the last 40 years. (January 2007)

Widespread deforestation threatens the biodiversity of the Amazon. As more of the rain forest is lost, fewer trees and plants remain to produce moisture and cloud cover and to remove greenhouse gases from the air. These changes are having a global impact.

**COMPARE ACROSS REGIONS**

## A Global Climate Challenge

Changes in the Amazon rain forest are party responsible for changes in climate identified by scientists. Climate change is a gradual shift in Earth's climate due to natural causes. In its history, Earth has undergone many climate changes.

Global warming, on the other hand, is a term used by some scientists to describe the rapid warming of Earth's surface observed over the last century. They argue that the heavy use of fuels that release carbon dioxide contributes to rising temperatures. Since the beginning of the 20th century, the average temperature of the globe has risen 1.4°F. While that number might seem small, scientists **acknowledge**, or recognize, that a change of even one or two degrees is cause for concern. Rapid changes in average temperature can destroy habitats and change ecosystems.

Deforestation is underway in the Amazon rain forest near Rondonia State, Brazil.

Protecting the Amazon and other rain forests around the world is critical to the health of the planet. Many countries in South America are working to protect the Amazon rain forest. Across the Atlantic, countries in the African rain forest of the Congo River Basin face similar challenges. Leaders of Cameroon, Democratic Republic of Congo, Guinea, and Ghana are cooperating to manage sustainable forestry operations and to protect easily-damaged wildlife habitats.

### Before You Move On
**Monitor Comprehension** What problems threaten the health of the Amazon rain forest?

**ONGOING ASSESSMENT**
## READING LAB  GeoJournal

1. **Summarize** What is the difference between climate change and global warming?
2. **Describe** In what ways do rain forests clean Earth's air?
3. **Interpret Graphs** According to the graph, what has been the general trend in deforestation since 2004? What do you think caused this trend?

SECTION 1.5  **199**

# SECTION 2 HISTORY
## 2.1 The Inca

**Main Idea** The Inca ruled a vast empire in a difficult, mountainous environment.

The Inca Empire in South America stretched along the Pacific coast. The empire included parts of present-day Colombia, Ecuador, Peru, Bolivia, Chile, and Argentina. From 1438 until the Spanish conquest in the 1530s, the Inca Empire was one of South America's largest and most advanced civilizations.

## Workings of the Empire

The Inca built the capital of their empire in Cuzco, in what is now Peru. Inca government and society were highly organized. The emperor—called the Sapa Inca—was considered a **descendant**, or a relative of, the sun god, Inti. At its high point, the empire included 80 provinces and about 12 million people.

Within Inca society, families were organized into groups based on **kinship**, or blood relationship, and common land ownership. Inca married within their kinship groups. They also worked together and shared land and resources. The leader of each family worked for the empire for a few months of the year as builders, farmers, craftsmen, or foot soldiers.

### INCA EMPIRE, c. 1520

> **Critical Viewing** The ancient ruins of Machu Picchu are located in Peru, at an elevation of 7,710 feet. Based on what is shown in the photo, what do you think might have been the challenges of building a city at high elevations?

## Achievements of the Inca

The Inca **utilized**, or made practical use of, their advanced engineering skills to adapt to the mountainous environment. For example, the Inca farmed on **terraced** fields, or flat fields cut into slopes or mountainsides. They also built irrigation canals to water their crops because the climate was arid.

The Inca adapted to their mountainous surroundings in other ways. They built suspension bridges using vines and wood. A **suspension bridge** is a bridge used to cross canyons or water. The Inca also built a system of roads that helped keep the empire unified.

Another example of the Inca's engineering skills is **Machu Picchu** (MAH choo PEE choo), built in the 1400s. The Inca built this complex city on a mountain by constructing giant walls, terraces, sloping ramps, and steep stairways.

**Visual Vocabulary Suspension bridges** are bridges used to cross canyons or water. This rebuilt suspension bridge in Peru is modeled after the bridges the Inca built.

Some archaeologists believe it served as a royal estate. Today it is a UNESCO World Heritage Site because of its historical and archaeological significance.

By the 1530s, the empire faced internal problems that included a weak economy and civil war. A much smaller but better-equipped Spanish force conquered the Inca in 1532. Today, the descendants of the Inca, the Quechua, live in the Andes of Peru, Ecuador, and Bolivia.

### Before You Move On
**Make Inferences** In what ways did the Inca maintain control of their empire?

---

**ONGOING ASSESSMENT**

## PHOTO LAB  GeoJournal

1. **Analyze Visuals** Based on the photos and text, in what ways did the Inca utilize advanced engineering skills in their empire?
2. **Describe** How would you describe the Inca Empire based on the photo of Machu Picchu?
3. **Monitor Comprehension** What factors made it possible for a small force of Spanish conquerors to overcome the large Inca Empire?

# Exploring Nasca Culture
## with Christina Conlee

**Main Idea** Artifacts and other archaeological discoveries provide insight into the Nasca culture.

myNGconnect.com
For more on Christina Conlee in the field today

Archaeologists study how people lived in the past. They examine **artifacts**, or objects left behind by past cultures. NG Expeditions Council Grantee Christina Conlee is an archaeologist who studies the Nasca of Peru. The Nasca lived on a desert plateau in southern Peru nearly 2,000 years ago. They left behind beautiful ceramic artifacts as well as the Nasca lines, which are a series of **geoglyphs**, or large, geometric designs and animal shapes drawn on the ground. The Nasca were one of the first complex societies in South America.

## A Surprising Find

In 2004, Conlee and her team were **excavating**, or carefully uncovering, what they thought was a house in La Tiza, Peru. She was surprised to unearth, instead, an ancient burial site. Between 2004 and 2006, Conlee and her team excavated nine burial sites. Artifacts collected at the sites included elaborate ceramics and copper objects. Conlee believes that the kinds of artifacts she found belonged to individuals with a high social rank in the community.

Some of the artifacts uncovered were not typical of traditional Nasca ceramics. The presence of copper objects, shell ornaments, and more elaborate tomb paintings suggested a different population group. Were they local Nasca or people from another culture?

## Ancient Migrations

Conlee was determined to find the answer. Through chemical analysis she performed on the bones, Conlee determined that the remains did indeed belong to the Nasca. However, her analysis also proved that some remains belonged to a group of Nasca rivals called the Wari.

The Wari were more powerful than the Nasca. Conlee believes when they moved into Nasca areas, they may have caused some Nasca to move away. Ultimately, the migration of the Wari may even have led to the end of Nasca culture. Conlee's studies and discoveries continue to reveal new insights about Nasca culture and their migration patterns.

### Before You Move On
**Make Inferences** What did archaeological discoveries reveal about ancient Nasca culture?

**Visual Vocabulary** Geoglyphs are geometric designs and animal shapes drawn on the ground. The dry climate in Peru helped preserve the Nasca geoglyphs.

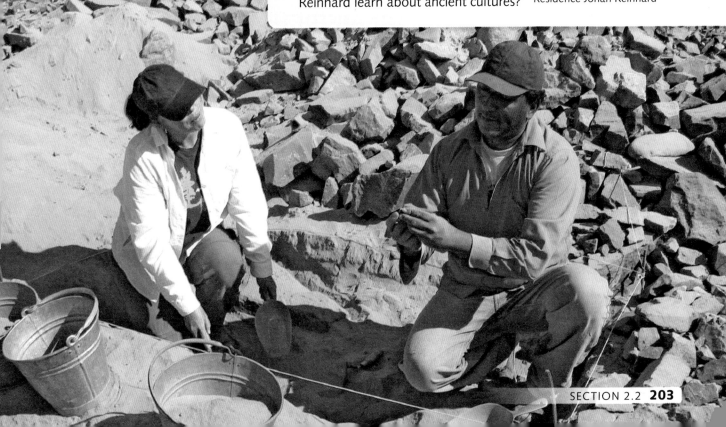

Christina Conlee and fellow archaeologist Aldo Noriega excavate a site in La Tiza, Peru.

### ONGOING ASSESSMENT
## VIEWING LAB

GeoJournal

1. **Analyze Visuals** Go to the **Digital Library** for an Explorer Video Clip about high-altitude archaeologist Johan Reinhard. What type of discoveries has he made?

2. **Describe** How do archaeologists such as Christina Conlee and Johan Reinhard learn about ancient cultures?

National Geographic Explorer-in-Residence Johan Reinhard

SECTION 2.2 **203**

# SECTION 2 HISTORY

## 2.3 People of the Lowlands

**TECHTREK**
myNGconnect.com For an online map and photos of the Yanomami

Maps and Graphs  Digital Library

**Main Idea** People have lived in the lowlands of South America for thousands of years.

People who lived in South America began to settle in low-lying areas of the continent more than five thousand years ago. The development of agriculture and a stable food supply allowed for larger groups of people to live there together.

### Lowland River Basins

The **lowland**, or low-lying, areas of South America include several fertile river basins. The lowlands include the Orinoco River and surrounding grasslands, the Amazon River Basin, and the Paraguay River Basin.

These rivers and basins contained abundant animal life and vegetation to support ancient people who lived there. For example, the Amazon River Basin was an ample source of fish. The Paraguay Basin, which includes Gran Chaco, a floodplain, provided fertile soils for growing food.

### Developing Agriculture

At first, lowland people lived as **nomads**, or people who move from place to place. As **hunter-gatherers**, or people who hunted animals and gathered plants and fruits for food, they moved to different locations as food became scarce.

As early as 3000 B.C., people began to farm the land and build villages in the lowlands. Establishing a stable food supply encouraged different groups to settle in villages rather than continue to move from place to place.

The **Guaraní** (GWAH rah NEE) people lived in the eastern and central lowlands on the Paraguay and Paraná Rivers. Before planting crops, the Guaraní cleared the land by cutting down and burning existing forest and vegetation. This agricultural technique is called **slash-and-burn**. Typically, Guaraní women grew corn and root vegetables such as sweet potatoes and cassava. In addition to farming, Guaraní men hunted and fished.

### SLASH-AND-BURN AGRICULTURE

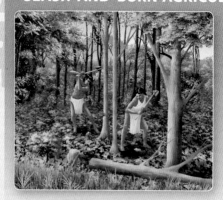

**1 Slash** Wooded areas and jungles are too thick to plant crops. Farmers slash, or cut down, trees.

**2 Burn** Fallen trees and foliage are burned to clear the land. Ash produced by the fires is used as fertilizer.

**3 Fertilize and Plant** Cleared land is fertilized with ash. Crops such as corn and sweet potatoes are planted.

Another group, the **Tupinambá** (too pee NAAM baa), settled near the mouth of the Amazon and southward along the Atlantic coast. They also cultivated crops using slash-and-burn agriculture. Because they lived in the Amazon River Basin and near the ocean, the Tupinambá also fished and hunted river mammals and turtles.

### The Yanomami

Today, indigenous groups continue to live in rain forests throughout the Amazon River Basin. The **Yanomami** (yaa noh MAA mee) are still hunter-gatherers who use the slash-and-burn technique to clear the land for farming. They live in villages but migrate to different areas to meet agricultural needs. Many have moved to the north-central lowlands, where they have found more fertile land.

**Before You Move On**
Make Inferences How did the physical features of the lowlands shape the lives of the people who lived there?

SOUTH AMERICA'S LOWLANDS

**4 Migrate** Groups move on to new locations after soil on cleared lands becomes less productive.

### ONGOING ASSESSMENT
## SPEAKING LAB  GeoJournal

1. **Turn and Talk** Turn to a partner and describe the process of slash-and-burn agriculture in the lowlands. Why did lowland groups engage in slash-and-burn agriculture?

2. **Create Charts** With a partner, make a chart comparing several aspects of nomadic and settled lifestyles. Discuss the characteristics of each lifestyle and what the advantages of each might be.

|  | NOMADIC | SETTLED |
|---|---|---|
| Movement | 1. Move frequently |  |
| Food Sources | 2. |  |

3. **Location** In what parts of the continent did the Guaraní and Tupinambá live? Where do the Yanomami live? Use the text to help you locate these areas on the map.

SECTION 2.3 **205**

# SECTION 2 HISTORY
## 2.4 The Spanish in South America

**Main Idea** The arrival of the Spanish in the 1500s shaped the history and culture of the South American continent.

In 1494, in order to avoid conflicts over exploration and settlement, Spain and Portugal signed a **treaty**, or an agreement between two or more countries. The **Treaty of Tordesillas** (tor duh SEE uhs) drew a line on a map that divided the newly discovered lands between the two countries. The treaty divided South America into two parts. The Spanish claimed lands west of the line, and Portugal claimed lands east of the line.

The Treaty of Tordesillas set the stage for the Spanish conquest of most of South America. Four years later, Christopher Columbus, sailing on behalf of Spain, landed on South America's northern coast.

### The Spanish Conquest

As they had in Mexico, Spanish conquistadors arrived in South America determined to expand Spain's empire and search for resources such as gold and silver. In 1533, **Francisco Pizarro**, with a small army of men, overthrew the Inca emperor, **Atahualpa** (AH tah WAHL pah). Pizarro founded the city of Lima, Peru, which became the center of the Spanish government and empire in South America. Other conquistadors explored and conquered Colombia, on the northern coast, and most of Chile, on the western coast. (See the map opposite.) The Spanish conquest of South America permanently **transformed**, or changed, much of the continent and its people.

▲ **Critical Viewing** *The Spaniard and the Inca Chief* by James McConnell. What stories does this painting tell about the Spanish and the Inca?

---

This gold mask is one example of the wealth the Spanish sought in South America.

Gold statues discovered in the mountains of Argentina illustrate the extent of the Inca Empire.

**1400**

**1438–1533** Inca Empire rules vast area of South America.

**1450**

**1494** Spain and Portugal sign the Treaty of Tordesillas.

**1498–1500** Columbus's third voyage reaches northern coast of South America.

## Impact on Native Populations

Deadly diseases that traveled with the Spanish to South America wiped out entire villages and native populations. Because these groups had no resistance to diseases such as smallpox, measles, and influenza, many died quickly.

The Spanish enslaved natives and forced them to work on plantations and ranches and in mines. Large numbers of enslaved native people died from the effects of harsh labor conditions.

Missionaries who arrived after the 1550s viewed South America as an opportunity to spread Christianity. The goal was to **convert**, or persuade native populations to change their religious beliefs. Some conversions were forced. Many native people began to practice the Catholic faith, and some blended aspects of Christianity with their own religious practices.

### Before You Move On
**Summarize** What impact did the Spanish have on the history and culture of South America?

### EXPLORERS IN SOUTH AMERICA

### ONGOING ASSESSMENT
## MAP LAB
*GeoJournal*

1. **Movement** According to the map, where were Spanish conquistadors most active in the 1520s and 1530s? Who made multiple expeditions?

2. **Evaluate** Locate the line of demarcation on the map. How would you describe the impact of that line on South America?

3. **Interpret Time Lines** About how many years passed between the signing of the Treaty of Tordesillas and the overthrow of the Inca?

Pedro Álvares Cabral, Portuguese navigator, lands on eastern coast of South America in 1500.

1500

1524 Pizarro reaches northwestern coast of South America.

1532 Spanish conquer Inca Empire.

1535 Spanish establish Lima, Peru, as center of empire in South America.

1541 Spanish establish Santiago, Chile.

1550

# SECTION 2 HISTORY
## 2.5 Brazil and the Slave Trade

**Main Idea** Portuguese colonization and the arrival of slaves from Africa influenced the history of Brazil.

In 1500, Portuguese navigator Pedro Álvares Cabral was headed to India with his fleet when he went off course. He landed on the southeastern coast of present-day Brazil. Cabral realized the area lay within the land allowed Portugal by the Treaty of Tordesillas and claimed it.

### Sugar and Slaves

Portuguese interest in Brazil was limited until the 1530s. Unlike the Spanish in Peru, the Portuguese did not conquer the native population and quickly take over the land. Instead, Portuguese colonization of Brazil took place over several decades.

Portuguese settlers discovered that these new lands contained natural resources valued in European markets. First, the Portuguese exported brazilwood, which was sought for its red color used to dye fabric. Then colonists realized that sugarcane, which grew abundantly in Brazil, was a more valuable crop. They built plantations and began exporting sugarcane and sugar products to Europe. The Portuguese tried enslaving natives to work the sugar plantations, but overwork

**Critical Viewing** This 1819 painting shows slaves working on a plantation in Brazil. What details do you notice?

and disease killed many of them. The Portuguese then turned to another source for labor: Africa.

Because of their earlier exploration, the Portuguese knew about the slave markets in Africa. By the mid-1500s, the Portuguese and other European countries were exporting African slaves across the Atlantic to South America and the Caribbean. The Portuguese were able to create a **monopoly**, or complete control, of the slave trade. A continuous trade in slaves began in Brazil in 1560 and lasted well into the 19th century.

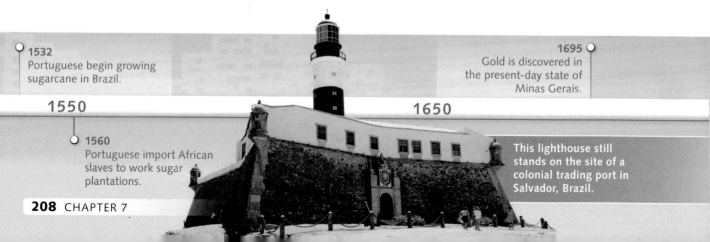

- 1532 Portuguese begin growing sugarcane in Brazil.
- 1560 Portuguese import African slaves to work sugar plantations.
- 1695 Gold is discovered in the present-day state of Minas Gerais.

This lighthouse still stands on the site of a colonial trading port in Salvador, Brazil.

## Portuguese Wealth

Brazil's abundant natural resources, combined with the African slave labor to extract them, made the Portuguese wealthy. As valuable natural resources like gold and diamonds were discovered, Portuguese demand for slave labor increased. Slaves were brought in to work on sugar and coffee plantations and in gold and diamond mines. Portuguese slave traders imported more than five million slaves from Africa to Brazil.

The Portuguese successfully put down **rebellions**, or revolts, against their rule until the early 1800s. Brazil finally declared independence from Portugal in 1822, but not everyone was free. Even though the slave trade had ended in 1850, slavery continued in Brazil for several decades until it was abolished in 1888.

### Before You Move On
**Monitor Comprehension** What influence did the slave trade have on Brazil's history?

### PORTUGUESE CLAIMS IN BRAZIL

## ONGOING ASSESSMENT
### READING LAB — GeoJournal

1. **Movement** Why did the Portuguese import slaves from Africa?
2. **Summarize** In what way did slaves contribute to Portuguese wealth?
3. **Interpret Maps** According to the map, where in South America did the Portuguese settle? What is the pattern of their settlement?
4. **Interpret Time Lines** How long did slavery exist in Brazil? What year did slavery end? Use the time line to form your answer.

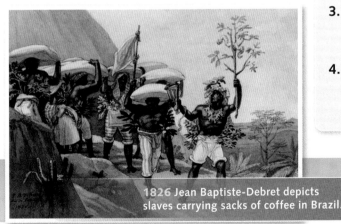
1826 Jean Baptiste-Debret depicts slaves carrying sacks of coffee in Brazil.

- **1725** Diamonds are discovered in Minas Gerais.
- **1750**
- **1822** Brazil declares its independence from Portugal.
- **1840s** Coffee becomes Brazil's top export.
- **1850**
- **1888** The Golden Law abolishes slavery in Brazil.

SECTION 2.5

# SECTION 2 DOCUMENT-BASED QUESTION

## 2.6 Simón Bolívar on Independence

Most of South America was ruled by the Spanish for more than 300 years. Making the **transition**, or change, from colonial rule to independence was difficult. Simón Bolívar led the revolution against the Spanish. Bolívar was born in 1783 to a wealthy family in Caracas, Venezuela. Both of his parents died when he was a child. After their death, his uncle made sure that he received an education and sent Bolívar to Europe. There, Bolívar learned new ideas about freedom and government. In 1810, he joined the independence movement in Venezuela. Present-day Bolivia was named in his honor.

### DOCUMENT 1
### The Letter from Jamaica (1815)

Bolívar wrote this letter in Jamaica while living in **exile**, or a state of absence from one's home country, during Venezuela's struggle for independence. Here he describes the need for revolution and independence from Spain.

> The hatred we feel for the Peninsula [Spain] is greater than the sea separating us from it; it would be easier to bring the two continents together than to reconcile [unite] the spirits and minds of the two countries.

**CONSTRUCTED RESPONSE**

1. What sea is Simón Bolívar referring to in this passage?

2. What two countries in particular is Bolívar discussing?

3. How would you describe the overall tone of Bolívar's words in this passage?

### DOCUMENT 2
### The Angostura Address (1819)

After returning from exile in Jamaica, Bolívar delivered this speech in Venezuela in 1819.

> The continuation of power in the same individual has frequently led to the demise [downfall] of democratic government.... A just zeal [enthusiasm for a cause] is the guarantee of republican freedom, and our citizens should properly fear that the same ruler who has long ruled them will wish to rule them forever.

**CONSTRUCTED RESPONSE**

4. According to Bolívar, what should citizens fear most in a ruler?

5. What does Bolívar argue leads to the "demise of democratic government"?

6. How does a "just zeal" guarantee freedom, according to Bolívar?

**A JUST ZEAL IS THE GUARANTEE OF REPUBLICAN FREEDOM.**
— SIMÓN BOLÍVAR

Source: Tito Salas, *Retrato Ecuestre del Libertador*, 1936.

## DOCUMENT 3

### Simón Bolívar, "The Liberator"

To **liberate** means to set someone or something free. Simón Bolívar earned the name "the Liberator" for his brave efforts against the Spanish during the struggle for independence. Bolívar is also often referred to as the "George Washington of South America."

**CONSTRUCTED RESPONSE**

7. In 1936, the Venezuelan government commissioned artist Tito Salas to paint this work for the National Pantheon, a monument built to honor national heroes. What details did Salas include to present Bolívar as a liberator?

### ONGOING ASSESSMENT
# WRITING LAB  GeoJournal

**DBQ Practice** By the early 1800s, many countries in South America began to press for independence from Spain. In what ways did Bolívar lead Venezuela's fight for independence?

**Step 1.** Review the colonial period of South America in Sections 2.4 and 2.5.

**Step 2.** On your own paper, take notes about the main ideas expressed in each document in this lesson.

> Document 1: The Letter from Jamaica
> Main Idea(s) _____
> Document 2: The Angostura Address
> Main Idea(s) _____
> Document 3: Painting of Bolívar
> Main Idea(s) _____

**Step 3.** Construct a topic sentence that answers this question: Why did Bolívar want to liberate South America from Spanish rule?

**Step 4.** Write a detailed paragraph that explains why Bolívar wanted self-rule and freedom for South America. Go to **Student Resources** for Guided Writing support.

# CHAPTER 7
# Review

## VOCABULARY

On your paper, write the vocabulary word that completes each of the following sentences.

1. People who lived in the South American lowlands practiced _____ agriculture.
2. The Portuguese had a _____ on the slave trade.
3. Archaeologists _____ artifacts to learn about past cultures.
4. Over 1,000 _____ feed into the Amazon River.
5. A _____ is an agreement between two or more countries.

## MAIN IDEAS

6. What are two extreme physical features on the South American continent? (Section 1.1)
7. What two factors help determine where people settle in South America? (Section 1.2)
8. In what ways does the Amazon River nourish the rain forest? (Section 1.3)
9. What unpredictable weather event impacts Peru and what are its effects? (Section 1.4)
10. How do rain forests protect the health of the planet? (Section 1.5)
11. How did the Inca adapt to their mountain environment? (Section 2.1)
12. In what ways have archaeologists contributed to an understanding of Nasca culture? (Section 2.2)
13. What natural resources helped the people of the lowlands thrive? (Section 2.3)
14. How did the arrival of the Spanish impact native populations? (Section 2.4)
15. In what ways did slave labor benefit the colonial Portuguese economy? (Section 2.5)
16. Why was Simón Bolívar called "the Liberator"? (Section 2.6)

## GEOGRAPHY

### ANALYZE THE ESSENTIAL QUESTION

How does elevation influence climate in South America?

### Critical Thinking: Compare and Contrast

17. Compare and contrast climate and elevation in the Andes and the Amazon River Basin.
18. Compare and contrast these two altitude zones: *tierra templada* and *tierra fría*.
19. Compare the ways in which currents and landforms influence climate in Chile.

### INTERPRET MAPS

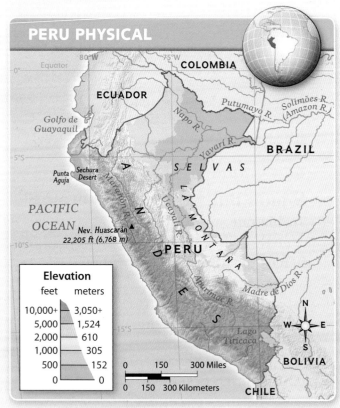

20. **Identify** What major mountain range lies in Peru, and what map element can help you determine its location?
21. **Describe Geographic Information** Where does Peru's largest area of lowlands lie in relation to the equator? Use the word "latitude" in your answer.

## HISTORY

### ANALYZE THE ESSENTIAL QUESTION
How did mountains, plateaus, and rivers shape the region's history?

**Critical Thinking: Draw Conclusions**

22. In what ways does Machu Picchu demonstrate the Inca's engineering skills?

23. Describe what archaeological evidence left behind on the desert plateau reveals about migration patterns of the Nasca culture.

24. How did people of the lowlands change from a nomadic to a settled way of life, and what role did the region's abundant rivers play in that change?

25. Which mountain resources in South America encouraged Spanish exploration, conquest, and colonization?

### INTERPRET PRIMARY SOURCES
Portuguese historian Pero de Magalhães Gândavo wrote *The Histories of Brazil* in 1576. Read his description of the valuable resources found in Brazil, and then answer the questions below.

> Certain Indians arrived in the Captaincy of Porto Seguro... giving news of the existence of green stones in a mountain range many leagues inland; and they were emeralds... and there were many other mountains of blue earth in which they [Indians] assured them there was much gold.

26. **Identify** According to this passage, what important minerals were found inland?

27. **Make Inferences** Based on your reading, how did the discovery of valuable resources in Brazil influence decisions about importing slaves from Africa?

## ACTIVE OPTIONS
**Synthesize** the Essential Questions by completing the activities below.

28. **Write Journal Entries** Describe South America from the perspective of a traveler in the region. Select two countries. In your journal, write two entries on each country: one on climate and elevation, and another on physical features, such as mountains and rivers. Use the following tips to help you write your journal entries. **Share your journal entries with a classmate or friend.**

    **Writing Tips**
    - Take notes before you begin to write.
    - Write an outline to help you organize details for the two countries you select.
    - Include as many details as possible.
    - Write in first-person narrative.

29. **Create Charts** Work in a group to make a four-column comparison chart that compares Brazil, Argentina, Venezuela, and Ecuador. Use online sources to gather the data. Compare your group's data with data found by other groups. What sources did you find most useful in your research?

|  | Brazil | Argentina | Venezuela | Ecuador |
| --- | --- | --- | --- | --- |
| Climate |  |  |  |  |
| Elevation |  |  |  |  |
| Population |  |  |  |  |
| Main Crops |  |  |  |  |
| Year of Independence |  |  |  |  |

REVIEW 213

# CHAPTER 8

# SOUTH AMERICA TODAY

## PREVIEW THE CHAPTER

**Essential Question** In what ways is South America culturally diverse?

### SECTION 1 • CULTURE

1.1 Indigenous Cultures .................................. 216

1.2 Language Diversity .................................... 218

1.3 Daily Life ...................................................... 220

**KEY VOCABULARY**
- mestizo
- roots
- language family
- topography
- immigrate
- cuisine

**ACADEMIC VOCABULARY**
predominant

**TERMS & NAMES**
- Aymara
- Quechua
- Guaraní
- Munduruku
- Creole
- Candomblé

**Essential Question** How is modern South America building its economies?

### SECTION 2 • GOVERNMENT & ECONOMICS

2.1 Comparing Governments ........................ 222

2.2 The Pampas Economy .............................. 224

2.3 Chile's Food Production ......................... 226

2.4 Products of Peru ....................................... 228

**KEY VOCABULARY**
- prosperous
- coup
- soybean
- fertilizer
- temperate
- Mediterranean climate
- export revenue
- profitable

**ACADEMIC VOCABULARY**
ruthless, diversify, erratic

**TERMS & NAMES**
- Dry Pampas
- Wet Pampas

**Essential Question** How has Brazil become an economic power?

### SECTION 3 • FOCUS ON BRAZIL

3.1 Brazil's Growing Economy ...................... 230

3.2 São Paulo ................................................... 232

3.3 Impact of the Olympics .......................... 234

**KEY VOCABULARY**
- steel
- ethanol
- biofuel
- megacity
- slum
- infrastructure
- venue

**ACADEMIC VOCABULARY**
foremost, impact

**TERMS & NAMES**
- São Paulo
- Rio de Janeiro

# TECHTREK
**FOR THIS CHAPTER**

- Student eEdition
- Maps and Graphs
- Interactive Whiteboard GeoActivities
- Digital Library

Go to **myNGconnect.com** for more on South America.

This cathedral in Brazil's capital city reflects the country's efforts to modernize.

# SECTION 1 CULTURE
## 1.1 Indigenous Cultures

TECHTREK
myNGconnect.com For photos of indigenous cultures in South America
Digital Library

**Main Idea** Indigenous cultures in South America maintain traditions in a modern world.

Indigenous groups have lived in South America for thousands of years. Their descendants still live and work in the region today.

### European Contact

As you have learned, the arrival of the Spanish, the Portuguese, and other Europeans during the colonial period changed life for the indigenous people in South America. Unfamiliar diseases and warfare with Europeans killed many native people and substantially reduced their populations.

European contact also introduced a new population group: the mestizo. Many people in South America are **mestizo**, or of mixed European and native ancestry. Some mestizos also have African **roots**, or cultural origins, because of the large numbers of slaves who were imported in the colonial period.

**Critical Viewing** Potato farmers work near the Aymara community of San Jose, Peru. Based on what you see in the photo, what is farming like for the Aymara?

216 CHAPTER 8

## Maintaining Traditions

The Aymara, Quechua, and Guaraní are the three largest indigenous groups in South America today. The **Aymara** (eye MAHR uh) live in the Andes of Peru and Bolivia. Today, the Aymara continue some of the traditions of their ancestors, such as speaking their native language, also called Aymara. The Aymara also continue to herd llamas and alpacas and grow crops such as potatoes and quinoa (KEEN wah), a grain that grows well in the mountains.

The **Quechua** (KEHCH wah) live in the Andes of Peru, Ecuador, and Bolivia. Like the Aymara, many Quechua farmers live in isolated mountain villages, far away from modern cities and lifestyles. Their religious practices are a blend of Catholicism and native beliefs. The Quechua have maintained traditions such as weaving and speaking their native language, Quechua.

The **Guaraní** (gwah rah NEE) live in Paraguay and are the main indigenous group in that country. Most people in Paraguay trace their roots to both Guaraní and Spanish ancestors. The Guaraní culture is represented in Paraguay's folk art and the Guaraní language.

The **Munduruku** (moon doo ROO koo), another important native group, live in Brazil. Their ancestors farmed, hunted, and fished in the rain forest. Although they live in relative isolation, one way the Munduruku adapt to the modern world is by selling forest products such as latex, a liquid substance harvested from native rubber trees.

### INDIGENOUS POPULATIONS OF SOUTH AMERICA (2007)

| Country | National Population | Indigenous Population | Indigenous Percentage |
|---|---|---|---|
| Argentina | 33,900,000 | 372,996 | 1.10 |
| Bolivia | 8,200,000 | 4,142,187 | 50.51 |
| Brazil | 155,300,000 | 254,453 | 0.16 |
| Chile | 14,000,000 | 989,745 | 7.07 |
| Colombia | 35,600,000 | 620,052 | 1.74 |
| Ecuador | 10,600,000 | 2,634,494 | 24.85 |
| French Guiana | 104,000 | 4,100 | 3.94 |
| Guyana | 806,000 | 45,500 | 5.65 |
| Paraguay | 4,800,000 | 94,456 | 1.97 |
| Peru | 22,900,000 | 8,793,295 | 38.40 |
| Suriname | 437,000 | 14,600 | 3.34 |
| Venezuela | 21,300,000 | 315,815 | 1.48 |

Source: International Union for Conservation of Nature, 2007

### Before You Move On

**Monitor Comprehension** What traditions do indigenous groups maintain today?

## ONGOING ASSESSMENT
## DATA LAB

1. **Analyze Data** According to the chart, in what countries do the highest and lowest numbers of indigenous people live? In what ways might low percentages impact indigenous populations?

2. **Make Inferences** Based on the chart and text, what percentage of people in Paraguay are indigenous? What other way might they identify themselves?

3. **Explain** What does "mestizo" mean? In what way does this cultural group reflect South America's colonial past?

# SECTION 1 CULTURE
## 1.2 Language Diversity

**Main Idea** South America has a rich diversity of languages.

As you know, South America is a land of varied cultures. It is also a land of diverse languages. The various languages that South Americans speak come from at least 50 **language families**, or groups of related languages.

### European Influences

The **predominant**, or main, languages in South America are Spanish and Portuguese. Other European languages spoken include French and Italian. Spanish, Portuguese, French, and Italian are Romance languages, or languages that come from Latin. During the 1800s, the land extending from Mexico through South America became known as Latin America. The widespread presence of the Romance languages helped to define Latin America as a culture region.

European presence in South America also brought about the development of Creole. **Creole** is a language that results when two other languages are combined together. During the colonial period, plantation workers created a common language that was a blend of European and non-European languages. In this way, groups who spoke different first languages found a way to communicate. Some people living on the Atlantic and Caribbean coasts of South America still speak Creole today.

Although European influence on languages in South America is strong, indigenous languages continue to thrive.

The large number of indigenous groups on the continent is one reason why more than 500 languages are spoken today. A second reason is because of the physical features, or **topography**, of the land and the density of its jungles. The mountain ranges and rain forests isolated indigenous groups and kept their languages separate from European languages.

### Indigenous Languages

In some countries, indigenous languages hold official status because of their widespread use. For example, along with Spanish, Quechua is considered an official language in Peru. In Paraguay, where Guaraní and Spanish are both official languages, more people speak and understand Guaraní than Spanish.

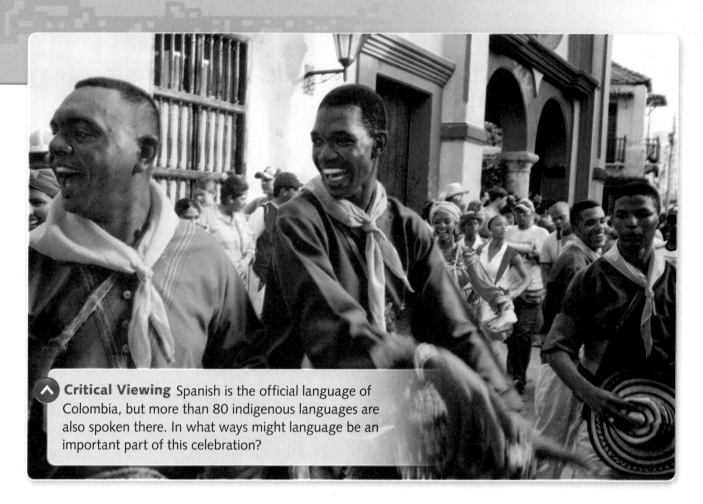

**Critical Viewing** Spanish is the official language of Colombia, but more than 80 indigenous languages are also spoken there. In what ways might language be an important part of this celebration?

Unlike Quecha and Guaraní, however, some indigenous languages in South America are spoken only by small groups of people in remote parts of the continent. Several of these languages—such as Kallawaya in Bolivia, Sáliva in Colombia, and Maka in Paraguay—are endangered, or in danger of vanishing completely.

National Geographic's Enduring Voices Project is working toward preserving these and other endangered languages. Enduring Voices works in language-diverse places, such as South America, to document and record indigenous languages. As this project captures the sounds and words of these languages, it is also helping to preserve the histories, songs, and stories of the cultures that speak them.

### Before You Move On
**Summarize** What factors have contributed to South America's language diversity?

## ONGOING ASSESSMENT
## LANGUAGE LAB  GeoJournal

1. **Location** How did the topography of the continent help preserve indigenous languages?
2. **Draw Conclusions** Why do you think people who speak endangered languages such as Kallawaya and Maka would be willing to work with the Enduring Voices Project?
3. **Interpret Maps** Based on the map, which indigenous language families are represented along the Tocantins River?
4. **Monitor Comprehension** What do you think might happen to some of the indigenous languages that are spoken by a small number of South Americans? Why?

SECTION 1.2

# SECTION 1 CULTURE
## 1.3 Daily Life

**Main Idea** South Americans' daily activities reflect aspects of their varied cultures.

Like most people around the world, South Americans' daily lives revolve around how they worship and celebrate, how they learn and play, and what they eat.

### Religious Practices

As you have read, Roman Catholicism became the main religion in South America during the colonial period. Spanish and Portuguese colonists were Roman Catholic, and they converted many native people to Catholicism. Roughly 80 percent of South Americans are Roman Catholic, with Brazil having the world's largest Catholic population.

After the Spanish and Portuguese, other Europeans **immigrated**, or moved permanently, to South America. Some of them were Protestants, or Christians who separated from the Catholic church. Today, most Protestants in the region live in Chile and the Guianas and in parts of Brazil, Bolivia, and Ecuador.

Other religious practices exist alongside Christianity in South America. In Brazil, a local religion called **Candomblé** combines African spiritual practices with Catholicism. The annual Carnival festival in Brazil mixes Roman Catholic practices with traditional African celebrations.

### School and Sports

Every country in South America provides public education. However, some children struggle to receive adequate schooling. These children may live in rural areas with few schools and may not attend school regularly. Other children may leave school at an early age to help earn money for their families. Though some barriers to education exist, the majority of people on the continent are able to read and write.

**Critical Viewing** A teacher conducts class in Peru. How would you describe the students, their classroom, and what they are doing?

Playing *futbol*—known in the United States as soccer—or watching a favorite team compete is a popular pastime in South America. Brazilian soccer teams frequently compete in, and often win, the World Cup, a global soccer tournament that occurs once every four years.

## Regional Food

South American **cuisine**, or the food that is characteristic of a particular place, varies in different parts of the continent. In coastal countries like Chile, seafood is often a main ingredient, such as in a stew called *paila marina* (pie yuh mah REE nah). In Uruguay and Argentina, where there is a good deal of grazing land, beef is often served. In the mountains of Peru, llamas provide a good source of meat. Stews with black beans, or *feijoadas* (fay ho AH daz), rice, and vegetables are common throughout South America.

### Before You Move On
**Monitor Comprehension** What daily activities reflect the various cultural elements of South America?

**Critical Viewing** This market is located in Santiago, Chile. Based on this photo, what kind of food can you infer might be important in Chilean cuisine?

## ONGOING ASSESSMENT
## WRITING LAB  GeoJournal

**Describe Culture** Imagine you and your family are living in a South American country for a short time. Write a blog entry describing daily life during your stay.

**Step 1** Outline your blog entry. Write a topic sentence for each paragraph you plan to write.

**Step 2** Use the research links at **Connect to NG** to research your country of choice. Include aspects of culture mentioned in the text.

**Step 3** Look for photos that might illustrate different aspects of daily life in that country.

**Step 4** Share your blog entry with classmates.

# SECTION 2 GOVERNMENT & ECONOMICS
## 2.1 Comparing Governments

**Main Idea** The governments of South America are moving toward democracy and strengthening their economies.

In the early 1800s, movements for independence in countries across South America brought an end to colonial rule. However, in many countries, real power remained in the hands of a few wealthy families. South American governments have gone through many changes since independence. Argentina, Peru, and Chile provide three examples of the challenges South American governments have struggled to overcome.

### Argentina

Argentina has faced economic and political challenges. After independence in 1816, dictators held power for several decades. In the 1850s, the country adopted a new constitution and entered into a **prosperous**, or economically strong, period that lasted until the late 1920s. In 1930, military officers staged a **coup**, or takeover, of the government. After another coup in 1943, Colonel Juan Domingo Perón gained support among workers.

In 1946, Perón was elected president. He raised wages for workers and established social and economic programs. Many of Perón's reforms were expensive, and corruption weakened his administration. In the 1950s, he was overthrown by military leaders who were dissatisfied with his leadership.

For decades, Argentina sought to move forward. Since the 1980s, democratically elected presidents have faced serious economic crises. Today, though, Argentina's constitutional government is stable, and, since recovering from a serious financial crisis in the early 2000s, its economy is one of the strongest in South America.

### Peru

Peru gained independence in 1821. For most of its history, Peru's government has shifted between democratically elected presidents and military rule. Many leaders—even those elected democratically—have favored wealthy landowners over ordinary citizens. Political instability made economic and social progress difficult in Peru.

In 2001, Peruvians elected their first Quechua president, Alejandro Toledo. Toledo had the support of the Quechua people and a large majority of Peru's poor population. His election demonstrated that Peru's government could represent all of its citizens.

**Critical Viewing** Former president Alejandro Toledo greets children in Peru. What does this photo lead you to think about Toledo's approach to leadership?

## Chile

Chile declared independence in 1818. Since then, Chile has mostly been a representative democracy. Like many other countries in South America, though, Chile has experienced rule by a dictator. In 1973, Salvador Allende's government was overthrown by the military. General Augusto Pinochet (peen oh SHAY) acted as a dictator in Chile for nearly two decades, from 1973 to 1990. Pinochet was **ruthless**, or cruel, and no one was allowed to disagree with his policies.

Chile returned to democratic rule in 1990. In 2006, the people of Chile elected its first female president, Michelle Bachelet Jeria. Bachelet's election was especially meaningful. Her father was killed during Pinochet's rule, and she and her mother were both imprisoned and exiled because they opposed Pinochet. As president, Bachelet helped ease poverty, expanded social reforms, and used profits from copper exports to create new employment opportunities.

### Before You Move On
**Summarize** In what ways are Argentina, Peru, and Chile alike and different in their movement toward democracy and strong economies?

---

**ONGOING ASSESSMENT**

## SPEAKING LAB  GeoJournal

1. **Turn and Talk** Turn to two other classmates and compare the similar challenges facing the governments of Argentina, Peru, and Chile since independence. Take notes while you compare. Include important dates and people. Ask each of your partners a question based on your discussion.

2. **Identify** Why was Michelle Bachelet's election significant? With a partner, discuss how her election was shaped by her family history.

---

Michelle Bachelet Jeria arrives at the presidential palace after her inauguration in Chile in 2006.

# SECTION 2 GOVERNMENT & ECONOMICS

## 2.2 The Pampas Economy

**Main Idea** The Pampas is a fertile region that contributes to the economic success of Argentina.

The wide, grassy Pampas is South America's major crop-growing region. The Pampas is also the grazing ground for another valuable export—cattle.

### An Agricultural Heart

The Pampas is a large plain that stretches across central Argentina from the Atlantic Ocean to the foothills of the Andes Mountains. The Pampas covers about 295,000 square miles in the northern half of the country, nearly one-quarter of Argentina's land area.

During the colonial period in the 1800s, the Spanish imported horses and cattle to the region. Spanish and mestizo cowboys, called gauchos, herded cattle and sheep on the Pampas, a practice continued by modern gauchos.

Cattle grazing takes place in the region's dry zone, or the **Dry Pampas**, located in the west. The humid zone, or the **Wet Pampas**, is in the east. Agriculture abounds in the Wet Pampas, where nearly 40 inches of rain falls on average each year. Together, the cattle industry and agriculture contribute greatly to the region's economic prosperity.

In the 1980s, a new crop became profitable in Argentina: **soybeans**, a type of bean grown for food and also for industrial products such as plastics, inks, and adhesives. Soybeans grow quickly and unlike crops such as wheat and corn, do not need as much **fertilizer**, a substance added to soil to enrich it. In response to demand, Argentines increased soybean production. Low production costs and high prices in global markets make soybeans a valuable crop in Argentina.

**Critical Viewing** This gaucho is tending sheep on the Pampas in Argentina. Based on the photo, how would you describe the work of a gaucho?

**① DRY PAMPAS** Argentines are among the world's top consumers of beef. Beef consumption per capita in Argentina is 119 pounds per year compared to 82 pounds per year in the United States.

**② SOYBEANS** These soybeans are grown in the Wet Pampas. Soybeans are used to make vegetable oil and livestock feed. Soybeans can also be cooked and eaten, and are a good source of protein.

**③ WET PAMPAS** This patchwork of crops is in the Wet Pampas, Argentina's main agricultural area. Soybeans, wheat, corn, flax, and alfalfa are some of the crops grown here.

### Before You Move On

**Summarize** How does the rich soil of the Pampas contribute to Argentina's economy?

### ONGOING ASSESSMENT
# READING LAB  GeoJournal

1. **Monitor Comprehension** What crop has become an important export in Argentina, and why is it profitable?
2. **Interpret Maps** How might Argentina's coastal location benefit its agriculture industry?

SECTION 2.2  **225**

SECTION 2 GOVERNMENT & ECONOMICS

# 2.3 Chile's Food Production

TECHTREK

myNGconnect.com For a map of Mediterranean climates and photos of agriculture

Maps and Graphs

Digital Library

**Main Idea** The mild climate in central Chile supports abundant agricultural exports.

Chile is a long, narrow country. It borders the Pacific Ocean and stretches more than 2,600 miles north and south. Central Chile enjoys a **temperate**, or mild, climate that supports an extensive agriculture industry.

## Mediterranean Climates

The temperate climate along the coasts of southern Europe and northern Africa on the Mediterranean Sea is called a **Mediterranean climate**. This climate is defined by hot, dry summers and mild, rainy winters. Mediterranean climates are found in southern Australia, southern and central California in the United States, southern South Africa, and central Chile.

One reason such widely scattered parts of the world have similar climates is because of latitude. Remember that latitude measures distance from the equator in degrees. Places with Mediterranean climates have similar latitudes either north or south of the equator. Generally, these latitudes measure between 30°S–40°S or 30°N–40°N. In addition to similar latitudes, places with Mediterranean climates lie on western coasts and have similar precipitation patterns because of their coastal positions.

## Chile's Agricultural Bounty

Because of Chile's location on the western coast of South America at nearly 30°S, farmers in central Chile can grow a rich mix of crops. Fruits such as grapes, peaches, and apples are grown for export in Chile's fertile valleys.

Two crops in particular show how Chile takes full advantage of its Mediterranean climate: grapes and olives. A thriving wine industry depends on the many varieties of grapes that grow in Chile's vineyards. Chile has also become a major exporter of olives and specialty

**Critical Viewing** Chile exports 40 percent of its fruit to the United States. What details in this photo tell you these peaches are imported from Chile?

## WORLD MEDITERRANEAN CLIMATES

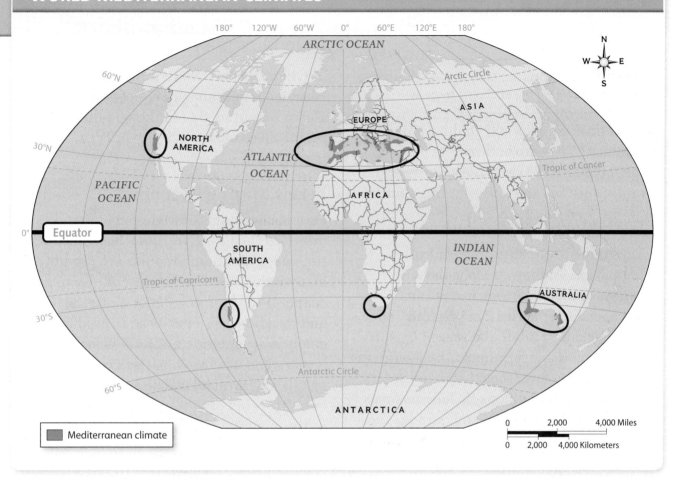

olive oils. Chilean wine and olive oil exports increased dramatically in the 1990s and early 2000s. Both wine and olive products are exports traditionally dominated by Mediterranean countries such as Italy, France, and Greece.

While agricultural production is a source of growth for Chile's economy, copper remains its most valuable export. The growth of agricultural exports since the late 1980s helped Chile **diversify**, or vary, its economy. Diverse economies depend on multiple industries—such as mining and agriculture—and tend to be stronger and more competitive.

### Before You Move On
**Summarize** What main agricultural exports are supported by Chile's Mediterranean climate?

---

**ONGOING ASSESSMENT**
## MAP LAB

1. **Interpret Maps** Look at the grid on the map of Mediterranean climates. Which Mediterranean climates are located south of the equator? At what latitude do they lie?

2. **Location** According to the map, where is the largest Mediterranean climate region located? Based on your reading, what might the weather be like during the winter months in that region?

3. **Summarize** Summarize the characteristics of a Mediterranean climate. Use the chart below to organize your summary.

| MEDITERRANEAN CLIMATES | |
|---|---|
| 1. | |
| 2. | |

SECTION 2.3

# SECTION 2 GOVERNMENT & ECONOMICS

## 2.4 Products of Peru

**Main Idea** Agriculture and mining industries in Peru are helping the economy to grow.

What do asparagus and gold have in common? They are both products of the mountains of Peru. Agriculture and mining are key industries in this country.

### High Mountain Agriculture

The Andes Mountains in Peru feature high peaks and low, steep valleys. The valleys are fertile in some areas, but often not suitable for agriculture. **Erratic**, or inconsistent, rainfall and rugged terrain make large-scale farming in the mountains difficult. The few crops that do grow well in these conditions include potatoes, wheat, corn, vegetables, and quinoa. These crops are generally grown for local people and are not exported. The limited cash crops grown for export include sugarcane, wheat, coffee, and asparagus, a vegetable valued in international markets.

Other economic activities in the Peruvian mountains include cattle and alpaca ranching. By far, though, the most profitable economic activity in Peru is mining.

### Peru's Mining Economy

Peru is a leading exporter of metals and minerals such as silver, zinc, lead, copper, tin, and gold. These products are used in many industries worldwide. For example, copper is made into wire for electrical and telephone systems. Lead is used for automobile batteries, and gold is made into jewelry and electronic parts. Exports of metals and minerals alone supply nearly two-thirds of Peru's total **export revenue**, or the money earned from exports.

Two developments helped Peru increase export revenues from mining so that it became more profitable than agriculture. The first was new government policies on mine ownership, which began in

> **Critical Viewing** This Peruvian farmer uses traditional methods to plow a field. What can you infer about the challenges of this type of farming?

**Critical Viewing** Copper is made into electrical wire and into tubes and pipes used for plumbing. Based on the photo, what can you infer about the importance of this export?

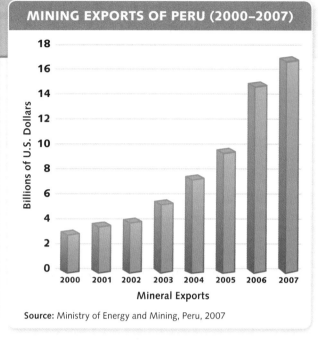

the early 1990s. These policies brought about private ownership of the mines. This shift allowed for investment in needed improvements and upgrades, which led to a rapid growth of the mining industry in Peru. Existing mines became more **profitable**, or financially successful, and new mines opened. Today the Yanacocha gold mine, which opened in Peru in 1993, is the largest gold mine in South America.

Peru's mining revenues also increased because of an upward trend in world prices for silver, gold, and other metals and minerals since the late 1990s. For 2012, the country's mining revenues reached $25 billion and were estimated to continue to increase to $30 billion by the end of 2013.

### Before You Move On
**Monitor Comprehension** What exports have helped Peru's economy to grow?

### ONGOING ASSESSMENT
## DATA LAB  GeoJournal

1. **Identify** Look at the bar graph. What happened to mining exports between the years 2000 and 2007?
2. **Interpret Graphs** What is the difference in mining revenues between 2000 and 2007, measured in dollars? What accounts for that difference?
3. **Summarize** What governmental policy change in the 1990s led to a growth of the mining industry between 2000 and 2007?

# SECTION 3 FOCUS ON BRAZIL
## 3.1 Brazil's Growing Economy

**Main Idea** Brazil is a leading industrial country and has a strong, diverse economy.

Brazil is the largest and most populous country in South America with many different exports. Brazil is an emerging force in the global economy.

### Diverse Products

Agriculture, forestry, ranching, and fishing all contribute to the country's economy. Brazil exports bananas, oranges, mangoes, cacao beans, soybeans, rice, cashew nuts, and pineapples. It is the **foremost**, or leading, global coffee producer and grows one-third of the world's total number of coffee beans. Brazil is also one of the leading exporters of sugarcane and raw sugar.

Forests in Brazil contain many raw materials, including timber, or wood prepared for use in construction. Wood from Brazilian forests is also used to make pulp for paper products. Mahogany, a rain forest hardwood, is a valuable export used to make fine furniture.

Cattle ranching is a big business in Brazil. Export products from cattle include leather and beef. Today, Brazil exports nearly as much beef as Australia and the United States.

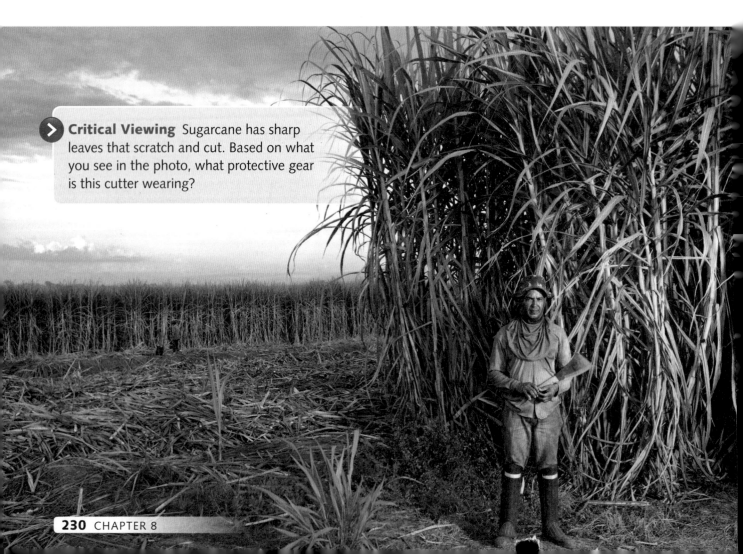

> **Critical Viewing** Sugarcane has sharp leaves that scratch and cut. Based on what you see in the photo, what protective gear is this cutter wearing?

Brazil's coastline stretches more than 4,600 miles along the Atlantic Ocean. New technology, facilities, and processes will allow Brazil to develop its commercial fishing industry.

## Mining and Manufacturing

Mining contributes to Brazil's growing economy. Brazil is a top producer and exporter of iron ore, bauxite (aluminum ore), gold, copper, and diamonds. Oil wells dot Brazil's coast.

In addition to its mineral production, Brazil's manufacturing industry is central to the country's economic strength. Brazil manufactures **steel**, a strong metal made from iron and other metals. Brazilian steel is used in automobiles, transportation equipment, and aircraft. Brazil also manufactures computers and electronic equipment.

## Fuel of the Future

Brazil has produced and exported sugarcane and sugar since the 1500s. For several decades, the country has been developing its sugar-based ethanol industry. **Ethanol** is a liquid removed from sugarcane or corn. It is mixed with gasoline to make an alternative fuel called a **biofuel**. Cars that run on biofuels use less gasoline. Because they are made from agricultural products, biofuels are a renewable source of energy. Today, Brazil is poised to be a global leader in the production of biofuel.

### Before You Move On
**Summarize** What factors enable Brazil to develop a diverse economy?

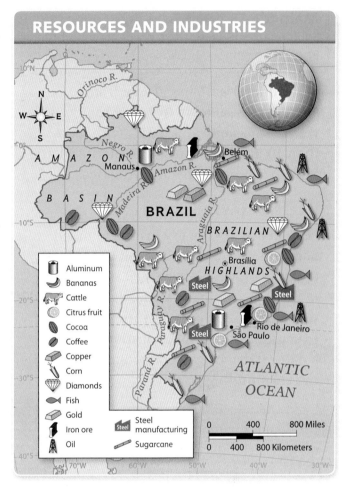

## ONGOING ASSESSMENT
## MAP LAB

1. **Interpret Maps** Study the map. What are two resources that provide fuel for Brazil? Where are these resources found?

2. **Draw Conclusions** Where are most cattle resources located in Brazil? Why might cattle not be concentrated in the Amazon Basin?

3. **Human-Environment Interaction** In what ways have Brazilians used natural resources to build a strong economy?

4. **Form and Support Opinions** What part of Brazil's economy do you think will be the most important in the future? Why?

# SECTION 3 FOCUS ON BRAZIL
## 3.2 São Paulo

TECHTREK
myNGconnect.com For a graph and photos of São Paulo, Brazil
Maps and Graphs | Digital Library

**Main Idea** São Paulo is the largest city in the Southern Hemisphere and a major contributor to Brazil's economy.

**São Paulo**, Brazil, has grown from a quiet mission town to one of the world's **megacities**, or cities with more than 10 million people. It is a cultural and industrial center of South America.

### Early Growth

Portuguese missionary priests founded São Paulo in 1554. They built a mission and school, which remained the focus of the town for many years. São Paulo was also a point of departure for military expeditions. Its hilltop location provided a natural defense and panoramic views of the surrounding area.

In the late 1600s, gold was discovered in the nearby mountains of the state of Minas Gerais. This resource proved extremely valuable, and by the mid-1700s, Brazil was producing nearly half of the world's supply of gold. Roughly 50 years later, gold deposits were mostly depleted. However, coffee production soon replaced gold mining as the main economic activity. By the mid-1800s, coffee had become a significant export crop. Wealth gained from coffee production transformed São Paulo and contributed to its rapid growth in industry and population.

In just 20 years between 1880 and 1900, the population of São Paulo jumped from 35,000 to 240,000. Some of the population growth came from rural to urban migration. However, much of the city's growth was the result of immigration from Asia and Europe.

### The Modern City

In the mid-1900s, São Paulo became the industrial center of Brazil. By the 1950s, the automotive industry was well developed. Jobs in the industry attracted workers from other parts of Brazil and other South American countries. São Paulo continues to manufacture and export one million cars each year.

In addition to manufacturing industries, tourism has contributed to the city's economic growth. Today, São Paulo attracts people from countries all over the world. Visitors to São Paulo enjoy the city's museums, its shopping district, its diverse collection of restaurants, and a bustling nightlife.

One result of the rapid expansion of São Paulo's economy is the explosive population growth since 1950. Many people who migrated there made their homes in **slums**, or overcrowded and poverty-stricken urban areas. Slums, or *favelas*, as they are called in Brazil, developed on the outskirts of the city. Slums are not unique to São Paulo, but are characteristic of large cities that experience rapid population growth.

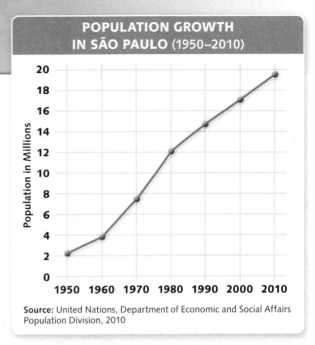

Source: United Nations, Department of Economic and Social Affairs Population Division, 2010

### Before You Move On
**Summarize** What industries are most important in São Paulo today?

These young people are part of a fashion event staged at a São Paulo train platform.

### ONGOING ASSESSMENT
## DATA LAB  GeoJournal

1. **Interpret Graphs** According to the graph, during what two decades did São Paulo experience the most rapid rate of growth? What accounts for that increase?

2. **Analyze Data** Using the data in the graph, how much larger in millions of people was São Paulo in 2010 than it was in 1950? How would you describe population growth in São Paulo since the 1990s?

3. **Make Inferences** In what ways might explosive population growth and the development of slums be related?

SECTION 3.2  233

# SECTION 3 FOCUS ON BRAZIL

## 3.3 Impact of the Olympics

**Main Idea** Brazil is hoping to maximize the economic and social impact of the 2016 Olympic Games.

Athens, Greece, hosted the first modern Olympics in 1896. Since that time, the International Olympic Committee (IOC) has decided which countries would host the games. Not once had the committee chosen a city in South America. That changed in 2009 when the IOC selected **Rio de Janeiro**, Brazil, as the host of the 2016 Olympics.

### Preparing the City

Rio has previous experience hosting international sports events. In 2007, Brazil hosted the Pan-American Games, which are similar to the Olympics but open only to countries in the Western Hemisphere. At that time, Brazil's government made improvements to the **infrastructure**, or the basic systems that a society needs, such as roads, bridges, and sewers. However, more improvements are needed to host the Olympics. Thirty-four competition **venues**, or locations for organized events, need to be built or updated. Efficient transportation to and from the venues is also important.

The Olympic Games last only a few weeks. However, many in Rio de Janeiro hope that one **impact**, or effect, will be long-term improvements in the health of their city. Overcrowded and often dangerous favelas surround Rio. As part of its Olympic preparations, the city tore down and rebuilt some of these locations. Areas that were once slums have become paved neighborhoods with running water, electricity, and gas—all of which were missing in the existing favelas.

**Critical Viewing** Brazilian soccer star Pélé (far right) celebrates winning the bid with the Rio Olympic committee. Based on this photo, what can you say about the committee's reaction?

## Hosting the Games

Benefits come to countries that host the Olympic Games. First, constructing buildings such as stadiums and the Olympic Village (where athletes stay) creates jobs. Ideally, these structures will continue to be used now that the games are over. Businesses such as hotels and restaurants benefit, too. Athletes and tourists from around the world flood a host city and spend money on lodging, food, and other goods. Billions more watch the Olympics on television. With such wide exposure, Rio de Janeiro should now earn a reputation as a thriving and world-class city.

### Before You Move On
**Make Inferences** In what ways did the 2016 Olympics change Rio?

## ONGOING ASSESSMENT
## PHOTO LAB  GeoJournal

1. **Analyze Visuals** Look at the photo of Rio de Janeiro. As a tourist visiting Rio, what impression might you have of the city?
2. **Describe** In what ways might the hills and coastline shown in the photo help make Rio an appealing location for the 2016 Games?
3. **Region** What makes the IOC's choice of Rio de Janeiro as an Olympic host a departure from previous choices?
4. **Form and Support Opinions** Consider what you learned from the text and photos. Write two or three sentences that explain whether or not you think the 2016 Games will have a lasting, positive impact in Rio. Support your opinion with facts and details. Go to **Student Resources** for Guided Writing support.

The Christ the Redeemer statue atop Corcovado Mountain overlooks Rio de Janeiro.

# CHAPTER 8
# Review

## VOCABULARY

For each pair of vocabulary words, write one sentence that explains the connection between the two words.

1. venue; infrastructure

> Olympic hosts often build new infrastructure such as roads and new competition venues for the events.

2. export revenue; profitable
3. Mediterranean climate; temperate
4. soybean; fertilizer
5. megacity; slum
6. steel; ethanol

## MAIN IDEAS

7. What are the three largest indigenous groups in South America? (Section 1.1)
8. Why are so many different languages spoken in South America? (Section 1.2)
9. In what ways has religion shaped the daily lives of South Americans? (Section 1.3)
10. What governmental changes have occurred in Argentina, Peru, and Chile since independence? (Section 2.1)
11. How does the Pampas region contribute to Argentina's economy? (Section 2.2)
12. What are the main characteristics of the Mediterranean climate in Chile? (Section 2.3)
13. What are Peru's most profitable exports? Why? (Section 2.4)
14. What exports and industries shape Brazil's economy? (Section 3.1)
15. In what way has population growth put pressure on São Paulo? (Section 3.2)
16. How is Rio de Janeiro preparing to host the 2016 Olympic Games? (Section 3.3)

## CULTURE

**ANALYZE THE ESSENTIAL QUESTION**

In what ways is South America culturally diverse?

**Critical Thinking: Summarize**

17. In what ways do governments in South America recognize indigenous languages?
18. How did the majority of South Americans become Roman Catholic?

## GOVERNMENT & ECONOMICS

**ANALYZE THE ESSENTIAL QUESTION**

How is modern South America building its economies?

**Critical Thinking: Find Main Ideas**

19. What industry has contributed to Peru's economic growth?
20. How does Chile's Mediterranean climate influence its economy?

**INTERPRET GRAPHS**

21. **Analyze Data** According to the graph, about when did Argentina and Brazil surpass the United States as soy exporters?
22. **Make Generalizations** Look at the graph. What decade represents the most rapid period of growth for soy exports overall?

## FOCUS ON BRAZIL

### ANALYZE THE ESSENTIAL QUESTION
How has Brazil become an economic power?

**Critical Thinking: Synthesize**

23. What factors point to Brazil's current and future economic strength?

24. In what ways does São Paulo reflect Brazil's economic potential?

### INTERPRET MAPS

25. **Location** According to the map, where are most cities located in Brazil? How would you explain this?

26. **Make Inferences** Locate Brasília, the current capital of Brazil, and Rio de Janeiro, its former capital. What might the transfer of capital city from Rio de Janeiro to Brasília in 1960 predict about future development in Brazil?

## ACTIVE OPTIONS

**Synthesize** the Essential Questions by completing the activities below.

27. **Create a Poster** Introduce a country in South America by creating a poster. Use the **Digital Library** to choose photos or create your own pictures that represent different aspects of that country, including its economy and cultural diversity. Include pictures of different people and languages, main exports, foods, sports, and celebrations. **Present your poster to the class.**

> **Poster Tips**
> - Create a catchy title for your poster.
> - Organize your pictures in categories.
> - Write labels explaining each picture on your poster.

28. **Conduct Internet Research** Work in a group to gather two new facts about the economies of Brazil, Argentina, Colombia, and Peru. Use online sources to search for your facts. Then create a chart to record your facts. Compare your findings with those of the other groups.

| | |
|---|---|
| Brazil | Fact 1: _____ |
| | Fact 2: _____ |
| Argentina | Fact 1: _____ |
| | Fact 2: _____ |
| Colombia | Fact 1: _____ |
| | Fact 2: _____ |
| Peru | Fact 1: _____ |
| | Fact 2: _____ |

COMPARE ACROSS REGIONS  UNIT 4

# Sports and the Olympics

As you have learned, the 2016 Olympics will be held in Rio de Janeiro, Brazil, which will mark the first time Brazil has hosted the games. In 2008, China hosted its first Olympic Games. Nearly 11,000 athletes from 204 countries competed in Beijing. Spectators around the world were awed by competing athletes who broke 40 world records and 130 Olympic records.

In any Olympics, the medal count draws the interest of many people. The top medal winners at the 2008 Games were the United States (110), China (100), and Russia (72). Exploring different factors that help determine high medal counts reveals patterns as well as surprises.

### Compare
- Australia
- Brazil
- China
- Jamaica
- United States

### POPULATION AND ECONOMICS

Countries that win medals generally have populations of at least one million. According to 2008 population statistics, the most populous country represented at the games was China, followed by India and the United States. However, high populations do not guarantee high medal counts. China and the United States won many medals at the 2008 Olympics, but India won just three. Other, less populous countries won more, including two of the least populous countries at the games, Slovenia and Jamaica.

Population is not the only factor in determining a high medal count. How strong a country is economically matters as well. The Gross Domestic Product (GDP), or the measure of the total value of goods and services provided in a country, is one indicator of the economic health of a country. Countries with higher GDPs have more resources for training and supporting athletes for international competitions. At the Beijing Games, the highest ranking GDPs, measured in U.S. billions of dollars, belonged to the United States, Japan, and China. However, countries with low GDPs were competitive with countries with higher GDPs. Zimbabwe, with one of the lowest GDPs, won four medals.

### HOME ADVANTAGE

One advantage for hosting the Olympic Games has been, historically, a higher medal yield. Host countries tend to win 30 to 40 percent more medals than in the previous Olympic cycle. At the Beijing Games, China won 100 medals. Four years earlier at the Summer Games held in Athens, Greece, China had won only 63 medals.

# OLYMPIC SNAPSHOT
(Beijing Olympics, Summer 2008)

**KEY** ⬤ = Total Medal Count  👤 = Population (millions)  $ = GDP (billions)

**BRAZIL**
- ⬤ = 15
- 👤 = 191.9
- $ = 1,621.27

**UNITED STATES**
- ⬤ = 110
- 👤 = 305
- $ = 14,195.03

**CHINA**
- ⬤ = 100
- 👤 = 1,327.7
- $ = 3,941.54

**AUSTRALIA**
- ⬤ = 46
- 👤 = 21.2
- $ = 1,046.79

**JAMAICA**
- ⬤ = 11
- 👤 = 2.7
- $ = 12.82

**Source:** The Wall Street Journal, © 2011

## ONGOING ASSESSMENT
### RESEARCH LAB  GeoJournal

1. **Explain** Which three countries have the highest GDPs and populations? How does Australia compare to these countries?
2. **Identify** Of the countries listed above, which has the lowest GDP? How does its medal count compare with Brazil's?
3. **Make Predictions** How many medals did Brazil win in 2008? Based on historical patterns of "home advantage," how many medals can you guess Brazil might win at the 2016 Games?

**Research and Draw Conclusions** Select two countries, each from a different part of the world. Research their performances at one of these Summer Olympic Games: Athens (2004), Sydney (2000), or Atlanta (1996). Compare your findings to the Olympic snapshot of the 2008 Beijing games above. In what ways do your countries compare to the top medal winners? What surprises did you find?

COMPARE ACROSS REGIONS

REVIEW & EXTEND **UNIT 4**

# Active Options

**TECHTREK**

myNGconnect.com For photos of Amazon rain forest life

 Digital Library

## ACTIVITY 1

**Goal:** Extend your understanding of life in the Amazon rain forest.

### Create a Photo Gallery

The Amazon rain forest is home to thousands of species of birds, land animals, and fish. Thousands of types of trees, plants, and insects live there, too. Do research and choose several examples of different Amazon plants, insects, and animals that interest you. Use the **Magazine Maker CD-ROM** to create a photo gallery of the plants, animals, and insects you selected. Write a description of each photo. Invite your friends to visit your virtual gallery.

Spider monkey, Amazon rain forest, Peru

## ACTIVITY 2

**Goal:** Learn more about South American history and culture.

### Prepare a Multimedia Presentation

Archaeologists have learned much about different cultures in South America. National Geographic grant recipient Christina Conlee explores artifacts left behind by the Nasca culture. Do online research to find information about them. Then select your favorite artifact and create a multimedia presentation about it. Include at least one photo of the artifact in your presentation. Write a label that describes the artifact. Create a title for your presentation.

## ACTIVITY 3

**Goal:** Research daily life in Latin America.

### Write a Feature Article

The culture region called Latin America includes Mexico, Central America, the islands of the Caribbean, and South America. Because of years of Spanish colonial influence, Latin America is united by common cultural practices. Using online research, write a feature article about aspects of daily life that unite Latin America as a culture region. Focus on food, sports, festivals, religion, and traditional dress.

# UNIT 5
# Explore EUROPE with NATIONAL GEOGRAPHIC

### MEET THE EXPLORER

**NATIONAL GEOGRAPHIC**

Some archaeological sites are underwater. Emerging Explorer Katy Croff Bell works with archaeologists in the Mediterranean and Black seas to help them figure out where to look for submerged secrets.

### INVESTIGATE GEOGRAPHY

The Alps are the highest and most extensive mountain range in Europe. They stretch across central Europe and are concentrated in France, Germany, Italy, Switzerland, and Austria. This is the Lauterbrunnen Valley in Oberland, Switzerland.

### STEP INTO HISTORY

The Colosseum in Rome is one of the Roman Empire's greatest architectural and engineering achievements. The arena, completed in A.D. 80, seated nearly 50,000 spectators who watched gladiator games, performances, and even mock naval battles.

ONLINE WORLD ATLAS

London, United Kingdom
3,673 miles
Washington, D.C.

Go to myNGconnect.com for maps of Europe.

CONNECT WITH THE CULTURE

Architect I.M. Pei's modern pyramid serves as an entrance into the Louvre in Paris, France. The museum holds some of the world's greatest art treasures.

# CHAPTER 9

# EUROPE
## GEOGRAPHY & HISTORY

## PREVIEW THE CHAPTER

**Essential Question** How did Europe's physical geography encourage interaction with other regions?

### SECTION 1 • GEOGRAPHY

1.1 Physical Geography .................. 246
1.2 A Long Coastline ...................... 248
1.3 Mountains, Rivers, and Plains ........ 250
1.4 Protecting the Mediterranean ........ 252
   NATIONAL GEOGRAPHIC Photo Gallery ........ 254

**KEY VOCABULARY**
- peninsula
- uplands
- polder
- bay
- fjord
- canal
- waterway
- ecosystem
- marine reserve

**ACADEMIC VOCABULARY**
navigable, erosion

**TERMS & NAMES**
- Northern European Plain
- Alps
- Danube River
- Rhine River

**Essential Question** How did European thought shape Western civilization?

### SECTION 2 • EARLY HISTORY

2.1 Roots of Democracy .................. 256
2.2 Classical Greece ...................... 258
2.3 The Republic of Rome ................ 260
2.4 The Roman Empire ................... 262
2.5 Middle Ages and Christianity ........ 264
2.6 Renaissance and Reformation ....... 266

**KEY VOCABULARY**
- democracy
- city-state
- golden age
- philosopher
- republic
- patrician
- plebeian
- barbarian
- aqueduct
- feudal system
- serf
- perspective
- indulgence

**ACADEMIC VOCABULARY**
aristocrat, veto

**TERMS & NAMES**
- Acropolis
- Alexander the Great
- Julius Caesar
- Augustus
- Christianity
- Middle Ages
- Crusades
- Renaissance
- Johannes Gutenberg
- Martin Luther
- Reformation
- Counter-Reformation

**Essential Question** How did Europe develop and extend its influence around the world?

### SECTION 3 • EMERGING EUROPE

3.1 Exploration and Colonization ........ 268
3.2 The Industrial Revolution ............ 270
3.3 The French Revolution ............... 272
   DOCUMENT-BASED QUESTION
3.4 Declarations of Rights ................ 274
3.5 Nationalism and World War I ........ 276
3.6 World War II and the Cold War ...... 278

**KEY VOCABULARY**
- navigation
- colony
- textile
- factory system
- radical
- guillotine
- natural rights
- apartheid
- nationalism
- trench
- reparations
- concentration camp

**ACADEMIC VOCABULARY**
convert, alliance

**TERMS & NAMES**
- Industrial Revolution
- Enlightenment
- John Locke
- Reign of Terror
- Napoleon Bonaparte
- Treaty of Versailles
- Great Depression
- Adolf Hitler
- Holocaust
- Iron Curtain
- Cold War
- Berlin Wall

244

# SECTION 1 GEOGRAPHY
## 1.1 Physical Geography

**TECHTREK**
myNGconnect.com For online maps of Europe and Visual Vocabulary

Maps and Graphs | Digital Library

### EUROPE PHYSICAL

**Main Idea** Europe is made up of several peninsulas with varied land regions and climates.

Europe is a "peninsula of peninsulas." A **peninsula** is a body of land surrounded on three sides by water.

## A Peninsula of Peninsulas

Europe forms the western peninsula of Eurasia, the landmass that includes Europe and Asia. In addition, Europe contains several smaller peninsulas, including the Italian, Scandinavian, and Iberian. Europe also consists of significant islands, including Great Britain, Ireland, Greenland, Iceland, Sicily, and Corsica.

Four land regions form Europe. The Western Uplands are made up of **uplands**, or hills, mountains, and plateaus, that stretch from the Scandinavian Peninsula to Spain and Portugal. The **Northern European Plain** is made up of lowlands that reach across northern Europe. The Central Uplands are hills, mountains, and plateaus at the center of Europe. The Alpine region consists of the **Alps** and several other mountain ranges.

## Varied Climates

Most of Europe lies within the humid temperate climate region. The North Atlantic Drift, an ocean current of warm water, keeps temperatures relatively mild. Winds also affect climate. The sirocco (shuh RAH koh) sometimes blows over the Mediterranean Sea and brings wet weather to southern Europe at different seasons. The mistral is a cold wind that sometimes blows through France and brings cold, dry weather to the country.

### CLIMATE

In general, a Mediterranean climate brings mild, rainy winters and hot, dry summers and supports a long growing season. Hardy plants grow best in this climate. In contrast, Eastern Europe has a humid continental climate with long, cold winters. Iceland, Greenland, and northern Scandinavia have a polar climate and a limited growing season.

### Before You Move On

**Monitor Comprehension** What are the main land regions and climates in Europe?

## ONGOING ASSESSMENT
### MAP LAB

1. **Interpret Maps** Study both maps in this lesson. Which climate regions are found on the Scandinavian Peninsula? Based on the climate, where do you think most of the peninsula's population is concentrated?

2. **Compare and Contrast** Use both maps to determine what places in Europe have the coldest climates. What geographic characteristics do these places have in common?

# SECTION 1 GEOGRAPHY
## 1.2 A Long Coastline

**TECHTREK**
myNGconnect.com For an online map and photos of Europe's coastal features and ports

Maps and Graphs | Digital Library

**Main Idea** Europe's long coastline helped to promote trade, industry, exploration, and settlement on the continent.

Europe has more than 24,000 miles of coastline. If you walked 25 miles a day along the continent's coasts, it would take more than four years to walk the entire distance. These extensive coastlines provided early Europeans with great access to oceans and seas.

### Trade and Industry

Europe's water access has benefited the continent in many ways. These benefits include the growth of trade and the development of industry.

Trade has been central to Europe's growth. The civilizations of ancient Greece and Rome flourished largely because of trade. Early sailors traveled to nearly every port on the roughly 2,500-mile-long Mediterranean Sea. They brought back from other lands goods and ideas, such as grains, olive oil, and new religions, that greatly influenced European culture.

Europe also developed several industries that depend on oceans and seas, including a fishing industry. In fact, Europeans have fished along their coastlines for thousands of years.

In lowland areas such as the Netherlands, the people created a way to drain water from the sea in order to increase their farming industry. They built dikes, or giant walls, to hold back the sea in order to create **polders**. Most of the low-lying land of a polder, which once was part of the seabed, was transformed into farms. Today, the Netherlands has about 3,000 polders.

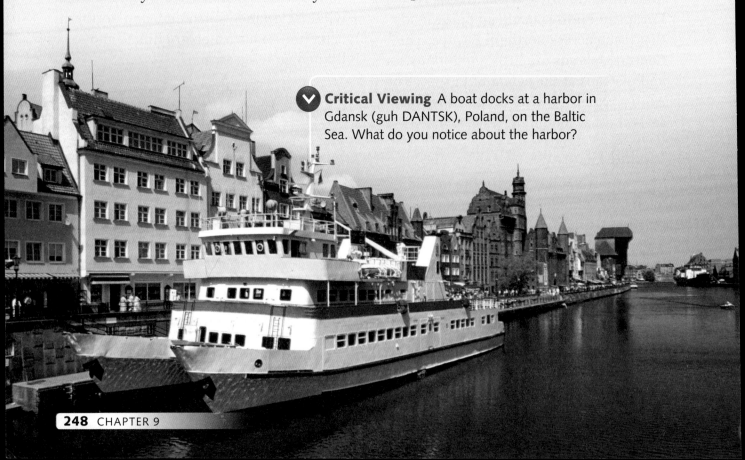

**Critical Viewing** A boat docks at a harbor in Gdansk (guh DANTSK), Poland, on the Baltic Sea. What do you notice about the harbor?

## EUROPE'S MAJOR LANDFORMS AND RIVERS

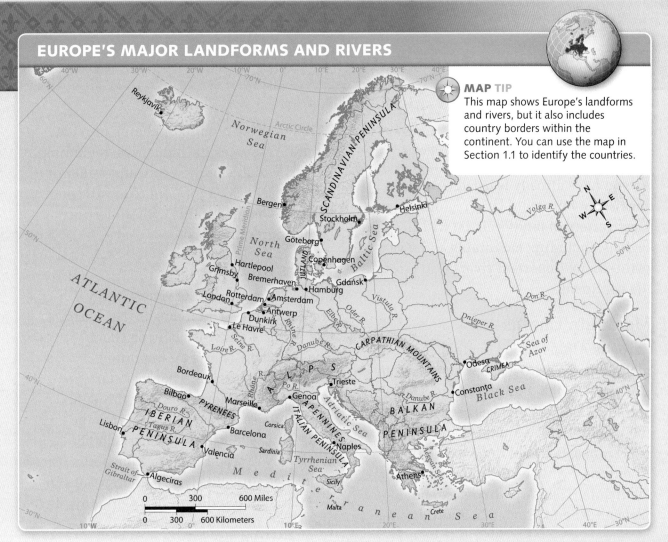

**MAP TIP** This map shows Europe's landforms and rivers, but it also includes country borders within the continent. You can use the map in Section 1.1 to identify the countries.

## Exploration and Settlement

The location of the continent near large bodies of water also encouraged exploration. In the 1400s, explorers helped European rulers obtain raw materials, spread religious beliefs, and build empires.

Over time, people settled around the ports where the ships docked. Towns often grew up near **bays**, which are bodies of water surrounded on three sides by land. Some of the towns, including Hamburg, Germany, became large cities as trade and industry expanded. In contrast, the deep and narrow bays of Norway, called **fjords** (fee ORDZ), did not encourage settlement.

### Before You Move On
**Summarize** In what ways has Europe benefited from its long coastline?

### ONGOING ASSESSMENT
## MAP LAB
GeoJournal

1. **Compare and Contrast** Use the map above and those in Section 1.1 to compare and contrast the Italian Peninsula with the Scandinavian Peninsula. What do the two have in common? What differences do you see?

2. **Analyze Cause and Effect** Complete the chart by writing one effect for each cause.

| CAUSE | EFFECT |
|---|---|
| Trade was conducted on Europe's oceans and seas. | Goods and ideas spread. |
| Industry developed on the coasts. | |
| Explorers traveled to new lands. | |
| Cities grew along the coasts. | |

SECTION 1.2  **249**

# SECTION 1 GEOGRAPHY

## 1.3 Mountains, Rivers, and Plains

myNGconnect.com For an online map and photos of Europe's landforms and natural resources

 Maps and Graphs

 Digital Library

**Main Idea** The landforms and resources in Europe support many economic activities.

As you have already learned, Europe consists of four main land regions. A great variety of landforms lie within these regions, including mountains and a vast plain. Many important rivers also cross the continent.

### Mountain Chains

Europe's Alpine region contains several mountain chains. The Alps stretch from Austria and Italy to Switzerland, Germany, and France. The Pyrenees are located to the west of the Alps and separate Spain and France. South of the Alps lie the Apennines, which run along the Italian Peninsula. The Carpathians extend through Poland, Romania, and Ukraine.

All of these mountain chains provide natural resources for industries, including forests, which supply wood, and mineral resources, such as iron ore. The valleys between the mountains contain fertile land for growing crops.

### Rivers and Plains

Europe has a wealth of rivers. Many are **navigable**, which means that boats and ships can travel easily on them. The **Danube River** is an important transportation route. The river starts in Germany and passes along or through ten countries before emptying into the Black Sea. The **Rhine River** is another vital body of water used to transport goods deep inland. The river originates in Switzerland, winds through Germany, and flows into the North Sea.

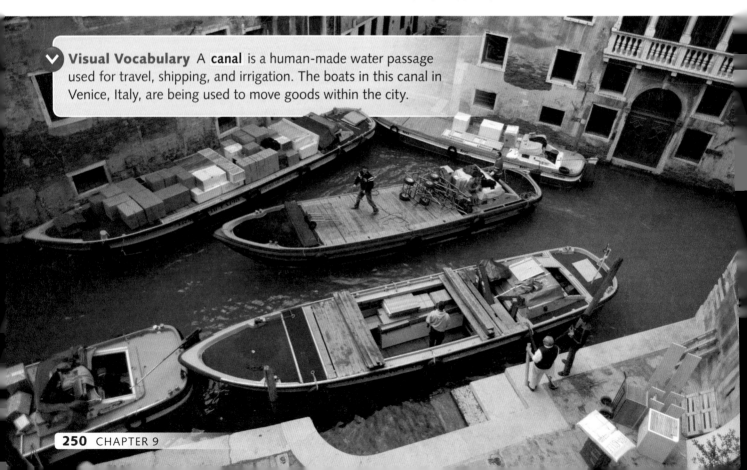

**Visual Vocabulary** A **canal** is a human-made water passage used for travel, shipping, and irrigation. The boats in this canal in Venice, Italy, are being used to move goods within the city.

## LAND USE AND NATURAL RESOURCES

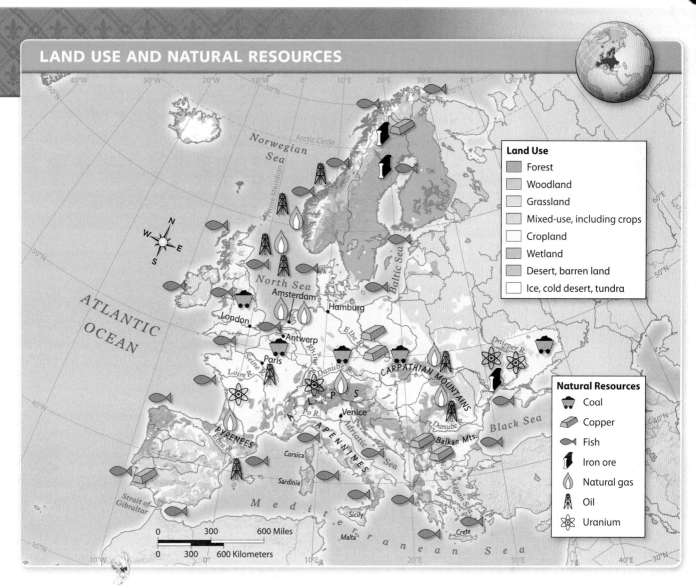

For centuries, Europeans have built canals. When linked together, canals and rivers form **waterways**, or navigable routes of travel and transport. The small country of the Netherlands has more than 3,000 miles of rivers and canals.

Many of the rivers in Europe cross the Northern European Plain. This vast lowland region stretches across France, Belgium, Germany, and Poland all the way to Russia. The fertile soil on the plain makes it ideal for growing crops, and thousands of farms are sprinkled throughout the region. The plain also contains some of the largest and most heavily populated cities, or urban centers, in Europe, including Paris, France.

### Before You Move On
**Summarize** What economic activities are supported by Europe's landforms and resources?

### ONGOING ASSESSMENT
## MAP LAB

1. **Interpret Maps** What natural resources are found in the Carpathian Mountains? What geographic challenges might workers deal with when they extract these resources?

2. **Draw Conclusions** Find the Danube River on the map. What natural resources are found along the river? What role might the Danube play in handling these resources?

3. **Make Inferences** What natural resources are found in the North Sea? What impact might they have on the countries bordering the sea?

SECTION 1.3 **251**

# 1.4 SECTION 1 GEOGRAPHY
## NATIONAL GEOGRAPHIC

**TECHTREK**
myNGconnect.com For a map, photos, and an Explorer Video Clip

Maps and Graphs | Digital Library

# Protecting the Mediterranean
### with Enric Sala

**Main Idea** Human activities have harmed the Mediterranean Sea's natural environment.

## Under the Sea

On June 4, 2010, National Geographic Explorer-in-Residence Enric Sala began an expedition: exploring the underwater world of the Mediterranean Sea. He wanted to find out how human activities have affected the sea's ecosystem.

An **ecosystem** is a community of living organisms and their natural environment. Three human activities have had an impact on the Mediterranean's ecosystem. One is overfishing, which occurs when people catch fish at a faster rate than the fish can reproduce. Another is pollution. The third is overdevelopment, which has occurred as coastal populations have grown. (See the chart on the next page.) Growing populations have added to the Mediterranean's pollution and the **erosion**, or wearing away, of its coastline.

myNGconnect.com
For more on Enric Sala in the field today

◀ **Critical Viewing** Sala swims with a sea turtle. What equipment is he using during this underwater exploration?

## MEDITERRANEAN EUROPE

Sala's work was inspired by Jacques Cousteau, the French underwater explorer. When he became an explorer himself, Sala sailed with Cousteau's son, Pierre-Yves Cousteau. They compared the condition of the Mediterranean Sea now with its condition 65 years earlier and found that the sea has been damaged. Sala concluded, "We have lost most of the large fish and the red coral because of centuries of exploitation [misuse]."

### Marine Reserves

In spite of the harm that has been done, Sala sees signs of hope. During their Mediterranean expedition, he and Cousteau visited the Scandola Natural Reserve, near Italy. Scandola is a **marine reserve**, or protected area where people are prohibited from fishing, swimming, or anchoring their boats. As a result, marine life is thriving at the reserve. "This marine reserve," Sala said, "has restored the richness that Jacques Cousteau showed us 65 years ago."

#### Before You Move On
**Monitor Comprehension** What can be done to protect the Mediterranean Sea from the human activities that have harmed its environment?

### POPULATION OF MEDITERRANEAN CITIES
(IN MILLIONS)

| City | 1960 | 2015 |
|---|---|---|
| Athens, Greece | 2.2 | 3.8 |
| Barcelona, Spain | 1.9 | 5.3 |
| Istanbul, Turkey | 1.74 | 14.1 |
| Marseille, France | 0.8 | 1.6 |
| Rome, Italy | 2.33 | 3.7 |

Source: UN, 2013; CIA World Factbook, 2016

### ONGOING ASSESSMENT
## DATA LAB  GeoJournal

1. **Interpret Charts** Based on the chart, which Mediterranean city had undergone the greatest growth by 2015? What impact did this growth probably have on the city's coastline?

2. **Make Inferences** According to the chart, which city had undergone the greatest growth by 2015? What does this number probably mean for the city's coastline?

3. **Turn and Talk** What could be done today to preserve fish populations for the future? Get together with a partner and come up with one or two specific suggestions. Share your ideas with the rest of the class.

# NATIONAL GEOGRAPHIC
# Photo Gallery • Scotland's Rugged Coast

For more photos from the National Geographic Photo Gallery, go to the **Digital Library** at myNGconnect.com.

The Carnival in Venice

Ancient Roman aqueduct

Musicians in Krakow, Poland

**Critical Viewing** These rocky pinnacles, or pointed formations, took shape thousands of years ago on a hill in Scotland called the Storr. The largest of the formations shown here is known as the Old Man of Storr.

Italy's Amalfi Coast

French pastries on display

The Parthenon in Athens

Armor from the 1600s

# SECTION 2 EARLY HISTORY
## 2.1 Roots of Democracy

**TECHTREK**
myNGconnect.com For an online map of ancient Greek city-states

Maps and Graphs

**Main Idea** The ideas out of which democracy grew first took root in ancient Athens.

The rulers of Athens laid the groundwork for **democracy**, a government in which people can influence law and vote for representatives. Democracy was one of the great achievements of Greek civilization.

### The Greek City-States

Greece lies on both the Balkan and Peloponnesus (pehl uh puh NEE suhs) peninsulas. The two are connected by an isthmus, or narrow strip of land. People first arrived in Greece around 50,000 B.C. Early civilizations developed between 1900 B.C. and 1400 B.C.

Around 800 B.C., several Greek city-states started to thrive. A **city-state** is an independent community that includes a city and its surrounding territory. The mountains on the peninsulas made transportation and communication difficult. As a result, each city-state developed independently. The two largest and most important Greek city-states were Athens and Sparta.

Each city-state established its own community and government. The earliest form of government in the city-states was a monarchy, in which a king or queen rules. Over time, a group of upper-class noblemen called **aristocrats** began to act as advisors to the king. In some city-states, the aristocrats set up a ruling council that served as the government. This council was a form of oligarchy (AHL ih gahr kee), in which a small group rules.

Around 650 B.C., tyrants in many of the city-states seized power away from the councils, took control of the government, and re-established one-person rule. Today, any harsh ruler may be called a tyrant. However, not all tyrants in ancient Greece were bad leaders. Some were fair and had the support of the Greek people.

### Democracy in Athens

Around 600 B.C. in Athens, a statesman named Solon controlled the government of the city-state. He established assemblies in which all the wealthy people of Athens—not just the aristocrats—made the laws.

Then in 508 B.C., a leader named Cleisthenes (KLIHS thuh neez) increased the people's power even more. He established a direct democracy. Under this government, all citizens voted directly for laws. However, only Athenian adult males were citizens and had the right to vote.

## Athens and Sparta

Democracy developed in Athens but not in all city-states. Sparta, Athens' rival, had an oligarchy ruled by a small group of warriors. They supervised a military training system for Spartan boys.

In 490 B.C., Athens and Sparta joined together to defeat the invading army of the Persian Empire under King Darius I. After that, however, the two city-states became fierce enemies.

### Before You Move On
**Summarize** What ancient Greek ideas served as the roots of modern democracy?

**Visual Vocabulary** The **Acropolis** of Athens is a rocky hill that once served as the city's fortress and contained its most important temples.

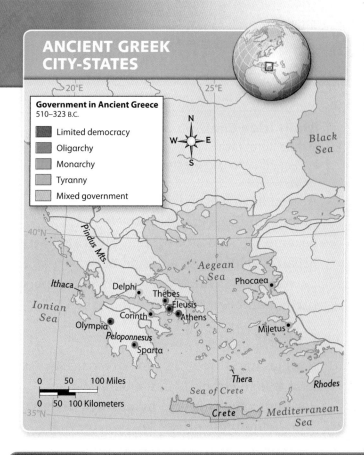

### ANCIENT GREEK CITY-STATES

## ONGOING ASSESSMENT
# READING LAB   GeoJournal

**Synthesize** Use the Greek roots in the chart to form a word in English that completes each of the following sentences:

a. The form of government that represents the people is called _____ (use your own paper).

b. People empowered to enforce a city's laws are the _____ (use your own paper).

c. The ruler of a kingdom is also called a _____ (use your own paper).

| SELECTED ENGLISH WORDS FORMED FROM GREEK ROOTS | | |
|---|---|---|
| **Greek Root** | **Meaning** | **English Word** |
| demos | people | democracy |
| polis | city-state | policy |
| aristo | best | aristocracy |
| monos | one | monarchy |
| oligo | few | oligarchy |

SECTION 2.1 **257**

# SECTION 2 EARLY HISTORY
## 2.2 Classical Greece

**TECHTREK**
myNGconnect.com For an online map and photos of Classical Greece

Maps and Graphs | Digital Library

**Main Idea** Greek ideas about democracy, architecture, philosophy, and science have had a lasting influence on Western culture.

As you have learned, democracy began in ancient Greece. In 461 B.C., Pericles became the leader of Athens. His rule began a **golden age**, a period of wealth and power during which democracy developed further and Greek culture flourished.

### Golden Age of Greece

Pericles had three goals for Greece. The first was to strengthen democracy. He accomplished this goal by paying citizens who held public office. This meant that even people who were not wealthy could afford to serve in government.

The leader's second goal was to expand the empire. Pericles built a strong navy and used it to increase Athens' power over the other Greek city-states.

Pericles' third goal was to make Athens more beautiful. He began rebuilding the city, including the Acropolis. Many of Athens' temples had been destroyed during the war with Persia. Pericles constructed a new temple called the Parthenon, dedicated to the goddess Athena for whom Athens was named.

### Greek Achievements

The golden age of Greece was a period of extraordinary achievements. Greek architects designed temples and theaters with graceful columns. **Philosophers**, people who examine questions about the universe, searched for the truth. Socrates (SAHK ruh teez) and his student, Plato (PLAY toh), were leading philosophers.

In the sciences, the mathematician Euclid (YOO klihd) developed the principles of geometry. The physician Hippocrates (heh PAH kruh teez) changed the practice of medicine by insisting that illnesses originated in the human body and were not caused by evil spirits.

### Greek Culture Spreads

Greece's golden age ended around 431 B.C., when war broke out between Athens and Sparta. The conflict, known as the Peloponnesian War, lasted 27 years and weakened both Athens and Sparta.

Statue of the goddess Athena

○ 1900–1400 B.C. Early Greek civilizations develop.

**1900 B.C.** — **1000 B.C.**

Early Greek gold lion's head

○ 800 B.C. Greek city-states begin to thrive.

**750 B.C.**

## ALEXANDER'S EMPIRE, 330 B.C.

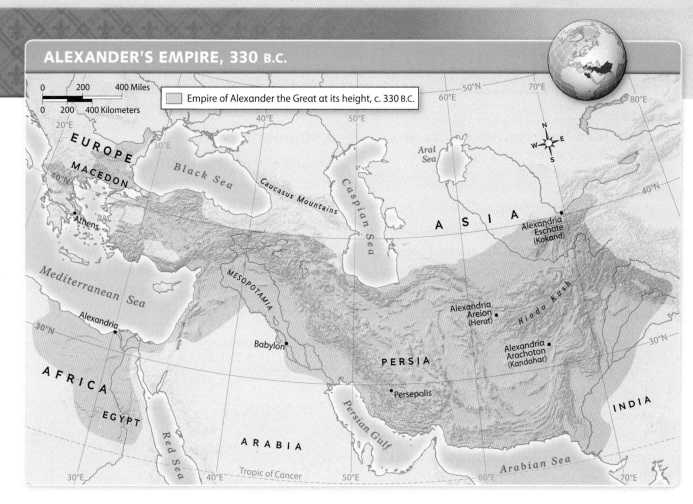

Around 340 B.C., King Philip II of Macedon took advantage of the weakened city-states and conquered Greece. In 334 B.C., Philip's son, **Alexander the Great**, became king and began to extend his father's empire. Alexander loved Greek culture and spread its ideas throughout the lands he conquered. Alexander died in 323 B.C. at the age of 33. The Greek ideas about democracy, science, and philosophy that he helped spread shaped the modern world.

### Before You Move On
**Monitor Comprehension** What Greek ideas have had a lasting influence on Western culture?

### ONGOING ASSESSMENT
## MAP LAB

1. **Draw Conclusions** Study the map. Across which continents did Alexander's empire spread? What helped him unite his vast empire?
2. **Make Inferences** Based on the map, what great empires did Alexander conquer? What do these conquests suggest about Alexander?

---

- **508 B.C.** Athens becomes a direct democracy.
- **490 B.C.** Greeks defeat the Persian Empire.
- **461 B.C.** Pericles becomes leader of Athens; Greek Golden Age begins.
- **431 B.C.** Peloponnesian War between Athens and Sparta begins.

 Pericles

- **340 B.C.** Philip II of Macedon conquers Greece.
- **334 B.C.** Alexander the Great begins to extend his father's empire.

500 B.C. — 250 B.C.

# SECTION 2 EARLY HISTORY
## 2.3 The Republic of Rome

**Main Idea** The Roman Republic created a form of government that Europe and the West would later follow.

Around 1000 B.C., the peninsula of Italy was dotted with hundreds of small villages. According to an ancient legend, two brothers named Romulus and Remus founded Rome in 753 B.C. The brothers were said to be the children of a god and to have been raised by a wolf.

### The Beginnings of Rome

Archaeologists actually believe that people known as the Latins founded Rome around 800 B.C. They came from a region of Italy called Latium and lived on Rome's seven steep hills, which provided protection from enemy attack. The Tiber River, which flows through Rome, provided water for farming and a route for trade. Over time, Rome developed into a wealthy city-state.

> **Critical Viewing** The Roman Forum contained the ancient city's most important buildings, including the Senate. In what ways does this photo reflect Rome's former glory?

### A Republic Forms

Around 600 B.C., the Etruscans, a people from northern Italy, conquered Rome. One Etruscan king named Tarquin was a brutal tyrant. In 509 B.C., the Romans rebelled against him, and Roman leaders began to create a republic. A **republic** is a form of government in which the people elect officials who govern according to law.

## GOVERNMENT OF THE ROMAN REPUBLIC

**EXECUTIVE**
- Two consuls
- Elected to one-year term
- Led the government and controlled the army

**LEGISLATIVE**

*Senate*
- 300 members made up of patricians
- Not elected; selected by the consuls to serve for life
- Made the laws and advised consuls

*Assembly*
- Made up of plebeians
- Elected tribunes as representatives
- Made the laws and selected consuls

**JUDICIAL**
- Eight judges
- Governed provinces
- Served for one year

**Legal Code**
- Twelve Tables
- Established rights and responsibilities of Roman citizens

Two main classes of people lived in Rome at this time. The **patricians** were mostly wealthy landowners. The **plebeians** were mostly farmers. At first, only the patricians could take part in government. They controlled the Senate and made laws.

In 490 B.C., plebeians gained the right to form an assembly and elect legislative representatives called tribunes. The assembly made laws and elected the consuls, the two executive officials who led the government for a year at a time. One consul could **veto**, or reject, a decision made by the other consul.

The judicial branch was made up of eight judges who served for one year. These judges oversaw the lower courts and governed the provinces.

Around 450 B.C., the government published the Twelve Tables. These were bronze tablets that set down the rights and responsibilities of Roman citizens. At this time, only adult male landowners born in Rome were citizens. Roman women were citizens but could not vote or hold office.

### The Roman Way

Citizens of Rome believed in values that were known as the Roman Way. These values included showing self-control, working hard, doing one's duty, and pledging loyalty to Rome. The Roman Way helped to unite all Roman citizens.

The Romans applied these values as the Republic began to conquer new lands and expand. During the second century B.C., Rome defeated the empire of Carthage in northern Africa. By 100 B.C., Rome controlled most of the lands around the Mediterranean. About this time, tensions began to grow between patricians and plebeians. These tensions triggered a war between the two groups. The war set the stage for the end of the Roman Republic—and the birth of the Roman Empire.

**Before You Move On**

**Summarize** What structures, laws, and values made up the government of the Roman Republic?

## ONGOING ASSESSMENT
## DATA LAB

1. **Interpret Charts** According to the chart, members of the Senate were selected by the consuls. However, the assembly elected the consuls. In what way did this arrangement help to control the Senate's power?

2. **Turn and Talk** Study the chart and consider what you know about the U.S. government. Then, with a partner, compare and contrast the two systems of government.

# SECTION 2 EARLY HISTORY
## 2.4 The Roman Empire

**Main Idea** The Roman Empire was one of the largest in history and left a legacy in technology and language.

As you have read, tensions grew between the patricians and plebeians not long after Rome defeated Carthage. The Roman soldiers had brought back great wealth from the conquered territory. They used their wealth to buy large plots of farmland. Small farmers could not compete with them, and the gap between rich and poor widened. In 88 B.C., war erupted between the patricians and plebeians.

### Creation of the Empire

After years of war, a general named **Julius Caesar** rose to power and became the sole ruler of Rome in 46 B.C. Caesar started projects to help the poor and tried to re-establish order in Rome, but he developed powerful enemies. In 44 B.C., a small group of senators stabbed him to death.

Octavian, Caesar's nephew, fought in the long, internal war for power that followed Caesar's death. Octavian won, but the war put an end to the Roman Republic. Calling himself **Augustus**, which means "honored one," Octavian became the ruler of the Roman Empire in 27 B.C. His rule began a period called the *Pax Romana*, which is Latin for "Roman peace."

### Rome's Decline

For about 500 years, the Roman Empire was the most powerful in the world and extended over three continents. Around A.D. 235, however, Rome had a series of poor rulers. In addition, German tribes, whom the Romans called **barbarians**, began invading the empire from the north.

In 330, Emperor Constantine moved the capital of the weakened empire from Rome to Byzantium, in present-day Turkey, and renamed the city Constantinople. He also made Christianity lawful throughout the empire. **Christianity**, which is based on Jesus' life and teachings as described in the Bible, began in the Roman Empire.

In 395, the empire was divided into an Eastern Empire and a Western Empire with two different emperors. In 476, invaders overthrew the last Roman emperor and ended the Western Empire.

### Rome's Legacy

The Western Empire fell, but Rome left the world a great legacy, or heritage. For example, Roman engineers built a network of roads that connected the empire. Many of the roads are still in use.

## THE ROMAN EMPIRE AT ITS HEIGHT, A.D. 200

**Expansion of the Roman Empire**
- Roman Republic in 264 B.C.
- Roman Empire at its height, c. A.D. 200

The engineers also developed the arch and used it to construct buildings and **aqueducts**, which carried water to parts of the empire. Latin, the language of Rome, became the basis for Romance languages, such as Spanish and Italian. Many English words have Latin roots.

### Before You Move On
**Summarize** Describe the Roman Empire's rise, fall, and legacy.

### ONGOING ASSESSMENT
# MAP LAB
GeoJournal

1. **Interpret Maps** According to the map, over which continents did the Roman Empire extend? What challenges might the size of the empire have presented to its rulers?
2. **Location** Find Byzantium on the map. Why do you think Constantine chose this location to become the capital of the Eastern Empire?

---

A.D. 300

○ **A.D. 330** Constantine moves the capital from Rome to Byzantium.

○ **A.D. 395** Empire is divided into an Eastern Empire and a Western Empire.

A.D. 500

○ **A.D. 476** Rome falls to invaders.

Painting of German invader

# SECTION 2 EARLY HISTORY
## 2.5 Middle Ages and Christianity

> **Main Idea** The Roman Catholic Church and the feudal system influenced Western Europe during the Middle Ages.

After the fall of the Roman Empire, Western Europe entered a period known as the **Middle Ages**, which lasted from about 500 to 1500. During this period, Western Europe consisted of numerous kingdoms. Castles, like those on the Rhine River in Germany, served as defensive fortresses. The Roman Catholic Church helped unite people during the Middle Ages, and the feudal system provided a social structure.

### The Roman Catholic Church

In 1054, Christianity officially divided into two parts: the Roman Catholic Church in Western Europe and the Eastern Orthodox Church in Eastern Europe. The Roman Catholic Church was the center of life for most people in Western Europe. It cared for the sick, provided education, and helped preserve books and learning.

The Church also played a leading role in government. It collected taxes, made its own laws, and waged wars. In 1096, the Church began a series of **Crusades**. These were military expeditions undertaken to take back holy lands in Southwest Asia from Muslim control. The Crusades cost many lives and ended in 1291.

### The Feudal System

The many kingdoms of Western Europe were often at war. From about 400 to 800, a German group called the Franks stopped the fighting and unified most of Western Europe. Their most important leader was Charlemagne (SHAHR luh mayn).

When Charlemagne died in 814, warfare between the kingdoms returned and Western Europe again became divided. To provide security for each kingdom, the feudal system developed. The **feudal system** was a social structure that was organized like a pyramid. At the top was a king who owned vast territory. Beneath the king were lords, powerful noblemen who owned land. The lords gave pieces of their land to vassals, who pledged their loyalty and service to the lord. Some vassals also served as knights, who were warriors on horseback.

Each lord lived on an estate called a manor, which functioned as a small village. **Serfs**, who farmed the lord's land in return for shelter and protection, were at the bottom of the pyramid. Serf families dwelt in small huts on the manor and gave most of the crops they grew to their lord.

### The Growth of Towns

In time, the growth of towns helped end the feudal system. Trade and businesses developed, and people began to leave the manors. A deadly disease called the bubonic plague, which swept through Europe in 1347, also weakened the feudal system. The plague killed millions and greatly reduced the workforce in the towns. Desperate for workers, employers offered higher wages. As a result, farmers left the country to seek the higher-paying jobs in the towns.

**Before You Move On**
**Summarize** In what ways did the Roman Catholic Church and the feudal system influence Western Europe during the Middle Ages?

# MANOR IN THE MIDDLE AGES

This model shows a simplified view of a feudal manor. Rolling fields and farmland lay outside its walls.

The church was the center of life on the manor.

The lord lived in relative safety and ease in his castle.

Serfs lived in tiny huts with dirt floors.

Guards protected the manor from rival lords.

## ONGOING ASSESSMENT
### VIEWING LAB — GeoJournal

1. **Interpret Visuals** What details in the model suggest the measures taken to protect those who lived in the manor?

2. **Make Inferences** Notice the position of the church in the manor. Why might it have been positioned near the lord's castle?

3. **Compare and Contrast** Based on the model and what you have read, in what ways did life probably differ for lords and serfs?

SECTION 2.5 **265**

# SECTION 2 EARLY HISTORY
## 2.6 Renaissance and Reformation

**Main Idea** Both the Renaissance and the Reformation brought great change to Europe.

You have learned that the growth of towns in Western Europe helped put an end to the feudal system. The Roman Catholic Church also began to lose some of its power at this time. As these key structures of the Middle Ages weakened, the **Renaissance** began to take hold. The Renaissance was a rebirth of art and learning that started in Italy in the 1300s and had spread through Europe by 1500.

### The Renaissance

Several other factors led to the Renaissance. Increased trade in the growing towns brought some Italian merchants great wealth. This wealth allowed them to buy the artists' work.

As you have learned, the Western part of the Roman Empire fell in 476. The Eastern part continued and became known as the Byzantine Empire. The fall of this empire in 1453 also advanced the Renaissance. Scholars from the empire came to Italy, bringing with them ancient writings of the Greeks and Romans. Studies of these works encouraged humanism, which focuses on human rather than religious values. In addition, a German inventor named **Johannes Gutenberg** developed a printing press in 1450 that printed many books in a short amount of time. Soon, more people in Europe had access to knowledge—including the new humanist ideas.

The result was an explosion of ideas expressed in art, architecture, and literature. Artists such as Leonardo da Vinci, Michelangelo (my kuhl AN juh loh), and Raphael (rah fee ELL) used **perspective** to make a painting look as if it had three dimensions. Architects used elements of ancient Greek and Roman design to create churches and buildings. Writers wrote in the everyday spoken language of their region. For example, Dante Alighieri (ah lah GYER ee) wrote *The Divine Comedy* in Italian, not Latin.

### The Reformation

Meanwhile, some people began looking more critically at the Church. **Martin Luther**, a monk in Germany, was shocked by the corrupt practices of some priests. To raise funds, they often sold **indulgences**, which relaxed the penalty for sin.

Gutenberg's printing press

Portrait of Dante

- **1300s** Renaissance begins in Italy.
- **1308** Dante starts writing *The Divine Comedy* in Italian.
- **1450s** Johannes Gutenberg develops the printing press.

**Critical Viewing** *The School of Athens* by Italian artist Raphael portrays important ancient Greek philosophers and scientists. Why do you think the artist chose to celebrate ancient Greece?

In 1517, Luther wrote the 95 Theses, in which he objected to such practices, and nailed them to a church door. Luther's actions started the **Reformation**, the movement to reform Christianity. Over time, people founded Protestant churches. The term comes from the word *protest*.

In response, the Church began a reform movement called the **Counter-Reformation** and placed more emphasis on faith and religious behavior. Nonetheless, the conflict between Catholics and Protestants would continue for the next 300 years.

**Before You Move On**
**Monitor Comprehension** What changes did the Renaissance and Reformation bring to European culture and society?

## ONGOING ASSESSMENT
### READING LAB  GeoJournal

1. **Summarize** What was the Renaissance?
2. **Analyze Cause and Effect** In what ways did the development of Gutenberg's printing press help spread humanist ideas?
3. **Draw Conclusions** What humanist ideas might have led people to look at the Roman Catholic Church more critically?

Michelangelo's *David*

1504 Michelangelo completes the sculpture *David*.

1517 Martin Luther nails his 95 Theses to the door of a church in Wittenberg, Germany.

1500                  1600

1497 Leonardo da Vinci finishes painting *The Last Supper*.

In this illustration, Martin Luther posts the 95 Theses.

# SECTION 3 EMERGING EUROPE

## 3.1 Exploration and Colonization

**TECHTREK**
myNGconnect.com For an online map of European colonization in Africa, Asia, and the Americas

Maps and Graphs

**Main Idea** To expand trade, Europeans explored Africa, Asia, and the Americas and established colonies on all three continents.

Around 1415, Prince Henry of Portugal decided that he would send explorers to Africa to establish new trade routes. Henry, who became known as Prince Henry the Navigator, founded a **navigation** school. The school taught sailors about mapmaking and shipbuilding and marked the beginning of the Age of Exploration.

### European Exploration

Portugal was the first of many European countries to sponsor voyages of exploration. Europeans wanted to find gold and establish trade with Asia to obtain spices, silk, and gems. They also wanted people in other lands to **convert**, or change their religion, to Christianity.

The voyages were filled with danger. Explorers often sailed for months in ships that were small and not always able to withstand strong storms at sea. The men also faced disease and attacks by native peoples. Furthermore, the explorers were traveling to unknown lands. Mapmakers often marked unexplored places with the phrase "Here be dragons."

Nonetheless, Portuguese explorers such as Bartolomeu Dias and Vasco da Gama sailed along the coast of Africa in the late 1400s to open up trade with Asia. Italian explorer Christopher Columbus uncovered a "new world"—the continents of North America and South America—in 1492. In the 1530s, Jacques Cartier (kahr TYAY) explored parts of North America for France. An Englishman, Sir Francis Drake, sailed around the world in 1577.

> **Critical Viewing** In this painting, Columbus and his crew land in North America as Native Americans arrive to meet the explorers in their canoes. What qualities must explorers have had to undertake their voyages?

# EARLY COLONIZATION OF AFRICA, ASIA, AND THE AMERICAS

## Establishing Colonies

In addition to trade, Europeans used the voyages of exploration to claim lands for their own countries. When explorers landed in a new place, they declared it a colony. A **colony** is an area controlled by a distant country. As you have learned, Spanish explorers claimed colonies in Mexico and South America. The French and the English also established colonies in North America. By 1650, European countries controlled parts of Africa and Asia as well.

European exploration and colonization resulted in a sharing of goods and ideas known as the Columbian Exchange. From the Americas, Europeans obtained new foods, such as potatoes, corn, and tomatoes. Europeans introduced wheat and barley to the Americas. They also introduced diseases like smallpox. The diseases killed millions of native peoples.

### Before You Move On
**Monitor Comprehension** What inspired Europeans to undertake voyages of exploration, and what did they gain as a result?

## ONGOING ASSESSMENT
### MAP LAB

1. **Interpret Maps** According to the map, where in Asia did France establish a large colony? Why was this location beneficial geographically?
2. **Identify Problems and Solutions** Study the map. Who was Spain's main rival for colonies in South America? What problems might have arisen from their rivalry?

SECTION 3.1

# SECTION 3 EMERGING EUROPE
## 3.2 The Industrial Revolution

**TECHTREK**
myNGconnect.com For an online map and images of the Industrial Revolution

 Maps and Graphs
 Digital Library

**Main Idea** The Industrial Revolution was an age of great developments in technology that changed how people worked and lived.

The Age of Exploration opened up trade around the world and brought great wealth to many western European countries. To increase this wealth, businesses looked for new ways to expand production. The result was the **Industrial Revolution**, a period when industry grew rapidly, and the production of machine-made goods greatly increased.

### The Revolution Begins

The Industrial Revolution started in Great Britain in the 1700s as a result of new inventions and technologies. The **textile** industry, which deals with the manufacturing of cloth, was the first to be transformed by the revolution. In 1769, textile manufacturers began using machines that were run using water from a stream. Then, around 1770, James Hargreaves invented the spinning jenny. This machine allowed workers to make cotton and wool yarn at a much faster rate.

Before these inventions, most people made cloth by hand in their homes. However, the new machines were too large and expensive to use in small houses. Instead, the machines were placed in factories, and workers manufactured the goods there. In these early factories, each person worked on a small part of the product. This way of producing goods is called the **factory system**.

At first, factories were powered by water. Then around 1776, James Watt developed the steam engine, which was powered by coal. As a result, coal became an important raw material, and Britain benefited from its rich deposits of the fuel.

In the late 1700s, the Industrial Revolution spread to the rest of Europe. France and Belgium became leading manufacturers of textiles. Germany built factories for processing iron. Railroad systems developed in the 1800s. In 1825, George Stephenson built the first railroad in England. By 1850, thousands of miles of tracks crossed Europe.

> **Critical Viewing** England's Iron Bridge, built in 1779, was the world's first arch bridge made of iron. Based on what you have read, what made this bridge possible?

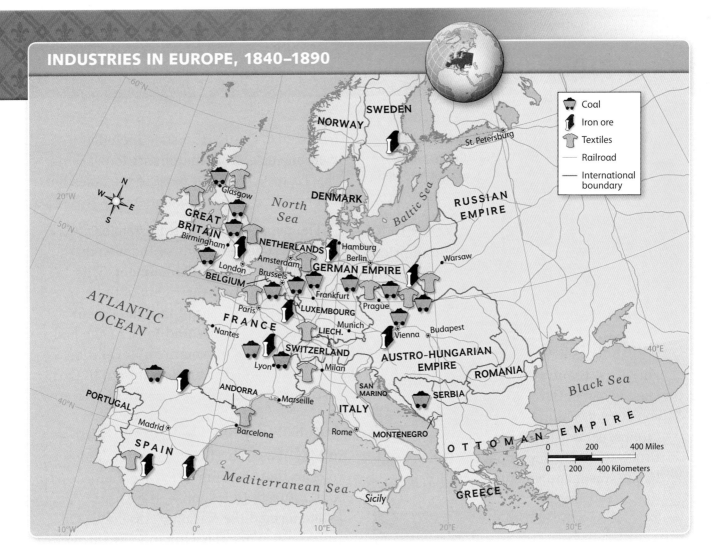

INDUSTRIES IN EUROPE, 1840–1890

## Impact of the Revolution

The Industrial Revolution had a tremendous impact on how people worked and lived. Cities grew rapidly because people migrated there for factory jobs. Standards of living rose, and a prosperous middle class grew.

However, factory workers often faced harsh conditions. Laborers worked as many as 16 hours a day. Child labor was common. Boys and girls as young as five years of age worked in factories and mines. Some were chained to their machines.

Many workers lived in small, crowded houses in neighborhoods where open sewers were common. Diseases spread quickly in these cramped buildings.

Over time, the workers' quality of life improved as sewer systems were created and other public health acts were passed.

### Before You Move On

**Summarize** In what ways did the Industrial Revolution change how people lived and worked?

### ONGOING ASSESSMENT
### MAP LAB
GeoJournal

1. **Interpret Maps** Where in Europe were most of the industries concentrated? What does this suggest about the economies of countries in other parts of Europe?

2. **Human-Environment Interaction** Which industry was the most widespread in Europe? Why was this such an important industry?

3. **Evaluate** Based on the map, which countries probably imported the fewest raw materials?

SECTION 3.2 **271**

# SECTION 3 EMERGING EUROPE
## 3.3 The French Revolution

**Main Idea** The late 1700s in France was a period of economic and political unrest, which led to the French Revolution and the rise of Napoleon.

By the summer of 1789, the French people had not yet benefited from the Industrial Revolution. Harvests were poor, and prices skyrocketed. On July 14, mobs attacked the Bastille, Paris's ancient prison. This action sparked the French Revolution.

### Roots of the Revolution

For years, France's lower and middle classes had suffered injustices. French society was composed of three large groups, called the Three Estates. The First Estate was made up of clergy. The Second Estate was made up of the nobility, or aristocrats. The Third Estate included everyone else, from merchants to peasants. The Third Estate paid most of the taxes but had no voice in government.

The people of the Third Estate began to call for change. Many of them were influenced by the **Enlightenment**. This movement stressed the rights of the individual. The ideas of Enlightenment thinkers like Voltaire and **John Locke** had helped inspire the American Revolution in 1776. The American Revolution, in part, inspired the revolution in France.

### The Revolution Begins

In May 1789, the Third Estate demanded reforms, but the king of France, Louis XVI, refused. In response, the Third Estate formed the National Assembly. On August 26, 1789, the assembly issued the *Declaration of the Rights of Man and of the Citizen.* This document guaranteed liberty, equality, and property to citizens. The assembly tried to form a new government in which Louis would share power with an elected legislature. However, he again refused to cooperate.

French citizens stormed the Bastille because they thought it held guns and gunpowder.

The guillotine was considered an efficient and painless method of execution.

**1785**

**1789** Mobs attack the Bastille.

**1790**

**1792** Jacobins seize power.

**1793** King Louis XVI and Marie Antoinette are executed; Reign of Terror begins.

**1794** Robespierre is executed, and the Reign of Terror ends.

**1795**

## The Radicals Take Over

Finally, in 1792, the Jacobins, a group of **radicals**, or extremists, seized power and formed the National Convention. The following year, the group executed Louis XVI and Marie Antoinette, his queen.

The violence soon got worse. Jacobin leader Maximilien Robespierre led a **Reign of Terror**. The Jacobins used a machine called the **guillotine** (GHEE uh teen) to cut off the heads of an estimated 40,000 people. In July 1794, the French finally turned on Robespierre and executed him.

## Napoleon's Rise

After five years of violence, the French were exhausted. France was at war with Prussia, Austria, and Britain, and the government was not ruling effectively.

A young general, **Napoleon Bonaparte**, saw his chance and overthrew the government. Over the next five years, Napoleon increased his powers. He then declared himself Emperor Napoleon I and set about conquering other European powers and building an empire. Britain and Prussia finally defeated him in 1815.

### Before You Move On
**Summarize** What led to the French Revolution and the rise of Napoleon?

**Critical Viewing** Marie Antoinette, shown here, was often accused of reckless spending. What details in this painting support this accusation?

### ONGOING ASSESSMENT
# SPEAKING LAB  *GeoJournal*

**Express Ideas Through Speech** Get together in a group and do research to prepare a panel discussion in which you will present the viewpoints of various figures from this section.

**Step 1** Decide who each person in your group will be. You might choose from King Louis XVI, Marie Antoinette, Maximilien Robespierre, and Napoleon or be a member of the Third Estate.

**Step 2** Come up with a few questions that your panel will discuss. The questions should focus on the French Revolution, the Reign of Terror, and Napoleon's rise.

**Step 3** Present the panel discussion before the class. At its conclusion, invite questions and answer them in character.

1800 — 1805 — 1810

○ 1799 Napoleon overthrows the French government.

○ 1804 Napoleon names himself Emperor.

Statue of Napoleon on horseback

1815 ○ Napoleon is defeated.

# SECTION 3 DOCUMENT-BASED QUESTION
## 3.4 Declarations of Rights

**TECHTREK**
myNGconnect.com For photos of the documents and Guided Writing
 Digital Library
 Student Resources

As you have learned, thinkers like John Locke and Voltaire led the Enlightenment. They asserted that people have natural rights, or rights that people possess at birth, such as life, liberty, and property. Two key documents describe these rights: the American Declaration of Independence and the French *Declaration of the Rights of Man and of the Citizen*. In 1993, Nelson Mandela of South Africa received the Nobel Peace Prize. In his speech at the ceremony, he explained that the rights detailed in the declarations are still important.

### DOCUMENT 1
*from the* **Declaration of Independence** (July 4, 1776)

> We hold these truths to be self-evident, that all men are created equal, that they are endowed [provided] by their Creator with certain unalienable [guaranteed] Rights, that among these are Life, Liberty, and the pursuit of Happiness; that, to secure these rights, Governments are instituted among Men, deriving their just powers from the consent of the governed.

This painting illustrates the signing of the Declaration of Independence.

**CONSTRUCTED RESPONSE**

1. What rights are citizens guaranteed?

### DOCUMENT 2
*from the* **Declaration of the Rights of Man and of the Citizen** (August 26, 1789)

> The representatives of the French people, organized as a National Assembly, ... have determined to set forth in a solemn declaration the natural, unalienable, and sacred rights of man. Articles:
>
> 1. Men are born and remain free and equal in rights. Social distinctions [classes] may be founded only upon the general good.
>
> 2. The aim of all political association is the preservation of the natural ... rights of man. These rights are liberty, property, security, and resistance to oppression.

**CONSTRUCTED RESPONSE**

2. Think about what you learned in Section 3.3 about the roots of the French Revolution. In what ways might the ideas in this document have inspired the French people to revolt?

# MEN ARE BORN AND REMAIN FREE AND EQUAL IN RIGHTS.
— DECLARATION OF THE RIGHTS OF MAN AND OF THE CITIZEN

Mandela and fellow Nobel recipient, F. W. de Klerk, were elected co-presidents of South Africa in 1994.

## DOCUMENT 3

### from **Nobel Lecture** by Nelson Mandela (December 10, 1993)

Nelson Mandela helped lead the struggle to end **apartheid** (uh PAHRT hyt) in South Africa. This system had denied black South Africans their rights. In recognition of his efforts, Mandela received the Nobel Peace Prize. The following excerpt is from his acceptance speech.

> The value of our shared reward will and must be measured by the joyful peace which will triumph, because [of] the humanity that bonds both black and white into one human race. . . .
>
> Thus shall we live, because we will have created a society which recognizes that all people are born equal, with each entitled in equal measure to life, liberty, prosperity, human rights, and good governance.

**CONSTRUCTED RESPONSE**

3. How do the rights Mandela discusses reflect those described in Documents 1 and 2?

## ONGOING ASSESSMENT
## WRITING LAB  GeoJournal

**DBQ Practice** Think about the ideas in the Declaration of Independence and the *Declaration of the Rights of Man and of the Citizen*. How did these ideas influence Nelson Mandela?

**Step 1.** Review your answers to Constructed Response questions 1, 2, and 3.

**Step 2.** On your own paper, jot down notes about the main ideas expressed in each document.

Document 1: Declaration of Independence
Main Idea(s) _____

Document 2: Declaration of the Rights of Man and of the Citizen
Main Idea(s) _____

Document 3: Nobel Lecture
Main Idea(s) _____

**Step 3.** Use your notes to construct a topic sentence that answers this question: How did the Declaration of Independence and the *Declaration of the Rights of Man and of the Citizen* influence Nelson Mandela?

**Step 4.** Write a paragraph that explains specific phrases and ideas in the Declaration of Independence and the *Declaration of the Rights of Man and of the Citizen*. Go to **Student Resources** for Guided Writing support.

# SECTION 3 EMERGING EUROPE
## 3.5 Nationalism and World War I

**Main Idea** Nationalism, new alliances, and growing tensions in Europe led to World War I.

After the French Revolution, the French people developed powerful feelings of nationalism. **Nationalism** is a strong sense of loyalty to one's country. During the 1800s, nationalism swept through Europe.

### Italy and Germany Unify

Nationalism led to unification efforts in Italy and Germany. In 1800, the Italian Peninsula was made up of separate city-states. In 1870, the states came together to form a unified Italy. Germany was also composed of many different states in the early 1800s. Beginning in 1865, Prussia, the most powerful German state, led the way to unification. Driven by nationalist feelings, Prussia fought to take control of other German states away from their non-German rulers. In 1871, the states came together as a united German Empire.

### Growing Tensions in Europe

By 1900, tensions had begun to grow among European powers. Nationalism had united some countries from within. However, nationalism also created fierce competition among rival countries. Mainly, the countries competed for raw materials and colonies in Africa and Asia. To strengthen their position, Britain, France, and Russia formed an **alliance**, or agreement to work toward a common goal, called the Triple Entente. The German Empire and Austria-Hungary formed an alliance known as the Central Powers.

These alliances were tested in June 1914, when Archduke Franz Ferdinand of Austria-Hungary was assassinated in Serbia by a nationalist from Bosnia-Herzegovina. The assassin belonged to a group that was unhappy with Austrian rule of Bosnia-Herzegovina and wanted to unite with Serbia. Immediately after the assassination, Austria-Hungary declared war on Serbia. Then, because Serbia was a Russian ally, Russia declared war on Austria-Hungary. Within weeks, much of Europe had been drawn into war.

### A Brutal War

The Great War, as it was called, dragged on for four brutal years. Both sides fought from **trenches**, or long ditches that protected soldiers from the enemy's gunfire. Both sides also used deadly technology, including machine guns, airplanes, tanks, and poison gas. German U-boats, or submarines, sank British ships.

1870 Italy unifies.

1870

1871 German states unite to form the German Empire.

Prussian prime minister Otto von Bismarck oversaw German unification.

1885

Illustration of Archduke Ferdinand's assassination

1900

1914 Archduke Ferdinand is assassinated; World War I begins.

In 1917, Germany seemed to gain an advantage when the Communist Party seized control of Russia's government and economy and made peace with Germany. That same year, the United States entered the war on the side of France and Britain. The fresh American troops helped turn the tide against Germany. In 1918, Germany surrendered to France, Britain, and the United States. By the time the war ended, ten million soldiers had died. About seven million civilians also lost their lives.

## Impact of the War

In 1919, Germany signed the **Treaty of Versailles**. Under this peace treaty, Germany was forced to pay several billion dollars in damages and accept full blame for the war. Many of Germany's territories were taken away, and new countries were formed, including Austria, Hungary, Czechoslovakia, Yugoslavia, and Turkey. The treaty angered and humiliated the German people and did little to ease tensions in Europe. These tensions would help lead the way to another world war in a little more than 20 years.

### Before You Move On

**Monitor Comprehension** In what ways did nationalism, new alliances, and growing tensions in Europe lead to World War I?

### ONGOING ASSESSMENT
# MAP LAB
GeoJournal

1. **Region** According to the map, which empires ruled much of Europe in 1914?
2. **Make Inferences** Note the countries that remained neutral in the war. What geographic factors might have encouraged their neutrality?

Poster illustrating the alliance of Britain, France, and Russia in 1915

Leaders who signed the treaty

- 1915
- 1917 United States enters the war.
- 1918 World War I ends.
- 1919 Treaty of Versailles is signed.
- 1930

French prime minister Georges Clemenceau

American president Woodrow Wilson

British prime minister David Lloyd George

# SECTION 3 EMERGING EUROPE

## 3.6 World War II and the Cold War

**Main Idea** After World War II was fought to defeat the Axis Powers, the Cold War developed between the democratic United States and the Communist Soviet Union.

At the end of World War I, Germany lost its military power. As you have read, the Treaty of Versailles also placed full blame for the war on Germany and forced it to pay **reparations**, or money to cover the losses suffered by the victors. The Great Depression, which began in 1929, further damaged Germany's economy. The **Great Depression** was a severe downturn in the world's economy. During this crisis, **Adolf Hitler** rose to power in Germany.

### World War II

Hitler became the leader of the National Socialist German Workers' Party, or the Nazis. In 1936, Hitler made an alliance with Italy. Germany also formed an alliance with Japan, where the military had seized power. Germany, Italy, and Japan formed the Axis Powers.

Germany's invasion of Poland in 1939 started World War II. Two of Poland's allies, Great Britain and France, declared war on Germany soon after the invasion. Germany responded by conquering Poland and then quickly took over most of Europe, including France.

In 1941, Japan attacked the United States at Pearl Harbor, Hawaii. As a result, the United States abandoned its neutrality and entered the war on the side of Britain and the Soviet Union. Together, they were known as the Allies. Over time, many other countries took sides and joined either the Allies or the Axis Powers.

After more than five years of war, Germany surrendered on May 8, 1945. Allied troops were stunned to find the Nazi **concentration camps** where six million Jews and other victims had been murdered. This mass slaughter was called the **Holocaust**. Japan continued to fight until the United States dropped atomic bombs on Hiroshima and Nagasaki. Japan surrendered on September 2, 1945.

### The Cold War

After World War II, the Soviet Union established Communist governments in Eastern Europe. Germany was divided into Communist East Germany and democratic West Germany. The imaginary boundary that separated Eastern and Western Europe was called the **Iron Curtain**. The division marked the beginning of the **Cold War**, a period of great tension between the United States and the Soviet Union.

To defend against possible attack, both sides forged military alliances. Western Europe and the United States formed NATO (North Atlantic Treaty Organization), while Communist Eastern Europe formed the Warsaw Pact. The two never directly waged war against each other during the course of the Cold War.

In the 1980s, many eastern European countries overthrew their Communist governments. In 1991, the Soviet Union itself collapsed. The Cold War ended, and democracy replaced communism throughout Eastern Europe.

**Before You Move On**
Make Inferences In what ways did World War II help lead to the Cold War?

## POST-WORLD WAR II EUROPE, 1950

**The Iron Curtain**
- NATO member countries
- Warsaw Pact member countries
- Neutral countries, non-Communist
- Neutral countries, Communist
- Iron Curtain

### Visual Vocabulary
The **Berlin Wall** divided Communist East Berlin from democratic West Berlin. It was torn down in 1989.

### ONGOING ASSESSMENT
## PHOTO LAB
GeoJournal

1. **Analyze Visuals** Study the photo. In what way was the Berlin Wall a solid symbol of the Iron Curtain?
2. **Evaluate** Soldiers patrolled the wall on East Berlin's side. How can you tell that this photo was taken on the West Berlin side?

# CHAPTER 9
# Review

## VOCABULARY

Match each word in the first column with its definition in the second column.

| WORD | DEFINITION |
|---|---|
| 1. ecosystem | a. sold to relax penalty for sin |
| 2. democracy | b. strong sense of loyalty to one's country |
| 3. plebeians | c. community of living organisms and their environment |
| 4. indulgence | d. area controlled by a distant country |
| 5. colony | e. common people |
| 6. nationalism | f. government of the people |

## MAIN IDEAS

7. What are some of the significant islands in Europe? (Section 1.2)
8. Where is much of Europe's farming industry located? (Section 1.3)
9. What events led to the development of democracy in ancient Greece? (Section 2.1)
10. How were plebeians represented in the Roman Republic? (Section 2.3)
11. In what way did the Roman Empire influence language? Why? (Section 2.4)
12. In what ways did the Roman Catholic Church serve as a unifying force in Western Europe? (Section 2.5)
13. What achievements in arts and literature did the Renaissance inspire? (Section 2.6)
14. How did the Industrial Revolution change Europe? (Section 3.2)
15. Why did the French people welcome Napoleon's rise to power? (Section 3.3)
16. Why did Russia declare war on Austria-Hungary in 1914? (Section 3.5)
17. What was the Cold War? (Section 3.6)

## GEOGRAPHY

### ANALYZE THE ESSENTIAL QUESTION

How did Europe's physical geography encourage interaction with other regions?

**Critical Thinking: Evaluate**

18. In what ways do rivers like the Danube make trade easier within Europe?
19. Why has trade been central to Europe's growth throughout its history?

## EARLY HISTORY

### ANALYZE THE ESSENTIAL QUESTION

How did European thought shape Western civilization?

**Critical Thinking: Draw Conclusions**

20. What elements of the democracy practiced in ancient Greece did the United States adopt?

### INTERPRET MAPS

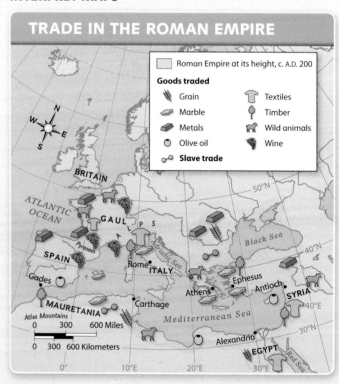

21. **Movement** From what part of the Empire did Rome obtain its grains? its textiles?

## EMERGING EUROPE

### ANALYZE THE ESSENTIAL QUESTION
How did Europe develop and extend its influence around the world?

### Critical Thinking: Make Inferences

22. In what way did improvements in navigation and shipbuilding lead Europe to establish colonies in other parts of the world?
23. Why did the Industrial Revolution make European leaders eager to establish colonies in the Americas and Asia?
24. What are some of the positive effects of nationalism? What are some negative effects?

### INTERPRET TABLES

| MILES OF RAILWAY TRACK IN SELECTED EUROPEAN COUNTRIES (1840–1880) | | | |
|---|---|---|---|
| | 1840 | 1860 | 1880 |
| Austria-Hungary | 144 | 4,543 | 18,507 |
| Belgium | 334 | 1,730 | 4,112 |
| France | 496 | 9,167 | 23,089 |
| Germany | 469 | 11,089 | 33,838 |
| Great Britain | 2,390 | 14,603 | 25,060 |
| Italy | 20 | 2,404 | 9,290 |
| Netherlands | 17 | 335 | 1,846 |
| Spain | 0 | 1,917 | 7,490 |

Source: Modern History Sourcebook

25. **Analyze Data** How much railway track did the Germans build between 1840 and 1880? What might account for this increase?
26. **Draw Conclusions** France and Spain are almost the same size. Note the difference in the extent of the railway system in each one in 1880. What does this suggest about the level of industrialization in each country?

## ACTIVE OPTIONS

**Synthesize** the Essential Questions by completing the activities below.

27. **Write Tour Notes** Suppose that you are going to lead a group of tourists on a trip on one of Europe's rivers. Select and research the river. You might choose the Danube, Rhine, Tiber, Rhone, Thames, or Seine. Then write notes for a guided tour of the river. Start by describing its location, size, and appearance. Next, point out important sites along the river and explain their historical significance. Finally, discuss the river's uses today. **Gather photos of the river and conduct the tour with your group of "tourists."**

> **Writing Tips**
> - Use language that appeals to the senses to help your tourists see and experience the river.
> - Include stories about the sites and historical events to hold your audience's interest.
> - Involve your audience by comparing the river to one that is familiar to them.

28. **Create a Slide Show** Prepare a slide show of famous European buildings, using **Connect to NG** or other online sources. Research and identify five buildings, such as the Pantheon in Rome, Italy. Write two sentences that explain the importance of each building to European history. Copy the chart below to help you organize your information.

| BUILDING | IMPORTANCE TO EUROPE |
|---|---|
| 1. | |
| 2. | |
| 3. | |
| 4. | |

REVIEW 281

# CHAPTER 10
# EUROPE TODAY

## PREVIEW THE CHAPTER

**Essential Question** How is the diversity of Europe reflected in its cultural achievements?

### SECTION 1 • CULTURE

1.1 Languages and Cultures ......................... 284

1.2 Art and Music ........................................ 286

1.3 Europe's Literary Heritage ..................... 288

1.4 Cuisines of Europe ................................ 290

**KEY VOCABULARY**
- dialect
- heritage
- perspective
- abstract
- troubadour
- opera
- genre
- epic poem
- novel
- staple
- cuisine

**ACADEMIC VOCABULARY**
cosmopolitan

**TERMS & NAMES**
- Romantic period
- impressionism
- Baroque period
- Classical period

**Essential Question** What are the costs and benefits of European unification?

### SECTION 2 • GOVERNMENT & ECONOMICS

2.1 The European Union ............................. 292

2.2 The Impact of the Euro ......................... 294

2.3 Democracy in Eastern Europe ................ 296

**GLOBAL ISSUES: MIGRATION**
2.4 Changing Demographics ....................... 298

**KEY VOCABULARY**
- tariff
- currency
- euro
- sovereignty
- eurozone
- consumer
- democratization
- privatization
- demographics
- aging population

**ACADEMIC VOCABULARY**
exchange, assimilate

**TERMS & NAMES**
- Common Market
- European Union (EU)
- Orange Revolution

Buildings with traditional red-tiled roofs line this square in Prague in the Czech Republic.

# SECTION 1 CULTURE

## 1.1 Languages and Cultures

**TECHTREK**
myNGconnect.com For photos reflecting European culture

Digital Library

**Main Idea** Europe has a great variety of languages, cultures, and cities.

Europe has more than a half billion people, yet they live in an area that is one-half the size of the United States. In addition, the continent of Europe contains more than 40 countries. The result is a great diversity, or wide variety, of languages and cultures.

### European Languages

Many of the languages spoken in Europe today fall into three language groups: Romance, Germanic, and Slavic. The Romance languages include French, Spanish, and Italian. The Germanic languages are spoken mostly in northern Europe and include German, Dutch, and English. Most people in Eastern Europe speak Slavic languages, such as Russian, Polish, and Bulgarian.

Some countries in Europe have more than one official language. Belgium, for example, has three: Dutch, French, and German. Even in countries with only one official language, people may speak different dialects. A **dialect** is a regional variety of a language. In Italy, for instance, people in Rome speak a dialect of Italian that differs from that in other cities.

### Cultural Traditions

Because Europe is composed of many different countries and ethnic groups, it has a rich cultural **heritage**, or tradition. Europe's cultural diversity is reflected in its religions and celebrations.

Christianity is the dominant religion in Europe. Today, about 45 percent of the continent's total population is Catholic. Protestantism is most common in Northern Europe.

**Critical Viewing** In this photo of the Palio, a centuries-old cultural tradition is honored as these horses race in Italy. What details in the photo convey the excitement of the race?

In recent years, Islam has become the fastest-growing religion in Europe. Immigrants from Turkey, North Africa, and Southwest Asia move to Europe and bring their Muslim faith with them.

Many of the holidays celebrated in Europe are rooted in religion. However, Europeans enjoy other kinds of festivals as well. One of the most colorful is the Palio, a horse race held each summer in Siena, Italy. In this race, which dates back to the Middle Ages, ten riders from ten of the city's neighborhoods compete.

## City Life

More than 70 percent of Europeans live in urban areas. In Belgium, over 95 percent of the people live in or near its cities. Most of Europe's cities are **cosmopolitan**, which means that they bring together many different cultures and influences. London is an example of a cosmopolitan city. Its restaurants and shops reflect the South Asian, Caribbean, and East Asian origins of some of its newer citizens.

Many European cities date back hundreds of years. As a result, they developed in ways very different from American cities. These cities are often smaller in area than those in the United States and have narrow, winding streets. Most people live in apartments rather than individual houses. For recreation, city dwellers visit their many parks. They also tend to use public transportation more often than most Americans.

### Before You Move On
**Make Inferences** What might be some of the advantages and disadvantages of Europe's great variety of languages and cultures?

### ONGOING ASSESSMENT
## PHOTO LAB  GeoJournal

1. **Analyze Visuals** Note the way the riders are dressed in the photo. What do you suppose their clothes represent?

2. **Draw Conclusions** Study the photo and recall what you have read about the Palio. How is this race an example of a cultural tradition?

3. **Turn and Talk** Get together with a partner and discuss the holidays and festivals that you celebrate. Take notes on your discussion and be prepared to share with the rest of the class.

# SECTION 1 CULTURE

## 1.2 Art and Music

**TECHTREK**
myNGconnect.com For photos of European art and samples of European music

Digital Library

**Main Idea** European art and music have developed over thousands of years.

Throughout the centuries, European art and music have changed to reflect different styles and beliefs.

### European Art

European art grew out of the artistic achievements of ancient Greece and Rome. Greek and Roman gods and goddesses were frequent subjects of the artists from these cultures, but they were portrayed to represent realistic human forms.

Much of the art of the Middle Ages reflected the influence of Christianity. The religious subjects were often presented as two-dimensional figures. During the Renaissance, artists used **perspective** to give their work greater depth. Although religious subjects were common, artists also painted portraits of people.

In the **Romantic period** of the early 1800s, artists moved away from religious themes to paint landscapes and other natural scenes that would convey emotion. **Impressionism** emerged in the late 1800s. Impressionist artists, such as Claude Monet, used light and color to capture a moment. By 1900, artists wanted to create a new form of art. These modern artists often worked in an **abstract** style, which emphasized form and color over realism.

**Critical Viewing** The *Mona Lisa*, by Italian Renaissance artist Leonardo da Vinci, is probably the most famous portrait of all time. What about this painting might account for its popularity?

**Critical Viewing** This painting, *Impression, Sunrise*, by French artist Claude Monet, gave the impressionist movement its name. What kind of mood is conveyed in this painting?

Most opera houses contain a stage, an orchestra pit, and several levels of balconies. The ceiling of the Paris Opéra, shown here, was painted by Russian-born artist Marc Chagall.

### European Music

Like art, European music began in ancient Greece and Rome. Musicians played on a few simple instruments and were often accompanied by singers.

During the Middle Ages, music was used in religious ceremonies. Singers called **troubadours** performed songs about knights and love. These songs influenced Renaissance music, when instruments such as the violin were introduced.

The new instruments helped inspire the complex rhythms in the music of the **Baroque period**, which lasted from about 1600 to 1750. **Opera**, which tells a story through words and music, was born then.

The **Classical** and Romantic periods followed the Baroque period and continued until about 1910. Composers from these two periods, such as Ludwig van Beethoven (BAY toh vuhn) of Germany, wrote works using instruments and techniques that are still used today.

### Before You Move On
**Monitor Comprehension** What styles and beliefs have influenced European art and music?

#### ONGOING ASSESSMENT
## LISTENING LAB  GeoJournal

1. **Analyze Audios** Listen to the music clip of Beethoven's Fifth Symphony in the **Digital Library**. Describe the music's mood. What instruments help to convey this mood?

2. **Form and Support Opinions** What do you think of the opening? Support your opinion by referring to specific details in the music.

# SECTION 1 CULTURE
## 1.3 Europe's Literary Heritage

**Main Idea** European literature has reflected new ways of thinking over the centuries.

Plays by the English playwright William Shakespeare (1564–1616) are performed almost every day. European writers such as Shakespeare have influenced literature for centuries. They wrote in many different **genres**, or forms of literature, including poetry, plays, and novels.

### Literary Origins

European literature began with the ancient Greeks and Romans. Around 800 B.C., the Greek poet Homer wrote the epic poems *The Iliad* and *The Odyssey*. An **epic poem** is a long poem that tells the adventures of a hero who is important to a particular nation or culture. Around 20 B.C., the Roman poet Virgil wrote *The Aeneid*, an epic poem about the founding of Rome.

One of the greatest writers of the late Middle Ages and early Renaissance was the Italian poet Dante (1265–1321). He wrote *The Divine Comedy*, an epic poem of over 14,000 lines that deals with the religious beliefs and politics of his time.

Many later works of the Renaissance focused on human behavior. Shakespeare explored this theme in plays such as *Hamlet*. Spanish writer Miguel de Cervantes (1547–1616) wrote what is considered the first modern novel, *Don Quixote* (kee HO tee). A **novel** is a long work of fiction, containing characters and a plot. The printing press, which Johannes Gutenberg developed in the 1450s, helped spread the popularity of these books.

### The 1700s and 1800s

In the mid-1700s, Enlightenment ideas about reason and government inspired the movement toward democracy. These ideas, in turn, led French and English writers of the time, such as Voltaire and John Locke, to explore the rights of the individual.

In the 1800s, writers of the Romantic period continued this exploration, with an emphasis on emotion and nature. For example, German author Johann Wolfgang von Goethe (GHER tuh) (1749–1832) wrote *The Sorrows of Young Werther*, a novel about a sensitive young artist.

Other writers of the 1800s took a much more realistic look at life. In novels such as *Sense and Sensibility*, British writer Jane Austen (1775–1817) used humor to examine women's role in society.

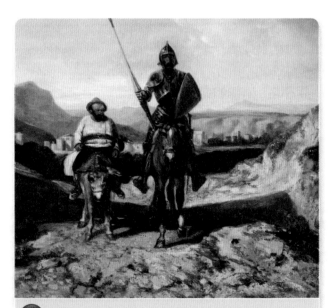

**Critical Viewing** Inspired by tales of knights, Don Quixote (right) goes to battle evil with his servant Sancho Panza (left). What details in this painting suggest that Cervantes' novel is a comedy?

**Critical Viewing** Austen's novels typically end with marriage, such as the one shown in this scene from a film adaptation of *Sense and Sensibility*. Based on the photo, how would you describe an English wedding in the 1800s?

Another British writer, Charles Dickens (1812–1870), commented on social issues, including poverty, in such novels as *Oliver Twist*. Norwegian playwright Henrik Ibsen (1828–1906) wrote plays, such as *A Doll's House*, which criticized the traditional role of husbands and wives at that time.

## Modern Literature

The two world wars had a great impact on the modern literature of the 20th century. Writers at this time reflected the sense that life was uncertain and unpredictable. Some rejected traditional genres and experimented with writing new forms of plays, poems, and novels.

Many modern writers examined the inner workings of the mind. In the novel *Ulysses*, for example, Irish writer James Joyce (1882–1941) focused on the thought processes of the main character over the course of a single day. Romanian playwright Eugene Ionesco (1909–1994) used ridiculous situations to comment on what he saw as the emptiness of life. Writers worldwide have been influenced by these authors.

### Before You Move On
**Summarize** What new ways of thinking has European literature reflected over the centuries?

### ONGOING ASSESSMENT
# WRITING LAB  GeoJournal

**Write Reports** Think about what you have learned about Europe's literary heritage. Then consider the following question: In what ways do beliefs and events influence literature? Select a writer mentioned in this lesson and write a report in which you answer this question.

**Step 1** Research to learn more about the writer and the period in which he or she lived.

**Step 2** Find out how the beliefs and events of the time influenced the writer.

**Step 3** Write a brief report in which you explain these influences. Support your ideas with specific references to one or two of the writer's works. Go to **Student Resources** for Guided Writing support.

# SECTION 1 CULTURE

## 1.4 Cuisines of Europe

**Main Idea** Landforms and climate have influenced the cooking traditions of Europe.

Throughout most of Europe, foods such as meat, bread, and cheese are **staples**, or basic parts of people's diets. However, the **cuisine**, or cooking traditions, of most European countries is largely determined by the landforms and climate of a particular region.

### Foods of Western Europe

The hot, dry climate in the Mediterranean countries of Spain, France, Italy, and Greece is ideal for growing olives, tomatoes, and garlic. As a result, these are key ingredients in their cuisines. Fish from the bodies of water surrounding the countries is also an important menu item.

The cuisines of France and Italy have influenced cooking throughout the world and are especially known for their sauces. French sauces are typically made of milk or cheese, while those of Italy are often tomato based.

People in western European countries with cooler climates often eat more filling fare. In Germany, Great Britain, and Ireland, potatoes grow well and are popular side dishes. In contrast with the light sauces of France and Italy, German cooks often serve heavier gravies.

In Scandinavian countries such as Sweden and Norway, the people often eat herring and other fish. Scandinavians also enjoy deer meat provided by the herding culture of these northern countries.

> **Critical Viewing** The French often enjoy long, relaxed meals with friends and family. What attitude toward life and food does this traditional way of eating suggest?

### Foods of Eastern Europe

Eastern Europe's cold climate results in a shorter growing season than that of Western Europe. In Russia, root vegetables such as turnips and beets are well adapted to the country's climate. A soup called *borscht*, which is made from beets, is a traditional offering on cold winter nights.

The fertile soil of Hungary allows Hungarian farmers to grow grains and potatoes. These crops are used to make a variety of breads and dumplings. A meat stew called *goulash* is Hungary's national dish. It is made with beef, potatoes, and vegetables, and seasoned with paprika. Paprika is a red spice that was brought to Hungary by the Turks in the 1500s.

Bread like this round loaf is the traditional centerpiece at a Ukrainian wedding.

Like Hungary, Ukraine has fertile soil and fields of wheat and other grains. The country is known for its bread. On special occasions, cooks prepare breads decorated with ornaments made of dough.

### Before You Move On
**Summarize** How have landforms and climate influenced European cooking traditions?

### ONGOING ASSESSMENT
## SPEAKING LAB  GeoJournal

**Turn and Talk** What is the traditional cuisine of the United States? What impact have dishes brought by immigrants from other countries had on how and what Americans eat? Get together in a small group and discuss these questions. Be prepared to share your ideas with other groups.

# SECTION 2 GOVERNMENT & ECONOMICS
## 2.1 The European Union

TECHTREK
myNGconnect.com For an online map of the European Union

Maps and Graphs

**Main Idea** The European Union was formed to unite Europe and benefit it economically.

In 1948, the United States established a program called the Marshall Plan to help Europe rebuild after World War II. To manage the U.S. aid money, European countries formed the Organization for European Economic Cooperation in 1948. As a result, European countries discovered that they could rebuild their countries faster when they worked together.

### The Common Market

In 1957, some European countries sought even closer economic ties. They formed the European Economic Community (EEC), which became known as the **Common Market**. The first countries to join were Belgium, France, Italy, Luxembourg, the Netherlands, and West Germany. Several more countries joined during the 1980s.

The Common Market pledged to create "an ever closer union among the European peoples." However, it was primarily formed to create a single market among the member nations. A single market is one in which a group of countries trades across its borders without restrictions or tariffs. A **tariff** is a tax paid on imports and exports.

### A United Europe

In 1992, the countries of the Common Market sought to extend their economic organization throughout Europe. They met in Maastricht in the Netherlands and signed the Treaty of Maastricht, which created the **European Union (EU)**. By 2016 the EU had 28 member nations with a total of more than 500 million people. When considered as a single economy, the EU is the largest in the world.

European Union flag

> **Critical Viewing** The EU flag, seen here in front of the organization's Parliament building in Strasbourg, France, is also the flag of Europe. The flag's circle of stars represents European unity. Why is this a fitting symbol for the EU?

## EUROPEAN UNION

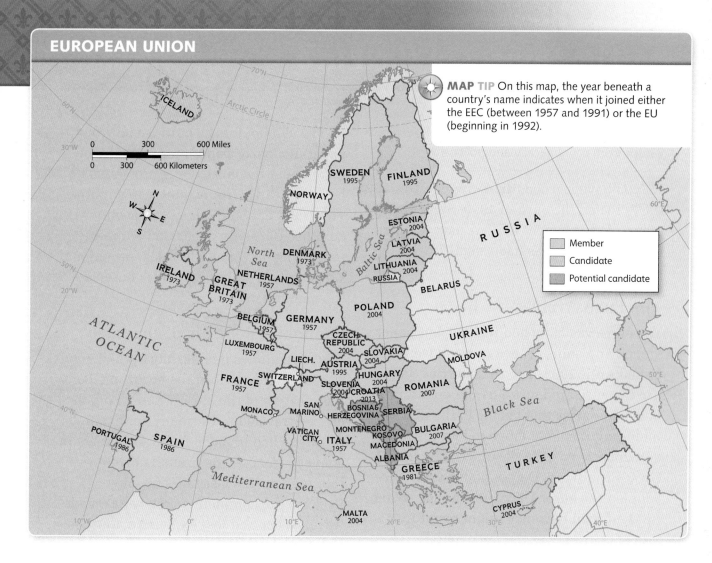

**MAP TIP** On this map, the year beneath a country's name indicates when it joined either the EEC (between 1957 and 1991) or the EU (beginning in 1992).

The EU has a government with an executive, a legislative, and a judicial branch. These branches propose, pass, and enforce the organization's policies and legislation. The EU also has agencies that direct economic policies. Through these agencies, the EU has eliminated tariffs among most member nations and founded a European bank. In 1999, the EU created a common **currency**, or form of money, called the **euro**. By 2016, 19 member nations had adopted this currency.

One of the requirements for joining the EU is having a stable democracy that respects human rights. Some countries that have applied for membership, such as Turkey, are still under review. However, other European countries, including Norway, have chosen not to join. Norway does not want to give up its **sovereignty**, or control over its own affairs.

### Before You Move On
**Monitor Comprehension** How has the European Union helped unite Europe?

### ONGOING ASSESSMENT
# MAP LAB
**GeoJournal**

1. **Interpret Maps** Study the map and recall what you have learned about Europe's geography. In what ways does Europe's geography encourage unity?

2. **Draw Conclusions** Based on the map, in what area are most of the members who joined in later years located? What political situation might have prevented them from joining earlier?

SECTION 2.1 **293**

# SECTION 2 GOVERNMENT & ECONOMICS
## 2.2 The Impact of the Euro

**Main Idea** The euro has helped to unify Europe both economically and politically.

As you have learned, the European Union (EU) created the euro in 1999. Since then, many of the member nations have adopted the euro, which has had a significant impact on Europe.

### The Euro Arrives

The euro was launched in 1999, but paper money and coins of the currency were not issued until 2002. The 19 countries that use the euro are known as the **eurozone**. Some countries that belong to the EU, including Romania, hope to join the eurozone soon. Other countries, including Great Britain and Denmark, have not adopted the euro. They believe that giving up their own currency might result in a loss of control over their economies.

The symbol for the euro is €. Euro paper money, or banknotes, is the same throughout the eurozone. Euro coins, however, differ from country to country. The front, or common, side of each coin has the same image and a number indicating its value. The back, or national, side shows a design that was chosen by the member nation.

### Economic Benefits of the Euro

The euro allows people, money, and goods to move freely within the eurozone. Before the creation of the euro, when a French citizen traveled to Germany, for example, he or she had to pay a fee to **exchange**, or convert, francs—the French currency—into marks—the German currency. Because the common currency has made travel easier and less expensive, tourism has increased within much of Europe.

The prices of fruit in this Italian market are in euros.

A single currency means lower fees for conducting business. As a result, trade has increased among European nations by an estimated 10 percent since 2002. The currency has also made costs easier to compare for companies within the eurozone. As a result, companies can import the least expensive products and then pass along the savings to **consumers**, the people who buy the goods.

## European Debt Crisis

In 2010, the unity of the eurozone was tested. Greece and Ireland—countries in the eurozone—were deeply in debt. To help them manage their debt, the other eurozone nations loaned the two countries money. In return, Greece and Ireland had to raise taxes and reduce spending. Economists continue to debate the consequences of these measures.

1-EURO COIN OF THREE EUROZONE COUNTRIES

**AUSTRIA** Mozart, Austrian composer

**GERMANY** Eagle, German symbol

**IRELAND** Harp, Irish symbol

**Before You Move On**
Summarize In what ways has the euro helped unify Europe?

### ONGOING ASSESSMENT
## VIEWING LAB  GeoJournal

1. **Compare and Contrast** Study the euros in the diagram. Note that the front of the coin is the same for all three countries shown. What elements appear on the front of each euro? What does each element represent?

2. **Make Inferences** Why might the countries have wanted their own design on the euro?

3. **Conduct Internet Research** Go online to find out what impact helping Greece and Ireland has had on the euro and the eurozone. Share your findings with the class.

# SECTION 2 GOVERNMENT & ECONOMICS

## 2.3 Democracy in Eastern Europe

**Main Idea** Eastern European countries have faced many challenges in their transition to democracy.

After World War II, many eastern European countries came under the control of the Soviet Union. The citizens of these countries lacked democratic freedoms and had a low standard of living. In 1981, Poland rebelled peacefully against its Communist government. By the late 1980s, similar rebellions had spread throughout Eastern Europe. Finally, in 1991, Russia and several other republics declared their independence, and the Soviet Union collapsed.

### The Road to Democracy

Since gaining their independence, Poland, Hungary, and the Czech Republic developed stable democratic governments In other countries, **democratization**, or the process of becoming a democracy, has been more difficult to achieve. In 1991, civil war broke out among ethnic groups in Yugoslavia. Over time, the country divided into several new democratic countries, including Serbia and Croatia.

Ukraine has also had setbacks. In 2004, the Ukrainian people staged the **Orange Revolution** and peacefully removed their prime minister, Viktor Yanukovych.

**Critical Viewing** Polish citizens shop and relax in a spacious mall in Warsaw. What does the shopping complex in this photo suggest about Poland's economy?

Yanukovych returned to politics to become prime minister again and then president in 2010. However, he was removed from power in 2014 after being accused of corruption. Petro Poroshenko was elected president and took office in June 2014.

## Rebuilding Economies

The former Communist countries of Eastern Europe also began to rebuild their economies. They changed from government-controlled economies to market economies. They accomplished this goal through **privatization**. That means that government-owned businesses became privately owned.

Eastern European countries have had mixed results since making the adjustment to a market economy. Poland has had the greatest success. It has a fast-growing economy and exports goods throughout Europe. Other countries have been slower to establish new businesses. They have also experienced rises in prices and unemployment.

The leaders of many eastern European countries wish to integrate with the rest of Europe. They want to join the European Union and NATO, a military alliance of democratic states in Europe and North America. While some citizens of eastern European countries believe that they were more secure under Communist leaders, others—particularly young people—disagree. They favor democracy and feel that this form of government can better help them solve their countries' problems.

### Before You Move On
**Summarize** What challenges have eastern European countries faced in their transition to democracy and a market economy?

---

**ONGOING ASSESSMENT**

## READING LAB  GeoJournal

1. **Identify Problems and Solutions** What problem did Ukraine face in its transition to democracy? In what way did the solution to the problem reflect democratization?

2. **Make Inferences** Why do you think younger eastern Europeans might be more willing than older people to support the democratic movements in their countries?

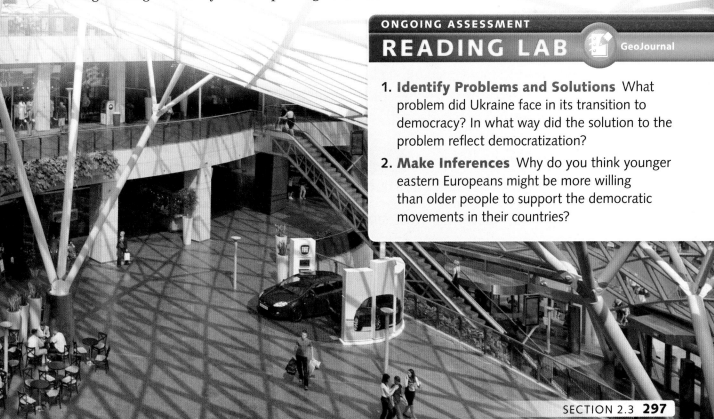

SECTION 2 GLOBAL ISSUES • Migration

# 2.4 Changing Demographics

**Maps and Graphs**  **Global Issues**

TECHTREK
myNGconnect.com For an online graph of Europe's changing demographics

**Main Idea** New immigrants are changing Europe.

Every May, people from Germany, Denmark, Hungary, Bulgaria, and other European countries come together to celebrate Europe's diversity on Europe Day. The celebrations reflect Europe's changing **demographics**, the characteristics or the profile of a human population. The population has become more diverse as people from Africa and Asia have immigrated to Europe.

**KEY VOCABULARY**

**demographics**, n., the characteristics of a human population, such as age, income, and education

**aging population**, n., a trend that occurs as the average age of a population rises

**ACADEMIC VOCABULARY**

**assimilate**, v., to be absorbed into a society's culture

## Migration Within Europe

For years, Europe has had an **aging population**. In other words, the average age of people on the continent has been rising. This trend has had a number of causes. For one thing, Europeans have been living longer because of better medical care. For another, most families are having fewer children. So today, senior citizens form a higher percentage of Europe's total population.

The trend created a need for more workers to replace the many senior citizens who were retiring. Workers were also needed to keep the economy strong and pay taxes to support such social services as education and health care. As a result, immigrants came to Europe to take the newly available jobs. People immigrated to Great Britain from former colonies, such as India, Pakistan, and Bangladesh. France also attracted immigrants from former colonies, especially Algeria and Morocco. In the 1970s, people from Turkey began coming to Germany for jobs. Germany now has approximately 2 million people of Turkish heritage. The fall of communism in Eastern Europe in the 1980s and 1990s also resulted in increased migration within Europe.

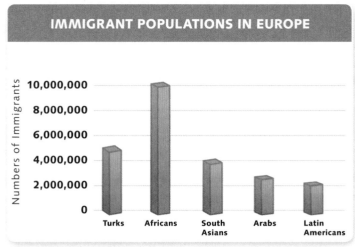

### Before You Move On
**Summarize** In what way did Europe's aging population create a need for immigrants?

298 CHAPTER 10

**Inspiring people to care about the planet**
National Geographic Society Mission

## Reasons for Immigration

Most immigrants have come to Europe to seek a better life. Some left their home countries for economic reasons—usually to find jobs. Others left for political reasons—to escape conflict in their countries or an unjust government. Once they have found work in Europe, many immigrants send money to relatives back home. Eastern Europeans have migrated to Western Europe for similar reasons. Membership within the European Union (EU) has made their migration easier.

## Challenges of Immigration

The global economic crisis and resulting unemployment in Europe led a number of young people from Spain and Portugal, for example, to emigrate to Brazil for jobs. Brazil needed workers in construction, engineering, and architecture to prepare for the 2014 World Cup of soccer and the 2016 Olympic Games. This reversed a migration trend.

The mix of different cultures has also created problems. Many immigrants are Muslims, with cultural and religious practices that differ from those of their Christian neighbors. Some Europeans would like to see the Muslim immigrants **assimilate**, or be absorbed into their society's culture. They believe the immigrants should adopt European traditions and values. Others believe a more multicultural approach is better. This approach encourages tolerance and embraces all cultures.

**COMPARE ACROSS REGIONS**

## Australia's Skilled Immigrants

Like Europe, Australia has an aging population. Many in the Australian government believe that immigration can help change that trend. As a result, each year the government identifies gaps in the country's workforce. It then determines the number of skilled immigrants who can come to Australia. Between 2010 and 2011 more than 127,000 immigrants arrived on this skilled migration program. Many came from Great Britain, India, and China.

Of course, Europe receives many more immigrants than Australia each year. Italy alone took in more than 558,000 immigrants in 2010. Many Europeans appreciate the cultural enrichment that immigrants bring to their countries. However, some believe that Europe, like Australia, should set immigration limits.

### Before You Move On
**Monitor Comprehension** What are the ways in which new immigrants are changing Europe?

**ONGOING ASSESSMENT**
## READING LAB  GeoJournal

1. **Interpret Graphs** According to the bar graph, from which two continents has the immigrant population in Europe primarily come?
2. **Make Inferences** Why might some immigrants resist assimilation?
3. **Identify Problems and Solutions** What has Australia done to solve some of the problems posed by immigration?

# CHAPTER 10
# Review

## VOCABULARY

Match each word in the first column with its meaning in the second column.

| WORD | DEFINITION |
|---|---|
| 1. abstract | a. tax on imports and exports |
| 2. dialect | b. control over one's own affairs |
| 3. sovereignty | c. art style emphasizing form and color |
| 4. tariff | d. absorb into another culture |
| 5. privatization | e. privately owned businesses |
| 6. assimilate | f. regional language |

## MAIN IDEAS

7. Why are there so many different dialects in Europe? (Section 1.1)
8. What is the fastest-growing religion in Europe? Why might that be so? (Section 1.1)
9. During the Romantic period, what themes did artists use in their paintings? (Section 1.2)
10. Why do you think the ancient Greeks and Romans celebrated historic events in epic poems? (Section 1.3)
11. How does the soup called *borscht* reflect Russia's geography? (Section 1.4)
12. What is one of the requirements for joining the European Union? (Section 2.1)
13. In what way has the euro helped increase trade among European nations? (Section 2.2)
14. Why did the Ukrainian people stage the Orange Revolution? (Section 2.3)
15. What factors have led people from other parts of the world to immigrate to Europe? (Section 2.4)

## CULTURE

### ANALYZE THE ESSENTIAL QUESTION

How is the diversity of Europe reflected in its cultural achievements?

**Critical Thinking: Draw Conclusions**

16. In what ways do the many dialects in Europe show its great diversity?
17. What aspects of ancient Greek and Roman art inspired Renaissance artists?
18. Why did pasta with tomato sauce develop in Italy rather than Russia?

**INTERPRET MAPS**

19. **Region** Where do most Protestants live in Europe? Where do most Catholics live?
20. **Make Inferences** Find the countries on the map in which a relatively small part of the population belongs to a different Christian denomination. What challenges might the people in the minority religion face?

## GOVERNMENT & ECONOMICS

### ANALYZE THE ESSENTIAL QUESTION

What are the costs and benefits of European unification?

**Critical Thinking: Analyze Cause and Effect**

21. In what way did the Marshall Plan help bring about the formation of the European Union?
22. What was the response of the eurozone countries when Greece and Ireland became deeply in debt in 2010?
23. What was the impact of the fall of communism on Eastern Europe? What was the impact on Western Europe?
24. What problems are caused by illegal immigration to Europe?

### INTERPRET CHARTS

**COST OF A TEN-MINUTE PHONE CALL TO THE U.S. IN EUROS (€)***

| Country | 1997 | 2012 |
| --- | --- | --- |
| Belgium | 7.50 | 2.17 |
| Czech Republic | 3.09 | 2.34 |
| Denmark | 7.41 | 2.72 |
| Ireland | 4.61 | 1.96 |
| Spain | 6.17 | 1.66 |
| France | 6.78 | 1.96 |
| United Kingdom | 3.50 | 2.19 |

* 1997 prices have been converted to euros
Source: Eurostat

25. **Analyze Data** According to the chart, how did the cost of making a telephone call change between 1997 and 2012?
26. **Analyze Cause and Effect** What move made by the European Union might have brought about the change in the cost of a telephone call?

## ACTIVE OPTIONS

**Synthesize** the Essential Questions by completing the activities below.

27. **Write a Speech** Suppose that you are the leader of a European country that has been invited to join the European Union. Write a speech that will persuade the citizens of your country to vote in favor of joining. **Deliver your speech to the class and ask the members of your audience to vote on whether they are in favor of joining the European Union.**

   **Writing Tips**
   - Take notes on three benefits that would result from joining the European Union.
   - Support each benefit using facts and statistics.
   - Address any concerns your audience might have about joining and explain why the advantages outweigh any disadvantages.

**TECHTREK** myNGconnect.com For photos of European art

28. **Create an Art Chart** Select three European works of art. You can choose Raphael's *School of Athens* or Leonardo da Vinci's *Mona Lisa* from the **Digital Library**, or you can search for other works online. Then research each piece to find out what period it is from and what theme, or subject matter, it represents. Copy the chart below to help you organize your information. Display the artwork and your findings on a poster. Be prepared to explain the relationship between each work's period and theme.

| WORK OF ART | PERIOD | THEME |
| --- | --- | --- |
|  |  |  |
|  |  |  |
|  |  |  |

# COMPARE ACROSS REGIONS UNIT 5
# World Languages

**TECHTREK**
myNGconnect.com For an online graph on world languages

Maps and Graphs

In this unit, you learned about the diversity of languages in Europe. Every culture in the world uses language to communicate. Scholars estimate that there are about 7,000 languages spoken today.

Most countries name one or more languages as official languages. An official language is the one used by a country's government. For example, French is the official language of France, and English and French are the official languages of Canada. Most countries have groups of people whose first language differs from the official language. It is estimated that at least half of the people in the world speak one or more languages in addition to their first language.

### Compare
- Africa
- Americas
- Asia
- Europe
- Oceania

## LIVING LANGUAGES

Although there are thousands of world languages, many are not widely spoken. In addition, sometimes the distinction, or difference, between a language and a dialect is not clear.

The following are the ten most widely spoken languages in the world in order of their ranking. Each is spoken as a native language by at least 100 million people. Some are official languages in widely different regions of the world. Some of the languages appear in only one region of the world.

1. Mandarin Chinese
2. Spanish
3. English
4. Hindi/Urdu
5. Arabic
6. Bengali
7. Portuguese
8. Russian
9. Japanese
10. German

## DYING LANGUAGES

Some languages are spoken by so few people that they are in danger of dying out. In fact, linguists, or people who study languages, estimate that one language dies every two weeks.

Languages die for different reasons. Some simply disappear with the death of the last speaker. Others fade more slowly as a dominant language replaces it. A few linguists are trying to document some of these languages to preserve the history and culture of the people who spoke them.

The graph at right shows the number of languages spoken on the world's continents, as well the names of a few of the languages spoken. Note that "Americas" consists of the number of languages spoken on the North American and South American continents. Compare the data in the graph and use it to answer the questions.

302 UNIT 5

# NUMBER OF LANGUAGES SPOKEN BY CONTINENT

Source: *Ethnologue*, 16th Edition, 2009

- EUROPE
- AFRICA
- ASIA
- AMERICAS
- OCEANIA

**2,322** (Asia): ARABIC, Lao, TAGALOG, hindi, PERSIAN, xiang, Bengali, PASHTU, tibetan, Thai, VIETNAMESE, NYAW, Zhuang, URDU, cantonese, HLAI, Korean, Jingpho, Kashmiri, DOGRI, MONGOLIAN, Wu, NICOBARESE, KUY, INDONESIAN, javanese, TAMIL, MIN, punjabi, CHAM, GAN, TURKMEN, TETUM, Saraiki, burmese, MON, TATAR, ARAMAIC, bodo, Kurdish, KAREN, Gondi, Hmong, Filipino, azeri, Dzongkha, HEBREW, Mandarin, Mizo, Japanese, Ainu, Sundanese, CEBUANO, KAZAKH

**2,110** (Africa): SWAHILI, Tswana, Arabic, Lingala, BERBER, kongo, somali, Kirundi, setswana, Kanuri, Rwanda-Rundi, BAMBARA, Tsonga, LUO, Shona, OROMO, Tigrinya, Zulu, VENDA, HAUSA, PEDI, FRENCH, SOTHO, FULA, CHICHEWA, Sango, Ibibio, SWAZI, UMBUNDU, Afrikaans, Xhosa, Spanish, GBE, Malagasy, TSHILUBA, yoruba, PORTUGUESE, Gikuyu, Twi, igbo, Ndebele, Sesotho, Kiswahili, Malinke, English, LUHYA

**1,250** (Oceania): ULITHIAN, Takuu, NAURUAN, Aranda, Nukuoro, ANUTA, PILENI, Mae, POHNPEIAN, Rapa, pitcairnese, TRUKESE, palauan, Kapingamarangi, bislama, yapese, CAROLINIAN, SAMOAN, HINDUSTANI, Chamorro, French, Tahitian, English, TUVALUAN, niuean, Tok Pisin, Sonsoralese, Marshallese, Maori, I-KIRIBATI, Tongan, Hiri Motu, FIJIAN

**993** (Americas): french, English, xinca, KEKCHI, CREE, PAPIAMENTO, CREOLE, ojibwe, ALEUT, navajo, quiche, Oneida, DANISH, tagish, HOPI, Spanish, ZUNI, QUECHUA, Aymara, nahua, HINDI, Sranan Tongo, MAM, INUIT, mayan, Haida, PORTUGUESE, DUTCH

**234** (Europe): SARDINIAN, Norwegian, BOSNIAN, greek, Scottish Gaelic, Irish, russian, DUTCH, BELARUSIAN, ITALIAN, Latvian, french, Danish, Czech, LATIN, POLISH

Speech bubble: Hola ▷ Guten tag ▷ **Hello** ▷ Kia ora ▷ Hujambo ▷ Ohayou

---

## ONGOING ASSESSMENT
### RESEARCH LAB  GeoJournal

1. **Compare and Contrast** On which continent are the most languages spoken? the fewest? What do these numbers suggest about the cultural unity of each continent?

2. **Analyze Data** Study the graph and the list of languages with the most native speakers. How many of the total languages in Europe are among those spoken most in the world?

**Research and Create Charts** Research to find out more about the people who speak Hindi and Portuguese. Create a chart for each language showing approximately how many people speak it and where it is spoken. Which language has more native speakers? Which language is an official language in more places? What might account for this?

COMPARE ACROSS REGIONS 303

REVIEW & EXTEND UNIT 5

# Active Options

**TECHTREK**

myNGconnect.com For photos of Renaissance art, nuclear power plants in Europe, and European cuisine

Digital Library

## ACTIVITY 1

**Goal:** Extend your understanding of Renaissance art.

### Write a Renaissance Arts Magazine

The Renaissance was a period of great artistic activity in Europe. Choose a city in Europe that was influenced by the Renaissance between 1400 and 1600. With a group, plan and publish a magazine showcasing that city's artistic achievements. Use the **Magazine Maker CD-ROM** to find photos and information on the following:

- art
- architecture
- literature
- fashion

Brunelleschi's dome atop the Cathedral of Florence in Italy

## ACTIVITY 2

**Goal:** Research the use of nuclear power in Europe.

### Create a Pro-and-Con Chart

Some European countries are planning to build nuclear plants, while others have chosen to close existing plants. Use online research links to create a pro-and-con chart that explains some of the advantages and disadvantages of nuclear power. Be prepared to present your chart and explain the issues.

## ACTIVITY 3

**Goal:** Learn about European culture through its music.

### Put Together a Playlist

Get together in a group and create a playlist of music from Europe—traditional, classical, or the music of today. Choose examples that express universal themes, or ideas and situations most everyone understands. Share your list and explain the themes. Tell where the music is from and whether it shows influences from other cultures.

# UNIT 6

# Explore Russia & the Eurasian Republics

## with NATIONAL GEOGRAPHIC

### MEET THE EXPLORER

**NATIONAL GEOGRAPHIC**

Tracing ancient trade routes, NG Fellow Fredrik Hiebert excavated a 4,000-year-old Silk Road city in Turkmenistan. He also searches for underwater settlements in the Black Sea.

### INVESTIGATE GEOGRAPHY

North of the Arctic Circle in Russia, reindeer herded by a nomadic clan charge across the tundra. Many groups in the Arctic rely heavily on the reindeer. The animals, also known as caribou, provide food, clothing, and shelter, as well as transportation.

### CONNECT WITH THE CULTURE

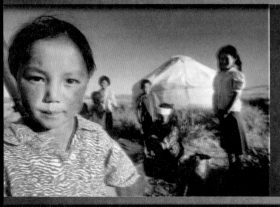

A young Mongolian Kazak family stands in front of a yurt, a shelter used by nomads. Kazaks are a nomadic, animal-herding people. Yurts allow them to live in harsh climates and give them the freedom to move and graze their animals in traditional tribal communities.

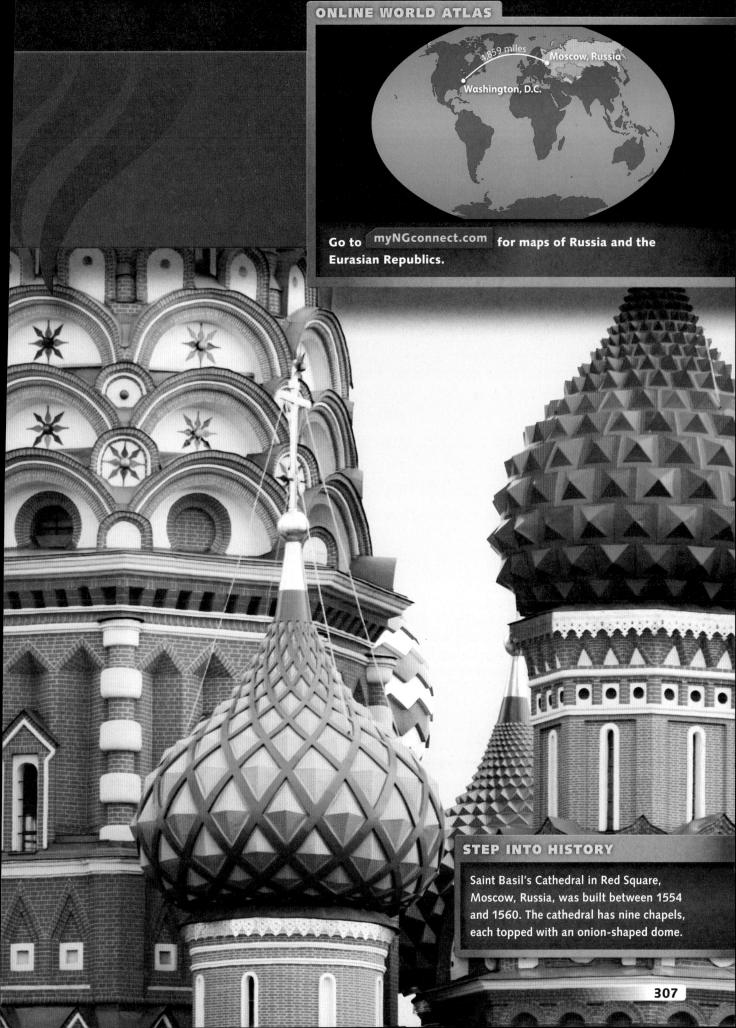

## ONLINE WORLD ATLAS

4,859 miles
Moscow, Russia
Washington, D.C.

Go to myNGconnect.com for maps of Russia and the Eurasian Republics.

## STEP INTO HISTORY

Saint Basil's Cathedral in Red Square, Moscow, Russia, was built between 1554 and 1560. The cathedral has nine chapels, each topped with an onion-shaped dome.

# CHAPTER 11

# Russia & the Eurasian Republics
# GEOGRAPHY & HISTORY

## PREVIEW THE CHAPTER

**Essential Question** How have size and extreme climates shaped Russia and the Eurasian republics?

### SECTION 1 • GEOGRAPHY

1.1 Physical Geography .................................. 310
1.2 Land of Extreme Climates ....................... 312
1.3 Natural Resources ................................... 314
1.4 Exploring Siberian Lakes ......................... 316
1.5 Central Asian Landscapes ....................... 318

**KEY VOCABULARY**
- landmass
- steppe
- permafrost
- tundra
- taiga
- nonrenewable fossil fuel
- peat
- hydroelectric power
- methane
- greenhouse gas
- semiarid
- arid
- pesticide

**ACADEMIC VOCABULARY**
isolate, remote

**TERMS & NAMES**
- Ural Mountains
- North Atlantic Drift
- Siberia
- Black Sea
- Caspian Sea
- Aral Sea

**Essential Question** How has geographic isolation influenced the region's history?

### SECTION 2 • HISTORY

2.1 Early History ............................................ 320
2.2 European or Asian? ................................. 322
2.3 Defending Against Invaders ................... 324
2.4 Serfdom to Industrialization .................. 326
2.5 The Soviet Union .................................... 328

**KEY VOCABULARY**
- state
- tribute
- czar
- reign
- secular
- invader
- scorched earth policy
- serf
- strike
- communism
- socialism
- collective farm
- propaganda

**ACADEMIC VOCABULARY**
expand, promote

**TERMS & NAMES**
- Slav
- Kievan Rus
- Genghis Khan
- Mongol Empire
- Silk Roads
- Peter the Great
- Catherine the Great
- Nazi Germany
- Industrial Revolution
- V. I. Lenin
- Bolshevik
- Russian Revolution
- Soviet Union
- Cold War
- Mikhail Gorbachev

Russian brown bear cubs

# SECTION 1 GEOGRAPHY
## 1.1 Physical Geography

RUSSIA AND THE EURASIAN REPUBLICS PHYSICAL

310 CHAPTER 11

**Main Idea** Russia and the Eurasian republics cover a huge area and contain a variety of geographic features.

Russia and the Eurasian republics take up about one-sixth of the land surface of the entire earth. The region's geographic features have limited its population.

## Huge Landmass

Russia is the largest country in the world in total area. Its **landmass**, or continuous extent of land, stretches almost 6,000 miles from east to west. The Eurasian republics lie south of Russia. The republics in the Caucasus Mountains are Armenia, Azerbaijan (ahz ur by JAHN), and Georgia. Those in Central Asia include Kazakhstan (kah zahk STAHN), Kyrgyzstan (kihr gih STAN), Tajikistan (tah jik ih STAN), Turkmenistan (turk MEN ih stan), and Uzbekistan (ooz BEK ih stan).

Plains, or large areas of level ground, cover much of the region. The relatively low **Ural Mountains** separate the Northern European Plain from the West Siberian Plain. In much of southwestern Russia and northern Kazakhstan, the very large plains are called **steppes**. Much of this land is good for agriculture and grazing.

## Natural Barriers

Russia has oceans on its northern and eastern borders and mountainous areas along much of its southern border. These natural barriers separate Russia from its neighbors. The deserts and mountains of Central Asia keep republics such as Kyrgyzstan and Tajikistan **isolated**, or cut off from other countries.

Because so much of Russia's coastline lies to the north of the Arctic Circle, few ports stay open to ships and trade all year long. Murmansk, located far to the north on Kola Peninsula, is one such port. An ocean current called the **North Atlantic Drift** warms the waters around Murmansk and keeps them ice-free most of the time.

### Before You Move On

**Monitor Comprehension** What are some of the key geographic features of this region?

### ONGOING ASSESSMENT
# MAP LAB

1. **Place** Find the Ural Mountains on the physical map. What is the elevation of the mountains? Locate the steppes. How does its elevation compare with that of the Ural Mountains?

2. **Interpret Maps** Use the population density map to determine how density differs west and far east of the Urals. Then study the physical map. What physical features might contribute to the population distribution?

SECTION 1.1 **311**

# SECTION 1 GEOGRAPHY

## 1.2 Land of Extreme Climates

**TECHTREK**

myNGconnect.com For a climate map and photos of extreme climates

Maps and Graphs  Digital Library

**Main Idea** The extreme climates of this region have an impact on where and how people live.

About half the land in Russia is so cold that it has **permafrost**, or permanently frozen ground, beneath it. Yet parts of Russia can also reach 100°F in summer, and large areas of Central Asia are desert. Because of these extremes, most of the population lives in the western part of the region, where the climate is not as harsh.

### Cold, Dark Winters

Latitude is an important factor in the climate of a region. The northern boundary of Russia is a coastal plain along the Arctic Ocean, with no natural barriers to keep out arctic winds. The high northern latitudes of Moscow and areas to its north help bring this region long, dark, snowy winters. St. Petersburg, for example, has a latitude of almost 60° N. For about one month each winter, there is has hardly any daylight in the city.

### Climate and Vegetation

Climate affects the types of vegetation that grow in different areas. **Tundra**, or flat land found in arctic and subarctic regions, exists in **Siberia**, which lies in central and eastern Russia. Here, only small plants can grow. Permafrost prevents most trees from growing because their roots can't spread deep under the ground.

**Critical Viewing** Snow and ice cover the ground throughout much of the long winter in the arctic city of Noril'sk. Based on the photo, what challenges might people encounter in the city's wintry weather?

312 CHAPTER 11

## CLIMATE

Just south of the tundra is the **taiga** (TY guh), or forest area. Mostly small evergreens like pines grow here. This area provides valuable timber resources.

Extremes in temperature and moisture make it hard for some areas to be used for agriculture. Much of the northern territory has short summers and, as a result, short growing seasons. The semiarid and desert areas are limited to herding and grazing. Farming is concentrated in the fertile soils of the western plains and steppes, along the **Black Sea**, the **Caspian Sea**, and in some river valleys.

### Before You Move On
**Summarize** In what ways do this region's extreme climates affect the people who live there?

## ONGOING ASSESSMENT
### MAP LAB

1. **Location** What is the climate like in the northeastern part of Russia? How does latitude affect the climate in this region?

2. **Interpret Maps** What areas in the region have arid or semiarid climates? What areas have a humid cold climate with dry winters?

3. **Pose and Answer Questions** Study the map and then create another climate question to ask a partner.

4. **Human-Environment Interaction** Look at the population map in Section 1.1 and then compare it with the climate map on this page. In what ways might climate determine where people live in the region?

# SECTION 1 GEOGRAPHY
## 1.3 Natural Resources

TECHTREK
myNGconnect.com For a resource map and photos of resources

 Maps and Graphs    Digital Library

**Main Idea** Russia and the Eurasian republics have plentiful natural resources, but many of them are in remote locations.

This region is among the world's richest in natural resources. These resources are important for the countries' economies.

### Energy Resources

Russia and the Eurasian republics have plentiful energy resources, especially oil and natural gas. Russia is also a leading coal producer. These resources are **nonrenewable fossil fuels**. They cannot reproduce quickly enough to keep pace with their use. Russia also has large amounts of **peat**, which is very old decayed plant material. Peat is burned like coal. In addition, some rivers provide **hydroelectric power**. Power production plants use the force of the rivers' water to generate electricity.

### Mineral Resources

The region contains large quantities of mineral resources that provide raw materials for factories and support industrial development. These resources include metallic ores, such as iron and aluminum, along with gold, copper, platinum, uranium, cobalt, manganese, and chrome.

Almost 20 percent of the world's reserves of iron ore are located in the region, with Russia and Kazakhstan among the main sources of this mineral. Iron ore is used to produce iron and steel, which are used in the construction of roads, railways, and buildings.

In 2010, huge reserves of minerals were found in nearby Afghanistan. Soon, this country might compete with Russia and the republics as a major producer of iron, copper, and other metals.

**Critical Viewing** Open-pit mines, like this one in Siberia, are dug when the diamonds are near the surface. What benefits might this mine bring to the town?

The mine is in the often snowy town of Mirny.

Stepped walls help prevent landslides.

A ramp allows trucks to carry away rocks and diamonds.

Water often pools at the bottom of the mine.

## NATURAL RESOURCES

## The Challenge of Location

Much of the region's resources are in **remote**, or hard to reach, locations. For example, Siberia contains oil fields, hydroelectric power sources, and minerals, such as nickel and gold. Many of these resources are located in the far-eastern, and coldest, parts of Siberia. The permafrost there makes it difficult to drill or mine for the natural resources and transport them to market. As a result, the resources in these areas of Siberia remain largely untouched.

### Before You Move On
**Summarize** What are some important natural resources of Russia and the Eurasian republics, and why are some of them hard to reach?

## ONGOING ASSESSMENT
### DATA LAB — GeoJournal

**Create Charts** You have learned that Russia and the Eurasian republics contain plentiful mineral resources that provide raw materials for factories. Identify the region's mineral resources in the map above. Then research to find out what consumer goods are made from these minerals. Create a chart like the one below to record your findings.

| MINERALS | GOODS |
| --- | --- |
| Iron | steel, medicine, magnets, auto parts, paper clips |
| Aluminum | kitchen utensils, drink cans, foil, airplane and car parts |

SECTION 1.3  **315**

# SECTION 1 GEOGRAPHIC
## NATIONAL GEOGRAPHIC

# Exploring Siberian Lakes
### with Katey Walter Anthony

**Main Idea** As permafrost thaws in Siberia, it releases methane gas into the atmosphere.

## Expedition to Siberia

Emerging Explorer Katey Walter Anthony first went to Siberia as a high school exchange student. Now she works with other scientists from Alaska and Russia at the Northeast Science Station in Cherskiy, Siberia,  to study how climate change is affecting the area—and possibly the entire world. Siberia's frigid climate makes the work difficult.

Anthony and some other scientists are concerned about Siberia's permafrost. They believe that because of global warming, the permafrost below its lakes is thawing, releasing carbon that was locked inside the frozen ground. Carbon is formed from dead prehistoric animals and the plants they ate. The carbon is then turned into methane, a colorless, odorless natural gas that can have a negative impact on the environment.

myNGconnect.com
For more on Katey Walter Anthony in the field today

Anthony checks for methane on a frozen lake.

ANTHONY'S EXPEDITIONS

## PROCESS OF METHANE RELEASE

- ❸ Methane is released into the atmosphere.
- ❷ Methane is released into the lake as the permafrost thaws.
- ❶ Organic matter is trapped in the permafrost.

Labels: soil, frozen lake, sediments, permafrost

**METHANE RELEASE** Warming temperatures thaw the permafrost and release methane gas. The methane heats the air as it is released into the atmosphere as a greenhouse gas.

## Methane and Climate Change

Methane is a greenhouse gas, a gas that traps the sun's heat over the earth. Anthony explains, "It's 25 times more powerful than carbon dioxide on 100-year time scales" and could have the most powerful effect of all on global warming. Siberian lakes could release about ten times as much methane as is in the atmosphere now. Some experts believe this could cause temperatures across the world to rise higher and faster.

To check for methane, Anthony chops holes into the ice on the lakes to collect gas samples that she brings back to the lab. Sometimes she wants to know right away what gases might be present, so she lights a match. If a giant flame shoots up, she knows she's found methane.

### Before You Move On
**Monitor Comprehension** What impact does methane have on the atmosphere?

## ONGOING ASSESSMENT
# VIEWING LAB  GeoJournal

1. **Analyze Visuals** Go to the **Digital Library** to view the video clip on Emerging Explorer Katey Walter Anthony. How might her work in Siberia be applied to other places in the world?

2. **Interpret Models** Study the model. What does the methane have to pass through to enter the atmosphere? Why does the gas penetrate this material so easily?

3. **Analyze Cause and Effect** List some causes and effects of methane release using a chart like the one below. Why do some scientists believe that increasing methane in Siberian lakes is both a cause and an effect of global warming?

| CAUSES | EFFECTS |
|--------|---------|
|        |         |
|        |         |

# SECTION 1 GEOGRAPHY
## 1.5 Central Asian Landscapes

myNGconnect.com For online maps and photos of Central Asia

Maps and Graphs  Digital Library

**Main Idea** Human activities have led to the shrinking of the Aral Sea, which has damaged the surrounding landscape.

As you have learned, methane gas is threatening Siberia's environment. Central Asia's landscape has also been damaged. Human activities have nearly destroyed one of its most important bodies of water.

## Adapting to Dry Conditions

Central Asia includes landforms such as deserts, mountains, forests, and steppes. Though the area doesn't experience the extreme cold of northern Russia, there are places in northern Kazakhstan that can reach 0°F in winter. In general, summers are hot and longer than in the northern parts of the region. Central Asia's temperatures vary so widely in part because it is not protected by a large body of water, which would help to keep temperatures moderate.

Large parts of Central Asia are **semiarid** or **arid**, meaning there is little or no rainfall. These dry lands are best suited for livestock grazing. Because of irrigation in some river valleys, however, farmers have also been able to grow crops such as cotton.

## The Shrinking Aral Sea

Major efforts to grow cotton led to the shrinking of the **Aral Sea** in Kazakhstan and Uzbekistan. The body of water is actually a salt-water lake. The rivers that fed the lake were redirected into canals for irrigation. The Aral Sea was once the fourth largest lake in the world, but now it is only a fraction of the size it was in 1960.

Pollution contributed to problems in the Aral Sea. Fertilizer and **pesticides**, which are chemicals that kill harmful insects and weeds, ran into it. As the lake shrank, salt and pesticides destroyed the habitat of many plants and animals and threatened human health. The lake's once-thriving fishing industry was also damaged. One resident of the area said, "My father and grandfather were fishermen in this town, but as you can see, the boats are now sitting in the middle of a desert."

In 2005, Kazakhstan, with the help of the World Bank, built a dam to save the North Aral Sea. That part has increased in size, and fishing has returned to the area. However, the southern part of the lake, in Uzbekistan, is almost completely gone.

### Before You Move On
**Monitor Comprehension** What human activities caused the Aral Sea to shrink, and what damage to the landscape has occurred as a result?

> **Critical Viewing** Camels walk past a stranded ship on land where the Aral Sea used to be. What does this picture suggest about the Aral Sea?

North Aral Sea, 1973

Aral Sea, 1973

This satellite image shows what the Aral Sea looked like in 1973. At this time, the North and South Aral seas were full of water.

North Aral Sea, 2009

Aral Sea, 2009

This satellite image shows what the Aral Sea looked like in 2009. The photo reveals that efforts to save the North Aral Sea have worked.

## ONGOING ASSESSMENT
## PHOTO LAB  GeoJournal

1. **Analyze Visuals** Study the photos above of the Aral Sea. How much time passed between the photo on the left and the photo on the right? What happened during that time period?

2. **Identify** What details in the photo below tell you that this part of the Aral Sea has been devastated for a long time?

3. **Human-Environment Interaction** Why did people redirect the Aral Sea? What steps have been taken to fix what happened as a result?

SECTION 1.5 **319**

# SECTION 2 HISTORY
## 2.1 Early History

TECHTREK
myNGconnect.com For an online map of the Silk Roads and photos of artifacts

Maps and Graphs    Digital Library

**Main Idea** Settlers and conquerors from Europe and Asia shaped the early history of Russia and the Eurasian republics.

Settlement of Russia by different groups dates back to around 1200 B.C. At that time, a people called the Cimmerians lived north of the Black Sea in what is now southern Ukraine. Over hundreds of years, many other groups ruled this region.

### Kievan Rus

The people whose culture had the most lasting influence on early Russia were the **Slavs** (slahvz). Some historians think they were farmers near the Black Sea around 700 B.C. or earlier. Others think they came from Poland around the A.D. 400s. By the A.D. 800s, the Slavs had built towns near rivers in Ukraine and western Russia.

In 862, Vikings from Scandinavia called the Varangian Russes (vahr ANG ee ehn ROOS ehz) took control of the Slavic town of Novgorod. Russia's name may have come from this tribe. About 20 years later, a Varangian prince captured Kiev (KEE ehf). He established a **state**, or defined territory with its own government, that came to be known as **Kievan Rus**.

### Mongol Rule and Trade

By the late 1100s, Kiev's power had declined. Struggles for control within the ruling family weakened the state.

In the early 1200s, **Genghis Khan** (JEHNG gihs KAHN) established the **Mongol Empire** in Central Asia. In 1240, his grandson, Batu Khan, extended the empire by taking over Kievan Rus and much of Russia.

Russian princes had to declare their loyalty and pay **tribute**, or taxes, to the Mongol ruler, the khan. Mongols selected one of the princes to serve as grand prince and represent Russian interests.

The empire isolated Russia from European influence for more than 200 years. However, Mongol rule kept Russia and Central Asia open to the East. The **Silk Roads**, ancient trade routes, carried goods and new ideas throughout the Mongol Empire. As you can see on the map on the next page, the Silk Roads connected Southwest Asia and Central Asia with China. Goods traded included gold, jade, and silk. Merv, in Turkmenistan, and Samarkand, in Uzbekistan, were important stops on the Silk Roads.

Gold horse and rider made by early settlers of Ukraine

c. 882 Varangian Russes found Kievan Rus.

1000 B.C.    A.D. 800    1000

1200 B.C. Cimmerians settle in southern Ukraine.

A.D. 800s Slavs build towns on the Dnieper River.

Ceiling of St. Sophia, Kievan Rus cathedral

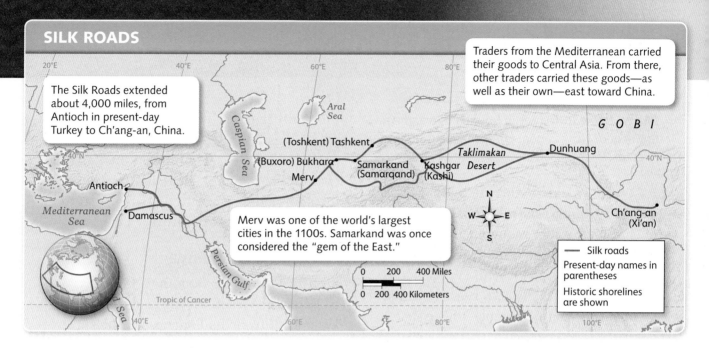

### SILK ROADS

The Silk Roads extended about 4,000 miles, from Antioch in present-day Turkey to Ch'ang-an, China.

Traders from the Mediterranean carried their goods to Central Asia. From there, other traders carried these goods—as well as their own—east toward China.

Merv was one of the world's largest cities in the 1100s. Samarkand was once considered the "gem of the East."

— Silk roads
Present-day names in parentheses
Historic shorelines are shown

## The Rise of Moscow

Around 1330, the Mongols allowed Grand Prince Ivan I of Moscow to collect tribute for them. Ivan kept some of the money for himself and used it to buy land and **expand** his territory. Moscow became stronger as Mongol rule weakened.

In 1380, Grand Prince Dmitry defeated the Mongols in battle. In 1480, Ivan III— also called Ivan the Great—refused to pay tribute to the Mongols, and they never demanded it again. This action ended Mongol rule in Russia. Ivan the Great's grandson, Ivan IV, expanded Russia and made it into an empire. He became Russia's first **czar** (zahr), or emperor.

### Before You Move On
**Summarize** How did settlers and conquerors shape the early history of Russia?

### ONGOING ASSESSMENT
## MAP LAB
GeoJournal

1. **Place** Why do you think the main route of the Silk Roads split in two between the cities of Kashgar and Dunhuang?
2. **Interpret Maps** Study the map. Why was Samarkand an important stop on the routes?
3. **Analyze Cause and Effect** What actions strengthened Moscow and what resulted from its rise? Use a chart like the one below to list the causes and effects.

| CAUSES | EFFECTS |
|--------|---------|
|        |         |
|        |         |

- 1237–1240 Mongols conquer Russia.
- 1380 Prince Dmitry defeats Mongols.
- 1480 Ivan III ends Mongol rule.

1200 — 1400

Mongol emperor Genghis Khan

Ivan the Great Bell Tower

# SECTION 2 HISTORY

## 2.2 European or Asian?

**TECHTREK**
myNGconnect.com For an online map and photos of historic buildings

 Maps and Graphs
 Digital Library

**Main Idea** Russia lies in both Europe and Asia but only began to adopt European ideas under the reign of Peter the Great.

As you have already learned, Russia was isolated from Western Europe for a couple of centuries under Mongol rule. This isolation continued under the Russian czars, beginning with Ivan IV. Then, starting in the 1600s, two czars began to bring a European influence to Russia.

### Spanning Europe and Asia

Geographically, Russia extends across the landmass of Europe and Asia, which is often called Eurasia. The Ural Mountains separate the continents. Many of the Russian people live in the European area west of the mountains.

In spite of lying partly in Europe, Russia did not develop a European culture. In the late 900s, Christianity became the main religion in Russia. However, while Western Europe adopted Roman Catholicism, Russia embraced the Eastern Orthodox branch of Christianity. The Russian people came to distrust western European ideas and culture.

### European Influence

Finally, Czar Peter Romanov, known as **Peter the Great**, recognized that Western Europe had surpassed Russia both economically and militarily. Peter, who ruled from 1682 to 1725, decided to modernize Russia. As a result, he became the first Russian czar to travel to Western Europe. From his travels, he brought back European ideas about how government and business should be run. He also introduced a Western style of architecture. Many buildings in St. Petersburg, including the Peterhof Palace shown below, reflect this style.

> **Critical Viewing** The Peterhof Palace is often called "the Russian Versailles," after the French palace of King Louis XIV. What words would you use to describe the palace?

**Catherine the Great**, empress from 1762 to 1796, also introduced European ideas to Russia, focusing on the arts and education. Western forms of entertainment, such as opera, became popular during her **reign**, or rule. In addition, Catherine founded many new schools and supported the idea of educating women. She also built hospitals and **promoted**, or encouraged, vaccinating against smallpox. Planners redesigned cities, and European architecture replaced older Russian styles.

The empress attempted to create a more **secular** country, one in which the church was less powerful. In fact, Catherine wanted to pass laws that would allow citizens to practice the religion of their choice. However, her attempt at reform was unsuccessful.

### Before You Move On
**Summarize** What European ideas did Russia adopt under Peter the Great?

### EUROPEAN AND ASIAN RUSSIA

 **MAP TIP** Map projections sometimes distort, or bend, the area shown. The Mollweide projection here makes Europe look somewhat mashed together but it is helpful in showing vast expanses of polar regions in small spaces.

### ONGOING ASSESSMENT
## READING LAB  GeoJournal

1. **Make Inferences** Based on the map, in what ways do you think Russia's location encouraged its isolationism?
2. **Synthesize** How was Russia under Peter and Catherine different than under Mongol rule?
3. **Interpret Maps** In which continent does most of Russia lie?
4. **Analyze Visuals** What impression of Peter the Great is this statue meant to convey? Explain your answer with specific references to the statue.

SECTION 2.2

# SECTION 2 HISTORY

## 2.3 Defending Against Invaders

**TECHTREK**
myNGconnect.com For an online map and photos of historical figures

 Maps and Graphs
 Digital Library

**Main Idea** For centuries, Russia used its geographic strengths—location, size, isolation, and climate—to defend against invaders.

Throughout history, Russians adapted to the challenges of distance and climate in their country. **Invaders**, or enemies entering by force, however, often had a hard time overcoming these challenges.

### Russia Builds an Empire

You have learned that Ivan IV became the first czar of Russia in 1547. Ivan began building Russia's empire by conquering territory from the Mongols. Eventually, Russia expanded across Siberia to the Pacific Ocean.

As Russia continued to grow under Peter the Great and Catherine the Great, the country was often forced to defend its empire against invaders. Enemies from Europe had to cope with Russia's vast expanse and harsh climate. They were rarely successful at invading the country. Two famous examples are the failed invasions of Napoleon and Hitler.

### Napoleon Invades Russia

Napoleon I became emperor of France in 1804, and his empire included much of Europe. Angered that Britain had defeated him in battle, he came up with a plan to limit Britain's trade with European countries. Czar Alexander I refused to go along with the plan. Alexander thought reducing trade with Britain would be bad for Russia's economy.

In revenge, Napoleon invaded Russia in the summer of 1812. The Russians followed a **scorched earth policy**, where troops retreat in front of the advancing army and destroy crops and other resources that might supply the enemy. When Napoleon arrived in Moscow, the city was nearly deserted. The Russians had burned it to destroy any supplies that might have helped the French troops.

By mid-October, Napoleon started his retreat. He knew the harsh Russian winter was coming. Soon, snow and extreme cold began to weaken the French army. Russian troops attacked the army as it fled. Of the roughly 420,000 troops that set out for Moscow, only about 10,000 survived.

Ivan IV stands before St. Basil's, which he had constructed.

**1500**

1547
Ivan IV is crowned first czar.

**1600**

1604–1613
Civil war and invasions

1613
Romanov rule begins. The family, which included Peter the Great and Catherine the Great, ruled until 1917.

**1700**

1703
St. Petersburg is founded.

Picture of Nicholas II, the last of the Romanovs

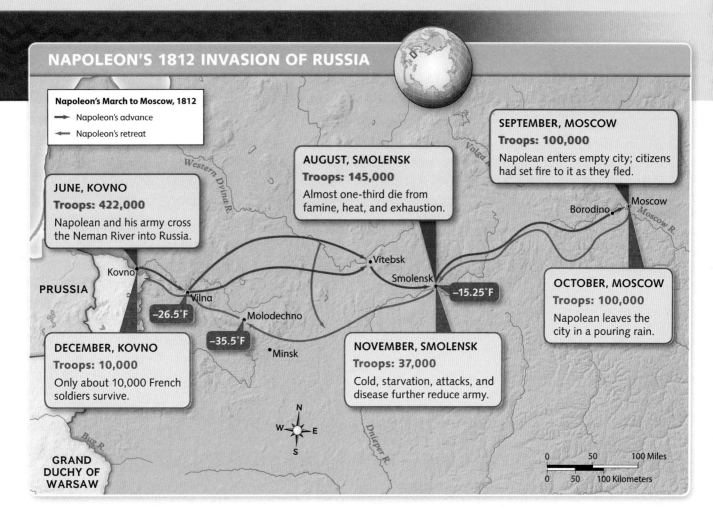

## Hitler Invades Russia

History repeated itself when Adolf Hitler sent troops from **Nazi Germany** to invade Russia during World War II. Frigid winter weather resulted in the deaths of many Nazi soldiers. Hitler was unable to take Moscow in 1941, or Stalingrad in 1943. Of the 300,000 Nazi troops that fought at Stalingrad, only about 5,000 came home.

### Before You Move On

**Monitor Comprehension** What role did Russia's geography play in defeating Napoleon and Hitler?

### ONGOING ASSESSMENT
### MAP LAB  GeoJournal

1. **Movement** Using the map, figure out how far Napoleon's army had to travel to get to Moscow from Kovno.
2. **Interpret Maps** What does the map tell you about the weather as the French troops retreated from Moscow?
3. **Compare and Contrast** In what ways were Nazi Germany's experiences in Russia similar to those of Napoleon?

- 1812 Napoleon defeated in Russia
- 1943 Nazi forces defeated at Stalingrad in World War II

1800 — 1900 — PRESENT

In this painting, Moscow burns as the French army battles Russian soldiers.

# SECTION 2 HISTORY

## 2.4 Serfdom to Industrialization

**Main Idea** Both peasants and industrial workers led difficult lives, but they played important roles in Russia's history.

For centuries, most Russian workers were peasants. They worked the land for wealthy landlords. The peasants were free to leave the estate as long as they paid their debts after the harvest.

### The Beginning of Serfdom

Ivan IV, also known as Ivan the Terrible, changed this system in the 1500s. He murdered many nobles and gave their lands to people who supported him. These people needed the peasants to farm their estates, so Ivan passed laws that tied the peasants to the land as **serfs**. Serfs had their own houses and a small plot of land to farm, but they had to pay the landlord rent. Serfs could not leave the estate without permission and had few rights.

In 1861, Czar Alexander II freed the serfs. He wanted these workers to be free to work in industry and help modernize Russia. As a result, he ended their legal ties to their landlords. Serfs could then become industrial workers for wages.

### Industrial Revolution in Russia

Under Peter the Great, early industries, such as shipbuilding and metalworking, had begun. Yet the **Industrial Revolution** didn't really begin in Russia until the 1890s. By 1913, it had become the fifth largest industrial nation in the world. As in other countries, however, the change from rural work to factory work was difficult. Most industrial workers and peasants were very poor.

**Critical Viewing** Serfs clear the land of large stones in this illustration. Based on the illustration, what can you conclude about the type of work serfs performed?

Many peasants starved because of poor harvests, and city factory workers were unhappy with their working conditions. Frequent worker **strikes**, or work stoppages, and protests gave way to political unrest. Revolutionary activist and politician **V. I. Lenin** led a political group called the **Bolsheviks**. They wanted workers to take over industry and the government. In February 1917 the Bolsheviks began the **Russian Revolution** and overthrew the czar. Lenin became leader of the new government.

### Before You Move On
**Summarize** What roles did serfs and industrial workers play in Russian history?

## RESOURCES AND INDUSTRIES IN WESTERN RUSSIA, 1900

Legend:
- Iron ore
- Coal
- Oil
- Textiles
- Metalworking
- Sugar production
- Railroads

### GROWTH OF KEY INDUSTRIES IN RUSSIA, 1890–1900

| Industry | Percentage growth |
|---|---|
| Coal | 131 |
| Cotton manufacturing | 76 |
| Manufactured iron | 116 |
| Oil production | 132 |
| Pig iron | 190 |

**Source:** Peter Stearns, *The Industrial Revolution in World History*, 1998

## ONGOING ASSESSMENT
# DATA LAB

1. **Interpret Graphs** According to the graph, which industry grew the least between 1890 and 1900? The most? Why do you think these industries grew at these rates?

2. **Synthesize** Based on the graph, the map, and the text, what conclusions can you draw about the Industrial Revolution in Russia?

3. **Movement** Examine the map shown above. Why do you think several railroad lines ran through Moscow?

SECTION 2.4

# SECTION 2 HISTORY
## 2.5 The Soviet Union

myNGconnect.com For photos of Soviet propaganda

Digital Library

**Main Idea** The Communist Soviet Union had a powerful central government that controlled the region from 1922 until 1991.

The Bolsheviks, who led the Russian Revolution, believed that a Communist form of government and a socialist economic system were the answers to the problems of the Industrial Revolution. Under **communism**, a single political party controls the government and economy. **Socialism** is a system in which the government controls economic resources. The Bolsheviks wanted to end private ownership of land and resources and establish a classless society.

### A Communist State

In 1922, Russia, Ukraine, Belarus, and the Transcaucasian republics of Armenia, Azerbaijan, and Georgia formed the Union of Soviet Socialist Republics (U.S.S.R.), also known as the **Soviet Union**. Later, other republics came under the control of the Soviet Union and the central Communist government in Moscow.

From 1927 to 1953, the Soviet people lived under the total command of Josef Stalin. Stalin's government isolated its citizens from contact with the West.

### The Cold War

After World War II, the Soviet Union and the United States were the two most powerful countries in the world. Tension and conflict arose between the two because of their very different political and economic systems. The conflict came to be known as the **Cold War** because the countries did not fight each other directly.

The Cold War led the United States and Soviet Union to develop nuclear weapons. The United States went to war in Korea and Vietnam during the 1950s and 1960s to try to prevent communism from spreading to these countries. The Cold War also resulted in a "space race." The Soviet Union won the race in 1957 when it launched its Sputnik satellite into space.

### A Controlled Economy

The Soviet Union also became an industrial leader and a world power, second only to the United States. The government owned most businesses and agriculture. On **collective farms**, workers produced a certain amount of food—determined by the government—and received a share of surplus crops. Still, the Soviet Union had trouble feeding all of its people.

A banner of Lenin overlooks Red Square as Soviet Russia celebrates the anniversary of the 1917 Russian Revolution.

- **1922** Soviet Union forms.
- **1924** Stalin becomes Soviet leader.
- **1941–1945** Soviet Union fights Germany in World War II.
- **1945** Cold War begins.

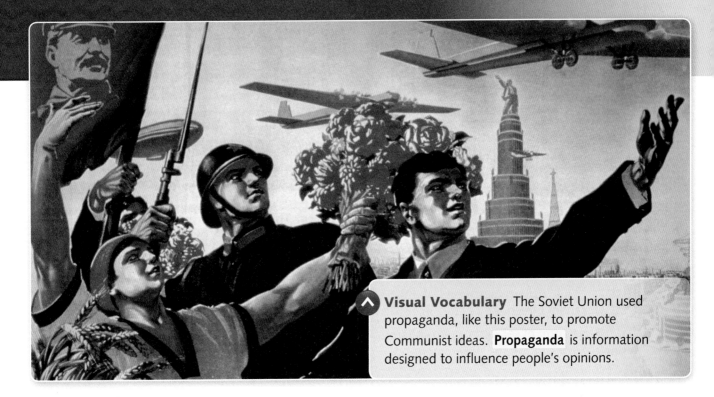

**Visual Vocabulary** The Soviet Union used propaganda, like this poster, to promote Communist ideas. **Propaganda** is information designed to influence people's opinions.

Quality of life was poor, too. People had guaranteed jobs, but their standard of living was much lower than that in Western countries. For example, they had little access to consumer goods. President **Mikhail Gorbachev** (mih KYL GAWR buh chawf) tried to reform and improve the economy. However, a movement toward adopting democratic forms of government was spreading across Eastern Europe. In 1991, the Soviet Union collapsed, and its republics gained their independence.

### Before You Move On
**Monitor Comprehension** In what ways did the Communist government of the Soviet Union control its republics?

## ONGOING ASSESSMENT
# VIEWING LAB  *GeoJournal*

1. **Analyze Visuals** In the propaganda poster, the figure in the back is holding a flag with Stalin's picture on it. What are the other figures holding? What might each figure represent?

2. **Make Inferences** Notice the attitude and gestures of the figures in the poster. How do you think the poster was supposed to make the Russian people feel?

3. **Draw Conclusions** How do you think the American people felt when the Soviet Union sent the first satellite into space? How might the space race have intensified the Cold War?

---

- **1957** Soviets launch Sputnik satellite.
- **U.S. president Ronald Reagan and Gorbachev**
- **1985** Gorbachev begins reform programs and works with President Reagan to end Cold War.
- **1975**
- **PRESENT**
- **1961** Soviets send first person into space.
- **1991** Soviet Union collapses.

# CHAPTER 11
# Review

## VOCABULARY

On your paper, write the vocabulary word that completes each of the following sentences.

1. _____ under the tundra keeps trees from growing there.
2. Some rivers provide _____, using the power of water to generate electricity.
3. _____ is information designed to influence people's opinions.
4. Czar Alexander I freed the _____ in 1861.
5. _____ is a form of socialism in which a single party controls the government and economy.

## MAIN IDEAS

6. What natural barriers separate Russia and the republics from their neighbors? What impact do the barriers have on these countries? (Section 1.1)
7. Why are Russian winters cold and dark in some places? (Section 1.2)
8. What are the two main types of natural resources in this region? (Section 1.3)
9. What does Katey Walter Anthony hope to learn by studying the presence of methane in Siberian lakes? (Section 1.4)
10. What type of agriculture is suited to the dry climate of Central Asia? (Section 1.5)
11. How was Kievan Rus founded? (Section 2.1)
12. What western European influences did Peter the Great and Catherine the Great introduce to Russia? (Section 2.2)
13. What factors helped defeat the Nazis in Russia? (Section 2.3)
14. In what ways were serfs different from peasants in Russia? (Section 2.4)
15. What methods did the government of the Soviet Union use to control the country's economy? (Section 2.5)

## GEOGRAPHY

### ANALYZE THE ESSENTIAL QUESTION

How have size and extreme climates shaped Russia and the Eurasian republics?

**Critical Thinking: Analyze Cause and Effect**

16. Why does Russia have few ice-free ports?
17. What impact do Russia's size and extreme climate have on its use of natural resources?
18. In what way did the climate in Central Asia contribute to the shrinking of the Aral Sea?

### INTERPRET MAPS

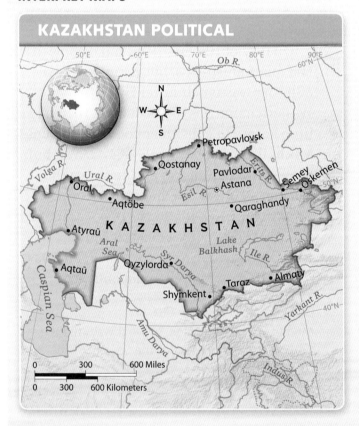

19. **Location** Which two cities are best located to take advantage of energy resources in the Caspian Sea?
20. **Make Inferences** Why do you think Kazakhstan moved its capital from Almaty to Astana in 1997?

## HISTORY

### ANALYZE THE ESSENTIAL QUESTION

How has geographic isolation influenced the region's history?

**Critical Thinking: Make Inferences**

21. What effect did the Mongols have on Russia's relationship with Europe?
22. During Napoleon's invasion of Russia, what impact might distance have had on the army's horses and supplies?
23. In what way might physical geography have prevented the Industrial Revolution from starting in Russia as early as it started in western European countries?

### INTERPRET TABLES

**TIME INDUSTRIAL LABORERS NEEDED TO WORK TO BUY SELECTED GOODS, 1986**

|  | Moscow | United States |
|---|---|---|
| Loaf of bread | 11 min* | 18 min |
| Liter of milk | 20 min | 4 min |
| Grapefruit | 112 min | 6 min |
| Chicken | 189 min | 18 min |
| Bus fare (2 miles) | 3 min* | 7 min |
| Postage stamp | 3 min | 2 min |
| Pair of jeans | 56 hrs | 4 hrs |
| Washing machine | 177 hrs | 48 hrs |

* price supports by government

Source: Radio Free Europe, published in the *New York Times*, June 28, 1987

24. **Compare and Contrast** What type of food must have had the biggest price difference? What food was probably more expensive in the United States than in Moscow?
25. **Make Generalizations** Based on the chart, what do you think life was like for industrial laborers in the Soviet Union in 1986?

## ACTIVE OPTIONS

**Synthesize** the Essential Questions by completing the activities below.

26. **Write an Email** Schools sometimes conduct exchange programs with schools in other countries. Given what you know about the geography, history, and culture of Russia, decide which area of Russia you would like to visit. Write an email to a classmate recommending a particular location in Russia using the tips below. **Send the email to a classmate and ask for feedback on the information provided.**

    > **Writing Tips**
    > - Organize your ideas under two or three main headings before you begin writing.
    > - Use a clear, straightforward style to present specific, useful, and interesting details in your email.
    > - Make sure you explain why you offer the advice that you do.

    Go to **Student Resources** for Guided Writing support.

27. **Gather and Share Information** Work in a group to gather two new facts about Russia and each of the republics. Use online sources to help you find the facts. Then create a chart to record your facts and share them in an oral presentation. Compare your findings with those of other groups.

    | Russia | Fact 1: _____ <br> Fact 2: _____ |
    |---|---|
    | Armenia | Fact 1: _____ <br> Fact 2: _____ |
    | Azerbaijan | Fact 1: _____ <br> Fact 2: _____ |

# CHAPTER 12
# Russia & the Eurasian Republics Today

## PREVIEW THE CHAPTER

**Essential Question** What features, such as size and climate, have influenced Russian culture?

### SECTION 1 • CULTURE

1.1 Climate and Culture .................................. 334
1.2 Trans-Siberian Railroad ........................... 336
1.3 St. Petersburg Today ................................ 338

**KEY VOCABULARY**
- culture
- nomad
- yurt
- terrain
- gauge
- port
- diplomacy

**ACADEMIC VOCABULARY**
enlist

**TERMS & NAMES**
- Trans-Siberian Railroad
- Hermitage Museum

**Essential Question** How have Russia and the Eurasian republics dealt with recent political, economic, and environmental challenges?

### SECTION 2 • GOVERNMENT & ECONOMICS

2.1 The Soviet Collapse ................................. 340
2.2 Russia's Government ............................... 342
2.3 Unlocking Energy Riches ......................... 344
   **GLOBAL ISSUES: POLLUTION**
2.4 After Chernobyl ....................................... 346

**KEY VOCABULARY**
- perestroika
- glasnost
- coup
- federal system
- proportional representation
- revenue
- pipeline
- radioactive
- fallout
- half-life
- contaminate

**ACADEMIC VOCABULARY**
autonomy, vulnerable

**TERMS & NAMES**
- Russification
- Kremlin
- Chernobyl

Uzbek women and girls from a mountain village wear a *rumol*, a traditional head scarf.

# SECTION 1 CULTURE

## 1.1 Climate and Culture

**TECHTREK**
myNGconnect.com For photos reflecting climate and culture in Russia and the republics

Digital Library

**Main Idea** The variety of climates in Russia and the Eurasian republics has a major impact on the region's cultures.

**Culture** refers to a group's unique way of life. Cultural traits include food—what people eat and how they obtain it—shelter, clothing, religion, and language. The first three traits are influenced by climate.

### Enduring the Cold in Siberia

About 30 different native peoples live in the tundra and taiga of Siberia in Russia. These include the Yakut, Nenets, Evenki, Chukchi, and Inuit (IHN yoo iht). All of these people have similar cultures but different languages.

People living in these cold climates, with their long winters and short, hot summers, traditionally dress in clothes made from reindeer fur. Some people, such as the Nenets, still live as **nomads**, moving from place to place according to the seasons in search of food. This way of life is changing, however, and today many native people live in towns or cities.

### Herding in Central Asia

Like the Nenets in Siberia, some herders in Kazakhstan and other parts of Central Asia also continue to lead nomadic lives. In the region's dry climate, this traditional culture focuses on herding sheep, camels, cattle, and horses. Animals provide food and milk as well as hides and wool for clothing and tents. Although herding continues to be important in this region, many herders now live in rural villages in houses made of mud bricks that have been baked by the sun.

### Farming in the Steppes

On the steppes of Russia and parts of Central Asia, the rich soil and moderate climate—cold winters and warm summers—are suited to agriculture. A peasant farming culture developed in this climate. In this culture, people grew grain crops, especially wheat, and raised livestock. They lived in settled villages. Later they worked for landlords on large estates and became serfs.

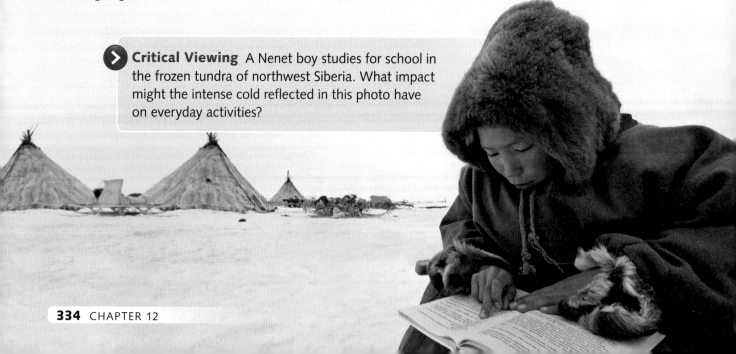

> **Critical Viewing** A Nenet boy studies for school in the frozen tundra of northwest Siberia. What impact might the intense cold reflected in this photo have on everyday activities?

## ADAPTING TO CLIMATE

**Cold**

**SIBERIA** Many native people of Siberia center their culture around reindeer. The warm, durable clothing worn by this family is made from reindeer skin. Reindeer meat and fat also provide nutritious food.

**Dry**

**CENTRAL ASIA** The animals the Kazakhs of Central Asia raise provide almost everything the herders need to live. Felt, or wool fibers pressed together, is used to make hats and traditional tents called yurts.

**Moderate**

**STEPPES** Life for farmers on the steppes revolves around the seasonal growing cycle. Farmers often use simple tools to grow their crops. Animals help with the work and provide meat and milk.

Today, large corporations or individuals own the farms. Farmers still grow wheat as in the past, and they grow other grain crops as well, including maize, or corn, and barley. Many of these grain products are exported to other countries. Beets and potatoes are also common crops, along with sunflowers, which are grown to make oil used in cooking.

### Before You Move On
**Summarize** What impact have the variety of climates in Russia and the Eurasian republics had on the region's cultures?

## ONGOING ASSESSMENT
# PHOTO LAB

1. **Analyze Visuals** The Central Asian yurt shown above is movable. Why might this be an important consideration for nomadic herders?

2. **Compare and Contrast** Study all three photos above. In what ways does the women's clothing reflect the climate in each region?

3. **Human-Environment Interaction** In what ways has climate influenced traditional culture in Siberia, Central Asia, and the steppes?

SECTION 1.1 **335**

# SECTION 1 CULTURE

## 1.2 Trans-Siberian Railroad

**Main Idea** The Trans-Siberian Railroad links the western and eastern parts of Russia.

The **Trans-Siberian Railroad** is the world's longest continuous railroad, spanning about 6,000 miles and crossing eight time zones. *Trans-Siberian* means "across Siberia." The railroad links Moscow with eastern Russia. It also carries people and goods to East Asia and Europe.

### Building the Railroad

Russia began to build the railroad in 1891. It was designed to connect Moscow with Vladivostok (vlah duh VAWH stock), a busy port on the Pacific Ocean. At that time, no reliable form of transportation linked this far eastern port with the European part of Russia.

Laborers began at both ends of the route and worked their way toward the center, but the climate and **terrain**, or physical features of the land, made progress very difficult. Workers had to lay tracks across long stretches of permafrost, and the route went through mountains, forests, rivers, and lakes. Materials—including explosives used to blast through rocks and cliffs—had to be brought thousands of miles to the work sites.

The construction project required many workers. As a result, thousands of Russian peasants, convicts, and soldiers were **enlisted**, or selected, to labor on the railroad. They were given only picks and shovels to do their difficult work. Human workers and horses hauled the heavy materials. The laborers worked long hours in extreme heat in the summer and extreme cold in the winter. They also had to deal with attacks by thieves and, occasionally, tigers. The main route, running from Moscow to Vladivostok, was finally completed in 1916.

**Critical Viewing** A Trans-Siberian Railroad train runs along the banks of Lake Baikal in Siberia. In this photo, the conductor looks out of the train as it curves around the lake. What challenges did the workers probably face while building this part of the railroad?

## ROUTES OF THE TRANS-SIBERIAN RAILROAD

### Effects of the Railroad

The railroad transformed Siberia and its traditional culture. Between 1891 and 1914, more than five million people immigrated to Siberia. New towns and cities grew up along the train's route. Soviet leaders began to industrialize Siberia and mine its plentiful raw materials. During World War II, the railroad moved Soviet troops and materials across Russia.

### The Railroad Today

Today, the railroad operates several more routes and has replaced all of the old steam engines with electric trains that carry passengers and freight. The railroad also plays an important role in the world economy. Container cargo, or goods packed in large steel boxes, travels from China and other parts of East Asia to Europe. One challenge is that the **gauge**, or width of the tracks, in Russia is wider than in Europe or China. As a result, containers need to be transferred to different trains at the borders. Still, shipping containers by land across Russia is much faster—and cheaper—than shipping by sea.

### Before You Move On

**Monitor Comprehension** In what ways has the Trans-Siberian Railroad helped to link Russia?

---

### ONGOING ASSESSMENT
## MAP LAB
 GeoJournal

1. **Interpret Maps** Use the map scale to determine the distance from Novosibirsk to Irkutsk on the Trans-Siberian Railroad.

2. **Movement** Trace the Trans-Siberian Railroad route to China. What benefit does it offer to manufacturers of consumer goods in China?

3. **Make Inferences** Why are there many tunnels and bridges along the route of the Trans-Siberian Railroad?

SECTION 1.2

# SECTION 1 CULTURE
## 1.3 St. Petersburg Today

**Main Idea** St. Petersburg is Russia's second largest city and a center of culture, industry, and trade.

St. Petersburg is located in northwest Russia on the Neva River. The river flows into the Gulf of Finland, which is at the most eastern part of the Baltic Sea. Because of its far northern location—at a latitude of about 60°N—St. Petersburg has long winter nights. In fact, for about one month a year, there's barely any daylight. For about three weeks in summer, the sky never gets completely dark. Special music and dance events are held during these "White Nights" to take advantage of the long days.

### Window to the West

St. Petersburg originally sat on isolated swampland, making it a poor place on which to build. Nevertheless, Peter the Great chose the site in 1703 to gain a **port**, or harbor, on the Baltic Sea for trade. He also wanted to create a modern city—for the times—that would resemble those in Western Europe and become Russia's "window to the West." Peter got his wish. He had St. Petersburg filled with islands, canals, and bridges like those in the western European cities of Amsterdam and Venice. He also built wide boulevards like those in Paris and London.

| FAST FACTS ON ST. PETERSBURG | |
|---|---|
| Population | 5 million people |
| Land Area | 550 square miles |
| Location | Average temperature in January: 21°F<br>Average temperature in July: 65°F<br>Latitude: 59° 57'N; longitude: 30° 19'E |
| Date Founded | 1703 |

St. Petersburg's historic architecture also reflects western European influences. Many of the buildings from the 1700s still survive, including the Winter Palace, **Hermitage Museum**, and summer palaces of several czars. St. Petersburg is also filled with many beautiful and historic Russian Orthodox churches and cathedrals in Western and Eastern style.

The city is Russia's cultural center. St. Petersburg boasts world-famous museums and ballet companies, such as the Kirov. It also contains many universities and theaters and the country's oldest music academy. In addition, St. Petersburg offers a variety of contemporary music, including jazz and rock.

### A Vibrant Economy

Manufacturing and construction are important industries in St. Petersburg. The city is also a center for trade and distribution of goods to and from Europe.

Above all, the local economy depends on tourism. More than three million visitors came to see the city's attractions in 2003 on its 300th anniversary. Many historic buildings that had been damaged during World War II were restored in time for the celebration. Vladimir Putin, Russia's president in 2003, is from St. Petersburg. He wanted the city to become a center of **diplomacy**, a place where international affairs could be conducted. Like Peter the Great, Putin wanted Russia to be more connected to the West.

### Before You Move On
**Summarize** In what ways is St. Petersburg a center of culture, industry, and trade?

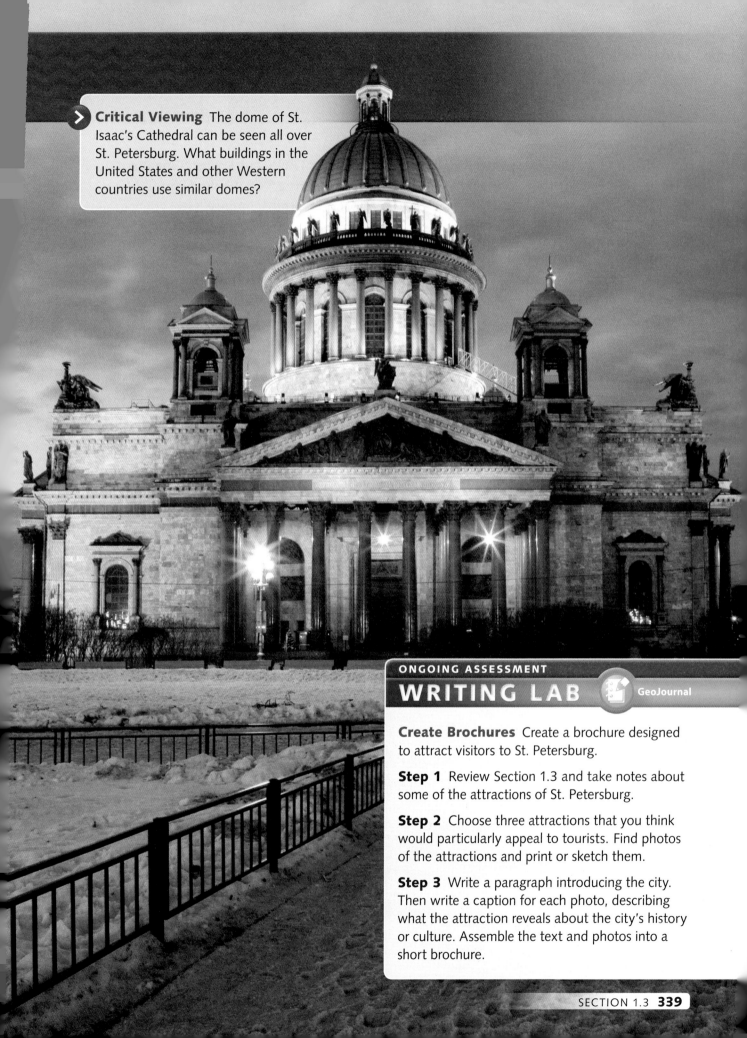

**Critical Viewing** The dome of St. Isaac's Cathedral can be seen all over St. Petersburg. What buildings in the United States and other Western countries use similar domes?

## ONGOING ASSESSMENT
## WRITING LAB  GeoJournal

**Create Brochures** Create a brochure designed to attract visitors to St. Petersburg.

**Step 1** Review Section 1.3 and take notes about some of the attractions of St. Petersburg.

**Step 2** Choose three attractions that you think would particularly appeal to tourists. Find photos of the attractions and print or sketch them.

**Step 3** Write a paragraph introducing the city. Then write a caption for each photo, describing what the attraction reveals about the city's history or culture. Assemble the text and photos into a short brochure.

# SECTION 2 GOVERNMENT & ECONOMICS
## 2.1 The Soviet Collapse

**Main Idea** Economic problems and people's desire for independence caused the Soviet Union to collapse in 1991.

As you have learned, the Soviet Union formed in 1922. During the 1970s and 1980s, the economy of the huge region slowed. Soviet money had almost no value outside the country. Stores had little food on their shelves and few goods to buy. In addition, people in the Eurasian republics had long resented the Soviet policy of **Russification**. Under this policy, the Soviet Union moved Russians into the republics and put them in charge. They also forced local people to learn the Russian language.

### Gorbachev Brings Reform

In 1985, Mikhail Gorbachev became head of the Soviet Communist Party and began to introduce reforms. He promoted a movement known as **perestroika** (pehr ih STROY kuh), which means "restructuring." Gorbachev wanted to restructure the economy. Less government control, he believed, would make the economy more effective.

After the economy had failed to improve by 1990, the government presented a new plan to change the economy in 500 days. According to this plan, the republics would have greater control over their economies, and state ownership of businesses would decline.

Gorbachev also introduced the policy of **glasnost** (GLAHS nuhst), or "openness," which encouraged people to speak openly about the government. This freedom of expression went beyond what Gorbachev had intended, however. People started to criticize and protest against the central government, and they began to demand even more freedom.

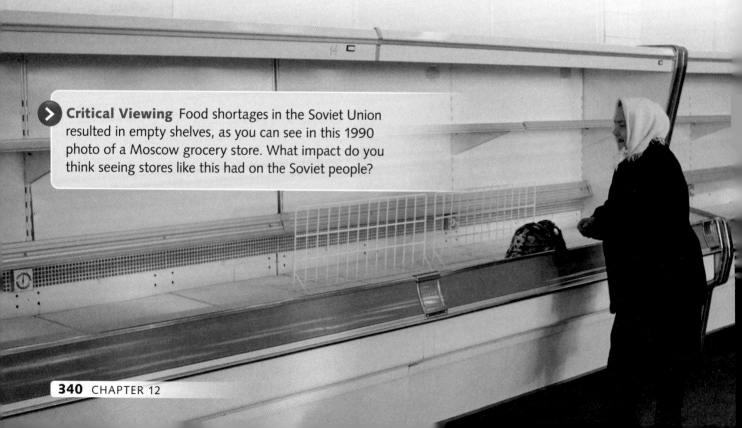

> **Critical Viewing** Food shortages in the Soviet Union resulted in empty shelves, as you can see in this 1990 photo of a Moscow grocery store. What impact do you think seeing stores like this had on the Soviet people?

## The Soviet Union Falls

After the collapse of the Berlin Wall in Germany in November 1989, the Communists no longer had control over Eastern Europe. People in many of the Soviet republics also wanted freedom from the central Communist government. Weakened by the economy, the government could no longer manage its vast empire. By the fall of 1990, all of the republics had declared their **autonomy**, or determination to govern themselves.

In August 1991, conservative Communists who opposed Gorbachev's political and economic reforms attempted a coup against him. A **coup** (KOO) is a sudden overthrow of the government by force. The coup failed, but Gorbachev knew he had lost power. On December 25, 1991, he resigned as president, and the Soviet Union dissolved. All of the republics became independent countries.

## Life After Independence

Some groups within the newly independent republics and Russia's own borders sought independence for themselves. In Azerbaijan, Armenians continue to fight over a part of the country where most of the people are ethnic Armenians. Within Russia, the Muslim republic of Chechnya (CHEHCH nee uh) has struggled to gain full independence.

The economic transition has been difficult as well. Less government control in Russia has led to a large gap between rich and poor, higher prices, unemployment, and an increase in corruption and organized crime. Some of the Eurasian republics still have centrally controlled economies and governments.

### Before You Move On
**Make Inferences** In what ways did attempts to reform the Soviet economy and political system contribute to the collapse of the Soviet Union?

---

**ONGOING ASSESSMENT**

## SPEAKING LAB  GeoJournal

**Turn and Talk** With a partner, discuss the causes and effects of the Soviet Union's collapse. Use a cause-and-effect chart like the one below to list and organize your ideas.

| CAUSES | EFFECTS |
|---|---|
| 1. | |
| 2. | |

# SECTION 2 GOVERNMENT & ECONOMICS
## 2.2 Russia's Government

**Main Idea** Russia's central government consists of three branches and, together with the president, holds most of the power.

Following the collapse of the Soviet Union, Russia adopted a new constitution. This set up a **federal system**, with a strong central government and local government units. The United States also has a federal system. Russia's government is more democratic than that of the former Soviet Union. Everyone 18 years and older can vote, and there are several political parties. However, the central government and the president hold most of the power.

### Presidential Power

In Russia, the president is the head of the executive branch and the most powerful government leader. Vladimir Putin served as president from 2000 to 2008. In 2008, Dmitry Medvedev (med VYED if) was elected president. He served one term.

Medvedev appointed Putin prime minister, so Putin was not far from power. In March 2012, Putin was reelected president. Opposition candidates and their parties in Russia are not strong, but there were protests to Putin's election.

### Legislators and Judges

There are two houses in the legislative branch. The lower house is the State Duma. Members are elected based on **proportional representation**. Under this system, a political party gets the same percentage of seats as the percentage of votes it received. A party must receive at least seven percent of the vote to get seats. Putin's party, the United Russia Party, holds about 64 percent of the seats.

The upper house of the legislative branch is the Federation Council. Executive and legislative leaders in each of the local government units appoint the members of this house. The State Duma is the more powerful legislative house. All bills must first be considered in the State Duma, even those proposed by the upper house.

> **Critical Viewing** The **Kremlin** is a historic complex of palaces and churches in the heart of Moscow. What feature in the photo suggests that the Kremlin once served as the city's fortress?

### COMPARE RUSSIAN AND U.S. GOVERNMENTS

| Russia | Government Branches | United States |
|---|---|---|
| • President, elected to a six-year term  Prime Minister, appointed<br>• Government Ministers, appointed | Executive | • President and Vice President, both elected to a four-year term<br>• Cabinet, appointed |
| • Federal Assembly:<br>  – Federation Council (166 members), appointed to a four-year term<br>  – State Duma (450 members), elected to a five-year term | Legislative | • Congress:<br>  – Senate (100 members), elected to a six-year term<br>  – House of Representatives (435 members), elected to a two-year term |
| • Constitutional Court Judges, appointed for life | Judicial | • Supreme Court Judges, appointed for life |

The highest court in Russia is the Constitutional Court. The Federation Council appoints these judges based on the president's recommendations. The judges are appointed for life. In general, the judicial branch in Russia is more **vulnerable**, or open, to political pressure than in most Western democracies. That means that officials in the executive and legislative branches are sometimes able to influence the judges.

## Central Control

The central government in Moscow still tries to control most levels of government. Those in power generally choose the people they want to be elected.

In 2000, then-president Putin reduced the number of local government units in Russia from 89 to 7 to increase central control of them. The president already exercised some power over these units because he nominated their governors.

### Before You Move On
**Summarize** What branches make up the central government, and in what ways do the government and the president exercise their power?

### ONGOING ASSESSMENT
## DATA LAB  GeoJournal

1. **Identify** Based on the chart, which officials in the Russian executive branch probably serve a role similar to that of the Cabinet members in the U.S. government?

2. **Compare and Contrast** Study the chart. In what way is the legislative branch in Russia similar to that in the United States? In what way is it different?

3. **Express Ideas Through Speech** Is the government in Russia more or less democratic than that in the United States? Think about which officials are elected or appointed and how power is handled. Discuss with a partner.

SECTION 2.2 **343**

# SECTION 2 GOVERNMENT & ECONOMICS

## 2.3 Unlocking Energy Riches

**TECHTREK**
myNGconnect.com For an online map and photos of the region's energy pipelines

 Maps and Graphs    Digital Library

**Main Idea** Oil and natural gas enrich the economies of Russia and several countries around the Caspian Sea.

In addition to reforming its government, Russia has worked to develop its economy. Oil and natural gas are its greatest sources of wealth. In fact, the country is the largest exporter of oil and natural gas in the world. These resources account for about two-thirds of the value of all Russia's exports and one-third of its **revenue**, or income.

### Siberian Boom Towns

About 70 percent of Russian oil comes from western Siberia. Since the end of the Soviet Union, the region has been booming, or growing rapidly. Workers from western Russia and Central Asia come to Siberia for good-paying jobs. They earn enough to buy apartments in cities such as Surgut. New suburbs have developed, and many people enjoy a higher standard of living. Oil wealth has also helped fund new airports, museums, and schools.

### Caspian Sea Riches

Some experts believe that the Caspian Sea may actually have more energy reserves than the Persian Gulf. Russia would like to control the energy from the Caspian. Russia, Turkmenistan, and Kazakhstan signed an agreement to build a pipeline to transport natural gas from the Caspian through Kazakhstan into Russia. From there, Russia exports the gas to Europe, but Russia's state-owned pipeline network does not control all of the region's exports.

New pipelines carry gas from Turkmenistan to China and Iran. Kazakhstan also uses tanker and rail transport, reducing its dependency on Russia.

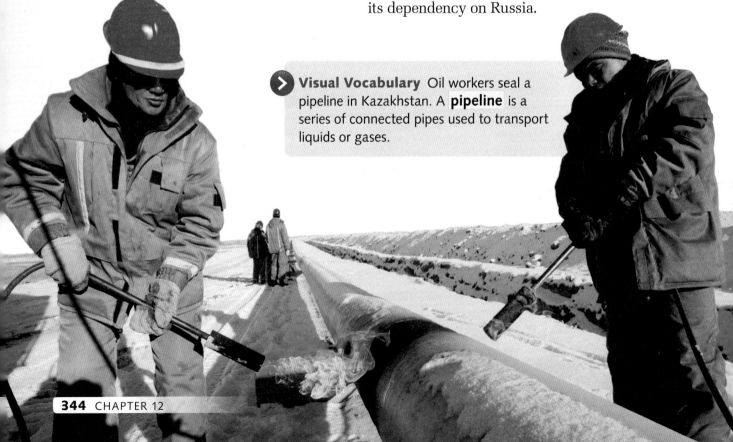

**Visual Vocabulary** Oil workers seal a pipeline in Kazakhstan. A **pipeline** is a series of connected pipes used to transport liquids or gases.

## OIL AND NATURAL GAS PIPELINES

Azerbaijan has large reserves of both oil and natural gas. It used to send the fuel through pipelines to Russia. Now it makes more money by trading mainly with the United States and by piping the fuel directly through Georgia and Turkey. By so doing, Azerbaijan does not have to share export profits with Russia.

For now, Russia gains a great deal of wealth from its exports of natural gas and oil. However, by relying too heavily on energy exports, the country risks using up these natural resources.

### Before You Move On
**Monitor Comprehension** In what ways have oil and natural gas enriched Russia and countries around the Caspian Sea?

### ONGOING ASSESSMENT
## MAP LAB

1. **Movement** About how far does natural gas travel by pipeline from Surgut in western Siberia to St. Petersburg in Russia? Why do you think pipelines might be a good way to transport the gas over long distances?

2. **Pose and Answer Questions** Come up with your own question about the information shown on the map. Then challenge a partner to answer your question.

3. **Draw Conclusions** What would Russia gain by controlling the energy from the Caspian Sea?

4. **Make Inferences** Why are countries in Central Asia developing pipelines to transport oil and natural gas directly to China and Iran?

# SECTION 2 GLOBAL ISSUES • Pollution
## 2.4 After Chernobyl

**Main Idea** The Chernobyl nuclear disaster severely damaged the environment in parts of Belarus, Ukraine, and western Russia.

On April 26, 1986, a nuclear reactor at a power plant in Chernobyl exploded and caught fire, resulting in the worst nuclear disaster in history. **Chernobyl** is in the country of Ukraine. At the time of the disaster, Ukraine was part of the Soviet Union. A **radioactive** cloud about 3,280 feet high spread over parts of Ukraine, Belarus, and Russia. Winds carried the **fallout** into parts of northern and central Europe. These radioactive materials have caused harm to humans, animals, and plants.

### Health Effects

Scientists know that it takes some radioactive materials a long time to break down and disappear. Each material has a particular **half-life**, the time needed for half of its atoms to decay and decrease. For example, the highly radioactive element cesium has a half-life of 30 years. That means that after 30 years, half of its atoms will still be radioactive.

Thirty people died within three months of the accident, most from being exposed to huge amounts of radioactive material. Over time, thousands of the people who helped clean up the disaster have developed health problems. Some of the children born to exposed parents carry the effects in their genes. In addition, radioactive iodine got into the milk of cows that grazed on **contaminated**, or infected, grass after the accident. This caused an increase in thyroid cancer, especially in children. Millions of people are still living on contaminated land. Officials cannot yet predict the long-term effects on people's health.

### Before You Move On
**Summarize** What are some of the health problems that occurred as a result of the Chernobyl nuclear disaster?

### KEY VOCABULARY

**radioactive**, adj., giving off energy caused by the breakdown of atoms

**fallout**, n., radioactive particles from a nuclear explosion that fall through the atmosphere

**half-life**, n., the time needed for half the atoms in a radioactive substance to decay and decrease

**contaminated**, adj., unfit for use because of the presence of unsafe elements

### CHERNOBYL DISASTER BY THE NUMBERS

**400**
Estimated number of atomic bombs it would take to equal the accident

**4,000**
Estimated number of people who may die from cancers caused by radiation exposure from the accident

**600,000**
Estimated number of people who received significant radiation exposure, including evacuees, residents, and those who helped clean up after the accident

**5 million**
Estimated number of people still living on contaminated areas in Ukraine, Belarus, and Russia

Source: UN Chernobyl Forum, 2006

## Environmental Damage

Belarus was hardest hit because of the direction the wind blew just after the accident. About 23 percent of the country was contaminated, including agricultural and forest land. In Ukraine, 7 percent of the land and 40 percent of its forests were contaminated. In Russia, the area bordering with Belarus was most affected.

Radiation is expected to remain in the soil for many years. Plants that grow in the region's forests are contaminated and so are the animals that feed on them. Many reindeer herds had to be killed shortly after the accident because they were so infected by the radiation.

The fallout has especially contaminated the fish in the rivers and lakes of Ukraine because these waters flow down from the site of the disaster. Some radiation that has gotten into groundwater sources will last for hundreds of years.

**COMPARE ACROSS REGIONS**

## France's Nuclear Program

In spite of what happened at Chernobyl, many countries have continued to develop nuclear energy. France, for example, has more than 55 nuclear plants that generate more than 75 percent of its electricity. Many of the French people welcome a new nuclear plant in their town because it brings jobs and prosperity to the area.

The French are aware of the dangers of nuclear energy plants, but they have little fear. Many of the plants offer tours, and advertisements help reinforce the idea that nuclear energy is a fact of life in France.

Few humans live in the 18-mile fenced area around the Chernobyl reactor now.

As a result, the people know that the plants are technologically more advanced and safer than the one in Ukraine. The French scientists and engineers who built them learned the lesson of Chernobyl.

### Before You Move On
**Monitor Comprehension** What has been learned from the Chernobyl disaster?

**ONGOING ASSESSMENT**
## READING LAB  GeoJournal

1. **Summarize** What damage did the Chernobyl disaster cause to the environment?
2. **Make Predictions** In what ways do you think the disaster will continue to affect people in Ukraine, Belarus, and Russia?
3. **Write Comparisons** Imagine that two new nuclear plants were being proposed: one in Russia and one in France. What do you think would be the reactions of the townspeople in each country? Write two sentences comparing their reactions. Make sure that you explain what led you to believe that the people in each country would feel that way. Go to **Student Resources** for Guided Writing support.

# CHAPTER 12
# Review

## VOCABULARY

For each pair of vocabulary words, write one sentence that explains the connection between the two words.

1. nomad; yurt

   *For centuries, nomads in Central Asia have traditionally lived in felt tents called yurts.*

2. perestroika; glasnost
3. federal system; proportional representation
4. radioactive; contaminated
5. fallout; half-life

## MAIN IDEAS

6. Describe some of the different types of climate in Russia and the Eurasian republics. (Section 1.1)
7. In what ways has Russia benefited from the Trans-Siberian Railroad? (Section 1.2)
8. Why is St. Petersburg called Russia's "window to the West"? (Section 1.3)
9. How did installing Russian officials in the republics help bring about the collapse of the Soviet Union? (Section 2.1)
10. In what way did Gorbachev's changes to the government contribute to the fall of the Soviet Union? (Section 2.1)
11. Which branch of Russia's central government has the most power? Which branch has the least power? (Section 2.2)
12. Why is western Siberia said to be "booming"? (Section 2.3)
13. What was the impact of the Chernobyl nuclear disaster on Ukraine, Belarus, and Russia? (Section 2.4)

## CULTURE

### ANALYZE THE ESSENTIAL QUESTION

What factors, such as size and climate, have influenced Russian culture?

### Critical Thinking: Make Inferences

14. What impact does climate have on nomadic herders in Siberia and Central Asia?
15. Why was it difficult to build the Trans-Siberian Railroad across Russia?
16. At what time of year might tourists prefer to visit St. Petersburg? Why?

### INTERPRET MAPS

LANGUAGES IN RUSSIA

17. **Region** What are the two major language families in Russia?
18. **Compare and Contrast** What do the languages spoken west and east of the Ural Mountains have in common? In what ways do they differ?

## GOVERNMENT & ECONOMICS

### ANALYZE THE ESSENTIAL QUESTION

How have Russia and the Eurasian republics dealt with recent political, economic, and environmental challenges?

**Critical Thinking: Make Generalizations**

19. What impact did the collapse of the Soviet Union have on Russia's role in the region?

20. Why did Russia's central government reduce the number of local government units?

21. In what ways do Russia's location and power help the country control energy resources in the region?

22. Why was Russia less affected by the Chernobyl nuclear disaster than Ukraine and Belarus?

**INTERPRET TABLES**

| NATURAL GAS RESERVES | | |
|---|---|---|
| Country | Amount (cu m)* | World Rank |
| Russia | 47.6 trillion | 1 |
| Iran | 33 trillion | 2 |
| Qatar | 25.2 trillion | 3 |
| Turkmenistan | 24.3 trillion | 4 |
| Saudi Arabia | 8.0 trillion | 5 |
| United States | 7.7 trillion | 6 |
| United Arab Emirates | 6.0 trillion | 7 |

Source: CIA World Factbook, 2012 (est.) / *cu m = cubic meters

23. **Analyze Data** According to the table, about how much more natural gas reserves does Russia have than the United States?

24. **Evaluate** What does this table suggest about where most of the world's natural gas reserves can be found?

25. **Draw Conclusions** Why do you think Ukraine imports natural gas from Russia?

## ACTIVE OPTIONS

**Synthesize** the Essential Questions by completing the activities below.

26. **Write Journal Entries** Write several journal entries from the point of view of a worker on the Trans-Siberian Railroad. Describe the difficulties of the work and the obstacles you encounter. Use the writing tips below to help you write your entries. **When you have finished, trade your entries with a partner and compare them.**

> **Writing Tips**
> - Include details on where you worked, the tools you used, and the conditions you endured.
> - Describe any unexpected experiences, such as an encounter with thieves or wild animals along the tracks.
> - Express your feelings about having been enlisted to carry out the work.

27. **Create Charts** Make a chart in which you compare the size, population, and major religions of Russia and the Eurasian republics. Include Kazakhstan, Kyrgyzstan, Tajikistan, Turkmenistan, and Uzbekistan in your chart. Use online sources to gather the data. With a partner, discuss what the biggest similarities and differences are between these countries.

| COMPARE COUNTRIES | | | |
|---|---|---|---|
| Country | Size | Population | Major Religions |
| Russia | | | |
| Armenia | | | |
| Azerbaijan | | | |
| Georgia | | | |

REVIEW **349**

COMPARE ACROSS REGIONS **UNIT 6**

# Natural Resources and Energy

**TECHTREK**
myNGconnect.com For an online graph on energy sources

Maps and Graphs

In this unit, you learned that Russia and many of the Eurasian republics have plentiful supplies of natural gas and oil. In fact, Russia has the world's largest supply of natural gas. Russia's Asian neighbor, China, is the world's biggest producer of coal. These energy sources are in great demand from countries all over the world.

The choices people make about energy are critical to a healthy planet. As you have learned, some energy sources pollute the environment as they burn. Many scientists believe that the carbon dioxide those sources give off contributes to an overall global temperature increase. To ease these problems, they would like to reduce the use of fossil fuels and develop nonfossil energy sources.

**Compare**
- Brazil
- China
- Mexico
- Russia
- United States

## FOSSIL FUELS

Fossil fuels are formed by buried plants and animals that have been dead for millions of years. This type of fuel is found in deposits beneath the earth's surface and must be burned to release its energy. These fuels supply about 85 percent of the world's energy. Fossil fuels include coal, oil, and natural gas.

These energy sources are classified as nonrenewable because they take millions of years to form and supplies are being used up faster than new ones can redevelop. Although natural gas is relatively clean—meaning that it doesn't create much pollution—fossil fuels, especially coal, tend to be dirty and release a great deal of harmful carbon dioxide into the air.

## NONFOSSIL FUELS

Nonfossil fuels are alternative sources of energy and include the following: hydroelectric power from water, nuclear power, wind power, solar energy from the sun, and biofuels made from vegetable oil.

Many of these energy sources are renewable because they can be quickly replenished. Some countries use significant amounts of nuclear energy and hydroelectric power. However, the other nonfossil fuels make up only two percent of the world's energy.

Russia and China use both fossil and nonfossil fuels. The graphs on the next page show the energy consumption in both countries. Compare the data in the graphs and use it to answer the questions.

**350** UNIT 6

# ENERGY CONSUMPTION PER COUNTRY

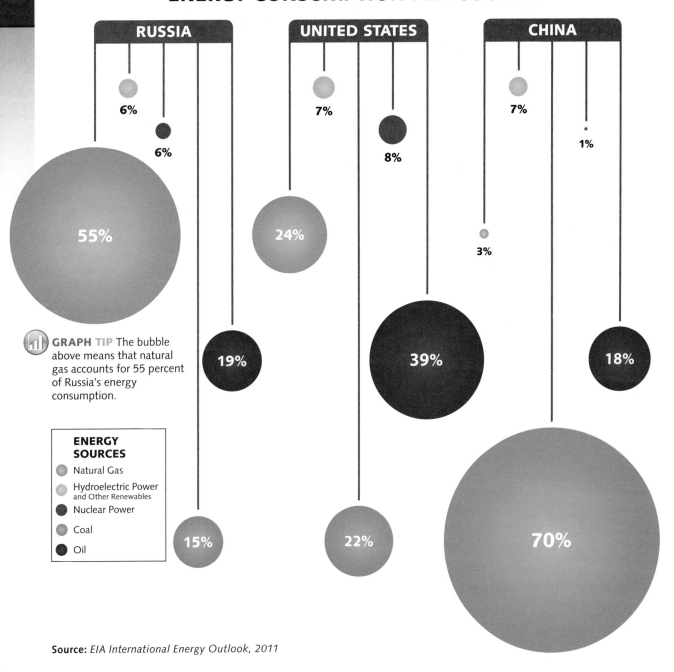

**GRAPH TIP** The bubble above means that natural gas accounts for 55 percent of Russia's energy consumption.

**ENERGY SOURCES**
- Natural Gas
- Hydroelectric Power and Other Renewables
- Nuclear Power
- Coal
- Oil

**Source:** EIA International Energy Outlook, 2011

## ONGOING ASSESSMENT
### RESEARCH LAB  GeoJournal

1. **Explain** What is the main source of energy in each country? Why might that be the case?
2. **Analyze Data** What is the energy consumption of nonfossil fuels in each country? What does this suggest about the development of alternative energy in Russia and China?

**Research and Make Comparisons** Research energy consumption in Brazil and Mexico and then compare with that in Russia and China. Which country appears to be investing more in nonfossil and renewable energy sources? What might this statistic suggest about its pollution levels?

COMPARE ACROSS REGIONS **351**

# REVIEW & EXTEND UNIT 6
# Active Options

**TECHTREK**
myNGconnect.com For photos of natural and cultural sites in Russia and the Eurasian republics and other UNESCO World Heritage sites

Digital Library

## ACTIVITY 1

**Goal:** Extend your understanding of animals native to Russia and the Eurasian republics.

### Design a Poster for Native Animals

Russia is home to both land and sea animals. Some of these animals, including those listed on the right, have been over-hunted or are considered undesirable neighbors. Research one of these animals and use the information to design a poster that will help viewers appreciate it. Consider the animal's habitat, surviving population, and efforts made to protect it.

- Blue whale
- Brown bear
- Gray wolf
- Polar bear
- Siberian tiger

Polar bears

## ACTIVITY 2

**Goal:** Research the natural and cultural heritage of Russia and the Eurasian republics.

### Write a World Heritage Guide

The United Nations Educational, Scientific and Cultural Organization (UNESCO) has identified hundreds of sites that form part of the world's natural and cultural heritage. A variety of these World Heritage sites are located in the Russia Federation and the Eurasian republics. Prepare a guide for students visiting the area. Choose five to ten UNESCO World Heritage sites in the region. Provide information for each site. If desired, use the **Magazine Maker CD-ROM** as a template.

## ACTIVITY 3

**Goal:** Learn about Russian culture through the country's folk tales.

### Hold a Russian Folk Tale Festival

Folk tales, such as "The Old Man and the Bear" and "Baba Yaga," are an age-old part of the Russian culture. With a group, find some Russian folk tales online and read them. Decide which ones you will present in the festival. Then decide how you will present the tales. You may each choose to retell a story, or you may decide as a group to act out one or more of the tales.

# UNIT 7
# Explore SUB-SAHARAN AFRICA
## with NATIONAL GEOGRAPHIC

### MEET THE EXPLORER

**NATIONAL GEOGRAPHIC**

Emerging Explorer Kakenya Ntaiya, educator and activist (shown at center), founded the first primary school for girls in her village. Her work provides African girls with an education and an opportunity to develop leadership skills.

### INVESTIGATE GEOGRAPHY

Giraffe mothers and their young run in Serengeti National Park, Tanzania. The Serengeti, connected to the Masai Mara National Park in neighboring Kenya, is the only place in Africa where large animal herds, including gnu, zebras, and gazelles, still make vast land migrations.

### STEP INTO HISTORY

This ancient Dogon village was built into the sandstone cliffs of Mali, West Africa. Today, most Dogon people are farmers, and practice the tradtional religion of their ancestors. Much of their social organization and cultural practices relates to this belief system.

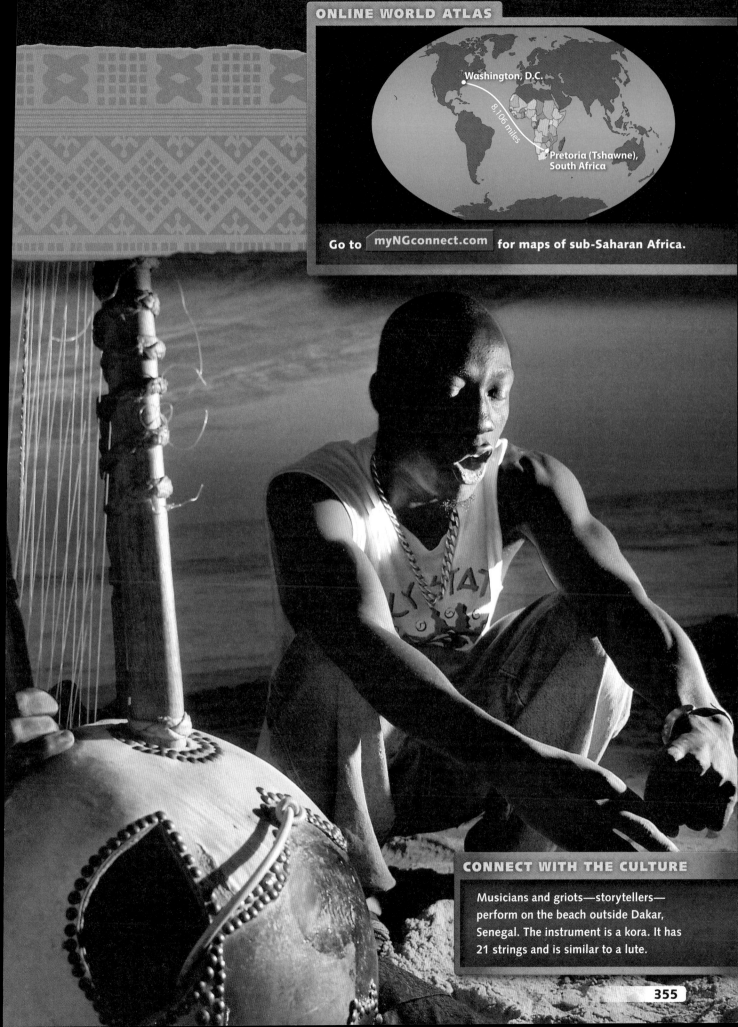

## ONLINE WORLD ATLAS

Go to myNGconnect.com for maps of sub-Saharan Africa.

## CONNECT WITH THE CULTURE

Musicians and griots—storytellers—perform on the beach outside Dakar, Senegal. The instrument is a kora. It has 21 strings and is similar to a lute.

CHAPTER 13

# Sub-Saharan Africa
## GEOGRAPHY & HISTORY

## PREVIEW THE CHAPTER

**Essential Question** How has the varied geography of sub-Saharan Africa affected people's lives?

### SECTION 1 • GEOGRAPHY

1.1 Physical Geography .................... 358
1.2 East Africa and the Rift Valley ................ 360
1.3 West Africa's Steppes .......................... 362
1.4 Rain Forests and Resources ...................... 364
1.5 Southern Plateaus and Basins ................. 366
1.6 Exploring Africa's Wildlife ....................... 368

**KEY VOCABULARY**
- basin
- savanna
- desertification
- rift valley
- pride
- interior
- deforestation
- transition zone
- highlands
- rain forest
- hydroelectric power
- landlocked
- escarpment
- habitat
- poaching
- nocturnal
- ecotourism

ACADEMIC VOCABULARY
dormant

**TERMS & NAMES**
- Sahel
- Kalahari
- Great Rift Valley
- Kilimanjaro
- Congo River
- Great Escarpment
- Zambezi River
- Okavango Delta

**Essential Question** How did trade networks and migration influence the development of African civilization?

### SECTION 2 • HISTORY

2.1 Bantu Migrations ...................... 370
2.2 Early States and Trade ............................ 372
2.3 Impact of the Slave Trade ....................... 374
2.4 Colonization to Independence ................. 376
   NATIONAL GEOGRAPHIC  Photo Gallery............ 378

**KEY VOCABULARY**
- agricultural revolution
- first language
- caravan
- *lingua franca*
- trans-Saharan
- alluvial
- city-state
- trans-Atlantic slave trade
- malnutrition
- imperialism
- colonialism
- missionary

ACADEMIC VOCABULARY
incentive

**TERMS & NAMES**
- Bantu
- Swahili
- Timbuktu
- Kongo
- Great Zimbabwe
- Middle Passage
- Berlin Conference
- Pan-Africanism
- Jomo Kenyatta
- Kwame Nkrumah

African elephants

# SECTION 1 GEOGRAPHY
## 1.1 Physical Geography

### SUB-SAHARAN AFRICA PHYSICAL

**Main Idea** Sub-Saharan Africa is divided into four parts, each with varied geographic features.

Sub-Saharan Africa lies south of the Sahara desert. The region extends from Senegal in western Africa to Ethiopia and Somalia in eastern Africa, and southward to the tip of the continent. Sub-Saharan Africa includes the island nation of Madagascar (ma duh GAS kahr).

## Africa's Continental Drift

As you have learned, the continents once belonged to one supercontinent called Pangaea. The movement of Earth's tectonic plates caused the continents to drift apart, and Africa became a separate continent. Plate tectonics also caused the formation of certain physical features on the African continent. Lakes, basins, and valleys were all formed as the plates moved. A **basin** is a region drained by a river system.

## Landforms and Water

Sub-Saharan Africa is divided into four parts: west, central, east, and south. West Africa includes savannas and much of the Sahel. The **Sahel** (saa HEL) is a semiarid grassland that separates the Sahara in the north from the tropical grasslands, or **savannas**, in the south. In the Sahel, one climate gradually changes to another. Parts of the Sahel, for example, are changing to desert. This process, called **desertification**, means there is less fertile land to grow food. Desertification is caused by many factors including climate change and overpopulation.

Central Africa's primary landform is rain forest, especially in the Congo Basin. In East Africa, **rift valleys**—deep valleys that formed when Earth's crust separated and broke apart—stretch from the Red Sea southward through Mozambique. Southern Africa features great plateaus and another major desert, the **Kalahari**. The Kalahari has limited surface water, but supports a variety of plants and wildlife.

Access to fresh water is an issue for sub-Saharan Africans due to limited water resources and poor sanitation. The situation is improving, however. For example, between 1990 and 2010, the percentage of people in Namibia with access to clean water increased from 57 percent to 93 percent as the government and individual communities began working together to solve the problem.

### Before You Move On
**Monitor Comprehension** What are the main geographic features of sub-Saharan Africa's four parts?

---

**ONGOING ASSESSMENT**
# MAP LAB  GeoJournal

1. **Human-Environment Interaction** According to the map, the approximate width of the Sahara is 3,000 miles from east to west. Why is the Sahara considered a natural boundary?

2. **Location** Use the map to point out two sources of fresh water in sub-Saharan Africa.

3. **Describe Geographic Information** Describe the causes of desertification and the problem it poses for sub-Saharan farming.

# SECTION 1 GEOGRAPHY

## 1.2 East Africa and the Rift Valley

**TECHTREK**

myNGconnect.com For online maps of East Africa and photos of the Great Rift Valley

 Maps and Graphs
 Digital Library

**Main Idea** East Africa is best known for the Great Rift Valley and the deep lakes found there.

East Africa is easy to spot on a map—it is shaped like a rhinoceros horn. In fact, the region is known as the Horn of Africa. The "horn" is formed by the countries of Somalia, Djibouti (ji BOO tee), Eritrea, and Ethiopia. Other countries in East Africa are Sudan, Kenya, Uganda, Rwanda, Burundi, and Tanzania.

### The Great Rift Valley

The most important physical feature of East Africa is the **Great Rift Valley**. This valley is part of a chain of valleys that stretch from southwest Asia to southern Africa. In some places the valley is 60 miles wide. Valley walls often rise more than 6,000 feet in height.

This chain of valleys was formed by tectonic plates that separated and created deep cracks, or rifts, in the earth's crust. The rift valleys have been forming for about 20 million years and continue to develop today.

Plate movements also created the freshwater lakes in the Great Rift Valley. As low spots and rifts developed, they slowly filled with rainwater to create lakes. West of Tanzania is Lake Tanganyika (TANG guhn YEE kuh). At 4,700 feet deep, it is the second deepest freshwater lake in the world.

### Plateaus and Savannas

East Africa sits mostly on plateaus. As the climate map shows, the two most substantial elevated areas are in Ethiopia and Kenya. The higher elevations of these areas means the temperatures are cooler, even though much of the area is on or near the equator.

In addition to the rift valleys and lakes, plate movements created volcanoes. Both **Kilimanjaro** (19,340 feet) in Tanzania and Mount Kenya (17,058 feet) in Kenya are **dormant**, or inactive, volcanoes. The volcanic soil around these mountains is fertile, meaning crops grow well.

> **Visual Vocabulary** A **pride**, or group, of lions moves through tall savanna grass in Kenya. A pride averages about 15 members.

360 CHAPTER 13

## EAST AFRICA POLITICAL

## EAST AFRICA CLIMATE

Both Tanzania and Kenya have vast savannas where wildlife such as lions, giraffes, and elephants roam freely or in protected reserves.

**Before You Move On**

**Summarize** Describe the physical characteristics of the Great Rift Valley.

### ONGOING ASSESSMENT
## MAP LAB

1. **Location** Using the maps, locate three sources of freshwater in East Africa.

2. **Interpret Maps** Based on the climate map, which countries have the largest variety of climate regions in East Africa? How does that information add to your understanding of this region?

3. **Make Inferences** How might climate, elevation, and access to water determine where people live in East Africa?

SECTION 1.2 **361**

# SECTION 1 GEOGRAPHY
## 1.3 West Africa's Steppes

**TECHTREK**
myNGconnect.com For an online map and photos of desertification

Maps and Graphs | Digital Library

> **Main Idea** West Africa's physical geography includes steppes and highlands as well as tropical coast and dry desert.

West Africa runs from the Atlantic coast south of the Sahara, eastward to the continent's **interior**, or area away from the coast. This part of Africa supports several different ways of life, depending on the landforms and climates found in each area.

### Steppes and Highlands

Semiarid steppes, or grasslands, define part of West Africa. As you know, the grassland between the Sahara desert and the tropical savannas is the Sahel. The Sahel runs through the middle of West Africa. The area has a short rainy season and, as a result, is very dry.

The growing population in West Africa has increased the demand for food crops. To create more cropland, West Africans have had to cut down forests, a practice called **deforestation**. Deforestation in West Africa has left the arid soil unprotected. Soil has eroded, or washed away, because of overuse. These conditions, along with climate change, contribute to desertification, an ongoing problem in sub-Saharan Africa.

**Highlands**, which are areas of higher mountainous land, are also found in West Africa. The Adamawa Highlands are on Nigeria's eastern border. The Futa Jallon Highlands rise in Senegal on its border with Guinea. This series of sandstone plateaus is marked by rugged canyons.

### Tropical Coast, Dry Desert

Coastal countries differ from interior countries in West Africa. Countries on the coast have more people and more cities than those in the interior. Coastal countries often have tropical climates. Some cities on the Gulf of Guinea get more than 80 inches of rain per year, and the coast of Côte d'Ivoire can get more than 10 feet per year. In contrast, the northern desert areas of Niger, an interior country, get less than 10 inches of rain per year.

> **Visual Vocabulary** A **transition zone** is an area between two geographic regions that has characteristics of both. The Sahel (shown here in Mali) is a transition zone between the Sahara in the north and the savannas in the south.

The largest cities in West Africa are located near or on the Atlantic coast. The coastal city of Lagos, Nigeria, is the largest city in West Africa and one of the largest in the world. Such coastal cities have many advantages. Adequate rainfall, fishing, and trade all make it easier for West African coastal cities to support growing populations.

The interior countries of West Africa are mostly covered by desert. Deforestation and desertification, along with the lack of water, make farming in the poor soil a constant struggle. About 80 percent of people in Chad make their living from subsistence farming and raising livestock. In 2003, Chad began exporting oil. This new economic resource may improve Chad's economy and make money available for improved farming technology.

### Before You Move On
**Monitor Comprehension** In what ways are West Africa's steppes different from the coasts?

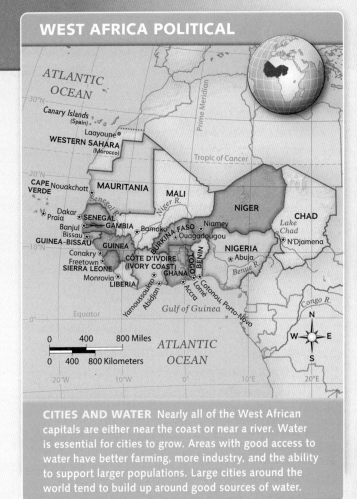

**CITIES AND WATER** Nearly all of the West African capitals are either near the coast or near a river. Water is essential for cities to grow. Areas with good access to water have better farming, more industry, and the ability to support larger populations. Large cities around the world tend to build up around good sources of water.

### ONGOING ASSESSMENT
## PHOTO LAB  GeoJournal

1. **Analyze Visuals** Based on the photo, what characteristics of the Sahel in Mali make it a transition zone?
2. **Location** What does the photo suggest about the challenges of living in this transition zone?
3. **Interpret Maps** What makes the capital of Nigeria different from other capitals in West Africa?

SECTION 1.3 **363**

### SECTION 1 GEOGRAPHY

# 1.4 Rain Forests and Resources

**TECHTREK**
myNGconnect.com For a land use map and photos of wildlife

 Maps and Graphs   Digital Library

**Main Idea** Central Africa is defined by the rain forests of the Congo River Basin and a variety of natural resources.

Central Africa is bordered by the Adamawa Highlands of West Africa and the Great Rift Valley of East Africa. On the north and south, Central Africa lies between the plateaus of the Sahel and Southern Africa.

## Rain Forest in the Congo Basin

The Congo Basin is the main geographic feature of Central Africa. The basin is located on the equator and surrounded by higher elevations. Within the basin is a **rain forest**, which is a forest with warm temperatures, plentiful rain, high humidity, and thick vegetation.

The rain forest in Central Africa is second in the world in size only to the South American rain forest. Vegetation grows so thick that sometimes sunlight does not reach the ground. The plants also make ground movement difficult, so most people live on the edges of the rain forest.

Spectacular wildlife also thrives in the rain forest. The okapi, related to the giraffe, lives in the Congo Basin. Some other rain forest animals include gorillas similar to the one shown here, leopards, and rhinoceroses.

The **Congo River** is a major waterway in Central Africa. Like the Amazon River of South America, the Congo is located in an equatorial area and flows into the Atlantic Ocean. Several large rivers in Central Africa feed into the Congo River, including the Ubangi, the Aruwimi, and the Lomami rivers.

## Resources of Central Africa

Central Africa includes the Central African Republic, Congo, Cameroon, São Tomé (sow too MAY) and Príncipe, Equatorial Guinea, and Gabon.

> **Critical Viewing** A western lowland gorilla walks in the forest in Democratic Republic of the Congo. Why might the rain forest be able to support so many plants and animals?

## CENTRAL AFRICA'S LAND USE

However, the largest, most populous country in Central Africa is the Democratic Republic of the Congo (DRC). The capital of the DRC is Kinshasa.

The DRC has a wealth of natural resources, including copper, forests, diamonds, and the Congo River itself. The river provides **hydroelectric power**, or electricity produced by a water source such as a river. The river is forced through turbines, or engines, and produces electricity for the DRC and other countries.

### Before You Move On
**Monitor Comprehension** Describe the rain forest and natural resources found in Central Africa.

## ONGOING ASSESSMENT
### MAP LAB

1. **Interpret Maps** How does the latitude of the Congo Basin help explain its climate and vegetation?

2. **Conduct Internet Research** Use the Internet to research the Amazon River Basin of South America. On a chart like the one below, compare and contrast the Congo River Basin and Amazon River Basin.

| | LOCATION | SIZE | RAINFALL |
|---|---|---|---|
| Congo River Basin | | | |
| Amazon River Basin | | | |

SECTION 1.4 **365**

# SECTION 1 GEOGRAPHY
## 1.5 Southern Plateaus and Basins

**TECHTREK**
myNGconnect.com For online maps and photos of Southern Africa

 Maps and Graphs   Digital Library

**Main Idea** Southern Africa's physical geography offers opportunities for economic development.

Southern Africa has fertile farmland, valuable natural resources, and abundant wildlife. The income from exporting natural resources gives Southern Africa the highest standard of living in sub-Saharan Africa.

### Basins and Plateaus

The Congo Basin extends into the Southern African countries of Angola and Zambia. From there, the land rises to a large plateau that spans most of Southern Africa. Six of the area's countries are **landlocked**, or have no direct access to a coast. Another major African basin, the Kalahari in Southern Africa, includes the Kalahari Desert. However, the basin also has areas rich with wildlife.

The plateau of Southern Africa is defined by the **Great Escarpment**. An **escarpment** is a steep slope. The Great Escarpment is the steep slope from the plateau down to the coastal plains of Southern Africa. The Great Escarpment is most dramatic in the countries of South Africa and Lesotho. It also extends northeast into Zimbabwe and northwest into Namibia and Angola.

The **Zambezi River** in Southern Africa collects water from the entire south-central part of Africa. The Zambezi flows through Angola, Zambia, along Zimbabwe's border, and through Mozambique (mo zam BEEK) to the Indian Ocean. The Kariba Dam on the Zambezi provides hydroelectric power. In fact, the countries of Zambia and Zimbabwe get most of their electricity from the dam.

**Critical Viewing** Miners near Johannesburg, South Africa, work deep underground to extract resources. Based on the photo, what are some of the dangers that miners might face?

366 CHAPTER 13

## Mining and Farming

A zone of mineral deposits winds through Zambia, Zimbabwe, and South Africa. Copper, gold, and diamonds are mined in this zone, and destined for jewelry and industrial uses. In fact, South Africa is one of the world's largest gold producers. Many people migrate there to work in the mines even though the work can be dangerous.

Southern Africa's temperate climate supports a variety of crops. For example, South Africa has many vineyards that thrive on its plateau, and Zimbabwe has tea plantations on its eastern escarpment. Angola produced nearly 20 percent of the world's coffee until 1975 when a civil war began. When the war ended in 2002, coffee production started again. Fruits, such as bananas, pineapples, and apples, are grown throughout Southern Africa. Corn, wheat, and other grains can also be found across the area.

South Africa, Botswana, and Namibia have the highest Gross Domestic Product in Southern Africa. Other countries are working to overcome factors that have weakened their economies, such as civil war and disease. For them, economic success is a goal for the future.

### Before You Move On

**Monitor Comprehension** In what ways do the area's resources and crops support its economy?

### ONGOING ASSESSMENT
# MAP LAB

1. **Compare and Contrast** According to the Gross Domestic Product map, which countries have the highest GDP? The lowest GDP?

2. **Interpret Maps** In what ways might South Africa's natural resources contribute to differences on the GDP map?

3. **Movement** What circumstances have led many workers to migrate to Southern Africa?

# SECTION 1 GEOGRAPHIC
## NATIONAL GEOGRAPHIC

**TECHTREK**
myNGconnect.com For an online map, photos, and an Explorer Video Clip

Maps and Graphs | Digital Library

# Exploring Africa's Wildlife
### with Dereck and Beverly Joubert

**Main Idea** Humans are working to protect endangered African big cats and their homes.

## Big Cats in Africa

Big cats, such as lions, cheetahs and leopards, are a vital part of African wildlife. Since the 1940s, the number of lions in Africa has been reduced from about 450,000 to 20,000—and humans caused most of this reduction. Hunting, movement into big cat **habitats** (natural homes), and **poaching** (illegal hunting) all contribute to fewer big cats in the wild.

## Working to Protect Habitats

"It seems like we were explorers from birth, wandering the wild Earth with a passion," said Dereck Joubert. The Jouberts, who are National Geographic Explorers-in-Residence, learned about wildlife on game reserves in Southern Africa. Today, they live in Botswana. "On our first trip to Botswana and the **Okavango** (oh kuh VAANG oh) **Delta**, in 1981, we felt we had come home," the Jouberts noted.

That same year, the Jouberts joined the Chobe Lion Research Institute in Botswana. They began an intensive study of lions,

myNGconnect.com
For more on Dereck and Beverly Joubert in the field today

## AFRICAN WILDLIFE RESERVES

☐ National Park, Wildlife Reserve
🐾 Chobe Lion Research Institute

## BIG CATS BY THE NUMBERS

**5**
Distance in miles from which a lion's roar can be heard

**12**
Average life span of a male lion in the wild

**23**
Distance in feet a cheetah can cover in one stride

**70**
A cheetah's top speed in miles per hour

**550**
Weight in pounds of a fully grown male lion

Source: Smithsonian National Zoo, National Geographic Society

---

which included adopting a **nocturnal**, or night-based, lifestyle. They worked during the night in the African wilderness for months at a time. Since leaving the Chobe Institute, the Jouberts have worked on their own in many different filmmaking, photography, and conservation projects.

The Jouberts have a mission: conserving big cat habitats. Protecting habitats also protects the biodiversity, or variety of life, in the habitat. The Jouberts' projects draw attention to the plight of big cats in Africa. Their knowledge of how big cats live helps other conservationists to develop habitat protection programs.

The Jouberts also support **ecotourism**, which is tourism that is focused on wildlife protection and responsible use of land and resources. Ecotourism teaches visitors about conservation issues. For the Jouberts, teaching is an essential part of their work. Ultimately, they believe that "We are part of a global community of lions and leopards, buffalo, dung beetles, snakes, trees, and ice caps, not somehow apart from it all."

### Before You Move On
**Summarize** What are some ways of protecting African big cats?

---

## ONGOING ASSESSMENT
# READING LAB  GeoJournal

1. **Make Inferences** Using what you know about sub-Saharan geography, what might explain the lack of wildlife reserves in the north?

2. **Interpret Data** What do the cheetah's speed and stride distance imply about its ability as a hunter?

# SECTION 2 HISTORY
## 2.1 Bantu Migrations

TECHTREK
myNGconnect.com For a map of the Bantu migrations and photos of Bantu culture

Maps and Graphs   Digital Library

**Main Idea** Bantu-speaking people migrated from West Africa and influenced the language and culture of the African continent.

As you have read, sub-Saharan Africa's varied physical features support a variety of crops today. About 10,000 years ago an agricultural revolution began in Central Africa. During the **agricultural revolution**, humans began to grow crops instead of gathering plants. Sometime after the agricultural revolution, around 2000 B.C., one of the greatest migrations in human history began—the **Bantu** migration. By A.D. 1000, Bantu peoples had spread from their Central African homeland south and east across sub-Saharan Africa.

### The Bantu People

People migrate for economic, political, religious, social, and environmental reasons. When groups of people move, cultures blend, new languages form, technology spreads—and sometimes conflict occurs.

Historians and anthropologists are not certain why the Bantu chose to migrate when and where they did. Whatever the reason, the Bantu people began to move and carried their cultural traits and skills along with them.

The Bantu had knowledge of iron working, and their use of iron weapons gave them an advantage over other tribal groups. As the Bantu people moved, they forced other groups to move or become absorbed into Bantu culture.

The many different groups that became part of Bantu culture kept much of their own culture as well. As a result, the nearly 85 million people who trace their history to the Bantu migrations share a very diverse culture today.

> **Critical Viewing** Bantu descendants pick tea in South Africa. One possible reason for the Bantu migration was to find better farmland. What does this photo suggest about farming methods in South Africa today?

## Bantu Languages

Today, people whose ancestors were Bantu exist in more than 400 ethnic groups, including the Zulu, the Swahili, and the Kikuyu. Original Bantu languages have evolved into more than 450 languages.

**Swahili** (also known as Kiswahwali) is one of the best-known of the surviving Bantu languages. For more than 5 million people, Swahili is the language they learn as children, or their **first language**. For 30 million other people, Swahili is their second language. Swahili is spoken mainly in East Africa and can be heard in several different forms, or dialects.

Swahili is heavily influenced by the Arabic language. Arab traders from northern Africa and Bantu-speaking people of eastern Africa met and exchanged goods over many centuries. Over time, Swahili became the language used to conduct trade.

Beginning in the early 19th century, Arab trade **caravans**—groups of merchants traveling together for safety—traveled farther into the interior of Africa, spreading Swahili to more people. Eventually, Swahili would be used by some Europeans who colonized parts of Africa where the language was spoken.

Today, many areas in Africa have two groups of people with different first languages. In these areas, Swahili is frequently spoken by both groups as a way to communicate. As a result, Swahili is the **lingua franca** (LEEN gwa FRAWN kah), or common language between multiple groups of people.

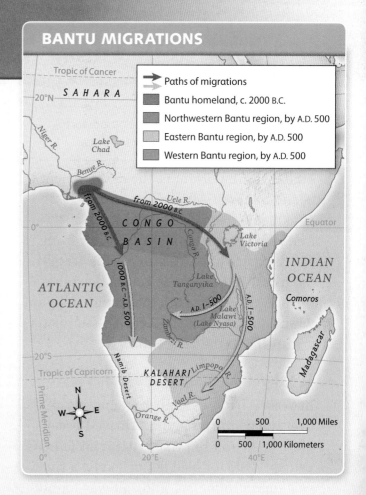

### Before You Move On

**Summarize** How did the Bantu influence the culture and language of the African continent?

### ONGOING ASSESSMENT
# MAP LAB
*GeoJournal*

1. **Interpret Maps** What landform in Africa might have influenced the Bantu people to migrate south instead of north?
2. **Movement** Based on the map, over what period of time did the Bantu migration take place?
3. **Draw Conclusions** How was the migration of the Bantu people supported by their use of metal-working technology?
4. **Make Inferences** Why might it be important for an area to have a *lingua franca*, or common language?

# SECTION 2 HISTORY
## 2.2 Early States and Trade

TECHTREK
myNGconnect.com For an online map of the African empires and and photos of Great Zimbabwe

 Maps and Graphs
 Digital Library

**Main Idea** Trade helped develop powerful states and empires in sub-Saharan Africa.

Trade was important in the development of sub-Saharan Africa. The promise of valuable trade goods brought Arab traders to a transportation corridor from North Africa to West Africa. This **trans-Saharan** trade, or trade across the Sahara, introduced Africans to the Islamic religion, which spread from the Arabian peninsula beginning in the 8th century A.D.

### West African Empires

In West Africa, a number of empires arose over the centuries and all thrived because of the gold and salt trade. **Alluvial** (a LOO vee ahl) gold, or gold deposited by a river, was found in forests, and salt was found in deserts. These goods were traded in the savanna between the forest and desert, where the empire of Ghana began.

Ghana gained wealth and power by taxing the gold and salt trade, controlling West Africa from A.D. 700 to the 1200s. Ghana declined, and the kingdom of Mali, led by King Sundiata, overtook Ghana. His great-nephew, Mansa Musa, continued to control trade, spread Islam, and make the city of **Timbuktu** a center of education.

Another empire that traded gold and salt was Songhai. Brought to its height by Askia Muhammad, Songhai prospered from the 900s to the 1400s. South of Songhai, the kingdom of Benin lasted from the 1200s to the 1800s. Benin actively traded with European countries such as Portugal and the Netherlands.

### East African Empires and States

The powerful East African empire of Aksum was located in present-day Ethiopia. Aksum flourished between A.D. 300 and 600. Adulis, its main port, was located on the Red Sea and served as a center of trade. Many **city-states** (independent states made up of a city and the territories depending on it), such as Mogadishu, formed on the coast of East Africa as trading grew.

### Other African States

Powerful states also arose in Central and Southern Africa. In Central Africa, the state of **Kongo** (different from the modern country of Congo) was founded in 1390. Kongo became known for its highly-organized government. Soon after its founding, the Portuguese arrived and became involved in many aspects of the state, including politics, trade, and religion.

Aksum Coins

300

○ A.D. 300
East African empire of Aksum begins growing to its height in present-day Ethiopia.

600

A.D. 700 ○
Ghana becomes center of gold and salt trade in West Africa.

900

Modern vendor cuts a slab of salt to sell in a Mali market.

## SUB-SAHARAN AFRICAN EMPIRES

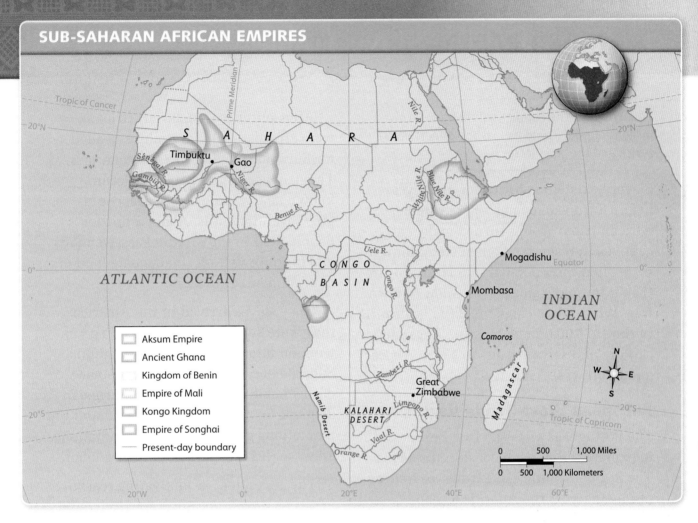

Between 1200 and 1450, the Shona people in Southern Africa built a walled city out of stone, called **Great Zimbabwe** (zim BAH bwe). *Zimbabwe* is a Shona word meaning "stone houses." The Shona traded gold, copper, and iron with places as far away as China and India.

### Before You Move On
**Summarize** In what ways did trade help develop the early states and kingdoms in Africa?

### ONGOING ASSESSMENT
## MAP LAB

1. **Place** Locate the city-state of Mogadishu on the map. Why would its location be beneficial for trade?
2. **Interpret Maps** Based on the map and the time line, which empire was oldest?
3. **Make Inferences** What might explain why the West African empires of Ghana, Mali, and Songhai overlapped?

**1200**

**1300s**
Mali empire in West Africa at its height

Mansa Musa ruled Mali for about 25 years.

**1500**

**1591**
Songhai empire comes to an end.

**1800**

SECTION 2.2 **373**

SECTION 2 HISTORY

# 2.3 Impact of the Slave Trade

myNGconnect.com For Guided Writing
Student Resources

**Main Idea** The European slave trade involved millions of people and had lasting effects on Africa and the Americas.

Slavery existed in Africa for many years before European contact. For example, African tribal groups turned male war captives into slaves. Women and children were often incorporated into families, and the children of some slaves could be born free. When Islam came to Africa beginning in the A.D. 700s, some Muslims began to capture and sell Africans to North Africa and Southwest Asia.

## European Slave Trade Begins

The Portuguese were the first Europeans to explore the African coast in the 1400s. The **trans-Atlantic slave trade**, or trading of slaves across the Atlantic Ocean, started around 1500. Enslaved people were brought to African coastal cities and held captive until sold. The Portuguese, Spanish, Dutch, French, and English all purchased slaves at African coastal ports.

After purchase, enslaved Africans were crowded onto large ships headed for European colonies in the Americas. This trip across the Atlantic Ocean, known as the **Middle Passage**, could take several months. About 2 million people died in the Middle Passage, many due to **malnutrition** (inadequate food or nourishment) or disease.

Once slaves arrived in the Americas, they were sold at auction, often to go to work on large farms called plantations. Sugar, tobacco, and cotton were some of the major plantation crops. European demand for these crops increased and the plantations got bigger. As the plantations grew, so did the demand for slave labor.

The **incentive**, or motivating reason, for slavery was profit. Europeans bought enslaved people in order to have a cheap and captive labor source. The plantation owners made more money because they did not have to pay the slaves.

> **Critical Viewing** This is a European ship used to transport enslaved Africans. What does this illustration suggest about conditions on the ships for enslaved Africans?

## Consequences of the Slave Trade

The trans-Atlantic slave trade lasted from the 1500s to the mid-1800s. Historians estimate that more than 12 million Africans were enslaved and shipped to the Western Hemisphere. The majority of slaves were sent to Brazil and the Caribbean.

People forced into slavery were generally young because they had a better chance of surviving the Middle Passage. Also, when they arrived at their destination, it was expected that young people would be able to work longer and harder in the fields.

| THE MIDDLE PASSAGE | | |
|---|---|---|
| | LEFT FROM AFRICA | ARRIVED IN THE AMERICAS |
| 1500–1600 | 277,506 | 199,285 |
| 1601–1700 | 1,875,631 | 1,522,677 |
| 1701–1800 | 6,494,619 | 5,609,869 |
| 1801–1867 | 3,873,580 | 3,370,825 |
| TOTAL | 12,521,336 | 10,702,656 |

Source: http://slavevoyages.org/tast/assessment/estimates.faces

Many Africans taken were male, and many were potential leaders in their community. Families were often torn apart. These losses weakened many African communities and completely destroyed others.

Millions of people in North America, the Caribbean, and South America are descendents of enslaved Africans. These people have shaped cultures in those regions by sharing their languages, customs, and traditions. The impact of the slave trade has lasted for centuries.

The journey across the Atlantic could take up to 90 days depending on the weather. Enslaved Africans were packed onto ships and chained below deck. They were only taken above deck for brief periods. Unable to stand up or move, many Africans died where they sat.

3.25 feet

### Before You Move On
**Summarize** How did the European slave trade develop and how did it change cultures?

### ONGOING ASSESSMENT
## DATA LAB  GeoJournal

1. **Interpret Charts** Based on the chart, how many people died during the Middle Passage?
2. **Analyze Data** During which time period did the greatest number of enslaved Africans die?
3. **Movement** How did the forced movement of Africans affect communities around the world?
4. **Write Reports** Research one of the following topics and then write a short report on how each affected sub-Saharan Africa: sugar plantations, the triangular trade, or slave ports. Go to **Student Resources** for Guided Writing support.

# SECTION 2 HISTORY
## 2.4 Colonization to Independence

TECHTREK
myNGconnect.com For online maps of colonialism and photos of African leaders
Maps and Graphs
Digital Library

**Main Idea** European powers colonized and ruled large parts of Africa until Africans began independence movements in the mid-1900s.

As you have learned, the Portuguese began exploring the African coast in the 1400s. By the 1500s, many European nations were seeking to control large parts of Africa.

### Imperialism and Colonialism

**Imperialism** is the practice of extending a nation's influence by controlling other territories. European imperialism in Africa began when Europeans started trading with slave merchants on the coast. Little by little, Europeans moved into African lands in search of profitable resources. Eventually, several European countries conquered African lands and established colonies. The practice of directly controlling and settling foreign territories is known as **colonialism**.

By the mid-1800s, European powers began to fight over their African colonies. They wanted more natural resources to fuel industrialization, or the transition to large-scale industries, in Europe. Because of advanced European weapons, there was little the Africans could do to stop them.

### Scramble for Africa

In 1884, Europeans held the **Berlin Conference** to settle their disputes about colonial claims in Africa. No Africans were invited to attend. Europeans at the conference divided Africa among themselves. By 1910, France, Germany, Belgium, Portugal, Italy, Spain, and Great Britain had established themselves as colonial powers in all parts of Africa.

Many Europeans believed that African culture and religion were inferior to those of Europe. They wanted to change African traditions. Europeans sent **missionaries**—people sent by a church to spread their religion among native populations—to convert Africans to Christianity.

### African Independence

In the early 1900s, **Pan-Africanism**, a movement to unify African people, grew among African leaders in London and other cities around the world. By the 1950s and 1960s, this nationalist movement had brought together many African leaders. **Jomo Kenyatta** (JOH moh ken YAA taa) of Kenya and **Kwame Nkrumah** (KWAA may en KROO mah) of Ghana helped to gain independence for their people.

A caravel ship used in trade voyages

1500 — 1600 — 1700

○ 1500s
European imperialism in Africa begins.

○ 1642
The Dutch take possession of Portuguese forts in West Africa.

## EUROPEAN COLONIES, 1938

## AFRICAN NATIONS, 2013

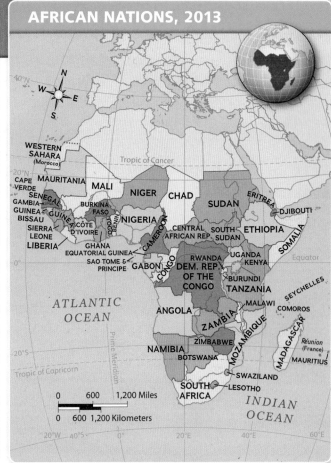

In 1963 the Organization of African Unity (OAU) was founded to promote Pan-Africanism. The OAU is known today as the African Union (AU). It continues to promote African unity and cooperation, but also functions as an economic group similar to the European Union.

### Before You Move On
**Make Inferences** What action by Africans helped bring colonialism to an end?

### ONGOING ASSESSMENT
## MAP LAB
*GeoJournal*

1. **Location** Based on the maps, which modern countries made up the colony of French Equatorial Africa?
2. **Interpret Maps** Looking at the maps, how would you describe the changes to Africa's internal borders from 1938 to today?
3. **Analyze Cause and Effect** Why did European countries colonize Africa? How were Africans affected by colonization?

This political cartoon (1892) suggests that Europe now controls the African continent.

1800

1884 — Berlin Conference divides Africa.

1900

1963 — Organization of African Unity (OAU) is founded.

2002 — African Union replaces OAU.

2000

Jomo Kenyatta

# NATIONAL GEOGRAPHIC
# Photo Gallery • Nigerian Salt Ponds

For more photos from the National Geographic Photo Gallery, go to the **Digital Library** at myNGconnect.com.

Mountain gorilla, Africa

Nigerian camel caravan

South African women

**Critical Viewing** These circular ponds contain a mixture of water and salty earth. The mud is moved to evaporation areas where salt is left behind. The color of the salt depends on where the mud came from.

Students in Botswana

African lions

Victoria Falls, Zimbabwe

Zulu village, South Africa

# CHAPTER 13
# Review

## VOCABULARY

For each pair of vocabulary words, write one sentence that explains the connection between the two words.

1. savanna; Sahel

> *The Sahel runs south of the Sahara and north of the savannas, forming a border between the two.*

2. ecotourism; habitats
3. Swahili; *lingua franca*
4. Middle Passage; trans-Atlantic slave trade
5. imperialism; Pan-Africanism

## MAIN IDEAS

6. What are the four regions of sub-Saharan Africa? (Section 1.1)
7. Which large landform in East Africa was created by the separation of tectonic plates? (Section 1.2)
8. Which region of West Africa supports more population, and why? (Section 1.3)
9. What is the major river in Central Africa, and what effect does it have on the region? (Section 1.4)
10. What are the two main types of landforms that make up Southern Africa? (Section 1.5)
11. What factors threaten the big cats of Africa? (Section 1.6)
12. From what African region did the Bantu people begin their migration? (Section 2.1)
13. How did gold and salt help Ghana develop as a kingdom? (Section 2.2)
14. What were some effects of slavery on communities in Africa? (Section 2.3)
15. What reasons did Europeans have for wanting to colonize Africa? (Section 2.4)

## GEOGRAPHY

### ANALYZE THE ESSENTIAL QUESTION

How has the varied geography of sub-Saharan Africa affected people's lives?

### Critical Thinking: Draw Conclusions

16. What might be some ways for Africans to combat the challenges of desertification?
17. How does not having access to a coastal port impact a country?
18. What advantages or challenges does physical geography create for economic development in Central Africa?

### INTERPRET TABLES

| IMPACT OF CHANGES IN LAND ||||||
|---|---|---|---|---|---|
| | Change | Environment | Human Health | Safety Issues | Politics & Economics |
| Desertification | | Loss of habitat; decline in variety of plant and animal life; increased soil erosion | Malnutrition, hunger | Wars over arable land and limited water resources | Poverty; decreased political and economic influence; population movement |
| Deforestation | | Decline in variety of plant and animal life; loss of habitat; reduced resources | Loss of potential new medical products | Increased landslides and flooding | Loss of forest products; loss of indigenous communities; loss of tourism opportunities |
| Soil Erosion | | Loss of soil and habitat; loss of farmland | Loss of food and water; hunger, malnutrition | Risk of flooding and landslides | Loss of property; reduced farm development |

Source: http://www.eoearth.org/article/Global_Environment_Outlook_(GEO-4):_Chapter_3#Introduction

19. **Summarize** How does desertification lead to loss of habitat?
20. **Form and Support Opinions** Think about the causes of desertification, deforestation, and soil erosion. Which one do you think would be easiest to solve? Explain.

## HISTORY

### ANALYZE THE ESSENTIAL QUESTION

How did trade networks and migration influence the development of African civilization?

### Critical Thinking: Analyze Cause and Effect

21. How can the effects of the Bantu migration be seen in Africa today?
22. In what ways did trade networks help build African kingdoms?

### INTERPRET MAPS

23. **Location** What climate zone covers most of the Sahara?
24. **Region** Think about the four areas of sub-Saharan Africa: West, East, Central, and Southern. What area has the largest number of different climate zones?
25. **Make Inferences** What is the relationship between the climate zones and the rain forests of the Congo Basin?

## ACTIVE OPTIONS

**Synthesize** the Essential Questions by completing the activities below.

26. **Create Time Lines** Review this chapter for information on the history of sub-Saharan African countries. Then use a computer or pen and paper to create a time line similar to those found in this book. **Post the time line in your classroom so others can add more entries.**

   **Time Line Tips**
   - Take notes before you begin.
   - Include important dates for a time line.
   - Include important rulers and events.
   - Find at least one photo that represents your country.

**Create Sketch Maps** Create a natural resources sketch map of sub-Saharan Africa. Use maps in this chapter and other online sources to gather information about natural resources in sub-Saharan Africa. Add them to an outline map similar to the one below. Use a map key to show the resources.

# CHAPTER 14

# Sub-Saharan Africa Today

## PREVIEW THE CHAPTER

**Essential Question** What historical and geographic factors have influenced the cultures of sub-Saharan Africa?

### SECTION 1 • CULTURE

1.1 Africa's Borders and Cultures .................. 384

1.2 African Music Goes Global ..................... 386

1.3 Kenya Modernizes ................................. 388

1.4 Exploring Traditional Cultures ............... 390

**KEY VOCABULARY**
- ethnic group
- transportation corridor
- griot
- oral tradition
- reserve
- domestic policy
- modernization
- literacy rate
- ethnobotanist
- medicinal plant

**ACADEMIC VOCABULARY**
evident

**TERMS & NAMES**
- African Union
- Youssou N'Dour
- Nairobi
- Jomo Kenyatta

**Essential Question** How have conflict and government instability slowed economic development in sub-Saharan Africa?

### SECTION 2 • GOVERNMENT & ECONOMICS

2.1 Prized Mineral Resources ....................... 392

2.2 Nigeria and Oil ..................................... 394

2.3 Agriculture and Food Supply .................. 396

**GLOBAL ISSUES: HEALTH**

2.4 Improving Public Health ........................ 398

2.5 Sudan and Somalia ................................ 400

2.6 Ending Apartheid .................................. 402

**KEY VOCABULARY**
- mineral
- commodity
- coup
- famine
- erosion
- legume
- microcredit
- epidemic
- pandemic
- vaccine
- infectious
- refugee
- clan
- failed state
- segregation
- apartheid
- homeland

**ACADEMIC VOCABULARY**
concentrated

**TERMS & NAMES**
- Lost Boys of Sudan
- African National Congress
- Nelson Mandela
- Steve Biko

# TECHTREK FOR THIS CHAPTER

**Student eEdition** | **Maps and Graphs** | **Interactive Whiteboard GeoActivities** | **Digital Library**

Go to myNGconnect.com for more on sub-Saharan Africa.

Traditional Masai in Tanzania herd a flock of goats into a pen.

# SECTION 1 CULTURE

## 1.1 Africa's Borders and Cultures

**Main Idea** European powers created new colonial borders that ignored the existing borders of traditional African cultures.

Africa is a continent of diverse cultures and countries. Today, about two thirds of Africans live in rural villages that have varied customs and languages. About 1,000 languages are spoken in Africa by different **ethnic groups**, which are groups of people who share a common culture, language, and sometimes racial heritage. Many Africans identify themselves first as members of a tribe rather than a country.

### The Impact of Colonialism

Before Europeans arrived, borders between African cultural groups developed by agreement or conflict. Natural features such as oceans, lakes, mountain ranges, and rivers often defined borders. Natural borders were easier to control than borders set by groups of people. Borders along bodies of water also provided **transportation corridors**, or routes to easily move people and goods from place to place.

When Europeans colonized the continent, they routinely ignored the borders of African cultural groups. Instead, colonial borders were set to address the European need for resources. To do so, some sub-Saharan African cultural groups were divided, forced to share territory with rivals, or both.

As African countries began to gain independence in the 1900s, they did not change their colonial borders. As a result, few sub-Saharan African countries share one common culture.

### Conflict and Cooperation

Cultural differences across Africa have often led to civil wars over political control, territory, and resources. Somalia's ethnic groups, for example, have never united to form a single nation. Military dictatorships often have been necessary to impose order.

> **Critical Viewing** Fishermen sun-dry fish caught in a lake near their village in Malawi. What can you infer about this fishing culture from the photo?

## COOPERATIVE INTERNATIONAL ORGANIZATIONS IN AFRICA

| Organization | Purpose |
| --- | --- |
| African Union | Peace-keeping; oppose colonization; promote unity; cooperate for economic development |
| Economic Community of West African States (ECOWAS) | Promote economy, industry, transportation, energy resources, agriculture, and natural resources |
| Southern African Development Community | Support local economy, transportation networks, and political interaction |
| Economic Community of Central African States (ECCAS) | Promote industry, transportation, communication, energy, natural resources, economy, tourism, education |
| East African Community (EAC) | Improve cooperation in transport and communication, industry, security, immigration, and economic matters |

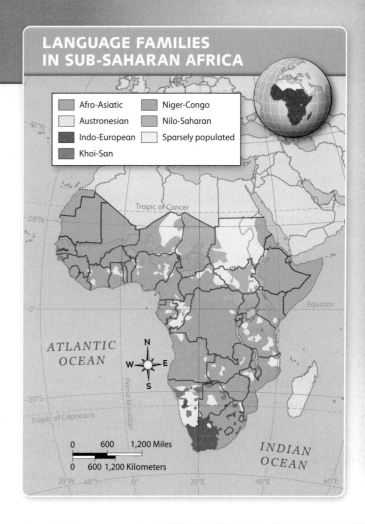

LANGUAGE FAMILIES IN SUB-SAHARAN AFRICA

- Afro-Asiatic
- Austronesian
- Indo-European
- Khoi-San
- Niger-Congo
- Nilo-Saharan
- Sparsely populated

To address these challenges, African countries have developed organizations (see chart above) such as the **African Union** and the East African Community (EAC). These organizations promote cooperation and economic progress within and between countries.

### Before You Move On
**Summarize** How were colonial borders different from sub-Saharan Africa's traditional borders?

## ONGOING ASSESSMENT
# MAP LAB

GeoJournal

1. **Interpret Maps** Locate country borders near areas with the Nilo-Saharan language group. What can you infer about the location of language groups and the borders that were created during colonization?

2. **Human-Environment Interaction** On the map, identify the oceans that surround Africa. How do natural features such as rivers, lakes, and oceans help countries control territory and methods of transportation?

3. **Make Generalizations** How did colonization lead to conflict between present-day cultures in Africa?

4. **Synthesize** What is one goal that is shared by the organizations shown in the chart?

SECTION 1.1

# SECTION 1 CULTURE

## 1.2 African Music Goes Global

**myNGconnect.com** For audio clips of African music and photos of musicians

Digital Library

**Main Idea** African music connects the people to their past and communicates their cultures to the world.

Music has always been a way for Africans to celebrate their cultures. Today, African music has become part of many cultures. Experts link African music to the development of jazz, blues, rock-and-roll, and gospel in the United States.

### A Wealth of Music

Each area of sub-Saharan Africa has distinctive music that reaches a wide range of audiences. Advancements in communication and transportation have helped to spread African music.

West African music is influenced by the stories and music of griots. **Griots** (GREE oh) are traditional storytellers. For centuries, they orally passed on the histories of West African cultures. Because they were not written down, these histories are part of the **oral tradition**, the practice of orally passing stories from one generation to the next.

Griots accompany their songs with harps, lutes, and drums. Their influence is **evident**, or clearly present, in West African music today.

Much West African music is a fusion, or blend, of griot music and other African music with global music. Mbalax (uhm BALAKS), for example, is a blend of griot percussion and songs with Afro-Cuban influences. **Youssou N'Dour** is a contemporary griot who plays mbalax music. His popularity has helped spread the music out of West Africa to audiences around the world.

In South Africa, music sometimes focused on politics. One example is protest music. Miriam Makeba and a number of other South African musicians left South Africa to protest government policies in the 1970s and 1980s—gaining global audiences for their music.

### Before You Move On

**Summarize** What are some examples of how African music connects the people to their past and communicates their culture to the world?

South African singer Lira performs in Soweto, South Africa.

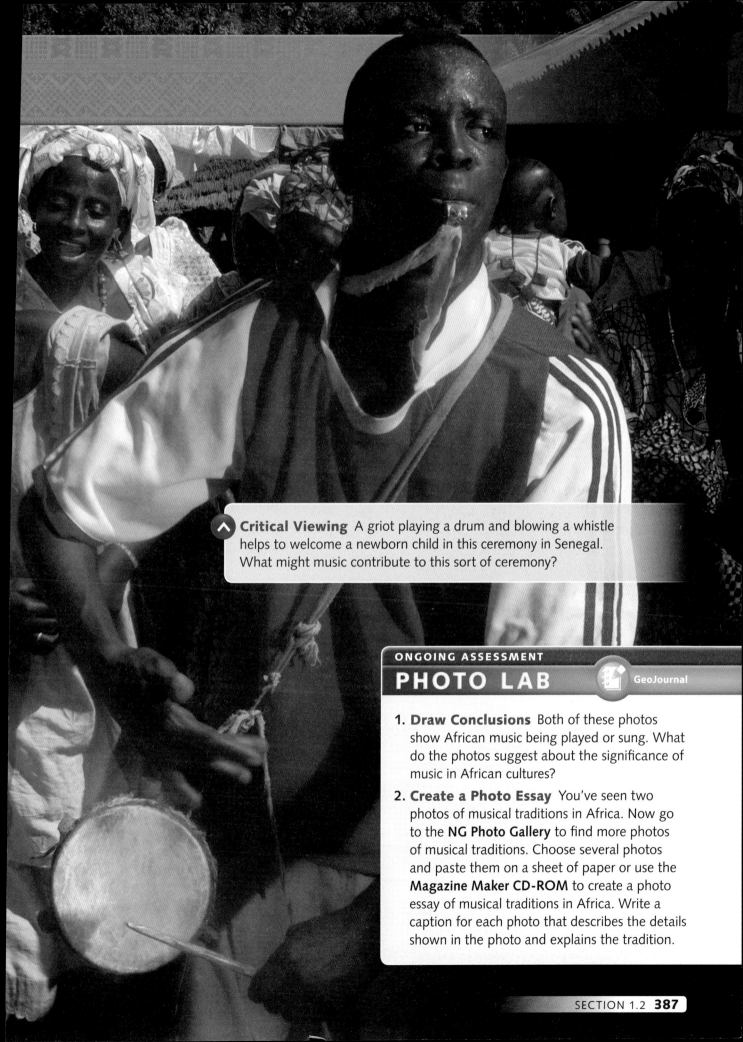

**Critical Viewing** A griot playing a drum and blowing a whistle helps to welcome a newborn child in this ceremony in Senegal. What might music contribute to this sort of ceremony?

## ONGOING ASSESSMENT
## PHOTO LAB   GeoJournal

1. **Draw Conclusions** Both of these photos show African music being played or sung. What do the photos suggest about the significance of music in African cultures?

2. **Create a Photo Essay** You've seen two photos of musical traditions in Africa. Now go to the **NG Photo Gallery** to find more photos of musical traditions. Choose several photos and paste them on a sheet of paper or use the **Magazine Maker CD-ROM** to create a photo essay of musical traditions in Africa. Write a caption for each photo that describes the details shown in the photo and explains the tradition.

# SECTION 1 CULTURE
## 1.3 Kenya Modernizes

**TECHTREK**
myNGconnect.com For an online map and photos of Kenya

Maps and Graphs | Digital Library

**Main Idea** Kenya is modernizing its economy and improving its standard of living.

Kenya's geographic features range from beaches, savannas, deserts, and farmland to snow-capped mountains such as Mount Kenya, for which the country is named. Large mammals such as the African elephant, giraffes, and lions roam Kenya's **reserves**, land set aside for the protection of animals. The diverse geography and wildlife are key to Kenya's economic future.

### Cultural Diversity

You can hear many languages and see an array of traditional clothing on a busy street in **Nairobi** (ny ROE bee), Kenya's capital. More than 40 ethnic groups live in Kenya. The largest are the Kikuyu (ki KOO yoo), Luhya, Luo, Kalenjin, and Kamba. This diversity makes Kenya a multicultural society.

As you've learned, cultural diversity has sometimes led to rivalry and conflict in Africa. Even so, diverse groups have been able to work toward common goals.

For example, the people of Kenya united in their struggle for independence from British rule. After independence in 1963, **Jomo Kenyatta**, a Kikuyu, became Kenya's first elected leader. To help all Kenyans pull together, he appointed members from different ethnic groups as his advisors.

### Modernization

Kenyatta and other leaders encouraged **domestic policies**, or government plans within a country, that led to economic growth and modernization. **Modernization** is the development of policies and actions designed to bring a country up to world standards in technology and other areas.

Education is an important factor in modernization for Kenya. Most children there attend free elementary schools. The adult **literacy rate**, or percentage of people who can read, soared from 32 percent in 1970 to 85 percent in 2003.

**Critical Viewing** Students work at Kibagare Good News Centre School in Nairobi, Kenya. What details in the photo remind you of a day in your classroom?

Nairobi, Kenya

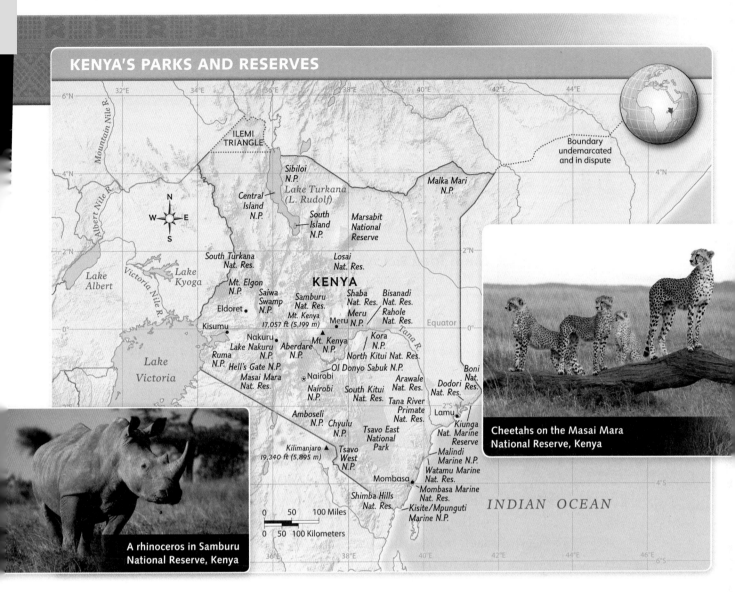

### KENYA'S PARKS AND RESERVES

A rhinoceros in Samburu National Reserve, Kenya

Cheetahs on the Masai Mara National Reserve, Kenya

Kenya's modernization is also evident in its growing tourist industry. The scenic beauty and unique wildlife found in Kenya's national parks and reserves attracts tourists. The government is determined to protect the parks and reserves because of the jobs and money they bring into the country.

In 2000, Kenya, Uganda, and Tanzania formed the East African Community. Today, Rwanda and Burundi are also EAC members. All five countries work together to encourage entrepreneurship and trade in the region.

### Before You Move On

**Summarize** What steps has Kenya taken to modernize?

### ONGOING ASSESSMENT
## MAP LAB

GeoJournal

1. **Interpret Maps** What does the number of national parks and reserves shown on the map indicate about their importance in Kenya?

2. **Place** In what ways are the national parks and reserves shown on the map part of the economic modernization of Kenya?

3. **Form and Support Opinions** What makes Kenya diverse? Explain an advantage or disadvantage of diversity in Kenya. Support your opinion with details from the map and from what you have read about Kenya.

4. **Turn and Talk** Think about Jomo Kenyatta's decision to include different ethnic groups in his council. Talk to a partner and discuss effective ways of bringing people together.

# 1.4 SECTION 1 GEOGRAPHIC NATIONAL GEOGRAPHIC

**TECHTREK**
myNGconnect.com For photos of Explorers at work and an Explorer Video Clip
Digital Library

## Exploring Traditional Cultures
### with Grace Gobbo and Wade Davis

**Main Idea** Preserving indigenous cultures and traditional ways of life benefits modern societies around the world.

### Cures in the Forest

When Emerging Explorer Grace Gobbo walks through a Tanzanian rain forest, she sees more than trees, flowers, and vines. She sees possible cures. As an **ethnobotanist**, she studies the relationship between cultures and plants.

Tanzanians have relied on traditional healers for centuries. Gobbo interviewed more than 80 traditional healers who rely on **medicinal plants**, or plants used to treat illnesses. Scientific research has confirmed that many of the plants these healers use do help treat illnesses. In fact, large drug companies are now using the same plants in many medicines.

However, many of the forests where these plants grow are at risk from logging and clear-cut farming. By teaching the value of medicinal plants, Gobbo hopes to inspire people to preserve their forests and maintain their traditional healing methods. "Nothing was written down," she explains. "The knowledge is literally dying out with the elders."

myNGconnect.com
For more on Grace Gobbo and Wade Davis in the field today

390 CHAPTER 14

### AFRICAN PLANTS AND USES

| Plant Latin name | Location | Medicinal Use |
|---|---|---|
| Tulbaghia violacea | Southern Africa | lowers blood pressure |
| Catharanthus roseus | Madagascar | treats rheumatism (joint or muscle disorder) |
| Peltophorum africanum | Southern Africa | relieves stomach problems |
| Strychnos madagascariensis | Southern Africa | relieves stomach problems |
| Harpagophytum procumbens | Southern Africa and Madagascar | reduces swelling |
| Sutherlandia frutescens | Southern Africa | treats cancer |
| Agapanthus praecox | Southern Africa | treats heart disease and chest pain |

## Preserving Culture

Like Gobbo, National Geographic Explorer-in-Residence Wade Davis is an ethnobotanist who has studied the use of native plants around the world. His current focus, however, is indigenous cultures around the globe—the customs and practices of people in their daily lives. Davis has lived among 15 cultural groups in the Americas, Asia, Africa, and the Arctic.

Davis has proposed that there is a cultural web of life, just as there is a biological connection between all forms of life. He believes that every culture is connected and that all indigenous cultures have contributed to the cultural web. Davis hopes to preserve this cultural diversity, just as Gobbo works to protect medicinal plants and knowledge of traditional healing in Tanzania.

Davis believes that the development of modern industry and technology has contributed to a decline in the number of indigenous cultures. As a result, the entire cultural web suffers from the loss of each culture. Davis further believes that, like the world's plants and animals, human cultures should be protected so that they can continue to contribute to humanity.

### Before You Move On
**Monitor Comprehension** Why are Gobbo and Davis working to preserve traditional cultures?

---

### ONGOING ASSESSMENT
# VIEWING LAB  GeoJournal

1. **Analyze Visuals** Go to the **Digital Library** to watch the Explorer Video Clip of Wade Davis. How are his contributions similar to and different from those of Grace Gobbo?

2. **Summarize** According to Wade Davis, why is preserving indigenous cultures important for humanity?

3. **Make Predictions** What image or idea from the video clip was most surprising to you? Write down two questions you still have.

National Geographic Explorer-in-Residence Wade Davis

SECTION 2 GOVERNMENT & ECONOMICS

# 2.1 Prized Mineral Resources

**TECHTREK**

myNGconnect.com For an online map and photos of mining in sub-Saharan Africa

Maps and Graphs   Digital Library

**Main Idea** Sub-Saharan Africa has mineral resources that could improve life for its people.

The mineral resources of sub-Saharan Africa have the potential to lift the economies of sub-Saharan countries. However, careful management is needed to maximize benefits for the African people.

## Mineral Riches

Sub-Saharan Africa has large deposits of gold, diamonds, and other **minerals**, which are inorganic solid substances formed through geological processes. Many of these minerals are exported to Europe, North America, and Asia where they are used to make automotive and electronic products.

Many sub-Saharan countries use mineral resources as **commodities**, which are materials or goods that can be bought, sold, or traded. These commodities can bring in enormous wealth, but economic progress has been slow. Over time, government corruption has taken much of the profits from mineral mining.

### SELECTED OIL AND DIAMOND EXPORTS SUB-SAHARAN AFRICA, 2005–2006

| Oil | | |
|---|---|---|
| Country | World Rank | Percentage of Country's Total Exports |
| Nigeria | 6 | 91.9 |
| Angola | 12 | 96.6 |
| Equatorial Guinea | 21 | 92.7 |
| Democratic Republic of the Congo | 22 | 89.6 |
| Sudan | 29 | 88.0 |
| Chad | 37 | 94.6 |

| Diamonds | | |
|---|---|---|
| Country | World Rank | Percentage of Country's Total Exports |
| South Africa | 5 | 6.9 |
| Botswana | 9 | 83.5 |
| Namibia | 14 | 43.5 |
| Angola | 16 | 2.4 |
| Democratic Republic of the Congo | 17 | 41.5 |
| Central African Republic | 32 | 36.8 |

Source: The World Bank

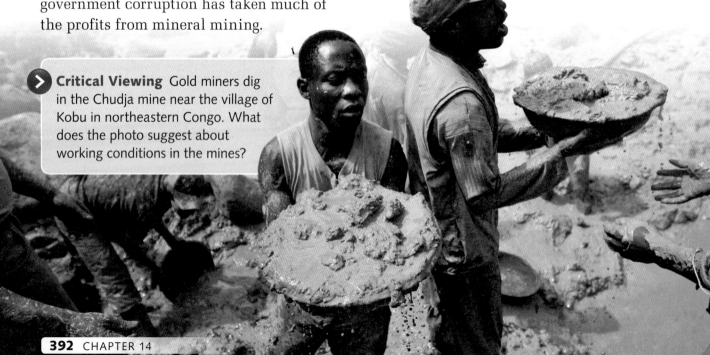

> **Critical Viewing** Gold miners dig in the Chudja mine near the village of Kobu in northeastern Congo. What does the photo suggest about working conditions in the mines?

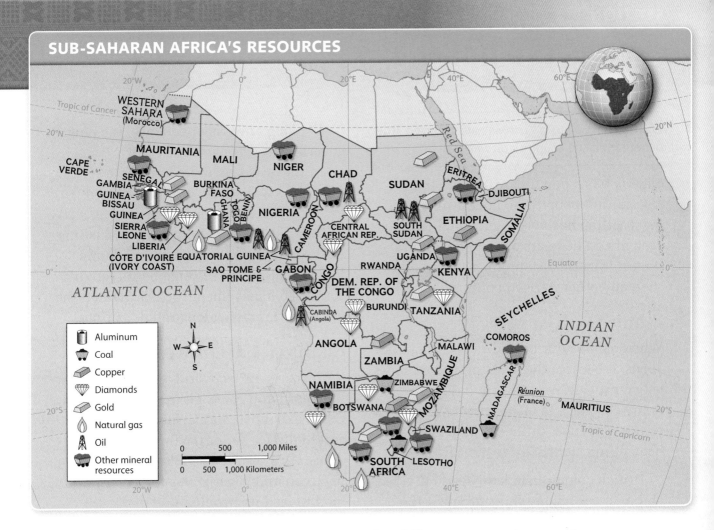

## SUB-SAHARAN AFRICA'S RESOURCES

### Economic Improvement

Today, unstable governments continue to challenge many sub-Saharan countries. For example, the Central African Republic has significant diamond resources as the chart shows, but a lack of infrastructure, smuggling, and political unrest have limited the development of its diamond mining industry. As a result, the country remains one of the poorest in the world.

Some countries such as South Africa, Namibia, and Tanzania have successfully used their mineral profits to build their economies. In Botswana, the government is a partner in the country's largest diamond mining company. It has invested the profits in education, infrastructure, and health care. Countries with resources that follow Botswana's example can improve the stability of their economies and improve the lives of their citizens.

### Before You Move On

**Monitor Comprehension** In what ways do sub-Saharan countries' mineral resources improve life for their people?

### ONGOING ASSESSMENT
## MAP LAB
GeoJournal

1. **Place** Based on the map, what mineral resources can be found in Tanzania?

2. **Compare and Contrast** Pick two countries on the map and compare their resources. What can you infer about their economies based on their resources?

3. **Interpret Charts** Based on the chart, what would happen to Angola's economy if its oil exports were disrupted?

SECTION 2.1 **393**

# SECTION 2 GOVERNMENT & ECONOMICS

## 2.2 Nigeria and Oil

**Main Idea** Nigeria's people have received little benefit from the country's oil wealth.

Nigeria has a population of more than 150 million people, the largest in Africa. Its coastline has major ports, and its river system and delta are among the world's largest. Nigeria also produces more oil than any other African country.

### Ethnic Conflict

When oil was discovered in Nigeria in the 1950s, it became the focus of the country's economy. However, Nigeria has faced many obstacles in attempting to use its oil riches. The most difficult challenge has been ethnic conflict.

British colonialism left Nigeria divided among three major ethnic groups—the Hausa-Fulani, the Igbo, and the Yoruba. There were also 250 smaller groups and at least two major religions. When Nigeria gained independence in 1960, these groups struggled for control.

Not long after independence, the new government was overthrown in a **coup**, or an illegal takeover by force. In a series of coups, rival ethnic groups replaced existing governments—often putting military leaders in charge. After Muslims in the north adopted *sharia*, or Islamic law, Christians in the area began moving south. This movement further divided the country, this time along religious lines.

### Challenges of Oil Wealth

In the early 1970s, world oil prices increased and the Nigerian economy grew. Cities expanded and farmers left their rural land for better paying jobs in the cities. As a result, agricultural production declined and Nigeria was forced to import food to feed its people.

The oil industry has added to the ethnic tension. Different groups struggle to control the oil and its profits. Conflict and corruption have prevented Nigerians from

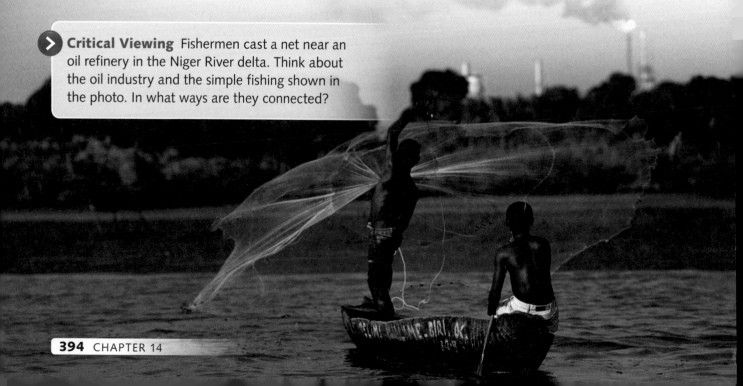

> **Critical Viewing** Fishermen cast a net near an oil refinery in the Niger River delta. Think about the oil industry and the simple fishing shown in the photo. In what ways are they connected?

sharing in the oil wealth. Most people living in the Niger River delta, where oil production is **concentrated**, or centered, live in poverty.

The delta region is also heavily polluted from oil spills. People in the delta have protested against the pollution and demanded that more of the oil profits be used to improve the area.

## Nigeria's Progress

Between 2007 and 2011, Nigeria made economic gains. Its leaders received favorable loans to repay the country's debt to other countries. Officials began putting in place economic reforms designed to improve Nigeria's infrastructure. Maintaining these improvements while resolving ethnic conflicts will be a challenge for Nigeria's future.

### Before You Move On
**Summarize** Why have Nigeria's people benefited so little from the country's oil wealth?

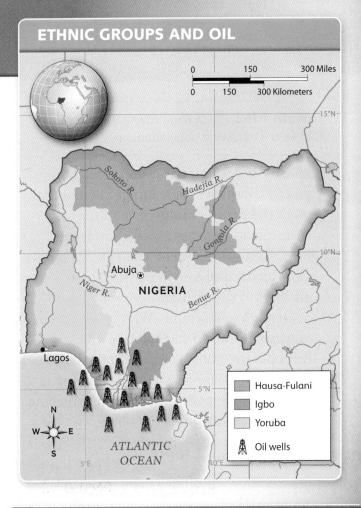

### ETHNIC GROUPS AND OIL

Legend:
- Hausa-Fulani
- Igbo
- Yoruba
- Oil wells

## ONGOING ASSESSMENT
### MAP LAB  GeoJournal

1. **Location** What ethnic group appears to have the most access to the country's oil industry?
2. **Interpret Maps** Identify the delta region on the map. What are some possible consequences of polluting this region?
3. **Analyze Cause and Effect** Use a chart similar to the one below to list effects related to Nigeria's oil industry.

| CAUSE | EFFECTS |
|---|---|
| Nigeria's oil industry | 1. |
| | 2. |
| | 3. |

SECTION 2.2  **395**

# SECTION 2 GOVERNMENT & ECONOMICS

## 2.3 Agriculture and Food Supply

**TECHTREK**
myNGconnect.com For photos of agriculture and Guided Writing
Student Resources | Digital Library

**Main Idea** Africa is improving its ability to feed its growing population.

Images of African children in need are all too common. Their hunger is real and often the result of famine. **Famine** is a widespread and sustained shortage of food. Natural disasters, such as droughts and floods, and armed conflicts can all disrupt the availability of food and contribute to a famine. Widely reported famines occurred in Sudan in 1998, Ethiopia in 2000, 2002, and 2003, Niger in 2005, and Somalia in 2011. Famine is a constant danger in Africa and threatens the health of millions of people.

### A Threatened Food Supply

Population growth has been rapid in some parts of Africa, but food production has not increased as rapidly. The average African eats 10 percent fewer calories today than he or she did 20 years ago. The number of undernourished people in sub-Saharan Africa increased from about 90 million in 1970 to 214 million in 2012. Many African children suffer from malnutrition, which is a lack of nutrients essential to good health. Because of hunger and malnutrition, African children typically have shorter life expectancies than other children around the world.

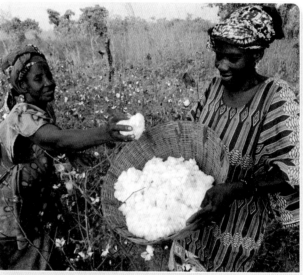

**Critical Viewing** Women harvest cotton in Mali. Based on the photo, how would you describe this Mali cotton farm?

**Visual Vocabulary** **Erosion** is the wearing away of Earth's surface by natural forces. The grid-like structures in the photo are fences made from branches. The fences keep sand from eroding onto land used for crops and grazing.

African food crops include corn, yams, and sorghum, which is a type of grain. Before colonization, Africans grew these crops with the goal of feeding everyone. After colonization, European settlers took the most fertile land, and farming shifted from food crops to cash crops for export. Cash crops such as coffee and cotton are grown because they can be sold for more money than food crops. The income from cash crops allows farmers to purchase more, but less food is grown to feed the African people.

Since independence, some land has been returned to Africans. However, African land can be easily overworked and overgrazed, exhausting the soil. Farmers also struggle with droughts, soil erosion, and desertification.

### New and Better Farming

Africans are working to improve agricultural practices. For example, farmers are moving animals from place to place to avoid overgrazing. They are also growing different crops on the same plot to avoid exhausting the soil. More **legumes**, which are peas or beans, are being planted. These add to the food supply and release helpful nutrients into the soil. Farmers are also enriching the soil by using fertilizer with animal and plant waste instead of chemicals.

**Microcredit**, or small loans, has helped poor farmers invest in land, tools, and seeds. Relief agencies have begun to provide free seeds and tools to farmers. Scientists hope to develop seeds that are more productive and drought resistant. These seeds could help farmers grow more food crops for their own use as well as cash crops for export. If successful, these efforts will feed Africans, extend life expectancies, and improve economies.

**Before You Move On**
**Summarize** What is Africa doing to improve its agriculture?

### ONGOING ASSESSMENT
## WRITING LAB  GeoJournal

1. **Summarize** What is the difference between food crops and cash crops?
2. **Analyze Cause and Effect** What is the relationship between reduced food production and life expectancy?
3. **Write Comparisons** Write a paragraph comparing African food crops and cash crops. Describe why is it important for African farmers to balance growing food crops with cash crops. Then work in groups and share your paragraphs. Go to **Student Resources** for Guided Writing support.

# SECTION 2 GLOBAL ISSUES • Health
## 2.4 Improving Public Health

**TECHTREK**
myNGconnect.com For data about public health in Africa

Global Issues

**Main Idea** National governments and international agencies are committed to improving health care in Africa by reducing the impact of diseases.

**KEY VOCABULARY**

**epidemic**, n., an outbreak of disease affecting a large number of people within a community at the same time

**pandemic**, n., an outbreak of disease occurring over a wide geographic area

**vaccine**, n., treatments to increase immunity to a particular disease

**infectious**, adj., capable of spreading rapidly to others

Nearly a half million people die of malaria each year, and 9 out of 10 of them are in Africa. Eighty-five percent of the victims are children under the age of five. Malaria is one of many chronic diseases that plague the continent.

### Climate, Poverty, and Disease

In 2012, 52 years was the estimated average life expectancy for sub-Saharan Africans. That is about 30 years shorter than the average in North America. One cause is the deadly toll taken by diseases such as malaria, sleeping sickness, and yellow fever.

Only 15 percent of the world's people live in Africa, but 90 percent of tropical disease cases occur there. The organisms that cause disease thrive in its tropical climates, as do the insects that carry the diseases.

For example, malaria is carried by a mosquito and causes flu-like symptoms. Severe forms can cause brain and nerve damage and death. Malaria probably began in ancient times as an **epidemic**—an outbreak limited to a particular community. Later, it became a **pandemic**, meaning it spread over a large area, in this case, across much of the world. Malaria can be prevented by draining or treating water where insects breed and by sleeping under mosquito nets treated with chemicals to kill insects.

In addition to the tropical climate, poverty contributes to the spread of disease. Poor Africans may lack mosquito nets, screened windows, and technology for draining insect breeding grounds. Poverty also raises the rates of other infections because of overcrowding, poor sanitation, and contaminated water.

### Before You Move On
**Summarize** What are some factors that contribute to the spread of disease in Africa?

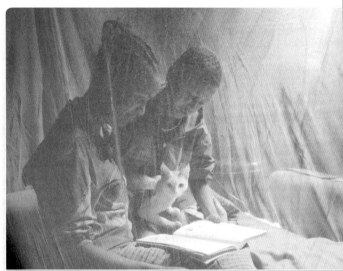
Students in Ethiopia read before school under the protection of a mosquito net.

398 CHAPTER 14

# Inspiring people to care about the planet
National Geographic Society Mission

**COMPARE ACROSS REGIONS**

## The Fight Against Disease

The fight against diseases such as malaria, HIV/AIDs, and yellow fever has become a global commitment. Many agencies offer medication to treat diseases and vaccines to prevent them. **Vaccines** are treatments designed to increase immunity, or resistance, to a particular disease. Scientists are trying to find new vaccines and cures for **infectious** diseases, or diseases that can spread rapidly.

Eliminating malaria is a global goal for the 21st century. Worldwide, there are an estimated 250 million cases every year. A non-governmental organization (NGO) called Malaria No More works to fight malaria in Africa. An NGO is a nonprofit volunteer group founded by citizens. Malaria No More hands out mosquito nets to protect people.

The World Health Organization (WHO) battles malaria around the world. The group's Mekong Malaria Program in Southeast Asia works with governments and NGOs to monitor disease outbreaks and treatment programs in the region.

A major source of funding for malaria treatment programs is the Bill and Melinda Gates Foundation. The foundation funds research for a vaccine and more effective anti-malarial drugs for people who have already been infected.

All of these programs reflect a renewed commitment on the part of national governments, private foundations, and NGOs to fight malaria and other diseases around the world.

### MALARIA BY THE NUMBERS

**37**
Percentage that malaria incidence fell globally between 2000 and 2015

**60**
Number of seconds between each death of an African child due to malaria

**97**
Countries worldwide reporting malaria cases in 2015

**2,414**
Estimated number of people worldwide who die every day from malaria

**801,000**
Estimated number of Africans who die every year from malaria

Source: World Health Organization, United Nations Children's Fund

### Before You Move On
**Monitor Comprehension** What are the main tools in the global fight against disease?

## ONGOING ASSESSMENT
## READING LAB  GeoJournal

1. **Analyze Cause and Effect** What is one cause of the short life expectancy among sub-Saharan Africans?
2. **Make Predictions** How might development of a vaccine against malaria affect the life expectancy in Africa?
3. **Make Inferences** How might you expect malaria to impact the economies of countries with large numbers of cases?

# SECTION 2 GOVERNMENT & ECONOMICS
## 2.5 Sudan and Somalia

**TECHTREK**
myNGconnect.com For online maps
Maps and Graphs

**Main Idea** Civil wars involving ethnic and religious groups in Sudan and Somalia have limited progress in these countries.

For years, Sudan and Somalia have been plagued by war, famine, and disease. Many people survived conflict only to die of starvation. In both countries, climate, history, and culture all combined to help create this difficult situation.

As you've learned, colonial rule in Africa left countries divided among different ethnic and religious groups. Sudan and Somalia also face a geographic division: The Sahel runs through both of these countries, dividing them between northern Africa and sub-Saharan Africa.

### Conflict in Sudan

Sudan is mostly desert in the north. Nomadic Muslim herders of Arabic descent live there. The south is swamp and savanna that was settled by farmers of African descent. Northerners are mainly Muslim, while southerners practice mainly Christianity or indigenous religions. Both north and south include a mix of ethnic groups.

These differences have resulted in human rights abuses. Since Sudan's independence in 1956, the government has oppressed Christians in the country. In Darfur, the government created a largely Arab militia in 2003. The militia attacked government protestors, but also attacked militia members' personal enemies. Almost 400,000 people were killed, and about 2.5 million people became **refugees**, or people who flee a place to find safety.

Neighboring countries such as Chad set up refugee camps for people fleeing Sudan. One camp housed more than 250,000 people in 2010. The camps have no permanent shelter or sewer system, often suffer food shortages, and rely on the host countries and relief agencies for help.

The refugees included a group known as the **Lost Boys of Sudan**. These young men were orphaned by the civil war and stuck together to escape the violence. They traveled to Ethiopia, then back to Sudan, and then to camps in Kenya. In 2001, the United States took in more than 3,500 of these young men. Many attended high school and college in the United States.

> **Critical Viewing**
> Eyl (AY uhl), shown here, is a town in Somalia's Puntland state. What might you infer from this photo about economic conditions in Eyl?

SUDAN POLITICAL

SOMALIA POLITICAL

## Somalia

Somalians have suffered through conflict among five major **clans**, or large, family-based units with loyalty to the group. Even within the clan, sub-clans may clash. As a result, Somalians have never united to form a single nation.

In 1991, clan-based groups overthrew a crumbling military government and battles raged between rival clans. This conflict disrupted farms that were already threatened by flood and drought. With no central government, some clans turned to piracy, which continues today. They attack and rob foreign ships or hold the ships and their crews for ransom.

Somalia is considered by many to be a **failed state**, a country in which government, economic institutions, and civil order have broken down. The regions of Somaliland and Puntland have claimed independence but have not received international recognition. In the 21st century, forces from the United Nations and the African Union have been trying to keep peace in the region.

### Before You Move On
**Summarize** How have conflicts limited progress in Sudan and Somalia?

### ONGOING ASSESSMENT
## MAP LAB
*GeoJournal*

1. **Interpret Maps** Using the map of Sudan, describe the geographic distribution of the main religions.

2. **Region** Using the map of Somalia, describe how it is divided. Based on your description, what challenges might Somalia face?

3. **Draw Conclusions** What is a failed state, and why might Sudan and Somalia both be considered failed states?

SECTION 2.5  **401**

# SECTION 2 GOVERNMENT & ECONOMICS
## 2.6 Ending Apartheid

TECHTREK
myNGconnect.com For photos of South Africa

Digital Library

**Main Idea** In 1994, South Africa moved from a minority, racist government to a democratically elected government.

In 1994, long lines of South Africans of all races stood for hours under a blazing hot sun. The majority of them waited to vote for the first time in their lives. This day marked the end of racial **segregation**, or separation by race, and the birth of a new democracy.

### Rich Resources Lure Colonists

South Africa stretches from warm tropics in the north to chilly waters in the south. Its geography and climate support a variety of crops, and it has large deposits of diamonds, gold, and other minerals.

In the 19th century, both Dutch and British colonists laid claim to these lands. The Dutch settlers, known as Boers or Afrikaners, formed their own republics. Boers had enslaved Africans and imported other people as laborers from Asia. Both the British and the Boers seized lands from Zulus, Xhosa, and other Africans. Even more Africans were forced to work for Boer and British colonists after the discovery of large gold and diamond deposits around 1870.

### The Beginning of Apartheid

In 1902, after a series of wars, the Boer territories became British colonies known as the Union of South Africa. The British colonial government divided South Africa into white and black areas. Only white Africans could vote. Most of the land, and all of the best land, was reserved for the white minority. Black Africans were left with the land that was less useful or productive. Cities were declared white, and black Africans could enter them only to work.

**Critical Viewing** A crowd in South Africa celebrates at a soccer match. What can you infer from the photo about how a sporting event might help bring people together?

In 1948, new laws created **apartheid** (uh PART hite), or the legal separation of races. In addition to being divided by race, South Africans could only own land in areas assigned to them. Black South Africans were required to carry identification papers at all times.

In 1970, the government made all black South Africans citizens of a homeland instead of citizens of South Africa. **Homelands** were supposed to be self-governing areas, but the national government actually controlled them. The homelands separated the races even more.

## The End of Apartheid

In the early 20th century, black Africans formed the **African National Congress** (ANC) to protest their treatment. Later, South Africa outlawed the ANC and imprisoned its leaders, including **Nelson Mandela**. In 1977, **Stephen Biko**, president of a student protest organization, was arrested and beaten to death. He and Mandela became symbols for the protest movement in South Africa.

In 1989, a new white president, F. W. de Klerk, began to change apartheid laws. He legalized the ANC and released its leaders, including Mandela. In 1994, voting rights were extended to all South Africans, and the people elected Mandela president.

Decades of apartheid left South Africa far from racial equality. The government has worked to give black South Africans access to better jobs and farmland, but a large economic gap remains between most blacks and whites in the country. In 2002, the government took control of

Nelson Mandela gestures to the crowd at a rally shortly after his release from prison.

the country's mineral resources, in part to make sure that black South Africans profited fairly. Progress is slow, but South Africa's ability to harness its resources has allowed it to develop the most prosperous economy in Africa.

### Before You Move On
**Summarize** How did South Africa change from minority rule to a more democratically elected government?

## ONGOING ASSESSMENT
## READING LAB

1. **Monitor Comprehension** What was apartheid and how did it begin?
2. **Movement** In what ways did colonization and apartheid remove Africans from their homes?
3. **Compare and Contrast** Give an example of how the government under apartheid limited peoples' rights and an example of how the government expanded rights after apartheid.

# CHAPTER 14
# Review

## VOCABULARY

For each pair of vocabulary words, write one sentence that explains the connection between the two words.

1. ethnic group; transportation corridor

   *Bodies of waters can act as transportation corridors between different African ethnic groups.*

2. griot; oral tradition
3. ethnobotanist; medicinal plant
4. commodity; mineral
5. famine; erosion
6. epidemic; pandemic

## MAIN IDEAS

7. What factors were ignored when the colonial boundaries of African countries were drawn? (Section 1.1)
8. How has Africa's music influenced music in the United States? (Section 1.2)
9. What challenges has Kenya faced in modernizing its economy? (Section 1.3)
10. How are Grace Gobbo and Wade Davis helping to preserve traditional cultures? (Section 1.4)
11. What circumstances have kept Africans from profiting from mineral resources? (Section 2.1)
12. How has population growth made it harder for Africa to feed its people? (Section 2.3)
13. What is being done to reduce the spread of malaria in Africa? (Section 2.4)
14. What are some of the causes of the conflict in Sudan and Somalia? (Section 2.5)
15. How has life improved in South Africa since the end of apartheid? (Section 2.6)

## CULTURE

### ANALYZE THE ESSENTIAL QUESTION

What historical and geographic factors have influenced the cultures of sub-Saharan Africa?

**Critical Thinking: Draw Conclusions**

16. Without the influence of colonialism, would African countries be as ethnically diverse as they are today? Why or why not?
17. In what ways have sub-Saharan Africa's geography, climate, and resources helped to shape the lifestyles of its people?

### INTERPRET MAPS

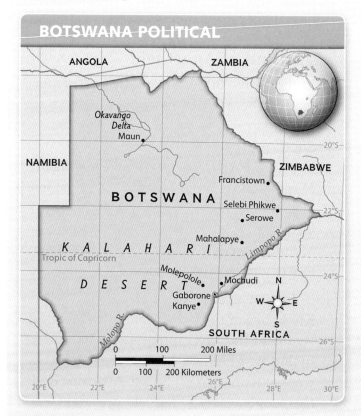

BOTSWANA POLITICAL

18. **Location** Which city can be found near 22°S latitude and 28°E longitude?
19. **Human-Environment Interaction** What type of geographic feature defines the irregular shape of Botswana's southern border?

## GOVERNMENT & ECONOMICS

### ANALYZE THE ESSENTIAL QUESTION

How have conflict and government instability slowed economic development in sub-Saharan Africa?

**Critical Thinking: Analyze Cause and Effect**

20. Describe some specific natural resources that have helped sub-Saharan African countries develop economically.
21. How did colonialism slow down Africa's economic development?
22. What effects did apartheid have on racial equality in South Africa?

### INTERPRET GRAPHS

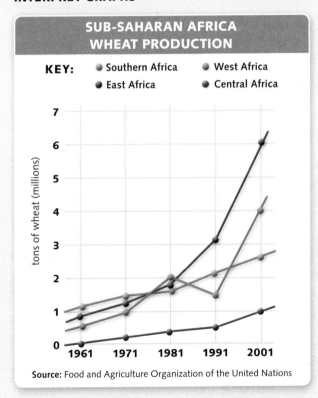

23. **Analyze Data** In which region of Africa did food production increase the most?
24. **Evaluate** What additional information is needed in order to determine growth in food production per capita?

## ACTIVE OPTIONS

**Synthesize** the Essential Questions by completing the activities below.

25. **Gather Information** Suppose it is 2001 and a group of Lost Boys from Sudan are coming to your school. With a group, talk about ways to welcome them. List words and daily routines they would need to know. Conduct research about their language. **Share your ideas orally with the class.**

> **Collaboration Tips**
> - Listen, take notes, and ask others to clarify what they say; respond to their questions as well.
> - Retell what you've heard to confirm your understanding.
> - Agree in advance on how you will explain your ideas to the class.

26. **Create a Chart** Make a three-column chart showing comparisons among three countries in sub-Saharan Africa. Be sure each country comes from a different area (West, East, Central, or Southern). Use online sources to gather the data for the categories shown below. Based on the data, which country has the most people? The fewest?

|  | Ghana (West Africa) | Uganda (East Africa) | Lesotho (South Africa) |
|---|---|---|---|
| Year Country Gained Independence |  |  |  |
| Population |  |  |  |
| Square Miles of Land |  |  |  |
| Main Economic Resources |  |  |  |

# Deserts of the World

COMPARE ACROSS REGIONS UNIT 7

Deserts make up about one fifth of the land on Earth. Geologically speaking, deserts as we know them are relatively young—forming within the last 65 million years. Deserts are dry lands that can lose more water through evaporation than they get from precipitation. Rainfall in deserts is usually less than 10 inches a year. Limited rainfall, low humidity, often high daytime temperatures, and winds all contribute to the desert's dryness.

Deserts sometimes cover large areas and extend beyond country borders. For example, the Sonoran Desert crosses the border between Mexico and the United States. The Kalahari Desert covers parts of three countries—Botswana, Namibia, and South Africa.

### Compare

- Botswana
- China
- Mexico
- Mongolia
- Namibia
- South Africa
- United States

## PHYSICAL CHARACTERISTICS

Many people think of a desert as a hot, sandy wasteland with no water and no life. In fact, there is an amazing diversity among deserts. Some are hot and sandy, but some have rainy seasons and can get very cold. While about one fourth of deserts are made of sand, the rest are composed of dirt, clay, rock, ice, and other materials. For example, Antarctica is a desert. Deserts can be flat or hilly, below sea level or in mountains.

## DESERT PLANTS AND ANIMALS

Deserts are home to a wide array of plants and animals that have adapted to the harsh climates. Desert plants, such as the saguaro cactus (below), can go long periods without water. Some plants have shallow root systems that cover a large area to gather as much water as possible. Many desert animals, such as the elf owl (below), are nocturnal, hunting for food after the sun goes down and the temperatures drop.

Namib Desert, Namibia

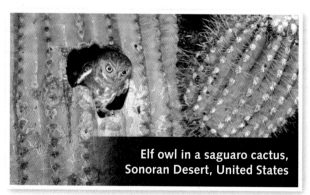

Elf owl in a saguaro cactus, Sonoran Desert, United States

# SELECTED DESERTS OF THE WORLD

**Location:**
United States, Mexico

**Approximate size:**
100,000 square miles

**Average yearly rainfall:**
4–12 inches

**Interesting animal species:**
Gila monster

**Examples of temperature ranges:**
120°F (summer)
32°F (winter)

**Location:**
Botswana, South Africa, Namibia

**Approximate size:**
220,000 square miles

**Average yearly rainfall:**
3–7.5 inches

**Interesting animal species:**
Meerkat

**Examples of temperature ranges:**
113°F (summer)
7°F (winter)

**Location:**
China, Mongolia

**Approximate size:**
500,000 square miles

**Average yearly rainfall:**
2–8 inches

**Interesting animal species:**
Bactrian Camels

**Examples of temperature ranges:**
113°F (summer)
-40°F (winter)

Source: World Wildlife Fund; worldatlas.com; gobidesert.org; Arizona-Sonoran Desert Museum

## ONGOING ASSESSMENT
### RESEARCH LAB  GeoJournal

1. **Analyze Data** Which desert has the widest range in temperatures?
2. **Compare and Contrast** How do the deserts compare in size?

**Research and Create Charts** Locate the Patagonian and Sahara deserts on a map. Do research to compare the deserts and use a chart like the one above to present your findings. Adjust the categories depending on the information you want to compare.

COMPARE ACROSS REGIONS 407

REVIEW & EXTEND UNIT 7

# Active Options

**TECHTREK**

myNGconnect.com For photos of endangered species and writing templates

 Digital Library    Student eEdition

### ACTIVITY 1

**Goal:** Extend your understanding of endangered animals.

## Write a Briefing

Do research and prepare a written briefing about one of the endangered sub-Saharan animals listed below. A briefing is a short summary of important information about a topic. Use the Internet to find reports or articles. Organize information from one or two of them. What are the latest and most important findings? Give your audience facts about the animal and about efforts to protect it.

- African elephant
- addax
- black rhinoceros
- mandrill (shown at right)
- pygmy hippopotamus
- lowland gorilla

mandrill

### ACTIVITY 2

**Goal:** Research sub-Saharan Africa's culture.

## Keep a Culture Journal

Congratulations! You have won a seven-day, all-expenses-paid tour to one of these countries: Kenya, Tanzania, South Africa, Botswana, Zimbabwe, or Zambia. The only condition is that you keep a daily journal describing what you see and do. Choose your country and begin your tour. Use the **Magazine Maker CD-ROM** to share your journal with friends back home.

### ACTIVITY 3

**Goal:** Review sub-Saharan Africa with a game.

## Play a Capital Game

With a group, look at a map and choose 25 countries in sub-Saharan Africa. Write the name of each country on a small card. On another small card, write the capital of each country. Now you're ready to play. Each player draws a country card and tries to identify the capital. If the player misses, the card is put at the bottom of the pile, and it is the next player's turn. If the player names the capital correctly, he or she keeps the card and it is the next player's turn. The one with the most cards wins. Vary the game by choosing a capital card and trying to name the country.

# UNIT 8

# Explore SOUTHWEST ASIA & NORTH AFRICA with NATIONAL GEOGRAPHIC

### MEET THE EXPLORER

**NATIONAL GEOGRAPHIC**

Emerging Explorer and urban planner Thomas Taha Rassam (TH) Culhane works in Cairo's poorest neighborhoods. He installs rooftop solar water heaters and household biogas systems using environmentally friendly technology.

### INVESTIGATE GEOGRAPHY

The Ramses temple, on the Nile River, was carved thousands of years ago in Egypt. Ancient Egypt was an oasis in the desert of northeastern Africa. Egypt depended on annual flooding of the river to support its society.

### STEP INTO HISTORY

Jerusalem is home to the Dome of the Rock (shown here), an Islamic and Jewish holy place. It also contains many sites of importance to Christianity, including the Church of the Holy Sepulchre.

## ONLINE WORLD ATLAS

Washington, D.C. — 5,810 miles — Cairo, Egypt

Go to myNGconnect.com for maps of Southwest Asia and North Africa.

### CONNECT WITH THE CULTURE

From the top of the world's tallest building, the Burj Khalifa in Dubai, dense development reflects the recent construction boom in the United Arab Emirates.

# CHAPTER 15

# SOUTHWEST ASIA & NORTH AFRICA GEOGRAPHY & HISTORY

## PREVIEW THE CHAPTER

**Essential Question** How have climate and location influenced the region in the past and today?

### SECTION 1 • GEOGRAPHY

1.1 Physical Geography .................... 414
1.2 Tigris and Euphrates Rivers ........... 416
1.3 The Arabian Peninsula ................. 418
1.4 Anatolian and Iranian Plateaus ....... 420

**KEY VOCABULARY**
- arid
- desertification
- alluvial plain
- silt
- irrigation
- petroleum
- nonrenewable
- fault
- oasis
- qanat

**ACADEMIC VOCABULARY**
reliable

**TERMS & NAMES**
- Sahara Desert
- Nile River
- Fertile Crescent
- Rub Al Khali
- North Anatolian Fault

**Essential Question** How did civilizations develop in Southwest Asia and North Africa?

### SECTION 2 • HISTORY

2.1 Mesopotamia ........................... 422
2.2 Birthplace of Three Religions ........ 424
2.3 Diffusion of Religions ................ 426
2.4 The Ottoman Empire .................. 428

**KEY VOCABULARY**
- cultural hearth
- agricultural revolution
- domesticate
- city-state
- cuneiform
- monotheistic
- messiah
- pilgrimage
- diffusion
- adherent
- sultan
- religious tolerance

**ACADEMIC VOCABULARY**
extent

**TERMS & NAMES**
- Ur
- Hammurabi
- Nebuchadnezzar
- Judaism
- Christianity
- Islam
- Hebrew Bible
- Christian Bible
- Qur'an
- Diaspora
- Constantine
- Ottoman Empire
- Byzantine Empire
- Osman
- Suleyman I

Arabian camel

**Essential Question** How did an advanced civilization develop in Egypt?

### SECTION 3 • FOCUS ON EGYPT

3.1 The Nile River Valley .................... 430

3.2 Egypt's Ancient Civilization .................... 432

3.3 The Great Pyramids .................... 434

**KEY VOCABULARY**
- floodplain
- hydroelectric power
- hieroglyphics
- dynasty
- pyramid
- pharaoh
- deity
- papyrus
- tomb
- sarcophagus

**ACADEMIC VOCABULARY**
regulate

**TERMS & NAMES**
- Aswan High Dam
- Hatshepsut
- Ramses II
- Re
- Isis
- Giza
- Great Pyramid of Khufu

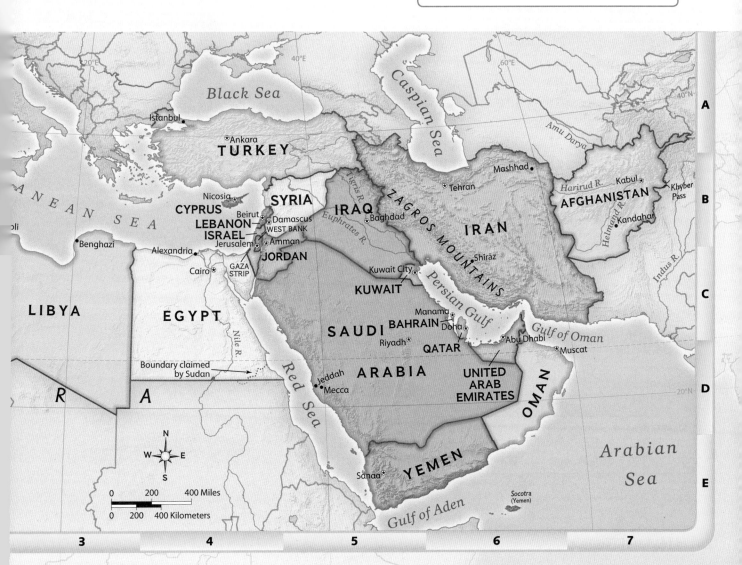

# SECTION 1 GEOGRAPHY

## 1.1 Physical Geography

**TECHTREK**
myNGconnect.com For online maps of Southwest Asia and North Africa and Visual Vocabulary
Maps and Graphs | Digital Library

### SOUTHWEST ASIA AND NORTH AFRICA PHYSICAL

**Visual Vocabulary**
Sahara Desert

**Elevation**

| feet | meters |
|---|---|
| 10,000+ | 3,050+ |
| 5,000 | 1,524 |
| 2,000 | 610 |
| 1,000 | 305 |
| 500 | 152 |
| 0 | 0 |
| Below sea level | |

**Visual Vocabulary**
Nile River

### SOUTHWEST ASIA AND NORTH AFRICA CLIMATE

**Climate Regions**

Dry
- Semiarid
- Arid

Humid Temperate
- No dry season
- Dry summer

Unclassified highlands

414 CHAPTER 15

**Main Idea** The expansive region of Southwest Asia and North Africa is hot, and water in the region is sometimes scarce.

Southwest Asia and North Africa span parts of the Asian and African continents. Two notable physical features located in this region are the world's longest river and its largest desert. Changes in climate in some places led to new patterns of migration and settlement.

## Physical Features and Climate

Much of Southwest Asia and North Africa is made up of vast, barren deserts. This desert area is **arid**, or very dry, with extremely high daytime temperatures. For example, most of the **Sahara Desert** and Arabian Peninsula receive fewer than four inches of rain per year. Daytime temperatures can climb to 130°F.

The region includes several different mountain ranges. The Atlas Mountains are located at the tip of northwest Africa. The Zagros Mountains in Iran stretch along the Persian Gulf, and the Taurus Mountains in Turkey border the Mediterranean Sea. The climate in these mountains is semi-arid and temperatures can fall below 0°F.

Three important rivers have historically supported life in this region, and they continue to do so today. The **Nile River** flows through Egypt, and the Tigris and Euphrates rivers flow through Syria and Iraq. Water is scarce in most of this region, except near rivers and coasts. Cultures have developed and grown along these waterways. Many of the world's oldest cities are located here.

## Desert Transformation

Today, the Sahara stretches 3,500 miles across North Africa from the Red Sea to the Atlantic Ocean. However, this broad area of the continent has not always been desert. The Sahara's climate has changed dramatically over time. Nearly 10,000 years ago, the world's largest desert was tropical grassland.

Around 5300 B.C., seasonal rains that watered the Sahara began shifting southward. Over time, the Sahara became a desert. The gradual transition from fertile to less productive land is called **desertification**. The drier climate prompted people living in the Sahara to migrate out of the desert. They settled in the Nile River Valley, where they found a **reliable**, or dependable, water source.

### Before You Move On

**Monitor Comprehension** In what ways have heat and water resources defined this region?

### ONGOING ASSESSMENT
## MAP LAB

1. **Location** On the physical map, locate the Nile, the Tigris, and the Euphrates rivers. Into which bodies of water do these rivers flow? What other major bodies of water can you infer might be important in the region?

2. **Make Inferences** Even though most of the region is hot and dry, variations exist. Based on the climate map, compare the climates of Lebanon and Yemen. Which country can you infer receives more rainfall?

3. **Describe Geographic Information** How did the climate of the Sahara change over time, and what caused the change?

# SECTION 1 GEOGRAPHY
## 1.2 Tigris and Euphrates Rivers

**TECHTREK**
myNGconnect.com For an online map and photos of the Tigris and Euphrates rivers

 Maps and Graphs
 Digital Library

**Main Idea** The Tigris and Euphrates rivers have supported life for thousands of years.

One of the world's earliest civilizations began between two rivers in a desert. The ancient name for this area is Mesopotamia, which means "land between the rivers."

### Two Rivers

As you have read, the climate of this region is hot and dry. However, the area surrounding and extending from the Tigris and Euphrates rivers is known as the **Fertile Crescent.** This fertile area arcs from the Mediterranean Sea to the Persian Gulf. Its rivers and floodplains were an ideal location for early agriculture.

The source of both the Tigris and Euphrates rivers is in the mountains of Turkey. The rivers flow through parts of Turkey, Syria, and Iraq. They join at Al Qurnah, Iraq, before emptying into the Persian Gulf.

Though the Tigris and the Euphrates flow in similar directions, each river has distinct features. As the longest river in southwestern Asia, the Euphrates River is about 1,740 miles long—about 600 miles shorter than the Mississippi River in the United States. After leaving eastern Turkey, the Euphrates flows southeast through Syria and across Iraq, and is fed by two major tributaries. Over its journey, the flow of the river slows down. As the Euphrates winds through hot desert land, much of its water evaporates.

The Tigris River is also a long river, about 1,180 miles in length. After leaving Turkey, it flows southeast through Iraq. Fed by several fast-moving tributaries, the Tigris carries more water than the Euphrates. The speed and unpredictable flow of the Tigris sometimes cause major floods. Though floods can be destructive, they have also enabled development of agriculture along the rivers.

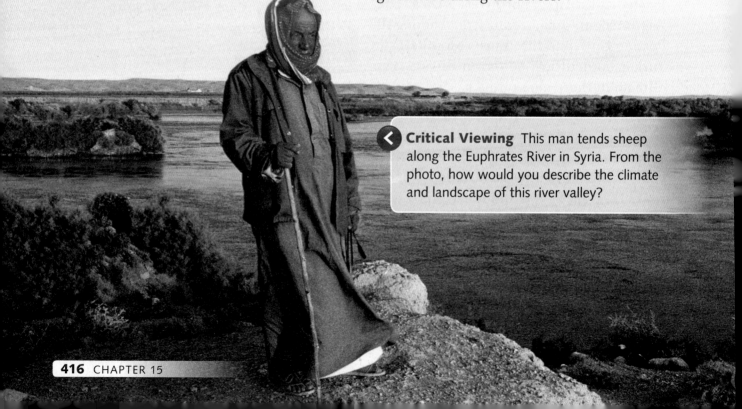

**Critical Viewing** This man tends sheep along the Euphrates River in Syria. From the photo, how would you describe the climate and landscape of this river valley?

## Fertile Land

The land between the rivers is an **alluvial plain**, which is a flat area of land located next to a stream or river that floods. Regular flooding deposits **silt**, or fine particles of soil, along the riverbanks. This silt makes the soil fertile.

Historically, farmers have relied on the natural flooding of the rivers to water their crops. However, farmers have also used irrigation. **Irrigation** is the process of redirecting water to crops using channels and ditches. When irrigated, crops in dry areas can grow over a wider area.

The Tigris and Euphrates rivers continue to support life. For example, most of Iraq's population lives between the two rivers and Iraq's capital, Baghdad, lies on the Tigris River. Frequent flooding used to overrun the city, but today Baghdad uses dams and embankments, or walls, to help control floods and regulate irrigation.

### Before You Move On
**Summarize** In what ways are the Tigris and Euphrates rivers important to this region?

### TIGRIS AND EUPHRATES RIVERS

### ONGOING ASSESSMENT
## MAP LAB  *GeoJournal*

1. **Interpret Maps** According to the map, where do the Tigris and Euphrates rivers join? Where is the mouth of these rivers?

2. **Location** Look at the map and identify where most cities in Iraq are located. What pattern do you notice?

3. **Draw Conclusions** In what ways has flooding of the Tigris and Euphrates rivers been both productive and destructive? Create a chart like the one below to help you answer the question.

| TIGRIS AND EUPHRATES RIVER FLOODS ||
|---|---|
| Productive | Destructive |
|  |  |
|  |  |

SECTION 1.2  417

# SECTION 1 GEOGRAPHY
## 1.3 The Arabian Peninsula

**TECHTREK**
myNGconnect.com For an online map and photos of the Arabian deserts

 Maps and Graphs     Digital Library

**Main Idea** The Arabian Peninsula is primarily desert and provides a large percentage of the world's petroleum.

Though the Arabian Peninsula is nearly surrounded by water, it is mostly desert. This peninsula's geologic history, or the way Earth developed over millions of years, provides the basis of its economic activity today.

### Sand and Heat

The Arabian Peninsula covers more than one million square miles. It is bordered on the west by the Red Sea, on the south by the Arabian Sea, and in the northeast by the Persian Gulf. Saudi Arabia is the largest country on the peninsula. Kuwait, Oman, Qatar, the United Arab Emirates, Yemen, Bahrain, and parts of Jordan and Iraq are also on the peninsula. Temperatures in this area often rise to 130°F, and very little rain falls.

Several deserts lie on the Arabian Peninsula. The Syrian Desert is located in the northern and central part of the peninsula. Another desert, the **Rub al Khali,** covers 250,000 square miles of southern Saudi Arabia—almost the size of the state of Texas. Its name, Rub al Khali, means "empty quarter." Except for small groups of nomads, almost no one lives in the Rub al Khali.

The coasts of the peninsula contrast with its desert interior. The western part of the peninsula along the Red Sea features mountain peaks, some as high as 9,000 feet. Fertile soil along the coasts even allows for some farming. For example, date palms, which can tolerate salty soils, grow abundantly along the Persian Gulf's coastal salt flats. Salt flats are soils with high concentrations of salt. Sand and gravel cover most of this area but a valuable resource lies under it.

> **Critical Viewing** The Matrah district in the Muscat Sultanate of Oman sits on the Gulf of Oman. Based on the photo, how would you describe the district?

## Oil

Large **petroleum**, or unrefined oil, reserves lie underneath the Arabian Peninsula. Petroleum develops from tiny animals and plants that died millions of years ago. Over time, heat and pressure change these organic materials into a new substance that moves through rock layers and collects in large deposits. Once extracted from the ground, petroleum can be refined into gasoline, diesel fuel, and other products.

Because it takes so long to form, petroleum is a **nonrenewable** natural resource. As you have read, a nonrenewable resource is a resource that cannot reproduce quickly enough to keep pace with its use.

Twenty-five percent of the world's known petroleum reserves are on the Arabian Peninsula. In fact, petroleum production is the most important industry on the peninsula. Most of the countries in the world depend on petroleum in one form or another for their energy needs. This dependence links the countries on the Arabian Peninsula to the rest of the world in critical ways.

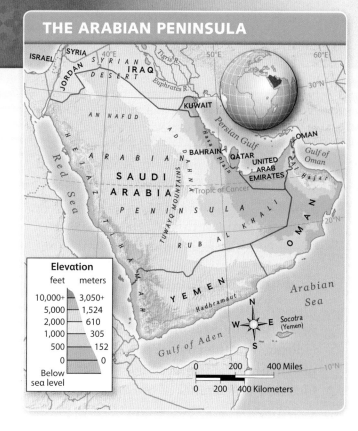

### Before You Move On
**Summarize** What role does the Arabian Peninsula play in the production of the world's petroleum?

### ONGOING ASSESSMENT
## PHOTO LAB  GeoJournal

1. **Analyze Visuals** Look at the photo and read the caption. Then locate approximately where this district lies on the map. Based on the photo and the map, how would you describe the physical features of this area? Consider elevation in your response.

2. **Make Inferences** Based on the photo and the text, what can you infer about patterns of settlement on the Arabian Peninsula?

3. **Human-Environment Interaction** In what ways does climate impact agriculture on the Arabian Peninsula?

SECTION 1.3  419

# SECTION 1 GEOGRAPHY

## 1.4 Anatolian and Iranian Plateaus

**TECHTREK**
myNGconnect.com For an online map and photos of Turkey and Iran

Maps and Graphs • Digital Library

> **Main Idea** The Anatolian and Iranian plateaus have a long history as a crossroads of trade.

Humans have lived on the Anatolian and Iranian plateaus for thousands of years. Though these plateaus are located in different parts of this region, they share characteristics and patterns of settlement.

### Plateaus and Mountains

Most of Turkey sits on a peninsula called Anatolia. The Anatolian Plateau  is located in the central part of this peninsula. The Iranian Plateau  is located in the center of present-day Iran.

Seismic activity formed both plateaus. The **North Anatolian Fault** runs east to west just south of the Black Sea. Activity along this **fault**, or fracture in Earth's crust, has caused many earthquakes. Scientists believe the North Anatolian Fault may have triggered a huge, long-lasting flood about 7,500 years ago, creating the Black Sea. Tectonic shifts also shaped the Iranian Plateau. When the Arabian and Eurasian plates pushed into each other, the Zagros Mountains formed.

The high mountains on each plateau help create a rain shadow. A rain shadow is a dry area on one side of a mountain range. Moisture from surrounding seas rises and condenses, but rain and snow fall only as the air rises up the side facing the moist winds. Land on the other side, sheltered from winds by the mountains, receives little precipitation.

### Trade and Settlement

The dryness of the plateaus has meant sparse settlement. However, the climate did not discourage travelers and traders. **Oases**, or fertile places in dry areas where water is found, dot the plateaus. Across the centuries, these oases were stops for trade caravans.

Throughout history, traders have crossed the Anatolian peninsula on their way between Europe and East Asia. Traders to and from ancient Greece

> **Critical Viewing** The Zagros Mountains border the Iranian Plateau. Based on what you see in the photo, what might be some challenges of living on this plateau?

## ANATOLIAN AND IRANIAN PLATEAUS

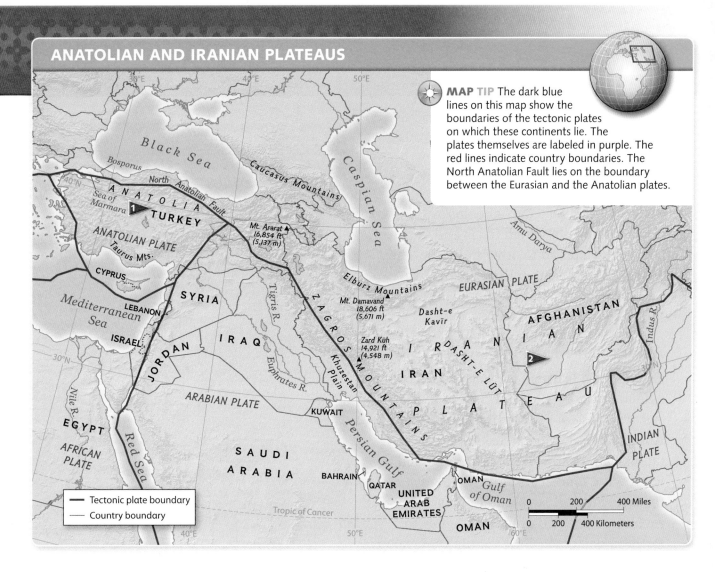

**MAP TIP** The dark blue lines on this map show the boundaries of the tectonic plates on which these continents lie. The plates themselves are labeled in purple. The red lines indicate country boundaries. The North Anatolian Fault lies on the boundary between the Eurasian and the Anatolian plates.

traveled along the Mediterranean coast. Some people moved inland, across the Taurus Mountains, and settled on the Anatolian Plateau. These settlements became stops on important trade routes.

Early human settlement took place on the Iranian Plateau, too. About 2,500 years ago, people invented a system to bring water into their arid lands. They built **qanats** (kuh NOTZ), or underground tunnels, to carry mountain waters to dry plains. This technology is still used today in Iran's capital, Tehran.

### Before You Move On
**Monitor Comprehension** In what way does the location of the Anatolian and Iranian plateaus make them an important crossroads?

### ONGOING ASSESSMENT
## MAP LAB   *GeoJournal*

1. **Interpret Maps** Locate the tectonic plate boundaries on the map. Which plates does the North Anatolian Fault border? On which plates do Turkey and Iran lie?

2. **Draw Conclusions** Locate the Zagros Mountains on the map. What conclusions can you draw about how these mountains formed?

3. **Place** What climate effect do the surrounding mountain ranges have on the Anatolian and Iranian plateaus?

4. **Explain** Over time, how did trade influence settlement on the plateaus?

SECTION 1.4  **421**

# SECTION 2 HISTORY
## 2.1 Mesopotamia

**TECHTREK**
myNGconnect.com For an online map and photos of Mesopotamian artifacts

Maps and Graphs | Digital Library

**Main Idea** Mesopotamia's early civilization contributed much to other cultures.

Mesopotamia is known as an early **cultural hearth**, or center of civilization from which ideas and technology spread to other cultures. The emergence of farming there more than 10,000 years ago allowed for advanced societies to grow.

### Agriculture Develops

As you have learned, the Fertile Crescent extends from the Persian Gulf to the eastern shore of the Mediterranean Sea. About 9500 B.C., people in this fertile land began to shift from gathering food to growing food. This shift is called the **agricultural revolution**. This revolution enabled groups of people to settle in one place and eventually develop advanced civilizations.

In addition to farming, Mesopotamians began to **domesticate** animals, or keep them as a source of animal labor and food. Farming villages grew into bigger settlements and then cities. Eventually, these cities unified into the world's first **city-states**, or independent political units.

**MESOPOTAMIA**

(map showing Black Sea, Caspian Sea, Taurus Mountains, Zagros Mountains, Mediterranean Sea, Tigris R., Euphrates R., Ninevah, Assur, Syrian Desert, Babylon, Umma, Lagash, Uruk, Ur, Arabian Desert, Persian Gulf; Legend: Fertile Crescent, Ancient coastline (about 5000 B.C.); Scale: 0–300 Miles / 0–300 Kilometers)

### Sumer

The first Sumerian city-states formed around 3500 B.C. Sumer was an ancient Mesopotamian region in what is now southeastern Iraq. Located in the lower valley of the Tigris and Euphrates rivers, **Ur** was one of the most important Sumerian city-states. Between 2800 and 1850 B.C., Ur was a center of trade.

9500 B.C. Agricultural revolution begins in Fertile Crescent.

Ancient Mesopotamian map with cuneiform text

1900 B.C. Amorites conquer Mesopotamia.

9500 B.C. | 3500 B.C. | 2500 B.C.

3500 B.C. City-states begin to develop in Sumer.

Golden helmet from Ur, a Mesopotamian city-state

422 CHAPTER 15

Its location on the Euphrates River helped it become a major port. Sea-traders from Ur connected Mesopotamia with people as far away as the Indus Valley in South Asia.

Sumerians made many significant cultural contributions. They used advanced mathematics and invented the first wheeled vehicles and the first codes of law. Sumerians also used written language to record their knowledge. Sumerian writing, called **cuneiform** (kyoo NEE uh form), is the earliest known form of writing. Clay tablets carved with cuneiform provide scientists with many details about daily life and culture in Sumer.

## Babylonia

Around 1900 B.C., the Amorites, a nomadic group from Arabia, conquered Mesopotamia. The Amorites adopted much of Sumerian culture and continued using cuneiform. Eventually, the conquered lands became known as Babylonia. The Babylonian Empire included all of southern Mesopotamia. During the 1700s B.C., King **Hammurabi** developed a code of law known as Hammurabi's Code.

After Hammurabi's death, outside invaders weakened the Babylonian Empire. Nearly 1,000 years after Hammurabi's rule, another strong ruler emerged. **Nebuchadnezzar** (nehb buh kuhd NEHZ uhr) was Babylonia's king from 605 to 562 B.C. His ambitious projects included rebuilding the port at Ur on the Persian Gulf and the creation of the Ishtar Gate in Babylon, the empire's capital. The Persians captured Babylonia in 539 B.C. Then, in 331 B.C., Alexander the Great from Macedonia conquered the Persians and Babylonia was never independent again.

### Before You Move On
**Summarize** In what ways did Mesopotamia's early civilizations contribute to other cultures?

### ONGOING ASSESSMENT
## SPEAKING LAB  GeoJournal

1. **Express Ideas Through Speech** What was the impact of the agricultural revolution? Work with a partner to develop an oral presentation that describes the agricultural revolution and how it impacted human development.

2. **Conduct Internet Research** Go online to research the cultural contributions of Mesopotamian civilizations. Discuss your findings with a partner.

---

- 1750 B.C. Babylonian Empire begins decline.
- Detail from Ishtar Gate of ancient Babylonia
- 331 B.C. Alexander the Great takes over Babylonia.

1500 B.C. | 500 B.C.

- 1792 B.C. Hammurabi becomes king of Babylon.
-  Sculpture of Hammurabi
- 539 B.C. Persians conquer Babylonia.

SECTION 2.1 **423**

SECTION 2 HISTORY

# 2.2 Birthplace of Three Religions

TECHTREK
myNGconnect.com For a map of the birthplace of three religions

Maps and Graphs

**Main Idea** Three of the world's most influential religions began in Southwest Asia.

**Judaism**, **Christianity**, and **Islam** all began in Southwest Asia. All three religions are monotheistic, meaning that they worship one God.

## Judaism

The Jewish people trace their ancestry to Abraham, who lived in southern Mesopotamia around 1800 B.C. According to the Hebrew Bible, God told Abraham to move his people to Canaan—where Lebanon and Israel are located today. Over hundreds of years, the descendants of Abraham developed a distinct religion, Judaism. In 1000 B.C., King David made Jerusalem ▷1 the capital of the Kingdom of Israel. Jews built and rebuilt the sacred First and Second Temples there. Jerusalem remains the holiest city for Jews today.

The Jewish belief in one God was new. At the time, most people believed in many gods. The establishment of monotheism marked a shift in religious practices. The **Hebrew Bible** is the sacred text of Judaism. It contains Jewish history and teaches how to lead a moral life.

## Christianity

Christianity also developed as a monotheistic religion. Among Jewish teachings was the expectation of a Messiah, a leader or savior. Jews believe the Messiah is yet to come. Christians, however, believe the Messiah was a Jew named Jesus. Born in Nazareth ▷2, Jesus began to preach in Galilee sometime around A.D. 30. Jesus drew many followers as he preached. Roman leaders who resented his popularity sentenced Jesus to death and he died in Jerusalem.

According to Christian literature, Jesus rose from the dead and told his followers to spread his message. The sacred book of Christianity is the **Christian Bible**, which includes the Old Testament, containing the Hebrew Bible, and the New Testament. The New Testament is about the life and teachings of Jesus.

The Dome of the Rock (far left) in Jerusalem was built on the site where Muslims believe Muhammad ascended into heaven. It was also built on top of the site of the First and Second Temples, and Judaism's holiest site, the Temple Mount. Just below the Dome, where people are standing and praying, is the Western Wall, Judaism's most accessible holy site.

## Islam

Another monotheistic religion that developed in this region was Islam. According to its followers, called Muslims, in the early 600s, a man from Mecca named Muhammad received revelations from Allah, the Arabic name for God. Muslims recognize Abraham, Jesus, and others as God's messengers, or prophets. For Muslims, Muhammad is the last prophet.

The holy book of Islam is the **Qur'an**, which contains the revelations Muhammad received until his death. Muslims fulfill religious duties known as the Five Pillars of Islam. One of these duties includes a **pilgrimage**, or journey, to Mecca, in Saudi Arabia, at least once in a lifetime.

**Before You Move On**
**Monitor Comprehension** What characteristics do these three religions share?

**BIRTHPLACE OF THREE RELIGIONS**
Judaism, Christianity, and Islam

**Critical Viewing** The Church of the Holy Sepulchre in Jerusalem was built on the site where Christians believe Jesus was buried. What details in the photo convey the religious nature of this setting?

## ONGOING ASSESSMENT
## READING LAB  GeoJournal

1. **Monitor Comprehension** Based on your reading, what belief is common to Judaism, Christianity, and Islam? How did this belief differ from previous beliefs?

2. **Identify** What is a prophet? Who are some important prophets of Judaism, Christianity, and Islam?

3. **Place** Locate Jerusalem, Nazareth, and Mecca on the map above. What do these ancient cities have in common in terms of their location and religious significance?

SECTION 2.2  **425**

# SECTION 2 HISTORY
## 2.3 Diffusion of Religions

**TECHTREK**
myNGconnect.com For an online map of the spread of religion and photos of religious architecture

Maps and Graphs | Digital Library

**Main Idea** Religions spread around the world through migration, missionaries, and trade.

In the years since becoming established as major religions, Judaism, Christianity, and Islam have spread thousands of miles from their birthplaces in Southwest Asia. The **diffusion**, or spread, of each religion happened in different ways.

### Migration, Missionaries, Trade

Judaism spread through the migration of the Jewish people. In A.D. 135, the Romans killed, enslaved, or exiled most of the Jewish population. While many Jews began to move from their homeland due to persecution, or discrimination because of their beliefs, a small group of Jews maintained a continuous presence. By the 1200s, they had migrated as far as Eastern Europe. The spread of Jews around the world is called the **Diaspora**. As Jews in Europe faced increasing persecution in the late 19th century, growing numbers returned to their ancient homeland. In 1948, the Jewish people established modern Israel there. Today, about 15 million Jews live throughout the world.

After the Roman Emperor **Constantine** legalized Christianity in 313, it became the official religion of the Roman Empire. As the empire expanded, so did Christianity. During the Middle Ages, European kings promoted Christianity in conquered lands. Beginning in the 1500s, predominantly Christian and European countries began to colonize the Americas, Africa, and other regions. Christian missionaries spread their faith. Today, Christianity has about 2.3 billion **adherents**, or followers.

Islam spread through the expansion of Muslim rule and through trade. By the 1500s, Islam had moved from the Arabian Peninsula throughout North Africa, Southwest Asia, Southeast Europe, and parts of India. Muslim traders spread Islam as they traveled and exchanged goods. Today, Islam has about 1.6 billion followers, more than any religion except for Christianity.

### Before You Move On
**Summarize** Compare and contrast the ways in which Judaism, Christianity, and Islam spread.

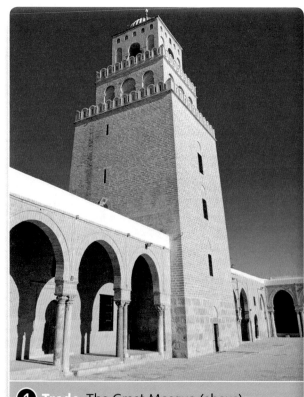

**1 Trade** The Great Mosque (above) in Kairouan, Tunisia, was founded in 670. Kairouan was an important stop for desert trade caravans. The Great Mosque continues to be an important holy site for Muslims today.

**2 Migration** In 1845, 37 Jewish immigrants from Germany founded the Temple Emanu-el synagogue in New York City. As the congregation grew, it constructed new buildings, including this one, built in 1928. Temple Emanu-el is the largest synagogue in the world.

## SELECTED EXAMPLES OF RELIGIOUS DIFFUSION

**3 Missionaries** Catholic missionaries arrived in present-day Bolivia in the 1540s to convert the Inca to Christianity. This cathedral in Sucre, Bolivia, was built in 1559. Christianity is the main religion in Bolivia and most of South America today.

### ONGOING ASSESSMENT
# VIEWING LAB

1. **Movement** Identify each location indicated on the map. What can you infer about how and where each religion spread?
2. **Evaluate** How does the religious community at Temple Emanu-el reflect the Diaspora?
3. **Explain** How did Christianity reach the Americas? How might colonization account for the number of Christian adherents today?

SECTION 2.3 **427**

# SECTION 2 HISTORY
## 2.4 The Ottoman Empire

**Main Idea** The Ottoman Empire was a powerful empire in Southwest Asia and North Africa that lasted more than five centuries.

The heart of the **Ottoman Empire** was located in what is present-day Turkey. As you have learned, Turkey is on the Anatolian Plateau, through which several trade routes crossed. Many groups had tried to control this area, including the Hittites, the Greeks, the Persians, and the Romans, who took control in A.D. 30.

### The Birth of the Ottoman Empire

When the Roman Empire split in 395, this area became part of the **Byzantine Empire**, the eastern part of the Roman Empire. The Byzantines ruled parts of Anatolia and Southeast Europe for 1,000 years. Turks from Central Asia began to invade and conquer parts of the Byzantine Empire in the 1300s. These Turks became known as Ottomans, after **Osman**, the name of their first leader.

In 1453, the Ottomans defeated the Byzantine Empire. They captured the city of Constantinople, renamed it Istanbul, and made it the capital of the empire. Istanbul became an important center of trade and wealth under Ottoman rule.

### The Empire at Its Height

The Ottoman Empire expanded its reach of power in the mid-1500s. Under the rule of **Suleyman I**, the Ottoman Empire stretched from present-day Hungary in Europe to the Persian Gulf and Red Sea in Asia. After Suleyman's reign, the empire continued to grow. It reached its farthest **extent**, or degree of spread, in the late 1600s.

> **Critical Viewing** Topkapi Palace in Istanbul served as a palace for Ottoman sultans from the mid-1400s to the early 1900s. Based on the photo, how might you describe this palace?

# THE OTTOMAN EMPIRE, 1683

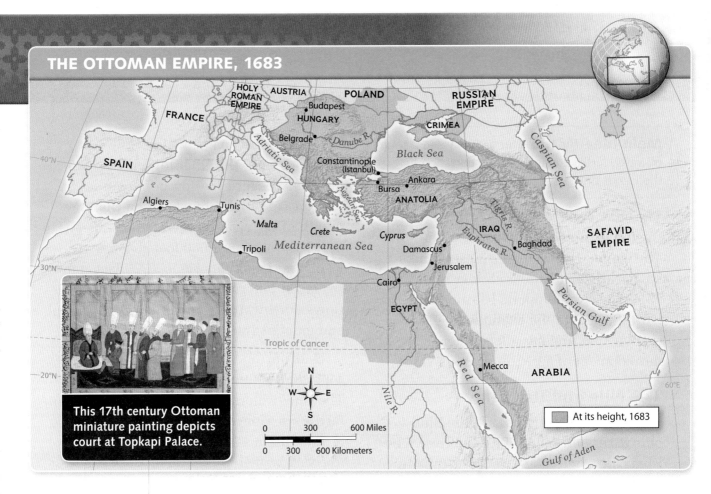

This 17th century Ottoman miniature painting depicts court at Topkapi Palace.

Most Ottoman wealth was gained through trade and taxation. The busiest trade routes in the region ran through the empire. The Ottomans controlled trade on rivers, as well as ports on important seas such as the Black Sea and the Mediterranean Sea. When the empire conquered other lands, Ottoman **sultans**, or leaders, appointed officials to collect taxes from their new subjects. Taxes were a major source of wealth in the Ottoman empire.

Because it covered such a vast area, the Ottoman Empire was composed of many different ethnic groups, including Turks, Greeks, Slavs, Arabs, and Armenians. The Ottomans were Muslim and they spread Islam throughout the empire. They were known for their **religious tolerance**. Existing religious groups maintained their own practices and communities within the empire.

Internal conflicts and wars with European countries began to weaken the empire by the late 1600s. The empire lasted into the 1900s, but after World War I, it lost most of its remaining territory. In 1923, Turkey, the last remnant of the Ottoman Empire, became a republic.

## Before You Move On

**Make Inferences** How did the Ottoman Empire grow so powerful and last so long?

### ONGOING ASSESSMENT
# MAP LAB

1. **Region** According to the map, which bodies of water did the Ottoman Empire control at its height? How did such control benefit the empire economically?

2. **Describe Geographic Information** Look at the map. Where is the capital of the Ottoman Empire? Why might that be a good location for the capital?

SECTION 2.4  **429**

# SECTION 3 FOCUS ON EGYPT
## 3.1 The Nile River Valley

**Main Idea** The Nile River has provided a source of water, fertile plains, and transportation for thousands of years.

The Nile is the longest river in the world, flowing through Africa for 4,132 miles. The United States at its widest point is still nearly 1,500 miles shorter than the Nile. The river supports life along its banks. In Egypt, for example, the Nile's regular flooding has allowed for the development of agriculture in its valley.

### The Nile and Its Valley

The Nile's main sources are the Blue Nile River and the White Nile River. The Blue Nile originates from Lake Tana in the Ethiopian highlands. The White Nile originates from Lake Victoria, which lies in Tanzania, Uganda, and part of Kenya.

Unlike most rivers, which flow east, west, or south, the Nile flows north. This happens because the southern sources of the river are higher in elevation than the mouth of the river on the Mediterranean Sea. Lake Tana is 6,000 feet above sea level, and Lake Victoria is 3,720 feet above sea level. The river drops in elevation as it flows northward through Sudan and Egypt and empties into the Mediterranean.

Each spring, snows melt at the river's sources, causing predictable flooding. The floodwaters deposit rich and fertile silt along the river's floodplain. A **floodplain** is the low-lying land next to rivers formed by sediment deposited by flooding. Historically, these favorable farming conditions supported permanent settlements and the development of an advanced civilization. The ebb and flow of the river allowed farmers to plan around its flood cycles.

**Critical Viewing** Palm trees grow along the Nile at Armana. How does this photo fit with the written description of the Nile River Valley?

NILE RIVER VALLEY

This satellite image captures the Nile River in Egypt at night. The majority of Egypt's population lives along the Nile.

## The Nile Valley Today

Use of the Nile River has changed since the 1960s, when construction of a major dam began. In 1970, the **Aswan High Dam** was completed. One reason for its creation was to generate **hydroelectric power**, a form of energy created using flowing water. The Aswan High Dam provides about 15 percent of Egypt's electric power. Another reason the dam was built was to **regulate**, or control, the flooding of the Nile. Man-made irrigation allows for crops to be grown year-round because farmers are no longer dependent on the natural flooding cycle.

Today, the Nile River Valley is a densely populated area. In fact, about 95 percent of Egyptians live along the Nile. Surrounded by desert, the river continues to support life and agriculture along its banks.

### Before You Move On
**Summarize** In what ways is the Nile River important to life in the region?

ONGOING ASSESSMENT
## PHOTO LAB

1. **Analyze Visuals** Based on the photo of the Nile River, what details demonstrate characteristics of a floodplain?
2. **Make Generalizations** Look at the satellite image above. What generalization can you make about settlement in the Nile Valley?
3. **Interpret Maps** What advantage might cities located along the Nile River have?
4. **Human-Environment Interaction** In what way did the human use of the Nile River change in 1970?

SECTION 3.1 **431**

# SECTION 3 FOCUS ON EGYPT
## 3.2 Egypt's Ancient Civilization

**Main Idea** Egypt's ancient civilization developed significantly during four main time periods.

As you have read, as the Sahara became a desert, people who lived there began to migrate to the Nile River Valley in search of a reliable source of water. They brought with them skills in pottery, metalworking, and agriculture. By around 3000 B.C., a writing system based on pictures called **hieroglyphics** (HY ruh GLIHF ihks) had developed. For the next several thousand years, ancient Egyptian civilization thrived along the Nile River.

### Old and Middle Kingdoms

Ancient Egyptian civilization is divided into four main periods, shown on the time line below. Around 3000 B.C., the kingdoms in Egypt unified under King Menes. His rule began the first **dynasty**, or series of rulers in the same family. Many dynasties followed. In the Early Dynastic period, agriculture and trade developed along the Nile. Egyptians began to build stone monuments called **pyramids**.

The Old Kingdom continued the growth of dynasties. Egypt expanded from the Nile Delta southward along the banks of the Nile. More advanced farming and trade practices enriched the Old Kingdom. Between 2150 and 2040 B.C., however, the annual floods of the Nile were not as strong, which led to a decrease in crops. This decrease created a long-term economic and political crisis in Egypt.

Menuhotep II's reign restored Egypt to stability and power in the 1900s B.C. and is considered the beginning of the Middle Kingdom. During this period, classic Egyptian arts, architecture, and literature flourished. In the 1600s B.C., however, Hyksos invaders from the East weakened and took over the Middle Kingdom.

### New Kingdom

Egyptians rebelled against the Hyksos and the defeat of the Hyksos ushered in the New Kingdom—ancient Egypt's greatest period of power and wealth. Its boundaries extended into the desert, to the Red Sea and along the eastern coast of the Mediterranean Sea.

During this period, Egyptians began to call their kings **pharaohs**. One New Kingdom pharaoh was **Hatshepsut** (hat SHEHP soot), a female pharaoh.

- 3200 B.C. Migration to Nile River Valley
- The Great Sphinx at Giza lies near King Khafre's pyramid.
- 2575–2150 B.C. Old Kingdom
- 1975–1640 B.C. Middle Kingdom

3000 B.C. — 2500 B.C. — 2000 B.C.

- 2950–2575 B.C. Early Dynastic Period
- Egyptian hieroglyphics

432 CHAPTER 15

During her reign, Egypt's economy was strengthened through trade. Another New Kingdom pharaoh, **Ramses II**, expanded Egypt's empire through conquest.

The New Kingdom began to decline about 1075 B.C. In 525 B.C., the Persian Empire conquered Egypt. In 332 B.C., Alexander the Great liberated Egypt from Persian rule. Alexander was a Macedonian king who overthrew many empires, including the Persians. Eventually, Egypt became a province of the Roman Empire.

## Egyptian Legacies

Ancient Egyptians made important contributions to culture, knowledge, and technology. Ancient Egyptian religion emphasized the belief in an afterlife, and was built around worship of **deities**, or gods. Two important deities were the sun god, **Re**, and the goddess, **Isis**.

Ancient Egyptians invented a paper-like material called **papyrus**. They used engineering skills to build cities and pyramids, and their mathematical and scientific observations endure today.

### Before You Move On
**Summarize** In what ways did ancient Egypt's civilization develop over its four main time periods?

**Critical Viewing** Artifacts like this statue from the tomb of Tutankhamen show how ancient Egyptians buried their royalty. What details from the photo indicate that King Tut is royalty?

## ONGOING ASSESSMENT
### READING LAB  GeoJournal

1. **Summarize** Ancient Egypt is frequently described as "the gift of the Nile." Based on your reading, summarize the details that might lead to this description.

2. **Movement** When and why did migration to the Nile River Valley occur?

3. **Interpret Time Lines** Based on the text and the time line, what explains the gaps in years between the Old, Middle, and New Kingdoms?

○ 1539–1075 B.C.
New Kingdom

Painting of the Egyptian goddess Isis

30 B.C. ○
Egypt becomes province of Roman Empire.

1500 B.C. — 1000 B.C. — 500 B.C.

332 B.C. ○
Alexander the Great conquers Egypt.

SECTION 3.2

# SECTION 3 FOCUS ON EGYPT
## 3.3 The Great Pyramids

**TechTrek**
myNGconnect.com For online photos of the Egyptian pyramids

Digital Library

**Main Idea** The ancient Egyptians built great tombs for their pharaohs.

The pyramids are among the enduring legacies of ancient Egypt. Archaeologists and engineers still marvel at how the Egyptians built the pyramids without modern machines and technology.

### Royal Tombs

Egyptian pyramids were built for two reasons. One purpose of the pyramids was to provide individual **tombs**, or burial places, for royalty. Secondly, kings and pharaohs demonstrated their power by having huge pyramids built for them. Ten pyramids were built at **Giza** alone.

Pyramids were holy sites, meant to transport the pharaohs to the afterlife. Ancient Egyptians believed the dead could enjoy earthly possessions, so they filled burial chambers with clothes, food, and furniture for that purpose.

### The Great Pyramid of Khufu

Built around 2550 B.C., the largest pyramid is the **Great Pyramid of Khufu,** which took 20 years to build. Initially, the pyramid stood 481 feet high. Over time, erosion has worn away the limestone and granite surface that had covered the outside of the pyramid. As a result, today the Great Pyramid is only 449 feet tall.

**Critical Viewing** Khufu's Great Pyramid (far right) is the oldest and tallest. It stands with the pyramids for Khafre (middle) and Menkaure (far left). What details in the photo convey the sizes of the pyramids?

The interior of the Great Pyramid has three separate burial chambers and several passageways. Because the Great Pyramid was looted, or burglarized, by thieves seeking valuable artifacts, only Khufu's **sarcophagus**, or coffin, remains in his burial chamber.

### Before You Move On
**Summarize** For what purposes did ancient Egyptians build the pyramids?

## ONGOING ASSESSMENT
## VIEWING LAB  GeoJournal

1. **Analyze Visuals** Based on the illustration, what does the Great Pyramid's construction reveal about ancient Egyptian engineering skills?
2. **Explain** Why might the King's Chamber be located in the center of the pyramid?
3. **Draw Conclusions** In what way did pyramid-builders try to prevent robberies? What conclusions can you draw about why guarding against theft was important?

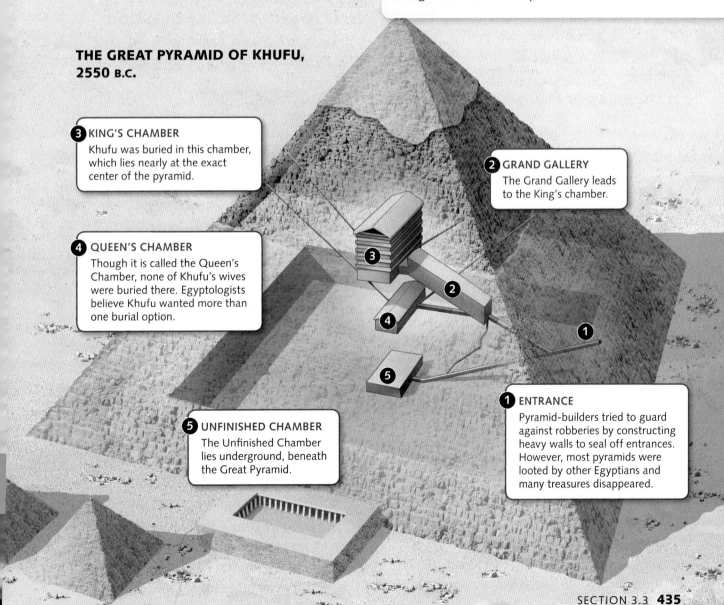

**THE GREAT PYRAMID OF KHUFU, 2550 B.C.**

**❸ KING'S CHAMBER**
Khufu was buried in this chamber, which lies nearly at the exact center of the pyramid.

**❹ QUEEN'S CHAMBER**
Though it is called the Queen's Chamber, none of Khufu's wives were buried there. Egyptologists believe Khufu wanted more than one burial option.

**❺ UNFINISHED CHAMBER**
The Unfinished Chamber lies underground, beneath the Great Pyramid.

**❷ GRAND GALLERY**
The Grand Gallery leads to the King's chamber.

**❶ ENTRANCE**
Pyramid-builders tried to guard against robberies by constructing heavy walls to seal off entrances. However, most pyramids were looted by other Egyptians and many treasures disappeared.

# CHAPTER 15
# Review

## VOCABULARY

For each pair of vocabulary words, write one sentence that explains the connection between the two words.

1. pharaoh; tomb

   *Pyramids were built as tombs for the pharaohs.*

2. hieroglyphics; papyrus
3. petroleum; nonrenewable
4. agricultural revolution; domesticate
5. desertification; arid

## MAIN IDEAS

6. Describe the climate in Southwest Asia and North Africa. (Section 1.1)
7. In what ways are the Tigris and Euphrates rivers important to the region? (Section 1.2)
8. What natural resource on the Arabian Peninsula contributes most to the economy? (Section 1.3)
9. What characteristics do the Anatolian and Iranian plateaus share? (Section 1.4)
10. Why did Mesopotamia become a cultural hearth? (Section 2.1)
11. What core belief do Judaism, Christianity, and Islam share? (Section 2.2)
12. How did Judaism, Christianity, and Islam spread? (Section 2.3)
13. In what ways did the Ottoman Empire gain wealth and power? (Section 2.4)
14. How did the Nile River help early civilizations form? (Section 3.1)
15. What contributions did ancient Egyptians make to human knowledge? (Section 3.2)
16. For what purpose did ancient Egyptians build the pyramids? (Section 3.3)

## GEOGRAPHY

### ANALYZE THE ESSENTIAL QUESTION

How have climate and location influenced the region in the past and today?

**Critical Thinking: Analyze Cause and Effect**

17. When the Sahara changed from grassland to desert, how did people living there respond?
18. In what ways did the flooding of rivers influence development in the region?

## HISTORY

### ANALYZE THE ESSENTIAL QUESTION

How did civilizations develop in Southwest Asia and North Africa?

**Critical Thinking: Summarize**

19. Summarize the ways in which European countries spread Christianity.
20. What circumstances and events led to the end of the Ottoman Empire?

### INTERPRET TABLES

**RELIGION IN SOUTHWEST ASIA AND NORTH AFRICA**

|  | Christianity | Islam | Judaism | Other |
|---|---|---|---|---|
| Egypt | 10.0% | 90.0% | — | — |
| Israel | 2.1% | 16.8% | 75.5% | 5.6% |
| Jordan | 6.0% | 92.0% | — | 2.0% |
| Lebanon | 39.0% | 59.7% | — | 1.3% |
| Morocco | 1.1% | 98.7% | 2.0% | — |
| Saudi Arabia | — | 100.0% | — | — |
| Turkey | — | 99.8% | — | 0.2% |

Source: CIA World Factbook, 2010

21. **Compare and Contrast** What country has the largest percentage of Christians?
22. **Make Inferences** Why does Israel have the largest Jewish population?

## FOCUS ON EGYPT

### ANALYZE THE ESSENTIAL QUESTION

How did an advanced civilization develop in Egypt?

**Critical Thinking: Draw Conclusions**

23. In what ways does the Nile River support life along its banks today?
24. Why is ancient Egypt referred to as "the gift of the Nile"?
25. In what ways do the pyramids demonstrate the advanced architectural and engineering skills of the ancient Egyptians?

### INTERPRET MAPS

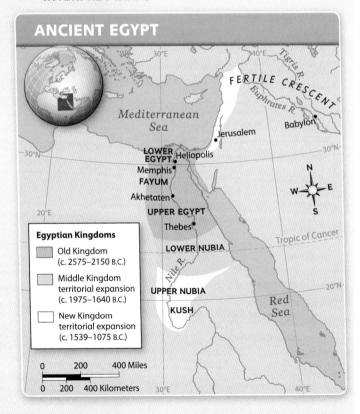

**ANCIENT EGYPT**

Egyptian Kingdoms
- Old Kingdom (c. 2575–2150 B.C.)
- Middle Kingdom territorial expansion (c. 1975–1640 B.C.)
- New Kingdom territorial expansion (c. 1539–1075 B.C.)

26. **Location** At its height, how far did ancient Egyptian kingdoms extend? Use the map scale to determine the north-south extent of the kingdoms.
27. **Explain** Which of Egypt's ancient kingdoms reached the furthest south? Based on what you know, what explains this expansion?

## ACTIVE OPTIONS

**Synthesize** the Essential Questions by completing the activities below.

28. **Write a Travel Article** Imagine you are a travel writer who has just returned from a trip to the region. Write a travel article that describes the history, physical features, and important cities, and which city has the most interesting history. Use the following tips to help you write your article. **Present your article to a small group of classmates.**

    **Writing Tips**
    - Before you write, make an outline that organizes your major points.
    - Include important dates and locations for specific events and points of interest.

    Go to **Student Resources** for Guided Writing support.

29. **Research Archaeological Sites** Many important artifacts from ancient civilizations have been found in Southwest Asia and North Africa. Conduct research, and prepare a chart that highlights two important archaeological sites in the region. Use online sources to gather your data. Organize your research with a chart similar to the one below.

|  | SITE #1 | SITE #2 |
|---|---|---|
| name and location of site | | |
| date of discovery | | |
| ancient civilization | | |
| important artifacts found there | | |
| what artifacts reveal about their civilization | | |

REVIEW **437**

# CHAPTER 16

# SOUTHWEST ASIA & NORTH AFRICA TODAY

## PREVIEW THE CHAPTER

**Essential Question** How have resources and migration shaped culture in Southwest Asia and North Africa?

### SECTION 1 • CULTURE

1.1 Migration and Trade .................................. 440
1.2 Istanbul: Bridging East and West ............ 442
1.3 Exploring Ancient Israel ........................... 444
1.4 Dubai: Desert City ..................................... 446

**KEY VOCABULARY**
- migration
- guest worker
- strait
- mosque
- breakwater
- tsunami
- sheikh
- emirate

**ACADEMIC VOCABULARY**
sprawl, interact

**TERMS & NAMES**
- Bedouin
- Tuareg
- Bosporus Strait
- Grand Bazaar
- Hagia Sophia
- Blue Mosque
- Dubai
- United Arab Emirates

**Essential Question** What forces have affected the development of modern countries in the region?

### SECTION 2 • GOVERNMENT & ECONOMICS

2.1 Comparing Governments ......................... 448
2.2 Oil and Wealth ........................................... 450
2.3 Tensions in Southwest Asia ..................... 452
2.4 Iraq's Problems and Promise .................. 454
2.5 Afghanistan: Moving Forward ................ 456
   **GLOBAL ISSUES: STANDARD OF LIVING**
2.6 Building Schools ....................................... 458

**KEY VOCABULARY**
- hereditary
- suffrage
- reserve
- petrochemical
- self-rule
- intifada
- coalition
- totalitarian
- extremist
- terrorist
- literate
- vocational

**ACADEMIC VOCABULARY**
distribute, disregard

**TERMS & NAMES**
- Knesset
- Jerusalem
- Palestine Liberation Organization (PLO)
- Shi'ite
- Sunni
- Kurd
- Saddam Hussein
- Taliban
- al-Qaeda

438 CHAPTER 16

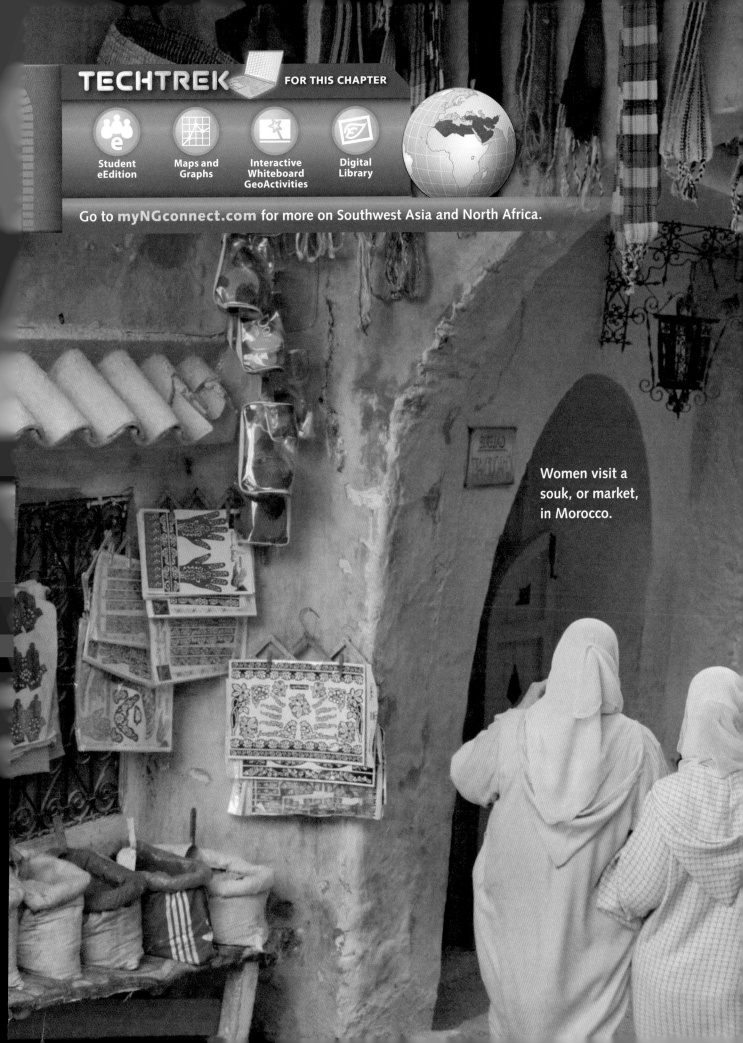

# TECHTREK FOR THIS CHAPTER

**Student eEdition** | **Maps and Graphs** | **Interactive Whiteboard GeoActivities** | **Digital Library**

Go to **myNGconnect.com** for more on Southwest Asia and North Africa.

Women visit a souk, or market, in Morocco.

# SECTION 1 CULTURE

## 1.1 Migration and Trade

TECHTREK
myNGconnect.com For photos of the Tuareg and Bedouin

Digital Library

**Main Idea** Trade and migration continue to define Southwest Asia and North Africa.

Migration, or the movement from one place to another, is a familiar process in Southwest Asia and North Africa. Three kinds of migration mark this region: movement for herding, movement for trade, and movement for job opportunities.

### Nomadic Herders

The **Bedouin** (BEHD u ihn) are a nomadic, Arabic-speaking people who live in the deserts of Saudi Arabia, Iraq, Syria, Israel, and Jordan. Most Bedouin trace their ancestry to the Arabian Peninsula. Strongly independent, Bedouin often identify themselves first as Bedouin rather than as citizens of a country.

The Bedouin move from place to place as they herd camels, sheep, goats, or cattle. Their migration patterns depend on the season and the needs of their herds.

### Salt Traders

As you have learned, the Sahara is a vast, hot desert and a difficult physical barrier to cross. However, for those who overcame the difficulties, trade across the Sahara could be very profitable. One group of traders, the **Tuareg** (TWAH rehg), historically dominated the trans-Saharan caravan trade—especially the salt trade. The Tuareg are a semi-nomadic people who live in various North African countries, including Algeria and Libya.

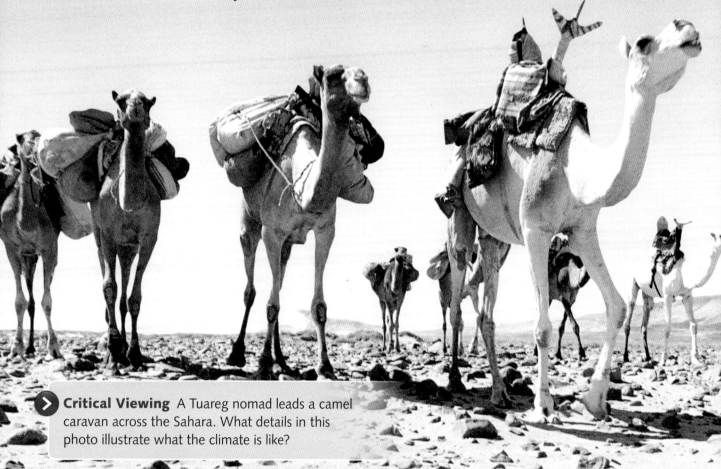

**Critical Viewing** A Tuareg nomad leads a camel caravan across the Sahara. What details in this photo illustrate what the climate is like?

Each winter, Tuareg traders travel in small caravans across a wide expanse of barren sand dunes in the Sahara. Caravans stop at oases along the way for rest and water. At these oases, the Tuareg trade goats for salt and millet, a grain, for dates, which are dried fruits of the date palm. After returning home, the Tuareg sell the salt and dates at markets for a profit.

This desert highway is located in the western Sahara.

## Guest Workers

Today's migration is prompted by a variety of different factors. Resource-rich countries such as Saudi Arabia draw millions of **guest workers**, or temporary laborers who migrate to work in another country. Companies employ guest workers to fill labor needs and shortages. Guest workers, seeking better wages than they might make in their home countries, work in industries such as oil, mining, and construction. In some cities, such as Dubai in the United Arab Emirates, guest workers represent a large percentage of the total population.

### Before You Move On
**Summarize** How would you describe the different types of migration in the region?

### ONGOING ASSESSMENT
# PHOTO LAB  GeoJournal

1. **Analyze Visuals** Based on the photo, how would you describe a caravan? What do people use to protect themselves in this environment?

2. **Compare and Contrast** Based on the photos of the caravan and the highway, compare and contrast modern and traditional methods of movement in this region. What might be benefits and drawbacks of both methods?

3. **Movement** What factors draw guest workers to this region today?

# SECTION 1 CULTURE

## 1.2 Istanbul: Bridging East and West

**TECHTREK**
myNGconnect.com For a map of the Bosporus Strait and photos of Istanbul

Maps and Graphs | Digital Library

**Main Idea** For centuries, Istanbul has been a thriving city at the crossroads of Europe and Asia.

As the only city in the world located on two continents, Istanbul ▶ is a cultural bridge between Europe and Asia. Istanbul has had several names. It began as the ancient city of Byzantium and became Constantinople in 330. When the Ottomans claimed the city as their new capital in 1453, Constantinople became known as Istanbul. Though it is no longer a capital, Istanbul is the cultural and industrial center of Turkey.

### Center of Trade and Culture

Istanbul sits on both sides of the **Bosporus Strait**. A **strait** is a narrow passage of water that connects larger bodies of water. The Bosporus Strait connects the Black Sea with the Sea of Marmara, the Aegean Sea, and, ultimately, the Mediterranean Sea. The Bosporus Strait is an important waterway that links Asia with Europe and North Africa.

Istanbul's location made this city an important center of trade for thousands of years. The **Grand Bazaar** stands as a reminder of Istanbul's history as a commerical hub. It has 5,000 shops and has been important for trade since the mid-1400s.

Istanbul is home to world-famous architecture. The Byzantines built the **Hagia Sophia** in the 500s as a Christian cathedral. In 1453, the building was converted into a **mosque**, or a Muslim place of worship. Another historical building is the **Blue Mosque**. Built in the early 1600s, it features six minarets, or tall, slender towers. The high ceilings of the Blue Mosque are decorated with more than 20,000 blue tiles that give the mosque its name.

People and cultures from all over the region converge in Istanbul. Most residents are Muslim Turks, but Istanbul is also home to many other ethnic and religious groups. Historically, diverse groups have practiced their own religions in this Muslim city.

### Rapid Growth

Today, more than 12 million people live in Istanbul, a hilly city that **sprawls**, or spreads out, for more than 90 square miles. Since the 1950s, rapid population growth has created a housing shortage. Many homes were built quickly and are not earthquake-safe, a real danger for a city that sits on the very active North Anatolian Fault.

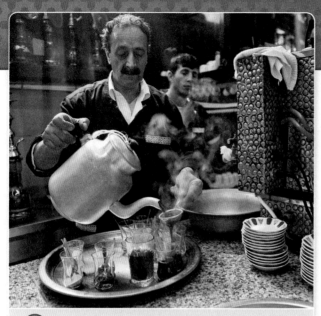

**Critical Viewing** This café is located near the Grand Bazaar. What details do you notice about how this tea is prepared and served?

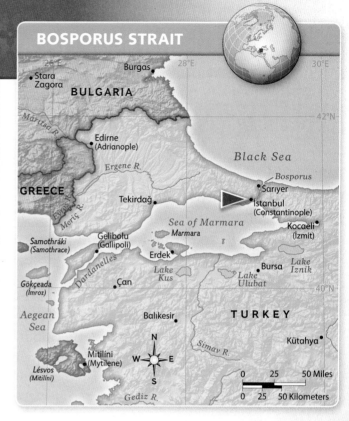

Determined to keep pace with growth, city leaders are modernizing infrastructure and services. For example, one major building project is the construction of a tunnel under the Bosporus Strait for high-speed commuter trains. As of its opening in October 2013, this modern—and earthquake-safe—tunnel has carried up to 1.5 million people back and forth across the Bosporus every day.

### Before You Move On
**Make Inferences** In what ways has Istanbul connected Europe and Asia?

**Critical Viewing** The Hagia Sophia (left) and the Blue Mosque (right) are located in Istanbul. How would you describe the city's architecture?

## ONGOING ASSESSMENT
# WRITING LAB

**Write Reports** Work with a partner to research and write a short report on Istanbul.

**Step 1** Make a three-column chart with the labels Location, Trade, and Culture.

**Step 2** Together, fill in the columns with details from the reading. Go to **myNGconnect.com** to add details and information to your chart.

**Step 3** Use the chart to help you write a two-paragraph report that describes the city as a hub of trade and culture, both historically and today.

## SECTION 1 GEOGRAPHIC

# Exploring Ancient Israel
## with Beverly Goodman

**Main Idea** Underwater exploration reveals clues about historical events in ancient Israel.

"Coastlines are the most dynamic natural environments on earth," says National Geographic Emerging Explorer Beverly Goodman. Goodman uses archaeology, geology, and anthropology to study how nature and people **interact**, or affect one another, on constantly changing coastlines.

### Herod's Harbor

In 2003, Beverly Goodman and a team of archaeologists were in Caesarea (say suh REE ah), Israel, to explore the ruins of the harbor built by Herod the Great, King of Judea, at the end of the first century B.C. Caesarea was a major port of trade between the Roman Empire and Asia. It was located on the coast of the Mediterranean Sea, in present-day Israel.

The harbor at Caesarea was one of the first harbors constructed on the open sea and had no peninsula or other natural protection. Instead, Herod's builders used concrete blocks to construct two huge **breakwaters**, or barriers, to protect the harbor. It was a gigantic structure, but for some reason it was heavily damaged and today lies in ruins on the seafloor.

**Critical Viewing** Caesarea was an important port city on the Mediterranean coast. How might the surrounding physical features make it difficult to preserve the ruins?

## An Ancient Tsunami

Initially, Goodman focused her work on understanding the construction of the ancient harbor. However, while diving, Goodman's team found layers of pottery, stone, and shells. This discovery would not have been unusual, except that the layer of shells was more than three feet thick. Goodman and her team determined that all the shells were deposited by a single, rapid, violent event. The discovery proved that a tsunami struck and destroyed the harbor sometime in the first or second century. A **tsunami** is a giant wave caused by an earthquake, volcano, or landslide.

Goodman is making similar excavations throughout the Mediterranean. She hopes to determine a pattern of tsunami activity across the region. Her findings may help the increasing number of people living on coasts anticipate tsunamis in the future.

Now an archaeological park open to visitors, Caesarea is located about 60 miles northwest of Jerusalem.

### Before You Move On

**Monitor Comprehension** In what way has underwater exploration led to an increased understanding of events in ancient Israel?

### ONGOING ASSESSMENT
## VIEWING LAB  GeoJournal

1. **Analyze Visuals** Watch the Explorer Video Clip and consider this question: What did you find interesting about Beverly Goodman's work?
2. **Pose and Answer Questions** After viewing the video clip, write two questions about what you saw. Ask a partner to answer your questions, and answer your partner's questions.
3. **Human-Environment Interaction** In what way does the construction of Herod's harbor demonstrate how people interacted with the environment to meet their needs?

# SECTION 1 CULTURE
## 1.4 Dubai: Desert City

**Main Idea** Dubai is a multicultural and rapidly developing city on the Persian Gulf.

In 1959, the **sheikh**, or Arab leader, of **Dubai** decided to turn a small fishing village into a modern city in the desert. Today, Dubai is one of the fastest growing cities in the world.

### Built on Sand

As both a city and a state, Dubai is one of seven **emirates**, or states, that make up the **United Arab Emirates** (UAE). The emirates are located on the Persian Gulf between Qatar, Saudi Arabia, and Oman. The UAE's location on the Persian Gulf makes it a prime trading center in this oil-rich part of the world.

Compared to other cities in the region, Dubai is a new city. Islands created out of sand from the Persian Gulf provide homes for wealthy residents and hotels for tourists. The Palm Jumeirah, the island featured below, is shaped like a palm tree. Just as Dubai extended its shoreline, it has also grown its skyline. In 2009, the tallest building in the world (shown at right) opened to tourists and businesses.

### Multicultural City

Dubai attracts tourists, investors, and workers from more than 150 countries. International business and tourism drive its economic growth. Businesses have moved here because they do not have to pay corporate taxes or income taxes. Dubai also has a well-developed international banking system. The global economic crisis hit Dubai and slowed its high growth rate. By 2013, tourism was helping its recovery.

Only about one in eight residents of Dubai are citizens of the United Arab Emirates. Guest workers from South Asian countries such as India represent more than 60 percent of the population. The official language of the emirate is Arabic, but people also speak Hindi, Urdu, English, and Bengali.

Because of its multicultural population, Dubai also shows a degree of religious tolerance. Located on the Arabian Peninsula, where the population is mostly Muslim, this city's Islamic mosques, Christian churches, and Hindu temples accommodate people of different faiths. Religious and ethnic conflicts among its residents are rare.

**Critical Viewing** The Palm Jumeirah is an island built out of sand from the Persian Gulf. In what ways might Dubai's sand islands extend its shoreline?

### Before You Move On
**Make Inferences** What factors contribute to Dubai's rapid growth?

### POPULATION OF DUBAI*

|  | Male | Female | Total |
|---|---|---|---|
| Employed Citizens | 30,725 | 11,053 | 41,778 |
| Employed Non-Citizens | 1,183,914 | 126,556 | 1,310,470 |
| Total Employed Persons | 1,214,639 | 137,609 | 1,352,248 |

*Aged 15 and older
**Source:** Dubai Statistics Centre—Labor Force Survey, 2009

**Critical Viewing** The Burj Khalifa stands 2,716.5 feet high and is the tallest building in the world. Based on what you can see in this photo, in what ways does the city contrast with its desert surroundings?

## ONGOING ASSESSMENT
# DATA LAB  GeoJournal

1. **Analyze Data** Look at the chart. Compare the total numbers of employed non-citizens with employed citizens in Dubai. What do you think explains the difference in numbers?

2. **Draw Conclusions** The total population of Dubai, employed and non-employed, is 1,570,923. What conclusions can you draw when you compare this number to the total number of employed citizens?

# SECTION 2 GOVERNMENT & ECONOMICS
## 2.1 Comparing Governments

**Main Idea** Countries in Southwest Asia and North Africa have varied types of governments.

Many governments in Southwest Asia and North Africa are monarchies. Other countries have established representative democracies. These varying governments must coexist as neighbors in the same region.

### Monarchies

Saudi Arabia is a monarchy ruled by a king. The position of king is **hereditary**, or passed on through family. The Council of Ministers helps the king govern but power rests with the monarch. Ministers, many of whom are from the royal family, are appointed by the king and can be dismissed at any time. Saudi kings govern by Islamic law and the country has no formal constitution. Opposition to those in power is not tolerated and political parties are banned. Only males at least 21 years of age have **suffrage**, or the right to vote. Women cannot vote. However, no national elections have yet been held.

Like Saudi Arabia, a hereditary monarch rules Jordan. Unlike Saudi Arabia, however, Jordan is a constitutional monarchy. In this form of government, monarchs must follow the constitution of the country and their power is not absolute. Jordan's legislative branch includes two parts: the National Assembly and the Chamber of Deputies. Monarchs appoint the members of the National Assembly but citizens elect the members of the Chamber of Deputies. All Jordanians 18 years of age and older can vote.

### Democracies

While countries such as Saudi Arabia and Jordan are monarchies, other countries in the region are democracies. Israel and Turkey are two examples.

Israel is a parliamentary democracy. In this type of democracy, instead of voting for individuals, citizens cast ballots for a particular party. Based on election results, each party is assigned a number of seats in the **Knesset** (kuh NEH set), or the Israeli parliament. The Knesset then elects a president and a prime minister. Israel does not have a formal constitution, but instead has a set of basic laws passed by the Knesset as its foundation.

After the end of Ottoman rule in 1923, Turkey became a republic. A republic is a form of democratic government in which representatives elected by the people hold power. Although it was established as a republic, only one party existed in Turkey for about 25 years. Since the 1950s, though, Turkey has been a multiparty republic. In 2007, voters in Turkey approved a constitutional amendment to establish direct presidential elections.

### Unrest in North Africa

Some governments in the region are technically democracies but have had a history of suppressing fair elections and limiting citizens' rights. Beginning in 2010, economic crises in these countries fueled political revolt.

In January 2011, several uprisings led to shifts in leadership. In Tunisia, President Ben Ali stepped down after days of protest

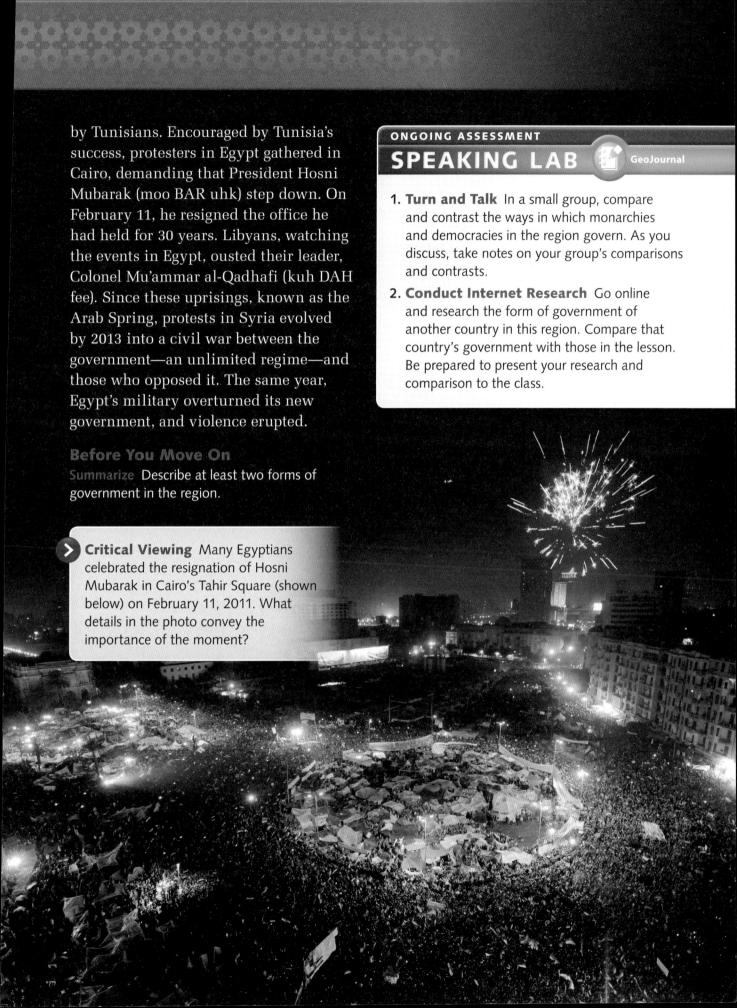

by Tunisians. Encouraged by Tunisia's success, protesters in Egypt gathered in Cairo, demanding that President Hosni Mubarak (moo BAR uhk) step down. On February 11, he resigned the office he had held for 30 years. Libyans, watching the events in Egypt, ousted their leader, Colonel Mu'ammar al-Qadhafi (kuh DAH fee). Since these uprisings, known as the Arab Spring, protests in Syria evolved by 2013 into a civil war between the government—an unlimited regime—and those who opposed it. The same year, Egypt's military overturned its new government, and violence erupted.

Before You Move On

Summarize Describe at least two forms of government in the region.

> Critical Viewing Many Egyptians celebrated the resignation of Hosni Mubarak in Cairo's Tahir Square (shown below) on February 11, 2011. What details in the photo convey the importance of the moment?

ONGOING ASSESSMENT

## SPEAKING LAB  GeoJournal

1. **Turn and Talk** In a small group, compare and contrast the ways in which monarchies and democracies in the region govern. As you discuss, take notes on your group's comparisons and contrasts.

2. **Conduct Internet Research** Go online and research the form of government of another country in this region. Compare that country's government with those in the lesson. Be prepared to present your research and comparison to the class.

# SECTION 2 GOVERNMENT & ECONOMICS
## 2.2 Oil and Wealth

**Main Idea** Oil-producing countries in Southwest Asia and North Africa possess important sources of energy and wealth.

Petroleum, or oil in its unrefined form, was discovered in the region in the 1900s. In fact, most of the world's known petroleum deposits are concentrated in Southwest Asia and North Africa. Much of the world, including the United States, depends on the region for this important energy source.

### An Oil-Rich Region

By extracting and exporting oil, several countries have grown wealthy. Among those countries, Saudi Arabia, Iran, Iraq, Kuwait, and the United Arab Emirates possess more than half of the world's known petroleum **reserves**, or future supply. About 20 percent of the world's known oil reserves are located in Saudi Arabia alone.

Even when oil prices drop, the major oil-exporters rank among the world's wealthiest countries. Some countries, such as Saudi Arabia, have used this wealth to build modern ports, airports, highways, and industrial plants. A few countries, including Kuwait and the UAE, have built industries that manufacture modern plastics and **petrochemicals**, or products made from petroleum.

Jobs in oil-rich countries have attracted guest workers. Some guest workers fill oil-producing and construction jobs. Others work in service industries or as domestic workers. Guest workers fill jobs that tend to be low-paying and temporary. In Saudi Arabia, guest workers comprise about 80 percent of the labor force; in Kuwait, they represent about 60 percent.

**Critical Viewing** This oil tanker is anchored in the Persian Gulf. What can you infer from the photo about the size of ships used to export oil?

## Oil Wealth and the Income Gap

Like oil itself, oil wealth is not evenly **distributed**, or spread out, within the region, or even within individual countries. Oil-rich countries include large areas of desert and remote villages. People who live in these underdeveloped rural areas tend not to benefit from oil wealth. Many rural residents have migrated to cities in the region seeking better economic opportunities.

However, the income gap is widening even within large, modern cities. For example, Riyadh, Saudia Arabia, is a rapidly-expanding city of more than four million people. Although it is the capital of one of the wealthiest countries in the world, housing is expensive. Many young Saudis are unemployed, and poverty is an increasing problem.

### Before You Move On
**Monitor Comprehension** In what ways has wealth gained from oil benefited and changed countries in the region?

### OIL RESERVES / PRODUCTION, TOP FIVE OIL-PRODUCING COUNTRIES

|  | Saudi Arabia | Iran | Iraq | Kuwait | UAE |
|---|---|---|---|---|---|
| Barrels of Reserves | 266.7 billion | 137.6 billion | 115 billion | 104 billion | 97.8 billion |
| Barrels Produced Per Day | 9.764 million | 4.172 million | 2.399 million | 2.494 million | 2.798 million |

Source: CIA World Factbook, 2009 estimates

### ONGOING ASSESSMENT
## MAP LAB

1. **Location** Look at the map. Near what geographic feature are most oil fields located? How might this feature benefit the countries as exporters?

2. **Interpret Maps** Of the top five oil-producing countries, which are located on the Arabian Peninsula? What other oil-producing countries are located there?

3. **Analyze Data** Which country in this region is second to Saudi Arabia in terms of oil production and reserves? According to the map, which countries produce less than 100,000 barrels per day?

# SECTION 2 GOVERNMENT & ECONOMICS

## 2.3 Tensions in Southwest Asia

**Main Idea** Israelis and Palestinians have struggled over issues of land, self-rule, and security for many years.

Current tensions in Southwest Asia have a long history. Complicated matters of land, security, and **self-rule**, or the government of a country by its own people, are critical issues for both Israelis and Palestinians.

### Founding Modern Israel

After World War I, many territories once ruled by the Ottomans became spheres of European rule. The area of present-day Jordan, Israel, the West Bank, and the Gaza Strip was placed under British control and named the British Mandate. Many Jews immigrated there, joining already established Jewish communities. An increased Jewish presence caused resentment among Palestinian Arabs.

The experience of the Holocaust during World War II prompted the United Nations (UN) to create a state for the Jewish people. In 1947, the UN voted to divide the British Mandate into two states: Arab and Jewish. Surrounding Arab countries and the Palestinian Arabs rejected the state offered by the UN. The Jews accepted the UN plan and in 1948 declared Israel an independent state.

Immediately, six Arab countries—Egypt, Iraq, Jordan, Syria, Saudi Arabia, and Lebanon—declared war against Israel. Before the war, many Palestinian Arabs fled to neighboring countries or to Arab towns in the West Bank. Israel won the war, and Egypt and Jordan took control of much of what the UN had proposed for an Arab State.

Israeli and Palestinian leaders meet to negotiate peace in 1993.

A series of Arab-Israeli wars followed for the next several decades. During these wars, the Arab states threatened or attacked Israel, and Israel defended itself. During the 1960s, Palestinian leaders created the **Palestine Liberation Organization (PLO)**. At the time, the PLO wanted to create a Palestinian state in place of Israel.

## Israelis and Palestinians Today

In 1987, Palestinians launched an **intifada**, or mass uprising. Palestinians protested—sometimes violently—against Israeli control of the Gaza Strip and the West Bank. In the 1990s, leaders of Israel, Arab countries, and the Palestinians began peace talks. Israel agreed to give the Palestinians self-rule in the Gaza Strip and the West Bank. Palestinians agreed to recognize Israel's right to exist and to renounce terrorism and violence. However, Palestinian terrorism against Israelis increased. Israel withdrew completely from the Gaza Strip and turned over control of much of the West Bank to the Palestinians.

Efforts toward peace stalled, however. Israelis and Palestinians could not agree on several issues—especially **Jerusalem**. Israel's capital is Jerusalem, but Palestinians also want to establish East Jerusalem as their capital. Tensions over the collapse of the peace process led to a second intifada in 2000. In response to the violence, in 2002, Israel began building a security barrier along the boundary between Israel and the West Bank. Permits are required to pass through the checkpoints along the barrier, which divides many workers from their workplaces and people from basic

**MAP TIP** Palestinians have self-rule in the Gaza Strip and limited rule in the West Bank. The permanent status of these areas is undecided.

services. However, violence against Israeli civilians has decreased by 90 percent.

Palestinians still desire a state. Israelis still desire security. World leaders are helping work toward a peaceful solution but progress is slow.

### Before You Move On
**Make Inferences** What issues divide Israelis and Palestinians?

## ONGOING ASSESSMENT
### MAP LAB

1. **Interpret Maps** Locate Israel on the map. Explain the significance of its location.
2. **Make Inferences** Find the Gaza Strip and the West Bank on the map. How might the location of these territories present challenges for Palestinian self-rule?
3. **Movement** In what ways is movement restricted among people who live in Israel and the West Bank?

SECTION 2.3 **453**

# SECTION 2 GOVERNMENT & ECONOMICS

## 2.4 Iraq's Problems and Promise

**Main Idea** Internal division and wars have caused great problems for Iraq, but a move toward democracy promises a better future.

For much of its modern history, Iraq has been torn by war and ruled by foreign rulers or dictators. In recent years, however, democracy—and with it, hope for the future—has begun to take root.

### Religious and Ethnic Divisions

Internal divisions have a long history in Iraq. When the Ottoman Empire ended after World War I, Great Britain established a monarchy in Iraq and defined the country's borders. The new territory brought together two distinct Arabic groups, **Shi'ite** (SHEE eyt) and **Sunni** (SOO nee) Muslims.

These two Muslim groups have been divided since the death of Muhammad, Islam's founder, in 632. The Shi'ites believe that the leaders of Islam should be descendants of Muhammad. The Sunnis believe that Islam's leaders should be chosen from those most qualified. About 75 percent of the world's Muslims are Sunni. However, Shi'ites represent about 60 percent of Iraq's population.

Iraq's newly established borders also included members of an ethnic group called the Kurds. The **Kurds** are Sunni Muslims but have a history, language, and culture that differs from their Arabic neighbors. Kurds represent 15 to 20 percent of Iraq's population. Because they are a minority ethnic group, the Kurds have at different times in their history experienced serious discrimination.

### War-Torn Nation

Iraq waged war in the 1980s with neighboring Iran partly as a result of the Sunni-Shi'ite divide. The largely Shi'ite Iran had overthrown its monarch in 1979 and established an Islamic government. Iraqi president **Saddam Hussein** was a Sunni Muslim who rose to power that same year. He feared that the Iranian Shi'ites would persuade the Shi'ites in Iraq to overthrow his government. The two countries fought for eight years, and the war ended with no clear winner.

In 1990, Iraq invaded Kuwait. Hussein claimed that Kuwait had been stealing Iraqi oil. In late 1990 and early 1991, the United States formed a **coalition**, or alliance, with other countries. Coalition countries drove Iraqi forces from Kuwait.

### Move Toward Democracy

In 2003, another coalition of countries invaded Iraq. The coalition believed that Iraq was concealing weapons of mass destruction. Although none were found, the coalition removed Hussein from power. He was tried and executed in 2006 in Iraq. Many Iraqis were relieved. As a **totalitarian** dictator, Hussein had ruled through terror.

In 2005, free elections were held in Iraq and legislators drafted a democratic constitution. Violence still occurs, but Iraqis are working to stabilize and rebuild their country.

**Before You Move On**
**Summarize** What factors do you think make Iraq's future look promising?

## IRAQ'S RELIGIOUS AND ETHNIC DIVISIONS

### ONGOING ASSESSMENT
# MAP LAB

 GeoJournal

1. **Interpret Maps** Locate the Tigris River. Which groups live along the river north of Baghdad? Which groups live along the river south of Baghdad? Which cities are located on the Tigris River?

2. **Location** Several different groups live in Iraq. According to the map, where do most Sunnis in Iraq live? Where do most Kurds live?

3. **Analyze Cause and Effect** Locate Iraq's Shi'ite population and Iraq's neighbor, Iran. How did the establishment of a Shi'ite government in Iran lead to the Iran-Iraq war?

**Critical Viewing** Iraqi women hold up ink-stained fingers indicating participation in provincial elections. What details in the photo suggest their feelings of pride?

SECTION 2.4 **455**

# SECTION 2 GOVERNMENT & ECONOMICS

## 2.5 Afghanistan: Moving Forward

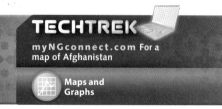

TECHTREK
myNGconnect.com For a map of Afghanistan

Maps and Graphs

**Main Idea** Afghanistan struggles to define itself and move forward in the modern political world.

Afghanistan is one of the world's poorest countries. After decades of war and upheaval, the country is trying to build a stable central government.

### Afghanistan Divided

The physical features and ethnic diversity of Afghanistan complicate efforts to unify the country. First, Afghanistan is landlocked and its mountains and deserts make travel in the country difficult. Second, about 80 percent of Afghanistan's population is rural. Widely scattered, isolated villages focus on local, rather than national, affairs. Third, Afghanistan's population includes several ethnic groups. Rural Afghans identify with their ethnic groups rather than as citizens of Afghanistan. Of Afghanistan's ethnic groups, Pashtun are most numerous, followed by Tajik, Hazara, and Uzbek.

In 1978, a Communist political party took power. This new government **disregarded**, or ignored, local ethnic and religious customs and forced policies of modernization. The speed and extent of change triggered rebellions throughout the country. Afghan leaders asked another communist government, the Soviet Union, for help. In 1979, the Soviets invaded Afghanistan. Islamic forces known as mujahedeen (moo ja heh DEEN) fought the Soviets for ten years.

After the Soviet Union withdrew from Afghanistan in 1989, the Afghan government fell, civil war tore the country apart, and locally armed leaders took control of much of the country. A group of Pashtuns known as the **Taliban** brought these local leaders under control and in 1996 seized the capital, Kabul.

### The Impact of Extremism

The Taliban are **extremists**, or people who hold rigid religious or political views. The Taliban immediately imposed their own brand of Islamic law. They destroyed non-Islamic works of art and enforced strict dress and behavior codes. The Taliban forbade women to work, go to school, or leave the house without a male relative.

Because of Taliban rule, Afghanistan became attractive to other extremist groups, including a group called **al-Qaeda**. On September 11, 2001, 19 al-Qaeda **terrorists**, or people who use violence to achieve political results, hijacked four U.S.

airplanes. They flew the planes into the towers of the World Trade Center in New York City and the Pentagon in Washington D.C. A plane headed for the White House crashed in a Pennsylvania field. In response, the United States and its allies bombed terrorist camps in Afghanistan. The Taliban were removed from power and a new government was established, but violent conflict continued.

Today, Afghanistan remains divided, but efforts to unify the country through elections continue. Most Afghan leaders believe that with political stability, economic development will follow.

### Before You Move On
**Make Inferences** What difficulties does Afghanistan face as it tries to move forward?

---

**ONGOING ASSESSMENT**
## READING LAB  GeoJournal

1. **Summarize** What factors have complicated efforts to unify Afghanistan?
2. **Analyze Cause and Effect** Why did the Soviet Union invade Afghanistan? What were the effects of the Soviet Union's withdrawal from Afghanistan ten years later?
3. **Movement** On the map, notice how much of Afghanistan is covered by mountains. How might Afghanistan's physical features influence settlement patterns?

**Critical Viewing** Afghan girls collect water outside Kabul, Afghanistan's capital. From what you can infer from the photo, in what ways is Kabul a modern city, and in what ways is it still developing?

# SECTION 2 GLOBAL ISSUES • Standard of Living
## 2.6 Building Schools

myNGconnect.com For photos of people building schools

Digital Library | Global Issues

**Main Idea** Building schools in Afghanistan and other developing countries is critical to the future.

**KEY VOCABULARY**

**literate,** adj., able to read and write

**vocational,** adj., related to a job or profession

For more than 30 years, Afghanistan has endured armed conflict, invasion, and instability. As a result, more than half of Afghan children do not attend school. Building schools in this country and in other parts of the developing world is a key part of improving children's lives and futures.

### Building Schools

According to the United Nations Children's Emergency Fund (UNICEF), only 28 percent of all Afghans are **literate**, or able to read and write. Lack of schools is one reason for low literacy rates in the war-torn country.

As you have read, when Taliban rulers took power in Afghanistan in the 1990s, they imposed their own form of Islamic law. Under Taliban rule, Afghan children, especially girls, had little or no access to education. In early 2002, just after the Taliban was pushed out, fewer than one million children in Afghanistan attended school. None of these students were girls.

Recognizing an immediate humanitarian need, multiple agencies worked with tribal leaders, military officials, and teachers in Afghanistan to build and re-open schools for Afghan children. By 2010, seven million students in Afghanistan attended school. Of these students, 37 percent were girls. New or re-built schools have been established in rural areas of Afghanistan and neighboring Pakistan, and literacy levels are climbing. In addition to teaching children to read and write, these schools offer health education and provide **vocational**, or job skills, training.

### Before You Move On
**Summarize** What factors motivated agencies to build or re-open schools in Afghanistan?

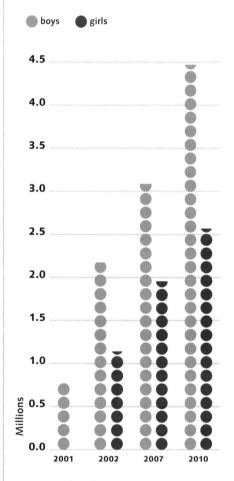

**SCHOOL ATTENDANCE* IN AFGHANISTAN**

*Equivalent of Grades 1–6

**Source:** Ministry of Education, Afghanistan, 2010; UNESCO Institute for Statistics, 2010

## Educating Girls

One challenge facing developing countries such as Afghanistan is that many girls do not attend school or they leave school at an early age. For example, on average, girls in Afghanistan attend school four fewer years than boys. Limited attendance at school results in lower rates of literacy.

Research shows that girls and women equipped with education can bring about positive change in their families and in their communities. Worldwide, a higher level of education for mothers is linked to better infant and child health. In addition, the number of years a girl spends in school is linked to her level of earnings as an adult. When women and girls earn income in the developing world, they usually invest 90 percent of that income back into their families.

**COMPARE ACROSS REGIONS**

## Schools for Africa

As one of the agencies working in Afghanistan, UNICEF recognizes the importance of building schools in war-torn and developing countries. Through its Schools for Africa program, UNICEF also partners with communities and local governments to build schools, improve classrooms, and increase access to education and opportunities for more than eight million children in at least 11 African countries.

One of these countries is Niger. Like Afghanistan, Niger is an extremely poor country. Only 38 percent of school-aged children attend school. Also like Afghanistan, school attendance rates

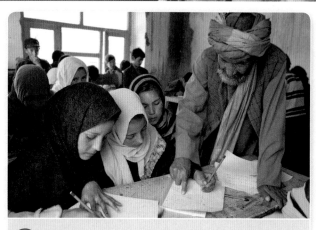

**Critical Viewing** Based on your school experiences and the past conflicts you read about, tell why this Afghan school's learning environment is important.

for girls are lower than for boys. Because of a lack of education, many girls marry and have children at a young age and tend to remain in poverty.

Schools for Africa is active in other sub-Saharan countries, including Malawi, Angola, and Rwanda. Through partnership with local residents, Schools for Africa aims to improve children's lives.

### Before You Move On
**Monitor Comprehension** In what ways does educating girls help strengthen families and communities?

**ONGOING ASSESSMENT**

## READING LAB  GeoJournal

1. **Summarize** What factors might explain the lack of educational opportunity in Afghanistan? How does this lack impact girls specifically?

2. **Interpret Graphs** Based on the graph and the text, what prediction might be made about future literacy levels in Afghanistan?

3. **Make Generalizations** In what ways are war and poverty linked to low literacy rates?

# CHAPTER 16
# Review

## VOCABULARY

For each pair of vocabulary words, write one sentence that explains the connection between the two words.

1. guest worker; migration

   *Modern migration in the region includes guest workers from many countries.*

2. breakwater; tsunami
3. petrochemical; reserve
4. sheikh; emirate
5. terrorist; extremist
6. literate; vocational

## MAIN IDEAS

7. In what ways have migration and trade shaped this region? (Section 1.1)
8. What factors make Istanbul a center of commerce? (Section 1.2)
9. How might underwater exploration of ancient sites help people living on coasts today? (Section 1.3)
10. What factors have encouraged the rapid economic growth of Dubai? (Section 1.4)
11. What are characterstics of two systems of government in the region? (Section 2.1)
12. What important resources do countries in this region possess? (Section 2.2)
13. What issues have led to violence among Israelis and Palestinians? (Section 2.3)
14. What challenges does Iraq face at home and internationally? (Section 2.4)
15. In what ways do ethnic and geographic factors divide Afghanistan? (Section 2.5)
16. Why is building schools in Afghanistan an important goal? (Section 2.6)

## CULTURE

### ANALYZE THE ESSENTIAL QUESTION

How have resources and migration shaped culture in Southwest Asia and North Africa?

**Critical Thinking: Make Generalizations**

17. In what ways has migration defined the population and culture of Dubai?
18. What factors have made Istanbul the cultural center of Turkey?
19. What characteristics do different patterns of migration in the region share?

### INTERPRET GRAPHS

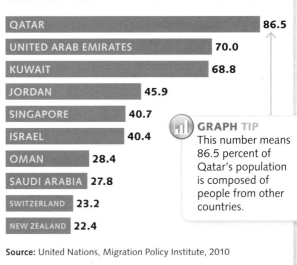

**INTERNATIONAL MIGRANTS IN SELECTED COUNTRIES, 2010**

Percentage of total population, countries with one million or more residents

| Country | % |
|---|---|
| QATAR | 86.5 |
| UNITED ARAB EMIRATES | 70.0 |
| KUWAIT | 68.8 |
| JORDAN | 45.9 |
| SINGAPORE | 40.7 |
| ISRAEL | 40.4 |
| OMAN | 28.4 |
| SAUDI ARABIA | 27.8 |
| SWITZERLAND | 23.2 |
| NEW ZEALAND | 22.4 |

Source: United Nations, Migration Policy Institute, 2010

**GRAPH TIP** This number means 86.5 percent of Qatar's population is composed of people from other countries.

20. **Explain** Of the countries selected for this graph, which of them are located on the Arabian Peninsula? What factors draw international migrants to countries located on the Arabian Peninsula?

21. **Make Inferences** Based on what you know, why might the United Arab Emirates have more international migrants than Israel?

## GOVERNMENT & ECONOMICS

### ANALYZE THE ESSENTIAL QUESTION

What forces have affected the development of modern countries in the region?

**Critical Thinking: Evaluate**

22. In what ways has conflict over land caused problems in Israel?

23. What are some ways in which Afghanistan is moving forward in the modern world?

24. What impact did the discovery of petroleum have on the culture of oil-rich countries?

### INTERPRET MAPS

GDP PER CAPITA, ARABIAN PENINSULA

25. **Make Inferences** What can you infer from the map about the economic significance of the oil industry?

26. **Compare** Compare the GDP per capita of Kuwait and Yemen. Based on what you know, what might explain the difference?

## ACTIVE OPTIONS

**Synthesize** the Essential Questions by completing the activities below.

27. **Write Discussion Questions** Explore the geographic, cultural, and historical factors that have shaped the region by writing discussion questions for a roundtable. As a group, choose three countries in this region and write discussion questions about each country. Elect a roundtable host, and decide which questions the host will ask. **Present your roundtable in class. Be prepared to answer questions from your audience.**

> **Writing Tips**
> - Research your country selection thoroughly.
> - Write questions that require more than a "yes" or "no" or one-word answer.

Go to **Student Resources** for Guided Writing support.

28. **Create Diagrams** Countries in this region have rich cultures. Working with a partner, choose two countries and do research to answer this question: Which cultural traits do these countries share, and which are unique to each country? Use online sources to gather your information. Use a Venn diagram to organize the information you find. Share your diagrams with other groups.

# COMPARE ACROSS REGIONS UNIT 8
# The World's Bread

myNGconnect.com For photos of bread

Digital Library

In Egypt, *baladi* and *pita* are breads eaten almost every day. All over the world, bread is a staple, or main food. Most cultures have one or more traditional forms of bread. Bread comes in a variety of shapes and sizes. It can be made from the ground flour of grains such as wheat, barley, rye, oats, or flaxseed.

Two main categories of bread are leavened and unleavened. Leavened breads are made with yeast, a microorganism used to make bread dough rise. Sandwich bread is an example of leavened bread. Unleavened breads—like many flatbreads eaten around the world—are made without yeast.

## Compare
- Egypt
- France
- India
- Mexico
- Peru
- Russia
- United States

### REGIONAL GRAINS, LOCAL BREADS

Types of bread vary around the world. Bread can be made from locally grown grains, or even vegetables. Flour made from barley is used to make Egyptian flatbreads. In Peru, potatoes provide a base for potato rolls. Both cornmeal and flour are used to make two different kinds of tortillas in Mexico and other Latin American countries.

How much processing goes into flours for bread varies by region. In highly industrialized countries such as the United States, wheat is processed to remove the rough outer layer of the wheat grain in order to produce a light, white flour. Because the bran has been removed, processed flours contain less fiber and are considered less nutritious.

### PROCESS AND DISTRIBUTION

In most regions of the world, making large quantities of bread follows these stages: The agriculture industry grows the grains. A processing plant manufactures flour from the grains and sells the flour to baking companies. Bakers bake the bread and sell it wholesale to grocery stores. **Wholesale** means goods are sold in large quantities for reduced prices. Grocery stores buy wholesale but sell retail. They are **retail** industries, meaning they sell goods in small quantities or single units directly to customers.

Whether yeasted breads such as baguettes in France or unleavened flatbreads such as naan in India, bread has fed people for centuries. Compare the regional breads on the opposite page.

# BREAD AROUND THE WORLD

The price of bread to consumers varies around the world. Some governments subsidize the price, or pay to reduce the cost, in order to make bread affordable to its poorest citizens. Elsewhere, grain prices directly sway the market. The prices per loaf shown here are estimates.

## 35¢
**Potato rolls** (Peru)

When wheat prices soared in 2008, the Peruvian government promoted bread made from locally grown potatoes.

## $1.37
**White bread** (United States)

Many Americans have switched to whole wheat loaves for health reasons, but heavily processed white bread is still popular with consumers.

## 89¢
**Brown bread** (Russia)

The Russian governnment hopes increased wheat plantings will continue to lower the cost of this staple.

## 1¢
**Pita** (Egypt)

A hollow, leavened flatbread, pita is the foundation of many meals. The Egyptian government subsidizes the price.

## $1.50
**Baguette** (France)

An icon of French cuisine, the price of baguettes has increased steadily. In 2002, a loaf cost about 70 cents.

Source: *National Geographic* October; 2008

## ONGOING ASSESSMENT
# RESEARCH LAB  GeoJournal

1. **Describe** What is the difference between leavened and unleavened bread?
2. **Make Generalizations** Which of the breads featured above is most expensive? Which is least expensive and why?
3. **Make Inferences** How might less-processed forms of bread make more sense nutritionally?

**Research and Create Charts** Research the breads of Mexico and India. Learn about different flours used to make breads in those countries, and determine whether their governments subsidize the price. Identify breads made both for everyday use and for special holidays or festivals. Create a chart organizing your information.

COMPARE ACROSS REGIONS **463**

REVIEW & EXTEND  UNIT 8

# Active Options

**TECHTREK**
myNGconnect.com For photos of items from the Silk Roads

Digital Library

## ACTIVITY 1

**Goal:** Extend your understanding of trade in the region.

### Draw a Map

For 2,000 years, the Silk Roads connected Asia, Europe, and Africa was a major route for trade and cultural exchange. Do research and draw a map that shows the path of the Silk Roads as they crossed Southwest Asia and North Africa. Include major cities and any other relevant detail.

Caravan crossing the Silk Roads, as detailed on a 14th century Spanish map

## ACTIVITY 2

**Goal:** Research ancient Egyptian culture.

### Write Field Notes

Conduct online research to take a virtual tour of the Valley of the Kings. Take notes on what you observe, and keep a written log of all the stops you make. Use the **Magazine Maker CD-ROM** to record in each log entry what you see there. Include any visuals that might let others see what you are experiencing.

## ACTIVITY 3

**Goal:** Learn more about recycling.

### Make a Decision

National Geographic Emerging Explorer Thomas Taha (TH) Rassam-Culhane helps people in Egypt and around the world think creatively about how to recycle used materials. With a small group, research online to learn about new ways to reuse materials. Then follow the process on page R19 to decide on a recycling project to propose to your school.

# UNIT 9
# Explore South Asia
## with NATIONAL GEOGRAPHIC

### MEET THE EXPLORER

Emerging Explorer Shafqat Hussain created Project Snow Leopard to save the endangered snow leopard species and boost the local economy in the mountain regions of Pakistan.

### INVESTIGATE GEOGRAPHY

Mount Everest, on the border of Nepal and China, is the ultimate climbing challenge. In 2006, 70-year-old Takao Arayama became the oldest to climb it. In 2010, 13-year-old Jordan Romero became the youngest. At 29,035 feet, Everest is the tallest mountain in the world.

### STEP INTO HISTORY

The Taj Mahal, a UNESCO World Heritage site, has attracted visitors to India since it was built in the 1600s. The white marble masterpiece was built by Emperor Shah Jahan as a memorial to his wife. The project took 20,000 workers 22 years to complete.

## ONLINE WORLD ATLAS

7,486 miles
Washington, D.C.
New Delhi, India

Go to myNGconnect.com for maps of South Asia.

### CONNECT WITH THE CULTURE

These women are participating in a wedding in Rajasthan, India. Weddings in South Asia are elaborate celebrations that may last for several days and include hundreds of guests.

# CHAPTER 17

# South Asia
## GEOGRAPHY & HISTORY

## PREVIEW THE CHAPTER

**Essential Question** How do South Asia's water systems affect how people in the region live?

### SECTION 1 • GEOGRAPHY

1.1 Physical Geography ................. 470

1.2 Exploring the Himalayas ......... 472

1.3 Living with Monsoons ............. 474

1.4 Resources and Land Use ......... 476

1.5 Conservation ........................... 478

**GLOBAL ISSUES: WATER RESOURCES**
1.6 South Asia's Water Crisis ........ 480

**KEY VOCABULARY**
- subcontinent
- plate
- delta
- elevation
- subsistence farmer
- monsoon
- evaporation
- drought
- arable
- famine
- sustainable
- conservation
- ecosystem
- pollution
- sanitation
- aquifer

**ACADEMIC VOCABULARY**
collide, reverse

**TERMS & NAMES**
- Himalaya Mountains
- Deccan Plateau
- Indus River
- Ganges Delta

**Essential Question** How have physical features, religion, and empires shaped South Asia's borders?

### SECTION 2 • HISTORY

2.1 Early Civilizations .................... 482

2.2 Historic Empires ...................... 484

2.3 Religion in South Asia ............. 486

2.4 Colonialism to Partition .......... 488

**DOCUMENT-BASED QUESTION**
2.5 Gandhi and the Bhagavad Gita ............... 490

**KEY VOCABULARY**
- isolation
- cultural hearth
- caste system
- empire
- tolerance
- deity
- reincarnation
- colonialism
- civil disobedience
- symbol

**ACADEMIC VOCABULARY**
isolation, displace

**TERMS & NAMES**
- Harappan
- Aryans
- Sanskrit
- Asoka
- Taj Mahal
- Hinduism
- Buddhism
- Jainism
- Sikhism
- Islam
- Vedas
- Mohandas Gandhi
- Partition
- Bhagavad Gita

# SECTION 1 GEOGRAPHY

## 1.1 Physical Geography

SOUTH ASIA PHYSICAL

**Main Idea** The key physical features of South Asia include mountains, rivers, and a delta.

On a map, South Asia looks like a diamond-shaped chunk of land that was shoved into Asia. At the center of this diamond is the Indian **subcontinent**, a separate region of the Asian continent.

## Continental Collision

South Asia is a separate **plate**, a rigid section of the earth's crust that can move independently. The plate is still moving. As a result, the **Himalaya Mountains** are being pushed higher. They are the highest mountain range in the world. The Himalayas were formed about 50 million years ago when the Indian plate **collided** with the Eurasian plate. This movement produced the world's tallest mountain, Mount Everest in Nepal. The Himalayas separate South Asia from the rest of Asia. Rainwater and snowmelt from these mountains form South Asia's rivers.

## Major River Systems

Much of South Asia's water drains into the Ganges River and the **Indus River**. These river systems provide drinking water and nourish farmland, and are considered holy by Hindus. In Bangladesh, the Brahmaputra River joins the Ganges, forming the low, fertile **Ganges Delta**. A **delta** is an area where a river deposits sediment as it empties into a body of water.

### Before You Move On
**Summarize** What are some key physical features of South Asia?

### ONGOING ASSESSMENT
## MAP LAB

1. **Location** Where is the Ganges River located?
2. **Identify** What is the climate like in northeast India near Dhaka and Kolkata?
3. **Draw Conclusions** The population in northeastern South Asia is very dense. What physical features might account for this?

SECTION 1.1 **471**

# 1.2 SECTION 1 GEOGRAPHY
## NATIONAL GEOGRAPHIC

**TECHTREK**
myNGconnect.com For a map of the Snowman Trek and an Explorer Video Clip

Maps and Graphs | Digital Library

# Exploring the Himalayas
### with Kira Salak

**Main Idea** South Asia's mountain systems create unique challenges for its people.

myNGconnect.com
For more on Kira Salak in the field today

## The Snowman Trek

The Snowman Trek involves hiking through some of the tallest and most isolated mountains in the world. Most of the passes between the mountain peaks are over 16,000 feet high. I eased comfortably into the first few days of the Trek, absorbing the beauty of the environment. On the third day we were introduced to the extreme **elevation**—the height above sea level—that we would experience the rest of the climb. It was nine hours, all uphill, to 11,800 feet.

We came to our first high mountain pass at Nyile La (roughly 16,000 feet). ▶1 This marked the first test for altitude sickness, a condition caused by a reduced level of oxygen in the air. As we climbed, the clouds parted to reveal many snow-topped peaks—the "rooftop of the world."

## Life in the Mountains

Later, we hiked to one of the most isolated villages, Thanza, ▶2 with an elevation of 13,700 feet. Only about 300 people live there. Most are **subsistence farmers**. They grow food for their families, but nothing to sell to anyone else. Food consists of a high-altitude rice, with meat and cheese from yaks, which are long-haired Tibetan oxen. Homes are built from what can be found. Walls are made from stone, and shingles from wood. Clay is the mortar that holds it together.

### BHUTAN: THE SNOWMAN TREK

As we left Thanza, all signs of human presence vanished. Even our trail disappeared from underfoot. When we were done with the Trek and arrived in Nikka Chhu village  we calculated that we'd each taken half a million steps and walked at least 216 miles.

### Before You Move On

**Monitor Comprehension** What key challenges do people living in South Asia's mountain systems face?

### ONGOING ASSESSMENT
## MAP LAB
*GeoJournal*

1. **Place** Based on the map and description of Bhutan, why do you think people built most of their towns where they did?
2. **Create Charts** Make a chart like this one. What geographic factors explain the differences between your life and life in Bhutan?

| NECESSITIES | MY LIFE | BHUTAN |
|---|---|---|
| shelter | | |
| access to water | | |
| access to food | | |
| level of fitness required | | |

SECTION 1.2  **473**

# SECTION 1 GEOGRAPHY
## 1.3 Living with Monsoons

**TECHTREK**
myNGconnect.com For online monsoon maps and photos of monsoons

Maps and Graphs | Digital Library

**Main Idea** Seasonal monsoons provide water for crops and bring fresh soil to farmland.

**Monsoons** are seasonal winds that bring intense rainfall during part of the year. These powerful wind patterns are an important feature of South Asia's climate.

### Summer: Wet Monsoons

From May to early October, winds blow northward from the Indian Ocean and bring heavy rain. Certain areas receive 100 inches of rain per year. During very wet years, these areas might receive over 300 inches.

The summer monsoon rains irrigate crops and fill reservoirs. The rains can also cause deadly floods and landslides. Despite these annual downpours, people forge ahead with their daily activities. They adapt by finding ways to save crops and navigate flooded streets.

### Winter: Dry Monsoons

From November to April, the winter monsoons **reverse** direction and blow southward. The air from this direction is usually dry. The dry monsoons do not produce as much rain as the wet monsoons do.

In fact, a very dry monsoon season can destroy crops, which threatens farmers' livelihoods and the economic well-being of the region. Because of this, people must carefully manage water they stored during the wet monsoons. They use this water to drink and to irrigate their crops. By the following June, after months of hot, dry weather, the heavy rain is welcome.

### Before You Move On
**Make Inferences** How do seasonal monsoons affect the region's economic development?

## SUMMER MONSOON
### Average Summer Precipitation

**WET MONSOONS** The wet monsoon winds pick up moisture from the Indian Ocean as they flow northwest. This moisture turns into heavy rain when it crashes into the mountains of the Indian subcontinent. The rain saturates the land and swells the rivers. This provides nourishment for crops, animals, and people alike.

## WINTER MONSOON
### Average Winter Precipitation

**DRY MONSOONS** The dry monsoons begin in the northeast and pick up moisture from the land as they move southwest. This is **evaporation**, the process of a liquid becoming a gas. If the wet monsoons do not produce enough rainfall, the dry monsoons can lead to a **drought**, or prolonged period without rain.

**Critical Viewing** These children don't seem bothered by their flooded schoolyard. What does this photo suggest about the children's life during monsoon season?

## ONGOING ASSESSMENT
# MAP LAB  GeoJournal

1. **Location** Locate Kolkata on the summer monsoon map. Why do you think Kolkata receives so much rain in the summer?

2. **Compare** Find Kolkata on the winter monsoon map. Does the city receive more or less rainfall in the winter? Why is there a difference in rainfall from summer to winter?

3. **Human-Environment Interaction** What might life be like for the people in Kolkata in summer? In winter?

SECTION 1.3 **475**

# SECTION 1 GEOGRAPHY

## 1.4 Resources and Land Use

**TECHTREK**
myNGconnect.com For an online resource map, chart, and Visual Vocabulary

 Maps and Graphs
 Digital Library

**Main Idea** The advances of the Green Revolution had both positive and negative effects.

South Asia's natural resources range from India's huge coal deposits to Sri Lanka's precious and semi-precious stones. The big story, however, is South Asia's **arable** land, or land that is suitable for farming, which feeds nearly 1.5 billion people.

### The Green Revolution

In spite of vast farmland, India often suffered terrible **famines**, or times when people starved because of a lack of food. During the 1950s and 1960s, India made slow progress in growing more food.

Then, in 1966, farmers started using new seeds to grow wheat. The seeds dramatically increased crop yield, or the amount of food grown on a unit of land. India's wheat production shot up from 10.3 million metric tons in 1960 to 20 million metric tons in 1970. Farmers also increased production of rice, fruits, sugarcane, and vegetables. This rapid and significant rise in food production was known as the Green Revolution.

However, the high-yield seeds required more fertilizer, irrigation, and pesticides, which are chemicals that kill diseases and insects. Fertilizers, pesticides, and irrigation cost money. As a result, the Green Revolution benefited mostly wealthy farmers. The Green Revolution also had negative effects on the environment. Rain washed fertilizers and pesticides into rivers, causing pollution.

### Sustainable Agriculture

Today, South Asian governments and farmers are adapting to the physical environment. They use new technologies and methods that are **sustainable**, or capable of being continued without long-term damage to the environment. More farmers are using natural fertilizers, such as manure, to enrich soils. They are also using crop rotation, in which they change by season the crops grown on a plot of land. These methods help to grow more food and also protect the environment.

**Before You Move On**
**Summarize** What have been the positive and negative effects of the farming methods used during the Green Revolution?

**Visual Vocabulary Arable** refers to land that is suitable for farming. This field is designed to let water soak into the ground. In India, more than 50 percent of the land is arable.

## LAND USE AND NATURAL RESOURCES

**Legend:**
- Copper
- Iron ore
- Coal
- Natural gas
- Oil
- Fish
- Forest
- Woodland
- Grassland
- Mixed-use, including crops
- Cropland
- Intensive cropland
- Wetland
- Desert, barren land
- Ice, cold desert, tundra

### GRAIN PRODUCTION AND POPULATION (1950–2012)

| | Food Grain Production (millions of metric tons) | Wheat Production (millions of metric tons) | Population (millions) |
|---|---|---|---|
| 1950 | 50.8 | 6.4 | 361 |
| 1960 | 82.0 | 10.3 | 439 |
| 1970 | 108.4 | 20.0 | 548 |
| 1980 | 129.6 | 31.8 | 683 |
| 1990 | 176.4 | 49.8 | 846 |
| 2000 | 201.6 | 76.3 | 1,000 |
| 2012 | 257.4 | 94.0 | 1,220 |

Sources: CIA Factbook; Indian National Science Academy

## ONGOING ASSESSMENT
### MAP LAB

1. **Location** What types of land use are most common around the main rivers in South Asia?

2. **Compare** Look at the physical map in Section 1.1 and compare it with the land use map on this page. Why do you think there are fewer natural resources in the northeast?

3. **Interpret Charts** How much wheat per person did India produce in 1950? In 2012? What happened to the amount of wheat grown?

4. **Human-Environment Interaction** How do physical features, climate, and land use in South Asia affect humans in the region?

SECTION 1.4  **477**

# SECTION 1 GEOGRAPHY
## 1.5 Conservation

**TECHTREK**
myNGconnect.com For an online population map and an Explorer Video Clip

 Maps and Graphs
 Digital Library

**Main Idea** People have contributed to the problems affecting the Ganges River.

Humans have a powerful impact on the land and the way it is used. One way they can protect the earth is through **conservation**, which is the protection of the environment. In South Asia, this includes protecting tigers in India and Bangladesh and preserving marine ecosystems in the Maldives. An **ecosystem** is a group of interacting organisms and their natural environment.

## Conserving the Ganges

In India, conservation of the rivers is a national issue. The Ganges River is a valuable natural resource. It is also considered sacred by Hindus, India's main religious group. However, a growing population and poor waste disposal have led to the contamination of the Ganges.

One source of pollution in the Ganges is raw sewage, including household wastes. Varanasi has three sewage plants that process 100 million liters of sewage a day. However, the city produces up to 300 million liters a day.

Another source of pollution comes from industry. In industrial centers such as Kanpur, big leather-producing tanneries flush waste containing arsenic, chromium, and mercury into the Ganges. Chromium can cause cancer, and mercury can damage the nervous system.

These pollutants also damage the river habitat, which threatens animals such as the river dolphin. Parts of the Ganges are so polluted that scientists refer to them as "dead." They can no longer support plant or animal life.

## Cleaning Up Pollution

Local non-governmental organizations (NGOs) are helping to reverse the damage. NGOs are nonprofit volunteer groups founded by citizens. For example, Rakesh Jaiswal founded EcoFriends to raise awareness of the condition of the Ganges at Kanpur. Jaiswal is hopeful that one day the Ganges will be clean again. Two Indian prime ministers, Rajiv Gandhi and Narendra Modi, launched aggressive programs to clean up the Ganges.

**Before You Move On**
**Monitor Comprehension** What are the main sources of pollution in the Ganges River?

**Critical Viewing** Pollution clogs the Ganges River. In what ways might pollution of the Ganges affect human life along its banks?

## POPULATION OF THE GANGES RIVER BASIN

**MAP TIP** This is a proportional symbol map. Each circle represents a different population size along the rivers. The larger the circle, the larger the population estimate.

**Populated places**
- More than 2 million
- 1 million – 2 million
- 500,000 – 1 million
- 200,000 – 500,000
- Less than 200,000

## ONGOING ASSESSMENT
# VIEWING LAB

*GeoJournal*

1. **Analyze Visuals** Go to the **Digital Library** to watch the video clip about pollution. What most interested you?

2. **Pose Questions** Draw this chart on your own paper. Write down three things you learned from the video clip and pose three questions you have after watching the clip.

National Geographic Emerging Explorer Alexandra Cousteau with Rakesh Jaiswal

| WHAT I LEARNED | QUESTIONS I STILL HAVE |
|---|---|
| 1. Kanpur is the 7th most polluted city in the world | |
| 2. | |

3. **Make Inferences** What connections can you make between the condition of the Ganges and population?

SECTION 1.5 **479**

# SECTION 1 GLOBAL ISSUES • Water Resources

## 1.6 South Asia's Water Crisis

**TECHTREK**
myNGconnect.com For data on South Asia's water crisis

Global Issues

**Main Idea** South Asia faces a water crisis because of pollution, water scarcity, and flooding.

### KEY VOCABULARY

**pollution**, n., making the environment dirty, foul, unclean, or contaminated

**sanitation**, n., measures such as sewers to protect public health

**aquifer**, n., a bed or layer beneath the surface of the earth that contains water

Several times a week, 12-year-old Somnath Dantoso drops a magnet attached to a fishing line into the Yamuna River. The river flows through New Delhi, India's capital city. He hauls rupees (India's coin money) out of the water, but he also hauls out garbage. The Yamuna is choking on **pollution**. It is loaded with food, tires, chemicals, sewage, and other waste. In fact, more than half of New Delhi's garbage ends up in the river. South Asia is facing a major water crisis.

### Polluted Water and Disease

Many other rivers in India are polluted. Water pollution is just as bad in other parts of South Asia. In Pakistan, 38.5 million people do not have safe drinking water. More than 50 million people lack proper **sanitation**, such as sewers to carry away dirty water. Rivers in many of Nepal's cities are so polluted that people cannot use them for drinking water.

In Bangladesh, millions of wells have arsenic, a poison, in the water. Experts believe that 20,000 Bangladeshis may die every year from drinking poisoned water. One United Nations scientist said that it could be "the largest mass poisoning of a population in history."

The pollution has severe consequences for the health of South Asians. Contaminated water kills 500,000 infants a year in South Asia and Southeast Asia. It can cause cancer and other deadly diseases.

### Before You Move On
**Monitor Comprehension** How does water pollution affect South Asia?

### POLLUTION BY THE NUMBERS

**55**
Percentage of New Delhi's people who are connected to the city's sewage system

**80**
Percentage of Yamuna River's pollution resulting from raw sewage

**500**
Millions of dollars that the Indian government has spent trying to clean up the Yamuna River

**855**
Length in miles of the Yamuna from the Himalayas to the Ganges

**1,815**
Growth of India's population every hour, 2012

**Sources:** Daniel Pepper, "India's Rivers Are Drowning in Pollution," *Fortune*, June 4, 2007; Rakesh Jaiswal, "India in Peril," *Smithsonian*, October 31, 2012

## Scarce Water and Drought

Water is growing scarce for the 1.5 billion people in South Asia. Because of increasing population and climate change, the region has had severe droughts recently. Elizabeth Kolbert wrote in *National Geographic* magazine:

> Since higher temperatures lead to increased evaporation, even areas that continue to receive the same amount of overall precipitation will become more prone to drought. (April 2009)

Bangladesh has already experienced severe drought. One reason is that upstream dams in India have reduced water flow into Bangladesh's rivers. In addition, the country's exploding population has taken millions of gallons of water from **aquifers**. Less water means less food, resulting in widespread hunger.

Water scarcity is also caused by shrinking glaciers in the Himalayas, which supply water to rivers and lakes in Asia. The glaciers used to build up in winter and then melt in spring, supplying water to rivers and lakes in Asia. According to some scientists, climate change is causing the glaciers to melt faster and shrink in size. These scientists fear that the glaciers will supply less water to rivers. People will have less water for drinking, sanitation, and irrigation.

While some areas will have droughts, other areas may experience more flooding. According to Kolbert, climate change is increasing air temperatures, and warmer air holds more moisture. As a result, the monsoons that hit South Asia every year may grow more severe.

**COMPARE ACROSS REGIONS**

## China and India's Solutions

South Asians are studying how other countries have worked to improve water quality. In China, for example, 70 percent of the rivers and lakes are polluted. In *National Geographic* magazine, Brook Larmer explained that the Chinese have formed thousands of organizations urging a reduction in water pollution. (May 2008)

India's government is also trying to clean up rivers and lakes. In 1986, it passed the Environmental Protection Act. The law allows the government to stop companies from polluting.

The people of South Asia are making significant contributions to their society by taking steps to clean up the region's rivers and lakes. Citizens are calling for companies to stop polluting and are pushing their governments to take action.

### Before You Move On
**Summarize** What is causing water scarcity in South Asia?

**ONGOING ASSESSMENT**

# READING LAB  GeoJournal

1. **Analyze Data** Analyze the data provided in "Pollution by the Numbers" and the map in Section 1.5. Which factors explain why the Ganges is more polluted than the Brahmaputra?
2. **Make Inferences** Based on the rate of India's population growth, what do you think might happen to the other figures shown?
3. **Write Comparisons** How are China and India dealing with water pollution? Write a sentence comparing their tactics.

# SECTION 2 HISTORY
## 2.1 Early Civilizations

**Main Idea** South Asia's ancient civilizations developed around the region's river systems.

The physical geography of a place often influences its history. Good farmland and the geographic **isolation** (separation) of South Asia made the Indus and Ganges river valleys **cultural hearths**, or centers of civilization from which ideas spread. Mountains and oceans provided natural boundaries that limited invasions. Fertile soil along the rivers provided arable farmland. Two major civilizations that thrived here were the Harappans and the Aryans.

### Harappan Civilization

The first urban civilization in South Asia was the **Harappan** (huh RA puhn) civilization. It developed along the Indus River in what is now Pakistan. The land was fertile and good for farming. As a result, people formed farming villages that grew into cities. The two greatest cities of this civilization were Mohenjo-Daro and Harappa. These cities were well planned and laid out in a grid pattern of straight streets. The cities also had brick houses, indoor plumbing, and a sewer system. These cities provide early examples of organized city planning that helped to support cultural growth.

The Harappan people developed advanced technologies and a system of measurement using weights and bricks that were a standard size. Among the ruins of these cities, archeologists have found stone seals with images of animals and script. Based on this evidence, scholars believe the Harappans may have developed a system of writing, but the scripts have not yet been translated.

After a period of prosperity, the Harappan civilization started to decline between 2000 and 1700 B.C. Historians believe some possible causes include the change of the Indus River's course, floods, and earthquakes.

### Vedic Period

According to many historians, around 1500 B.C., waves of new people began crossing the Hindu Kush into India. The migrants were a collection of tribes called **Aryans**, meaning "noble ones." Some historians, though, believe that other groups within South Asia began mixing with one another. The period following the height of the Harappan civilization is known as the Vedic period, named for religious texts called the Vedas. These were recorded in **Sanskrit**, the basis of many modern languages in South Asia. The Vedas are the holiest book of Hinduism, the major religion of India today.

- 2600–2500 B.C. Emergence of the Harappan, or Indus Valley, civilization
- 2000–1700 B.C. Decline of the Harappan civilization

3000 B.C. — 2500 B.C. — 2000 B.C.

- 2500 B.C. A sculpture fragment of a priest-king from the site of Mohenjo-Daro
- 2000 B.C. Beginning of Aryan migration into the Indus Valley

## ANCIENT RIVER VALLEY CIVILIZATIONS

Some historians believe that people migrated to South Asia following the routes shown in the map.

Vedic society was organized into four social categories called varnas, based on one's interests. There were four groups:

Brahmins: to teach
Kshatriyas: to fight and rule
Vaisyas: to own land and trade
Sudras: to serve

By the end of the Vedic period, another social grouping called Jates emerged. Jatis were based on occupation and later became birth-based. Jatis grew more rigid, and Europeans later called this a **caste system.** Eventually, Indians of all religions identified with one or another of the castes. The caste system is not as prominent in modern India.

### Before You Move On
**Monitor Comprehension** Where did South Asia's ancient civilizations develop?

### ONGOING ASSESSMENT
## LANGUAGE LAB  GeoJournal

1. **Compare** Read aloud the Sanskrit words in the chart. Compare them to the English words.

| SELECTED INDO-EUROPEAN LANGUAGES | | | |
|---|---|---|---|
| Sanskrit | pitar | matar | dvi |
| English | father | mother | two |
| Greek | patéras | matros | dyo |
| Latin | pater | mater | duo |
| Spanish | padre | madre | dos |

2. **Draw Conclusions** A language family is a group of languages that come from a common ancestor. After looking at the chart, what can you conclude about languages in the same language family?

3. **Interpret Time Lines** During what period of time were the Vedas written?

- 1500–1200 B.C. Vedas written

1500 B.C.

- 1500 B.C. Aryan migration into northern India

1500 B.C. Pages from the Vedas, sacred text in Sanskrit

500 B.C.

# SECTION 2 HISTORY
## 2.2 Historic Empires

**TECHTREK**
myNGconnect.com For an online map on empires and photos of artifacts and architecture

Maps and Graphs | Digital Library

**Main Idea** Three South Asian empires made significant cultural contributions in religion, science, and the arts.

The Mauryan, Gupta, and Mughal empires (an **empire** is the land or people ruled by one leader) dominated the history of South Asia between 321 B.C. and A.D. 1858. Like the Harappan and Aryan empires, these later empires had the advantage of protective mountains and arable land. The Mauryans, Guptas, and Mughals also helped spread three major religions in the region—Buddhism, Hinduism, and Islam.

### The Mauryan Empire

The Mauryan (MOWR yuhn) Empire was founded in the Ganges River valley in 321 B.C. The Mauryans built an efficient and organized government, which allowed them to run the empire. They also had a standing, or permanent, army.

The leader **Asoka** ruled for nearly 40 years and brought the empire to its height around 250 B.C. As the Mauryan empire grew and flourished, Asoka turned away from conquest in favor of more peaceful policies. He studied Buddhist nonviolent teachings and built many stupas, which are Buddhist religious structures.

Buddhism began in India, but gained more followers in east and southeast Asia as it spread. After Asoka's death, the Mauryan Empire declined.

### The Gupta Empire

The Gupta (GUP tuh) Empire began around A.D. 321 in the fertile Ganges River valley. Gupta leaders practiced Hinduism, which became the major religion of South Asia.

Gupta artists and scientists created lasting cultural contributions. Advances in metal working, literature, mathematics (including the development of the decimal), and astronomy were part of this legacy. Eventually, invasions weakened the Guptas and by A.D. 540, their reign was over.

### The Mughal Empire

One thousand years later, the Mughal (MOO guhl) Empire was established in 1526 by Babur. Mughal rulers practiced Islam and came from Central Asia. Islam became a unifying force in South Asia and grew into a large religious minority.

Akbar the Great came to power in 1556 and ruled for 49 years. Akbar expanded the empire and practiced religious **tolerance**, or respect for others' beliefs.

**321 B.C.–185 B.C. Mauryan Empire**
Asoka's Pillar, c. 273 B.C., inscribed with his laws. The lion is a symbol of India.

**A.D. 321–A.D. 540 Gupta Empire**
Gold coin from Gupta Empire, c. A.D. 321

500 B.C. — A.D. 1 — A.D. 500

- c. 250 B.C. Height of the Mauryan Empire
- 321 B.C. Chandragupta Maurya founds the Mauryan Empire.
- 400 Height of the Gupta Empire
- 321 Chandra Gupta I founds the Gupta Empire.

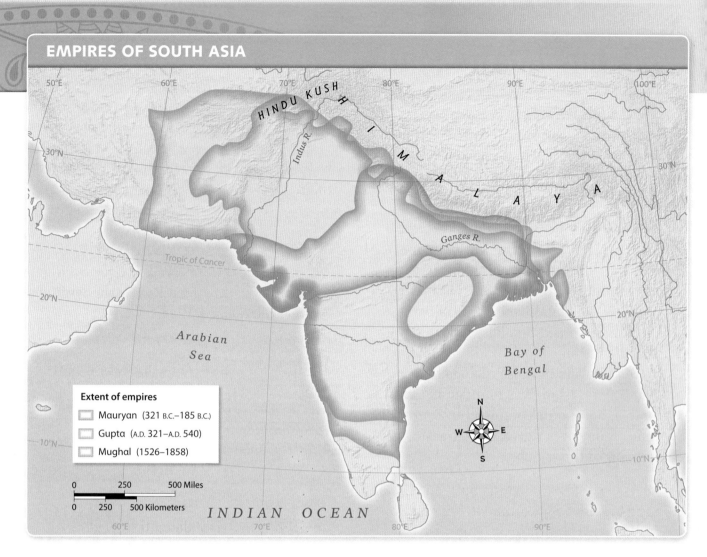

### EMPIRES OF SOUTH ASIA

**Extent of empires**
- Mauryan (321 B.C.–185 B.C.)
- Gupta (A.D. 321–A.D. 540)
- Mughal (1526–1858)

Mughal leaders developed a large empire through military conquest. Cultural contributions of Mughal artists included the architecture of the **Taj Mahal** and detailed paintings called miniatures. The empire came to an end in 1858 when the British took control of the territory.

### Before You Move On
**Monitor Comprehension** What cultural contributions did the three empires make?

## ONGOING ASSESSMENT
# MAP LAB

1. **Movement** What physical features influenced the empires' ability to control their territory?
2. **Make Generalizations** Based on the map and what you know about arable lands, what geographic features were vital for the success of these empires?

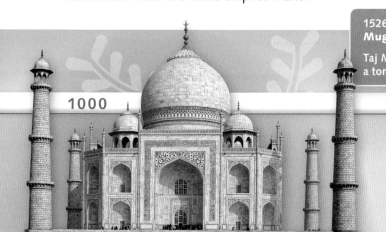

**1526–1858**
**Mughal Empire**
Taj Mahal, 1600s, built as a tomb for Shah Jahan's wife

1000　　　　　　　　　　1500　　　　　　　　　　PRESENT

● 1700
Height of the Mughal Empire

● 1526
Babur founds the Mughal Empire.

SECTION 2.2 **485**

# SECTION 2 HISTORY
## 2.3 Religion in South Asia

**Main Idea** Religion was an important part of South Asia's history. It is still important to the culture today.

Religion is an important part of South Asia's history because it shaped borders and cultures. Two of the world's main religions—**Hinduism** and **Buddhism**—were founded in India along with two other religions, **Jainism** and **Sikhism**. Buddhism spread into East Asia and Southeast Asia. Today, Buddhism has more followers in those regions than in South Asia. **Islam** was founded in Saudi Arabia, but it quickly spread to South Asia. Today, it is the second-largest religion in the region.

**SIKHISM** Sikhism began in India in the late A.D. 1400s. It was created to combine aspects of Hinduism and Islam. Sikhs believe in one god, truthful living, and equality of humankind. Today, most Sikhs live in the Indian state of Punjab.

**Followers in South Asia:** Approximately 1 percent of the population

**JAINISM** Jainism developed in the 7th century B.C. Jains believe in *ahimsa*, or nonviolence towards all living things. Most Jains live in northwestern India.

**Followers in South Asia:** Approximately 1 percent of the population

**BUDDHISM** Buddhism was founded in India around 525 B.C. by a prince named Siddhartha Gautama. Gautama left his royal life to find a solution to human suffering. After six years, he discovered and began teaching the Four Noble Truths: 1) Suffering is a part of life; 2) Selfishness is a cause of suffering; 3) It is possible to move beyond suffering; and 4) There is a path that leads to this point.

People who followed Gautama's teachings called him the Buddha, or "Enlightened One." They recorded his teachings in a set of books called the *Tripitaka*, or "Three Baskets." During the next few centuries, Buddhism spread throughout Asia. In South Asia, it is the main religion of Bhutan and Sri Lanka. You can learn more about Buddhism in the chapters on East Asia.

**Followers in South Asia:** Approximately 2 percent of the population

**HINDUISM** Hinduism is the world's third largest religion and developed over thousands of years. Today it is the main religion in India and Nepal, with significant Hindu population in Bangladesh, Sri Lanka, and Pakistan. Hindus worship many **deities** as the representation of one supreme being, often called Brahman. Hindus also believe in **reincarnation**, or the rebirth of the soul. After death, a person's soul is reborn into another physical life. The kind of life is determined by the soul's *karma*, or actions during a previous life. If the soul has lived a good life, it is reborn into a better life. If the soul has lived an evil life, it is reborn into a worse life. This process continues until the soul lives a perfect life.

These beliefs and others are recorded in many different texts. The most important are the **Vedas**, the *Puranas*, the *Ramayana*, and the *Mahabharata*.

**Followers in South Asia:** Approximately 63 percent of the population

## SOUTH ASIA'S RELIGIONS

 **ISLAM** Islam arrived with Muslim traders in the early A.D. 700s. Today it is the main religion of Pakistan and Bangladesh. It is also the largest minority religion in India. You can learn more about Islam in the chapters on Southwest Asia.

**Followers in South Asia:** Approximately 30 percent of the population

### Before You Move On
**Summarize** Why is religion important to South Asia's history?

### ONGOING ASSESSMENT
## DATA LAB          GeoJournal

1. **Create Graphs** Using the map and data in this section, make a circle graph that shows the percentages of followers of the religions in South Asia. How does the graph help you understand the distribution of religions across South Asia?

2. **Pose and Answer Questions** With a partner, ask and answer three questions about South Asia's religions.

3. **Location** In which two South Asian countries is Islam the dominant religion? What South Asian country separates these two Islamic countries from each other?

SECTION 2.3  **487**

# SECTION 2 HISTORY
## 2.4 Colonialism to Partition

**Main Idea** Colonialism, independence, and conflict set new national borders in South Asia.

In 1600, during the Mughal Empire, the British established the East India Company, beginning an extensive trading relationship with countries in East and South Asia. As the Mughal Empire declined, the East India Company was able to take control over parts of India. Eventually, this power grew into **colonialism**, the control by one power over a dependent area or people.

### Colonialism Limits Growth

In order to profit from India's rich natural resources, the British used trade practices that favored Britain. They shipped India's raw materials back to England and forced India to import British goods. The British also tried to prevent Indian manufacturers from producing certain goods that British manufacturers made, so that only British goods could be sold. This crippled India's economic growth for more than 100 years.

In 1857, Indians unsuccessfully rebelled against British control. The East India Company was disbanded in 1858, but the British established direct rule over India called the *raj*. British colonization had a lasting impact on the region's borders.

### Seeking Independence

In 1885, Hindus formed the Indian National Congress (INC). Muslims formed the Muslim League in 1906. Both groups opposed British rule. India's movement for independence grew in the 1930s under the leadership of lawyer **Mohandas Gandhi** (GAHN dee). Gandhi campaigned for **civil disobedience**, or the nonviolent disobeying of laws, against the British.

India also faced conflict between Hindus and Muslims. Many Muslims called for their own land. In July 1947, the British Parliament passed the India Independence Act. British India would be divided into two countries: majority-Hindu India and majority-Muslim Pakistan, East and West. The part of Pakistan known as East Pakistan became Bangladesh in 1971.

### Partition Creates Boundaries

Before **Partition**, or division, people scrambled to decide where to live. Many Muslims in India moved to Pakistan, and many Pakistani Hindus migrated to India. For hundreds of thousands of Indians, the movement was not peaceful. In August of 1947 they were **displaced** (forced to leave their homes) or even killed in the religious conflicts that followed.

Silk textiles are a thriving industry in South Asia.

1850

1900

○ 1857 First Indian revolt for Independence

○ 1885 Indian National Congress Party leads India's movement for independence.

○ 1906 Muslim League leads India's Muslims in movement for their own nation.

In 1948, Gandhi was assassinated. The boundaries set by Partition have strained relations in South Asia to the present.

## Conflict Over Kashmir

Partition established the present-day borders of South Asia. After Partition, India and Pakistan began a violent conflict over a region known as Kashmir. Both India and Pakistan believed the region belonged to them.

In 1949, a cease-fire was called, and Kashmir was divided. India maintains control over the portion called Jammu and Kashmir; Pakistan controls the portion known as Gilgit-Baltistan. However, Indian and Pakistani soldiers stand guard along the disputed border, and tension remains at a high level.

Part of the continuing conflict centers around the control of water for drinking and irrigation. Many rivers, including the Indus, begin in Kashmir, and maintaining control of the water is important to both countries. In fact, both countries have threatened to use nuclear weapons to resolve the dispute.

**Before You Move On**
**Summarize** What impact did British rule in India have on borders in South Asia?

### CONFLICT OVER KASHMIR

## ONGOING ASSESSMENT
# MAP LAB  GeoJournal

1. **Interpret Maps and Time Lines** On this map, striped areas represent conflict between two countries. Which three countries disagree about their borders? Based on the time line, when was Kashmir divided?

2. **Location** How does the presence of rivers factor into the cause of the conflict in Kashmir?

3. **Turn and Talk** What physical features of this region might make it hard to control borders? Turn to your classmate and identify details from this map, or other physical maps in the chapter, that might help answer this question. Take notes to share with the class.

○ 1930
Gandhi continues his civil disobedience campaigns.

○ 1947
India gains independence. Partition occurs.

○ 1949
Kashmir divided between India and Pakistan

1950

Mohandas Gandhi

PRESENT

# SECTION 2 DOCUMENT-BASED QUESTION

## 2.5 Gandhi and the Bhagavad Gita

Every religion has sacred text, and in Hinduism the most important may be the **Bhagavad Gita** (bah guh vahd GEE tah). It is a poem, part of the larger *Mahabharata* (muh hah BAH ruh tuh), which makes up many of Hinduism's stories, traditions, and beliefs. It is an example of literature that has spread beyond Indian society and conveys religious themes. The Bhagavad Gita was written around the first or second century A.D. in Sanskrit, an ancient language of South Asia. The poem is still loved by millions of Hindus.

### DOCUMENT 1
### Gandhi on the Bhagavad Gita

> When disappointments stare me in the face, and when I see not one ray of light... I turn to the Bhagavad Gita and find a verse to comfort me, and I immediately begin to smile in the midst of overwhelming sorrow. My life has been full of external tragedies, and if they have not left any visible and indelible [permanent] effect on me, I owe it to the teaching of the Bhagavad Gita.

Gandhi with his granddaughters

**CONSTRUCTED RESPONSE**

1. Think about what you learned in Section 2.4. What specific "external tragedies" might Gandhi be referring to?

### DOCUMENT 2
### *from* the Bhagavad Gita
translated by Ranchor Prime

The Bhagavad Gita includes a conversation between the deity Krishna and a young warrior, Arjuna. Krishna explains to Arjuna his responsibilities as a warrior. Gandhi especially liked this passage:

> One without self-control cannot have a clear intelligence or a steady mind. An unsteady mind finds no peace, and without peace where is joy? As a strong wind sweeps away a boat on the water, so the mind dwelling on even one of the senses carries away the intelligence.

**CONSTRUCTED RESPONSE**

2. According to this passage, humans should not allow emotions to overcome logic. Think about what you have learned about how Hinduism unifies its followers. Explain how this work of literature transcends, or goes past, the boundaries of societies to express a universal theme.

# WHEN DISAPPOINTMENT STARES ME IN THE FACE AND ALL ALONE I SEE NOT ONE RAY OF LIGHT, I GO BACK TO THE BHAGAVAD GITA.

— GANDHI

### DOCUMENT 3

## Vishnu, Preserver of the Universe

Hindus believe Krishna, the narrator of the Bhagavad Gita, is a human form of Vishnu, a Hindu deity. Every element in this image of Vishnu is a **symbol** (something that stands for another thing) of an idea in Hinduism. For example, the crown symbolizes Vishnu's highest authority.

**CONSTRUCTED RESPONSE**

3. The lotus flower in his left hand shows that Vishnu is spiritually perfect. How does this symbol reflect the qualities described in Documents 1 and 2?

**ONGOING ASSESSMENT**

## WRITING LAB

**DBQ Practice** Think about the Bhagavad Gita and the picture of Vishnu. Why would these ideas and symbols be important to Gandhi and what he wanted for Hindus?

**Step 1.** Review Gandhi's influence on history in Section 2.4 and the ideas of Hinduism in Section 2.3.

**Step 2.** On your own paper, jot down notes about the main ideas expressed in each document.

> Document 1: Quotation from Gandhi
> Main Idea(s) _____
>
> Document 2: Excerpt from the Bhagavad Gita
> Main Idea(s) _____
>
> Document 3: Image of Vishnu
> Main Idea(s) _____

**Step 3.** Construct a topic sentence that answers this question: Which ideas and symbols from the Bhagavad Gita and the image of Vishnu would be important to Gandhi?

**Step 4.** Write a detailed paragraph that explains why each idea or symbol in your topic sentence was important to Gandhi and the history of India. Go to **Student Resources** for Guided Writing support.

# CHAPTER 17
# Review

## VOCABULARY

For each pair of vocabulary words, write one sentence that explains the connection between the two words.

1. subcontinent; plate

> *The Indian subcontinent is on a different plate than the rest of Asia.*

2. evaporation; drought
3. arable; sustainable
4. conservation; ecosystem
5. deity; reincarnation
6. mythology; symbol

## MAIN IDEAS

7. How were the Himalaya Mountains formed? (Section 1.1)
8. How do people living at high elevations in Bhutan feed themselves? (Section 1.2)
9. What are the two kinds of monsoons in India, and what are the effects of each? (Section 1.3)
10. What were two results of the Green Revolution in India during the 1960s? (Section 1.4)
11. What factors contributed to the pollution of the Ganges River? (Section 1.5)
12. What are the consequences of water pollution in South Asia? (Section 1.6)
13. How did Hinduism develop in India? (Sections 2.1, 2.3)
14. What events led to India's independence from Britain? (Section 2.4)
15. How did Islam spread to India? (Section 2.3)

## GEOGRAPHY

### ANALYZE THE ESSENTIAL QUESTION

How do South Asia's water systems affect how people in the region live?

**Critical Thinking: Analyze Cause and Effect**

16. What is the geographic connection between snowmelt in the Himalayas and South Asia's extensive river systems?
17. What are the effects of the wet and dry monsoon seasons on India's farmers?
18. What geographic factors make the Ganges Delta one of the primary agricultural regions of the Indian subcontinent?

### INTERPRET CHARTS

**THE GREEN REVOLUTION AND INDIAN VILLAGES (1955–1986)**

| | 1955 | 1965 | 1975 | 1986 |
|---|---|---|---|---|
| Percentage of farmland irrigated | 60 | 80 | 100 | 100 |
| Use of farm machinery per village | none | none | 4 tractors | 9 tractors |
| Population per village | 876 | *N/A | N/A | 1,076 |
| Average size of farms | 4.8 acres | N/A | 5.7 acres | N/A |

*Not available
Source: Brace, Steve. *India*. Des Plaines, IL: Heinemann Library, 1999, p. 29

19. **Make Generalizations** What happened to the amount of land irrigated in the average Indian village between 1955 and 1975?
20. **Summarize** What was the trend in the use of farm machinery? What do you think explains this trend?

# HISTORY

## ANALYZE THE ESSENTIAL QUESTION

How have physical features, religion, and empires shaped South Asia's borders?

**Critical Thinking: Summarize**

21. How did the Mauryan and Mughal empires bring religious diversity to India?
22. How did the Aryan migrations help to create cultural unity in the Indian subcontinent?
23. What religious differences led to boundaries between India and Pakistan?

**INTERPRET MAPS**

24. **Movement** What physical barrier did the Aryans have to cross to migrate into India?
25. **Make Inferences** Why might the Aryans' path of migration have avoided the western region of India?

## ACTIVE OPTIONS

**Synthesize** the Essential Questions by completing the activities below.

26. **Write a Speech** Imagine that you live in a small village in South Asia today. You must persuade the village government to support sustainable agriculture. Write and deliver a one-minute speech to the village about its benefits. **Share your speech with the class.**

> **Writing Tips**
> - Make sure your introduction is lively enough to catch the reader's attention.
> - Present clear, strong evidence in support of your opinion.
> - End with a strong conclusion that summarizes your position.

27. **Create Charts** Make a three-column chart showing comparisons between India, Pakistan, and Bangladesh. Use online resources to gather the data. Then answer this question: What conclusions can you draw from the data recorded?

|  | India | Pakistan | Bangladesh |
|---|---|---|---|
| Year country was formed |  |  |  |
| Population |  |  |  |
| Square miles of land |  |  |  |
| Main crops |  |  |  |
| Form of Government |  |  |  |
| Religion of majority of population |  |  |  |

# CHAPTER 18

# South Asia TODAY

## PREVIEW THE CHAPTER

**Essential Question** How is diversity reflected in South Asia's cultures?

### SECTION 1 • CULTURE

1.1 Hinduism Today .................................. 496

1.2 Changing Traditions .......................... 498

1.3 Daily Life ............................................ 500

1.4 Popular Culture ................................. 502

**KEY VOCABULARY**
- karma
- pilgrimage
- henna
- sari
- literacy rate
- shalwar-kameez
- cricket
- popular culture

**ACADEMIC VOCABULARY**
discrimination

**TERMS & NAMES**
- Krishna
- Bollywood

---

**Essential Question** Why has India experienced an economic boom?

### SECTION 2 • FOCUS ON INDIA

2.1 The Largest Democracy ..................... 504

2.2 Economic Growth .............................. 506

2.3 The Golden Quadrilateral ................. 508

**KEY VOCABULARY**
- democracy
- federal republic
- infrastructure
- outsourcing
- developed nations
- developing nations
- microlending
- modernization

**ACADEMIC VOCABULARY**
accommodate, establish

**TERMS & NAMES**
- Parliament
- Golden Quadrilateral

---

**Essential Question** What are some effects of South Asia's rapid changes?

### SECTION 3 • GOVERNMENT & ECONOMICS

3.1 The Impact of Urbanization ............. 510

3.2 Pakistan's Changing Government ... 512

3.3 Fighting Poverty in Bangladesh ...... 514

**KEY VOCABULARY**
- slum
- push-pull factors
- military dictatorship
- overpopulation
- cyclone

**ACADEMIC VOCABULARY**
deny

**TERMS & NAMES**
- Mohammed Ali Jinnah
- Benazir Bhutto

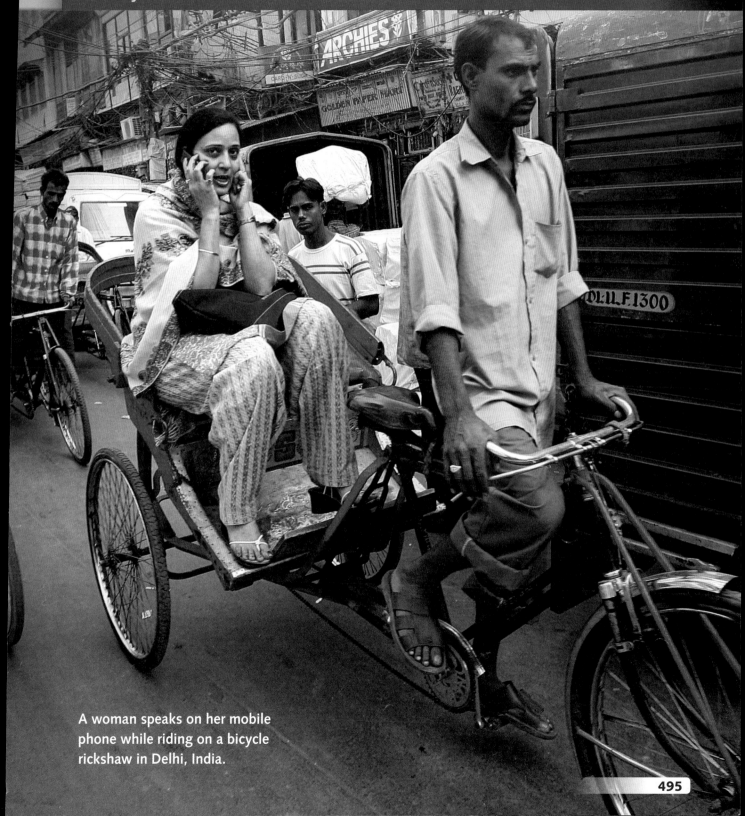

# TECHTREK FOR THIS CHAPTER

- Student eEdition
- Maps and Graphs
- Interactive Whiteboard GeoActivities
- Digital Library

Go to **myNGconnect.com** for more on South Asia.

A woman speaks on her mobile phone while riding on a bicycle rickshaw in Delhi, India.

# SECTION 1 CULTURE

## 1.1 Hinduism Today

**Main Idea** Hinduism unifies its followers through a wide variety of religious beliefs.

Approximately 63 percent of South Asians identify themselves as Hindus. In India, the majority is even larger. Hindus make up over 80 percent of the population, more than 930 million people. Hinduism in India is more than a religion—it is a way of life shared by a diverse population.

▲ **Critical Viewing** This tenth-century painting shows Krishna as the protector of cows. What details in the painting tell you that cows are valued by Hindus?

### Basic Beliefs

Hindus believe that everything in the universe is part of one spiritual force called Brahman. To become one with Brahman, every soul must go through reincarnation, in which it is reborn into different forms of life. A soul's **karma**, or actions during a life, determines the soul's form in the next life. If a soul has good karma, it will be reborn into a higher state, such as a human. If the soul has bad karma, it will be reborn into a lower state, such as a plant.

These beliefs are recorded in sacred texts called the Vedas, which date back to Aryan civilization. Hindu texts, such as the *Mahabharata* and the *Ramayana*, teach Hindu beliefs in the form of epic poems. As you've learned, the Bhagavad Gita is part of the *Mahabharata*.

### Hindu Deities and Sacred Spaces

According to Hindus, there are many gods and goddesses, but all come from Brahman. The three most important deities are Brahma, the creator of the universe; Vishnu, the preserver of the universe; and Shiva, the destroyer of the universe. **Krishna**, an avatar of Vishnu, is a popular Hindu deity. He is often pictured playing his flute.

Hindus believe that the Ganges River is sacred. Millions of Hindus make **pilgrimages**, or religious journeys, to the city of Varanasi to worship on the *ghats*, or stone steps and platforms along the river. Hindus also believe that bathing in the water gives them better karma.

## The Caste System

In addition to Hinduism, the Aryan migration into India also brought the caste system. This system divided society into different social groups, or varnas, based on a person's occupation. As you have learned, there were four original varnas in the caste system: the *Brahmans* (priests and scholars); the *Kshatriyas* (rulers and warriors); the *Vaisyas* (merchants and professionals); and the *Sudras* (artisans, laborers, and servants).

Eventually, an unofficial fifth group was created to include those outside the varna system. These "Untouchables" performed the lowest jobs in Indian society, such as tanning animal skins and collecting garbage. Today, this group prefers to call itself *Dalit*.

The rules of the caste system were very strict. People could not marry out of their caste or even eat with people of a higher caste. It was nearly impossible to move into a different caste. In the last 50 years, however, these rules have become less rigid. India's constitution now forbids **discrimination**, or unfair treatment, against members of any caste and many groups have formed to struggle against such discrimination.

### Before You Move On
**Monitor Comprehension** What religious beliefs unite Hindus?

## ONGOING ASSESSMENT
### READING LAB — GeoJournal

1. **Summarize** Name and describe the three most important Hindu deities.

2. **Compare and Contrast** Think about other world religions that you've learned about. How are they similar to or different from Hinduism? Copy the graphic organizer and use it to compare and contrast religions.

3. **Write a Response** Why is Hinduism so important to India? Think about the size of the Hindu population and Hinduism's part in Indian history. Write two to three sentences explaining Hinduism's importance.

▼ **Critical Viewing** How is this Hindu temple similar to other religious buildings you have seen?

# SECTION 1 CULTURE
## 1.2 Changing Traditions

**TECHTREK**
myNGconnect.com For photos of South Asian traditions

Digital Library

**Main Idea** In South Asia, traditions blend with modern practices.

Many groups have helped shape South Asia's culture, from the Aryans to the British. Some aspects of the region's culture are very old and some are modern. These photographs show the ancient cultural traditions practiced by people in South Asia today.

### Traditional Decoration

Since ancient times, Indian women and girls have decorated their hands with fancy patterns for special occasions, such as weddings or festivals. They use a reddish powder called **henna** to create the intricate designs (shown at right). Each design has a special meaning: A flower can mean happiness; a square might mean honesty; and a triangle might represent creativity.

### Before You Move On
**Make Inferences** How do traditions blend with modern practices in South Asia?

> **Critical Viewing** What traditional and modern elements can you identify in this photo?

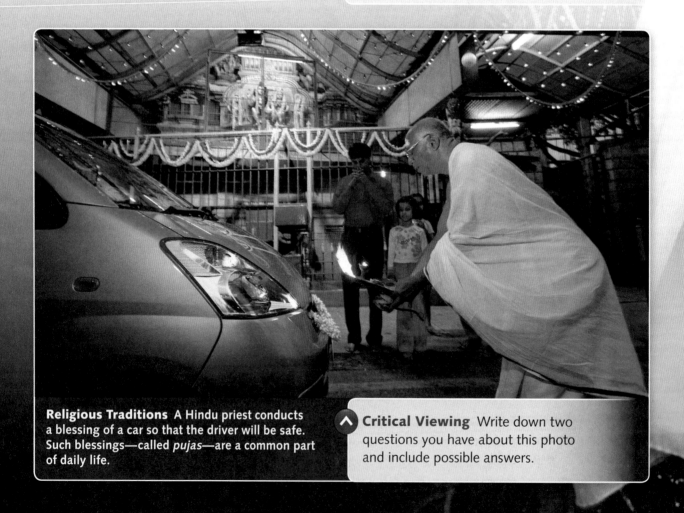

**Religious Traditions** A Hindu priest conducts a blessing of a car so that the driver will be safe. Such blessings—called *pujas*—are a common part of daily life.

**Critical Viewing** Write down two questions you have about this photo and include possible answers.

**Visual Vocabulary** A **sari** is a traditional Indian garment for women worn wrapped around the body. Many saris are silk.

**Visual Vocabulary** **Henna** is a reddish powder used to create designs on skin. Henna is used for special occasions.

## ONGOING ASSESSMENT
## PHOTO LAB
*GeoJournal*

1. **Create a Photo Essay** Look at the two photos of traditions in South Asia. Now go to the **Digital Library** to find more photos of traditions. Choose several photos and print them out. Then paste them on paper or use the **Magazine Maker CD-ROM** to create a photo essay of traditions in South Asia. Write a caption for each photo that explains the tradition and how it is a part of modern life.

2. **Make Inferences** Many South Asians wear Western-style clothing. Why might wearing saris continue to be popular today?

# SECTION 1 CULTURE
## 1.3 Daily Life

**Main Idea** Some aspects of culture in South Asia show how tradition is important in everyday life.

Modern practices have spread to many parts of South Asia. However, people continue to observe early traditions.

### Schools

Most countries in South Asia offer some level of free public education like the public schools in the United States. India's schools require attendance until age 14. In Bhutan, formal eduation has only been in place since 1950. However, education is definitely a shared value in the region. Schools are set up wherever there is space: mountain huts, river boats, desert tents, and even train platforms.

The **literacy rate** (the percentage of people who can read and write) for 15-to-24-year-olds ranges from 74 percent in Bangladesh to 96 percent in the Maldives. Rising literacy rates indicate that the region is developing.

### Clothing

South Asia's climate extremes determine what people wear. For example, Bhutanese herders wear coats and yak-hair hats to protect themselves from the cold, while Sri Lankan fishermen dress lightly because of the heat.

Many men and women wear a **shalwar-kameez** (see photo opposite), a long shirt with loose-fitting pants. Once considered a traditional form of Muslim dress, the shalwar-kameez is now worn by Muslims and Hindus alike. Some women wear a sari, a single piece of cloth (often silk) wrapped to form a long dress. Western clothes are increasingly popular too.

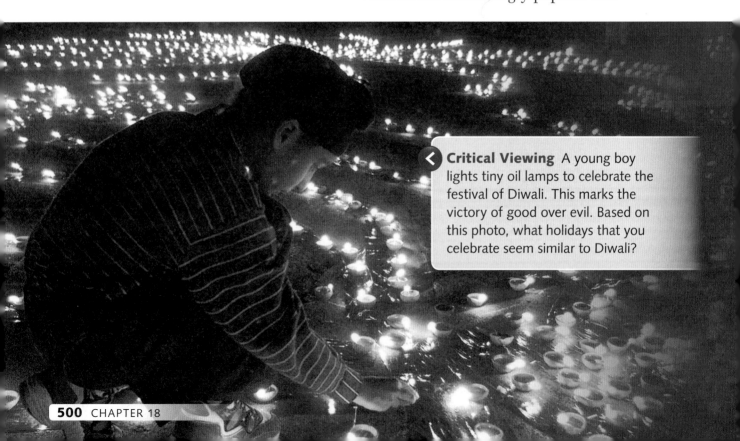

**Critical Viewing** A young boy lights tiny oil lamps to celebrate the festival of Diwali. This marks the victory of good over evil. Based on this photo, what holidays that you celebrate seem similar to Diwali?

## Food

South Asia's cuisine—the food commonly served in a region—is richly spiced and often vegetarian. Cows are sacred to Hindus, so many Indians do not eat meat. Pakistan and India serve curry dishes with potatoes, eggplant, and okra. In Bhutan, yaks provide meat and milk, which is used to make butter and cheese. Each region prepares its own mixture of spices, or *masala*, to use in sauces. Rice is often the center of a meal, especially in Bangladesh and eastern and southern India. *Chapati*, a flat wheat bread, is common in northern India. *Dal*, a blended stew of peas, beans, or lentils, is a frequent side dish.

**Visual Vocabulary** shalwar-kameez

**Visual Vocabulary** cricket

hockey, golf, and soccer are also popular. India has professional cricket and soccer leagues that draw many enthusiastic fans.

### Before You Move On
**Summarize** What are three ways in which tradition is part of everyday life in South Asia?

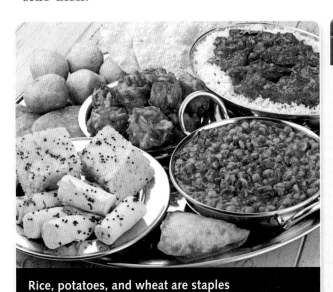

Rice, potatoes, and wheat are staples of South Asian cuisine.

## Sports

Traditional sports like kabaddi (described as a combination of wrestling and rugby) have existed in India for nearly 4,000 years. Western sports such as **cricket** (a team game similar to baseball), field

## ONGOING ASSESSMENT
# WRITING LAB  GeoJournal

1. **Describe Culture** What is your culture like? Create a culture diagram to describe it
   **Step 1** Cut a large paper circle. Write "My Culture" in the center, and divide it into four sections with the labels Schools, Clothing, Food, Sports.
   **Step 2** Find photos from magazines or online that represent each category of U.S. culture. Paste photos in the appropriate sections.
   **Step 3** Turn the circle over and write a detailed paragraph that describes your culture in each category.

2. **Compare** How is South Asia's daily life different from daily life in the United States? How is it the same? Use your diagram to compare. Discuss your ideas with a classmate.

SECTION 1.3 **501**

# SECTION 1 CULTURE
## 1.4 Popular Culture

**TECHTREK**
myNGconnect.com
For samples of Indian music

Digital Library

**Main Idea** Music and movies help unify the diverse peoples and cultures of South Asia.

Music and movies are a lively part of South Asia's popular culture, which consists of the arts, music, and other elements of everyday life in a region.

### Unifying a Culture

South Asia has a long musical history, ranging from religious music to blockbuster movie soundtracks. Traditional music echoes in today's spectacles of Bollywood movies. With wide availability and appeal, music and movies foster the development of a common culture.

### Music

Young people in South Asia listen to Bollywood music, Indian rock, and Western pop. Many also enjoy Bhangra, a form of folk dance and music from the Punjab region of India and Pakistan. Some modern Indian musicians are fusing Indian and Western styles to create a new sound that keeps Indian musical traditions alive.

**Critical Viewing** Young people in New Delhi dance to a wide variety of music. What influences of Western popular culture do you see in the photograph?

The musician on the left plays traditional music near the Ganges River at Varanasi. The DJ on the right plays music in Mumbai. Classical Indian music is based on religious themes and can be heard in both traditional and contemporary music.

## Movies

India's film industry began in Bombay (Mumbai) and is called **"Bollywood"** after its American counterpart, Hollywood. Many Bollywood movies have themes based on Hindu stories, feature musical numbers, and use the Hindi language. India produces more feature films than any other country in the world, and their international audience is growing.

The southern city of Chennai has its own film industry that produces films in the Tamil language. This film center is called "Kollywood" after the Kodambakkam district of Chennai.

Cast members from the 2008 hit movie *Slumdog Millionaire* pose at the Academy Awards. Movies like *Slumdog Millionaire* showed that films about India are often popular in Western culture.

### Before You Move On
**Make Inferences** How do music and movies unify people of South Asia?

▲ **Critical Viewing** Actors perform a scene in a Bollywood movie. Based on details in this scene, how does this movie compare to movies you have seen?

### ONGOING ASSESSMENT
## SPEAKING LAB  GeoJournal

**Turn and Talk** Do music and movies bring people together in the United States? Think about your answer, and then work with a group to develop a response that can be presented orally to the class. Your first sentence will answer the question. Each person in the group adds one sentence for support. Be prepared to present your response to the class.

# SECTION 2 FOCUS ON INDIA
## 2.1 The Largest Democracy

TECHTREK
myNGconnect.com For a chart of India's governmental structure

Digital Library

**Main Idea** The democratic government of India faces many challenges in governing more than one billion citizens.

In 2009, 64-year-old Meira Kumar became the Speaker of the lower house of India's **Parliament**. She was the first woman elected to this position—a milestone for the world's largest democracy. India's Parliament is the legislative branch of the government. It has two houses like the U.S. Congress.

### Governing One Billion People

India has passed many milestones since its independence. One of the first was crafting a new constitution in 1949 to create a **democracy**, or a government in which citizens make decisions either directly or indirectly through elected representatives. India's constitution established a **federal republic**, which is a democratic form of government in which voters elect representatives and the central government shares power with the states. The United States is also a federal republic.

India's government has three branches: the legislative, the executive, and the judicial. The legislative branch has two houses, the *Lok Sabha* (similar to the U.S. House of Representatives) and the *Rajya Sabha* (similar to the U.S. Senate). Members of the *Lok Sabha* are elected every five years, and those of the *Rajya Sabha* serve six-year terms.

Parliament and the states' legislatures elect the president. The president appoints justices to the Supreme Court. However, the leader with the most power is the prime minister, who is head of the political party with the most members in the *Lok Sabha*. The prime minister heads the Council of Ministers, who operate the government.

### India's Politics Today

In 2009, India held its national elections. Voting took place in phases over a one-month period and brought Meira Kumar into the headlines. Kumar promised to work toward a "casteless" society in India.

**Critical Viewing** The Presidential Palace in Delhi was designed by British architect Edwin Lutyens, who combined classical architecture with Indian decorative features. How does the architecture of this building convey its importance?

## INDIA'S GOVERNMENTAL STRUCTURE

**LEGISLATIVE (Parliament)**
- Rajya Sabha — Upper House: House of States (250 members)
- Lok Sabha — Lower House: House of the People (545 members)

**EXECUTIVE**
- President
- Prime Minister
- Council of Ministers

**JUDICIAL**
- Supreme Court
  - Chief Justice
  - 25 associate justices

The Indian government is tackling major challenges, such as India's fast-growing population. By 2030, India is expected to pass China as the world's most populous nation. The government is already working to improve its infrastructure in order to **accommodate,** or make room for, that growth. **Infrastructure** includes the basic systems that a society needs, such as roads, bridges, and sewers. You will read about one of India's massive road projects, the Golden Quadrilateral, later in this chapter. In addition, because of the shortage of fossil fuels in India, the government is increasing its investment in nuclear power to produce electricity.

Another challenge for the government is reaching an agreement with Pakistan over control of largely Muslim Kashmir. This is a goal most observers believe is important to the region's future.

### Before You Move On
**Make Inferences** What is a major challenge facing India in governing its citizens?

## ONGOING ASSESSMENT
# READING LAB  GeoJournal

1. **Compare and Contrast** Name two ways in which the government of India is similar to the government of the United States. Name one way they are different.
2. **Analyze Visuals** How is India's government organized?
3. **Make Inferences** Why do you think the Prime Minister of India has the most power?

SECTION 2.1

# SECTION 2 FOCUS ON INDIA

## 2.2 Economic Growth

TECHTREK
myNGconnect.com For photos of economic growth
Digital Library

**Main Idea** Many factors have contributed to India's rapid economic growth.

If you have a computer problem and call for technical help, your call may be answered in the southern Indian city of Bangalore. Bangalore is the leading center of India's telecommunications industry. Companies there handle support services for U.S. computer and software companies. **Outsourcing**, the shifting of jobs to workers outside of a company, has been a big part of India's economic growth.

### A Growing Economy

The value of the goods and services produced in a country divided by the number of people in that country is called the per capita Gross Domestic Product (GDP). Countries with a high per capita GDP are known as **developed nations**, and countries with a low per capita GDP are called **developing nations**. Developed nations have healthier, more educated people, consume more goods and services, use more energy, and employ more people in manufacturing and service industries than developing nations. Developing nations have a lower standard of living and less developed infrastructure.

Most South Asian countries are developing nations. But some economists see India as an emerging market, with a high growth rate and goods and services that compete in global trade. India's GDP was growing quickly until the global financial crisis. Between 2008 and 2013, the country's economy slowed. Nevertheless, its estimated middle class of more than 300 million people nearly matches the entire U.S. population.

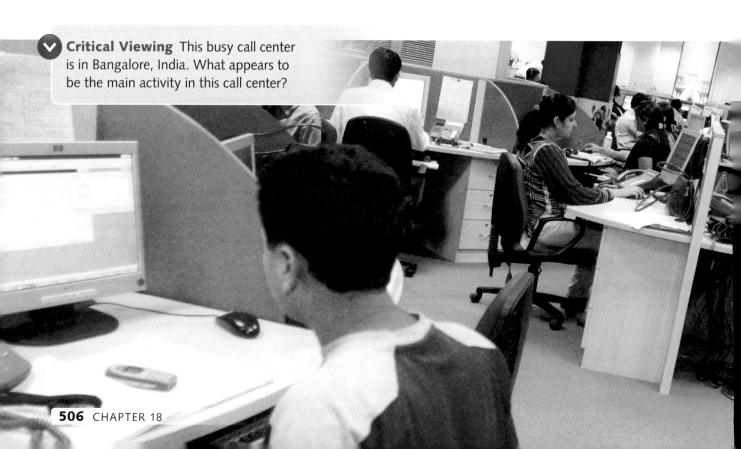

**Critical Viewing** This busy call center is in Bangalore, India. What appears to be the main activity in this call center?

## Factors Affecting Growth

The teaching of English—the language of international commerce—is a legacy of British colonialism that has worked in India's favor. A common language, a stable democratic government, and improvements in infrastructure have encouraged foreign investment.

What does India's growth mean to the average person? With nearly 70 percent of the people trying to exist on only two dollars a day, it means slow but improving access to more consumer goods and more opportunities. **Microlending**, the practice of making small loans to people starting their own businesses, is one way that the country's growing prosperity reaches even its poorest citizens.

### Before You Move On
**Monitor Comprehension** What economic factors contributed to rapid growth in India?

### COMPARISON OF SELECTED CONSUMER GOODS, SOUTH ASIA AND THE UNITED STATES

| | Televisions | Cell Phone Subscriptions | Personal Computers | Total Number of Automobiles |
|---|---|---|---|---|
| | (per thousand people) | | | |
| Bangladesh | 85 | 566 | 22 | 185,000 |
| Bhutan | 33 | 656 | 16 | 29,189 |
| India | 83 | 720 | 33 | 13,673,611 |
| Maldives | 40 | 530 | 152 | 3,917 |
| Nepal | 10 | 438 | 5 | 93,266 |
| Pakistan | 56 | 616 | 5 | 2,331,900 |
| Sri Lanka | 142 | 871 | 35 | 410,282 |
| U.S.A. | 893 | 1059 | 755 | 134,880,000 |

Between 2003 and 2011     Source: Encyclopedia Britannica

### ONGOING ASSESSMENT
## DATA LAB  GeoJournal

1. **Identify** Based on the chart, which South Asian country has the most consumer goods overall?
2. **Location** How do you explain the Maldives having the most cell phone subscriptions in South Asia?
3. **Evaluate** Based on what you know, what factors allow the United States to have so many consumer goods?

# SECTION 2 FOCUS ON INDIA
## 2.3 The Golden Quadrilateral

TECHTREK
myNGconnect.com For online maps and photos of India's infrastructure

Maps and Graphs • Digital Library

**Main Idea** India is modernizing its roadways to support its economic growth.

In the mid-1990s, India's prime minister complained, "Our roads don't have a few potholes. Our potholes have a few roads." Since then, the country has invested large amounts of money to improve its roads.

### A Road to the 21st Century

In 1998, the prime minister announced that India would build a 3,633-mile superhighway called the **Golden Quadrilateral** (GQ) to connect four major cities: Delhi, Mumbai, Chennai, and Kolkata. The GQ is an example of India's efforts to improve its infrastructure. **Modernization** efforts like the GQ bring countries up to present standards in technology and other areas. The GQ will help spread economic growth from India's bustling cities to its thousands of poor villages.

When India gained independence in 1947, drivers shared unpaved roads with cattle and slow-moving tractors. In 1998, the government **established** (put into place) the National Highways Authority to build thousands of miles of new roads, including the Golden Quadrilateral, at a cost of more than $30 billion. It is the largest public works project in the country's history. With the GQ's advanced technology, road sensors will automatically notify crews when it needs repairing. Providing faster and more reliable transportation is one step toward modernizing India's infrastructure and preparing for a profitable future.

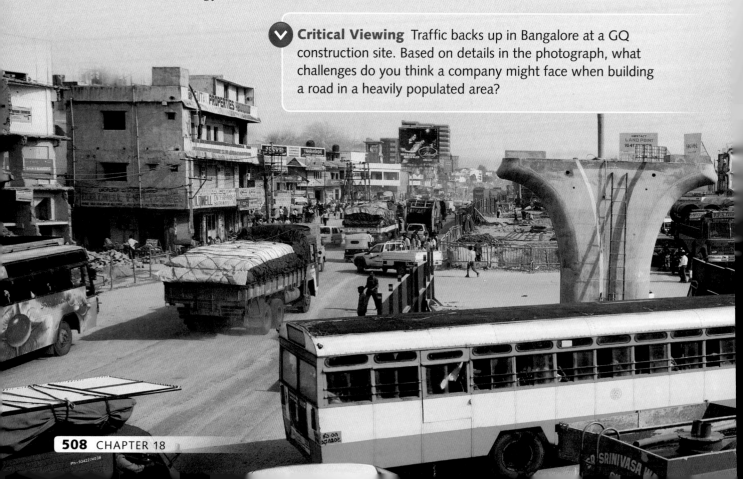

**Critical Viewing** Traffic backs up in Bangalore at a GQ construction site. Based on details in the photograph, what challenges do you think a company might face when building a road in a heavily populated area?

## The Impact of the GQ

India's investment in the GQ has already had a tremendous impact on its economy—and on people's lives. For example, farmers who could sell their crops only in nearby towns can now use the GQ to carry their products to Chennai and sell them for higher prices.

As India builds better roads, more Indians are purchasing automobiles and trucks—a clear sign of economic growth. In fact, recently India's economy has grown at an annual rate of 9 percent—a rate second only to China's growth. For strong growth to continue, India will need to keep investing in major projects like the Golden Quadrilateral.

### Before You Move On
**Monitor Comprehension** How have improvements in infrastructure helped India's economic growth?

INDIA'S HIGHWAYS AND THE GOLDEN QUADRILATERAL

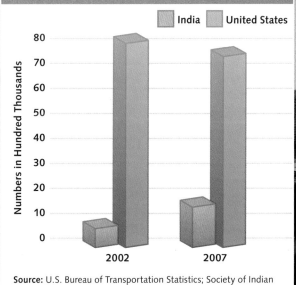

SALES OF PASSENGER VEHICLES IN INDIA AND THE UNITED STATES

**Source:** U.S. Bureau of Transportation Statistics; Society of Indian Automobile Manufacturers

## ONGOING ASSESSMENT
### DATA LAB

1. **Interpret Graphs** According to the graph, how were sales of vehicles in India in 2007 different from sales in 2002?

2. **Interpret Graphs** How did the difference between Indian and U.S. sales change during the period shown on the graph? What might explain the change?

3. **Movement** Based on the map and what you know from your reading, how would you explain the location of three of the four major cities connected by the GQ?

# SECTION 3 GOVERNMENT & ECONOMICS

## 3.1 The Impact of Urbanization

TECHTREK
myNGconnect.com For photos of urban areas
Digital Library

**Main Idea** South Asia's cities are facing the challenges of rapid growth.

Dharavi is a slum in Mumbai, India. A **slum** is an area in a city that has crowded, unclean housing, poor sanitation, and bad living conditions. Residents of Dharavi often walked a mile to get water, but now some have their own water taps. It's a small sign of progress in an area where improvement comes slowly.

### Growing Cities

Since independence, cities in India have exploded in population. In 1951, about 62 million Indians lived in urban areas—cities and their surrounding communities. In 2001, nearly 300 million people lived in cities. Thirty-five cities have more than a million residents. People have flocked to India's cities because of a number of push-pull factors. **Push-pull factors** are the reasons why people migrate. "Push" factors cause people to leave a place, and "pull" factors draw them to another place.

One key factor pushing people to leave the countryside is poverty. Factors pulling people to cities include job opportunities and better education.

India's cities have gradually divided into small wealthy sections and vast slums. The slums have very poor infrastructure and lack clean water and electricity. Education and health-care services are inadequate. Permits to build new structures are hard to get. City centers are also jammed with growing numbers of automobiles, creating traffic congestion and adding to air pollution—which causes thousands of deaths each year. Despite these problems, the jobs and educational opportunities in the cities do help some people to improve their lives.

Pakistan faces similar issues. There, nearly 36 percent of Pakistan's population live in cities. The city of Karachi (kuh RAH chee) is approaching 24 million people—about 13 percent of Pakistan's entire population.

**Critical Viewing** Slums of Mumbai stand amidst new construction. What does this photo suggest about the differences between the way wealthy and poor people live?

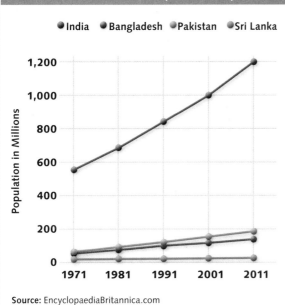

POPULATION GROWTH IN SOUTH ASIA (1971–2011)

Source: EncyclopaediaBritannica.com

**Critical Viewing** This crowded city street is in Karachi, Pakistan. What can you infer about life in a Pakistani city from this photo?

## Pakistan's Solutions

As in India, people are leaving rural areas in Pakistan and moving to the cities. With urban populations increasing at a rate of 3 percent per year, Pakistan's infrastructure has trouble keeping up. This leads to overcrowding in slums.

One solution to overcrowding has been the development of "secondary" cities around major urban centers. For example, Gujranwala is a secondary city to Lahore. Secondary cities provide rural migrants with an affordable place to settle and find employment. The secondary cities also relieve strain on big cities such as Lahore. They do this by slowing the flow of people into the main city. Other South Asian cities, particularly in Bangladesh and Sri Lanka, face similar pressures.

### Before You Move On
**Make Inferences** What are some solutions to the challenges of rapid growth in South Asian cities?

### ONGOING ASSESSMENT
# DATA LAB  GeoJournal

1. **Analyze Data** About how much did Bangladesh's population grow from 1971 to 2011? How much did Pakistan's grow?

2. **Make Inferences** Think about the total land area of India, Pakistan, and Sri Lanka, their arable land, and the populations indicated on the graph. Based on what you know, explain the rates of population growth in each country.

SECTION 3.1 **511**

# SECTION 3 GOVERNMENT & ECONOMICS

## 3.2 Pakistan's Changing Government

**Main Idea** Pakistan's government has alternated between democracy and military dictatorships.

Pakistan was created in 1947 as an Islamic country. Since then, it has alternated between civilian governments and **military dictatorships**, in which the army runs the government and **denies**, or refuses to recognize, citizens' rights. The changes in Pakistan's government have limited social and economic progress in the country.

### A Lack of National Unity

One cause of Pakistan's changes in government has been a lack of unity. The country is divided into four provinces: Punjab, Sindh, Balochistan, and the North-West Frontier Province. Each province has its own tribal groups and languages. No leader has been able to pull the provinces together to form a unified nation.

The lack of unity was a problem from the start. In 1947, the country had two parts. West Pakistan was west of India. East Pakistan was east of India, more than 1,000 miles away from West Pakistan. Most leaders came from West Pakistan, which troubled the people of East Pakistan. In 1971, East Pakistan declared independence and became Bangladesh.

### Military Dictatorships

Another cause of Pakistan's changes in government has been inconsistent leadership. **Mohammed Ali Jinnah**, who led the founding of Pakistan, believed in democracy. When he died in 1948, the country lacked a strong leader. Pakistanis grew angry about slow economic development, and unrest between different groups of Pakistanis increased. In 1958, the military took over the government and established a military dictatorship.

After a civil war in 1971, Zulfikar Bhutto was elected to lead Pakistan. He promised to take steps to end poverty. However, in 1977, the military took control of Pakistan again, and Bhutto was executed in 1979.

> **Critical Viewing** These women are getting ready to vote at a polling center in Pakistan. What can you infer about their attitude toward voting?

### Benazir Bhutto's Election

Elections were held again in 1988 after the death of the military leader. Pakistanis elected Bhutto's daughter, **Benazir Bhutto**, as prime minister. Her opponents accused her of corruption. She denied the charges, but there were many threats on her life, and she fled Pakistan in 1999.

With the military in power again, Pakistan experienced moments of prosperity. Yet it also continued to deny people's rights. In 2007, the military promised elections, and Benazir Bhutto returned to Pakistan. However, she was assassinated later that year. Bhutto's husband, Asif Ali Zardari, was elected president in 2008.

### Continuing Challenges

Pakistan's relationship with its larger neighbor, India, continues to be strained, particularly over the region of Kashmir. The two countries are frequently at a stand-off over the area. Violence regularly erupts, and war is a very real threat. Difficulty controlling the border with its eastern neighbor, Afghanistan, also continues to put a great deal of pressure on Pakistan's government.

Pakistan's internal problems show that successful countries need strong leaders who can bring different groups together. They also demonstrate the importance of having an effective government that makes sure the country's constitution is enforced, guaranteeing rights to all of its citizens.

**Before You Move On**
Make Inferences Explain why the military has taken power in Pakistan at various times.

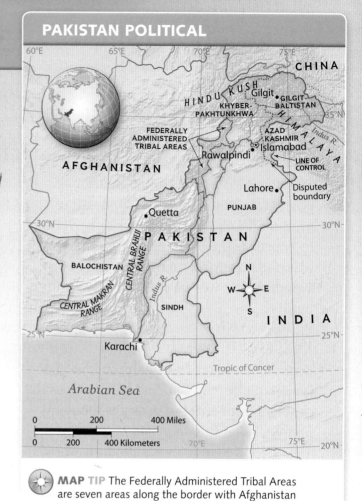

**MAP TIP** The Federally Administered Tribal Areas are seven areas along the border with Afghanistan that are inhabited by mainly Pashtun tribes. These areas are loosely governed by regulations put in place by the British *raj*.

1. **Analyze Cause and Effect** What circumstances have caused Pakistan to alternate between democratic governments and military governments?

2. **Location** Find the five major cities in Pakistan on the map. How might geographic factors explain their location?

3. **Make Inferences** Why would it be difficult to govern the four provinces of Pakistan?

4. **Evaluate** Why has it been difficult for Pakistan to move forward as a country?

# SECTION 3 GOVERNMENT & ECONOMICS

## 3.3 Fighting Poverty in Bangladesh

**Main Idea** Bangladesh is one of the poorest countries in the world, but it is making progress toward improving its economy.

The biggest problem in Bangladesh is poverty. In 2009, the gross domestic product (GDP) per capita was only $1,500. In contrast, U.S. GDP per capita that year was $46,000. More than eight out of ten Bangladeshis live on less than two dollars per day.

### Poverty in Bangladesh

Economists identify three problems that help explain why Bangladesh is poor. One problem is **overpopulation**, or too many people living in one place. In 2010, over 156 million people in Bangladesh were living on only about 55,600 square miles of land. That is roughly equivalent to half of the U.S. population living in an area the size of the state of Illinois.

The second problem facing Bangladesh is natural disasters. Rivers like the Ganges surge over their banks during summer monsoons. Many **cyclones**—the name for hurricanes in South Asia—also hit Bangladesh. Floods and cyclones destroy crops, adding to the problems of an uneven food supply and poverty.

Lack of education is a third problem. Only 48 percent of Bangladeshis can read and write. In contrast, 99 percent of Americans can read and write. In addition, about 8 million Bangladeshi children work in factories and at other jobs. Because they work for wages, they are not able to go to school.

### Toward a Brighter Future

When Bangladesh became an independent nation in 1971, it had a military government like Pakistan's. Since 1991, though, it has had a democratically elected government. The government has improved the economy by encouraging exports and building new industries.

It is not easy to find products to export. Bangladeshi farmers grow rice, wheat, tea, and sugarcane, but those crops must feed the country's large population, and there is usually nothing left over to export. However, the country does export huge amounts of jute. Jute is a strong fiber used in making ropes, carpets, types of paper, and other products.

**Critical Viewing** Garments are a major export in Bangladesh. What conditions shown in this photo of a factory help explain why the garment industry has been successful?

# BANGLADESH RESOURCES AND INDUSTRIES

Bangladesh has also built several industries. New factories produce paper, cement, and steel. (Locate them on the resources map above.)

Garment manufacturing—shown in the photo at left—has been particularly successful. In fact, garments are the country's largest export. Most of the employees in the garment industry are women, and today the clothing they produce is sold in countries around the world. This economic progress gives hope for a brighter future for Bangladesh.

### Before You Move On
**Summarize** What are two strategies that Bangladesh has used to develop its economy?

## ONGOING ASSESSMENT
## MAP LAB
*GeoJournal*

1. **Human-Environment Interaction** What natural resource appears most frequently on the map? How would this natural resource contribute to building industries in Bangladesh?

2. **Write a Paragraph** Pick one resource on the map. Write a paragraph that explains where that resource is located in Bangladesh and why it might be important to Bangladesh's economy. Share your paragraph with classmates.

3. **Make Inferences** How has a lack of education interfered with economic development in Bangladesh?

4. **Make Generalizations** Why do you think that the garment industry has grown so quickly in Bangladesh?

SECTION 3.3  **515**

# CHAPTER 18
# Review

## VOCABULARY

For each pair of vocabulary words, write one sentence that explains the connection between the two words.

1. pilgrimage; karma

   *Hindus believe that a pilgrimage to the Ganges River will give them good karma.*

2. shalwar-kameez; sari
3. democracy; federal republic
4. infrastructure; modernization
5. developed nations; developing nations

## MAIN IDEAS

6. In what ways does Hinduism unify South Asia? (Section 1.1)
7. For what purpose do Indian women use henna? (Section 1.2)
8. How do foods differ from region to region in South Asia? (Section 1.3)
9. How do music and movies foster a common culture in South Asia? (Section 1.4)
10. Name two challenges facing the government of India today. (Section 2.1)
11. How has outsourcing contributed to rapid growth? (Section 2.2)
12. What is the Golden Quadrilateral, and what cities will it connect? (Section 2.3)
13. What problems are caused by rapid growth in South Asia's cities? (Section 3.1)
14. What are two causes of Pakistan's many changes in government? (Section 3.2)
15. What three factors have contributed to poverty in Bangladesh? (Section 3.3)

## CULTURE

### ANALYZE THE ESSENTIAL QUESTION

How is diversity reflected in South Asia's cultures?

**Critical Thinking: Draw Conclusions**

16. What is one basic belief of Hinduism? How does it influence Indian culture?
17. Why do you think established traditions have continued in South Asia today?
18. In what ways is South Asian food similar to and different from American food?
19. How does South Asia's music reflect the region's diversity?

### INTERPRET MAPS

**LANGUAGES OF SOUTH ASIA**

- Indo-European
- Dravidian
- Tibeto-Burman
- Munda
- Mon-Khmer
- Burushaski

20. **Region** According to the map, what is the major language family in South Asia?
21. **Make Inferences** Why might there be so many different language families in South Asia?

## FOCUS ON INDIA

### ANALYZE THE ESSENTIAL QUESTION
Why has India experienced an economic boom?

**Critical Thinking: Make Generalizations**

22. In what ways is India's government similar to and different from the U.S. government?
23. If India's economy keeps expanding, what effect might that have on the other countries of South Asia?
24. How will the Golden Quadrilateral help to build India's economy?

## GOVERNMENT & ECONOMICS

### ANALYZE THE ESSENTIAL QUESTION
What are some effects of South Asia's rapid changes?

**Critical Thinking: Analyze Cause and Effect**

25. How do Pakistan's secondary cities help relieve strain on the major cities?
26. How has Pakistan's unstable government affected its citizens?

### INTERPRET CHARTS

| LITERACY RATE AND GDP PER CAPITA IN BANGLADESH (1981–2011) | | | | |
|---|---|---|---|---|
| | 1981 | 1991 | 2001 | 2011 |
| Adult Literacy Rate (% people) | 29% | 35% | 47% | 57% |
| GDP Per Capita ($) | $319 | $514 | $664 | $747 |

Source: The World Bank

27. **Analyze Data** What is the relationship between Bangladesh's adult literacy rate and GDP per capita?
28. **Evaluate** How would you explain the relationship between these two economic indicators? Why might GDP per capita affect the literacy rate?

## ACTIVE OPTIONS

**Synthesize** the Essential Questions by completing the activities below.

29. **Write a Letter** Describe in a letter to a friend the different things that bring South Asians together. Also describe in your letter what you think the future holds for South Asia as a region and for specific countries in that region. Use the following tips to help you write your letter. **Read your letter to a classmate.**

   **Writing Tips**
   - Take notes on your ideas about what unifies South Asia.
   - Use an informal tone and style to describe those ideas in your letter.
   - Be sure to include colorful details to keep your friend's interest.

myNGconnect.com For photos of South Asia

30. **Create Visuals** Make a slide show of South Asia photos from the **Digital Library** or other online sources. Pick one photo for each section in the chapter and give it a title. Then write a couple of sentences explaining how the photo represents South Asia. Copy the chart below to help you organize your information.

| Photo | How It Represents South Asia |
|---|---|
| Section 1 | |
| Section 2 | |

# COMPARE ACROSS REGIONS UNIT 9
# Economic Growth

myNGconnect.com For an online infographic about economic growth

Student Resources

Many factors—geography and climate, natural resources, the freedom of the markets to compete, and government policies—play a complex part in whether a country's economy can grow.

A government might control all or a large part of how goods and services are created and then how they are sold or provided to citizens. Some monarchies, totalitarian regimes, and societies that have communist and socialist economies are like that. Or a country might have a free-market economy, in which the government has a smaller role, such as to protect consumers. When India and Hungary became more free-market based in the 1990s, their economic growth increased.

### Compare
- India
- Hungary
- Vietnam
- South Africa

## INDIA

Think about the factors of production: 1) natural resources, 2) labor or workers, 3) capital (factories, computers, tractors, for example), and 4) people who start new businesses, such as entrepreneurs. Factors of production are needed to produce goods and services. They influence economic systems.

One factor, labor, means having enough skilled workers. India's large labor force has had to become skilled in industry and technology for the country to become a fast-growing world market.

India also has had to increase its capital—the buildings, machinery, and materials needed for growth. In 2013, India was in the top ten countries with the largest GDP. However, its GNI, Gross National Income per capita (per person), remained low.

## HUNGARY

Under communism, Hungary's economy was controlled by the government. Entrepreneurs could not start their own businesses. New businesses that can freely compete are one benefit of a free enterprise system. Another is freely buying and selling capital at fair prices. Both of these factors encourage growth.

Today, with a free-market system, Hungary's economy is one of the most prosperous in Eastern Europe. The country has had a high literacy rate, and it has a lower percentage of people living in poverty than in the United States. But like India and many other nations, Hungary suffered a slowdown because of the global economic crisis, which began between 2007 and 2008.

# ECONOMIC INDICATORS FOR INDIA AND HUNGARY
(Data Comparison)

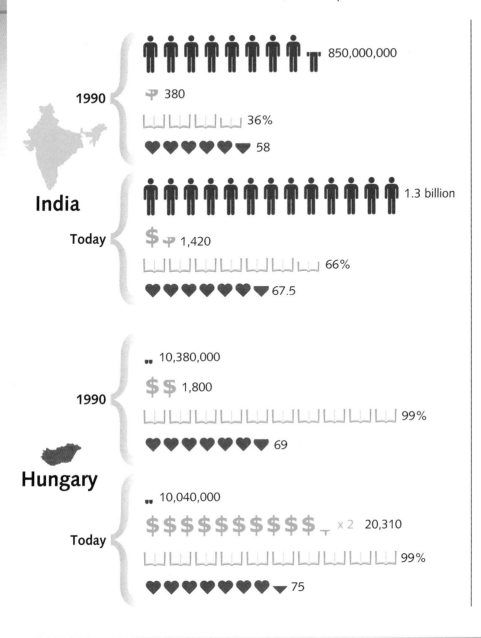

**KEY**
- Population: 100,000,000 people
- GNI: $1,000
- Adult Literacy Rate: 10%
- Life Expectancy: 10 years

**U.S. Benchmark**
Population: 313,914,040
GNI: $48,620
Adult Literacy Rate: 98%
Life Expectancy: 79 years

**Sources:** World Bank, Encyclopedia Brittanica, CIA World Factbook

## ONGOING ASSESSMENT
### RESEARCH LAB — GeoJournal

1. **Compare and Contrast** Which measurement increased the most in both countries between 1990 and today?
2. **Analyze Data** How might population growth affect economic growth in India and Hungary?

**Research and Create Graphs** Research the emerging markets of Vietnam in Southeast Asia and South Africa in sub-Saharan Africa. Create a graph for each country, showing its most recent key emerging market measurements. Then compare the data for all four countries—India, Hungary, Vietnam, and South Africa. Based on the key measurements, which country's economy do you think might emerge as the strongest?

COMPARE ACROSS REGIONS 519

REVIEW & EXTEND **UNIT 9**

# Active Options

**TECHTREK**
myNGconnect.com For photos of South Asia's culture and Guided Writing

 Digital Library   Student Resources

## ACTIVITY 1

**Goal:** Extend your understanding of endangered animals.

### Write About Endangered Species

National Geographic explorer Shafqat Hussain created Project Snow Leopard to protect the endangered snow leopard species in South Asia. With a partner, research one of the following endangered animals from the region:

- Bengal tiger
- Indian rhinoceros
- Ganges River dolphin
- Indus River dolphin
- Indian elephant
- Lion-tailed macaque

Create a presentation that tells about the animal, why the animal is endangered, and why it is important to protect the animal. Present your findings to the class. **Go to Student Resources** for Guided Writing support.

snow leopard

## ACTIVITY 2

**Goal:** Research South Asia's culture.

### Create *South Asia Today* Magazine

You and your classmates are on the staff of the magazine *South Asia Today*. Work together and use the **Magazine Maker CD-ROM** to create next month's issue. Go to the **Digital Library** to find images and information on:

- clothing
- festivals
- food
- movies
- music
- schools
- shopping
- sports

## ACTIVITY 3

**Goal:** Review South Asia with a game.

### Host a Geography Bee

Write a question about South Asia and its answer on a card. Your teacher will collect the questions and ask them to the class. Stand to answer the question. The last person standing wins!

**520** UNIT 9

NATIONAL GEOGRAPHIC
World Cultures and Geography

GEO

# UNIT 10

# Explore East Asia
## with NATIONAL GEOGRAPHIC

### MEET THE EXPLORER

**NATIONAL GEOGRAPHIC**

Using advanced tools such as satellite imagery and remote sensors, Emerging Explorer Albert Yu-Min Lin searches for the tomb of Genghis Khan. His search leaves the land undisturbed and respects Mongolian beliefs and tradition.

### INVESTIGATE GEOGRAPHY

The upper levels of a temple overlook snowcapped Mount Fuji in central Japan. Mount Fuji is a volcano that has been dormant for over 300 years. A shrine on its peak (12,388 feet) attracts thousands of climbers every summer.

### CONNECT WITH THE CULTURE

With a population of over 1.3 billion, China is the world's most populous country. A large and growing population creates challenges, such as polluting auto traffic and overcrowded public transportation. Bicycle use helps to ease the problem.

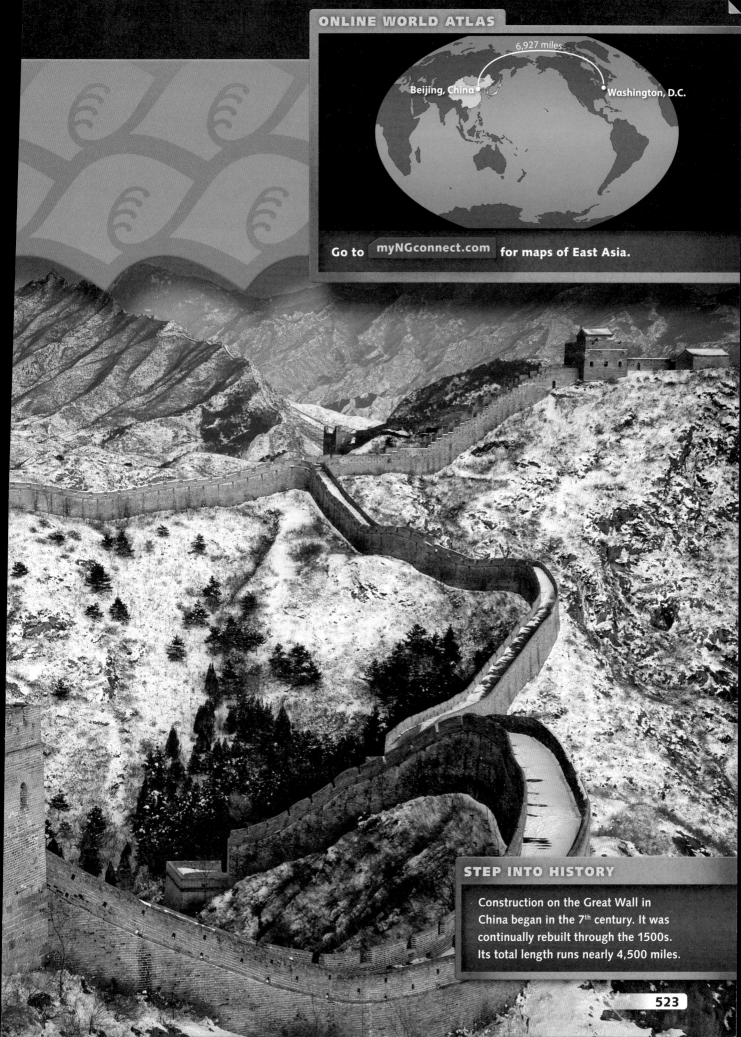

## ONLINE WORLD ATLAS

6,927 miles
Beijing, China — Washington, D.C.

Go to myNGconnect.com for maps of East Asia.

## STEP INTO HISTORY

Construction on the Great Wall in China began in the 7th century. It was continually rebuilt through the 1500s. Its total length runs nearly 4,500 miles.

# CHAPTER 19

# East Asia
## GEOGRAPHY & HISTORY

## PREVIEW THE CHAPTER

**Essential Question** How did geographic factors affect population distribution?

### SECTION 1 • GEOGRAPHY

| | |
|---|---|
| 1.1 Physical Geography | 526 |
| 1.2 China's Rivers and Plains | 528 |
| 1.3 The Island Arc of Japan | 530 |
| 1.4 The Korean Peninsula | 532 |
| 1.5 Mongolia's Desert Landscape | 534 |
| 1.6 Tracking Aquatic Wildlife | 536 |

**KEY VOCABULARY**
- mainland
- basin
- alluvium
- loess
- archipelago
- eruption
- typhoon
- demilitarized zone (DMZ)
- steppe
- semiarid
- ger
- animal-borne
- carapace

**ACADEMIC VOCABULARY**
erode, refuge

**TERMS & NAMES**
- Kunlun Mountains
- Tibetan Plateau
- Chang Jiang
- Huang He
- North China Plain
- Ring of Fire
- Kanto Plain
- Gobi desert

---

**Essential Question** What influences, beliefs, and encounters helped shape China?

### SECTION 2 • CHINA'S HISTORY

| | |
|---|---|
| 2.1 Early Dynasties | 538 |
| NATIONAL GEOGRAPHIC Photo Gallery | 540 |
| 2.2 Confucianism | 542 |
| 2.3 Silk Roads and Trade | 544 |
| 2.4 Exploration and Isolation | 546 |
| 2.5 Communist Revolution | 548 |

**KEY VOCABULARY**
- dynasty
- dynastic cycle
- empire
- terra cotta
- ethical system
- moral
- caravan
- maritime
- tribute
- expedition
- occupy
- collective farm
- elite

**ACADEMIC VOCABULARY**
ban, barter

**TERMS & NAMES**
- Shang
- Zhou
- Qin
- Shi Huangdi
- Great Wall of China
- Han
- Silk Roads
- Confucianism
- Zheng He
- Jiang Jieshi
- Mao Zedong
- Cultural Revolution

---

**Essential Question** What factors had an impact on the histories of Japan and Korea?

### SECTION 3 • HISTORY OF JAPAN & KOREA

| | |
|---|---|
| 3.1 Japanese Samurai | 550 |
| 3.2 Japan Industrializes | 552 |
| 3.3 Korea's Early History | 554 |
| 3.4 The Korean War | 556 |

**KEY VOCABULARY**
- samurai
- shogun
- competitive
- zaibatsu
- rivalry
- celadon
- retreat
- armistice

**ACADEMIC VOCABULARY**
dissolve, seize

**TERMS & NAMES**
- Silla
- Koguryo
- Paekche
- Koryo
- Choson
- 38th parallel
- Cold War

524 CHAPTER 19

# SECTION 1 GEOGRAPHY
## 1.1 Physical Geography

EAST ASIA PHYSICAL

## CLIMATE

## POPULATION DENSITY

**Main Idea** East Asia's landforms, bodies of water, and climate influence where people live.

East Asia includes China and Mongolia, which stretch across the **mainland**, or continental landmass, of Asia. The region also contains the island countries of Japan and Taiwan as well as North Korea and South Korea on the Korean Peninsula.

## Landforms and Water

High mountains and plateaus, such as the **Kunlun Mountains** and **Tibetan Plateau**, cover much of East Asia. The Gobi desert covers parts of Mongolia and China.

Major rivers, such as the **Chang Jiang** (chahng jyahng) in China, flow through East Asia. River **basins**, the low areas drained by these rivers, support large populations. Many East Asians also live near coastal areas and on low-lying plains near the Pacific Ocean. People throughout the region depend on the Pacific for fish, trade, and transportation.

## Climate

East Asia's climate ranges from tropical, in parts of China, South Korea, and Japan, to desert, in much of Mongolia. Monsoons greatly influence the region's climate. These seasonal winds bring hot, rainy summers and cool, dry winters to some parts of East Asia.

### Before You Move On

**Summarize** In what ways do landforms, bodies of water, and climate help determine where people live in East Asia?

### ONGOING ASSESSMENT
# MAP LAB

1. **Interpret Maps** Use the physical map to locate the Tibetan Plateau on the population density and climate maps. Why do you think few people live on the plateau?

2. **Compare and Contrast** Compare the climate of inland East Asia with that of its coastal areas. What role might climate play in the population distribution of these areas?

# SECTION 1 GEOGRAPHY

## 1.2 China's Rivers and Plains

**TECHTREK**
myNGconnect.com For an online map and photos of China's rivers and cities
Maps and Graphs | Digital Library

**Main Idea** China's main river systems support large populations.

China's two largest river systems begin in the melting snow and ice of the country's western mountains and plateaus. As the waters flow down, they **erode**, or wear away, the land and deposit alluvium in river basins. **Alluvium** is soil carried by flowing water and is ideal for farming.

### River Systems

The Chang Jiang, or Yangtze, is China's longest river. It passes through the Sichuan (sheesh whan) Basin and the Chang Jiang Plain. China's second longest river, the **Huang He** (hwahng huh), flows through the **North China Plain**. The Huang He is often called the Yellow River because it flows through an arid region that contains **loess** (less), or fine yellowish soil. The wind carries the loess and deposits it in the river.

China's river valleys and fertile plains provide rich agricultural land for growing rice and wheat. They also support many cities. The capital, Beijing (bay ghing), is in the North China Plain. Shanghai, the country's largest city, is at the mouth of the Chang Jiang on the East China Sea.

### Controlling the Rivers

Frequent flooding of the rivers leaves behind deposits of fertile soil. Over the centuries, however, the flooding has also caused millions of deaths. Chinese engineers have built barriers such as dams and levees to hold the water back.

To connect China's major rivers, which flow from west to east, the ancient Chinese built a structure called the Grand Canal, which flows from north to south.

> **Critical Viewing** Boats on a branch of the Chang Jiang sail past skyscrapers in Shanghai. Based on the photo, how do the people in the city use the river?

## CHINA'S RIVERS AND CITIES

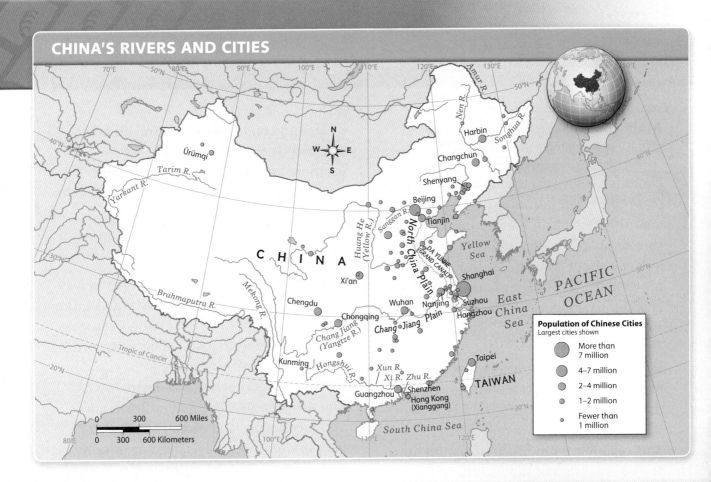

The roughly 1,100-mile-long canal—the longest in the world—links Beijing in the North China Plain to Hangzhou, just south of Shanghai. Barges travel along the canal, carrying bulk goods such as coal and gravel.

### Before You Move On
**Monitor Comprehension** In what ways do China's rivers support large populations?

### ONGOING ASSESSMENT
## MAP LAB

1. **Location** Study the map. How does the Grand Canal promote the transport of goods in China?

2. **Evaluate** Use the map above and the physical map in Section 1.1 to identify the landform where many of China's large cities are located. Why do you think the cities developed there?

# SECTION 1 GEOGRAPHY

## 1.3 The Island Arc of Japan

TECHTREK
myNGconnect.com For an online earthquake map and photos of Japan

Maps and Graphs | Digital Library

> **Main Idea** Japan is a mountainous island country with limited natural resources.

Japan is a closely related group of islands, or **archipelago** (ahr kuh PEH luh goh). The country consists of four main islands and thousands of smaller ones in an arc that extends about 1,400 miles in length. From north to south, the four main islands are Hokkaido, Honshu, Shikoku, and Kyushu.

### A Mountainous Land

Mountains cover almost three-fourths of Japan and run like a spine down the middle of the four main islands. The Japanese islands were formed by the top part of a mountain range thrusting up from the bottom of the Pacific Ocean.

Japan is on the **Ring of Fire**, an area rimmed by the Pacific Ocean where earthquakes and volcanic **eruptions**, or blasts, frequently occur. Every year there are about 1,500 earthquakes and thousands of eruptions from the country's active volcanoes.

On March 11, 2011, northern Japan was rocked by a 9.0-magnitude earthquake, the strongest recorded in the country's history. The earthquake triggered a tsunami, with huge waves sweeping away everything in their path and devastating entire towns. The disaster destroyed countless homes and businesses, and more than 19,000 people were counted as dead or missing. Rebuilding will take years.

> **Critical Viewing** A tsunami wave washes over a street in the town of Miyako. What details in the photo help convey the size of the wave?

The tsunami severely damaged several nuclear power plants. By 2013, it was clear that actions taken to protect the public had prevented radiation exposure. But concerns about contaminated water still exist. Robots, instead of human workers, gather contamination data.

## A Crowded Country

Because so much of Japan is covered with mountains and forests, the people crowd on its plains. About 80 percent of Japan's nearly 130 million people live on the plains of Honshu.

The **Kanto Plain** lies to the east of the Japanese Alps, which cross the central part of Honshu. The flat land of the Kanto Plain is good for agriculture and industry. Tokyo, Japan's capital and largest city, is located there.

## Limited Natural Resources

Japan's most important natural resource is fish, caught in the Pacific Ocean and Sea of Japan. The country's fishing industry is one of the largest in the world. The seas are also essential to trade, which is vital to Japan's economy.

Because Japan has only small quantities of mineral resources, the country must import raw materials such as iron ore and lead. Japan also imports most of its energy resources, including petroleum and coal. Japan uses these imports in its industries and then exports finished products, such as automobiles and electronics.

### Before You Move On
**Summarize** How do Japan's mountains and limited natural resources affect life in Japan?

### EARTHQUAKE 2011

**MAP TIP** The map shows the location, or epicenter, of Japan's 2011 earthquake and its larger aftershocks, a series of quakes that occur after the main one.

## ONGOING ASSESSMENT
# MAP LAB
GeoJournal

1. **Location** Locate the area hit by the earthquake on the inset map. Why might this area be prone to earthquake activity?
2. **Make Inferences** Study the map of Japan's 2011 earthquake. Based on the map, between which two major cities do you think most of the devastation occurred?

# SECTION 1 GEOGRAPHY

## 1.4 The Korean Peninsula

**Main Idea** North Korea and South Korea have a similar geography, but these two countries on the Korean Peninsula remain divided.

The Korean Peninsula was divided into two countries after World War II. The two countries are the Democratic People's Republic of Korea, or North Korea, and the Republic of Korea, or South Korea. North Korea has a Communist government, while South Korea is democratic.

### Geography and Climate

Like Japan, the Korean Peninsula is largely mountainous, with coastal plains and river valleys where most of the people live. With a population of about 50 million, South Korea has more than twice as many people as North Korea and is much more densely populated.

In South Korea, the southern and western plains are important areas for farming. The country's climate supports agriculture, with hot, humid summers and cold, dry winters. South Korea's two main industrial cities—Seoul, the capital, and Busan—are also located on the plains.

North Korea is more mountainous than South Korea. As a result, North Korea has less agricultural land than its neighbor. In general, North Korea's climate is colder and wetter than South Korea's, but **typhoons**, or hurricanes, can bring heavy rain and flooding to both countries.

### Natural Resources and Industries

South Korea's natural resources include iron ore, copper, and coal. However, because these resources are limited, South Korea imports much of the raw materials it needs for industrial production. South Korea has a highly industrialized economy. Shipbuilding and steel are among the country's major industries, along with the manufacture of electronics and automobiles.

North Korea's natural resources include coal, magnesite, and iron ore. Unlike South Korea, many of North Korea's industries produce military equipment.

> **Critical Viewing** Steel is a major industry in South Korea. What details in the photo suggest that this plant in Busan is thriving?

## A Divided Land

From 1950 to 1953, the two Koreas fought in a conflict known as the Korean War. North Korea initiated the war by invading South Korea with the goal of taking over the peninsula. At the end of the war, the two countries remained divided. You will learn more about the Korean War in Section 3.

After the war, a **demilitarized zone (DMZ)**, or neutral area, was created between North Korea and South Korea. The DMZ is about 150 miles long and 2.5 miles wide. Soldiers still patrol the land, which has become a wilderness.

Over time, the isolation of the DMZ has created a **refuge**, or safe place, for large numbers of wildlife. Animal species, such as rare cranes, tigers, and bears live there. Scientists once thought that these animals had died out on the Korean Peninsula.

### Before You Move On
**Monitor Comprehension** In what ways are North Korea and South Korea alike and different?

### LAND USE AND NATURAL RESOURCES

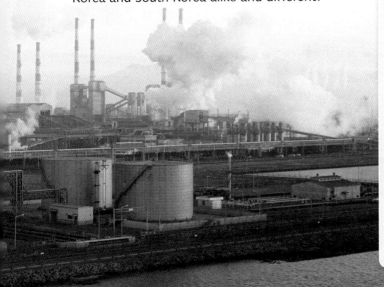

## ONGOING ASSESSMENT
# WRITING LAB  *GeoJournal*

**Compare and Contrast** Write a paragraph in which you compare and contrast North Korea and South Korea. Use a Venn diagram like the one below to organize information about each country's geography, climate, natural resources, and industries. Exchange your finished paragraph with a partner and discuss your ideas. Go to **Student Resources** for Guided Writing.

SECTION 1.4 **533**

# SECTION 1 GEOGRAPHY

## 1.5 Mongolia's Desert Landscape

**Main Idea** Mongolians have adapted to life in a harsh, dry environment.

Mongolia is a landlocked country located between northern China and eastern Russia. The **Gobi desert**, the largest desert in Asia, covers much of southern Mongolia and extends into China.

### High and Dry

Mountains, plateaus, **steppes**, or dry grassland, and desert make up the physical geography of Mongolia. All of the country's land is 1,700 feet or more above sea level.

Most of the **semiarid**, or somewhat dry, land gets fewer than 20 inches of rain, and much of it falls in July and August. In the Gobi desert, only about seven inches of rain fall every year. Some parts of the Gobi receive no rain at all. *Gobi* means "place without water." The extreme dryness and strong winds that sometimes blow through the desert result in blinding dust storms.

Summers in Mongolia are generally short and hot, while winters are long and cold. In the Gobi desert, temperatures can reach 113°F in the summer and -40°F in the winter. In addition, temperatures can rise and fall in the desert by as much as 60 degrees within the same day.

### People and Environment

In Mongolia's harsh environment, not much livable space is available. Nearly half of the country's roughly three million people live in urban areas. About one-third of the population lives in the capital city of Ulaanbaatar (oo lahn BAH tawr).

Less than one percent of the land in Mongolia is arable. As a result, most of the people in rural areas are livestock herders. Many lead a nomadic life. They move from place to place with their sheep or goats, looking for good pastureland. Some herders live in **gers**. In Central Asia, these tents are called yurts.

> **Visual Vocabulary** In this photo, a nomadic Mongolian woman stands outside her ger on the Gobi desert. A **ger** is a portable tent made of felt.

## MONGOLIA PHYSICAL

### Preserving the Desert

As you have seen, the Mongolian people have learned to adapt to their desert environment. They also want to preserve it. The Gobi is expanding by a process called desertification, in which fertile land changes to desert. Much of this expansion is a result of human activities, including allowing livestock to overgraze pastureland. The Mongolians are trying to stop the process by reducing the number of livestock that graze the pastureland and by cutting down fewer trees.

Parts of the Gobi desert have also been turned into nature reserves and national parks. One of the parks in the southern part of the Gobi contains land of great interest to paleontologists, scientists who study prehistoric life. The area is one of the richest sources for dinosaur fossils in the world. In fact, the first dinosaur eggs were discovered here in the 1920s. The Gobi's harsh climate and remote location have helped protect and preserve the fossils for millions of centuries.

### Before You Move On
**Monitor Comprehension** How have Mongolians adapted to their harsh, dry environment?

---

**ONGOING ASSESSMENT**

# MAP LAB

1. **Region** Study the physical map of Mongolia. What does the map tell you about sources of water in the country?

2. **Synthesize** Compare the map on this page with the climate and population maps in Section 1.1. In what ways do climate and land features explain the population density of Mongolia?

3. **Draw Conclusions** Why do you think most of Mongolia's cities are located in the north?

# 1.6 SECTION 1 GEOGRAPHY
## NATIONAL GEOGRAPHIC

**TECHTREK**
myNGconnect.com For photos of an Explorer's work and an Explorer Video Clip

Digital Library

# Tracking Aquatic Wildlife
### with Katsufumi Sato

> **Main Idea** Animal-borne recorders help scientists study how animals behave and learn how to protect them.

## Tracking Technology

Since the 1990s, scientists have been attaching electronic data recorders to animals to gather information about everything from how fast the animal moves to how much it eats. These recorders are **animal-borne**, which means that the devices are carried by the animals themselves. Many of the recorders include cameras that take photos and videos of what the animal sees under natural conditions.

Emerging Explorer Katsufumi Sato is continually looking for ways to improve the technology, noting "We always work to perfect the instruments . . . and find better ways to attach and retrieve them." For example, when he first started studying sea turtles, he used a harness to attach the recorder. However, the harness slowed the turtle down as it swam. Eventually, Sato figured out how to glue the instrument to the turtle's **carapace**, or shell, to make it lighter and less awkward for the animal to carry.

myNGconnect.com
For more on Katsufumi Sato in the field today

## Unexpected Findings

Sato tracks sea creatures of all kinds. One of his research subjects is the cormorant, a large seabird that dives underwater to catch fish in its bill. For years, Japanese fishermen have been convinced that cormorants steal some of the fish in their nets. As a result, the fishermen have hunted and killed the birds. By attaching tiny animal-borne recorders to the birds, Sato hopes to learn the exact amount and type of fish the birds actually eat. The cormorants may not be as destructive as the fishermen believe.

Sato has also studied the loggerhead sea turtle. Many scientists thought that these turtles were dying because they ate the plastic bags that litter the ocean, mistaking them for jellyfish. The scientists believed that the turtles swallowed the bags and died of suffocation or starvation. However, animal-borne cameras proved that the turtles actually swim away from the plastic bags, recognizing that they are not food. This finding told scientists that they need to continue researching to learn what is really causing the turtles' declining numbers.

In another surprising study, Sato discovered that king penguins regulate how much air they inhale before a dive. They inhale more air before a deep dive and less air before a shallower dive. This behavior suggests that the penguins may be capable of planning their actions.

### Before You Move On
**Make Inferences** How might animal-borne recorders be used to save animals?

---

## ONGOING ASSESSMENT
## VIEWING LAB  GeoJournal

1. **Analyze Visuals** Go to the **Digital Library** to view the video clip on Katsufumi Sato. What findings described in the video surprised you?
2. **Pose and Answer Questions** Get together with a partner and discuss the video clip. Then each of you should write a question about Sato's work. Exchange questions with your partner and take turns answering them.
3. **Draw Conclusions** In what ways might Sato's work help protect endangered species?

---

◀ **Critical Viewing** King penguins, similar to those Sato studied, gather on a bay north of Antarctica. What data might one of Sato's devices record about the penguins in this environment?

SECTION 1.6  **537**

# SECTION 2 CHINA'S HISTORY

## 2.1 Early Dynasties

**TECHTREK**
myNGconnect.com For photos of ancient Chinese artifacts

Digital Library

> **Main Idea** Powerful families ruled and shaped ancient China for about 2,000 years.

In the 2100s B.C., farming settlements developed along the Huang He in eastern China. Over time, some of these settlements grew into cities and marked the beginning of an early civilization. Because Chinese culture developed from this early society, China is said to have the world's longest continuous civilization.

### Shang and Zhou Dynasties

Around 1766 B.C., kings from the **Shang** family took over some of the cities along the Huang He. The Shang established a **dynasty**, a series of rulers from the same family. Eventually, the Shang ruled much of the area along the North China Plain.

Shang society was largely agricultural, but the leaders also built large walled cities. The society was structured, with nobles at the top and peasants at the bottom. The Shang used horse-drawn chariots to defend themselves against invaders. They also developed a system of writing, which helped unify their lands.

The **Zhou** (joh) people often fought with the Shang. Around 1050 B.C., they defeated the Shang and established their own dynasty. Throughout much of their long rule—longer than any other dynasty in Chinese history—the Zhou rulers waged war against invaders from the north and west. The Zhou also fought among themselves. Zhou kings had placed lords in charge of different parts of the region. These lords battled each other to acquire more land. The continual warfare led to great disorder in Chinese society.

Chinese rulers believed that the length of their dynasty was determined by the gods worshiped in ancient China. The pattern in the rise and fall of dynasties came to be known as the **dynastic cycle**, which is shown on the next page.

### The Qin Dynasty

The **Qin** (chihn) Dynasty gained control of China in 221 B.C. *China* is thought to come from the name of this dynasty. Qin ruler **Shi Huangdi** (shee hwahng dee) strengthened the central government and expanded the lands under his control. He brought these lands together to form an **empire**, a group of states ruled by a single strong ruler, and became China's first emperor. Shi Huangdi unified his empire by building a system of roads and standardizing the Chinese currency.

○ 1766 B.C.
The Shang establish a dynasty on the North China Plain.

1600 B.C.

Shang Dynasty oracle bone used to pose questions to the gods

Bronze vessel from the Zhou Dynasty

900 B.C.

○ 1050 B.C.
The Zhou Dynasty begins its long rule.

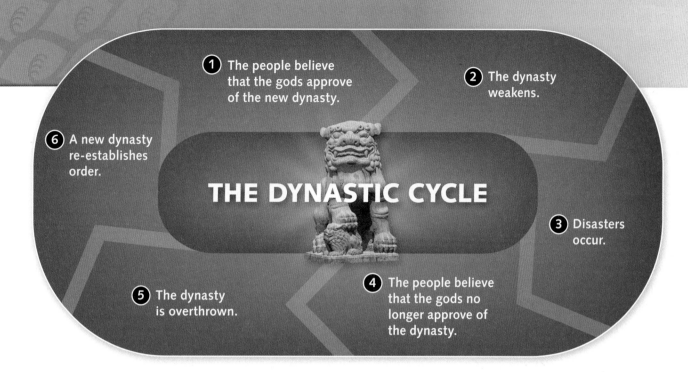

Shi Huangdi also began construction on the **Great Wall of China** to protect against invaders from the north. Thousands were forced to labor on the huge project. Later rulers continued to expand the wall until it stretched about 4,500 miles.

Qin rule ended four years after Shi Huangdi's death in 210 B.C. Archaeologists discovered thousands of **terra cotta**, or baked clay, warriors buried near his tomb in 1974. Experts believe these life-sized statues were created to guard his tomb.

## The Han Dynasty

The **Han** Dynasty came to power in China in 206 B.C. and lasted until A.D. 220. Han leaders expanded the empire and established a strong central government. During this time, China began to trade with Europe and Central Asia along routes known as the **Silk Roads**. The goods and ideas traded helped create a prosperous civilization and an advanced culture in China. Many Chinese still call themselves "the people of the Han." You will learn more about the Silk Roads in Section 2.3.

### Before You Move On
**Monitor Comprehension** How were the ancient Chinese dynasties alike, and how did they differ?

A terra cotta warrior found in Shi Huangdi's tomb

○ 221 B.C.
The Qin Dynasty forms an empire.

**200 B.C.**

206 B.C. ○
The Han Dynasty expands the Chinese Empire.

### ONGOING ASSESSMENT
## READING LAB

1. **Make Inferences** The Qin Dynasty ended shortly after the death of Shi Huangdi. What does that fact suggest about the ruler that followed him?

2. **Form and Support Opinions** Which dynasty do you think had the greatest impact on Chinese society? Explain your reasons.

3. **Evaluate** Why did the Chinese people probably accept the overthrow of an old dynasty and the rise of a new one?

# NATIONAL GEOGRAPHIC
## Photo Gallery • Terra Cotta Army

For more photos from the National Geographic Photo Gallery, go to the **Digital Library** at myNGconnect.com.

Great Wall of China

Tea farm, China

Tibet's Zar Gama Pass

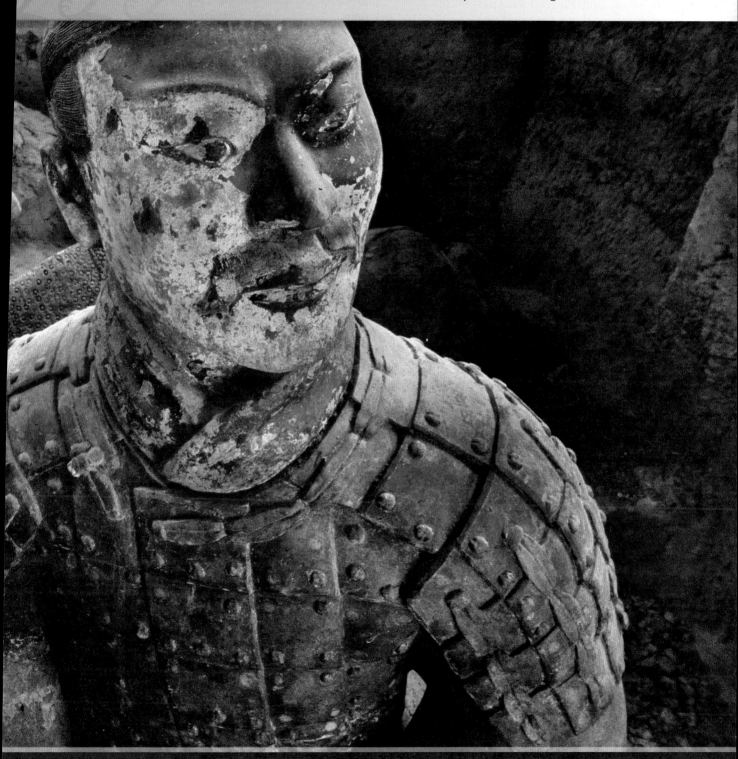

**Critical Viewing** More than 8,000 life-size terra cotta, or baked clay, soldiers were found in Xi'an, China, in 1974. The soldiers had been buried next to the tomb of Emperor Shi Huangdi around 210 B.C.

Nomads in Tibet

Forbidden City, Beijing

Kyoto's Golden Temple

Gobi desert

# SECTION 2 CHINA'S HISTORY

## 2.2 Confucianism

TECHTREK
myNGconnect.com For photos of Confucian temples

Digital Library

**Main Idea** The teachings of Confucius have influenced Chinese society for more than 2,000 years.

As you have read, the Chinese people endured long periods of conflict and disorder during the Zhou Dynasty. Confucius, a Chinese scholar and teacher born during that period, wanted to bring peace to his country. He developed ideas about the proper conduct of rulers and subjects and taught these ideas to others. His teachings form the basis of an ethical system called **Confucianism**. An **ethical system** teaches **moral**, or right, behavior. For more than 2,000 years, Confucius' teachings guided Chinese thought.

### Teachings

Confucius' teachings are based on five relationships: father and son; elder brother and younger brother; husband and wife; friend and friend; and ruler and subject. Confucius taught that if these relationships were conducted respectfully, peace would be restored to society. The teachings below illustrate Confucius' emphasis on the importance of social relationships and study.

> Don't worry if people don't recognize your merits; worry that you may not recognize theirs. (Analects 1.16)
>
> To study without thinking is futile [useless]. To think without studying is dangerous. (Analects 2.15)
>
> What you do not wish for yourself, do not do to others. (Analects 15.24)

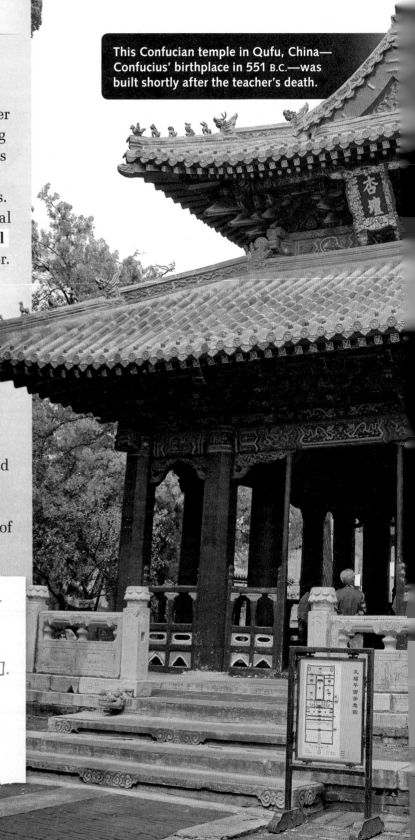

This Confucian temple in Qufu, China—Confucius' birthplace in 551 B.C.—was built shortly after the teacher's death.

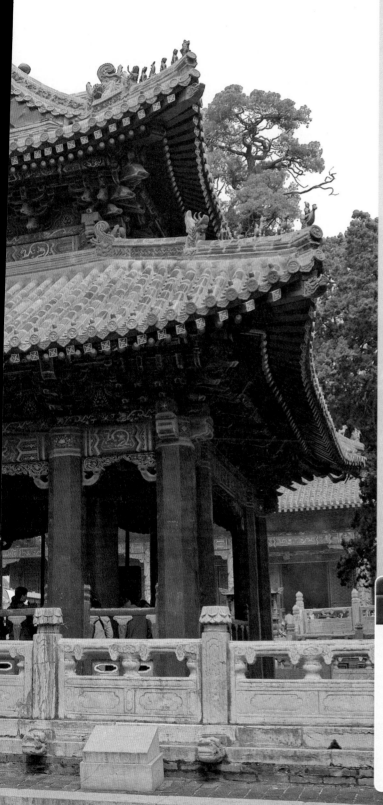

## Legacy

Confucius died believing that he had been a failure. In his lifetime, his teachings did not change Chinese society. However, his students wrote down his teachings and gathered them into a book called the *Analects* for future generations to read. In time, Han Dynasty rulers adopted Confucius' teachings and used them to select government officials based on their accomplishments rather than wealth.

Confucianism continued to influence Chinese society until the Communist Party seized power in 1949. The Communists **banned**, or outlawed, Confucianism because they considered it a religion. The government ended its ban in 1977. Since then, Confucianism has regained much of its influence. It is taught in schools, and many leaders have rediscovered Confucius' wisdom. Some officials have begun to use the teachings to guide their work and behavior.

### Before You Move On
**Summarize** What are the basic teachings of Confucianism?

### ONGOING ASSESSMENT
## PHOTO LAB  GeoJournal

1. **Analyze Visuals** What mood do you think the temple in the photo is meant to inspire?
2. **Make Inferences** Study the photo. Why do you think so many people visit the temple?
3. **Analyze Primary Sources** Reread the third sample of Confucius' teachings. What Confucian values does this teaching reflect?

# SECTION 2 CHINA'S HISTORY
## 2.3 Silk Roads and Trade

**TECHTREK**
myNGconnect.com For an online map of the Silk Roads and photos of trade items

 Maps and Graphs   Digital Library

**Main Idea** Goods were traded and ideas spread on the Silk Roads, which connected China with much of the world.

You have learned that the Silk Roads were a series of trade routes that began during the Han Dynasty. The routes connected China with Europe, India, Central Asia, and North Africa. The Silk Roads got their name from the trade of silk cloth, which was made only in China at that time.

### Trade Routes

The main overland route began in the city of Chang'an, the capital of ancient China. The route split in two to go around the Taklimakan Desert. It divided again to avoid the highest peaks of the Hindu Kush. Trade routes stretched west to Central Asia and Africa and south to India. Most of the overland traders traveled with **caravans**, or groups, with camels—which were well suited to the difficult terrain and dry climate.

The Silk Roads also included **maritime**, or sea, routes. From Nanjing, traders carrying Chinese goods traveled to Japan. From the ancient cities of Antioch and Tyre, trade continued to Rome.

### Goods and Ideas

The overland roads covered about 4,000 miles, but few traders traveled the routes from end to end. Most Chinese traders were forbidden to go beyond their country's borders. As a result, they mainly exchanged their silk, jade, and spices with Central Asian nomads and traders from India. These, in turn, traded with merchants from the Mediterranean. Market towns sprang up along the routes, and some became major cities.

> **Critical Viewing** Camels can travel for long periods without water. Why do you think that ability might have been useful on the Silk Roads?

# TRADE ROUTES OF THE SILK ROADS

Traders carried a variety of goods. Indian traders sold gems and sweet-smelling woods such as cedar. European goods included glass, pearls, and wool. Since there was not a single currency used by all the traders, they **bartered**, or exchanged goods without using money. For example, in exchange for their silk, Chinese traders might have received purple dyes from Mediterranean traders.

Government officials and missionaries also traveled on the routes. Their ideas spread along with the traders' goods. For instance, the Silk Roads helped spread Buddhism from India, where it began, to China and eventually to Korea and Japan.

## Before You Move On
**Monitor Comprehension** How did the Silk Roads promote the exchange of goods and ideas?

### ONGOING ASSESSMENT
# MAP LAB
GeoJournal

1. **Movement** Based on the map, what route might traders have taken from Dunhuang, China, to Delhi, India?
2. **Make Inferences** Study the map. Explain why Kashgar developed into a thriving city.
3. **Create Sketch Maps** Use the map and the information in the text to create a thematic map showing the activity on the Silk Roads. Use symbols to represent different goods and ideas.

SECTION 2.3 **545**

# SECTION 2 CHINA'S HISTORY
## 2.4 Exploration and Isolation

**TECHTREK**
myNGconnect.com For an online map and photos representing Zheng He's voyages

 Maps and Graphs
 Digital Library

**Main Idea** Zheng He made seven voyages of exploration that were intended to impress foreigners with China's wealth and power.

China dominated East Asia during the Ming Dynasty, which began in 1368. Ming emperor Yongle (yung lu) decided that he wanted to show the world China's wealth and power and take control of maritime trade. He also wanted other countries to pay him tribute, money demonstrating their recognition of his power.

### Zheng He

To carry out this mission, Yongle ordered Admiral **Zheng He** (jung huh) to lead seven expeditions, or voyages, between 1405 and 1433. Zheng He's fleet was much larger than those of European explorers such as Christopher Columbus, who would set out almost 100 years later. Zheng He commanded thousands of men and sailed with hundreds of treasure ships designed to carry huge stores of trade goods.

Shortly after Zheng He's last expedition, China withdrew into isolation. This isolation continued until the 1800s, when European powers began controlling China's economy.

**Before You Move On**
Summarize In what ways did Zheng He's expeditions demonstrate China's wealth and power?

**5** 1417–1419
Zheng's treasure fleet visited the Arabian Peninsula and, for the first time, Africa. In Aden, the sultan presented exotic gifts such as zebras, lions, and ostriches.

**6** 1421–1422
Zheng He's fleet returned foreign ambassadors to their native countries after stays of several years in China.

**7** 1431–1433
The last voyage marked the end of China's age of exploration. Historians believe that Zheng died on the return trip and was buried at sea.

> **Critical Viewing** This illustration compares Zheng He's large ship with that of a European explorer. What impression might Zheng's ship have made when it arrived in a foreign port?

# SECTION 2 CHINA'S HISTORY
## 2.5 Communist Revolution

TECHTREK
myNGconnect.com For an online map and photos of Communist China

Maps and Graphs  Digital Library

**Main Idea** The Chinese Communist Party under Mao Zedong significantly changed life in the People's Republic of China.

As you have read, China fell under the control of European powers in the 1800s. The Chinese Nationalist Party wanted to end this foreign control and modernize China. The party overthrew the ruling government and established the Republic of China in 1912. Over time, the Chinese people began to turn to the Communist Party, which arose in Shanghai in 1921. Many young people in particular believed that the Communists would be better able to modernize China. The Nationalists and Communists began a long struggle for power. By 1930, the two groups were locked in a bloody civil war.

### Civil War

In 1933, Nationalist leader **Jiang Jieshi** (jee ahng jee shee) gathered a large army and attacked the Communists in their mountain base in Jiangxi (jee ahng shee) Province. Knowing they were outnumbered, the Communists fled to their base in Shaanxi (shahn shee) Province. Between 1934 and 1935, **Mao Zedong** (MOW dzuh dahng) led about 100,000 Communists over rough terrain on a series of marches that came to be known as the Long March. Tens of thousands of Communists died during the marches.

In 1937, the Japanese invaded China and **occupied**, or took possession of, parts of the country. The Nationalists and Communists set aside their differences to fight their common enemy. The two groups remained united against Japan throughout World War II. After the war, the Nationalists and Communists started fighting again. On October 1, 1949, the Communists declared victory and established the People's Republic of China on the mainland. The Nationalists fled and established a government in Taiwan, an island off the coast of the mainland.

### Chairman Mao

As Chairman of the Communist Party, Mao became the leader of China and established a totalitarian dictatorship. The government seized farmland and forced the peasants to work on **collective farms**. Between 200 and 300 families lived and worked on each collective farm, which was strictly supervised by the government. The government also took over all businesses and industries. By 1957, industrial output had greatly increased.

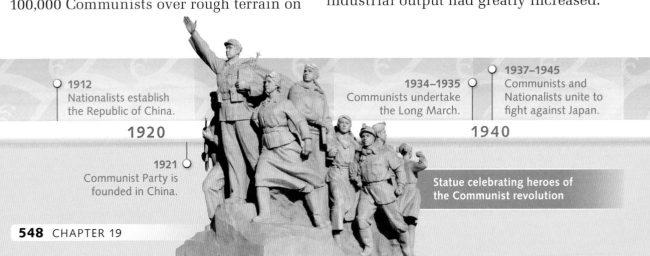

- 1912 Nationalists establish the Republic of China.
- 1921 Communist Party is founded in China.
- 1934–1935 Communists undertake the Long March.
- 1937–1945 Communists and Nationalists unite to fight against Japan.

Statue celebrating heroes of the Communist revolution

Encouraged by this initial success, Mao set a plan in motion the following year called the Great Leap Forward to make China's economy grow even faster. The plan failed. The government's poor management of China's industries slowed economic growth, and crop failures severely reduced agricultural yields. After an estimated 20 million people had died of hunger, Mao ended the plan in 1961.

Five years later, Mao began a new plan called the **Cultural Revolution** to remove what he considered anti-Communist elements from China. To help carry out the plan, young people belonging to military groups known as the Red Guards attacked anyone whom they considered **elite**, or superior. They particularly targeted teachers and intellectuals. Thousands of people were killed in the violence.

After Mao died in 1976, the situation slowly began to change in China. Today, the country is still Communist, but reforms begun in the 1980s have resulted in great economic growth and improved the lives of many Chinese people.

### Before You Move On
**Summarize** What happened in China as a result of Communist control under Mao Zedong?

## THE LONG MARCH

## ONGOING ASSESSMENT
# MAP LAB
GeoJournal

1. **Movement** Trace the route on the map from Jiangxi to Shaanxi Province. What directions did the route follow?

2. **Describe** Use the map on this page and the physical map in Section 1.1 to describe the terrain encountered by those who marched from Jiangxi to Shaanxi Province.

3. **Interpret Time Lines** For how many years did Mao Zedong rule China?

Mao Zedong

**1949** Communists establish the People's Republic of China with Mao Zedong as their leader.

**1958–1961** Mao begins the Great Leap Forward.

1960

**1966** The Cultural Revolution begins.

**1976** Mao dies.

**1980s** Reforms begin that eventually lead to great economic growth.

1980

SECTION 2.5 **549**

# SECTION 3 HISTORY OF JAPAN & KOREA

## 3.1 Japanese Samurai

> **Main Idea** Samurai were part of a feudal system that continued in Japan for about 700 years.

Emperors ruled Japan for hundreds of years, but in the 1000s, they began to lose authority. Landowning lords who belonged to wealthy families expanded their estates and grew increasingly powerful. They formed private armies and hired warriors known as **samurai** (SAM uh ry) to protect their estates.

### Samurai Code and Culture

Samurai means "one who serves." The relationship between lords and samurai was part of a Japanese feudal system similar to the one that arose in Europe during the Middle Ages. Like the knights of the Middle Ages, the samurai developed a code of behavior based on the values of honor, bravery, and loyalty to their lords.

The samurai were not only skilled warriors but also developed their own culture. They wrote poetry and created fine ink paintings. The samurai values of loyalty and hard work and the warriors' artistic interests influenced Japanese society, art, and literature.

### Shogun Rule

Powerful families battled each other for control of Japan during the 1100s. In 1192, a lord from the powerful Minamoto family won and received the title of **shogun**, or "military ruler," from the emperor. The shogun came to hold the true power in Japan. The Minamoto established a dynasty of shoguns that ruled Japan until the 1300s. After the Minamoto lost power, rival families fought for control. Japan became divided into many warring states and experienced a long period of disorder.

Finally, in 1603, a shogun named Tokugawa Ieyasu (ee yeh yah soo) defeated his rivals. He united Japan and moved the capital from Kyoto to Edo, which is present-day Tokyo. The Tokugawa shoguns restored peace. However, they also isolated Japan from the rest of the world. In 1868, this isolation ended when the Japanese people persuaded the shogun to resign. Imperial rule was restored, and the new emperor put an end to the samurai class.

### Before You Move On

**Monitor Comprehension** What role did the samurai play in Japan's feudal system?

Samurai armor

**Critical Viewing** This painting shows samurai on horseback during a battle. What details in the painting illustrate the samurai's skill and bravery?

## ONGOING ASSESSMENT
## PHOTO LAB   GeoJournal

1. **Analyze Visuals** Study the photo of samurai armor. What words would you use to describe the warrior's equipment?

2. **Compare and Contrast** Compare the samurai in the photo and in the painting. Which details are similar? Which are different?

3. **Write Reports** Think about another group of warriors that you know about from history, comics, or movies such as *Star Wars*. In what way does their code compare with that of the samurai? Write a report comparing the two.

# SECTION 3 HISTORY OF JAPAN & KOREA
## 3.2 Japan Industrializes

**Main Idea** Japan transformed itself into an industrialized economy beginning in 1868.

As you have learned, the emperor was restored to power in Japan in 1868. The restoration came about because many of the Japanese people wanted to end Japan's isolation. They wanted to make the country **competitive** with, or able to challenge, the newly industrialized Western nations. After the last of the Tokugawa shoguns resigned, a new era called *Meiji* (may jee), meaning "enlightened rule," began.

### Building on Western Models

For hundreds of years, Japan's economy had been based on agriculture and fishing. The Meiji government looked to Western nations as models for a new industrial economy. Many foreign experts came to Japan to instruct the Japanese in English, engineering, and science. British advisers helped the Japanese government design railroad and communication systems and set up new industries.

At first, the government invested directly in coal and zinc mines and large-scale industries, especially those needed to build a modern military force. These industries included shipyards and weapons factories. The government also developed factories to produce textiles and silk. The government planned to export these goods so the country could acquire raw materials. Japan turned to China and Korea as sources for these raw materials and as potential markets for the finished products. By the 1880s, however, the government could not afford to continue

This woodblock print by the Meiji-era artist known as Hiroshige III illustrates a Tokyo railway station and train built in 1872 with the help of Western advisers and engineers.

this level of investment in the country's industries and sold the industries to private investors.

### Large Businesses Develop

Over time, business in Japan became concentrated in several large, family-controlled organizations called **zaibatsu** (zeye BAHT sue). Each zaibatsu owned a number of different businesses, such as manufacturing, transportation, banking, insurance, trade, and real estate.

Some of the zaibatsu were rooted in the shogun era. For example, the Mitsui family had been successful textile merchants under the Tokugawa shoguns. This family expanded first into banking and, in time, owned more than 270 companies.

Other zaibatsu began during the Meiji era. Mitsubishi, for instance, started as a large shipping firm. The family eventually established financial services and real estate businesses and industries such as oil, steel, and shipbuilding.

The zaibatsu were largely **dissolved**, or broken up, after World War II ended in 1945. Businesses owned by the zaibatsu were **seized**, or taken control of, and reorganized into smaller holdings.

However, the value the zaibatsu placed on hard work and thoroughness continues to influence Japanese industry today.

### Before You Move On
**Summarize** By what means did Japan transform into an industrialized economy beginning in 1868?

## ONGOING ASSESSMENT
# READING LAB  GeoJournal

1. **Make Inferences** Why do you think Japan turned to Western nations for help in becoming industrialized? In what way was this a reversal of Japanese policy during the Tokugawa period?
2. **Analyze Cause and Effect** What impact do you suppose industrialization had on Japan's economy and people?
3. **Synthesize** Some former samurai were able to adapt to and find roles in the newly industrialized society. Based on what you have learned about these warriors, what qualities might have made them good employees?

▼ **Critical Viewing** In this 1935 photo, Japanese emperor Hirohito (front, left) inspects a weapons factory. The black and white photo was taken before color photography had been developed. Black and white photos can convey mood, lighting, and texture more powerfully than color photos. What other advantages might black and white photography have over color photography? What might be some of its disadvantages?

# SECTION 3 HISTORY OF JAPAN & KOREA
## 3.3 Korea's Early History

**Main Idea** A series of kingdoms and dynasties arose in Korea, developing a distinct culture.

In Section 2, you learned about the early dynasties that came to power in China. In 108 B.C., China's Han Dynasty expanded into Korea and took control of an area in the northwest. At that time, the Korean people belonged to scattered tribes. Eventually, groups of tribes united and formed three kingdoms on the peninsula: **Silla** (SIHL uh), **Koguryo** (koh gur YOO), and **Paekche** (PAHK CHAY).

### The Three Kingdoms

Silla, the first kingdom to form, arose around 57 B.C. in the southeastern part of the peninsula. Koguryo emerged about 20 years later when several tribes united in the northern part of the peninsula and parts of eastern Manchuria in China. Paekche formed around 18 B.C. in the southwest. At first, Koguryo was the largest and most powerful of the three.

The three kingdoms were rivals and often invaded each other's territory. In spite of their **rivalry**, or opposition, the kingdoms had similar cultures, and their people spoke the same language. The kingdoms produced items such as leather goods, tools, and woolen clothing, which they exported to China. In return, the kingdoms received porcelain, paper, silk, and weapons. The kingdoms also absorbed some ideas from China, including its writing system, Confucianism, and Buddhism. Eventually, Buddhism spread from the Korean kingdoms to Japan.

In A.D. 660, the Tang Dynasty in China joined with Silla to defeat the other two kingdoms. Shortly after, Silla succeeded in driving out the Chinese. By 668, the Silla kingdom ruled the entire peninsula.

### Koryo and Choson Dynasties

By 935, Silla had weakened, and its kingdom was overthrown. The Korean Peninsula came under the control of the **Koryo** Dynasty, which ruled for more than 450 years. The Koryo Dynasty modeled its government after that of China. Like the early kingdoms, Koryo was strongly influenced by Confucianism.

Silla crown

Hunting scene in Koguryo wall painting

- 108 B.C. Chinese Empire expands into Korea.
- 57 B.C. Silla arises in the southeastern part of the peninsula.
- 37 B.C. Koguryo develops in the north.
- 18 B.C. Paekche forms in the southwest.

100 B.C. — 1 B.C.

A distinctly Korean culture developed during the Koryo period. Artists created **celadon** pottery, with its characteristic green glaze. They also carved all of the Buddhist scriptures onto more than 80,000 wooden blocks for printing. The blocks are called the Tripitaka Koreana. Invading Mongols destroyed the blocks in 1232, but a new set was re-created within 20 years. The temple of Haeinsa in South Korea is home to the set today and has been named a UNESCO World Heritage Site.

In 1392, the Koryo Dynasty was defeated and replaced by the **Choson** Dynasty, which lasted for 518 years. During this time, Korea continued to adopt elements of Chinese culture. Korean culture also flourished at this time, with developments in architecture, science, and technology. The Choson Dynasty ended in 1910 when Korea came under the control of Japan. Japanese occupation continued until 1945, when World War II ended.

### Before You Move On
**Summarize** How were the Korean dynasties similar and different?

### EARLY KINGDOMS, A.D. 500s

### ONGOING ASSESSMENT
## MAP LAB  *GeoJournal*

1. **Interpret Maps** On the map, use your finger to trace the outline of the Koguryo Kingdom. What parts of present-day countries did the kingdom consist of?
2. **Draw Conclusions** Remember that Silla conquered Koguryo and Paekche in 668. How would the size of Silla at its greatest extent have compared with that of the Koryo and Choson dynasties, as shown on the map?

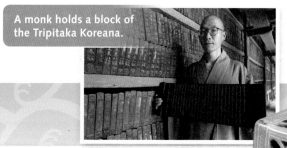
A monk holds a block of the Tripitaka Koreana.

Celadon pillow

A.D. 500

A.D. 668 — Silla controls the entire peninsula.

A.D. 935 — Koryo Dynasty overthrows Silla.

A.D. 1392 — Choson Dynasty begins.

1500

SECTION 3.3 555

# SECTION 3  HISTORY OF JAPAN & KOREA

## 3.4 The Korean War

myNGconnect.com For an online map and photos of the Korean War

Maps and Graphs  Digital Library

**Main Idea** After three years of war, North Korea and South Korea remained divided.

When the Japanese occupation ended in 1945, the Korean Peninsula was divided along the 38° North latitude line—usually referred to as the **38th parallel**. The United States held the land south of that line, and the Soviet Union occupied the land to the north. The United States and the Soviet Union came to be locked in a **Cold War**. This means that they did not fight one another directly in battle. Instead, they supported opposing groups in wars that took place in other parts of the world. One such war occurred in Korea.

### Fighting Begins

In 1947, the United Nations called for free elections that would create one government for Korea. The elections were held, but the Soviet Union stepped in and established a Communist government in the north. The following year, Communist North Korea and democratic South Korea were created, with the 38th parallel serving as the border.

On June 25, 1950, North Korea attacked South Korea. The United Nations called for an international force to come to South Korea's aid. The United States supplied most of the troops and placed them under the command of General Douglas MacArthur, who had been an important military leader during World War II.

The fighting wore on for several years. At one point, the North Koreans pushed deep into South Korea and captured most of the peninsula. Then the UN forces made a surprise landing at Inchon and took back the South Korean capital of Seoul. As the North Koreans **retreated**, or drew back, the UN forces pushed north and captured the North Korean capital of Pyongyang. The course of the war shifted from one side to the other through 1952, but in the end, little territory was gained or lost.

### Impact of the War

As many as four million soldiers and civilians were killed in the war. Much of the Korean Peninsula was damaged by bombs dropped by jet aircraft.

General MacArthur lands at Inchon.

**JUNE 1950** North Korea invades South Korea.

**OCTOBER 1950** UN forces capture Pyongyang.

1950

**SEPTEMBER 1950** North Korea captures most of the peninsula; MacArthur's forces land at Inchon and take back Seoul.

Korean civilians particularly suffered during the war. Half of all industries and a third of all homes were destroyed. In addition, many people died of starvation.

### The War Ends

In 1953, the UN forces and North Korea signed an **armistice**, or agreement to stop fighting, but a treaty was never signed. An area set near the 38th parallel, called the demilitarized zone (DMZ), still divides the two countries today. Since 1953, North Korean troops have guarded one side of the zone, while South Korean and American troops have guarded the other.

Today, Communist North Korea is largely isolated from the rest of the world, while South Korea is a democracy with a global, market economy. Over the years, the two countries have discussed the possibility of uniting. However, political differences and the threat of North Korea's nuclear arms program have made this unification increasingly unlikely.

### Before You Move On
**Make Inferences** What, if anything, did the Korean War accomplish?

Statues of American soldiers in the Korean War Veterans War Memorial in Washington, D.C.

## ONGOING ASSESSMENT
## MAP LAB
GeoJournal

1. **Movement** Study the map. Which country's forces moved further into enemy territory?
2. **Interpret Maps** Based on the map, why do you think it was easy for Chinese forces to rapidly come to North Korea's aid?
3. **Analyze Time Lines** Based on the time line captions, how would you characterize the course of the war?

- **JANUARY 1951** Chinese forces occupy Seoul.
- **MARCH 1951** UN forces retake Seoul.

1952

South Korean soldiers patrol the DMZ.

- **JULY 1953** An armistice is signed, but North and South Korea remain divided.

1954

SECTION 3.4

# CHAPTER 19
# Review

## VOCABULARY

On your own paper, write the vocabulary word that completes each of the following sentences.

1. _____ is yellow silt that blows from the Gobi desert onto the Huang He.
2. Tropical hurricanes called _____ bring heavy rains to North Korea.
3. The rise and fall of dynasties in China followed a pattern known as the _____.
4. Confucianism is considered an _____, which teaches right behavior.
5. Japanese warriors known as _____ were similar to the knights of the European Middle Ages.

## MAIN IDEAS

6. Where do most East Asians live? Why? (Section 1.1)
7. What purpose does the Grand Canal in China serve? (Section 1.2)
8. What are the main natural resources of North Korea and South Korea? (Section 1.4)
9. What steps has Mongolia taken toward preserving the Gobi desert? (Section 1.5)
10. Who was Shi Huangdi and what were his accomplishments? (Section 2.1)
11. Why did the Chinese people embrace Confucianism? (Section 2.2)
12. What were the Silk Roads? (Section 2.3)
13. In what ways did Mao Zedong control life in China? (Section 2.5)
14. What qualities were valued under the samurai code of behavior? (Section 3.1)
15. Why did Japan begin to industrialize after 1868? (Section 3.2)
16. What happened at the conclusion of the Korean War? (Section 3.4)

## GEOGRAPHY

**ANALYZE THE ESSENTIAL QUESTION**

How did geographic factors affect population distribution?

**Critical Thinking: Analyze Cause and Effect**

17. Why is eastern China more densely populated than western China?
18. Why do many of Japan's people live on the island of Honshu?
19. How has Mongolia's harsh environment affected the ways people live and work?

**INTERPRET TABLES**

| RIVERS IN CHINA | |
| --- | --- |
| River | Length (miles) |
| Chang Jiang | 3,900 |
| Huang He | 3,395 |
| Xi Jiang | 1,250 |
| Yalu | 490 |

Source: CIA World Factbook

20. **Analyze Data** About how many times bigger is the Chang Jiang than the Yalu?
21. **Draw Conclusions** Based on the table, what conclusions can you draw about rivers in China?

## CHINA'S HISTORY

**ANALYZE THE ESSENTIAL QUESTION**

What influences, beliefs, and encounters helped shape China?

**Critical Thinking: Make Generalizations**

22. What impact has Confucianism had on Chinese government and society?
23. What did China gain from its trade on the Silk Roads and its voyages of exploration?

### INTERPRET MAPS

24. **Interpret Maps** Study the map. Why do you think Buddhism spread to China before it came to Korea and Japan?

25. **Make Predictions** What other region shown on the map was most likely influenced by Buddhism?

### HISTORY OF JAPAN & KOREA

**ANALYZE THE ESSENTIAL QUESTION**

What factors had an impact on the histories of Japan and Korea?

**Critical Thinking: Make Inferences**

26. Why do you think Japan's rulers chose to isolate the country once they had restored peace in the early 1600s?

27. In what ways has China influenced Korea throughout its history?

### ACTIVE OPTIONS

**Synthesize** the Essential Questions by completing the activities below.

28. **Write a Speech** Choose a historical figure from the chapter that particularly interests you and write a speech from the figure's point of view. To get information for your speech, do some research online on the historical figure. Then use the tips below to help you write the speech. **Once you have finished writing the speech, deliver it to your class.**

   **Writing Tips**
   - Describe some of the historical figure's beliefs and accomplishments.
   - Include brief stories from the person's life to hold your audience's interest.
   - Use a tone that will help convey the person's personality. For example, for Shi Huangdi, you might use a boastful tone. For Confucius, you would probably use a more modest tone.

29. **Create a Poster** Choose two countries of East Asia and compare and contrast their geography, history, and situation today. Use online sources to help you gather your data. Then organize your ideas in a Venn diagram like the one below. Finally, create your poster, using photos to illustrate your ideas.

CHAPTER 20

# East Asia TODAY

## PREVIEW THE CHAPTER

**Essential Question** How do traditions and modernization create a unique way of life in East Asia?

### SECTION 1 • CULTURE

1.1 Religious Traditions .................... 562
1.2 China's Inventions ...................... 564
1.3 Shanghai's Rapid Growth ............ 566
     NATIONAL GEOGRAPHIC Photo Gallery ........... 568
1.4 Japanese Anime .......................... 570
1.5 Bullet Trains ............................... 572

**KEY VOCABULARY**
- meditation
- animism
- polytheism
- monotheism
- porcelain
- movable type
- multinational corporation
- economic globalization
- anime
- manga
- bullet train
- magnetic levitation

**ACADEMIC VOCABULARY**
aerodynamic

**TERMS & NAMES**
- Daoism
- Shinto
- Special Economic Zone

**Essential Question** What problems does East Asia face today, and what are its opportunities?

### SECTION 2 • GOVERNMENT & ECONOMICS

2.1 China's Economy Today ............. 574
     **GLOBAL ISSUES: POPULATION GROWTH**
2.2 China's Population Policy .......... 576
2.3 China's Three Gorges Dam ........ 578
2.4 Republic of China (Taiwan) ....... 580
2.5 Japan's Economic Future ........... 582
2.6 Comparing North and South Korea ......... 584
     NATIONAL GEOGRAPHIC Photo Gallery ........... 586

**KEY VOCABULARY**
- entrepreneur
- gross domestic product (GDP)
- one-child policy
- fertility rate
- gorge
- reservoir
- martial law
- pagoda
- capital
- free trade
- drought
- famine

**ACADEMIC VOCABULARY**
comply, controversy

**TERMS & NAMES**
- Three Gorges Dam
- Lost Decade

Young Japanese women dressed in traditional kimonos meet on a snowy day in Nagano, Japan.

# SECTION 1 CULTURE
## 1.1 Religious Traditions

**Main Idea** Buddhism and other religions spread throughout East Asia.

As you have learned, Buddhism began in India, but it came to have a far greater influence in East Asia. Missionaries traveling on the Silk Roads during the Han Dynasty first brought Buddhist teachings to China. Over time, Buddhism blended with other religious traditions throughout East Asia.

### Blending Beliefs

Before Buddhism spread to East Asia, many people in the region practiced Confucianism and **Daoism**. Like Confucianism, Daoism is an ethical system. It stresses the harmony between people and nature. As Buddhism gained popularity, many East Asians combined elements of all three traditions.

Buddhism focuses on helping people end their physical and mental suffering by teaching them to give up worldly possessions. According to Buddhist teachings, one way to achieve this goal is through meditation. **Meditation** is the practice of using concentration to quiet and control one's thoughts.

In Mongolia, Buddhism mixed with **animism**, a belief that everything has a soul, including objects in nature. The native religion of Japan, **Shinto**, is similar to animism. The followers of this religion worship the spirits of their ancestors. They believe that these spirits exist in natural forces, including trees, rocks, and rivers.

> **Critical Viewing** Prayer flags flutter in the wind at a Buddhist temple in China. How might this temple's setting aid meditation?

## Absorbing New Beliefs

People who practice animism and Shinto worship many different gods. Belief in more than one god is called **polytheism** (PAHL ee thee ihz uhm). Eventually, religions that teach belief in one god, such as Christianity and Islam, came to East Asia. Belief in one god is called **monotheism** (MAHN uh thee ihz uhm). Muslim traders in northwestern China helped spread Islam in the 700s. Today, it is the dominant religion in that part of China. Christian missionaries began to come to East Asia in the 1600s to spread their religion. Today, about 30 percent of South Koreans are Christians.

Throughout East Asia's history, governments supported different religions and philosophies. Han rulers, for example, promoted the ideas of Confucianism as a model for government and society. When Communist leaders came to power in China and North Korea, they banned the practice of religion, emphasizing Communist philosophy instead. Since the 1970s, however, there has been more religious tolerance in China.

### Before You Move On
**Monitor Comprehension** How were religions spread and absorbed in East Asia?

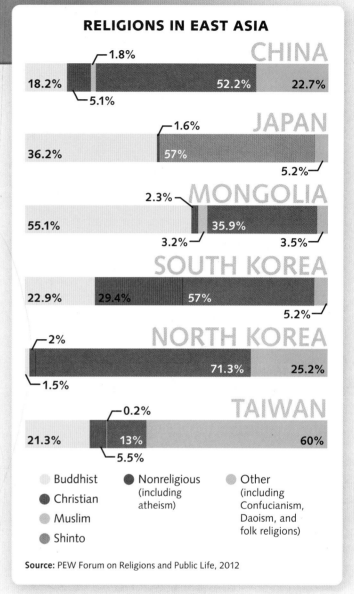

**RELIGIONS IN EAST ASIA**

CHINA: 18.2%, 5.1%, 1.8%, 52.2%, 22.7%
JAPAN: 36.2%, 1.6%, 57%, 5.2%
MONGOLIA: 55.1%, 2.3%, 3.2%, 35.9%, 3.5%
SOUTH KOREA: 22.9%, 29.4%, 57%, 5.2%
NORTH KOREA: 2%, 1.5%, 71.3%, 25.2%
TAIWAN: 21.3%, 5.5%, 0.2%, 13%, 60%

- Buddhist
- Christian
- Muslim
- Shinto
- Nonreligious (including atheism)
- Other (including Confucianism, Daoism, and folk religions)

**Source:** PEW Forum on Religions and Public Life, 2012

### ONGOING ASSESSMENT
# DATA LAB

1. **Interpret Graphs** On the graph, notice the high percentages of "Other" in China, Mongolia, and Taiwan. What probably accounts for the numbers in this category?

2. **Analyze Data** Why do so many people in China consider themselves nonreligious?

3. **Draw Conclusions** As you have learned, many East Asians combine and practice more than one religion. What does this suggest about the representations in the graphs?

# SECTION 1 CULTURE
## 1.2 China's Inventions

**TECHTREK**
myNGconnect.com For photos of Chinese inventions
Digital Library

**Main Idea** Chinese inventions introduced products and technologies that continue to be used today.

The Tang Dynasty (618–907) and Song Dynasty (960–1279) in China were periods of great technological advancements. New inventions transformed life in China and spread throughout much of the world. Many of these inventions are part of modern life.

### Porcelain and Gunpowder

In the 700s, the Chinese developed porcelain, a type of strong ceramic pottery. They kept the technology used in making it a closely guarded secret for hundreds of years. Porcelain became a valuable trade good that was—and is—exported to many parts of the world. Because of its close tie to Chinese culture porcelain is often referred to as "china."

During the 800s, the Chinese invented a very different product: gunpowder. Gunpowder was first used for fireworks. Within 300 years, however, the Chinese were using it to shoot weapons. Fireworks, meanwhile, became a popular feature of Chinese celebrations because they were believed to frighten away evil spirits. Today, China is the world's largest manufacturer and exporter of fireworks.

### Printing and Paper Money

The Chinese invented paper around A.D. 100. By the 700s, they began printing books using wooden blocks upon which text had been carved. Before block printing, people had reproduced books by copying each one by hand.

In the 1040s, the Chinese advanced printing further with the invention of movable type. Using this technique, individual characters carved on clay or metal blocks were placed in a frame to form a page of text. The blocks could be moved around to create many different pages. The Chinese used their printed books to spread the teachings of Confucius and Buddha.

The first paper money appeared in China in the 800s, but it did not become widely used until the 1020s. During the Song Dynasty, paper currency replaced the heavy metal coins that merchants and buyers previously carried. Paper money made trade easier and greatly expanded China's economy. However, people did not trust the new currency at first. After all, unlike the coins formerly used in trade, the paper itself had no value.

**Critical Viewing** This porcelain teapot was made during the Song Dynasty. Why do you think porcelain became so desirable?

## Magnetic Compass

Like paper, the compass was another early Chinese invention, dating from the 200s B.C. The compass from this period was used to position buildings and furniture in ways believed to bring good luck to the owners. By the 1100s, the Chinese had developed a magnetic compass for navigation by floating a magnetized needle in a bowl of water. They had learned that the needle always pointed in the north-south direction. With this floating compass, the Chinese explorer Zheng He traveled throughout Asia and Africa, demonstrating China's power and spreading its inventions.

### Before You Move On
Make Inferences In what ways have Chinese inventions influenced today's society?

**ONGOING ASSESSMENT**

**SPEAKING LAB** GeoJournal

**Turn and Talk** In your opinion, which of the inventions described in this lesson has had the greatest impact on society? Get together with a partner and discuss, using reasons to support your opinion. Consider the invention's positive and negative effects. Organize your ideas in a paragraph and be prepared to share them.

> **Critical Viewing** Fireworks mark the opening of the 2008 Summer Olympics in Beijing. What aspects of this photo help explain why fireworks are so popular today?

# SECTION 1 CULTURE
## 1.3 Shanghai's Rapid Growth

**TECHTREK**
myNGconnect.com For photos reflecting Shanghai's growth

Digital Library

> **Critical Viewing** People crowd Nanjing Road, a busy shopping district in Shanghai. What details in this photo suggest that Shanghai is a thriving center of business?

**Main Idea** Shanghai's growth has resulted in a building and population boom.

Shanghai's location near the mouth of the Chang Jiang has attracted settlers since ancient times. Over the centuries, it became an important port city and busy trade center. Today, with a population of around 20 million, Shanghai has grown into China's largest city. It is also the country's center for finance and business.

## Factors Fueling Growth

Shanghai's rapid growth began in the 1990s, when economic reforms led to new investments in the city. The economy of much of China is controlled by the central government. In 1990, however, the government allowed an area in Shanghai called Pudong to become a **Special Economic Zone**. This is an area that has the ability to develop a market economy with no government control of business.

As a result, Shanghai began to attract multinational corporations. A **multinational corporation** is a large company that is based in one country but establishes offices, branches, or plants in several others. Shanghai now hosts hundreds of multinational corporations. They have contributed to bringing economic globalization to China. **Economic globalization** occurs when economic activities are conducted across national borders.

## Building and Population Boom

To make room for these corporations, the city witnessed a building boom in the 2000s. Thousands of new skyscrapers were constructed. One of the most important is the Shanghai World Financial Center, which opened in Pudong in 2008. It is the tallest building in China and home to the Shanghai Stock Exchange.

As Shanghai's economy has grown, so has its population. Since the 1990s, about four million people from China's rural areas have come to Shanghai seeking jobs. They now make up about 25 percent of the workforce. The city center became so crowded that about a half million households were relocated to the suburbs. The exploding growth in population also resulted in higher levels of pollution.

To address these problems, city planners are building several new towns around Shanghai's urban center. Each one will house about 500,000 people. To protect its environment, Shanghai has also funded efforts to use buses and taxis that run on cleaner fuel and has moved most of its factories out of the city center.

### Before You Move On
**Monitor Comprehension** What factors fueled Shanghai's growth, and what happened in the city as a result?

## ONGOING ASSESSMENT
### READING LAB  GeoJournal

1. **Make Inferences** What economic reasons might have led Chinese leaders to allow Pudong to become a Special Economic Zone?
2. **Analyze Causes and Effects** What have been some of the effects of Shanghai's rapid economic and population growth?
3. **Make Predictions** Describe what you think Shanghai might be like ten years from now.

# NATIONAL GEOGRAPHIC
# Photo Gallery • Shanghai's Skyscrapers

For more photos from the National Geographic Photo Gallery, go to the **Digital Library** at myNGconnect.com.

Bullet train passes Mt. Fuji

Toy factory, China

Silla palace, South Korea

**Critical Viewing** The Oriental Pearl Tower and other skyscrapers line the banks of the Huangpu River in Shanghai. The Oriental Pearl, which is a television tower, features the revolving restaurant shown in the photo.

Japanese Kabuki drama

Buddhist monastery, Tibet

Japanese tea ceremony

Celebration in Taiwan

569

# SECTION 1 CULTURE

## 1.4 Japanese Anime

**TECHTREK**
myNGconnect.com For images of Japanese anime and manga

Digital Library

> **Main Idea** Anime reflects both Western and Japanese cultural influences and has become popular around the world.

**Anime** is a style of animation, or cartoon, developed in Japan. The characters and settings in this art form are either drawn by hand or computer-generated. Anime is closely related to **manga**, or Japanese comic books. Manga stories are told using a series of panels. Many anime films are based on the art and stories in popular manga. In fact, an anime film may even use manga panels to introduce a story.

### Beginnings

Although anime films were produced in the early 1900s, the style and techniques used today didn't develop until the 1960s. Around that time, manga artist and animator Osamu Tezuka began to adapt some of the techniques used by Walt Disney in his own films. For example, Disney cartoons such as *Bambi* inspired Tezuka to draw characters with large eyes, which became a typical anime feature.

Anime uses elements from Western-style cartoons, but it also draws on Japanese culture. Ancient Japanese myths, Shinto, and Buddhism have all influenced many of the cartoons. For example, Shinto nature spirits populate the world depicted in *Spirited Away*, an anime film made in 2001 by director Hayao Miyazaki.

### Popularity

Anime was a popular art form in Japan from the beginning, but Western audiences did not embrace it until the 1980s. Since then, anime has continued to gain fans—and respect—around the world.

One reason for anime's popularity is its broad range of subjects. Anime is used to tell stories in many different forms, including fantasy and science fiction. As a result, it appeals to all ages and both genders. Anime is also big business. In Japan, the art form and products related to anime earn more than $5 billion a year.

**Before You Move On**
**Summarize** In what ways does anime reflect Western and Japanese cultural influences, and why is it so popular?

▲ **Critical Viewing** The artist in the top photo is drawing a panel, like that shown at bottom, for an anime film. What skills are probably needed to create anime?

This image shows a scene from Miyazaki's *Castle in the Sky*, in which a young boy and girl search for a floating castle.

Lighting and colors convey an exciting mood.

The boy and girl have round, expressive eyes like many anime characters.

The camera angle in this scene shows perspective and depth.

### ONGOING ASSESSMENT
## PHOTO LAB
GeoJournal

1. **Analyze Primary Sources** Examine the anime image above. Who might be the intended audience for the film? What details in the image support your ideas?
2. **Analyze Visuals** What personality traits do you detect in the characters shown above?

SECTION 1.4  **571**

# SECTION 1 CULTURE
## 1.5 Bullet Trains

**Main Idea** Bullet trains and other high-speed rail services have transformed travel in much of East Asia.

Japan pioneered the use of high-speed rail in 1964, with trains that traveled at speeds of more than 125 miles per hour (mph). They earned the nickname **bullet trains** because of their appearance and speed. Each train has a rounded nose like that of an airplane and a sleek, **aerodynamic** design. This means that the train moves with little resistance from the wind. The technology has spread, and now bullet trains connect cities in countries throughout East Asia.

> **Critical Viewing** A Japanese bullet train arrives in a station in Tokyo. Based on the photo, how would you describe the location of the station?

### High-Speed Rail Spreads

The first bullet trains linked Tokyo with Osaka and allowed people to travel the 320-mile route in four hours instead of six. Today, the system connects Tokyo with all major cities on the main island of Honshu. Additional lines are being built on Kyushu as well. The trains on these lines travel at speeds up to 186 mph. The newer trains are also quieter and more energy efficient.

By the 2000s, bullet trains were running throughout much of East Asia. In 2004, South Korea established a high-speed rail system, connecting Seoul with major industrial cities and ports. In 2007, a high-speed rail system began in Taiwan and connected Taipei with cities in the southwestern part of the country.

# EAST ASIA HIGH-SPEED RAIL ROUTES, 2010

The Chinese began developing high-speed rail in the 1990s. By 2009, China operated the fastest trains in the world, with speeds up to 245 mph. China has also taken advantage of another high-speed train technology: the magnetic levitation train, or Maglev for short. A **magnetic levitation** train rides on a cushion of air over tracks laid with many powerful magnets. Today, Maglev trains run between the Shanghai airport and Pudong, while achieving a top speed of 268 mph.

## Benefits

The greatest benefit of high-speed trains may be the time passengers save by riding them. However, the trains have also given an economic boost to once-remote rural locations that are now linked to cities and towns. In addition, bullet trains produce less pollution than traditional trains. In fact, the Maglev is almost pollution-free.

### Before You Move On

**Summarize** In what ways have high-speed rail services transformed travel in East Asia?

## ONGOING ASSESSMENT
# MAP LAB

1. **Movement** On the map, study the rail routes in Japan. What generalization can you make about the routes?

2. **Region** According to the map, in what part of the country are most of China's routes concentrated? Why do you think that is so?

3. **Make Inferences** Why might people prefer to travel from one city to another on a bullet train rather than on a plane or by car?

SECTION 1.5 **573**

# SECTION 2 GOVERNMENT & ECONOMICS

## 2.1 China's Economy Today

**TECHTREK**
myNGconnect.com For online graphs of China's GDP and photos of Chinese industry

 Maps and Graphs   Digital Library

**Main Idea** Since 1979, China has become a leader in the global economy.

As you have learned, the Chinese government established a Special Economic Zone, or SEZ, in Shanghai in 1990. You may remember that an SEZ is an area in which businesses are free to develop without government control. China began to allow the creation of SEZs in the early 1980s. As a result, China has had one of the world's fastest-growing economies, and Chinese people today enjoy a higher standard of living.

### New Strategy for Growth

China limited its SEZs to cities and provinces along the Pacific Coast because its leaders believed that this location would promote and facilitate overseas trade. The first SEZ was established in 1980 in Shenzhen, which at the time was a small fishing village just north of Hong Kong. Since then, Shenzhen has become a thriving city with about ten million people. Other important SEZs are in Xiamen and the province of Hainan.

Chinese leaders encouraged trade and foreign investment in these SEZs by offering lower taxes and fewer regulations on imports and exports. The lower costs also attracted **entrepreneurs**, people who start up new businesses. In addition, the lower wages paid to Chinese workers persuaded regional companies and multinational corporations to open branches in the country.

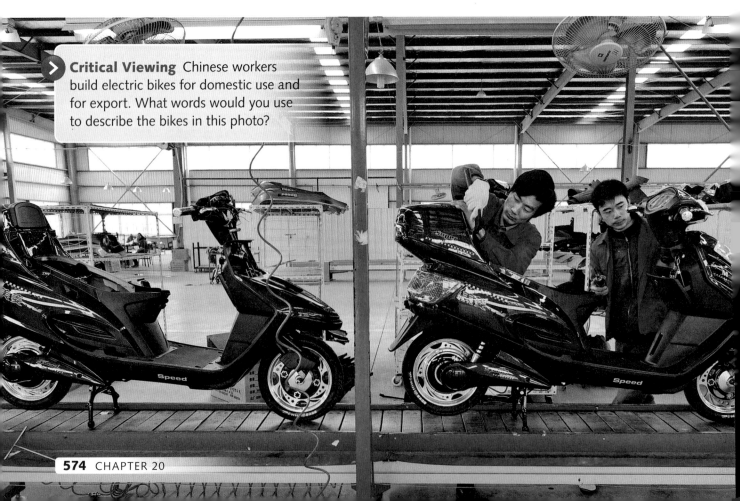

> **Critical Viewing** Chinese workers build electric bikes for domestic use and for export. What words would you use to describe the bikes in this photo?

China's strategies have worked. In 2013, China had the world's third largest economy after the United States and the European Union. Economists measure the size of a country's economy based on **gross domestic product (GDP)**, the total value of all the goods and services produced in a country in a given year. China's GDP increases about ten percent per year.

## A Diverse Economy

China's growing economy is powered by industry, services, and agriculture. The country's main industries include mining and iron and steel manufacturing. Globally, China is most known for its exports of clothing and textiles, electronics, and toys. China exports more than any other single country in the world. About 20 percent of its exports go to the United States.

Inside China's borders, service industries provide activities such as banking, insurance, trade, communication, education, health care, recreation, and transportation. As China welcomes more and more visitors, tourism is also becoming a large part of its economy.

Agriculture contributes little to China's GDP, but it employs 35 percent of the country's workers. Only about 15 percent of Chinese land is arable, but the country produces most of the food it needs for its large population.

### Before You Move On
**Summarize** What steps has China taken to grow and diversify its economy?

CHINA'S GDP BY SECTOR
- 10.1% Agriculture
- 44.6% Services
- 45.3% Industry

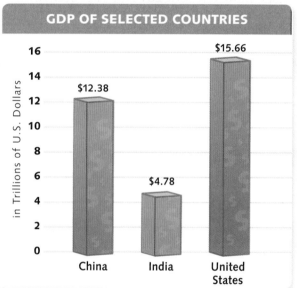

GDP OF SELECTED COUNTRIES
- China: $12.38
- India: $4.78
- United States: $15.66
(in Trillions of U.S. Dollars)

Source: CIA World Factbook 2013

## ONGOING ASSESSMENT
### DATA LAB — GeoJournal

1. **Draw Conclusions** According to the pie chart, how much of China's GDP is supplied by agriculture? Since 35 percent of the Chinese people work in agriculture, what conclusions can you draw about their earnings?

2. **Analyze Data** Study the bar graph and compare China's GDP with that of India. What do these numbers suggest about the relative rate of each country's GDP growth?

3. **Make Inferences** Based on what you have read, why are Chinese-made consumer goods sold at low prices in countries such as the United States?

SECTION 2.1

# SECTION 1 GLOBAL ISSUES • Population Growth

## 2.2 China's Population Policy

**Main Idea** China's one-child policy has slowed the population growth rate, but it is also bringing great change to Chinese society.

**KEY VOCABULARY**

**one-child policy**, n., law that restricts urban Chinese families to one child

**fertility rate**, n., average number of children born per woman

**ACADEMIC VOCABULARY**

**comply**, v., follow or obey

With about 1.3 billion people, China has the largest population in the world. For many years, Communist leaders encouraged the country's population to grow because they believed that having more people—and workers—would strengthen China's economy. In time, however, it became difficult to provide food, housing, education, and jobs for the country's citizens.

## Slowing Population Growth

To curb their rapidly growing population, the government introduced the **one-child policy** in 1979. This law limited families living in urban areas to one child. Those who **complied**, or followed, the policy were rewarded with more food, improved housing, and better education and job opportunities for that child. Those who did not comply had to pay large fines. As a result, the rate of population growth declined to about 0.65 percent per year by 2010, which is about half of what it was in 1979. There are some exceptions to the policy. For example, minority populations and rural families may have two or more children. In 2013, the government began to loosen restrictions, and, in 2015, the policy began to be phased out.

### Before You Move On

**Monitor Comprehension** What policies did China put in place to decrease population growth?

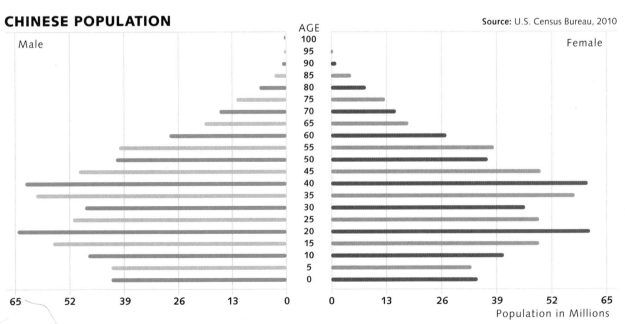

**CHINESE POPULATION**

Source: U.S. Census Bureau, 2010

## Effects of China's Policy

In general, China's one-child policy has worked. The **fertility rate**, or the average number of children born per woman, dropped from about six in the 1950s to about two by 1995. The Chinese government claimed that, since it was put in place, the policy has helped reduce the number of births in China by 400 million. According to the government, the policy has also improved the standard of living for many Chinese people.

The policy's biggest long-term impact will be on the age distribution of China's population. With fewer children being born, the elderly will soon outnumber young people in China. Traditionally, daughters and daughters-in-law have been the primary caretakers of elderly parents. However, if a single adult child cannot take care of his or her parents, the seniors may have to depend on retirement funds or charity for support. In response, the Chinese government has begun to create community services to help the elderly who live alone.

Three generations of a family pass a poster promoting population control.

**COMPARE ACROSS REGIONS**

## India's Population Growth

Like China, India has a huge population. With about 1.2 billion people, India is second only to China in population—for now. Scientists predict that India's population will exceed China's by 2030. In contrast with China, about 50 percent of India's population is under the age of 25. While this statistic guarantees that India will have a large workforce, the young population will also put a strain on the country's schools and other resources.

In spite of this growth, India does not enforce strict population policies. Instead, the country encourages the advancement and education of its women. India's leaders believe that educated women are more likely to have smaller families. The government also offers cash bonuses to women in rural areas who wait to start a family and agree to have fewer children.

### Before You Move On
**Summarize** What have been the effects of China's one-child policy?

## ONGOING ASSESSMENT
## READING LAB  GeoJournal

1. **Make Generalizations** Based on what you read, what is the connection between overpopulation and standard of living?
2. **Interpret Graphs** Study the population pyramid. If these numbers hold, which groups will be the largest in 20 years? What impact might this situation have on China's economy?
3. **Form and Support Opinions** In your opinion, which country's policies have a better chance of working, China's or India's? Why?

# SECTION 2 GOVERNMENT & ECONOMICS

## 2.3 China's Three Gorges Dam

**TECHTREK**
myNGconnect.com For photos of the Three Gorges Dam and Guided Writing

**Main Idea** The Three Gorges Dam in China is the largest dam in the world and has both benefits and drawbacks.

The **Three Gorges Dam** on the Chang Jiang River is nearly 1.3 miles wide and 600 feet high. The dam was built to prevent flooding along the eastern part of the river between the cities of Chongqing and Wuhan. It was also designed to generate hydroelectric power, promote economic development in the region, and ease navigation along the Chang Jiang.

## Building a Giant Dam

Geographic factors determined where the dam would be built. The Three Gorges is an area where the Chang Jiang is very narrow—only about 350 feet wide as the river flows between steep cliffs. The area is named for this geographic feature. A **gorge** is a deep, narrow passage surrounded by steep cliffs. When snows melt in the spring or heavy rains fall during summer monsoons, the river can rise very quickly in the narrow area. Over the years, this seasonal rise has caused major flooding and millions of deaths.

As work on the dam began in 1994, a large artificial lake called a **reservoir** formed behind it to store the water being held back. This reservoir is about 500 feet deep, and it is longer than Lake Superior in the United States, the largest natural body of fresh water in the world. During dry periods, the dam will slowly release water from the reservoir. This raises the level of the river downstream to allow large ships to travel to China's interior.

Work on the dam was completed in 2006, but it did not become fully operational until 2011. The dam is a major engineering achievement. However, it has stirred **controversy**, or debate, both within China and abroad.

**Critical Viewing** The Three Gorges Dam has been called China's Great Wall across the Chang Jiang. Based on this photo, what feelings might the dam inspire in the Chinese people?

## Benefits and Drawbacks

The Chinese government expects to reap great benefits from the dam, which has already prevented serious flooding along the Chang Jiang. Improving conditions for commercial shipping along the river should increase economic development in the region. In addition, the government predicts that the hydroelectric power created by the dam will provide about 10 percent of China's total energy.

Critics, however, claim that the drawbacks of the dam, which cost an estimated $25 billion, outweigh the benefits. To create the reservoir, a large expanse of land had to be cleared. About 1.5 million people were displaced from their cities and villages, and more than 1,000 historic and cultural sites are now under water. The critics also fear that the dam may cause landslides and destroy the habitat of endangered animals. Many argue that building several smaller dams and using newer technology for producing energy would have been a better solution.

### Before You Move On
**Summarize** Why did China build the Three Gorges Dam?

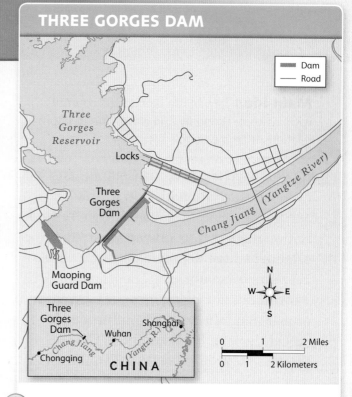

**MAP TIP**
The smaller map shows where the Three Gorges Dam is located on the Chang Jiang. Notice that the dam is west of Shanghai in eastern China.

## ONGOING ASSESSMENT
# WRITING LAB  GeoJournal

**Form and Support Opinions** Do you think that building the Three Gorges Dam was a good idea? Use a chart like the one below to list the dam's benefits and drawbacks. Then write a brief paragraph in which you state and support your opinion with evidence. You may want to do additional research to learn more about the dam. Go to **Student Resources** for Guided Writing Support.

| BENEFITS | DRAWBACKS |
|---|---|
| 1. | |
| 2. | |

# SECTION 2 GOVERNMENT & ECONOMICS
## 2.4 Republic of China (Taiwan)

**TECHTREK**
myNGconnect.com For an online map and photos of Taiwan

 Maps and Graphs
 Digital Library

**Main Idea** Taiwan is a democratic country with a strong economy and is working to improve relations with the People's Republic of China.

As you have learned, the Communists defeated the Nationalists in China in 1949. After their defeat, about two million Nationalists fled to Taiwan where they established the Republic of China, Taiwan's official name. The island of Taiwan lies about 100 miles off the coast of mainland China, which is officially called the People's Republic of China. Since the war between the Communists and Nationalists never formally ended, there are still tensions between the two neighboring countries.

### A Democratic Government

In 1949, the Nationalists placed Taiwan under **martial law**—a government maintained by military power—which continued into the 1980s. During their long, one-party rule, Nationalist leaders claimed that they had authority over all of China, including the mainland.

Beginning in 1987, more political parties were allowed to take part in elections, and martial law was lifted. As a result, people on the island experienced more democratic freedom. However, the country's political status, or legal position, has remained uncertain.

> **Visual Vocabulary** A large pagoda rises on a mountain in Taiwan. Most often found in Asia, a **pagoda** is a multistoried structure that is usually used for religious purposes.

^ **Critical Viewing** Motorcyclists wait at an intersection during rush hour in Taipei. What conclusions can you draw about Taiwan's capital based on this photo?

During the Cold War, many countries regarded Taiwan as the legitimate government of China. Today, these countries recognize the government of the People's Republic of China in Beijing but unofficially treat Taiwan as an independent state. Mainland China considers Taiwan one of its provinces. It has chosen not to challenge Taiwan, however, as long as the island country does not declare its independence.

## A Strong Economy

During the 1960s and 1970s, Taiwan began to develop a market economy. Today, it has the 20th largest economy in the world with a gross domestic product (GDP) of about $900 billion.

Exports have fueled the growth of Taiwan's GDP. Important exports include electronics and machinery. Because Taiwan depends so much on exports, its economy declines when world demand for its goods decreases. Farming in Taiwan is limited because about two-thirds of the country is covered in mountains.

Taiwan wants to strengthen economic ties with mainland China, its leading export and import partner. Chinese investors have been able to invest directly in businesses in Taiwan, and Taiwanese financial firms have opened on mainland China. This improved economic cooperation may also lead to better diplomatic relations between the two.

### Before You Move On

**Monitor Comprehension** What steps has Taiwan taken to improve its government, economy, and relations with China?

Source: CIA World Factbook 2010

### ONGOING ASSESSMENT
## DATA LAB
GeoJournal

1. **Make Inferences** Why does agriculture account for so little of Taiwan's GDP?
2. **Interpret Graphs** Based on the Export Partners graph, where are most of Taiwan's partners located? Why do you think this is so?
3. **Synthesize** As you have read, much of Taiwan's economy depends on its exports. Which sector of the GDP probably provides most of those exports?

SECTION 2.4 **581**

# SECTION 2 GOVERNMENT & ECONOMICS

## 2.5 Japan's Economic Future

**TECHTREK**

myNGconnect.com For a population graph of Japan and photos of its industries

Maps and Graphs | Digital Library

**Main Idea** Japan's economy has declined sharply since the 1990s, and the country faces many challenges.

From the 1960s until the late 1980s, Japan had the strongest economy in East Asia. It was second only to the United States in gross domestic product (GDP) worldwide. However, by 1990, the country's economic growth had begun to slow down. Many businesses had borrowed huge amounts of money and were deeply in debt. With little **capital**, or money for investment, production declined. The situation continued throughout the 1990s, a period sometimes called Japan's **Lost Decade**.

In the 2000s, Japan's economy began to recover. However, its progress was halted in 2008 by the worldwide economic downturn, which led to a decrease in the demand for Japanese exports. Many industries also suffered as a result of the 2011 earthquake and tsunami.

### Economy Today

Today, Japan's GDP is the third largest in the world, behind the United States and China. About 77 percent of Japan's economy is invested in service industries and about 22 percent in manufacturing.

Because Japan has few natural resources, the country must import most of the raw materials needed for its many industries. Japan's economy depends heavily on exported goods produced by these industries, including cars, computers, and other electronics.

Japan is still among the world's top five exporters. However, it has been less aggressive than many other Asian countries in promoting free trade with its neighbors. **Free trade** is trade that does not impose tariffs, or taxes on imports. These taxes can make Japan's exports more expensive than those of other Asian countries with free-trade agreements.

> **Critical Viewing** A technician inspects flat-screen televisions in a factory in Osaka. How do the televisions in this photo compare with those made in America?

582 CHAPTER 20

## Challenges Ahead

In addition to competition from other countries, Japan's aging population may have a harmful effect on the country's economy. In 2050, Japan's population is expected to be about 20 percent smaller than it was in 2009. Furthermore, about one-third of the people in 2050 are expected to be over age 65. These statistics suggest that Japan will have a shortage of workers as well as of entrepreneurs, since these risk takers are generally younger.

Currently, Japanese workers are required to retire at age 60. In addition, men hold most of the higher paying jobs in most industries. Japan could increase its workforce by encouraging older workers to keep working and by providing more economic opportunities for women. The country could also relax the limits it places on foreign investments in Japanese companies. Foreign investors could bring in new capital and ideas that might help make Japan become more competitive.

### JAPAN'S POPULATION (2009)

■ = 1 million people

| Age Groups | Population (millions) |
|---|---|
| 0–9 | 11.1 |
| 10–19 | 12 |
| 20–29 | 14.4 |
| 30–39 | 18.3 |
| 40–49 | 16.4 |
| 50–59 | 16.9 |
| 60–69 | 17.8 |
| 70–79 | 12.7 |
| 80–89 | 6.6 |
| 90+ | 1.3 |

Source: Population Census, Statistics Bureau of Japan

Japan's economy also faced challenges following the 2011 earthquake and tsunami. After the disaster, manufacturing fell sharply and power supplies decreased when several of Japan's nuclear power plants became permanently disabled. The cost of repairing the damage caused by the natural disaster was estimated at about $300 billion. Some analysts feared that the disaster could put a halt to Japan's economic recovery.

### Before You Move On
**Summarize** What were the causes and effects of Japan's economic decline?

## ONGOING ASSESSMENT
# DATA LAB  GeoJournal

1. **Make Predictions** According to the graph, how many Japanese people were in the 0–9 age group in 2009? What will this statistic mean for Japan in 20 years?

2. **Interpret Graphs** Study the graph. What were Japan's four largest age groups in 2009? Why might these statistics have a negative impact on Japan's economy in 20 years?

3. **Draw Conclusions** What burdens might Japan's aging population place on young Japanese workers?

# SECTION 2 GOVERNMENT & ECONOMICS

## 2.6 Comparing North and South Korea

**Main Idea** North Korea and South Korea have different government and economic systems and different ways of life.

As you have already learned, the Korean Peninsula was divided into North Korea and South Korea after World War II. Since then, these neighbors have developed in very different ways.

### North Korea

North Korea, which is officially called the Democratic People's Republic of Korea, has a Communist government that follows a policy of self-reliance. This means that the country depends mostly on its own efforts and abilities. The policy has kept North Korea isolated from most other countries.

The North Korean government controls most of the country's economy. The country's isolation policy has restricted trade, but it does conduct a limited amount of commerce, particularly with China. Much of its industry is devoted to the production of military equipment.

Only about 18 percent of North Korea's land is arable, and the climate does not support a long growing season. In the 1990s, floods and **drought**, a long period of extremely dry weather, greatly decreased food production. The resulting **famine**, or extreme food shortage, is believed to have caused as many as three million deaths.

### South Korea

South Korea, which is officially called the Republic of Korea, was ruled largely by repressive military leaders until 1987. Since then, the country has developed into a successful democracy, with free elections and multiple political parties.

Unlike North Korea, South Korea did not isolate itself from the rest of the world. It accepted grants and loans from countries such as the United States and Japan. As a result, South Korea's market economy grew quickly beginning in the 1960s. Today, it is one of the largest in the world.

South Korea's economy depends heavily on manufacturing. Shipbuilding has long been a major industry. In fact, South Korea is the world's leading builder of ships. The country also produces automobiles and electronics. Like Japan, South Korea has few natural resources and relies largely on the goods it exports to fuel its economy.

### Reunification

North Korea and South Korea have had a tense relationship since the end of the Korean War in 1953. In the early 2000s, meetings between the countries' leaders suggested that relations were improving. Attempts were made to encourage interaction between the two countries. However, North Korea's development of nuclear weapons and its military attack on a South Korean island in 2010 stirred up hostilities once again. Since then, there has been little hope that the two countries would reunite soon.

### Before You Move On

**Make Inferences** What impact have the different governments and economic systems in North Korea and South Korea had on their people?

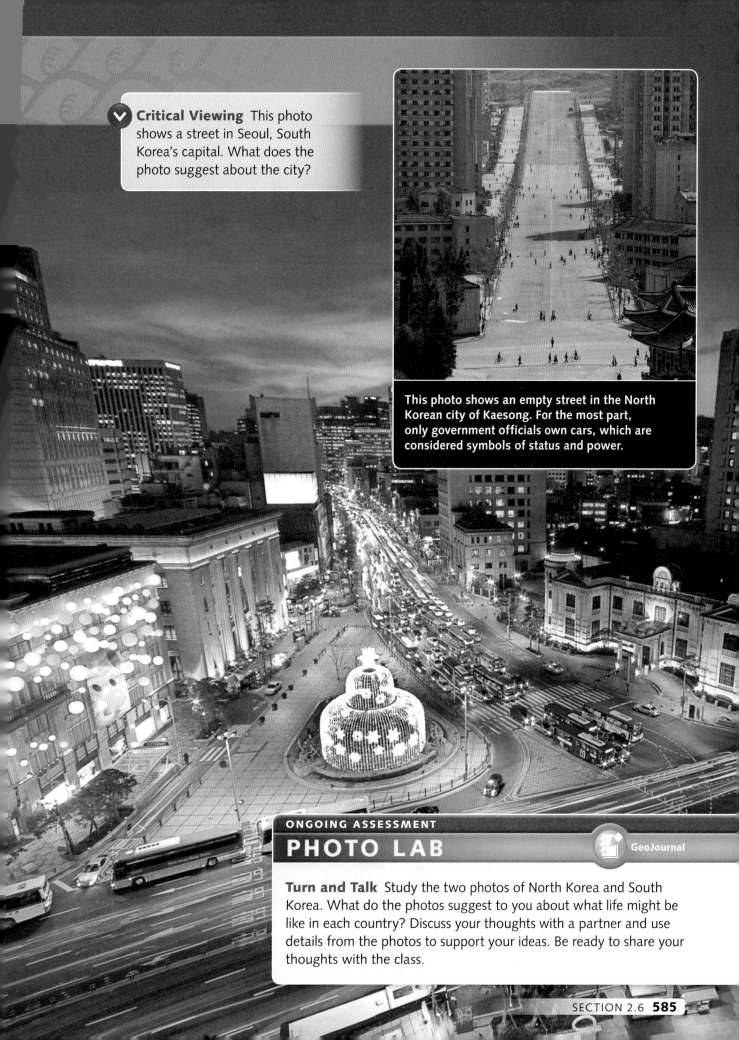

**Critical Viewing** This photo shows a street in Seoul, South Korea's capital. What does the photo suggest about the city?

This photo shows an empty street in the North Korean city of Kaesong. For the most part, only government officials own cars, which are considered symbols of status and power.

ONGOING ASSESSMENT
# PHOTO LAB
GeoJournal

**Turn and Talk** Study the two photos of North Korea and South Korea. What do the photos suggest to you about what life might be like in each country? Discuss your thoughts with a partner and use details from the photos to support your ideas. Be ready to share your thoughts with the class.

# NATIONAL GEOGRAPHIC
## Photo Gallery • Buddhas in South Korea

For more photos from the National Geographic Photo Gallery, go to the **Digital Library** at myNGconnect.com.

Capsule hotel, Japan

Japanese rice cakes

Hong Kong street market

**Critical Viewing** Rows of golden Buddhas welcome visitors to a South Korean temple. Buddhism spread from India to China in the A.D. 100s. Chinese monks eventually introduced the religion to the Korean Peninsula.

Shinto torii gate

Kabuki theater actor

Crowded subway, Japan

The Bund, Shanghai

# CHAPTER 20
# Review

## VOCABULARY

For each pair of vocabulary words, write one sentence that explains the connection between the two words.

1. monotheism; polytheism

> *Belief in one god is called monotheism, while belief in more than one god is called polytheism.*

2. anime; manga
3. gorge; reservoir
4. free trade; capital
5. famine; drought

## MAIN IDEAS

6. What are East Asia's religious traditions and in what manner did they arise in the region? (Section 1.1)
7. How did the Chinese use the inventions of movable type and paper money? (Section 1.2)
8. What factors helped Shanghai become China's largest city? (Section 1.3)
9. Why did Japan develop a high-speed rail system in the 1960s? (Section 1.5)
10. Where did China encourage most economic growth after 1979? (Section 2.1)
11. What did China hope to accomplish with its one-child population policy? (Section 2.2)
12. Why are some people critical of the Three Gorges Dam? (Section 2.3)
13. What is Taiwan's relationship with China? (Section 2.4)
14. In what way is Japan's population related to its economic future? (Section 2.5)
15. How do the governments and economic systems of North Korea and South Korea differ? (Section 2.6)

## CULTURE

### ANALYZE THE ESSENTIAL QUESTION

How do traditions and modernization create a unique way of life in East Asia?

**Critical Thinking: Draw Conclusions**

16. Why did many East Asians blend new religious ideas with traditional beliefs?
17. Why did China allow an area of Shanghai to become a Special Economic Zone?
18. In what ways does high-speed rail travel promote economic development in East Asia?

### INTERPRET MAPS

19. **Region** Which language is spoken in much of China? What might be the disadvantages for those who don't speak the language?
20. **Compare and Contrast** Which East Asian countries are most united by a common language? What does this fact suggest about the people of these countries?

## GOVERNMENT & ECONOMICS

### ANALYZE THE ESSENTIAL QUESTION

What problems does East Asia face today, and what are its opportunities?

**Critical Thinking: Compare and Contrast**

21. Compare the economic opportunities in China's coastal cities and rural areas. Which area provides greater opportunities? Why?

22. In what ways are the economies of Japan, South Korea, and Taiwan similar?

23. What do North Korea and South Korea have in common? Do you think these similarities might help the two countries unite some day? Explain why or why not.

### INTERPRET CHARTS

#### LIFE IN THE TWO KOREAS

|  | North Korea | South Korea |
|---|---|---|
| GDP (valued at U.S. prices) | $40 billion | $1.611 trillion |
| agriculture | 23% | 3% |
| industry | 48% | 39% |
| services | 29% | 58% |
| GDP per person | $1,800 | $32,400 |
| Life expectancy at birth | 69.5 years | 79.5 years |
| Infant mortality | 25 deaths/ 1,000 live births | 4 deaths/ 1,000 live births |
| Literacy | 99% | 97.9% |

Source: CIA World Factbook 2010

24. **Draw Conclusions** Based on the statistics in the chart under GDP per person, what conclusions can you draw about the standard of living in North Korea and South Korea?

25. **Make Inferences** Study the statistics on infant mortality. What do these numbers suggest about health care in each country?

## ACTIVE OPTIONS

**Synthesize** the Essential Questions by completing the activities below.

26. **Create a Web Page** Make a home page for a Web site about East Asia today. Write a short introduction to the region and include links to each country. Use visuals, including photos, maps, charts, and graphs, that illustrate East Asia's culture and economic endeavors. Use the tips below to help you prepare your Web page. **Display your Web page and invite your classmates to "click" on the links. Be prepared to summarize what your viewers would find on each link.**

> **Writing Tips**
> - Use bullet points to summarize cultural and economic topics in East Asia today.
> - Provide labels for your links that will capture your audience's attention.
> - Include captions with your visuals that are brief and informative.

27. **Design an Anime Storyboard** Get together in a small group to design a storyboard for a scene in an anime film. First, brainstorm to come up with ideas for the scene's characters and plot. You can use online sources to learn more about anime. Then sketch and write dialogue for each panel of the scene. Use a chart like the one below to organize your ideas.

| PANEL 1 | PANEL 2 | PANEL 3 |
|---|---|---|
| Characters: | Characters: | Characters: |
| Action: | Action: | Action: |
| Dialogue: | Dialogue: | Dialogue: |

COMPARE ACROSS REGIONS  UNIT 10

# Mobile Access

myNGconnect.com For an online graph on mobile phone use

Maps and Graphs

Perhaps no other manufactured product since the personal computer has had so broad a global impact as the mobile phone and its cousin the smart phone. These devices play a part in economics, politics, civic participation, and the spread of culture—in activities that go far beyond taking photos, texting, or checking email.

In developing countries not so long ago, people had little to no access to global information. Today, those who cannot afford computers for Internet connections buy smart phones instead, which are less expensive.

## Compare

- Argentina
- Bangladesh
- Brazil
- China
- India
- Mongolia
- Nepal
- South Korea
- United States
- Venezuela

### CHINA AND MEDIA

China has the largest world population, and it has the most mobile phone subscribers. The BBC reported in 2013 that 464 million Chinese, or 44% of the population, used smartphone or wireless devices for Internet access. (In 2002, the percentage was 4.6%.) China's communist government, however, controls and censors media. It requires media providers to do the censoring, too. A blog post criticizing the government can be removed in minutes.

Chinese artist Ai Weiwei (EYE way way) used a blog to gather the names of children who died in poorly constructed schools during the 2008 earthquake in Sichuan province. The government had built the schools. It blocked the artist's Internet access on his computer. So he and those helping him used cell phones instead. The 9,000 names they gathered became part of an art project called "Remembering." A child's backpack, one for each victim, was mounted on long outdoor walls at a German museum.

### OTHER USES AND BENEFITS

Brazil's mobile phone and Internet usage tripled from 2007 to 2012. Mobile phone and smartphone use is on the rise not only in South America, East Asia, and South Asia but in countries in remote regions. In 2013, there were reportedly 4.5 billion mobile phones in the developing world. Developing countries are using resources to install mobile networks. This helps small businesses quickly keep up with market news using phones. Some studies show that cell phone use can increase a country's GDP.

Thousands of uses have been developed through mobile apps. These software applications are downloaded for banking, games, movies, social media, music, car rentals, and finding businesses, as well as for education, health information, government information, and public safety weather alerts.

# CELL PHONE USAGE*

## NUMBER OF PEOPLE USING CELL PHONES

● East Asia ● South Asia ● U.S. Benchmark

**GRAPH TIP** The number at the top of each bar indicates the country's world ranking in terms of cell phone usage.

- China — 1st (~540 million)
- South Korea — 20th
- Mongolia — 116th
- India — 2nd (~360 million)
- Bangladesh — 25th
- Nepal — 131st
- United States — 3rd (~255 million)

*Sources: UN Database, 2007

## PERCENTAGE OF POPULATION

These graphs show the percentage of the country's population that uses cell phones.

China 42%  South Korea 89%

Mongolia 61%

India 32%  Bangladesh 23%

Nepal 4%

United States 84%

---

## ONGOING ASSESSMENT
# RESEARCH LAB  GeoJournal

1. **Analyze Data** Study the numbers for China and India. Why do you suppose these countries have the most cell phone subscribers?

2. **Draw Conclusions** Note that Mongolia has more than 2.5 million people while Nepal has more than 29 million. What conclusions can you draw about economic development in Mongolia and Nepal based on their cell phone usage?

**Research and Compare** Research cell phone usage in the South American countries of Brazil, Argentina, and Venezuela. Use graphs similar to those above to compare the numbers of cell phone subscribers in those countries. Then compare your findings with the statistics shown here. Which South American country has the most cell phone use? What might this statistic suggest about economic development in the country?

COMPARE ACROSS REGIONS  **591**

REVIEW & EXTEND **UNIT 10**

# Active Options

**TECHTREK**
myNGconnect.com For photos of cherry blossoms in Japan and Korean kimchi

Digital Library

## ACTIVITY 1

**Goal:** Learn about the importance of cherry blossoms in Japan.

### Celebrate the Cherry Blossom

Every year, the Japanese people hold festivals to celebrate the cherry blossom. In Japanese culture, the beauty of the blossoms and the brief duration of their bloom symbolize the fleeting nature of life. The blossoms are often used in manga and anime. Research to learn more about the importance of cherry blossoms in Japan. Then choose one of the following ways to celebrate the flower:

- Write a haiku in honor of the cherry blossom. A haiku is a three-line poem with five syllables in the first line, seven in the second line, and five in the third line. The poem does not rhyme.
- Sketch or paint a cherry blossom.
- Find and display examples of cherry blossoms in Japanese art.
- Research and report on three other places outside Japan that hold cherry blossom festivals.

Anime illustration of cherry blossoms

## ACTIVITY 2

**Goal:** Extend your knowledge of China's culture.

### Create a Chinese Culture Magazine

Learn more about China's rich cultural heritage. Get together in a small group to design a magazine page on one part of Chinese culture. You might focus on the country's art, music, food, or martial arts. Research your topic and use the **Magazine Maker CD-ROM** to create your page. Combine the pages created by different groups into a magazine.

## ACTIVITY 3

**Goal:** Make predictions based on prior knowledge.

### I Predict!

What's the next stage for bullet trains or mobile phones? Predict discoveries in science and technology that could affect culture or politics. How might technology improve health? Explain the basis for your predictions.

# UNIT 11

# Explore Southeast Asia
## with NATIONAL GEOGRAPHIC

### MEET THE EXPLORER

**NATIONAL GEOGRAPHIC**

Emerging Explorer Jenny Daltry searches for unknown species of snakes, frogs, and crocodiles in unexplored corners of South Asia and Southeast Asia. Her work helps conserve these animals' habitats. Here, she inspects the fangs of a snake.

### INVESTIGATE GEOGRAPHY

Mount Merapi is a volcanic mountain peak located near the center of the densely populated island of Java, Indonesia. It is the most active of the country's volcanoes. Its ash creates fertile soil, luring farmers in spite of the dangers.

### STEP INTO HISTORY

A Buddhist monk prays at a statue in the Angkor Wat temple complex, Cambodia. Angkor Wat was originally built as a Hindu worship center in the 12th century. It remains the largest religious structure in the world.

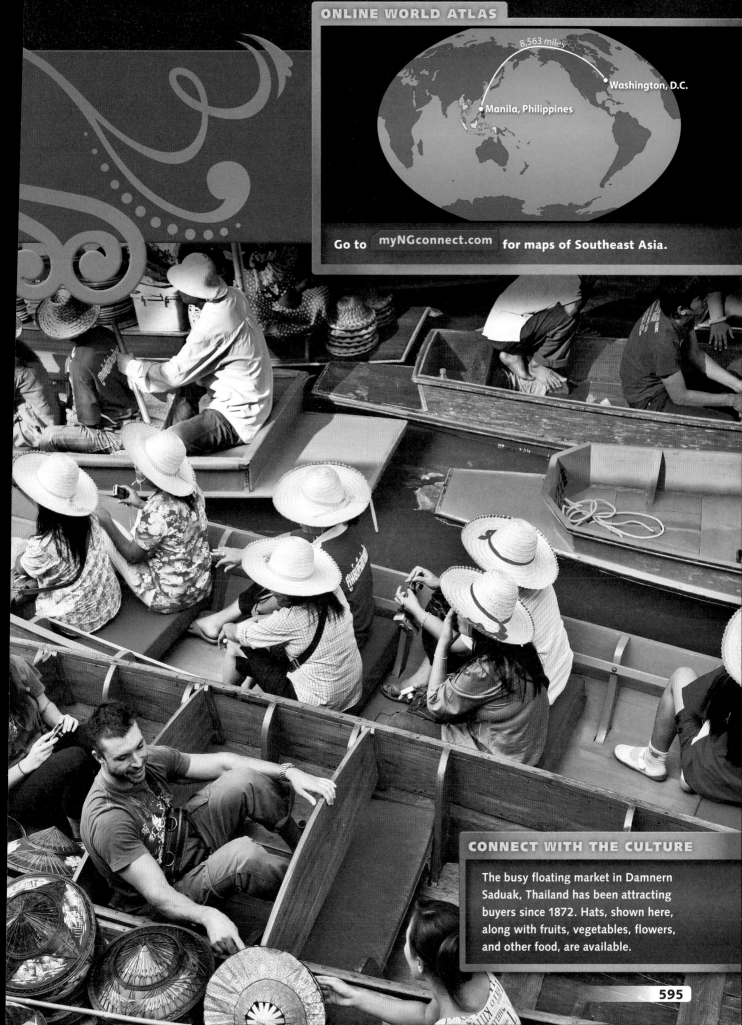

ONLINE WORLD ATLAS

8,563 miles
Washington, D.C.
Manila, Philippines

Go to myNGconnect.com for maps of Southeast Asia.

CONNECT WITH THE CULTURE

The busy floating market in Damnern Saduak, Thailand has been attracting buyers since 1872. Hats, shown here, along with fruits, vegetables, flowers, and other food, are available.

# CHAPTER 21

# Southeast Asia
## GEOGRAPHY & HISTORY

## PREVIEW THE CHAPTER

**Essential Question** What are the geographic conditions that divide Southeast Asia into many different parts?

### SECTION 1 • GEOGRAPHY

1.1 Physical Geography ................................. 598

1.2 Parallel Rivers ........................................... 600

1.3 The Malay Peninsula .............................. 602

1.4 Island Nations .......................................... 604

1.5 Discovering New Species ........................ 606

**KEY VOCABULARY**
- land bridge
- landlocked
- typhoon
- tsunami
- subsistence fishing
- ecologist
- bauxite
- biodiversity
- dynamic
- dormant
- zoologist
- wallaby

**ACADEMIC VOCABULARY**
enhance

**TERMS & NAMES**
- Ring of Fire
- Mekong River
- Chao Phraya River
- Irrawaddy River
- Malay Peninsula
- Foja Mountains

**Essential Question** How have physical barriers in Southeast Asia influenced its history?

### SECTION 2 • HISTORY

2.1 Ancient Valley Kingdoms ........................ 608

2.2 Trade and Colonialism ........................... 610

2.3 Indonesia and the Philippines ................ 612

**DOCUMENT-BASED QUESTION**

2.4 The Vietnam War ..................................... 614

**KEY VOCABULARY**
- complex
- bas-relief
- monopoly
- colonialism
- fossil
- commerce
- launch
- resistance

**ACADEMIC VOCABULARY**
transform

**TERMS & NAMES**
- Khmer Empire
- Angkor Wat
- Borobudur
- Dutch East India Company
- Manila
- Emilio Aguinaldo
- Ho Chi Minh

596 CHAPTER 21

# SECTION 1 GEOGRAPHY
## 1.1 Physical Geography

**TECHTREK**
myNGconnect.com For a physical map of Southeast Asia and Visual Vocabulary

Maps and Graphs | Digital Library

SOUTHEAST ASIA PHYSICAL

**Main Idea** Southeast Asia is a mountainous region with both mainland and island countries.

Southeast Asia has two kinds of countries: mainland and island. Indonesia and the Philippines are islands that were once connected by **land bridges**, strips of land connecting two land masses. Glaciers that melted over 6,000 years ago caused the sea level to rise, which separated these land masses. Malaysia is unique in that it includes land on the Asian continent as well as the island of Borneo.

## Mainland Countries

The region's mainland countries are part of the Asian continent and include Myanmar, Thailand, Cambodia, Vietnam, and Laos. This cluster is linked by a long coastline. Only Laos is **landlocked**, or surrounded by land on all sides. Elevation is generally higher in the northern and eastern coast of mainland Southeast Asia. Mountains hold the source of several of the region's major rivers, which people rely on for transportation, food, drinking water, and irrigation.

The people of Myanmar, Laos, Vietnam, and Cambodia live mostly in small villages in the mountains or near waterways. However, many of the region's river deltas are densely populated. Bangkok, the country's most developed and densely populated city, is in the delta of the Chao Phraya River. Sediment deposits from this and other rivers created the fertile soil of Thailand's Central Plain, which is ideal for growing rice.

Southeast Asia generally has a tropical climate, although temperatures vary based on elevation and distance from the ocean. Mainland countries receive rainfall from May to September, the wet monsoon season. **Typhoons**, fierce tropical storms with heavy rains and high winds, often strike during this time. The rest of the year is the dry monsoon season.

## Island Countries

Southeast Asia's island countries include Indonesia and the Philippines. They sit on the **Ring of Fire**, a volcanic zone around the Pacific Ocean where the plates that make up the earth's crust meet. Both countries have many active volcanoes. Undersea earthquakes can cause a **tsunami** (soo NAH mee), a giant ocean wave with enormous power. In 2004, an earthquake-caused tsunami near Sumatra killed at least 225,000 people. Its effects were felt as far away as East Africa.

### Before You Move On
**Monitor Comprehension** How did islands form in the region?

---

### ONGOING ASSESSMENT
## MAP LAB

1. **Categorize** Use the map and text to explain the differences between the two types of countries that make up Southeast Asia.
2. **Make Inferences** Find Laos on the map. What difficulties might it face as a result of being landlocked? What might be a benefit?
3. **Human-Environment Interaction** Why are the region's river deltas densely populated?

SECTION 1.1

# SECTION 1 GEOGRAPHY
## 1.2 Parallel Rivers

**Main Idea** River systems in Southeast Asia support life in many ways.

Three parallel rivers run through mainland Southeast Asia: the Mekong (may KONG), the Chao Phraya (chow PRY uh), and the Irrawaddy. They begin in the highlands and flow south through valleys between mountains. As they approach the sea, they divide into a triangular shape made up of smaller streams. These river deltas are composed of silt, or fertile soil the rivers carried from upstream.

### The Mekong River

At 2,600 miles, the **Mekong River** is the longest in Southeast Asia. It runs through the middle of the mainland and forms part of the borders of Myanmar, Laos, and Thailand. The mouth of the river, where it empties into the South China Sea, is in Vietnam near Ho Chi Minh City.

The Mekong Delta covers nearly 25,000 square miles, about the size of West Virginia. The densely populated delta is a rich rice-growing region. Some countries in the region are working to harness the river's power to produce hydroelectricity.

### The Chao Phraya River

The **Chao Phraya River**  is the most important river in Thailand. It is used to irrigate rice fields and serves as a major transportation route through the country. The capital city of Bangkok is located along its banks.

### The Irrawaddy River

The **Irrawaddy River**  is about half as long as the Mekong. It also supports rice farming and is used as a transportation network. As a result of the soil carried by the river and dumped at its mouth, the delta is growing by about 165 feet a year.

**Visual Vocabulary** An **ecologist** is a scientist who studies the relationship between organisms and their environments. National Geographic Fellow Zeb Hogan is an aquatic ecologist working here in the waters of the Mekong River.

## RIVERS AND POPULATION DENSITY

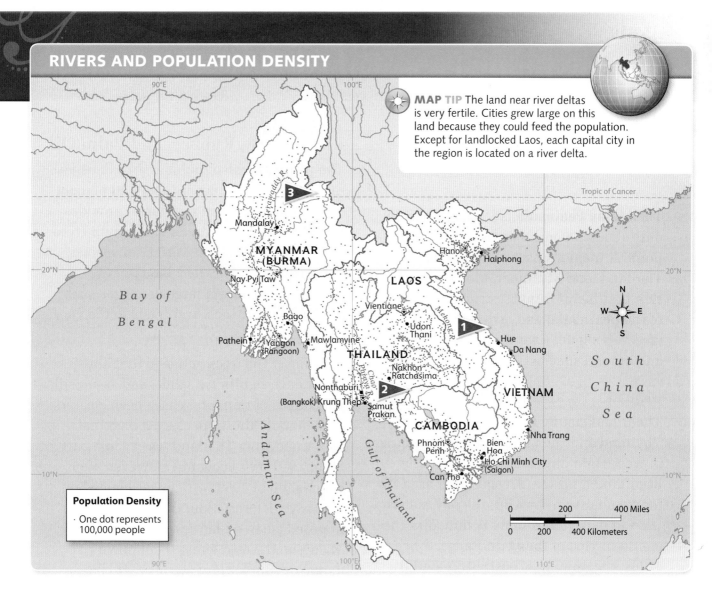

MAP TIP The land near river deltas is very fertile. Cities grew large on this land because they could feed the population. Except for landlocked Laos, each capital city in the region is located on a river delta.

**Population Density**
· One dot represents 100,000 people

In the rainy season, the Irrawaddy can rise more than 30 feet. Ports must have two areas for docking, one for each season. Farmers have adapted by storing water during the rainy season and releasing it onto their fields in the dry season.

## Challenges of the Rivers

The rivers of Southeast Asia are used by many people for **subsistence fishing**, catching just enough fish to live on. Countries in the region must work together to control the threat of overfishing. **Ecologist** Zeb Hogan is part of a program that buys live fish from local fishermen for study. Efforts like this can protect endangered fish while still allowing local people to make a living.

Dams built along the rivers help control water levels, but they also sometimes interfere with transportation and disrupt river environments.

### Before You Move On
**Summarize** What do river systems provide to support life in the region?

## ONGOING ASSESSMENT
### READING LAB  GeoJournal

1. **Monitor Comprehension** How are dams a threat to the region's rivers?
2. **Make Inferences** Why is water level a challenge for people who depend on the rivers?
3. **Region** Trace the course of the Mekong River on the map. What countries share this river?

# SECTION 1 GEOGRAPHY
## 1.3 The Malay Peninsula

**Main Idea** The mountains of the Malay Peninsula are rich in mineral resources and valuable rain forest land.

The **Malay Peninsula** is long and narrow, only about 200 miles across at its widest point. It includes parts of Malaysia, Thailand, and Myanmar. Mountain ranges rich in mineral resources run the length of the peninsula, and lush rain forest provides a habitat for thousands of plant and animal species.

### Mountains and Mining

The Bilaukataung Range mountains of Thailand and the Main Range mountains of western Malaysia have traditionally been mined for tin, a metal often used in food containers. **Bauxite**, the raw material used to make aluminum, is mined in the southern part of the Main Range. Since the 1970s, the number of easily accessed tin and bauxite deposits has been shrinking, causing a steady decline in mining.

### Rain Forest, Rubber, and Palm Oil

The peninsula also includes an extensive rain forest, which covers about 40 percent of the land area. The rain forest provides ideal habitat to hundreds of different trees and other plants. This variety of species in an ecosytem is called **biodiversity**. Animals range from large creatures such as elephants, rhinos, and tigers to the very small deer mouse.

Some of the trees native to the rain forest have significant value, and as a result large areas of rain forest have been cleared to plant only those species. At one time, teak wood from Thailand was a large part of the country's economy. However, after a landslide in 1989, which was blamed on excessive deforestation, the government imposed a ban on harvesting teak. In Malaysia, the rain forest is cut down to make room for large farms of rubber and palm oil trees. Palm oil is used with machinery, to make soap, and for cooking.

**Critical Viewing** This land in Malaysia is being prepared for a palm oil farm. What can you infer was present before the land was cleared?

Villlage on a hillside in Cameron Highlands, Malaysia

Critics claim that after depleting its mineral resources, Malaysia is now destroying its forests. To clear the land, farmers first cut the trees and then burn anything leftover. Fertilizers are then applied, permanently changing the soil and making it impossible for the rain forest to regrow quickly.

The environmental damage has an economic impact. Beginning in the early 2000s, rain forest tourism became an important part of the peninsula's economy. As a result governments are working to find a balance between land development and rain forest conservation.

### Before You Move On
**Summarize** How do environmental concerns conflict with the production of palm oil?

### ONGOING ASSESSMENT
## MAP LAB  GeoJournal

1. **Region** According to the map, which country has all four of the climates shown?
2. **Make Generalizations** Look at the physical map. Which type of land makes up most of the Malay Peninsula?
3. **Make Inferences** What effect might tourism have on the clearing of rain forest land?

SECTION 1.3  **603**

# SECTION 1 GEOGRAPHY
## 1.4 Island Nations

**TECHTREK**
myNGconnect.com For maps and photos of the island nations of Southeast Asia

 Maps and Graphs   Digital Library

**Main Idea** Geographic conditions on the islands affect settlement in Southeast Asia.

Five countries of Southeast Asia are islands or groups of islands: Indonesia, Singapore, Brunei, East Timor, and the Philippines. A part of Malyasia is on the island of Borneo. Mountains and water barriers among the islands have given rise to isolated cultures with distinct features.

### Volcanic Activity

Located where the Eurasian, Indian, Philippine, and Australian plates come together, the islands of Southeast Asia are in a **dynamic**, or continuously changing, geographic zone. These plates are constantly—yet slowly—moving into and over each other. Most islands in the region were formed by the crashing together of these plates. The collisions gradually formed small land masses with high volcanic mountains that slope downward toward coastal plains.

Although many of the volcanoes above ground are no longer active, some that are **dormant**, or inactive for long periods of time, can suddenly erupt. For example, Mount Sinabung on Sumatra in Indonesia had been quiet for 400 years before erupting in 2010. Although many farmers stayed with their farms, tens of thousands of people fled. Volcanic eruptions can destroy villages, but the ash also creates the fertile, nutrient-rich soil that allows for successful farming. The humid climate of the islands **enhances**, or improves, the quality of agriculture. As a result, many crops can be cultivated year-round.

### Indonesia

Indonesia is the giant of the region in both land area and population. It is nearly three times the size of Myanmar, the next largest. In population, Indonesia is more than three times as large as the Philippines, and it has more volcanoes than any other country in the world.

> **Critical Viewing** This village sits less than two miles from Mount Batur, an active volcano in Indonesia. What might be some advantages and disadvantages of living in this location?

## TECTONIC PLATES AND VOLCANOES

## POPULATION DENSITY

Indonesia is made up of thousands of islands. The five largest islands in size are Sumatra, Java, Borneo, Sulawesi, and New Guinea. Though Java is the smallest of the five, it is the most populous, with more than half of Indonesia's 250 million people. Four of Indonesia's five largest cities are on Java, including the capital, Jakarta. With the exception of the largest cities, most urban areas are more like large towns, each with its own local culture.

## The Philippines

Settlement patterns in the Philippines are similar to those in Indonesia. The greatest concentration of people is on lowland plains, which provide soil made fertile by volcanic eruptions. Like Indonesia, the Philippines has an extremely large capital city. This city, Manila, has more than 11 million people.

The Philippines also has many small rural settlements that subsist on fishing or rice farming. Houses near the ocean are built on columns made of timber to allow for changing tides and boat traffic.

### Before You Move On
**Summarize** In what ways has geography of the island nations affected life in Southeast Asia?

### ONGOING ASSESSMENT
# MAP LAB

GeoJournal

1. **Place** Based on the map of tectonic plates, which large island seems to be unaffected by active volcanoes or earthquakes? Why might this be?

2. **Compare** Based on the Population Density map, how does the population density of Borneo compare to that of Java? What might be the cause of this difference in population?

SECTION 1.4 **605**

# SECTION 1 GEOGRAPHY
## NATIONAL GEOGRAPHIC

**TECHTREK**
myNGconnect.com For a map of New Guinea, photos, and an Explorer Video Clip

Maps and Graphs | Digital Library

# Discovering New Species
## with Kristofer Helgen

**Main Idea** The unexplored Foja Mountains of Indonesia may be home to unidentified plant and animal species.

## Undiscovered Species

Today knowledge of animal and plant life is extensive and well documented. However, National Geographic Emerging Explorer Kristofer Helgen is a **zoologist**, a scientist who studies animals, and he knows there are hundreds of species that have yet to be discovered.

## The Foja Mountains

The **Foja Mountains** of Indonesia have wildlife not found in any other region. The area is also largely unexplored. Helgen described the Foja (FOY ya) Mountains as "one of the few places in the world with no villages, no roads, no human population at all." Human presence is not part of life in this rain forest. "The animals there just don't know people," he said. "That's a very, very rare thing in this day and age."

In 2005, Helgen took part in a research trip to this remote area. The team included Helgen, who studies mammals, and other experts on plants, butterflies, reptiles, and birds.

The researchers found 20 new frog species, 5 new kinds of butterfly, and several new plant species. Helgen himself found new rat and mouse species. He also discovered a unique type of **wallaby**, a small relative of the kangaroo, which he hopes to name as a new species.

myNGconnect.com
For more on Kristofer Helgen in the field today

New species of blossom bat

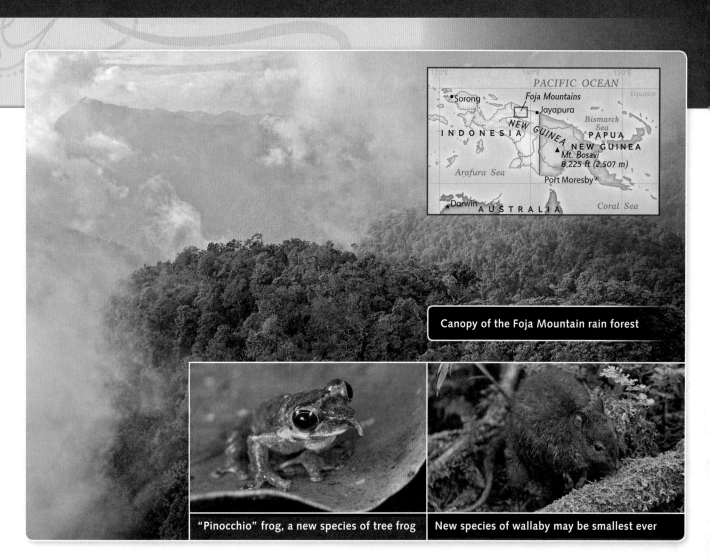

Canopy of the Foja Mountain rain forest

"Pinocchio" frog, a new species of tree frog

New species of wallaby may be smallest ever

## Scientific Opportunity

The research conducted in the Foja Mountains provides a chance for scientists to learn more about the diversity of animal and plant life. Because they have not been explored, the mountains also offer new scientists an opportunity to become trained in the discovery of new species. Students can join experienced scientists in future trips to the region. They will learn about working in the field even as they help make new discoveries.

### Before You Move On
**Make Inferences** Why might there be many undiscovered species of plants and animals in the Foja Mountains?

### ONGOING ASSESSMENT
## VIEWING LAB  GeoJournal

1. **Analyze Visuals** Go to the **Digital Library** and view the Explorer Video Clip about Kristofer Helgen. How would you describe the land?
2. **Place** What in the video explains why scientists are excited about this area?
3. **Make Inferences** How do you think animals that have never encountered humans would behave the first time they see a person? Why?
4. **Create Charts** Make a chart listing the kinds of new species that might be found in a place like the Foja Mountains. Remember that the region is home to a rain forest.

| PLANTS | ANIMALS |
|---|---|
|  |  |
|  |  |

SECTION 1.5

# SECTION 2 HISTORY
## 2.1 Ancient Valley Kingdoms

TECHTREK
myNGconnect.com For photos of ancient Southeast Asian kingdoms
Digital Library

**Main Idea** The development of Southeast Asia was influenced by nearby powers because of its important location for trade.

The location of Southeast Asia between the Pacific and Indian Oceans meant its surrounding waterways were on important trade routes. Two powerful civilizations, China to the north and India to the west, heavily infuenced the cultural direction of the region. The impact came through military force and invasion, as well as through trade.

### Mainland Empires

Chinese culture first came to Southeast Asia in 111 B.C., when the Chinese invaded and conquered part of what today is Vietnam. India had already established trade before the Chinese arrived. This trade had a strong influence on the area's religious practices.

By the A.D. 700s, Buddhist and Hindu empires were competing for influence and power in other parts of the region. The largest and longest lasting was the **Khmer Empire** of Cambodia. Centered along the Mekong River valley, the Khmer (kuh MAYR) Empire covered much of Southeast Asia and lasted from the A.D. 800s to the 1430s.

At the peak of the empire's success in the 1100s, its ruler, King Suryavarman II, built the massive Hindu temple **Angkor Wat** in the capital city.

Angkor Wat, Cambodia

- 111 B.C. China conquers Vietnam; Buddhism reaches Southeast Asia.
- c. A.D. 780–850 Sailendra rulers of Java build Borobudur temple.
- c. A.D. 890 Khmer Empire sets capital at Angkor.

100 B.C. — A.D. 600 — 800

- A.D. 600s–c. 1100 Srivijaya Empire of Sumatra

Borobudur temple, Java, Indonesia

608 CHAPTER 21

This religious **complex**, or set of interconnected buildings, was dedicated to the Hindu god Vishnu and served as a tomb for the king. Beautiful **bas-reliefs**, or sculptures that slightly project from a flat background, cover the walls with scenes from Hindu stories. Eventually forces from modern-day Thailand conquered the city and the complex fell into ruin.

In A.D. 939 the people of Vietnam broke from China and established the independent kingdom of Dai Viet. Though influenced by Chinese culture, Vietnam had its own cultural traits. For instance, women in Vietnam had higher social standing than women in China. Eventually Dai Viet grew weak and was reconquered by China in 1407.

## Island Empires

Modern Indonesia was also home to powerful empires. The earliest was Srivijaya (sree vi JY ah), which arose on southern Sumatra in the A.D. 600s. This kingdom controlled the Strait of Malacca and was therefore able to control trade from South Asia to China. It was known throughout Asia as a center of trade and also as a center of Buddhist study. The empire declined around A.D. 1100.

A second Indonesian power was the Sailendra dynasty, which arose in Java and flourished from about A.D. 780 to 850. Sailendran rulers built another famous temple complex, **Borobudur.** Each of the temple's three levels symbolizes a step toward enlightenment, the ultimate spiritual goal of Buddhism.

Another trading kingdom arose on Java around A.D. 1300. It is named for its capital city of Majapahit (mah jah PAH heet) in eastern Java. It gained power through the control of trade. However, by the 1500s, other powers had replaced it.

### Before You Move On
**Summarize** How did Chinese and Indian empires influence life in the region?

## ONGOING ASSESSMENT
### VIEWING LAB  GeoJournal

1. **Interpret Time Lines** Based on the time line, how long was Vietnam able to maintain its independence from China?
2. **Make Generalizations** Look at the photo of Angkor Wat. How would you describe the building skill of the Khmer Empire? Support your answer with details from the photo.
3. **Compare and Contrast** Based on the photos, how are the Borobudur temple and Angkor Wat similar and different?

- A.D. 939 — Vietnam gains independence from China; Dai Viet kingdom begins.
- A.D. 1113–1150 — Reign of Khmer king Suryavarman II, who built Angkor Wat
- A.D. 1407 — China conquers Vietnam again.
- A.D. 1290s–c. 1500 — Majapahit empire in Java

Angkor Wat bas-relief sculpture

SECTION 2.1

# SECTION 2 HISTORY
## 2.2 Trade and Colonialism

**Main Idea** The development of the spice trade in Southeast Asia led to colonization of the region.

As you have learned, trade with India and China brought the influence of these cultures to Southeast Asia. European influence arrived in the 1500s, as merchants hoped to establish a spice trade **monopoly**, or complete control of the market. Spices found in the region, such as cinnamon, nutmeg, and black pepper, could be sold for a high profit in Europe. Traders from Spain and Portugal came first, but the Netherlands' **Dutch East India Company** dominated the region for many years. This success established a strong Dutch influence in Indonesia.

### European Control

From the 1600s to the 1800s, Europeans tried to gain an economic hold on Southeast Asia. By 1850, through a combination of alliances, favorable trade agreements, and even military force, the majority of the region was ruled by European powers. (See the time line below.) Only Thailand and parts of the Philippines were independent. Britain, France, Spain, and the Netherlands controlled the rest.

The economic motives that led European powers to Southeast Asia changed the region. Increased production and an ongoing demand for goods strengthened the region's economy. However, trade and wealth, once in the hands of indigenous kingdoms, were now held by distant economic powers. **Colonialism**—one country ruling and developing trade in another country for its own benefit—continued in Southeast Asia well into the 20th century.

### Before You Move On
**Monitor Comprehension** How did development of the spice trade lead to colonization?

### SPAIN

**1521**
Ferdinand Magellan lands at the Philippines and claims them for Spain.

**1565**
Spain makes first settlement on the Philippine islands.

**1571**
Spain captures site of Manila.

**1830s**
Spain opens Manila to trade.

**1892**
Filipinos begin movement aimed at independence from Spain.

**1898**
United States defeats Spain in the Spanish-American War and wins control of Philippines.

### NETHERLANDS

**1619**
Dutch East India Company makes base at Batavia (modern Jakarta).

**1641**
The Netherlands captures Malacca from Portugal and becomes a major power in spice trade.

**1824**
The Netherlands and Britain reach agreement on control of Java and Sumatra (to Dutch) and Singapore and Malacca (to British).

**1825–1839**
The Netherlands fights to put down revolts on Java.

**1860**
The Netherlands and Portugal sign treaty to divide Timor between them.

### GREAT BRITAIN

**1781**
Britain captures Sumatra from Dutch.

**1786–1809**
Britain gains control of Malaya trade.

**1819**
Britain founds Singapore, which becomes a major port city.

**1824–1826**
Britain controls western Burma.

**1886**
Britain completes control of Burma in Third Burmese War.

**1888**
Britain wins southern Burma.

**1888**
Britain gains control of northern Borneo.

## SOUTHEAST ASIA UNDER COLONIAL RULE, c. 1895

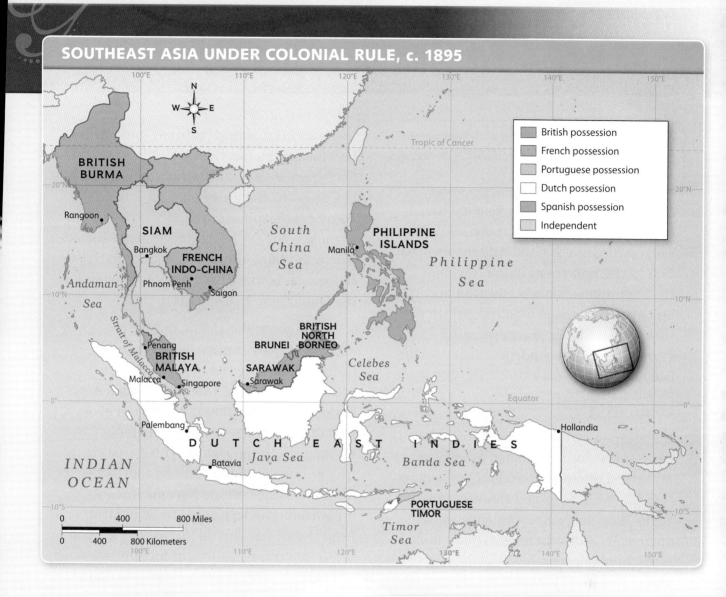

### FRANCE

**1644**
France forms French East India Company.

**1789**
French East India Company disbanded during French Revolution.

**1858**
France captures Saigon, Vietnam.

**1863**
France seizes Cambodia.

**1887**
France creates Indo-Chinese Union (Cambodia, Vietnam).

**1896**
France and Britain agree to allow Siam to remain independent to separate their colonies.

### ONGOING ASSESSMENT
## MAP LAB

1. **Region** Based on the map, which European countries held the most territory in the region around 1895?

2. **Movement** Think about the importance of waterways in influencing trade in this region. Which European country was in the best position to control trade? Support your answer with evidence from the map.

3. **Interpret Time Lines** Based on the dates and events, what do you think was the attitude of the people of Southeast Asia toward European control? Why do you think so?

SECTION 2.2 **611**

# SECTION 2 HISTORY

## 2.3 Indonesia and the Philippines

TECHTREK
myNGconnect.com For photos of Indonesia and the Philippines
Digital Library

**Main Idea** Indonesia and the Philippines are island countries that have faced similar challenges in becoming independent.

As you know, Southeast Asia has a long history of diversity because of its geographic location and unique resources. The island nations of Indonesia and the Philippines have been influenced and even controlled by other cultures throughout history. However, after long struggles for independence, each one has become its own nation.

### Indonesia

Indonesia may have been home to the earliest species of humans. **Fossils**, or preserved remains, found on Java suggest that human life existed there as early as 1.7 million years ago. Evidence shows that ancient societies there used hand tools, made implements out of metals, and wove cloth. They also traveled the sea as early as 2500 B.C., possibly to trade with other areas of Asia and beyond.

As its civilization matured, Indonesia became an intersection for trade in the East. Part of the country became known to Europe as the Spice Islands, for the many exotic spices that were a strong attraction for explorers and traders. For example, trade in nutmeg, a spice native to Indonesia, became extremely profitable for the Dutch who had settled there.

Throughout the 1800s, the Dutch expanded their control. Some revolts occurred, but the Dutch were able to maintain power. In 1830, they began a system that required all villages to give part of their crops to the government for export. As the Dutch gained wealth, Indonesians suffered. During the 1900s their resistance efforts became more organized. The Japanese seized control from the Dutch during World War II. As these two countries fought for control, Indonesians continued to resist. At the same time, a strong sense of national identity was developing. The country finally won independence in 1949.

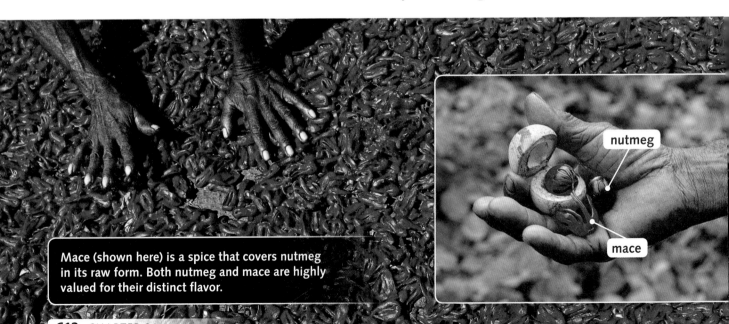

Mace (shown here) is a spice that covers nutmeg in its raw form. Both nutmeg and mace are highly valued for their distinct flavor.

**Critical Viewing** Emilio Aguinaldo (front, center) and members of the assembly of the First Philippine Republic, 1899. In early photography, subjects were required to sit for long periods to capture an image. How might this explain the expressions on these men's faces?

## The Philippines

When the Spanish seized control of **Manila** in 1571, they made it the capital of their new colony. Manila was and still is the economic, political, and cultural center of the Philippines. Trade with China led many Chinese people to settle in Manila. They became the major force in **commerce**, or the business of trading goods and services.

In the 1800s, Spain's economic power began to fade and Manila became open to trade with more countries. This allowed some Filipinos to gain wealth and influence they had not known before.

**Emilio Aguinaldo** was a key leader in the movement for Filipino independence. He fought alongside the United States when it defeated Spain in the Spanish-American War in 1898, and thought the islands would become independent. However, after the war, the United States kept the islands as its own colony.

In late 1941, during World War II, Japan attacked the United States at Pearl Harbor, then attacked the Philippines. The outnumbered U.S. forces stationed on the islands were forced to leave. Later in the war, the United States took back control and remained in power until 1946, when the Philippines were granted independence. Today, it is a stable democracy.

### Before You Move On
**Summarize** What challenges did Indonesia and the Philippines face in gaining independence?

### ONGOING ASSESSMENT
## READING LAB  *GeoJournal*

1. **Location** Why was Southeast Asia so important to Europeans?
2. **Make Inferences** Why were Indonesians able to develop a strong national identity?
3. **Draw Conclusions** Was trade or conquest more important in shaping these countries? Why?

## SECTION 2 DOCUMENT-BASED QUESTION
# 2.4 The Vietnam War

myNGconnect.com For photos of the war in Vietnam and Guided Wrting

Digital Library

Student Resources

In 1954 Vietnam was divided into two parts, North and South. President **Ho Chi Minh** of North Vietnam established a Communist government. War erupted in 1959 when he sent aid to overthrow the government in South Vietnam and create one country. The United States supported South Vietnam—it feared that defeat by the North would spread communism to other countries. U.S. forces fought there from 1964 to 1973. In 1975, South Vietnam surrendered. The reunited country became the Socialist Republic of Vietnam.

### DOCUMENT 1
### President Lyndon Johnson on U.S. Policy

In 1965, Johnson explained reasons for the involvement of the United States in the fighting in Vietnam:

> We are ... there to strengthen world order. Around the globe ... are people whose well-being rests, in part, on the belief that they can count on us if they are attacked. To leave Viet-Nam ... would shake the confidence of all these people in the value of an American commitment. ... Let no one think for a moment that retreat from Viet-Nam would bring an end to conflict. The battle would be renewed in one country and then another. The central lesson of our time is that the appetite of aggression is never satisfied.

**CONSTRUCTED RESPONSE**

1. How are Johnson's two reasons for fighting the Vietnam War related?

### DOCUMENT 2
### Letter from Ho Chi Minh

In 1967, Johnson sent Ho Chi Minh a letter suggesting that the two sides begin peace talks. Here is Ho Chi Minh's response:

> The Vietnamese people have never done any harm to the United States. But ... the United States Government has constantly intervened in Viet-Nam, it has **launched** [started] and intensified the war of aggression in South Viet-Nam for the purpose of prolonging the division of Viet-Nam and of **transforming** [remaking] South Viet-Nam into an American neo-colony and an American military base. ...
>
> The Vietnamese people ... are determined to continue their **resistance** [opposition] until they have won real independence.

**CONSTRUCTED RESPONSE**

2. How does Ho Chi Minh respond to the suggestion of peace talks?

> THE VIETNAMESE PEOPLE . . . ARE DETERMINED TO CONTINUE THEIR RESISTANCE UNTIL THEY HAVE WON REAL INDEPENDENCE.
>
> — HO CHI MINH

### DOCUMENT 3
### Photo of Warfare

The physical geography and climate of Vietnam presented challenges to soldiers. These challenges might be especially great for soldiers who were not native to the region and were not accustomed to heavy rains and dense plant life of the jungle.

**CONSTRUCTED RESPONSE**

3. What does the photo suggest about the conditions the soldiers faced in the war in Vietnam?

## ONGOING ASSESSMENT
## WRITING LAB  GeoJournal

**DBQ Practice** Think about how North Vietnam compares in size and power to the United States. What factors might have allowed North Vietnam a chance to win the war?

**Step 1.** Think about Johnson's determination to fight in Vietnam, Ho Chi Minh's statements, and what the photo shows about the war in Vietnam.

**Step 2.** On your own paper, jot down notes about the main idea expressed in each document.

```
Document 1: Excerpt: Johnson's Speech
Main Idea(s) _____
Document 2: Excerpt: Ho Chi Minh's Letter
Main Idea(s) _____
Document 3: Photo of Warfare
Main Idea(s) _____
```

**Step 3.** Construct a topic sentence that answers this question: What factors might have affected North Vietnam's ability to fight and win a war against South Vietnam and the United States?

**Step 4.** Write a detailed paragraph that answers the question above, using evidence from the documents. Go to **Student Resources** for Guided Writing support.

SECTION 2.4  615

# CHAPTER 21
# Review

## VOCABULARY

Use the following vocabulary words in a sentence that shows understanding of each term's meaning.

1. landlocked

   *All countries in Southeast Asia have some coastline except Laos, making it the region's only landlocked country.*

2. subsistence fishing
3. dynamic
4. dormant
5. commerce
6. launch

## MAIN IDEAS

7. How does the climate of the two parts of Southeast Asia differ? (Section 1.1)
8. Why are the upper reaches of the region's rivers less populated? (Section 1.2)
9. Where do you think most people live on mainland Malaysia? Why? (Section 1.3)
10. How do Manila and Java show that population in the Philippines and Indonesia tends to cluster? (Section 1.4)
11. What has allowed some species in the region to have gone undiscovered? (Section 1.5)
12. Why did trade play such a great role in the development of island kingdoms? (Section 2.1)
13. Which region of the world practiced colonialism in Southeast Asia and why? (Section 2.2)
14. What struggles did Indonesia and the Philippines face in developing their own culture? (Section 2.3)
15. What were North Vietnam and South Vietnam fighting about? (Section 2.4)

## GEOGRAPHY

### ANALYZE THE ESSENTIAL QUESTION

What are the geographic conditions that divide Southeast Asia into many different parts?

**Focus Skill: Analyze Cause and Effect**

16. Why are rivers and seas so important in this region?
17. Why are island countries like Indonesia and the Philippines good for farming?
18. In what way has Laos's lack of a coastline impacted its ability to engage in international trade?

### INTERPRET MAPS

19. **Summarize** What is the advantage of the location of the mainland portion of Malaysia?
20. **Make Inferences** Look at the locator map. In what ways has Malaysia's location helped it establish trading partners?

## HISTORY

### ANALYZE THE ESSENTIAL QUESTION

How have physical barriers in Southeast Asia influenced its history?

**Focus Skill: Draw Conclusions**

21. Why do you think one great empire never arose to unite all of Southeast Asia?

22. Which other cultural regions do you think had the most influence on Southeast Asia? Explain your reasons.

23. What made it difficult for Europeans to establish complete control of the countries in this region?

### INTERPRET MAPS

24. **Region** How much of Southeast Asia did the Khmer Empire control at its greatest point? In what year was that the case?

25. **Make Inferences** The kingdom of Dai Viet existed along the South China Sea. What factors would have allowed it to remain independent of the Khmer Empire?

## ACTIVE OPTIONS

**Synthesize** the Essential Questions by completing the activities below.

26. **Write a Press Release** You work for an art museum that is staging an exhibition of art and objects from Angkor Wat. Write a 3- or 4-paragraph press release to announce the exhibition. Include information about how long the exhibition will run and the museum's schedule. Be sure to word the press release in a way that would attract visitors to the exhibition. Use these tips to prepare your press release. **Read your press release aloud with a partner to evaluate each other's ideas.**

> **Writing Tips**
> - Make sure you include details that will attract visitors.
> - Clearly explain what kinds of objects the exhibition will include.
> - Remember to use vivid, appealing language to describe the objects.
> - Be sure to give the dates and times of the exhibition.

27. **Create a Time Line** Use information in the lessons and online research to gather your facts about five key dates in the history of two countries in the region. Then construct a time line showing the dates for both countries and why they are significant. Illustrate the time line with photos connected to some of the events.

|         | COUNTRY 1 | COUNTRY 2 |
|---------|-----------|-----------|
| Event 1 |           |           |
| Event 2 |           |           |
| Event 3 |           |           |
| Event 4 |           |           |
| Event 5 |           |           |

REVIEW

# CHAPTER 22

# Southeast Asia
## TODAY

## PREVIEW THE CHAPTER

**Essential Question** How have local traditions and outside influences shaped cultures in Southeast Asia?

### SECTION 1 • CULTURE

1.1 Religious Traditions .................................. 620
1.2 Thailand Today ......................................... 622
1.3 Regional Languages ................................. 624
1.4 Saving the Elephant ................................. 626

**KEY VOCABULARY**
- prehistoric
- ritual
- attribute
- wat
- monk
- metropolitan area
- dialect
- adapt
- language diffusion
- poach
- restore
- domesticate

**ACADEMIC VOCABULARY**
predominant

**TERMS & NAMES**
- Java
- Sumatra
- Bali
- Cardamom Mountains

**Essential Question** How are Southeast Asia's governments trying to unify their countries?

### SECTION 2 • GOVERNMENT & ECONOMICS

2.1 Governing Fragmented Countries ............ 628
2.2 Migration Within Indonesia ..................... 630
   **GLOBAL ISSUES: GLOBALIZATION**
2.3 Singapore's Growth ................................. 632
2.4 Malaysia and New Media ......................... 634

**KEY VOCABULARY**
- fragmented country
- remittance
- relocate
- trend
- port
- industrialize
- multinational corporation
- emergence
- reliable

**ACADEMIC VOCABULARY**
potential

**TERMS & NAMES**
- Madura
- inner islands
- outer islands
- Malaysia

618 CHAPTER 22

Tourists and locals use rickshaws and motorbikes to get around in Hanoi, Vietnam.

# SECTION 1 CULTURE

## 1.1 Religious Traditions

**TECHTREK**
myNGconnect.com For photos of religious traditions in Southeast Asia

Digital Library

**Main Idea** Religions in Southeast Asia have been shaped by both local traditions and outside influences.

Religious practices in Southeast Asia today blend many influences from several centuries. The **predominant**, or most common, belief systems have shifted in many parts of the region.

### Traditional Religion

The traditional religion in Southeast Asia is animism, the belief that spirits exist in animals, plants, objects, and places. These spirits are believed to influence people's lives. Many historians think that animism began in **prehistoric** times before there were written histories. Animists perform **rituals**—formal regularly repeated actions—to please spirits so they bring good fortune to their human families or villages. Many small, tribal groups in the region still practice forms of animism.

### Outside Influences

Other cultures entering Southeast Asia through trade or conquest brought their traditions with them. The Chinese brought Buddhism when they conquered Vietnam in 111 B.C. They also introduced a philosophy, or system of thought, called Confucianism, which became an important part of Vietnam's local religions. Beginning in the A.D. 100s, traders from India spread Buddhism and Hinduism across the region. By the 400s, Buddhism had taken hold on **Java** and later spread to **Sumatra**. In the 1100s, Cambodians built Angkor Wat to worship a Hindu god.

Arab traders carried Islam to Southeast Asia during the 1300s, where it spread from Malaysia to parts of Indonesia. Later, Europeans spread Christianity. Spain brought Roman Catholicism to the Philippines in the 1500s. The French introduced it to the mainland in the 1700s.

> **Critical Viewing** Muslim women pray at a mosque in Jakarta, Indonesia. What does the number of people suggest about the presence of Islam in Indonesia?

Children attend a prayer service in a Catholic school in Makassar, Indonesia.

Forest roots devour ruins at Ta Prohm, a Buddhist temple at Angkor, Cambodia.

## Religion Today

Over the centuries, the predominant religion of a country sometimes shifted depending on the beliefs of the ruling power. Today, Southeast Asia has a mix of religions. Buddhism is most prominent in mainland countries. Around 95 percent of the people in Thailand are Buddhists, and many Buddhist holy days are national holidays. Buddhism is also predominant in Myanmar. Islam is the main religion of Indonesia, which is the most populous Muslim country in the world. In Malaysia, about three out of five people are Muslim. Most people living in the Philippines and East Timor continue to practice Roman Catholicism, introduced by Europeans.

While some religions dominate each country, the region as a whole has religious diversity. For example, **Bali**, an island in Indonesia, is largely Hindu. Islam has many followers in the southern Philippines. Diversity is also found in many of the ancient, local traditions that are still practiced in each country today.

### Before You Move On
**Monitor Comprehension** Which religions were brought to the region by outside cultures?

### ONGOING ASSESSMENT
## PHOTO LAB  GeoJournal

1. **Compare and Contrast** What similarities and differences can you see in the photos?
2. **Synthesize** Taken together, what do the photos show about religion in Southeast Asia?
3. **Summarize** How has contact with other countries influenced religion in the region?

## SECTION 1 CULTURE
## 1.2 Thailand Today

**TECHTREK**
myNGconnect.com For photos of modern Thailand
Digital Library

**Main Idea** Thai culture today reflects traditional foundations and modern influences.

Modern Thai culture includes regional traditions and other global infuences that blend into a unique Thai identity.

### Classical Architecture

One **attribute**, or specific quality, of Thai culture is its remarkable architecture. Traditional buildings have steeply slanted roofs designed to shed the heavy monsoon rains. Many are built on legs to keep them high off the ground during the monsoon floods. The most important buildings in Thai architecture are **wats**, or Buddhist temples, influenced by designs from India, the Khmer empire, and China.

### Buddhist Monks

As you have learned, Buddhism is the dominant religion in Thailand. Almost every village has a wat with a community of **monks**, men who devote themselves to religious work. Buddhist monks wear orange or yellow robes, live simply, and focus on religious practices such as meditation and other rituals.

Most young men traditionally became monks for at least three months during one rainy season. However, as more and more young people migrate away from rural communities and attend non-religious schools, young men are making shorter commitments to religious life, or sometimes none at all.

**Visual Vocabulary** A **wat** is a Buddhist temple. Building on Marble Wat (above) started in 1900, during a period of rapid growth in the capital city of Bangkok.

The Chao Phraya River cuts through downtown Bangkok at dusk.

## Modern Influences

About four out of five young men and women now work in cities, especially Bangkok's large metropolitan area, the populated location that includes the city limits and surrounding area. Many still identify with their villages even though they mostly live and work in cities.

Urban life has also changed people's clothing, food, and entertainment. Most people in Thailand now wear Western-style clothing. Many urban women buy prepared food at local stores on their way home from work instead of cooking. Most homes now have televisions and other modern conveniences.

Young people in urban areas also turn to the Internet as a source of news, entertainment, and communication.

### Before You Move On
**Monitor Comprehension** What traditional and modern influences can be seen in Thai culture today?

**Critical Viewing** Suvarnabhumi Airport in Bangkok opened in 2006. Based on the photo, what can you say about the design for the airport?

### ONGOING ASSESSMENT
## PHOTO LAB — GeoJournal

1. **Make Inferences** What does the large photo show about life in modern Thailand?
2. **Analyze Visuals** What does each photo reveal about Thai culture? Use details from the photos to explain your answer.
3. **Human-Environment Interaction** How was traditional Thai architecture well suited to its environment?

# SECTION 1 CULTURE
## 1.3 Regional Languages

**TECHTREK**
myNGconnect.com For a graphic of languages in Southeast Asia

Student Resources

**Main Idea** Geographically isolated cultures and large historic migrations have created a diversity of languages in the region.

As you have learned, language is an important part of any culture. People use language to express their ideas, values, and history. Like religion, language can bring people together or it can divide them. The people of Southeast Asia speak hundreds of different languages.

### Native Languages

Each country in the region has a dominant native language. Generally the name of this language reflects the name of the country or its largest ethnic group. The dominant language is the official language used by government, business, education, and the media. In countries such as Indonesia and Malaysia, a common language has helped unify geographically fragmented areas. In Myanmar, the diversity of minority languages has made unifying the country more difficult.

Many people in Southeast Asia speak a **dialect**, a regional variation of a main language. Speakers of different dialects often belong to small groups that live in isolated communities. Some of these dialects exist only in oral form, and many dialects and ethnic languages are in danger of disappearing. People learn the official language as a way to **adapt**, or adjust to common practices, especially as cultures become more globally connected. When older speakers of an isolated language die, that language may disappear, resulting in a loss of traditional culture.

**Critical Viewing** Many people in rural villages in Vietnam speak a native language or a dialect of a main language. How might speaking a dialect influence this woman's business?

### Language Migration

The region's location has historically attracted diverse people to trade. This movement led to **language diffusion**, or the spread of languages from their original homes. Traders had to find a common language in order to communicate. Malay served this purpose for early traders from Arab countries and different parts of China. Today, English is often used as the common language.

Immigrants also brought their languages to Southeast Asia. For example, Chinese is the dominant language in Singapore. Many speakers of Chinese dialects live in Malaysia and Brunei and in major cities throughout the region. Immigrants from India brought various Indian languages to Malaysia, Singapore, and Myanmar.

**SPOKEN LANGUAGES IN SELECTED SOUTHEAST ASIAN COUNTRIES**

**CAMBODIA** — Official language: Khmer
Total Living Languages: 23
Other Asian Languages: Chinese, Vietnamese
Western Languages: English, French

**PHILIPPINES** — Official language: Filipino and English
Total Living Languages: 171
Other Asian Languages: Chinese, Languages of India and Southeast Asia
Western Languages: English

**INDONESIA** — Official language: Bahasa Indonesia
Total Living Languages: 719
Other Asian Languages: Chinese, Vietnamese
Western Languages: English, Dutch

**SINGAPORE** — Official languages: Malay, Mandarin Chinese, and English
Total Living Languages: 21
Other Asian Languages: Mandarin Chinese, Tamil
Western Languages: English

**LAOS** — Official language: Lao
Total Living Languages: 84
Other Asian Languages: Chinese, Vietnamese
Western Languages: English, French

**THAILAND** — Official language: Thai
Total Living Languages: 74
Other Asian Languages: Mandarin Chinese, Tamil
Western Languages: English

**MALAYSIA** — Official language: Bahasa Malaysia
Total Living Languages: 137
Other Asian Languages: Chinese, Languages of India and Southeast Asia
Western Languages: English

**VIETNAM** — Official language: Vietnamese
Total Living Languages: 106
Other Asian Languages: Khmer, Chinese
Western Languages: English, French

Sources: CIA World Factbook, www.ethnologue.com

European countries began establishing colonies in Southeast Asia in the 1500s. As the British, Dutch, French, and Spanish gained control of various countries, they used their languages to rule and do business. The United States controlled the Philippines for a time after Spanish rule and made English common there. As the countries of Southeast Asia gained independence in the 20th century, their governments chose dominant native languages to be the official languages. However, many people still speak a Western language as a second language.

## Before You Move On

**Summarize** In what ways is language use changing as cultures become more connected?

### ONGOING ASSESSMENT
### LANGUAGE LAB  GeoJournal

1. **Movement** According to the chart, what East Asian language is most widespread in these Southeast Asian countries?

2. **Draw Conclusions** Based on the chart, what Western language is a Southeast Asian business person most likely to learn? Explain.

3. **Identify Problems and Solutions** What might a country do to save endangered languages spoken by its people?

# SECTION 1 CULTURE

## 1.4 Saving the Elephant

**TECHTREK**
myNGconnect.com For photos of elephants in Southeast Asia

 Student Resources  Digital Library

**Main Idea** Some countries in Southeast Asia are working to protect the endangered Asian elephant.

Although smaller than their African cousins, Asian elephants are huge and awe-inspiring creatures. Some have been trained to use their enormous strength for the benefit of people. However, most elephants live in the wild. As the human populations of the region increase, their growing numbers increasingly threaten the Asian elephant population.

### Asian Elephants

As recently as 1900, it is estimated 80,000 Asian elephants may have been living in the wild. Today, their population is thought to range from 30,000 to 50,000.

Human behavior, such as **poaching**, or illegal hunting of a wild animal, is one reason for that population loss. People kill male Asian elephants for their ivory tusks. Ivory is highly valued for its beauty and hard texture. An international agreement banned trade in ivory in 1989, but it still continues illegally.

A bigger problem for Asian elephants is their loss of habitat. These huge animals need large areas of rain forest to find food, but humans have cleared much of the land for alternative uses, such as logging and iron mining. Land is also cleared for growing crops such as coffee. Once crops are planted, some elephants wander in from the remaining forests to eat them, and farmers trying to protect their crops sometimes kill those raiding elephants.

### Protecting Elephants

Many countries in Southeast Asia have tried to **restore**, or bring back, the wild Asian elephant population. In Cambodia's **Cardamom Mountains**, for example, many conservationists have begun to use modern technology, such as electric fences that run off solar power, to keep elephants confined to protected places. Other methods are more basic. Hammocks hung near crops make the elephants think humans are in the fields, so they stay away.

> **Critical Viewing** Asian elephants look for food in Sumatra. Based on the photo, how would you describe their habitat?

**Visual Vocabulary** Many elephants are **domesticated**, or trained to work with humans. Domesticated elephants can provide service or entertainment.

## ASIAN ELEPHANTS BY THE NUMBERS

**11+**
Height in feet

**12,000+**
Weight in pounds

**300**
Pounds of food (plants, grain) consumed in one day

**80,000**
Population estimate, wild, beginning of 20th century

**30,000+**
Population estimate, wild

**16,000**
Population in captivity (protected)

Sources: World Wildlife Federation, Fauna & Flora International, U.S. Fish & Wildlife Service, 2013

The results have been dramatic. The elephant-human interaction that leads to population loss was reduced so much that from 2005 to 2010, no elephants were killed anywhere in Cambodia. These efforts can lead to long-term protection of this endangered species.

### Before You Move On
**Monitor Comprehension** What efforts have been made to protect the wild Asian elephant in Southeast Asia?

### ONGOING ASSESSMENT
## READING LAB  GeoJournal

1. **Region** Which country has successfully reduced elephant killings?
2. **Interpret Charts** How has the population of Asian elephants changed over time?
3. **Human-Environment Interaction** What human activities threaten Asian elephants? Why are they a threat?

# SECTION 2 GOVERNMENT & ECONOMICS

## 2.1 Governing Fragmented Countries

**TECHTREK**
myNGconnect.com For photos from fragmented countries

Digital Library

**Main Idea** Geographic and ethnic divisions make it difficult for some countries in Southeast Asia to become unified.

In Southeast Asia, Indonesia, Malaysia, and the Philippines face challenges in forming unified countries. All three are **fragmented countries**, or countries that are physically divided into separate parts, such as a chain of islands. The three countries also have diverse ethnic groups.

### Indonesia

Indonesia's 17,000 islands span across 3,200 miles and are more different than they are similar. Java, for example, is densely populated and urbanized. Sumatra, on the other hand, is rural and contains large plantations. The country includes more than 300 ethnic groups and more than 700 languages are spoken.

To meet the complex challenges of fragmentation, Indonesia's government has tried to create a sense of nationhood. The country's motto, a saying that guides them, is "Diversity in unity." However, unity is not always easy to achieve. For example, there is conflict between the majority Malays and minority Chinese, and groups in northern Sumatra and Borneo have recently tried to gain independence. The government has focused on improving people's standard of living so that these groups will see the advantages of staying part of Indonesia.

**Critical Viewing** Perdana Putra is the Malaysian prime minister's palace. What other buildings that you have seen have onion domes similar to these?

## Malaysia

Malaysia includes both mainland and island areas. The mainland section is on the Malay Peninsula, and the island section is part of the island of Borneo. The challenge to Malaysia's government is to unify two parts of a country that are separated by several hundred miles of ocean. About half the total population is Malay, and their numbers dominate mainland Malaysia. The country also has sizable Chinese and Indian minorities, and these groups have generally achieved economic success. However, government policies have typically favored Malays, which has created tension between the two groups.

The Malaysian government has tried to achieve unity in several ways. Foremost is emphasis on economic growth. The country has made strong progress toward becoming a developed nation. This economic growth has helped ease some of the tensions among ethnic groups.

## The Philippines

Like Indonesia, the Philippines consists of thousands of islands, most of which are less than a square mile in size. Its wide variety of ethnicities includes Malays, Chinese, Japanese, Arab, and Spanish. Many Americans have immigrated to the country as well. The country has blended these groups into its own Filipino culture. The widespread use of Filipino, one of the country's official languages, helps to form the national identity. The other official language, English, is also widely spoken.

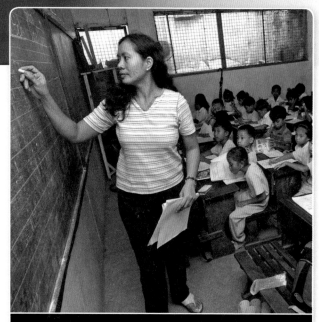

Filipino teacher, Leonora Jusay, gives a lesson to the 59 students in her class. Education is underfunded but school is widely attended.

Although the Philippine economy has grown, nearly a third of all people are poor. Due to a lack of jobs, a few million people have left the Philippines to find work in other countries. They send a share of their earnings back to the Philippines as **remittances** to help their families at home.

### Before You Move On
**Make Inferences** Why would geographic and ethnic divisions make it difficult for a country to come together?

### ONGOING ASSESSMENT
## READING LAB  GeoJournal

1. **Analyze Cause and Effect** What are the causes and effects of fragmentation in Indonesia, Malaysia, and the Philippines?
2. **Identify Problems and Solutions** What could the governments of these countries do to try to improve unity?
3. **Make Inferences** In what ways might widespread school attendance in the Philippines' positively impact the country?

# SECTION 2 GOVERNMENT & ECONOMICS

## 2.2 Migration Within Indonesia

**Main Idea** Efforts to bring unity to Indonesia's islands through relocation have had mixed results.

Indonesia is the fourth most populous country in the world. However, the majority of the people live on just a few of Indonesia's many islands. Living on the remote islands isolates citizens from the greater population, and the country remains fragmented.

### Relocation Policy

The Dutch, who had colonized the area, recognized the problem of unifying the vast chain of islands. In the 1800s, they began **relocating**, or moving, individuals and families from Java, **Madura**, and Bali—called the **inner islands**—to the surrounding and less-central islands, the **outer islands**. After Indonesia won independence in 1949, the Indonesian government continued relocating people.

Currently, more than half of Indonesia's people live on the island of Java—an island with a small percentage of the country's total land area. The government hopes that continuing to spread out the Javanese people will help to unify the country. Indonesians native to Java speak the official language, and their presence on the outer islands can help spread the official language to those places where it is infrequently heard. A common language can help unify a fragmented country.

### Effects of Internal Migration

So far, however, the practice of moving people among the islands has had unintended results. New arrivals came into conflict with native people, altering their way of life. Modern farming practices clashed with traditional land use and sometimes damaged the environment.

> **Critical Viewing** The outer islands, where this farmer works in a rice field, are less populated than the inner islands. What can you infer from the photo about life on the outer islands?

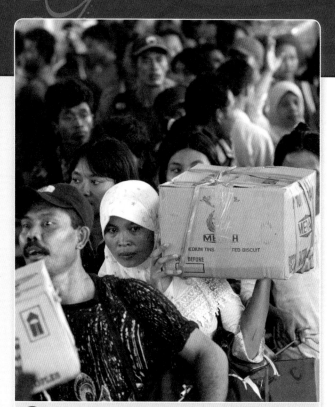

**Critical Viewing** Indonesians return to Jakarta after visiting their homes on other islands. What does the photo suggest about the movement from rural to urban areas like Jakarta?

The government hoped the new settlers could build successful farms, but many had trouble supporting themselves. Some ended up abandoning their new homes to return to the inner islands. As a result, Java and Bali remain crowded, making the government's relocation program ineffective. Crowding on Java and Bali grew even worse due to other **trends**, or changes over time. Indonesians living on the outer islands migrated there to flee rural poverty and find work on busier islands. Decades after the program began, Java and Bali are far more densely populated than Indonesia's other islands.

### Before You Move On
**Summarize** How have government policies and economic factors determined migration within Indonesia?

## INDONESIA: ISLAND POPULATIONS

**INNER ISLANDS:**
Java + Madura + Bali
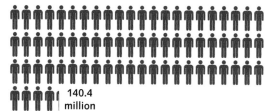
140.4 million

**OUTER ISLANDS:**
Sumatra

50.6 million

Sulawesi

17.3 million

Borneo
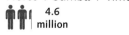
13.7 million

Flores + Sumba + Timor

4.6 million

Lombok + Sumbawa

4.4 million

New Guinea

3.6 million

Moluccas
2.5 million

 equals 2 million people

**Source:** *Europa World Year Book*, 2012; populations for 2010

### ONGOING ASSESSMENT
## DATA LAB
*GeoJournal*

1. **Interpret Graphs** According to the graph, which is the most populous outer island? Which is the least populous?

2. **Synthesize** When measuring population density, what does higher and lower population density indicate?

3. **Place** Look at the map of Indonesia in Section 1.1 of the previous chapter. Compare the sizes of the islands to the population figures here. Which of the outer islands do you think would have the highest population density? Why?

# SECTION 2 GLOBAL ISSUES • Globalization

## 2.3 Singapore's Growth

**TECHTREK**
myNGconnect.com For photos of Singapore and Guided Writing

 Digital Library
 Student Resources

**Main Idea** Singapore has grown economically due to its geographic location and economic policies.

The British established the modern port of Singapore in the early 1800s in an effort to compete with the Dutch in trade. Located just off the southern tip of the Malay Peninsula, the island has a perfect location near the shipping routes that link the Indian and Pacific oceans. This location gave it the **potential**, or possibility, of becoming a great **port**, a place where ships can exchange cargo. Today, the tiny island country is one of the world's busiest ports—and a strong economic power.

### KEY VOCABULARY

**port**, n., a place where ships can exchange cargo

**industrialize**, v., to develop manufacturing

**multinational corporation**, n., large business that has operations in many different countries

### ACADEMIC VOCABULARY

**potential**, n., possibility or promise

## Building Success

In 1963, Malaysia gained independence from Britain. Singapore was part of this new country. However, conflict arose between the majority Chinese population of Singapore and the Malays of the rest of Malaysia. In 1965, to reduce the tension, the government offered Singapore its independence, and Singapore took it.

Singapore thrived because of its prime location. It served as the main transit point for sending raw materials such as timber, rubber, rice, and petroleum from Southeast Asia to other parts of the world. Manufactured goods from the United States and Europe came into the port and were shipped to other ports in Southeast Asia. Cars and machinery were shipped into the city from the west to be distributed around the region.

Prime Minister Lee Kuan Yew led Singapore from 1959 to 1990. He emphasized the country's role as an important port and led the drive to **industrialize**, or develop manufacturing. However, the government strictly controlled life in Singapore. Streets were kept clean, and there was very little crime.

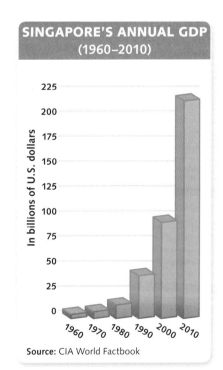

**SINGAPORE'S ANNUAL GDP (1960–2010)**

Source: CIA World Factbook

### Before You Move On

**Monitor Comprehension** What geographic assets have helped Singapore be part of the global economy?

632 CHAPTER 22

## Keeping Pace

Singapore today has a stable free market economy that relies on electronics exports and service businesses. It has one of the world's highest per capita GDP and is home to many **multinational corporations**. The global financial crisis only mildly affected its booming economy.

Singapore's leaders have set goals based largely on economic success. They invest heavily in infrastructure improvements to gain countrywide access to the most current technologies available. To attract foreign investment, Singapore offers low tax rates and other economic incentives. Economic policy focuses on key growth industries such as telecommunications and other technologies. Because these industries rely on educated, highly skilled workers, the country emphasizes improving the level of education in the workforce.

### COMPARE ACROSS REGIONS

As in Singapore, a large sector of the workforce in India is highly educated and trained in the newest technologies. Economic policy offers incentives to foreign investors, and many multinational corporations have set up operations there. However, India continues to face many challenges as it works to manage its economic growth. The poverty level is higher and the education level is lower than those of Singapore. India also struggles to improve its infrastructure to keep up with a global economy.

**Critical Viewing** This late-night street scene shows the business district in Singapore. Which details in the photo are typical of a prosperous city?

Greater stability has allowed Singapore to take advantage of changes happening elsewhere. When the British handed Hong Kong back to China in 1999, some business owners were concerned that Chinese rule would limit their freedom and prosperity. Singapore welcomed business owners who chose to move their companies there, which boosted economic growth.

### Before You Move On
**Summarize** How have Singapore and India's participation in the global economy differed?

### ONGOING ASSESSMENT
## WRITING LAB  GeoJournal

1. **Make Inferences** Describe how you think the government's position on taxation has affected economic development.
2. **Write Analyses** How has Singapore been affected by globalization? How has its economy been affected by other countries? Go to **Student Resources** for Guided Writing support.

# SECTION 2 GOVERNMENT & ECONOMICS
## 2.4 Malaysia and New Media

**TECHTREK**
myNGconnect.com For graphs and photos of communication in Malaysia

 Maps and Graphs   Digital Library

**Main Idea** Access to new media sources is changing the strict control on information formerly held by the government in Malaysia.

Since gaining independence in 1963, **Malaysia** has enjoyed prosperity and calm. The government has concentrated on building the economy. However, as in nearby Singapore, Malaysia has limited the freedoms its people can enjoy.

### Controlling Information

One of these limits is government control of access to information. Freedom of the press is limited. Newspapers must obtain licenses from the government to operate, and the government can cancel those licenses at any time. Similar laws place restrictions on companies that want to run radio or television stations. Also, the government withholds information from media.

Tough laws punish media outlets that criticize the government. Members of the ruling party own many major newspapers. Even independently owned newspapers usually do not criticize the government.

### New Media

The restrictions on the flow of information may be starting to loosen. Nearly two-thirds of Malaysians can now connect to the Internet. New media, like the Internet, give people access to new sources of information that the government may have more difficulty controlling.

Officially the government promises limitless Internet use. However, interference is not uncommon. Many Web site operators focusing on Malaysian news have been arrested multiple times for government criticism. Other journalists have been targeted as well.

> **Critical Viewing** A young woman uses her laptop in a public space in Malaysia. In what ways does the photo show both government control and freedom?

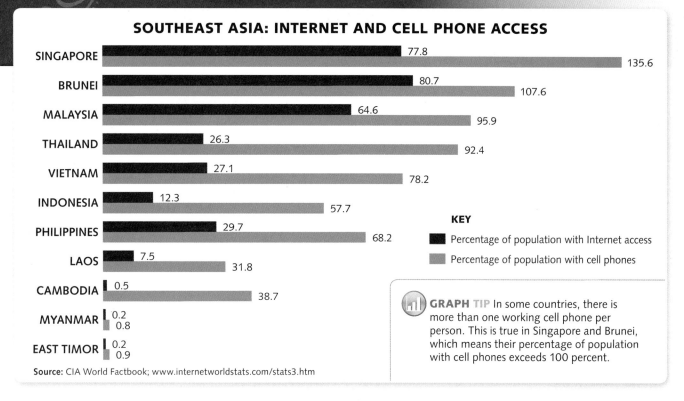

In other instances, police raided Web site headquarters and seized computers in order to find specific people who wrote articles critical of the government.

Many Malaysians believed that until the **emergence**, or arrival, of the Internet the government controlled and manipulated information available to citizens. During elections in 2008, Web sites became the leading source of news for people in the country.

Web sites were outside of government control. They also were completely open to opinions and ideas from anyone in the country. Any citizen could file stories or post videos to the sites. Because of such openness, Web sites may have influenced the 2008 elections. As a result of that vote, the ruling party lost 58 seats in the national legislature. It was the worst result for the party in more than 40 years.

New media technologies give people access to information that is difficult for people in power to control. Poll results found that older citizens trust newspapers and television news and are less likely to embrace additional sources of media. However, among voters in their twenties and thirties, only a small minority trusted traditional media, while more than 60 percent said online news sources were **reliable**, or trustworthy.

## Before You Move On
**Summarize** How has access to new media changed the strict government control of information in Malaysia?

## ONGOING ASSESSMENT
# DATA LAB

1. **Analyze Data** How do percentages for Internet access and cell phone use in Malaysia compare to those of other countries in the region?
2. **Make Inferences** Why do you think the government tries to keep such tight control of the news in Malaysia?
3. **Make Predictions** What do the poll results suggest about the way traditional media will be accepted in the future? Why?

# CHAPTER 22
# Review

## VOCABULARY

For each vocabulary word, write one sentence that explains its meaning and relates it to the content of the chapter.

1. prehistoric

> *Many religions in Southeast Asia are prehistoric, meaning they existed before history was written down.*

2. metropolitan area
3. dialect
4. domesticate
5. relocate
6. multinational corporation

## MAIN IDEAS

7. How has colonial history shaped religion in the region? (Section 1.1)
8. How does modern Thailand reflect both tradition and external influences? (Section 1.2)
9. In what way has globalization changed language use in Southeast Asia? (Section 1.3)
10. What methods can help protect the wild Asian elephant population in Southeast Asia? (Section 1.4)
11. How has fragmentation been a problem for Indonesia, Malaysia, and the Philippines? (Section 2.1)
12. What have been the results of Indonesia's policy of moving people to new areas? (Section 2.2)
13. How has Singapore built its economy? (Sections 2.3)
14. How have new media changed politics in Malaysia? (Section 2.4)

## CULTURE

### ANALYZE THE ESSENTIAL QUESTION

How have local traditions and outside influences shaped cultures in Southeast Asia?

**Focus Skill: Make Generalizations**

15. How have outside influences increased the diversity of Southeast Asia?
16. What difficulties do the countries of Southeast Asia face in trying to maintain traditional culture in the modern world?

### INTERPRET TABLES

**PERCENTAGES OF ETHNIC GROUPS IN SELECTED SOUTHEAST ASIAN COUNTRIES**

| Country | | |
|---|---|---|
| Indonesia | Javanese: 41% <br> Sundanese: 15% <br> Madurese: 3% | Minangkabau: 3% <br> Other: 38.4% |
| Laos | Lao: 55% <br> Khmou: 11% | Hmong: 8% <br> Other: 26% |
| Malaysia | Malay: 50% <br> Chinese: 24% | Indian: 7% <br> Other: 19% |
| Philippines | Tagalog: 28% <br> Cebuano: 13% <br> Ilocano: 9% | Bisaya/Binisaya: 8% <br> Hiligaynon Ilonggo: 8% <br> Other: 35% |
| Singapore | Chinese: 77% <br> Malay: 14% | Indian: 8% <br> Other: 1% |
| Thailand | Thai: 75% <br> Chinese: 14% | Other: 11% |
| Vietnam | Kinh (Viet): 86% | Other: 14% |

**Source:** CIA World Factbook

17. **Analyze Data** Which three countries have a single dominant ethnic group? What is the group in each case?
18. **Make Generalizations** Would you expect countries with one dominant ethnic group to have a single official national language? Why or why not?

## GOVERNMENT & ECONOMICS

### ANALYZE THE ESSENTIAL QUESTION

How are Southeast Asia's governments trying to unify their countries?

**Focus Skill: Summarize**

19. What economic and political concerns led Indonesia to adopt a policy of moving people from the inner islands to the outer islands?

20. Why did Singapore split from Malaysia and become independent?

21. How is the policy of Malaysia's government to limit freedom of the press related to the issue of fragmentation?

### INTERPRET MAPS

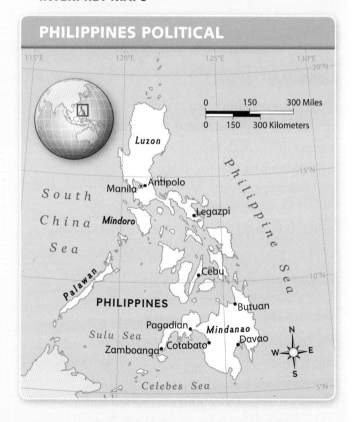

22. **Interpret Maps** Why is it appropriate to call the Philippines a fragmented country?

23. **Draw Conclusions** What transportation improvements might the government invest in to help unify the country? Explain.

## ACTIVE OPTIONS

**Synthesize** the Essential Questions by completing the activities below.

24. **Write a Culture Article** Introduce a friend to Southeast Asia today by writing an article about one country in the region. How do religious practices and regional languages shape culture in that country? Describe traditional art and music or the creative expressions of people who live there today. **Share your article with your friend.**

> **Writing Tips**
> - Research and take notes about the culture of your country.
> - Keep your paragraphs tightly focused on one topic.
> - Include details about the country's religions, art, music, or other aspects of culture.
> - Be sure you have an interesting introduction, smooth transitions between paragraphs, and a conclusion that summarizes your article.

25. **Create Graphs** Use online research to make a bar graph showing the per capita gross domestic product (GDP) of Cambodia, Malaysia, and the Philippines. Write a paragraph explaining which countries' economies might be affected by fragmentation. Use the example below as a guide for your graph.

COMPARE ACROSS REGIONS UNIT 11

# Endangered Species

myNGconnect.com For an online graph on endangered species

Maps and Graphs

Endangered species of both plants and animals can be found in every region of the world. A species is endangered when it runs the risk of becoming extinct, or disappearing completely from the world. Over time, various species have not survived. In fact, historically there have been five mass exinctions, in which a large number of existing species died out.

Today, the rate of extinction for plant and animal species has become hundreds of times faster than what scientists have observed through the fossil record. Many scientists believe the earth is currently in the midst of a sixth mass extinction. Those same scientists believe the main cause to be the destruction of habitat.

### Compare Ranges of Endangered Big Cats

- Asiatic Lion
- Cheetah
- Iberian Lynx
- Jaguar
- Tiger

### CAUSES OF EXTINCTION

Mining, logging, and clearing of forests for grazing cattle and growing crops all greatly change the natural landscape, or habitat, on which most species depend. Other development, such as the building of dams, highways, and housing, can increasingly divide animal populations into smaller, less diverse pockets. Some groups, like the big cats shown at right, face additional threats from hunters fearful of the animals' ability to harm people and livestock.

Climate change also can seriously stress a species' population and push it to extinction. For example, climate change can alter the amount of rain that falls, which affects plant growth and changes the food available in the habitat. When that happens quickly, species have difficulty adapting.

### CONSERVATION

Species and their ecosystems contribute much to the health and well being of humans. A diversity of plants and animals, and the habitats in which they are found, provide fertile soils, medicines, clean air and water, fibers, building materials, and food.

The International Union for Conservation of Animals (IUCN) is one organization trying to help countries around the world find solutions for balancing human needs and environmental challenges. To help target the habitats and animals in need of support, the IUCN maintains a database called the "Red List of Endangered Species." The list identifies animal species as near threatened, vulnerable, endangered, and critically endangered, as shown on the diagram at right.

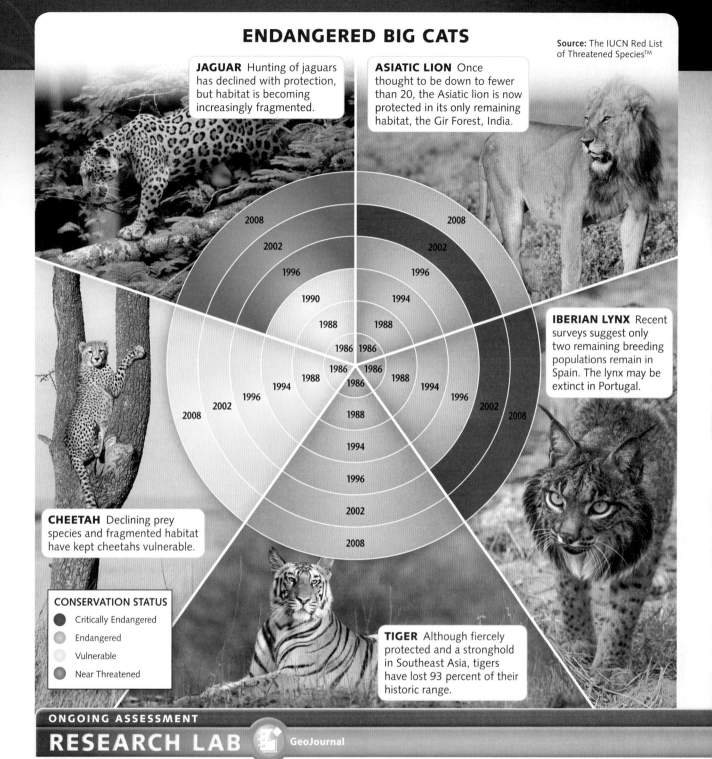

# ENDANGERED BIG CATS

Source: The IUCN Red List of Threatened Species™

**JAGUAR** Hunting of jaguars has declined with protection, but habitat is becoming increasingly fragmented.

**ASIATIC LION** Once thought to be down to fewer than 20, the Asiatic lion is now protected in its only remaining habitat, the Gir Forest, India.

**IBERIAN LYNX** Recent surveys suggest only two remaining breeding populations remain in Spain. The lynx may be extinct in Portugal.

**CHEETAH** Declining prey species and fragmented habitat have kept cheetahs vulnerable.

**TIGER** Although fiercely protected and a stronghold in Southeast Asia, tigers have lost 93 percent of their historic range.

**CONSERVATION STATUS**
- Critically Endangered
- Endangered
- Vulnerable
- Near Threatened

## ONGOING ASSESSMENT
# RESEARCH LAB  GeoJournal

1. **Analyze Data** Look at the diagram for the Asiatic lion. What happened between 2002 and 2008? How can you explain this?

2. **Draw Conclusions** Look at the diagram for the jaguar and the Iberian lynx. Think about where these cats live and the population density in those ranges. Why might this help explain why their levels of endangerment have changed?

**Research and Compare** Use the IUCN Red List of Threatened Species™ to research and compare the levels of endangerment within an animal group such as antelopes, seals, bats, bears, big apes, dolphins or otters. Identify the region and how humans are affecting the populations. What is being done to conserve the species? Create a table to describe your findings.

COMPARE ACROSS REGIONS

# REVIEW & EXTEND UNIT 11
## Active Options

**TECHTREK**
myNGconnect.com For photos of fruit

Digital Library

dragon fruit

### ACTIVITY 1
**Goal:** Learn about unusual and healthy food.

#### Make a Recommendation

Southeast Asia is home to a rich variety of plants that includes a wide assortment of fruits. Several of the more exotic fruits are listed below. With a partner, use the list below and online sources to research fruit from the region. Recommend three fruits that your classmates might want to try. Show what the fruits look like and give reasons for recommending those three.

- ciku
- dragon fruit
- durian
- jackfruit
- langsat
- longan
- mangosteen
- rambutan
- salak
- sapodilla
- soursap
- star apple

### ACTIVITY 2
**Goal:** Research a Southeast Asian city.

#### Culture Search

On a map or globe, locate each of the capital cities below, giving its coordinates. Then research one of the cities to learn what aspects of its cultural life interest you. Use the **Magazine Maker CD-ROM** to organize and illustrate your findings. Include places in the city you'd like to visit and facts to interest your readers.

- Bangkok, Thailand
- Kuala Lumpur, Malaysia
- Phnom Penh, Cambodia
- Singapore, Singapore

### ACTIVITY 3
**Goal:** Extend knowledge of Southeast Asian wildlife through research and drama.

#### Set the Stage

Some of the world's oldest rain forests are in Southeast Asia. With a group, compose a one-act play set deep in the rain forest. Focus the drama on the region's rain forest issues, and include in your play one of the animals found there, such as the Komodo dragon, flying snake, or orangutan. Perform the play for the class.

# UNIT 12

## EXPLORE
## AUSTRALIA, THE PACIFIC REALM & ANTARCTICA
### WITH NATIONAL GEOGRAPHIC

### MEET THE EXPLORER

National Geographic Fellow Elizabeth Lindsey helps conserve knowledge and traditions of Polynesian cultures. Her work focuses on such elements as Micronesian chants and traditional navigation methods that do not use any instruments.

### STEP INTO HISTORY

Easter Island, a UNESCO World Heritage site, is located 2,200 miles west of Chile. It is known for its enormous stone statues, standing from 10 to 40 feet tall. More than 600 statues remain on the island.

### CONNECT WITH THE CULTURE

The opera house in Sydney, New South Wales, Australia, is one of the most famous buildings in the world. Sydney, now one of the largest ports in the South Pacific, began as a convict settlement in the 18th century.

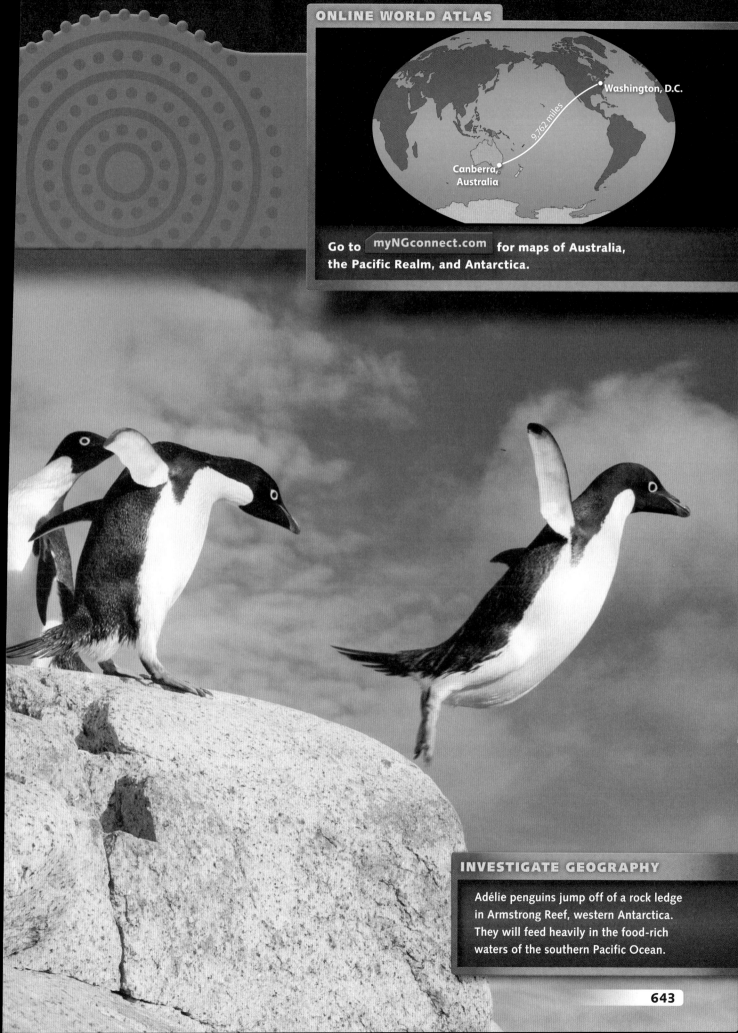

## ONLINE WORLD ATLAS

Go to myNGconnect.com for maps of Australia, the Pacific Realm, and Antarctica.

### INVESTIGATE GEOGRAPHY

Adélie penguins jump off of a rock ledge in Armstrong Reef, western Antarctica. They will feed heavily in the food-rich waters of the southern Pacific Ocean.

# CHAPTER 23

# AUSTRALIA, THE PACIFIC REALM & ANTARCTICA
# GEOGRAPHY & HISTORY

## PREVIEW THE CHAPTER

**Essential Question** How did geographic isolation influence the development of this region?

### SECTION 1 • GEOGRAPHY

1.1 Physical Geography .................................. 646

1.2 Indigenous Plants and Animals ............... 648

1.3 Biological Hitchhikers ............................. 650

1.4 The Pacific Islands .................................. 652

**GLOBAL ISSUES: HABITAT PRESERVATION**

1.5 Saving the Reefs ..................................... 654

**KEY VOCABULARY**
- coral reef
- atoll
- indigenous
- marsupial
- invasive species
- extinct
- feral
- hotspot
- coral island
- marine
- exoskeleton

**ACADEMIC VOCABULARY**
preserve

**TERMS & NAMES**
- Southern Hemisphere
- Southern Alps
- South Pole
- New Guinea
- Kingman Reef
- Great Barrier Reef

**Essential Question** How did geographic isolation shape the history of Australia and the Pacific Realm?

### SECTION 2 • HISTORY

2.1 Indigenous Populations ........................... 656

2.2 Seagoing Societies .................................. 658

2.3 From Convicts to Colonists ..................... 660

2.4 Exploring Vanishing Languages .............. 662

**KEY VOCABULARY**
- land bridge
- clan
- pictograph
- seafarer
- outrigger canoe
- navigation
- convict
- assisted migration
- generation
- linguist

**ACADEMIC VOCABULARY**
transport, convey

**TERMS & NAMES**
- Aborigine
- James Cook
- New South Wales
- Commonwealth of Australia
- Enduring Voices Project

644 CHAPTER 23

## SECTION 1 GEOGRAPHY
# 1.1 Physical Geography

### TECHTREK
myNGconnect.com For maps of Australia, the Pacific Realm, and Antarctica and Visual Vocabulary

 Maps and Graphs   Digital Library

### AUSTRALIA, THE PACIFIC REALM, AND ANTARCTICA PHYSICAL

**Main Idea** This region lies in the Pacific Ocean and is geographically isolated from other parts of the world.

Australia, much of the Pacific Realm, and Antarctica are located in the **Southern Hemisphere**. The region is spread across a vast ocean area.

## Australia

Australia is the only country in the world that is also a continent. Australia is sometimes called the "island continent" because it is surrounded by water. However, the climate is mostly dry. In fact, nearly 20 percent of the land mass of Australia is classified as desert. Few people live in the inland parts of the continent. Instead, most people live along the coasts where rainfall is plentiful.

## The Pacific Realm and Antarctica

The Pacific Realm is a large area of the Pacific Ocean made up of thousands of small islands and coral reefs. **Coral reefs** are rock-like structures built by layers of coral organisms. Coral reefs thrive in warm ocean waters between latitudes of 30°N and 30°S. **Atolls**—ring-shaped reefs, islands, or chains of islands made of coral—also dot this area of the Pacific.

New Zealand, located southeast of Australia, is made up of two major islands, North Island and South Island. New Zealand's mountain range, the **Southern Alps,** includes peaks that rise more than 12,000 feet.

Antarctica is centered on the **South Pole,** the southernmost point of Earth's axis. A thick layer of ice covers almost all

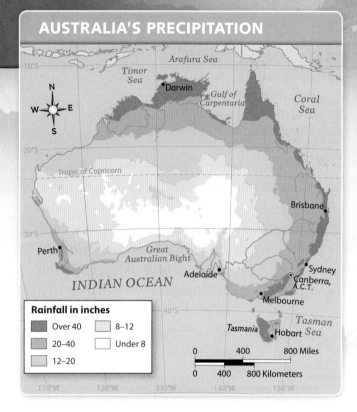

the continent's mountains, valleys, and islands. Antarctica has the coldest climate on Earth and is the continent with the highest average elevation.

### Before You Move On
**Monitor Comprehension** How is this region isolated from other parts of the world?

### ONGOING ASSESSMENT
## MAP LAB
GeoJournal

1. **Human-Environment Interaction** Based on the precipitation map, which part of Australia might produce few crops and why?
2. **Interpret Maps** Between what latitudes do the Pacific islands lie? Explain how their location is ideal for the formation of coral reefs.
3. **Compare and Contrast** Examine the physical map of the region. In what ways are Australia and Antarctica similar and different? Use a chart like this one to answer the question.

| SIMILARITIES | DIFFERENCES |
|---|---|
|  |  |

SECTION 1.1

# SECTION 1 GEOGRAPHY

## 1.2 Indigenous Plants and Animals

**Main Idea** Many plants and animals are found exclusively in this region.

Australia, New Zealand, and the Pacific islands have some of the world's most unusual indigenous plants and animals. **Indigenous** plants and animals are native to the region in which they are found. Because of the region's physical isolation, these species developed with little outside influence. The plants and animals that developed here are unlike those found elsewhere.

### Plant Species

Thousands of wildflowers, shrubs, and trees are native to the region. Among them are eucalyptus (YOO kuh LIHP tuhs) trees, whose leaves produce a fragrant oil used in medicines. These leaves are the only food that Australia's much-loved koalas eat. Another common plant is the acacia, with more than 700 varieties found in Australia. Most acacia have brightly colored flowers.

### Native Animals

Animals native to this region include **marsupials** (mar SOO pee uhlz), which are mammals whose females carry their babies in a pouch. Four types of marsupials live in this region. Kangaroos are the biggest marsupials. Other marsupials include koalas, wallabies, wombats, and Tasmanian devils.

Other indigenous mammals of Australia and **New Guinea** include the platypus and the echidna (ih KID neh). These mammals are the only mammals that lay eggs. The platypus and the echidna have beaks, another bird-like feature not typically found in mammals.

The region is home to many kinds of birds, including New Zealand's kiwi and kakapo, which do not fly. The cassowary (KAS eh ware ee) and emu, also flightless, stand five to six feet tall. Australia's kookaburra, a large kingfisher, is famous for its call, which sounds like laughter.

### Before You Move On

**Summarize** What plant and animal species are indigenous to the region?

*Cassowaries grow as high as five feet tall and can be aggressive.*

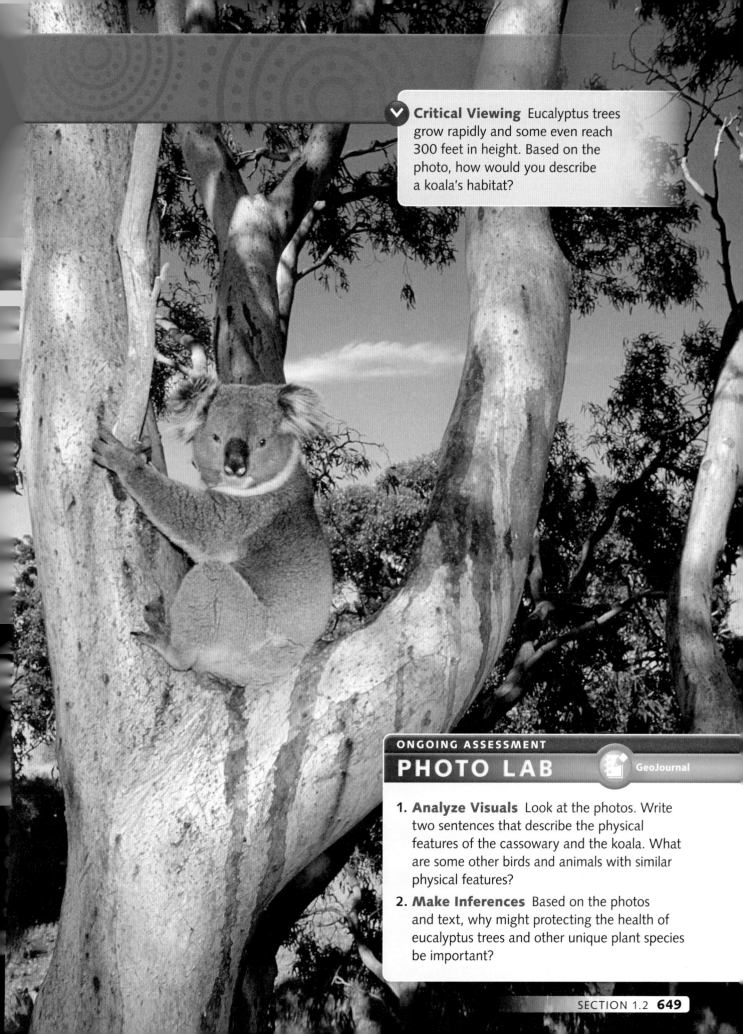

**Critical Viewing** Eucalyptus trees grow rapidly and some even reach 300 feet in height. Based on the photo, how would you describe a koala's habitat?

## ONGOING ASSESSMENT
# PHOTO LAB     GeoJournal

1. **Analyze Visuals** Look at the photos. Write two sentences that describe the physical features of the cassowary and the koala. What are some other birds and animals with similar physical features?

2. **Make Inferences** Based on the photos and text, why might protecting the health of eucalyptus trees and other unique plant species be important?

SECTION 1.2

# SECTION 1 GEOGRAPHY
## 1.3 Biological Hitchhikers

**TECHTREK**
myNGconnect.com For a map and photos of the region's invasive species

Maps and Graphs | Digital Library

**Main Idea** The introduction of non-native plant and animal species has changed the natural habitats of Australia and New Zealand.

Plants and animals, and their habitats, developed as they did because of the physical isolation of the region. The introduction of new plants and animals to Australia and New Zealand has had negative consequences.

### Invasive Species

Many non-native species were introduced to these countries on purpose—before people understood what problems might occur. New species also have arrived by accident. Planes and ships delivering produce and packages bring in hidden plant and animal "hitchhikers" that then invade natural habitats. Non-native plants and animals that disturb the habitats of native life forms are called **invasive species**. Because of invasive species, some native plants and animals have become **extinct**, meaning no more living members of that group exist. In fact, 19 different species of small mammals have become extinct in Australia since the introduction of non-native mammals such as cats, foxes, and rabbits.

**Visual Vocabulary** Invasive species are non-native plants and animals that disturb the habitats of native life forms. These rabbits in Australia are an invasive species.

### Dingoes, Rabbits, and Other Pests

Dingoes have lived in Australia for many years, but they were not indigenous to the island continent. Sea travelers from Asia brought dingoes with them about 3,000 years ago. Some dingoes escaped, became

**Critical Viewing** Dingoes are hunters. Based on the photo, what other animals do dingoes look like?

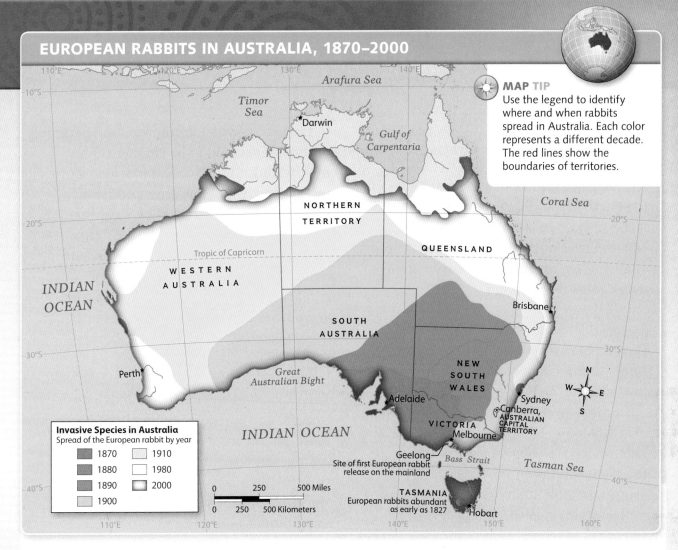

# EUROPEAN RABBITS IN AUSTRALIA, 1870–2000

**MAP TIP** Use the legend to identify where and when rabbits spread in Australia. Each color represents a different decade. The red lines show the boundaries of territories.

**Invasive Species in Australia**
Spread of the European rabbit by year
- 1870
- 1880
- 1890
- 1900
- 1910
- 1980
- 2000

wild, and preyed on native kangaroos and wallabies. Today, dingoes continue to threaten native animals and livestock.

When Europeans began to settle in Australia in the 1800s, they brought along rabbits, which quickly became **feral**, or wild, and spread across the continent. Feral rabbits have caused serious damage to crops and grazing lands. In the 1950s, the Australian government tried to wipe out feral rabbits but was not successful.

Non-native plants have also done damage. In fact, in New Zealand, more types of invasive plants thrive in the wild than do native plants. One climbing plant called "old man's beard" has spread widely since the 1940s. It smothers native foliage, and can even bring down large trees.

## Before You Move On

**Summarize** In what ways have invasive species impacted Australia and New Zealand?

## ONGOING ASSESSMENT
# MAP LAB

1. **Interpret Maps** Look at the map. In what direction have rabbits moved across Australia? By what decade was the island of Tasmania overrun with rabbits?

2. **Interpret Maps** Based on the map, where were rabbits first released in Australia? Using the map scale, determine approximately how far rabbits spread between the years 1870 and 2000.

3. **Movement** What are invasive species? In what ways do human migrations encourage their spread?

SECTION 1.3  **651**

## SECTION 1 GEOGRAPHY
# 1.4 The Pacific Islands

**Main Idea** Physical processes helped form the Pacific islands over a long period of time.

Northeast of Australia and New Zealand, the Pacific islands are a group of 20,000 to 30,000 islands scattered across millions of square miles of the Pacific Ocean. These islands were formed by geologic changes over thousands of years.

### High Islands, Low Islands

Geographers divide the Pacific islands into two main categories: high islands and low islands. High islands are formed by volcanic activity. As tectonic plates move over **hotspots**, or unusually hot parts of Earth's mantle, magma rises up through the plates and produces a volcanic eruption. Cooled molten material forms underwater volcanic cones. This process repeats, and over time, these volcanic cones emerge above water as islands.

Low islands also form over long periods of time. They tend to be smaller than high islands and sit only a few feet above sea level. Low islands are **coral islands**, which are islands created by a gradual buildup of the skeletons of corals and other tiny marine animals. Coral islands often sit on top of coral reefs. Many coral reefs form around the base of high volcanic islands. Over time these reefs become atolls.

### Climate, Agriculture, and Fishing

The climate in the Pacific islands is tropical and warm throughout the year. Most islands have a wet season and a dry season, but precipitation on the islands varies. Warm ocean air cools at higher elevations on high, mountainous islands and produces rain. High islands receive plentiful rainfall. Pineapples, sugarcane, and mangoes grow well in their rich

## PACIFIC ISLAND FORMATION

### HIGH ISLAND FORMATION

 **1** Magma rises from deep inside Earth's mantle.

 **2** Cooled molten material forms underwater volcanic cones.

 **3** Volcanic cones emerge from the water as high islands.

### LOW ISLAND FORMATION

 **1** Coral reefs form on the outer edges of high islands.

 **2** High islands appear to sink as erosion wears them away.

 **3** Over time, only the surrounding coral reef remains.

volcanic soil. Low islands, with no landforms to "catch" moisture, receive less rainfall. Though they are surrounded by ocean, drought is common on low islands. As a result, people on low islands depend less on agriculture and more on fishing.

### Before You Move On
**Monitor Comprehension** How did the two types of Pacific islands form?

### ONGOING ASSESSMENT
## VIEWING LAB  GeoJournal

1. **Sequence Events** Based on the illustration above, describe in your own words the sequence of events in high and low island formation. Include the words *atoll* and *coral reef* in your answer.

2. **Analyze Visuals** Based on the illustration and the text, in what way is molten material involved in island formation?

3. **Human-Environment Interaction** Why might farming on high islands be more successful than on low islands?

◀ **Critical Viewing** Kingman Reef lies under an atoll in the Line Islands and is one of the few unspoiled coral reefs remaining in the world. What details do you notice about life on this reef?

SECTION 1.4 **653**

# SECTION 1  GLOBAL ISSUES • Habitat Preservation

## 1.5 Saving the Reefs

**TECHTREK**
myNGconnect.com For photos of Australia's Great Barrier Reef

 Digital Library
 Global Issues

**Main Idea** The Great Barrier Reef is a marine ecosystem that is threatened by human activities.

Each year, many people visit Australia's **Great Barrier Reef**. In order to protect this ecosystem and others like it around the world, Australia and other countries are balancing tourism with preservation.

**KEY VOCABULARY**

**marine**, adj., sea based

**exoskeleton**, n., a hard, external covering that provides protection

**ACADEMIC VOCABULARY**

**preserve**, v., to protect

### The Great Barrier Reef

The Great Barrier Reef is the world's largest coral reef. Locate the Reef on the map on page 646. As you know, a coral reef is a complex **marine**, or sea-based, ecosystem. Small marine animals called corals secrete, or produce and discharge, calcium carbonate. This hardens into an **exoskeleton**, or a hard, external covering that provides protection. Over time, colonies of these corals form coral reefs that support marine life, from small fish to predators such as sharks.

The Great Barrier Reef stretches 1,250 miles and consists of around 2,900 separate reefs. It is home to one of the most diverse collections of plants and animals on Earth. Around 2,000 species of fish, 350 corals, and 4,000 mollusks, or soft-bodied marine animals with exterior shells, live here. In recent years, scientists have discovered hundreds of new animal species. Because of the diversity and extent of life on the reef, it is sometimes called "the rain forest of the sea."

The Australian government made the Great Barrier Reef a national park in 1975. By doing so, it reduced the potential negative impact of tourism through controls on tourist activities in the park. Visitors to the Great Barrier Reef can scuba dive, snorkel, view coral reefs from glass-bottomed boats, and swim with dolphins.

**Before You Move On**

**Summarize** What kinds of marine life does the Great Barrier Reef support?

### GREAT BARRIER REEF BY THE NUMBERS

**1.99 million**
Number of annual tourists, 2012

**72°–84°F**
Ideal water temperature range for corals

**10,000**
Average age of corals in years

**2,900**
Number of coral reefs that make up the Great Barrier Reef system

**900**
Number of islands in the Great Barrier Reef system

**600**
Species of starfish and sea urchins on the Great Barrier Reef

**30**
Species of whales and dolphins observed on the Great Barrier Reef

Sources: Great Barrier Reef Marine Park Authority, 2012

## Science at Work

Today, the Great Barrier Reef's survival is threatened. The reef's ecosystem depends on a specific temperature range for reef-building corals to thrive. These corals depend on algae, small oxygen-producing organisms, for nourishment. Even small increases in water temperatures of 1°–2°F can prompt the algae to leave the corals, resulting in the corals' death. This process is called "coral bleaching" because it causes the corals to turn white.

In addition to increased water temperatures, pollution and overfishing have also caused problems for corals and other marine life on the reef. In fact, if steps toward preservation are not taken, scientists estimate that corals on the Great Barrier Reef could be extinct by 2050.

Experts are working to **preserve**, or protect, the reef. One group of scientists is trying to identify all the different species living in the reefs. Another group is focused on the effects of rising water temperatures on corals and reef life. A deeper understanding of the reef could lead to new methods of saving it.

**COMPARE ACROSS REGIONS**

## Reef Preservation

Like the Great Barrier Reef, the Belize Barrier Reef—the second largest reef system in the world—attracts thousands of tourists each year. Belize is a small country in Central America that borders the Caribbean Sea. Mangrove trees, an important part of the reef's habitat, have been cut down. Sand and coral from the ocean floor have also been disturbed, causing the loss of sea life. As tourism has grown, the reef has also been overfished.

As in Australia, local communities in Belize are actively preserving their reef. For the reef to be saved, the Belize government must work more closely with the tourism industry to control human activities in this fragile part of the world.

### Before You Move On
**Monitor Comprehension** How do increasing water temperatures threaten coral reefs?

**Critical Viewing** White tip sharks like this one live on the Great Barrier Reef. How might their survival depend on the protection of the reef?

**ONGOING ASSESSMENT**
## READING LAB

1. **Summarize** In what ways have human beings contributed to the problems facing coral reefs?
2. **Make Inferences** What might Belize learn from Australia about reef preservation?
3. **Draw Conclusions** The Great Barrier Reef Marine Park keeps track of numbers of tourists to the reef. Why might it be doing this?

# SECTION 2 HISTORY
# 2.1 Indigenous Populations

TECHTREK
myNGconnect.com For photos of indigenous people and Australian rock art
Digital Library

**Main Idea** The languages and culture of Australia's indigenous people developed in isolation on the continent.

About 50,000 years ago, humans migrated to present-day Australia from Asia. They may have arrived by sea, or by **land bridge**, a narrow and temporary link between two land masses. The **Aborigines** (AB uh RIHJ uh neez) were Australia's first developed human culture. Most scientists believe this culture dates from 30,000 B.C.

## Populating the Continent

Australia's Aborigines are considered the oldest continuous human culture in the world. In fact, the phrase *ab origine* means "from the beginning." Because of Australia's isolated location, Aboriginal culture and languages developed apart from other human groups. Over time, more than 200 different languages developed. These languages were unlike any others in the world.

Like other prehistoric peoples, Aborigines were nomadic hunters and gatherers. To adapt to the continent's dry climate, they conserved water and carried it with them as they moved. They lived in distinct groups of larger **clans**, or family units. Each clan had a strong attachment to the land on which it hunted and gathered food. In addition to providing a source of food, the land held spiritual meaning for the Aborigines.

**Critical Viewing** Uluru (oo LOO roo) is located in Australia's Northern Territory. Based on what you see in the photo, what aspects of Uluru might inspire Aborigines to view it as sacred?

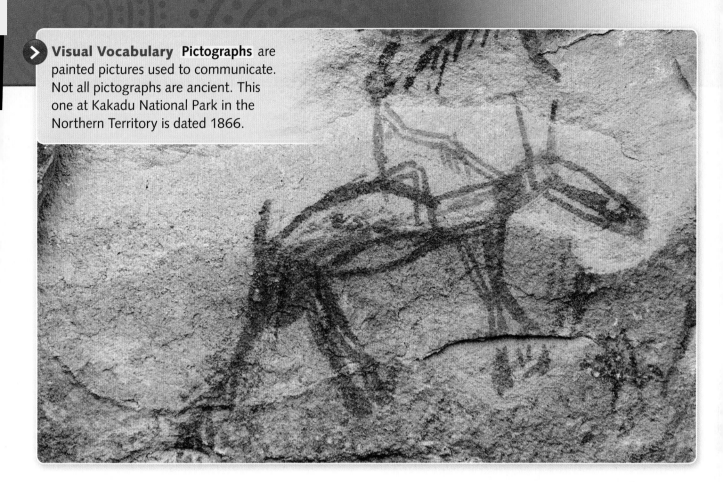

**Visual Vocabulary** **Pictographs** are painted pictures used to communicate. Not all pictographs are ancient. This one at Kakadu National Park in the Northern Territory is dated 1866.

Aborigines have considered the huge, red rock called Uluru, pictured at left, a sacred site for thousands of years. Rock art painted on cave walls at Uluru and other sites reveals much about Aborigines' beliefs. Some **pictographs** depict elements from the natural world while others portray religious ceremonies and important historical events.

## Conflicts with Europeans

By 1788, when Europeans arrived in Australia, between 500,000 and one million Aborigines lived on the continent. Though Aborigines had established complex societies, most European settlers viewed their way of life as uncivilized.

Conflicts arose when Europeans seized Aborigines' lands. Separated from their land, Aborigines were not able to hunt, and many fell into poverty. Many died of disease because they did not have immunity or resistance to European illnesses. By 1921, they totaled only about 62,000. Today, Aborigines represent less than 3 percent of Australia's population, or 548,370 citizens.

### Before You Move On
**Summarize** In what ways did Australia's location determine the way Aboriginal languages and culture developed?

### ONGOING ASSESSMENT
## PHOTO LAB

1. **Analyze Visuals** Based on the photo of rock art, what can you infer about how Aborigines used their surroundings to record events?
2. **Make Inferences** Uluru is a popular tourist site in Australia. Based on the photo and text, what might you infer about how Aborigines would feel about Uluru as a tourist site?
3. **Movement** In what ways did conflicts over land cause problems for Aborigines?

# SECTION 2 HISTORY
## 2.2 Seagoing Societies

**Main Idea** The first people to inhabit the Pacific Realm developed specific skills to adapt to life on the islands.

Several thousand years ago, ancient **seafarers**, or sea travelers, set sail from Southeast Asia, southern China, and Taiwan. They traveled over wide expanses of ocean and settled on islands in the Pacific Ocean.

### Living on the Ocean

The islanders built canoes with attached floats, called **outrigger canoes**. These early settlers were experts at **navigation**, or the process of determining location and routes, as they traveled between islands. Seafarers used the patterns of stars to map their course. They also studied the flight patterns of birds, which indicated where land was located. Seafarers created stick charts, an early form of maps, from sticks, shells, and twine. Shells on the charts indicated islands and different shapes of sticks represented ocean currents.

**Critical Viewing** This stick chart is a type of map used by Pacific seafarers. How does this map differ from maps you use?

Over time, different cultures developed. On larger islands, inhabitants practiced advanced farming methods. On smaller islands, people supported themselves through fishing. Mountainous land on some islands kept groups of people isolated from one another and cultures developed independently. However, on

**Visual Vocabulary** An **outrigger canoe** is a boat with an attached float that helps it balance. These boys use paddles to guide their outrigger canoe in Papua New Guinea.

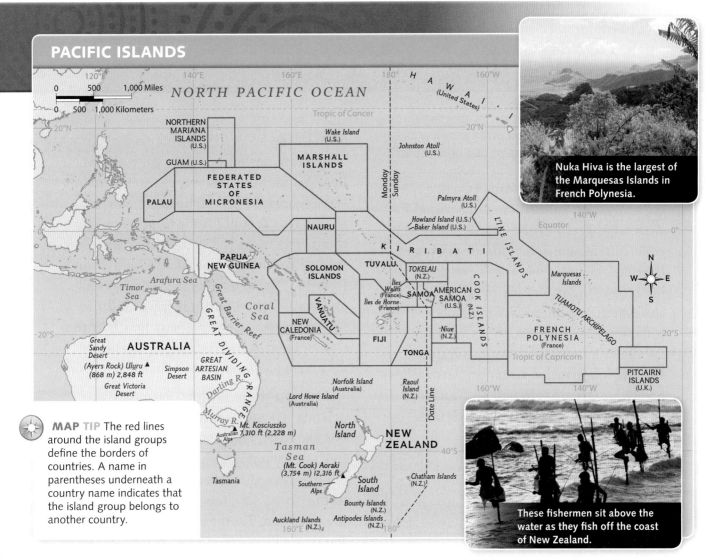

### PACIFIC ISLANDS

**MAP TIP** The red lines around the island groups define the borders of countries. A name in parentheses underneath a country name indicates that the island group belongs to another country.

Nuka Hiva is the largest of the Marquesas Islands in French Polynesia.

These fishermen sit above the water as they fish off the coast of New Zealand.

islands with few land barriers, people shared a similar language and culture. Sometimes groups of people moved from one island to another. In the 1100s, for example, the Maori people migrated from eastern Pacific islands to the North Island of New Zealand.

## Western Influence

For the most part, seafaring societies remained isolated from Western contact. A few Europeans explored the Pacific islands as early as the 1500s, but most islanders did not encounter Europeans until the 1700s. Some Europeans traveled to this region in search of wealth or to spread Christianity. Other Europeans, such as British navigator **James Cook**, charted new territories.

During multiple voyages in the 1770s, Cook mapped islands in the Pacific, including New Zealand. In fact, the Cook Islands, which lie northeast of New Zealand, are named for James Cook.

### Before You Move On
**Summarize** In what ways did the first people in the Pacific Realm adapt to the islands?

### ONGOING ASSESSMENT
## MAP LAB  GeoJournal

1. **Interpret Maps** The Pacific islands cover a broad expanse of ocean. In what way are the Pacific islands represented on this map?

2. **Identify** In the 1800s, most Pacific islands became colonies of Western countries. Based on the map, which islands remain possessions of other countries? How can you tell?

SECTION 2.2

# SECTION 2 HISTORY
## 2.3 From Convicts to Colonists

**Main Idea** Convicts and other immigrants to Australia led the development of a new country.

Seventeen years after James Cook explored Australia's coast, British ships sailed for Australia. From 1788 to 1868, the British government sent **convicts**, or prisoners, to overseas colonies in order to decrease overcrowding in British prisons. When the American colonies established independence in 1776, the government needed a new place to send convicts. British colonization began in Australia as a way for the government to **transport**, or ship, its convicts to prisons overseas.

### Settling the Colonies

Most transported convicts were young, healthy, and unmarried. Only 20 percent were women. Those who were married sometimes brought their families with them. The majority of convicts came from the working classes in England and Ireland. The government used convicts' skills and relied on their labor to build the colonies, including **New South Wales**. Convicts built roads, bridges, buildings, and farms as they served their sentences.

Though these settlers were British prisoners, they had certain freedoms. For example, many were allowed to live in their own homes and run their own businesses. At the end of their prison terms, freed convicts often remained in the colonies. Their contributions helped the Australian colonies to grow.

In 1831, the British set up a program for **assisted migration**. This program encouraged people to move to Australia by giving them financial help. Twenty years later, gold was discovered in Australia. This discovery drew many settlers to the continent including immigrants from China and the surrounding Pacific islands. An increased population and more available workers led to the growth

Photograph of a replica of James Cook's ship *Endeavour*

1750 — 1770 James Cook explores Australian coast. — 1787 First Fleet sets sail for Australia. — 1788 Convict transportation begins; British establish colony at New South Wales. — 1800 — 1831 British policy of assisted migration begins.

of industries, such as manufacturing and mining. As more of the continent was explored, some settlers began sheep farming. Wool became an important product and export.

Australia's population quickly grew from 400,000 in 1850 to more than one million by 1860. Population growth and European settlement pushed Aborigines off lands that had belonged to their ancestors for many years.

## Becoming A Country

In the 1890s, many Australians began to think that the colonies would benefit by becoming a single country. Colonists wrote a constitution and submitted it to the British government. In 1901, Great Britain approved the constitution, and the **Commonwealth of Australia** came into being. Though Australians now had their own government, they were still part of the British Empire. Today, Australia remains part of the British Commonwealth. Australia recognizes the British monarch as its head of state, but it is an independent country.

**Before You Move On**
Make Inferences In what ways did convicts lead the development of Australia?

**Critical Viewing** Thomas Gosse painted *Founding of the Settlement of Port Jackson at Botany Bay in New South Wales* in 1799. What does his painting lead you to think about the settlement of Port Jackson?

## ONGOING ASSESSMENT
### READING LAB — GeoJournal

1. **Summarize** According to your reading, why did the British begin to colonize Australia? What discovery in the 1800s attracted more immigrants to the continent?
2. **Movement** What migration patterns to and within the continent shaped Australia's history?
3. **Interpret Time Lines** Examine the time line. How many years did Great Britain transport convicts to Australia? Based on your reading, what other way did the British encourage migration to the continent?

Australian flag, designed in 1901

- 1851 Gold is discovered in Australia.
- 1868 Convict transportation is abolished.
- 1901 Australian colonies become Commonwealth of Australia.
- 1890s Australian colonies begin effort to become a single nation.

Illustration of gold miner in New South Wales

## 2.4 SECTION 2 HISTORY — NATIONAL GEOGRAPHIC

# Exploring Vanishing Languages
### with David Harrison and Greg Anderson

> **Main Idea** Many languages spoken by indigenous people are endangered but efforts are underway to revitalize them.

### The Importance of Language

Nearly 7,000 languages are spoken in the world today. Language is a vital part of a people's identity. Each one tells a great deal about the people who speak it. Language **conveys**, or communicates, what is important to a culture and helps pass along history, stories, and songs from one generation to the next. A **generation** is a group of individuals who are born and live about the same time. Older generations share knowledge, like language, with younger generations.

Today, many languages are disappearing. Languages spoken by indigenous people are especially threatened. Many languages have no written form and are spoken only by a small group of people. As the children of these people are exposed to other languages, they often stop speaking their native language. Older members of the group are then the only ones left who remember the language. With no written record, the indigenous language disappears.

myNGconnect.com
For more on David Harrison and Greg Anderson in the field today

David Harrison and Greg Anderson work with Charlie Mangulda in Australia.

By 2100, more than half the world's languages may no longer exist. Some experts believe that on average a language is lost every two weeks. As these languages vanish, a wealth of information about history and culture goes with them. Because indigenous people often use language to share knowledge about native plants and animals, when their languages vanish, information about the natural world also disappears.

## The Enduring Voices Project

National Geographic Fellows David Harrison and Greg Anderson are **linguists**, or language scientists. They founded National Geographic's **Enduring Voices Project** to study and document languages at risk. They work in many regions of the world. In 2010, in the foothills of the Himalayas in India, Harrison and Anderson documented a language previously unknown to researchers. Australia and Papua New Guinea are also among the places they identify as language hotspots because so many languages there are endangered. Over 800 languages are spoken in Papua New Guinea alone.

Harrison and Anderson use a variety of methods in their work. One is to interview the remaining speakers of the languages and make video and sound recordings. In addition to documenting the sounds and meanings of words, Harrison and Anderson gather valuable information about each group's history and culture. By studying these languages, they are also recording knowledge about the cultures.

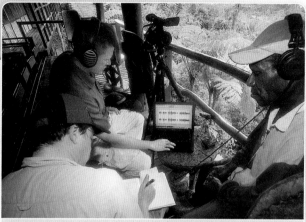

**Critical Viewing** Harrison and Anderson interview Felix Andi in Papua New Guinea. Based on what you see here, what kinds of technology help linguists document endangered languages?

Another method these linguists use is language revitalization. This involves supporting community-led efforts to bring vanishing languages back to life by teaching them to younger generations. Harrison and Anderson believe that educating young people is key to the survival of their languages.

### Before You Move On
**Summarize** Why are many languages spoken by indigenous people endangered?

### ONGOING ASSESSMENT
## LANGUAGE LAB  GeoJournal

1. **Explain** Why is it important to revitalize indigenous languages?

2. **Make Inferences** Some words from indigenous languages have been adopted into English, such as *dingo* and *kangaroo*. Why might this be? Think about the special features of these animals before you answer.

3. **Analyze Visuals** Watch the Explorer Video Clip. Name at least two things you learned about the Enduring Voices Project. Then pose two questions you might ask David Harrison and Greg Anderson.

# CHAPTER 23
# Review

## VOCABULARY

For each pair of vocabulary words, write one sentence that explains the connection between the two words.

1. indigenous; marsupial

   *The kangaroo, which is indigenous to Australia, is a marsupial.*

2. invasive species; extinct
3. coral reef; atoll
4. seafarer; outrigger canoe
5. convict; assisted migration
6. linguist; generation

## MAIN IDEAS

7. In what ways are Australia, the Pacific Realm, and Antarctica similar? (Section 1.1)
8. Why does the region have so many plants and animals unlike anywhere else in the world? (Section 1.2)
9. How were invasive species introduced in the region? (Section 1.3)
10. What two geologic processes created the Pacific islands? (Section 1.4)
11. In what ways are coral reefs around the world endangered? (Section 1.5)
12. How did migration and location shape the history of the Aborigines? (Section 2.1)
13. Why might Western influences be limited in the Pacific islands? (Section 2.2)
14. What groups helped develop the continent of Australia? (Section 2.3)
15. What steps are linguists taking to preserve vanishing languages? (Section 2.4)

## GEOGRAPHY

### ANALYZE THE ESSENTIAL QUESTION

How did geographic isolation influence the development of this region?

**Critical Thinking: Summarize**

16. Why is Australia called an "island continent"?
17. How did indigenous plants and animals develop in this region?
18. In what ways do invasive species threaten the indigenous plants and animals in the region?
19. What physical features make some Pacific islands better suited for farming than others?

### INTERPRET MAPS

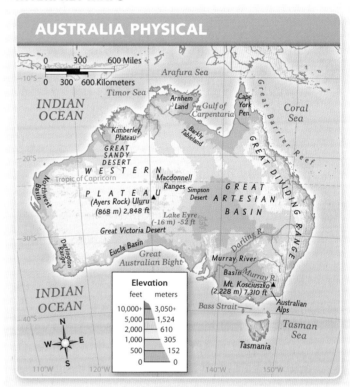

20. **Identify** What are the names of Australia's deserts? In what part of the continent do they lie?
21. **Location** In what sea is the Great Barrier Reef located? How is the name of the sea related to the reef?

## HISTORY

### ANALYZE THE ESSENTIAL QUESTION

How did geographic isolation shape the history of Australia and the Pacific Realm?

**Critical Thinking: Draw Conclusions**

22. In what ways did conflicts over land impact Aborigines in Australia?

23. What methods did ancient seafarers use to deal with the challenges of life on isolated islands in the Pacific Ocean?

24. Why might the British government have chosen Australia as a place to transport convicts?

25. On mountainous Pacific islands, how did physical features play a role in the establishment of multiple languages?

### INTERPRET TABLES

| COMPARING LANGUAGE USE IN AUSTRALIA AND PAPUA NEW GUINEA | | |
|---|---|---|
| | Australia | Papua New Guinea |
| Percentage of people speaking English | 78.5 | 2.0 |
| Number of indigenous languages spoken | 145 | 860 |

**Sources:** CIA World Factbook, 2010; Australian Government of Foreign Affairs and Trade 2010; BBC, 2010

26. **Compare and Contrast** Based on this table, in which country does the majority of the population speak English?

27. **Make Generalizations** Based on this table and what you have read, in which country—Australia or Papua New Guinea—did European settlers have a larger influence on indigenous populations?

## ACTIVE OPTIONS

**Synthesize** the Essential Questions by completing the activites below.

28. **Write a Letter** Imagine that you are a settler in 1851 who has arrived in Australia to search for gold. Write a letter to a family member back in England, describing what you observed when you arrived. Explain whether you think this person should join you. **Send your letter to a friend and ask if it painted a clear picture of Australia.**

    **Writing Tips**
    - Address your letter to a particular person.
    - Draw on what you have read about Australia to describe what the country is like.

    Go to **Student Resources** for Guided Writing support.

29. **Create a Visual Presentation** Use the **Digital Library** or other online sources to create a visual presentation of this region. Choose a photo to go with each of the following statements. Describe your photos with additional research you conduct.

    - Australia is home to many unusual plants and animals.
    - Thousands of living creatures inhabit the Great Barrier Reef.
    - Some Pacific islanders still practice a traditional way of life.
    - Researchers work to revitalize endangered indigenous languages.

CHAPTER 24

# AUSTRALIA, THE PACIFIC REALM & ANTARCTICA TODAY

## PREVIEW THE CHAPTER

**Essential Question** How are Australia, the Pacific Realm, and Antarctica becoming connected to the rest of the world?

### SECTION 1 • CULTURE

1.1 From Ranch to City .................................. 668

1.2 Immigration to Australia ........................ 670

1.3 Polynesian Cultures ................................. 672

1.4 Human Footprint in Antarctica .............. 674

**KEY VOCABULARY**
- urban
- salinization
- immigration
- labor force
- oral tradition
- emigrate
- ice shelf
- scientific station

**ACADEMIC VOCABULARY**
deplete, maintain

**TERMS & NAMES**
- Sydney
- Polynesian Triangle
- Cook Islands
- Charles Wilkes
- Richard Byrd
- Antarctic Treaty

**Essential Question** What new economic patterns are emerging in the region?

### SECTION 2 • GOVERNMENT & ECONOMICS

2.1 New Trade Patterns ................................. 676

2.2 Rights for Indigenous People ................. 678

2.3 Adventure Tourism ................................. 680

2.4 New Zealand Today ................................. 682

**KEY VOCABULARY**
- alliance
- free trade agreement
- reserve
- assimilation
- adventure tourism
- glacier
- crevasse
- renewable energy
- geothermal energy

**ACADEMIC VOCABULARY**
reclaim, lucrative

**TERMS & NAMES**
- Papua New Guinea
- Samoa
- Asia-Pacific Economic Cooperation (APEC)

666 CHAPTER 24

Surfers study the waves at Piha, on the western coast of North Island, New Zealand.

SECTION 1 CULTURE

# 1.1 From Ranch to City

**TECHTREK**
myNGconnect.com For a map and photos of ranches and cities in Australia

 Maps and Graphs
 Digital Library

**Main Idea** Most Australians live in coastal cities, and many have migrated from rural areas.

Today, Australia is one of the world's most **urban** countries—most people live in cities or nearby suburbs. Lately, even farmers and ranchers whose families have lived in rural areas for generations are choosing to relocate. Economic pressures in farming and ranching communities are pushing people toward cities.

## A Dry Climate Turns Drier

Much of Australia's climate is arid. About 60 percent of the continent receives fewer than 10 inches of rain per year. The dry climate of Australia's interior makes farming and ranching difficult. Farmers have mainly depended on irrigation, rather than rainfall, to water their crops. Sheep and cattle ranchers, whose livestock graze in dry areas, also depend on irrigation.

However, irrigation comes with problems of its own. Years of continuous irrigation have helped cause **salinization**, which occurs when salt accumulates in soil. Too much irrigation into normally dry land raises the natural water table and causes salts to rise to the surface. Crops cannot grow in this salty soil. Low rainfall and high evaporation rates make soil in the interior of Australia prone to salinization.

In 2002, the longest drought in Australia's history began. Increased demands **depleted**, or drained, major irrigation sources such as the Murray and Darling rivers in New South Wales. The drought caused severe water shortages and the amount of water allowed for irrigation was greatly reduced. Farmers could not water their crops and ranchers sold entire herds of sheep and cattle at auction.

**Critical Viewing** Ranchers in New South Wales herd sheep for auction. What can you infer about the conditions in which these ranchers work?

## The Pull to the Cities

In Australia's interior, few alternatives to farming and ranching exist. Faced with the combination of drought and poor soil, farmers and ranchers have had to make hard choices. Many are moving to the cities on the coasts.

Families are also drawn to urban areas because of more plentiful goods and services, including better schools for their children. Today, nearly 75 percent of the population lives in one of five coastal cities: **Sydney**, Melbourne, Brisbane, Perth, and Adelaide.

### Before You Move On
**Summarize** Why do most Australians live in or near cities?

### POPULATION DENSITY

**Population Density**
· One dot represents 25,000 people

### ONGOING ASSESSMENT
## MAP LAB  GeoJournal

1. **Location** Use the map to identify the locations of Australia's five largest cities. Which of those cities is most isolated?
2. **Interpret Maps** Locate South Australia and New South Wales on the map. In which state do more people live? What advantage might New South Wales have over South Australia, in terms of access to fresh water?

**Critical Viewing** Sydney is located on Australia's southeastern coast. Based on what you see in the photo, what might Sydney offer that might not be available in Australia's interior?

# SECTION 1 CULTURE

## 1.2 Immigration to Australia

TECHTREK
myNGconnect.com For photos of Australia's immigrant communities
Digital Library

**Main Idea** Australia includes people who have come from a wide variety of countries and cultures.

Since the end of World War II, the population of Australia has changed dramatically. A marked increase in **immigration**, or moving permanently from one country to another, has diversified the country.

### Postwar Immigration

After World War II, refugees from war-torn Europe sought new lands to call home. At the same time, the Australian government wanted to increase population and expand its available **labor force**, or the number of people available to work. The government loosened immigration restrictions, increasing the number of immigrants permitted entry. Both Australia and its new settlers benefited from the decision.

Between 1950 and 1960, Australia's labor force increased by 60 percent due to immigration. New arrivals found jobs in postwar construction projects and other industries. Immigrants from Poland, Germany, Hungary, Greece, Italy, and other European countries changed the ethnic makeup of Australian cities. Some immigrants did not speak English and introduced their native languages to Australia. Many immigrants opened businesses that sold food and goods from their native countries.

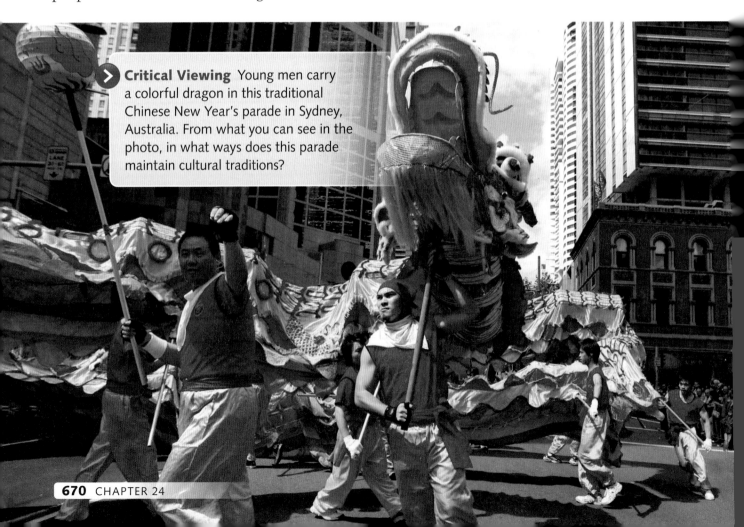

**Critical Viewing** Young men carry a colorful dragon in this traditional Chinese New Year's parade in Sydney, Australia. From what you can see in the photo, in what ways does this parade maintain cultural traditions?

## Australia's Diversity

In the 1970s, new groups began to immigrate to Australia. For example, many refugees from the war in Vietnam fled by boat to the continent. Over the next few decades, immigrants from other Asian countries, including Japan, China, and India, relocated to Australia in search of job opportunities. Currently, about 33 percent of Australia's population is of Asian descent.

Many immigrants from New Zealand, the Pacific islands, Canada, and the United States have also moved to Australia since the 1970s. Unlike the early years of settlement, fewer of those who resettle in Australia each year come from Great Britain. Today, nearly one-quarter of Australians were born in other countries. English remains the official language, but more than 100 languages are spoken, including Chinese, Greek, and Italian.

Diversity is especially reflected in Australia's cities. Nearly 40 percent of new immigrants now settle in Sydney. As Australia's largest city, Sydney is an attractive point of entry into the country. Melbourne, Australia's second largest city, boasts the largest Greek population in the world outside Greece. Brisbane and Perth receive large numbers of immigrants from Asia and Pacific island countries.

### Before You Move On
**Monitor Comprehension** In what ways has immigration to Australia changed over the years?

## ONGOING ASSESSMENT
### DATA LAB                                    GeoJournal

1. **Movement** What percentage of immigrants to Australia arrived from the Asian countries included on the graph? Which two Asian countries represent the largest percentage of immigrants to Australia and why?

2. **Make Generalizations** Look at the data above. In what way does Australia's colonial past continue to be reflected in the number of people who immigrate to the country?

3. **Explain** What world events in the 20th century prompted different waves of immigration to Australia? What were different groups of immigrants seeking?

# SECTION 1 CULTURE

## 1.3 Polynesian Cultures

**Main Idea** Polynesian people maintain their cultural traditions while migrating to urban areas.

Polynesia spans a vast area in the South Pacific known as the **Polynesian Triangle**. This triangle captures a distinct Polynesian culture that includes New Zealand, Easter Island, French Polynesia, Samoa, the Cook Islands, and hundreds of other islands. Many Polynesians try to **maintain**, or preserve and carry on, their traditional cultures.

### Traditional Polynesia

Historically, maintaining cultural practices was one function of the Polynesian family. Several generations of a single family lived together. Older members of the family taught younger members the arts of canoe-building, fishing, and navigation.

Polynesians have maintained traditional cultural practices and beliefs through the **oral tradition**, or passing stories by word of mouth from generation to generation. Polynesian stories explain how the universe was created and celebrate the connection between humans and the natural world.

Storytelling also involves the use of traditional Polynesian languages. In the 1970s, in response to fears that some languages were disappearing, many Polynesian schools established language programs that taught only traditional languages. Today, knowledge and use of the Maori, Samoan, Tongan, and Tahitian languages is more widespread.

In addition to storytelling, Polynesians have rich traditions in dance, music, and visual art forms such as woodcarving and basket-weaving. These art forms have provided ways for Polynesians to maintain and pass on their cultures. Dance, music, and visual arts are important elements of traditional Polynesian festivals, such as the Festival of Pacific Arts, held every four years in a different Polynesian country.

### Polynesia Today

Like Australia, Polynesia has become increasingly urban since the 1960s. A lack of economic opportunities has pushed some Polynesians to migrate from rural villages to urban areas. Cities that have grown because of rural migration include Apia in Samoa, Pago Pago in American Samoa, and Nuku'alofa in Tonga.

Many Polynesians have **emigrated**, or left their home country, to live in other countries, including New Zealand and the United States. Between 1970 and 2000, emigration from Samoa to New Zealand doubled. Samoans emigrated to seek jobs and educational opportunities.

The **Cook Islands** provide examples of both rural to urban migration and emigration from Polynesia. Because of economic pressures, Cook Islanders have migrated from the countryside to the cities. Today, nearly 75 percent of Cook Islanders live in cities. Many others have emigrated to New Zealand to seek better economic futures.

### Before You Move On
**Summarize** In what ways have Polynesian people maintained cultural traditions?

## ONGOING ASSESSMENT
## PHOTO LAB

1. **Analyze Visuals** How would you describe the vegetation in the photo? How might you describe the climate?
2. **Make Inferences** Inferring from the photo, what physical features of this island do you notice? Why might those features make this work difficult?
3. **Explain** What are different ways in which Polynesians migrate? What are different reasons for migration? Use a chart like the one below to organize your ideas.

| POLYNESIAN CULTURES ||
|---|---|
| Form of Migration | Reasons Why |
|  |  |
|  |  |

Men on Olosega Island in American Samoa harvest leaves of the panandus plant, which will be dried and woven into mats.

# SECTION 1 CULTURE

## 1.4 Human Footprint in Antarctica

**TECHTREK**
myNGconnect.com For an online map and photos of Antarctica

Maps and Graphs | Digital Library

**Main Idea** People all over the world have been interested in Antarctica for hundreds of years.

Antarctica lies on the South Pole and has one of the most extreme climates on Earth. No humans live there permanently, but people have left their "footprint" through exploration and research.

### Early Explorations

As early as 1773, Captain James Cook, a British naval captain, began looking for a southern continent. Ice blocked his southward journey. In 1820, British and Russian explorers were the first to report sighting land. American **Charles Wilkes** surveyed a 1,500-mile stretch of coastline in the late 1830s. His observations confirmed that Antarctica is a continent. Later explorers investigated Antarctica's thick sheets of ice and the ranges of the Transantarctic Mountains. They also explored the continent's coastal **ice shelves**, or floating sheets of ice that are attached to a landmass.

Twentieth century explorers made further discoveries. One dramatic event involved the race to the South Pole. In 1911, Robert Falcon Scott, from Great Britain, and Roald Amundsen, from Norway, each led expeditions to the South Pole. Amundsen's team reached the South Pole first. Though Scott's expedition

> **Critical Viewing** Researchers return to their station at the South Pole in Antarctica. Based on what you can see here, what preparations do people make in order to work in this extreme environment?

also reached the South Pole, none of its members survived the months-long journey back to base camp.

From 1929 through the 1940s, National Geographic supported American naval officer **Richard Byrd** as he explored the continent, both on land and by air. He studied its ice, rocks, and minerals and made observations about Earth's magnetism. Byrd's team also photographed Antarctica's coastline and discovered 26 new islands.

## An International Treaty

By 1958, scientists from several different countries had built more than 50 **scientific stations**, or places to carry out research. Seven of these countries claimed sections of Antarctica as their own. However, some countries, including the United States, did not recognize the claims as legal. Then in 1959, 12 countries signed the **Antarctic Treaty**. According to the treaty, the continent can be used only for peaceful purposes. Any scientific discoveries must be shared internationally.

Today, Antarctica continues to serve as a center of research and global cooperation. One study, called the International Polar Year 2007–2008,

involved scientists from more than 60 countries. Working together on more than 200 research projects, scientists collected data on thinning ice sheets in order to determine their effect on global sea levels. Studying changes in Antarctica helps scientists understand changes in other parts of the world.

### Before You Move On
**Make Inferences** In what ways have people expressed interest in Antarctica?

### ONGOING ASSESSMENT
## MAP LAB
*GeoJournal*

1. **Location** According to the map, which oceans surround Antarctica? From what central point do international claims originate?
2. **Interpret Maps** Based on the map, what country claims the largest area of Antarctica? What might explain the sizes of the South American and Australian claims?
3. **Explain** Why can no single country claim Antarctica as its own?

# SECTION 2 GOVERNMENT & ECONOMICS

## 2.1 New Trade Patterns

**TECHTREK**
myNGconnect.com For a map and photos of exported goods from the region

Maps and Graphs | Digital Library

**Main Idea** Trade is a source of economic growth for Australia, New Zealand, and the Pacific Realm.

For many years, geographic isolation and other factors made it difficult for Australia, New Zealand, and Pacific island countries to export goods to countries in other parts of the world. As the global economy expands, countries in this region are developing new trade partners.

### Trade Within the Region

Australia gained independence from Great Britain in the early 1900s. It continued to rely on Great Britain as its sole trading partner for several decades afterward. By the late 1900s, however, Australia began to increase trade with other countries.

As the largest and most populous country in this region, Australia serves as an economic and trade anchor, or hub, for surrounding countries. Australia, New Zealand, Papua New Guinea, and the other countries in the Pacific Realm have built trade partnerships with one another.

These trade partnerships benefit the entire region economically. For example, **Papua New Guinea** exports gold and coffee to Australia. **Samoa** exports cocoa beans to New Zealand. Regional economic development has strengthened ties among countries and has given the region a stronger global presence.

### New Global Partnerships

This region's increased participation in the world economy is due in part to an organization called the **Asia-Pacific Economic Cooperation (APEC)**. Founded in 1989, APEC's goal is to strengthen economic ties and trade **alliances**, or partnerships, among its members. Three member countries—Australia, New Zealand, and Papua New Guinea—are located in this region. Other APEC member countries border the Pacific Ocean, including China, Japan, Mexico, the United States, Chile, and Peru. APEC alliances help these countries compete in the global economy.

> **Critical Viewing** Bales of wool in this New Zealand warehouse await export. From the photo, what can you infer this man is doing and why?

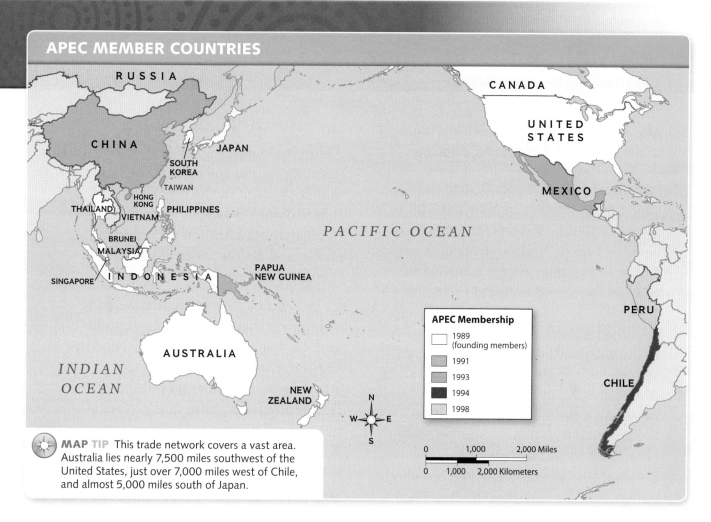

**MAP TIP** This trade network covers a vast area. Australia lies nearly 7,500 miles southwest of the United States, just over 7,000 miles west of Chile, and almost 5,000 miles south of Japan.

One of APEC's main achievements has been to establish free trade agreements among member countries. **Free trade agreements** are treaties between two or more countries that encourage trade by limiting tariffs, or taxes, on that trade. Since 1989, APEC has helped form more than 30 free trade agreements.

### Before You Move On
**Monitor Comprehension** In what ways have trade alliances helped this region become competitive in the global economy?

### ONGOING ASSESSMENT
## MAP LAB  GeoJournal

1. **Interpret Maps** Of the founding APEC member countries represented on the map, which are located in North America? Which APEC countries are located in Asia?

2. **Make Inferences** According to the map, when did Australia, New Zealand, and Papua New Guinea join APEC? Based on your reading, in what ways might APEC membership benefit their economies?

3. **Evaluate** In what ways might limiting tariffs on trade between countries help strengthen trade alliances?

SECTION 2.1 **677**

## SECTION 2 GOVERNMENT & ECONOMICS
# 2.2 Rights for Indigenous People

TECHTREK
myNGconnect.com For photos of Maori people and art
Digital Library

**Main Idea** Indigenous people in the region are working toward social and political equality.

As you have read, direct British rule ended in Australia and New Zealand in the early 1900s. However, the effects of colonial rule on indigenous people remain. Today, indigenous groups continue to address social and political problems.

### Aborigines in Australia

In the 18th and 19th centuries, British settlers took lands away from Australia's Aborigines. Many Aborigines died in violent conflicts with Europeans. Others were forced to live on **reserves**, or land set aside for native people. Colonial British rulers enacted laws and policies that promoted, and often forced, assimilation. **Assimilation** is a process in which a minority group is pressured to give up its cultural practices and be absorbed into another society's culture.

One way in which the British tried to force Aborigines to assimilate was by separating Aboriginal children from their parents. Children were placed in mission schools and homes. This practice took place from the early 1900s until the 1960s. An estimated 100,000 children were separated from their families, most of them permanently. They are referred to as the "Stolen Generation."

In the 1960s, the Aborigines' struggle to gain basic civil rights intensified. By 1962, Aborigines were able to vote in national elections for the first time. Five years later, they were finally included as citizens in the Australian census. The Aboriginal Land Rights Act passed in 1976 gave Aborigines the right to **reclaim**, or take back, land in the Northern Territory that had once belonged to their ancestors. In 2008, the Australian government apologized for the years of unjust treatment toward the Aborigines.

Today, Aborigines are still working to improve their lives. Unemployment and illiteracy rates are high. Especially in rural communities, access to adequate health care and education is unpredictable. After many years of hardship, Aborigines continue to press for social and political equality.

**Critical Viewing** Lowitja O'Donoghue, a member of the Stolen Generation, and Australia's Prime Minister Kevin Rudd meet in 2008. What does the photo lead you to think about indigenous peoples' reaction to the government's apology?

**678** CHAPTER 24

## Maori in New Zealand

The Maori (MOW ree) are the indigenous people of New Zealand. They arrived from Polynesia in the 1300s and were New Zealand's only inhabitants until the British arrived in the late 1700s. Like the Aborigines in Australia, the Maori struggled with the British for control of their land. Although the 1840 Treaty of Waitangi (WY tahng ee) granted the Maori legal and land rights, it was largely ignored for more than 100 years. The Maori people have organized politically, and, like Australia's Aborigines, have had some success achieving civil rights and reclaiming lands.

### Before You Move On
**Make Inferences** In what ways are indigenous people in the region regaining social and political equality?

## ONGOING ASSESSMENT
### READING LAB  GeoJournal

1. **Summarize** What is assimilation, and how did the British try to force this process on Aborigines in Australia?
2. **Make Inferences** In 1967, the Australian government counted Aborigines in the national census for the first time. What can you infer about how this inclusion might help Aborigines gain social and political equality?
3. **Compare** In what ways were the experiences of the Maori and the Aborigines similar?

▼ **Critical Viewing** Maori protesters gather in New Zealand to demand that the Treaty of Waitangi be followed. What can you infer about the strategy these protesters are using?

# SECTION 2 GOVERNMENT & ECONOMICS

## 2.3 Adventure Tourism

TECHTREK
myNGconnect.com For photos of adventure tourism
Digital Library

**Main Idea** Tourism that involves physical activity and adventure is a major industry in the region.

Australia, the Pacific Realm, and Antarctica are frequent destinations for adventure tourism. **Adventure tourism** takes travelers to remote areas and involves active enjoyment of the physical environment.

### Adventure Activities

New Zealand and Australia are popular destinations for adventure tourists. In New Zealand, rock climbing, mountain biking, and cave exploration are a few of the ways people enjoy the rugged terrain. Adventurers explore New Zealand's interior rivers and lakes by rafting, kayaking, and canoeing. Whale-watching expeditions provide opportunities to see the many different species that inhabit the waters off New Zealand's coast.

People seeking active enjoyment of unusual physical environments also travel to Australia—particularly to its Great Barrier Reef, which draws millions of tourists every year. In 2009, more than two million people visited this natural wonder. Activities on the Great Barrier Reef include snorkeling, scuba diving, fishing, and exploring reef waters in glass-bottomed boats.

In Antarctica, hikers trek across **glaciers**, which are large masses of ice and packed snow. Expert guides help hikers cross dramatic **crevasses**, or deep, open cracks in glaciers. Visitors to Antarctica can also observe emperor penguins, leopard seals, and other Antarctic birds and animals in their natural habitat.

### Economic Benefits

In recent years, adventure tourism has become a **lucrative**, or profitable, industry. People work as expert adventure guides and as drivers on tours. Others find jobs in hotels, stores, airports, and restaurants or on cruise ships that regularly carry thousands of people to the region.

In the Pacific islands, even those not directly involved with adventure tourism often benefit from it. Some local residents make a living creating and selling traditional baskets, mats, and masks. Other islanders earn income by entertaining guests with traditional feasts. Some critics worry that tourism as an industry might devalue, or trivialize, islanders' cultures and ways of life. However, those who support the industry focus on the opportunity to bring more income into their economies.

**Before You Move On**
Monitor Comprehension In what ways is this region an ideal destination for adventure tourism?

ONGOING ASSESSMENT
## PHOTO LAB

GeoJournal

1. **Human-Environment Interaction** Review each photo. In what ways do adventure tourists interact with the environment in this region?
2. **Make Generalizations** From what you can infer from the photos, what kinds of personal traits might adventure tourists share?

680 CHAPTER 24

**Visual Vocabulary** A **crevasse** is a deep, open crack in a glacier. With the help of a guide, this group is crossing a crevasse on Franz Josef Glacier, South Island, New Zealand.

**Critical Viewing** Paramotoring, a form of motorized flying, is another popular activity on South Island. What kinds of scenery might this person be able to enjoy while flying?

Scuba divers use an underwater platform to work on diving skills in the waters of Australia's Great Barrier Reef.

# SECTION 2 GOVERNMENT & ECONOMICS

## 2.4 New Zealand Today

**TECHTREK**
myNGconnect.com For a map and photos of New Zealand today

 Maps and Graphs   Digital Library

**Main Idea** New Zealand protects its culture and environment while developing new uses for natural resources.

For some countries, the desire to improve their economies sometimes means choosing between commercial uses of the land and protecting their cultures and natural environments. New Zealand has been able to balance these concerns.

### Preserving Maori Culture

As you have read, the Maori were New Zealand's first inhabitants. During the British colonial period, many Maori groups lost ancestral lands. Reclaiming lands that once belonged to their ancestors continues to be a key issue for the Maori. In addition to protecting land for future generations, the Maori people are embracing traditional customs in order to preserve their culture. They have renewed a focus on traditional arts and ceremonies as well as their native language. Maori-speaking elders and parents worked with the New Zealand government to establish language immersion schools for Maori children. These schools use the Maori language exclusively throughout the school day. In 1987, the government made the Maori language one of New Zealand's official languages.

### Energy for the Future

New Zealand has also made a special effort to preserve its natural heritage. The government has established many national parks and wildlife preserves. In fact, nearly 30 percent of the land is protected by the government.

Another way New Zealand protects its environment is by producing energy from solar, water, and wind sources. This kind of energy is not depleted by its use and is called **renewable energy**. One-third of energy consumed in New Zealand comes from renewable sources and that percentage is expected to increase.

One source of renewable energy in New Zealand is geothermal energy. **Geothermal energy** is heat energy from within the earth that can be turned into electricity. Geothermal reservoirs of hot water and steam are common in New Zealand

**Critical Viewing** This sculpture in Auckland City is a modern representation of a traditional Maori entry gate. Based on the photo, in what ways does New Zealand balance tradition with modernity?

because the country lies in an active volcanic zone. Geothermal energy is low-cost and non-polluting and provides at least 13 percent of New Zealand's electricity needs.

New Zealand is considered an environment-friendly country because of its foward-thinking development of renewable energy sources. Through the establishment of wind farms, geothermal plants, and hydroelectric power stations, New Zealand is forging a new path for sustainable energy production.

### Before You Move On
**Summarize** In what ways is New Zealand preserving its culture and its environment?

### ONGOING ASSESSMENT
## MAP LAB
*GeoJournal*

1. **Interpret Maps** Geothermal energy zones are categorized by temperature. In what zone is the Taupo Geothermal Region? Based on the map, how might geothermal sites on South Island be described?

2. **Make Predictions** Most people live on New Zealand's North Island. What cities might benefit most from future development of geothermal energy plants?

**Critical Viewing** This geothermal pool is located near Rotorua. What can you infer from the photo about the temperature of the water?

SECTION 2.4 **683**

# CHAPTER 24
# Review

## VOCABULARY

On your paper, write the vocabulary word that completes each of the following sentences.

1. One reason farmers move to urban areas is because _____ is ruining the soil.
2. Polynesians pass along stories and history through their _____.
3. APEC member countries have _____ with other member countries.
4. In Antarctica, many different countries have set up _____ to conduct research.
5. _____ provides 13 percent of New Zealand's energy needs.

## MAIN IDEAS

6. What factors have caused many Australian farmers and ranchers to move to urban areas? (Section 1.1)
7. In what ways have populations of immigrants to Australia changed over the years? (Section 1.2)
8. What cultural traditions are Polynesians preserving? (Section 1.3)
9. What does the Antarctic Treaty state about the use of Antarctica? (Section 1.4)
10. In what way have Australia and the Pacific islands strengthened their presence in the global economy? (Section 2.1)
11. What are many indigenous people trying to reclaim? (Section 2.2)
12. In what ways does adventure tourism highlight the region's diverse physical features? (Sections 2.3)
13. Name three things the New Zealand government has done to protect traditional culture and preserve the environment. (Section 2.4)

## CULTURE

### ANALYZE THE ESSENTIAL QUESTION

How are Australia, the Pacific Realm, and Antarctica becoming more connected to the rest of the world?

### Critical Thinking: Make Inferences

14. In what ways might Australia's long drought impact trade with other countries?
15. What characteristics attract immigrants from around the world to Australia?
16. How might rural life change as a result of current migration patterns?

### INTERPRET MAPS

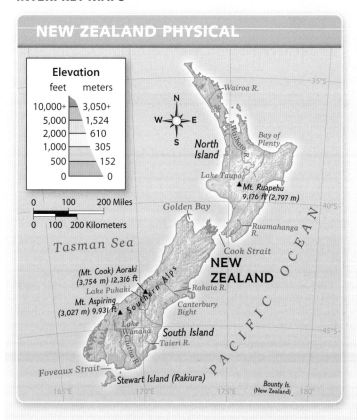

17. **Place** What two main islands make up New Zealand? Which island has higher elevations?
18. **Region** Unlike Australia, New Zealand does not have extensive deserts. Instead, what physical feature can you infer from the map legend covers much of New Zealand?

## GOVERNMENT & ECONOMICS

### ANALYZE THE ESSENTIAL QUESTION
What new economic patterns are emerging in the region?

### Critical Thinking: Draw Conclusions
19. How might new trade partnerships among different countries benefit the entire region?
20. What energy sources in New Zealand might represent the energy of the future?
21. In what ways might adventure tourism benefit countries in the region economically?

### INTERPRET GRAPHS

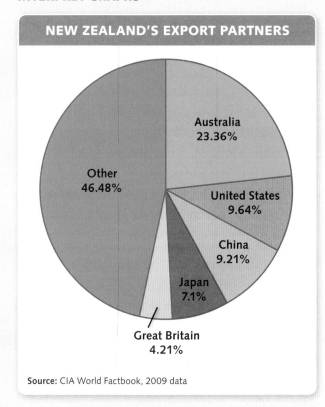

**NEW ZEALAND'S EXPORT PARTNERS**
- Australia 23.36%
- Other 46.48%
- United States 9.64%
- China 9.21%
- Japan 7.1%
- Great Britain 4.21%

Source: CIA World Factbook, 2009 data

22. **Analyze Data** What percentage of New Zealand's exports go to Australia? How does that number compare with Japan?
23. **Identify** Of the export partners shown on the pie chart, which are APEC members? Why might Australia represent nearly one-quarter of New Zealand's exports?

## ACTIVE OPTIONS
**Synthesize** the Essential Questions by completing the activities below.

24. **Write a Travelogue** Imagine that you are a magazine writer trying to persuade your readers to visit one of the places you learned about in this chapter. Write two paragraphs that describe the features of the place and why it is a good choice for a vacation. **Share your paragraphs with a partner or a friend.**

    **Writing Tips**
    - Describe how your readers might travel to the location, and how long the trip might take.
    - Include specific details about important places travelers should make a point to visit.
    - Highlight at least one lesser-known attraction.

    Go to **Student Resources** for Guided Writing support.

25. **Create Charts** Make a chart like the one below to compare information about three Pacific islands. Use online sources to gather the data. Then, with a partner, create three questions to ask other students.

|  | PAPUA NEW GUINEA | SAMOA | FRENCH POLYNESIA |
|---|---|---|---|
| Population |  |  |  |
| Urban Population |  |  |  |
| Square miles of land |  |  |  |
| Exports |  |  |  |
| Imports |  |  |  |

COMPARE ACROSS REGIONS UNIT 12

# Pastures and Crops

myNGconnect.com For a map of grassland use around the world

Maps and Graphs

Australia's grasslands have been used for many years to support both ranching and agriculture. Countries with large, open grasslands use them for similar purposes, including grazing livestock and growing a variety of crops.

## Compare
- Argentina
- Australia
- China
- India
- New Zealand
- Ukraine
- United States

Decisions about land use are influenced by many factors, such as climate, crop yield, and sustainability. In most countries, mixed use of grasslands is more common than devoting land solely to ranching or agriculture. Comparing two countries, Australia and Argentina, reveals similarities in how these two countries use grasslands.

### GRAZING

Grasslands provide grazing, or feeding, grounds for ranch livestock. Cattle, horses, and sheep are the most common ranch livestock. Other livestock on ranches around the world include goats, alpacas, emus, ostriches, bison, deer, and elk.

Ranching industries have been in place in Australia and Argentina since the 1880s. Cattle were not native to either continent. European settlers imported cattle during the colonial periods.

Australia and Argentina are top beef producers today. Australia's beef industry is export-oriented, while much of Argentina's beef is consumed domestically. Each country faces challenges with its beef industry. Drought on pasture lands in Australia prompted ranchers to reduce herds. In Argentina, beef producers struggle with government policies that limit how much beef can be exported.

### GROWING

Though their climates differ, Australia and Argentina are both top exporters of wheat. Most of Australia's wheat production is located in the southern and southwestern parts of the country. Argentina's fertile Pampas is a productive center of wheat cultivation.

As you can see on the map, both countries are located in the Southern Hemisphere. Their growing season occurs opposite of Northern Hemisphere countries. This enables them to make wheat sales on the global market during the Northern Hemisphere's off-season.

Around the world, demand for food production is increasing as the population climbs. In order to increase grain production, both Australia and Argentina are beginning to use marginal, or less desirable, lands to plant more wheat and other crops.

# Grassland Use Around the World
Major Crops and Livestock of Selected Countries

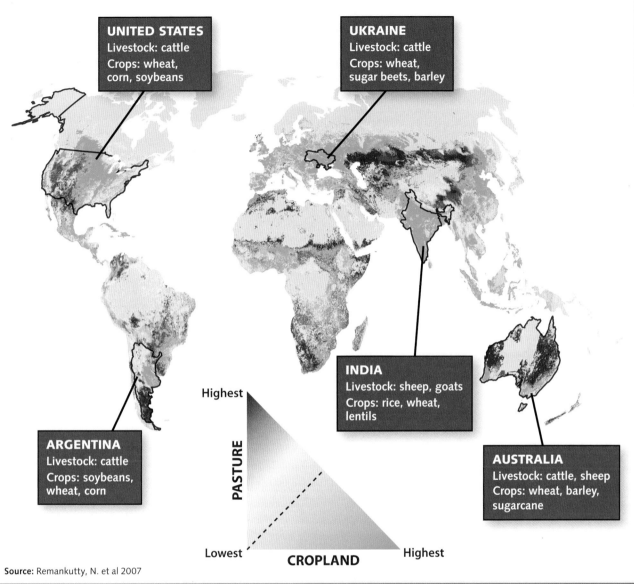

**UNITED STATES**
Livestock: cattle
Crops: wheat, corn, soybeans

**UKRAINE**
Livestock: cattle
Crops: wheat, sugar beets, barley

**ARGENTINA**
Livestock: cattle
Crops: soybeans, wheat, corn

**INDIA**
Livestock: sheep, goats
Crops: rice, wheat, lentils

**AUSTRALIA**
Livestock: cattle, sheep
Crops: wheat, barley, sugarcane

Source: Remankutty, N. et al 2007

## ONGOING ASSESSMENT
## RESEARCH LAB  GeoJournal

1. **Identify** Look at the map. What crop do all five countries shown above grow?
2. **Compare and Contrast** Find India and Australia on the map. Which country devotes more land use to crops, and which country devotes more to grazing? What climate factors might account for these choices?
3. **Make Inferences** Which country do you think produces more beef, Australia or Ukraine? Why do you think so?

**Research and Write a Report** Consider the similarities of Australia and Argentina in their use of grasslands. Research other countries including New Zealand and China to compare their use of grasslands. Write a report to present your findings. Consider these topics to guide your research:
- climate
- location of grasslands
- typical livestock
- crops grown for export

COMPARE ACROSS REGIONS

REVIEW & EXTEND **UNIT 12**

# Active Options

**TECHTREK**

myNGconnect.com For photos of animal life in the region

Digital Library

### ACTIVITY 1

**Goal:** Extend your knowledge about Antarctica.

## Prepare a Guidebook

You are leading a group going on a seven-day summer cruise to Antarctica. Do research and use the **Magazine Maker CD-ROM** to prepare a guidebook for your group that gives some facts about the continent and describes what visitors might expect to see on the cruise. Include examples of marine life and sea birds that inhabit the waters surrounding Antarctica.

leopard seal

### ACTIVITY 2

**Goal:** Research indigenous animal species in Australia.

## Write a Report

The Tasmanian devil and the Australian dingo are animals found only in Australia. Research one of these animals online. Write a report that details its habitat, what it eats, and what threats it might face. Include photos of your animal from the **Digital Library** with your written report.

### ACTIVITY 3

**Goal:** Learn more about constellations in the Southern Hemisphere.

## Identify the Stars

National Geographic Fellow Elizabeth Kapu'uwailani Lindsey studies the traditions of Pacific island seafarers who relied on their knowledge of the stars, waves, and birds to guide them. Many of the stars they observed in the Southern Hemisphere are not visible in the night skies in North America. Conduct research and identify five constellations of stars visible only in the Southern Hemisphere. Explain what shape each constellation forms and how that shape relates to its name.

# REFERENCE HANDBOOK

Skills Handbook ..................................................................... R2

Economics and Government Handbook ............................... R41

World Religions Handbook ................................................... R51

English Glossary ................................................................... R59

Spanish Glossary .................................................................. R78

Index ..................................................................................... R98

# Skills Handbook

| | |
|---|---|
| Find Main Idea and Details | R3 |
| Take Notes and Outline | R4 |
| Summarize | R5 |
| Sequence Events | R6 |
| Categorize | R7 |
| Describe Geographic Information | R8 |
| Express Ideas Through Speech | R9 |
| Write Outlines for Reports and Comparisons | R10 |
| Write Journal Entries | R11 |
| Pose and Answer Questions | R12 |
| Make Predictions | R13 |
| Compare and Contrast | R14 |
| Analyze Cause and Effect | R15 |
| Make Inferences | R16 |
| Draw Conclusions | R17 |
| Make Generalizations | R18 |
| Form and Support Opinions | R19 |
| Identify Problems and Solutions | R20 |
| Analyze Data | R21 |
| Distinguish Fact and Opinion | R22 |
| Evaluate | R23 |
| Synthesize | R24 |
| Analyze Primary and Secondary Sources | R25 |
| Analyze Visuals | R26 |
| Interpret Physical Maps | R27 |
| Interpret Political Maps | R28 |
| Create Sketch Maps | R29 |
| Create Charts and Graphs | R30 |
| Interpret Charts | R31 |
| Interpret Graphs | R32 |
| Create and Interpret Time Lines | R33 |
| Build Models | R34 |
| Create Databases | R36 |
| Create Graphic Organizers | R37 |
| Conduct Internet Research | R38 |
| Evaluate Internet Sources | R39 |
| Create Multimedia Presentations | R40 |

# Find Main Idea and Details

Every book, paragraph, or passage has a **main idea**, which is the sentence or sentences that state the subject of the text. A main idea can sometimes be unstated, or *implied*. In that case, the **details** in the text provide clues about the main idea. To find a main idea and details, follow the steps at right.

**Step 1** Look for a stated main idea in the first and last sentences of the paragraph. If no main idea is clear, look for details that offer clues about what the implied main idea is.

**Step 2** Next, find details that support and clarify the main idea. If the main idea is in the first sentence, supporting details follow it. If the main idea is in the last sentence, the details come before it.

## GUIDED MODEL

### Habitat Loss and Restoration

**Ⓐ** The loss of habitats can destroy an entire ecosystem. An ecosystem is a community of plants and animals and their habitat. **Ⓑ** <u>Earth has many different ecosystems that interact with each other.</u> **Ⓑ** <u>The destruction of one affects all the other ecosystems.</u> **Ⓑ** <u>For example, many scientists believe the destruction of rain forest habitats has led to global climate change.</u>

**TIP** When an author doesn't directly state the main idea of a passage, it's up to you to figure out the **implied main idea**. To do this, ask yourself, "What do the details of the passage have in common?" Find the connection between the details, put it into your own words, and you have the implied main idea.

**Step 1** Look for a stated main idea.

Read the first and last sentences in the paragraph. Do they state the main idea? If the main idea is not obvious, look for details that offer clues about what the main idea is. Some details explain the main idea. Other details give examples of the main idea.

**MAIN IDEA Ⓐ** The loss of habitats can destroy an entire ecosystem. In this example, the stated main idea is the first sentence of the paragraph.

**Step 2** Once you have figured out the main idea, find details that support it.

If the main idea is in the first sentence, it is usually followed by details that support and explain it. If the main idea is in the last sentence, the supporting details come before it. In the example at left, the details follow the main idea.

**DETAIL Ⓑ** Earth has ecosystems that interact with each other.

**DETAIL Ⓑ** The destruction of one affects all the other ecosystems.

**DETAIL Ⓑ** Many scientists believe the destruction of rain forest habitats has led to global climate change.

## APPLY THE SKILL

Turn to *Europe Geography & History*, Section 2.2, "Classical Greece." Read the "Golden Age of Greece" passage. Identify the main idea and details. Record them in a web like the one at right.

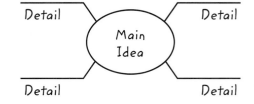

# Take Notes and Outline

**Taking notes** while you read helps you understand and remember important information, including facts, ideas, and details. Many writers organize their notes in an **outline**. To take notes, follow the steps shown at right.

**Step 1** Read the title to learn the topic of the passage.

**Step 2** Write down the main ideas and important details of the passage.

**Step 3** Summarize the main ideas and details in your own words.

**Step 4** Search for key words and write down the words and their definitions.

**Step 5** To save time and space, use abbreviations.

### GUIDED MODEL

**Ⓐ Economic Indicators**

**Ⓑ** A country's economic strength can be measured by several indicators. One is **Ⓒ gross domestic product** (GDP), or the total value of the goods and services that a country produces. **Ⓒ** Other indicators include **GDP** per capita (or per person), income, basic literacy rate, and life expectancy.

**Ⓑ** Economies are also placed in one of three categories. Countries with high GDPs, such as the United States, are **Ⓒ more developed countries**. Countries with lower GDPs are **Ⓒ less developed countries**.

**TIP** Outlines list the main ideas and supporting details in the order in which they best make sense. You can use time order, order of importance, or the order in which they appear in the text. Once the outline is complete, you can use it as a guide for your writing.

**Step 1** Read the title to learn the topic of the passage.
**TITLE Ⓐ** Economic Indicators

**Step 2** Write down the main ideas and important details of the passage.
**MAIN IDEAS Ⓑ** A country's economic strength can be measured by several indicators. Economies are also placed in one of three categories.

**Step 3** Summarize the main ideas and details in your own words.
*You can measure a country's economic strength using indicators, such as GDP, GDP per capita, income, basic literacy rate, and life expectancy. Countries can be placed in categories, such as as more developed, less developed, and newly industrialized.*

**Step 4** Search for key words and write down the words and their definitions.
**SAMPLE KEY WORD: Ⓒ Gross domestic product** is the total value of the goods and services a country produces.

**Step 5** To save time and space, use abbreviations. Here, for example, use "N" for *north* and **GDP** for *gross domestic product*.

### APPLY THE SKILL

Turn to *Europe Today*, Section 1.1, "Languages and Cultures." Read the "European Languages" passage. Take notes about what you read. Then organize your notes using an outline like the example shown at right.

I. Main Language Groups in Europe
  A. Romance
    1. French
    2. Spanish
    3.
  B. Germanic

# Summarize

When you **summarize**, you restate text in your own words and shorten it. A summary includes only the most important information and details. You can summarize a paragraph, a chapter, or a whole book. To summarize, follow the steps shown at right.

**Step 1** Read the text looking for the most important information. Watch for topic sentences that provide the main ideas.

**Step 2** Restate each main idea in your own words.

**Step 3** Write your summary of the text using your own words and including only the most important information.

## GUIDED MODEL

### Foods of Eastern Europe

**A** Eastern Europe's cold climate causes a shorter growing season than that in Western Europe. In Russia, root vegetables such as turnips and beets are well adapted to the country's climate. A soup called borscht made from beets is a traditional dish on winter nights.

**B** The fertile soil of Hungary allows Hungarian farmers to grow grains and potatoes. These crops are used to make a variety of breads. A meat stew called goulash is Hungary's national dish. It is made with beef, potatoes, and vegetables and seasoned with a spice called paprika.

**TIP** Try using a chart to record and organize the main idea and details you want to summarize. You can then use your notes to write your summary.

**Step 1** Read the text looking for the most important information. Watch for topic sentences that provide the main ideas.

**TOPIC SENTENCE A** Eastern Europe's cold climate causes a shorter growing season than that in Western Europe.

**TOPIC SENTENCE B** The fertile soil of Hungary allows Hungarian farmers to grow grains and potatoes.

**Step 2** Restate each main idea in your own words.

**RESTATED A** The climate of Eastern Europe is cold, so it has a shorter growing season than Western Europe.

**RESTATED B** Hungarian farmers grow grains and potatoes.

**Step 3** Write your summary of the text using your own words and including only the most important information.

**SUMMARY:** Eastern Europe has a cold climate, so it has a shorter growing season than Western Europe. Certain crops grow well there, such as potatoes and grains in Hungary.

### APPLY THE SKILL

Turn to *Europe Today*, Section 1.1, "Languages and Cultures." Read the "Cultural Traditions" passage. Identify the main ideas and important information, and restate them in your own words. Then write a summary of the passage using the topic sentence shown here.

**Cultural Traditions**

Europe's cultural traditions reflect the region's ethnic diversity. _____
_____
_____

## Sequence Events

When you **sequence events**, you put them in order based on when they occurred in time. Thinking about events in time order helps you understand how they relate to each other. To sequence events, follow the steps shown at right.

**Step 1** Look for time clue words and phrases such as names of months and days, or words such as *before, after, finally, a year later,* or *lasted* that help you sequence events.

**Step 2** Look for dates in the text and match them to events.

### GUIDED MODEL

### Greek Culture Spreads

Greece's golden age ended around **B** <u>431 B.C.</u>, when war broke out between Athens and Sparta. The conflict, known as the Peloponnesian War, **A** <u>lasted 27 years</u>. The war weakened both Athens and Sparta.

Around **B** <u>340 B.C.</u>, King Philip II of Macedon took advantage of their weakness and conquered Greece. In **B** <u>334 B.C.</u>, Philip's son, Alexander the Great, began to extend the Macedonian Empire. Alexander's love of Greece led him to spread its culture throughout his empire. Alexander died in **B** <u>323 B.C.</u> at the age of 33.

**TIP** A **time line** is a visual tool that can be useful for sequencing events. Time lines often move from left to right, listing events from the earliest to the latest.

**Step 1** Look for time clue words and phrases.
TIME CLUES **A** lasted 27 years

**Step 2** Look for specific dates in the text.
Be sure to read the text carefully because the dates in a paragraph may not always be listed in time order. Always match an event with its date. A chart like the one below can be a useful way to sequence dates and events.

SAMPLE DATE **B** 431 B.C., when Greece's golden age ends and the Peloponnesian War begins

| DATE IN TEXT | EVENT |
|---|---|
| 431 B.C. | Greece's golden age ends; Peloponnesian War breaks out between Athens and Sparta. |
| 340 B.C. | King Philip of Macedon conquers Greece. |
| 334 B.C. | Alexander the Great starts extending the Macedonian Empire. |
| 323 B.C. | Alexander the Great dies. |

### APPLY THE SKILL

Turn to *Europe Geography & History*, Section 2.3, "The Republic of Rome." Read the "A Republic Forms" passage. Identify the dates and events that occurred on each date. Use a time line like the one below to sequence the events.

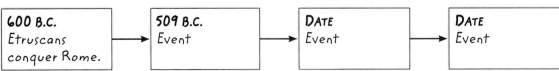

**R6** SKILLS HANDBOOK

# Categorize

When you **categorize**, you sort things into groups, or categories. Almost everything can be categorized, including objects, ideas, people, and information. Categorizing is important because it helps you recognize data patterns and trends. To categorize, follow the steps shown at right.

**Step 1** Read the title and text and ask yourself what the passage is about to determine how the information can be categorized.

**Step 2** Look for clue words to help you categorize information.

**Step 3** Decide what the categories will be.

**Step 4** Sort the information from the passage into the categories.

### GUIDED MODEL

**Ⓐ Political and Physical Maps**

Cartographers, or mapmakers, create Ⓑ different kinds of maps for many Ⓑ different purposes.

Ⓐ A political map shows features that humans have created, such as countries, states, and cities. These features are labeled, and lines show boundaries, such as country borders.

Ⓐ A physical map shows features of physical geography. It includes landforms, such as mountains, plains, deserts, and bodies of water. A physical map also shows elevation and relief. Elevation is the height of a physical feature above sea level. Relief is the change in elevation from one place to another.

**TIP** Every day, we put information into categories, such as healthy food versus junk food or fiction versus nonfiction. Knowing how information is categorized helps you understand how things are related and organize ideas.

**Step 1** Read the title and text and ask yourself what the passage is about to determine how the information can be categorized.
**THE PASSAGE IS ABOUT** Ⓐ the features of two types of maps.

**Step 2** Look for clue words to help you categorize information.
**CLUE WORDS** Ⓑ different kinds of maps, different purposes

**Step 3** Decide what the categories will be.
**THE CATEGORIES** political maps and physical maps

**Step 4** Sort the information into the categories.
Use a chart like the one below to help you organize the information.

| TYPE OF MAP | PURPOSES |
|---|---|
| Political | Shows man-made features, such as countries, states, cities, and boundaries |
| Physical | Shows features of physical geography, such as mountains, plains, deserts, and bodies of water; also shows elevation and relief |

### APPLY THE SKILL

Turn to *Europe Today,* Section 1.2, "Art and Music." Read the section carefully. Determine how categorizing can help you understand and organize the information about European art. Then create a chart like the one at right and categorize the information.

| EUROPEAN ART ||
|---|---|
| PERIOD | CHARACTERISTICS |
| Middle Ages | religious subjects; two-dimensional |
| Romantic Period | landscapes, nature, conveys emotion |

**SKILLS HANDBOOK**

# Describe Geographic Information

When you read text or study a chart or graph about a geographic subject, you take in information. One way to enhance your understanding of **geographic information** is to describe it. You can describe geographic information by following the steps shown at right.

**Step 1** Read the title of a passage, chart, or other visual to find out what geographic information it contains.

**Step 2** Read the passage or study the visual and identify its main ideas or topics.

**Step 3** Describe the information by summarizing it or responding to questions.

### GUIDED MODEL

**Ⓐ The Geography of Europe**

Ⓑ Europe forms the western peninsula of Eurasia, the landmass that includes Europe and Asia. In addition, Europe contains several peninsulas and significant islands.

Four land regions form Europe. The Western Uplands are hills, mountains, and plateaus that stretch from the Scandinavian Peninsula to Spain and Portugal. The Northern European Plain is made up of lowlands that reach across northern Europe. The Central Uplands are at the center of Europe. The Alpine region consists of the Alps and several other mountain ranges.

**TIP** You can use the steps above to describe the geographic information in a chart. Just imagine that the column headings are the main ideas and the information in the rows are the supporting details.

**Step 1** Read the title of a passage, chart, or other visual to find out what geographic information it contains.

Ⓐ *From its title, I see that this passage is about the geography of Europe.*

**Step 2** Read the passage or study the visual and identify its main ideas or topics.

**MAIN IDEAS** Ⓑ *Europe is a peninsula on Eurasia. It is formed by four land regions.*

**Step 3** Describe the information by summarizing it or responding to questions.

**SUMMARY** *Europe is a peninsula and contains several peninsulas and islands. It is formed by four land regions: the Western Uplands, the Northern European Plain, the Central Uplands, and the Alpine region.*

**QUESTIONS**

- *What landforms does Europe contain?*
- *What land regions make up Europe?*

### APPLY THE SKILL

Turn to *Europe Geography & History*, Section 1.3, "Mountain Chains." Use the information in this section to describe the location and natural resources of Europe's mountains. You might use questions like those shown at right to help you describe the information.

- What mountain chains are part of the Alpine region?
- What natural resources do Europe's mountain chains provide?

# Express Ideas Through Speech

When you **express ideas through speech**, you say what you are thinking using effective language. The purpose of your speech may be to narrate events, explain information, or convince your audience of your point of view. You can express ideas through speech by following the steps at right.

**Step 1** Determine the topic you will be presenting and the purpose of your speech. Do research to gather information using a wide variety of reliable sources.

**Step 2** Organize your ideas and create speaking notes and any visual aids you may need.

**Step 3** Practice speaking honestly, respectfully, clearly, and appropriately.

## GUIDED MODEL

> **A** <u>European Literature</u>
> **B** Ancient Greeks and Romans (focus on journeys, heroes, and adventures)
> • Iliad and Odyssey (Homer)
> • Aeneid (Virgil)
> Middle Ages (focus on religious beliefs and politics of the time)
> • Divine Comedy (Dante)
> Renaissance (focus on human experiences)
> • Hamlet (Shakespeare)
> • Don Quixote (Cervantes)
> —first modern novel

**TIP** When you express ideas through speech, it is important to hold your audience's attention. Vary your volume, speaking rate, and sentence structure to add interest to your presentation and emphasize meaning. Your body language is also important. Gestures, posture, facial expressions, and eye contact will engage your audience and enhance your speech.

**Step 1** Determine the topic and purpose of your speech and do research to gather the information you need.

**A** The topic of this student's speech is the history of European literature. The purpose is to explain the main periods of European literature.

**Step 2** Organize your ideas and create speaking notes and any visual aids you may need.

Notes and visual aids will help you stay on topic, remember your points, and emphasize your message.

**B** Well-organized notes will help you if you lose your place while speaking or forget a point you wanted to make. Keep your notes simple—just list headings and key points and important people and events.

**Step 3** Practice speaking honestly, respectfully, clearly, and appropriately.

Consider your audience and subject matter. Speak honestly and respectfully and use language that is appropriate to your audience.

## APPLY THE SKILL

Select a topic from the book, research it, and prepare speaking notes for a presentation about it like those shown at right. Then express your ideas through speech to an audience.

> <u>World War I</u>
> Began 1914 (explain causes)
> • nationalism
> • alliances
> • Archduke Ferdinand

# Write Outlines for Reports and Comparisons

**Writing an outline** is a useful way to synthesize, organize, and summarize information before you write a report or comparison. A good outline lists the most important ideas and supporting details in order, either chronologically or by level of importance. To write an outline, follow the steps shown at right.

**Step 1** Give your outline a title.

**Step 2** Determine and record the main ideas in logical order using Roman numerals.

**Step 3** Determine and record the ideas that support and explain the main ideas.

**Step 4** Determine and record specific details below the appropriate supporting ideas.

## GUIDED MODEL

**A** <u>Earth's Water</u>
**B** I. Fresh Water
   **C** A. Sources
      **D** 1. Rivers, water flowing downward
      2. Lakes, large inland bodies of water
      3. Streams, brooks, and creeks
   B. Uses
      1. Drinking and cooking
      2. Irrigating crops
II. Salt Water
   A. Oceans
      1. Large bodies of salt water
      2. Atlantic, Pacific, Indian, Arctic
      3. Ocean currents and their effects
   B. Seas
      1. Small bodies of salt water enclosed partly or completely by land
      2. Red, Caspian, Arabian

**Step 1** Give your outline a title.

**A** If you are writing an outline to organize information for a report or comparison, the outline title will probably become the title of your paper.

**Step 2** Determine and record the topics in logical order using Roman numerals.

**B** Each topic will probably make up at least one paragraph in your paper.

**Step 3** Determine and record the main ideas that support and explain each topic.

**C** Label these using indented capital letters below the topics. Not all topics need the same number of main ideas.

**Step 4** Determine and record specific details below the appropriate main ideas.

**D** Label these details using more deeply indented numerals below the main ideas. Not every main idea will have the same number of details.

**TIP** When you write an outline for a comparison, you can list all of the features of one subject under one topic and all of the features of the other subject under another topic. You can also use the features as topics and list facts about both subjects under each one.

## APPLY THE SKILL

Turn to *Europe Geography and History*, Section 3.1, "Exploration and Colonization." Then organize the information from the passage in an outline like the one shown at right. Use your outline to write a report.

<u>European Exploration</u>
I. Portuguese explorations
   A. Reasons
      1. Find gold
      2. Establish trade

# Write Journal Entries

When you **write journal entries** in response to something you have read, you record your reactions, make notes, ask and answer questions, and respond to new information in an informal written format. To use written journal entries to respond to a question or writing prompt, follow these steps.

**Step 1** Use what you have learned and read about the topic to answer the question.

**Step 2** Respond and react in a personal way to the question.

**Step 3** Record any related questions, comments, or notes you might have.

## GUIDED MODEL

QUESTION:
What challenges might immigrants moving to Europe face today?

JOURNAL ENTRY:

**A** Some immigrants may feel pressure to assimilate into European society and fear they will lose their identity and culture. Those belonging to non-Christian religions may feel discriminated against simply for having different religious beliefs.

**B** When my family moved, I hated that students at my new school had to wear a uniform. I refused to wear the jacket for the first week and got in trouble. I felt angry that I wasn't allowed to dress the way I wanted. I think I know a little bit about how the immigrants living in Europe feel.

**C** NOTE: Look up countries in Europe with large immigrant populations on the CIA World Factbook Web site.

**Step 1** Use what you have learned and read about the topic to answer the question.
  **A** The journal writer answers the question using facts from the book and his or her prior knowledge.

**Step 2** Respond and react in a personal way to the question.
  **B** The journal writer relates to the topic on a personal level and provides insight into the question based on personal experience.

**Step 3** Record any related questions, comments, or notes you might have.
  **C** The journal writer writes himself or herself a note to look up more information about the topic online.

**TIP** Journals can be used to record the steps in a research project (such as your observations during an experiment). You can also use a journal to write your thoughts and questions as you read a book or carry out a project.

## APPLY THE SKILL

Turn to *Europe Today*, Section 1.1, "Languages and Cultures." Then write a journal entry to respond to the following question: What is life like for Europeans who live in cities today? The beginning of a sample entry is shown at right.

Most Europeans today live in cities that reflect the cultures and ethnic origins of immigrant populations.

# Pose and Answer Questions

**Pose questions** about what you have read to check your understanding, keep track of details, and identify points you may have missed. To **answer questions**, review what you have read, search for answers in the text, and use your prior knowledge. To pose and answer questions, follow these steps.

**Step 1** Read the passage and identify information that is important or confusing.

**Step 2** Pose questions about key concepts or confusing information using question words, such as *who, what, when, where, why,* and *how.*

**Step 3** Answer the questions using the text and what you already know.

## GUIDED MODEL

### Establishing Colonies

**A** European exploration and colonization resulted in an exchange of goods and ideas known as the Columbian Exchange. From the Americas, Europeans obtained new foods, such as potatoes, corn, and tomatoes. Europeans introduced wheat and barley to the Americas. They also introduced diseases like smallpox. The diseases killed millions of native peoples.

**TIP** A two-column chart is a helpful way to record your questions and answers. Be sure to include the page number for each answer you found in the text in case you need to review or reread a passage or section.

**Step 1** Read the passage and determine which information is important or confusing.

**A** The most important topic seems to be the result of European exploration in the colonies. I'm confused about what the Columbian Exchange is.

**Step 2** Pose questions about key concepts or confusing information.

**My Questions**
1. **What** is the Columbian Exchange?
2. **Who** were killed by European diseases?
3. **Which** foods did Europeans obtain from the Americas? **Which** foods did Europeans introduce to the Americas?

**Step 3** Answer your questions using the text and what you already know.

Don't forget to examine charts, graphs, maps, and photographs as you answer.

**Answers**
1. The Columbian Exchange is the exchange of goods and ideas between Europeans and the areas they explored and colonized.
2. European diseases killed native peoples in the Americas.
3. Europeans obtained potatoes, corn, and tomatoes; the Americas obtained wheat and barley.

### APPLY THE SKILL

Turn to *Human & Physical Geography,* Section 1.1, "Earth's Rotation and Revolution." Read "Revolution and Rotation" and pose questions about information in the passage that confuses you. Record your questions in a chart like the one at right. Use the text and diagram to answer your questions.

| MY QUESTIONS | ANSWERS |
| --- | --- |
| What process creates the four seasons? | the rotation and tilt of Earth |

R12 SKILLS HANDBOOK

# Make Predictions

When you **make predictions,** you think about the events described in a passage or selection, use your prior knowledge, and guess or predict what will happen next. Making predictions as you read can help you understand and remember what you have read. To make predictions, follow these steps.

**Step 1** Preview the passage or selection to anticipate what it is about.

**Step 2** Use your personal knowledge. Ask yourself what you know about the topic.

**Step 3** As you read, make predictions about what will happen next.

**Step 4** Confirm or revise your predictions as you continue to read.

## GUIDED MODEL

### Ⓐ European Music

European music began in ancient Greece and Rome. Musicians played on Ⓒ simple instruments. During the Ⓑ Middle Ages, music was used in religious ceremonies. Also, singers called troubadours sang about knights and love. These songs influenced Ⓑ Renaissance music, when the Ⓒ violin was introduced.

The new instruments helped inspire the complex music of the Ⓑ Baroque period (1600–1750). Opera, which tells a story through words and music, was born then. The Ⓑ Classical and Romantic periods followed the Baroque and continued until about 1910.

**Step 1** Preview the passage or selection to anticipate what it is about.

*Ⓐ The title tells me that this passage is about European music. By skimming ahead, I see the names of many different types of music, Ⓑ musical periods, and Ⓒ words relating to instruments.*

**Step 2** Use your personal knowledge.

*I know that the Romantic period in art took place in the early 1800s. I have listened to Baroque music before.*

**Step 3** As you read, make predictions about what will happen next.

*I predict that this text will describe how music evolved in Europe.*

**Step 4** Confirm or revise your predictions as you continue to read.

*My prediction was correct. The passage describes how music evolved from the Middle Ages to the present.*

**TIP** Use a prediction chart to take notes as you read. A prediction chart allows you to record predictions, state whether they were correct, and explain why.

| PREDICTION | CORRECT? | EVIDENCE |
| --- | --- | --- |
| This passage will explain music I am familiar with. | Yes | It describes what opera is and when it developed. |

### APPLY THE SKILL

Turn to *Human & Physical Geography,* Section 1.2, "Earth's Complex Structure." Read the "Earth's Layers" text. Then use a prediction chart like the one shown at right to record and analyze your predictions.

| PREDICTION | CORRECT? | EVIDENCE |
| --- | --- | --- |
| This passage will describe the high temperatures deep within the earth. | | |

# Compare and Contrast

When you **compare** two or more things, you examine the similarities and differences between them. When you **contrast** things, you focus on only their differences. To compare and contrast, follow the steps shown at right.

**Step 1** Determine what the subject of the passage or paragraph is.

**Step 2** Identify two or more ideas, examples, or features relating to the subject that can be compared and contrasted.

**Step 3** Search for clue words that indicate similarities (comparing) and differences (contrasting).

## GUIDED MODEL

### Ⓐ Varied Climates

Most of Europe lies within the humid temperate climate region. The North Atlantic Drift, an ocean current of warm water, keeps temperatures relatively warm. Winds also affect climate. The sirocco sometimes blows over the Mediterranean Sea and brings humid weather to southern Europe at different seasons. The mistral is a cold wind in winter that blows through France, bringing cold, dry weather.

Ⓒ In general, Ⓑ <u>a Mediterranean climate brings cool, wet winters and hot, dry summers. This climate supports a long growing season.</u> Ⓒ <u>In contrast,</u> Ⓑ <u>Eastern Europe has long, cold winters.</u> Greenland, northern Scandinavia, and Iceland have polar climates and a limited growing season.

**Step 1** Determine what the subject of the passage or paragraph is.
**SUBJECT Ⓐ** the climates of Europe

**Step 2** Identify features of the subjects that can be compared and contrasted.
**FEATURE Ⓑ** climates of southern and eastern European countries

**Step 3** Search the passage for clue words that indicate similarities and differences.
Those sentences will help you compare. Then look for clue words that indicate how the two aspects are different. Those sentences will help you contrast.
**CLUE WORDS Ⓒ** in general (comparing); in contrast (contrasting)

**TIP** A Y-Chart and a Venn diagram, shown below, are useful graphic organizers for comparing and contrasting two topics. In a Y-Chart, list unique information on the branches and shared characteristics in the straight section. In a Venn diagram, list unique characteristics in the left and right sides and common characteristics in the overlapping area.

## APPLY THE SKILL

Turn to *Human & Physical Geography*, Section 2.4, "Natural Resources." Read the "Categories of Resources" passage. Compare and contrast renewable resources and nonrenewable resources using a Y-Chart or Venn diagram.

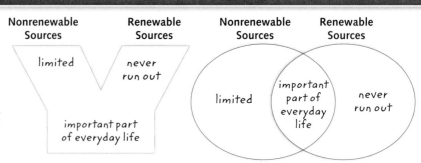

# Analyze Cause and Effect

A **cause** is an event or action that makes something else happen. The **effect** is an event that happens as a result of the cause. Analyzing cause-and-effect relationships can help you understand how events are related. To analyze cause and effect, follow the steps shown at right.

**Step 1** Determine the cause of an event. Look for signal words that show cause, such as *because, due to, since,* and *therefore.*

**Step 2** Determine the effect that results from the cause. Look for signal words such as *led to, consequently,* and *as a result.*

**Step 3** Look for a chain of causes and effects. An effect may be the cause of another action or event.

## GUIDED MODEL

### Rome's Decline

For about 500 years, the Roman Empire was the most powerful in the world and extended over three continents. **Ⓐ** Beginning around A.D. 235, however, Rome had a series of poor rulers. **Ⓐ** In addition, German tribes began invading the empire.

**Ⓑ** As a result, in 312, Emperor Constantine moved the capital from Rome to Byzantium, in present-day Turkey, and renamed the city Constantinople. In 395, the empire was divided into an Eastern Empire and a Western Empire with two different emperors. **Ⓒ** Because Rome had been severely weakened, invaders overthrew the last Roman emperor and ended the Western Empire in 476.

**Step 1** Determine the cause.

Ask yourself why the event took place. Notice that an event may have more than one cause.

**CAUSE(S)**

**Ⓐ** Beginning around A.D. 235, Rome had a series of poor rulers.

**Ⓐ** German tribes began invading the empire.

**Step 2** Determine the effect.

Ask yourself what happened as a result of the event or events.

**EFFECT Ⓑ** As a result, in 312, Emperor Constantine moved the capital from Rome to Byzantium.

**Step 3** Look for a chain of causes and effects.

An effect may cause another event.

**CAUSE/EFFECT Ⓒ** Because Rome had been severely weakened, invaders overthrew the last Roman emperor and ended the Western Empire in 476.

**TIP** Test whether events have a cause-and-effect relationship by using this construction: "Because [insert cause], [insert effect] happened." If the construction does not work, one event did not lead to the other.

## APPLY THE SKILL

Turn to *Human & Physical Geography*, Section 3.5, "Protecting Human Rights." Read "The Impact of Human Rights" passage. Identify a cause-and effect-relationship in the passage. Record the cause(s) and effect(s) in a graphic organizer like the one at right.

| CAUSE | EFFECT |
|---|---|
| Declaration of Human Rights was developed | World pressured South Africa to grant human rights to non-whites |

# Make Inferences

**Inferences** are conclusions or interpretations a reader makes from information that a writer does not state directly. When you make inferences, you use common sense and your own experiences to figure out what the writer means. To make inferences, follow the steps shown at right.

**Step 1** Read the text looking for facts and ideas.

**Step 2** Think about what the writer does not say but wants you to understand.

**Step 3** Reread the text and use what you know to make an inference.

## GUIDED MODEL

### Aryan Migration

Many historians believe a group of nomadic herders called the Aryans migrated from Central Asia into the Indus Valley around 2000 B.C. From there, they moved into northern India. **A** <u>Their language, Sanskrit, became the basis of many modern languages in South Asia.</u> The Aryans recorded religious teachings in Sanskrit in sacred texts called the Vedas. **B** <u>The Aryans established the beginnings of Hinduism, the major religion of India today.</u>

**TIP** Use a two-column chart to keep track of inferences. In the left column, write details, quotations, examples, statistics, and other facts. In the right column, write the inference that you draw from each fact. Note that an inference can be based on one fact or several facts.

**Step 1** Read the text looking for facts and ideas.

You must know the facts before you can make inferences. Ask yourself, "What facts does the writer state directly in this text?"

**FACT A** The Aryan language, Sanskrit, became the basis of many modern languages in South Asia.

**FACT B** The early religion of the Aryans established the beginnings of Hinduism, the major religion of India today.

**Step 2** Think about what the writer does not say but wants you to understand.

Ask yourself:
- How do these facts connect with what I already know?
- How does this information help me better understand Hinduism?

**Step 3** Reread the text and use what you know to make an inference.

**INFERENCE A** Many modern languages have ancient origins.

**INFERENCE B** Modern religions are influenced by ancient religions.

### APPLY THE SKILL

Turn to *Europe Geography & History*, Section 1.2, "A Long Coastline." Read the "Exploration and Settlement" text. Then make two inferences about how Europe's location near large bodies of water affected its people. Use a two-column chart like the one at right to record the facts you found and the inferences you made.

| FACTS | INFERENCES |
|-------|------------|
|       |            |

# Draw Conclusions

When you **draw conclusions**, you make a judgment based on what you have read. You analyze the facts, make inferences, and use your own experiences to form your judgment. To draw a conclusion, follow the steps shown at right.

**Step 1** Read the passage to identify the facts.
**Step 2** Make inferences based on the facts.
**Step 3** Use the inferences you have made and your own experiences and common sense to draw a conclusion.

## GUIDED MODEL

**Ⓐ** The euro was launched in 1999, but paper money and coins of the currency were not issued until 2002. **Ⓐ** By 2016, 19 of the 28 European Union (EU) nations had adopted the euro. The countries using the euro are known as the eurozone. **Ⓐ** Some EU countries, including Romania, hope to join the eurozone soon. Other countries, including Great Britain and Denmark, have not adopted the euro. They believe that giving up their own currency might result in a loss of control over their economies.

**TIP** Use a diagram to organize the facts you have identified, the inferences you have made, and your conclusion. A diagram can help you clarify your thinking.

**Step 1** Read the passage to identify the facts.
**FACTS Ⓐ** The euro was introduced to Europe in 1999. By 2016, 19 out of 28 EU nations used it. Some countries hope to join the eurozone. Others, including Great Britain and Denmark, have not adopted the euro because they think it would result in a loss of control over their economies.

**Step 2** Make inferences based on the facts.
**INFERENCES** *I see that most of the EU countries have adopted the euro, and at least one more wants to adopt it. However, some countries have chosen not to adopt the currency. That must mean that there are both advantages and disadvantages to using the euro.*

**Step 3** Use the inferences you have made and your own experience and common sense to draw a conclusion.
**CONCLUSION:** *In the future, more European countries will probably adopt the euro and become part of the eurozone. However, other countries will not adopt the euro because there are disadvantages to doing so.*

## APPLY THE SKILL

Turn to *Human & Physical Geography*, Section 2.3, "Extreme Weather." Read the "Scientific Solutions" passage and draw conclusions about the solutions scientists have come up with to combat the dangers of extreme weather. Write a few sentences summarizing your conclusion. A sample sentence is shown at right.

*Scientists are making great progress in predicting extreme weather* _____
_____
_____

# Make Generalizations

**Generalizations** are broad statements that apply to a set of ideas, a group of people, or a series of events. You make generalizations based on information you have read or heard. You can also draw on personal experiences. To make generalizations, follow the steps shown at right.

**Step 1** Look for the overall theme or message of the selection.

**Step 2** Find information in the passage that supports the theme.

**Step 3** Draw on your personal knowledge.

**Step 4** Make a generalization about the topic and put it into sentence form.

## GUIDED MODEL

### Life in the Ocean

Dr. Sylvia Earle has led more than 60 diving expeditions to explore marine life. During these dives, she has seen the incredible variety of ocean life— **B** "More than 30 major divisions of animals are known, from sponges and jellies to many kinds of beautiful worms and mollusks."

**A** Dr. Earle has also seen how people have harmed the oceans. **B** "Taking too much wildlife out of the sea is one way," she claims. "Putting garbage, toxic chemicals, and other wastes in is another." Dr. Earle has witnessed a huge drop in the number of fish in the sea. She has also noted pollution's impact on coral reefs.

**Step 1** Look for the overall theme.
   COMMON THEME **A** Humans harm the diverse marine life in the oceans.

**Step 2** Find information that supports the theme.
   SUPPORTING INFORMATION **B** Dr. Sylvia Earle, an ocean expert, states that people harm ocean life by taking too much wildlife out of the sea and putting garbage, toxic chemicals, and other wastes into it. Dr. Earle has witnessed a huge drop in the number of fish in the sea and pollution's impact on coral reefs.

**Step 3** Draw on your personal knowledge.
   *I know that living in polluted water is bad for living things. I watched a television program on how toxic waste in the oceans has caused the extinction of some species.*

**Step 4** Make a generalization about the topic and put it into sentence form.
   GENERALIZATION *If we don't stop polluting the waters and taking too much wildlife from the oceans, many species could become extinct.*

**TIP** You can also make generalizations about information from multiple sources. Determine what information the sources have in common. Then form a generalization that all of the sources would support.

## APPLY THE SKILL

Turn to *Europe Geography & History*, Section 2.5, "Middle Ages and Christianity." Read "The Feudal System" passage. Then make generalizations about the feudal system's role in Europe, like the one at right, using the information in the passage and what you may already know.

> The feudal system provided security for Europeans during the Middle Ages.

# Form and Support Opinions

When you **form an opinion**, you determine and assess the importance and significance of something. While an opinion is a person's own judgment, not a fact, sound opinions must be supported with examples and facts.

**Step 1** Read the passage. Look for reliable information about the subject, including facts, statistics, and quotations.

**Step 2** Form your own opinion about the subject.

**Step 3** Find facts to support your opinion.

## GUIDED MODEL

### Germany in World War II

**Ⓐ** Hitler became the leader of the National Socialist German Workers' Party, or the Nazis. He forged alliances with Italy and Japan.

**Ⓐ** Germany's invasion of Poland in 1939 started World War II. Great Britain and France, the Allies, declared war on Germany soon after, but Germany conquered Poland and quickly defeated most of Europe. In 1941, the United States entered the war on the side of the Allies.

**Ⓐ** Finally, in May 1945, Germany surrendered. Near the end of the war, Allied troops were stunned to find Nazi concentration camps where six million Jews and other victims had been murdered. This mass slaughter was called the Holocaust.

**Step 1** Read the passage. Look for reliable information about the subject.
**SUBJECT** Germany in World War II
**IMPORTANT INFORMATION ABOUT THE SUBJECT Ⓐ** Germany, Italy, and Japan formed an alliance. Germany's invasion of Poland in 1939 started World War II. Germany soon defeated most of Europe. In 1945, Germany surrendered and Allied troops discovered evidence of the Holocaust.

**Step 2** Form your own opinion.
Decide the subject's importance or wider meaning. What do you believe about the subject?
**MY OPINION** *Hitler and the Nazi Party's actions in the 1930s and 1940s dramatically violated human rights.*

**Step 3** Find facts to support your opinion.
If you cannot find facts to support your opinion, you must revise it.
**FACTS THAT SUPPORT MY OPINION** *Hitler was a powerful leader of the Nazis. They murdered six million Jews and other victims during the Holocaust.*

**TIP** To distinguish facts from opinions, ask yourself, "Can this sentence be proven?" If evidence can prove a statement, it is a fact. If the sentence cannot be proven because it expresses a thought or emotion, it is an opinion.

## APPLY THE SKILL

Turn to *Europe Geography & History,* Section 2.2, "Classical Greece." Read the "Greek Achievements" passage. Provide an opinion about what you have read and list facts to support your opinion. Use a chart like the one at right to organize your ideas.

| OPINION | EVIDENCE |
|---|---|
| Greece's golden age was a period of extraordinary achievements. | |

# Identify Problems and Solutions

Throughout history, people have faced problems and learned how to solve them. When you **identify problems,** you find the difficulties people faced. When you **identify solutions,** you learn how people tried to fix their problems. To identify problems and solutions, follow the steps shown at right.

**Step 1** Read the text and determine what problem or problems people faced.

**Step 2** Determine what caused the problems. There may be multiple causes.

**Step 3** Identify the solutions people used to resolve, or fix, the problems.

**Step 4** Determine if the solutions succeeded. Ask yourself, "Was the problem solved?"

## GUIDED MODEL

### Impact of the Revolution

The Industrial Revolution had a tremendous impact on how people worked and lived. Cities grew rapidly because people migrated to them for factory jobs. Standards of living rose, and a prosperous middle class grew.

**Ⓐ** However, factory workers often faced harsh conditions. **Ⓑ** Laborers worked as many as 14 hours a day. Child labor was common. Children as young as ten years old worked in factories and mines.

**Ⓑ** Many workers lived in small, crowded houses in neighborhoods where open sewers were common. Diseases spread quickly in these buildings. **Ⓒ** Over time, the workers' quality of life improved as public health acts were passed to provide sewage systems and larger buildings.

**Step 1** Read the passage and determine what problem people faced.
**PROBLEM Ⓐ** Factory workers faced harsh conditions during the Industrial Revolution.

**Step 2** Determine what caused the problem.
**CAUSES Ⓑ** long work hours, child labor, crowded lodgings, open sewers

**Step 3** Identify the solutions people used to resolve the problem.
**SOLUTIONS Ⓒ** Larger buildings for workers to live in were built, and sewage systems for improved sanitation were installed.

**Step 4** Determine if the solutions succeeded.
**SUCCESS?** Yes. Laws regulated working and living conditions, which are now much cleaner and safer. Sewers reduced disease. Children are not permitted to work in factories.

**TIP** A chart like the one below is a useful tool for recording information about a problem and solution. Organize your chart to have separate rows for a description of the problem, its cause, the solutions, and the success of the solution.

## APPLY THE SKILL

Turn to *Human & Physical Geography*, Section 2.5, "Habitat Preservation." Read the "Natural Habitats" passage and the "Habitat Loss and Restoration" passage. Identify the problems and solutions using a chart like the one at right.

| | |
|---|---|
| What was the problem? | |
| What were the causes? | |
| What was the solution? | |
| Did the solution work? | |

# Analyze Data

Information, or data, can be collected in charts, databases, graphs, diagrams, models, and maps. Regardless of its format, you can **analyze data** to draw conclusions, make comparisons, identify trends, and improve your understanding of the information. To analyze data, follow the steps at right.

**Step 1** Identify the data source, which could be a chart, graph, database, model, diagram, or map. Read its title to determine what information it contains.

**Step 2** Read any headings or subheadings to see how the information is organized.

**Step 3** Study the data to answer questions, draw conclusions, make comparisons, and identify trends.

## GUIDED MODEL

### A ECONOMIC INDICATORS OF SELECTED COUNTRIES

| B COUNTRY | GDP PER CAPITA (U.S. $) | LITERACY RATE (percent) |
|---|---|---|
| B Afghanistan | 1,000 | 28.0 |
| Brazil | 12,000 | 88.6 |
| China | 9,100 | 92.2 |
| Ethiopia | 1,200 | 42.7 |
| Germany | 39,100 | 99.0 |
| Mexico | 15,300 | 86.1 |
| United States | 49,800 | 99.0 |

Source: The World Bank; United Nations

**TIP** Data can be represented in ways other than charts and graphs. When analyzing data from a model, diagram, or map, examine all callouts, symbols, illustrations, and text boxes. Important data may be found where you least expect it.

**Step 1** Identify the data source and determine what information it contains.

A *This data source is a chart. Its title is "Economic Indicators of Selected Countries." That tells me it includes economic data on various countries.*

**Step 2** Read any headings or subheadings to see how the information is organized.

In a chart, columns go up and down, and rows go from left to right. Other data sources, such as graphs, models, and diagrams, organize information in different ways.

*This chart gives two types of economic indicator data recorded for each country.* B *The columns record the economic indicator data.* B *The rows list information for each country.*

**Step 3** Study the data to answer questions, draw conclusions, make comparisons, and identify trends.

*Using the data in this chart, I can compare the data in the literacy rate column to the data in the GDP per capita column. I can draw the conclusion that a high literacy rate is tied to a high GDP per capita.*

### APPLY THE SKILL

Turn to *Human & Physical Geography,* Section 1.2, "Earth's Complex Structure." Examine the "Tectonic Plate Movements" diagram. Then analyze the data in the diagram to answer the following questions.

1. What type of data does this diagram provide?
2. What type of plates are shown?
3. What are the four types of plate movements?
4. What is subduction?
5. Compare divergence and convergence.
6. Based on your analysis of the data in the diagram, what conclusions can you draw about the movement of Earth's plates?

# Distinguish Fact and Opinion

It is important to **distinguish fact from opinion** to separate someone's personal beliefs from concepts or events that are known to be true. To distinguish facts from opinions, follow the steps shown at right.

**Step 1** Read the passage and identify facts: information that can be proven to be true.

**Step 2** Identify opinions: claims or feelings about a topic or statements of personal beliefs.

**Step 3** Ask yourself, "Can this statement be checked to see if it is true?" If the answer is yes, you should decide how or where you can verify it.

## GUIDED MODEL

### Hosting the Games

Benefits come to countries that host the Olympic Games. **Ⓐ** First, constructing buildings such as stadiums and the Olympic Village (where athletes stay) creates jobs. Ideally, these structures will continue to be used now that the games are over. **Ⓐ** Businesses such as hotels and restaurants benefit, too. Athletes and tourists from around the world flood a host city and spend money on lodging, food, and other goods. Billions more watch the Olympics on television. **Ⓑ** With such wide exposure, Rio de Janeiro should now earn a reputation as a thriving and world-class city.

**TIP** When you identify personal beliefs in a historical text, you should also check for bias or a writer's prejudiced point of view. Look for words, phrases, or statements that reflect the positive or negative opinions of a particular group, social class, or political party.

**Step 1** Read the passage and identify information that can be proven to be true.

**FACTS Ⓐ** Constructing buildings such as stadiums and the Olympic Village create jobs. Businesses such as hotels and restaurants benefit, too. Athletes and tourists from around the world flood a host city and spend money on lodging, food, and other goods.

**Step 2** Identify opinions.

Look for personal beliefs, judgments, and feelings. Key words such as "ideally" and "should" are clues that a statement is an opinion.

**OPINIONS Ⓑ** With such wide exposure, Rio de Janeiro should now earn a reputation as a thriving and world-class city.

**Step 3** Ask yourself, "Can this statement be checked to see if it is true?" If the answer is yes, decide how and where to verify it.

**STATEMENT** Billions watch the Olympics on television.

**WHERE TO CHECK** encyclopedias, history books, reliable online sources

**FACT OR OPINION** fact

## APPLY THE SKILL

Turn to *Europe Geography & History*, Section 1.4, "Protecting the Mediterranean." Read the "Marine Reserves" passage. Use a chart like the one at right to record statements of fact and opinion and decide where you could verify it. After you have checked the statement, record whether it is a fact or opinion.

| |
|---|
| Statement: Scandola is a marine reserve that prohibits people from fishing, swimming, or anchoring their boats. |
| Where to check: |
| Fact or opinion: |

# Evaluate

When you read an informational text, you must **evaluate**, or assess what you have read. Sometimes you evaluate a passage to determine if its claims are believable. You may also evaluate to understand someone's actions. To evaluate something that took place, follow the steps shown at right.

**Step 1** Identify the action or event you want to evaluate.

**Step 2** Gather evidence about the positive impact of the action or event.

**Step 3** Gather evidence about the negative impact of the action or event.

**Step 4** Decide if the evidence is adequate.

**Step 5** Form your evaluation of the action.

## GUIDED MODEL

**Ⓐ** Eastern European countries have had mixed results since changing to a market economy. **Ⓑ** Poland has had the greatest success. It has a fast-growing economy and exports goods throughout Europe. **Ⓒ** Other countries have been slower to establish new businesses and become competitive. They have also experienced rises in prices and unemployment.

**Ⓑ** The leaders of many eastern European countries wish to integrate with the rest of Europe. They want to join the European Union and NATO, a military alliance of democratic states in Europe and North America.

**TIP** Making a list of the positive and negative outcomes of a decision, event, or action can help you evaluate. Read the passage and list the positives and negatives. Then review your list and form your evaluation.

**Step 1** Identify the action or event.

Ⓐ the impact of Eastern Europe's adoption of a market economy

**Step 2** Gather evidence about the positive impact of the action or event.

Ⓑ Poland's economy has grown quickly. Poland now exports goods throughout Europe. It has caused many eastern European leaders to reach out to the rest of Europe.

**Step 3** Gather evidence about the negative impact of the action or event.

Ⓒ Some countries have been slow to establish new businesses and become competitive. They have experienced rises in prices and unemployment.

**Step 4** Decide if the evidence is adequate.

*There is evidence to support both the positive and negative effects of the action. The evidence seems factual.*

**Step 5** Form your evaluation of the action.

*Changing to a market economy has had negative effects on some eastern European countries, However, it has been successful in others and has encouraged Eastern Europe to deepen connections with the rest of the world.*

## APPLY THE SKILL

Turn to *Europe Geography & History*, Section 2.5, "Middle Ages and Christianity." Read "The Growth of Towns" passage. Record evidence from the passage and determine whether it is positive or negative using a chart like the one at right. Then write a brief evaluation of the impact of the growth of towns on Europe.

| EVIDENCE | POSITIVE/NEGATIVE |
|---|---|
| Trade and business developed. | |

## Synthesize

When you read, you take in information, details, clues, and concepts. When you **synthesize**, you combine all of that data to form an overall understanding of what you have read. To synthesize, follow the steps shown at right.

**Step 1** Look for solid, factual evidence.
**Step 2** Look for explanations that connect facts.
**Step 3** Think about what you have experienced or already know about the topic.
**Step 4** Use evidence, explanations, and your prior knowledge to form a general understanding of what you have read.

### GUIDED MODEL

**Immigrants in Australia**

**A** Like Europe, Australia has an aging population. **B** The Australian government believes that immigration can help change that trend. As a result, each year the government identifies gaps in the country's workforce. It then determines the number of skilled immigrants that can come to Australia. Between 2010 and 2011, more than 127,000 immigrants came on this skilled migration program. Most of Australia's immigrants came from Great Britain, India, and China.

**Step 1** Look for factual evidence.
Identifying the facts will help you base your synthesis on reliable evidence.

**FACT A** Like Europe, Australia has an aging population.

**Step 2** Look for explanations that connect facts.
In this passage, the facts have a problem-solution connection.

**EXPLANATION B** Australia believes that immigration can help change its aging population trend.

**Step 3** Think about what you already know.
*I know that immigration made the United States diverse. I assume that the cultural diversity of Australia has grown due to that country's skilled migration program.*

**Step 4** Use evidence, explanations, and your prior knowledge to form a general understanding of what you have read.
*Because of its aging population, Australia has a government-regulated immigration policy to draw workers from other countries. Through this program, Australia is both solving the problem of its aging population and increasing its cultural diversity.*

**TIP** To synthesize, you must be able to determine what information is the most important. Once you have identified the most important facts, organize them, find explanations for them, and fit them in with what you already know. Synthesis occurs as you extract the most important information from a passage and give it personal meaning.

### APPLY THE SKILL

Turn to *Europe Geography & History*, Section 2.3, "The Republic of Rome." Read "The Roman Way" passage. Use a chart like the one at right to organize the evidence. Then write a brief synthesis statement about the importance of Roman values in the development of Rome.

| | |
|---|---|
| Evidence: | Romans applied the values of self-control, working hard, doing one's duty, and pledging loyalty to Rome. |
| Supporting explanation: | |
| Synthesis: | |

# Analyze Primary and Secondary Sources

**Primary sources** are materials written or provided by people who have had personal experience with an event. **Secondary sources** are materials written by people who did not witness or experience an event directly. To analyze primary and secondary sources, follow the steps at right.

**Step 1** Identify whether the material is a primary or secondary source.

**Step 2** Determine the quality and credibility of the primary or secondary source.

**Step 3** Determine the main idea of the secondary source material.

**Step 4** Identify the author of the primary source and his or her main point.

## GUIDED MODEL

### Mandela's Inspiration

**Ⓐ** Enlightenment thinkers asserted that **Ⓑ** people have natural rights, such as life, liberty, and property. They inspired Nelson Mandela in his struggle to end apartheid in South Africa. For his efforts, Mandela received the 1993 Nobel Peace Prize. The following is from his speech.

**Ⓒ** *The value of our shared reward will and must be measured by the joyful peace which will triumph.*

*Thus shall we live, because we will have created a society which recognizes that all people are . . . entitled . . . to life, liberty, prosperity, human rights, and good governance.*

**Ⓓ** —Nelson Mandela, 1993

**TIP** When you analyze a primary or secondary source, separate facts from opinions. These opinions may reflect the author's bias and the beliefs and ideas of his or her time.

**Step 1** Identify whether the material is a primary or secondary source.

Most textbook passages are secondary sources. An observation from an expert or eyewitness is likely to be a primary source. The secondary source material in this textbook passage is marked with **Ⓐ**.

**Step 2** Determine the main idea of the secondary source material.

**MAIN IDEA Ⓑ** People have natural rights that all humans should enjoy.

**Step 3** Determine the quality and credibility of the primary or secondary source.

The primary source in this passage is marked with **Ⓒ**. Ask yourself, "Is the author a reliable source of information? *Nelson Mandela is a political activist and Nobel Peace Prize winner. He is a credible source. The secondary source is my textbook, which is also a reliable source.*

**Step 4** Identify the author of the primary source and his or her main point.

**AUTHOR Ⓓ** Nelson Mandela

**MAIN IDEA Ⓑ** People have natural rights that all humans should enjoy.

### APPLY THE SKILL

Turn to *Europe Geography & History*, Section 1.4, "Protecting the Mediterranean." Read the "Under the Sea" passage. Determine what parts of the passage are primary and secondary sources. Then use a chart like the one at right to analyze the passage.

| |
|---|
| Quality of the source: |
| Main idea of secondary source material: |
| Author of primary source material: |
| Information provided by primary source: |

SKILLS HANDBOOK

# Analyze Visuals

Visuals, such as charts, graphs, maps, photos, artwork, and diagrams, illustrate ideas within a text. When you **analyze visuals**, you determine what information is being presented and how it relates to other information about that topic. To analyze visuals, follow the steps shown at right.

**Step 1** Study the visual and determine what information it provides.

**Step 2** Determine how the information in the visual relates to other information provided in a text.

**Step 3** Analyze the information presented in the visual to enhance your understanding.

## GUIDED MODEL

### Solstices and Equinoxes

The moment at which summer and winter start is called a **B solstice**. June 20 or 21 is the summer solstice in the Northern Hemisphere. On December 21 or 22, the Northern Hemisphere has its winter solstice.

The beginning of spring and autumn is called an **B equinox**. In the Northern Hemisphere, the spring equinox occurs around March 21, and the autumn equinox occurs around September 23.

**Step 1** Study the visual and determine what information it provides.

Examine any titles, labels, or captions for clues.

**A** *This diagram is titled "Earth's Four Seasons: Northern Hemisphere." The dates on the visual tell me the position of Earth in relationship to the sun throughout the year.*

**Step 2** Determine how the information in the visual relates to other information provided in a text.

**B** *The summer and winter solstices and spring and autumn equinoxes are described in the text. The visual illustrates the information from the text.*

**Step 3** Analyze the information presented in the visual to enhance your understanding.

Ask yourself: What information does the visual show? Why was this visual used? Does the visual represent information accurately? What does the visual show that is not explained in the text?

**TIP** Look for patterns and connections between items and information in a visual and the surrounding text. Study colors and symbols to understand what they represent.

### APPLY THE SKILL

Turn to *Human & Physical Geography*, Section 1.3, "Earth's Landforms." Analyze the visual and answer the following questions.

1. What type of information does this visual provide?
2. Which text passage does this visual enhance?
3. Use the information in the related text passage and the visual to explain what the continental shelf is and where it is located.
4. What does the visual show you about continental rise that is not explained in the passage?

# Interpret Physical Maps

**Physical Maps** provide information about the earth's physical features such as lakes, rivers, and mountains. You can learn about elevation, or relief, and absolute and relative location by studying physical maps. To read a physical map, follow the steps at right.

**Step 1** Read the title of the map.
**Step 2** Use the map legend.
**Step 3** Use the map scale to measure distance.
**Step 4** Use the compass rose or directional pointer to determine direction.
**Step 5** Use the latitude and longitude gridlines to determine the region's location on Earth.

## GUIDED MODEL

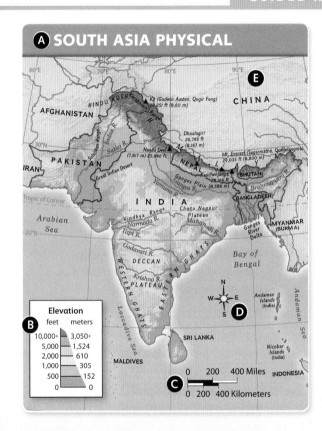

**Step 1** Read the title of the map.
**A** Read the title of the map to find out what type of map it is and what kind of information the map presents.

**Step 2** Use the map legend.
**B** A map legend explains the symbols used on the map. On a physical map, the legend usually provides information about physical features, such as mountains. The map also has a color-coded elevation scale to show how far above sea level each area is.

**Step 3** Use the map scale.
**C** Use the map scale to help you determine the distances between points on the map.

**Step 4** Determine direction.
**D** Use the compass rose or directional pointer to help you determine direction on the map.

**Step 5** Determine latitude and longitude.
**E** Examine the numbered gridlines on the map. The horizontal lines represent latitude. The vertical lines represent longitude.

**TIP** Making a chart is a good way to record important information from a physical map. Use a chart to record the map's title, legend information, scale information, latitude and longitude, and location.

## APPLY THE SKILL

Turn to *Europe Geography & History*, Section 1.1, "Physical Geography." Interpret the map to answer the following questions.

1. What is the map about?
2. What type of map is this and what region does it represent?
3. What is the highest elevation in the country of Romania?
4. What is the approximate distance between the easternmost point in France and the westernmost point in Italy, in both miles and kilometers?

R27

# Interpret Political Maps

**Political maps** provide information about human-made features such as cities, capitals, and borders between countries. Unlike physical maps, political maps do not focus on the physical features of a country or region. To read a political map, follow the steps shown at right.

**Step 1** Read the title of the map.
**Step 2** Use the map legend.
**Step 3** Use the map scale to measure distance.
**Step 4** Use the compass rose or directional pointer to determine direction.
**Step 5** Use the latitude and longitude gridlines to determine the region's location on Earth.

## GUIDED MODEL

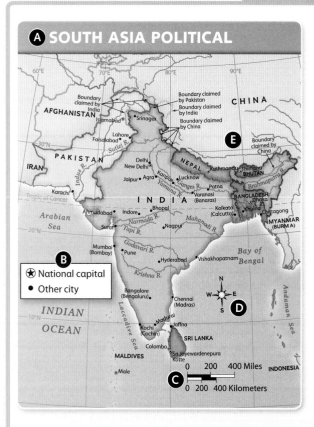

**Step 1** Read the title of the map.
Ⓐ Read the title of the map to find out what type of map it is and what information the map represents.

**Step 2** Use the map legend.
Ⓑ A map legend explains the symbols used on the map. Symbols for capital cities and major cities are usually found in the legend of a political map. On this map, New Delhi, India's capital, is represented by a star.

**Step 3** Read the map scale.
Ⓒ Use the map scale to help you determine the distances between cities or countries on the map.

**Step 4** Determine direction.
Ⓓ Use the compass rose or directional pointer to determine direction on the map.

**Step 5** Determine latitude and longitude.
Ⓔ Examine the numbered gridlines that intersect over the map. The horizontal lines represent latitude, and the vertical lines represent longitude. Latitude and longitude lines can help you establish the location of the countries or major cities shown.

**TIP** On many political maps, countries or states are shown in different colors. This makes it easy to distinguish their borders. Cities are often designated by a dot of varying sizes based on the city's population.

## APPLY THE SKILL

Turn to the chapter introduction for *Europe Geography & History*. Locate the Europe Political map next to the Preview the Chapter page. Interpret the map and answer the following questions.

1. What type of map is this and what region does it represent?
2. What is the capital city of Spain?
3. What is the approximate distance between the easternmost and westernmost coasts of Iceland?
4. Where is Berlin located in relation to Warsaw?

# Create Sketch Maps

In social studies textbooks, maps are a common visual aid. However, sometimes it is helpful to draw your own map to help you understand a place better or to visualize features described in the text. To **create a sketch map** of a country, follow the steps shown at right.

**Step 1** Determine which map you are going to draw and give it a title.

**Step 2** Sketch the outline of the location.

**Step 3** Add important political and physical features to your map.

**Step 4** Add a compass rose to your map.

**Step 5** If appropriate, sketch and label surrounding countries or regions.

## GUIDED MODEL

**Step 1** Determine which map you are going to draw and give it a title.

**Ⓐ** Study existing maps and text describing that geographical location.

**Step 2** Sketch the outline of the location.

Ask yourself, "What shape is this location? What borders it?" Remember that sketch maps do not have to be perfect. Everyone's sketch map will look different.

**Step 3** Add important political and physical features to your map.

**Ⓑ** This student chose to draw and label the Ebro River in Spain.

**Step 4** Add a compass rose to your map. **Ⓒ**

**Step 5** If appropriate, sketch and label surrounding countries or regions.

**Ⓓ** On this sketch map, the country of Portugal has been identified.

**TIP** Sketch your map on graph paper so that you can draw your location and its surrounding area to scale. Also, use a pencil when you sketch so that you can easily erase any mistakes.

## APPLY THE SKILL

Examine the Europe Political map next to the Preview the Chapter page in *Europe Geography & History*. Then examine the Europe Physical map and read the passage "A Peninsula of Peninsulas" in Section 1.1, "Physical Geography." Use a chart like the one at right to help you create a sketch map of Ireland.

| Map title: |
|---|
| Capital: |
| Major cities: |
| Surrounding countries: |

# Create Charts and Graphs

To organize information, it is useful to **create charts** and **graphs.** Charts simplify and summarize information. Graphs present numerical information. To create charts and graphs, follow the steps shown at right.

**Step 1** Determine whether you should use a chart or graph to represent your data.

**Step 2** Give your chart or graph a title to tell what kind of information it shows.

**Step 3** Create your chart or graph using appropriate labels for the data.

## GUIDED MODEL

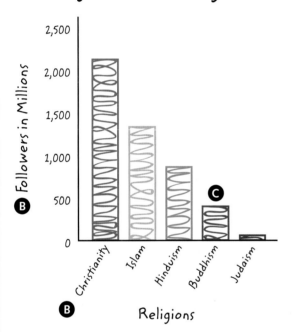

**Step 1** Determine whether you should use a chart or graph to represent your data.

*I want to represent the number of people who practice the major world religions: Christianity, Islam, Hinduism, Buddhism, and Judaism. The data is numerical, so I will create a graph. I will use a bar graph to compare the data and show which religion has the most followers and which religion has the least.*

**Step 2** Give your chart or graph a title to tell what kind of information it shows.

**Ⓐ** *I will call the bar graph "Major World Religions."*

**Step 3** Create your chart or graph using appropriate labels for the data.

**Ⓑ** *My horizontal axis will represent the five major world religions. My vertical axis will represent the number of followers in millions. I will label the axes accordingly.*

**Ⓒ** *I will record my data on the bar graph using a different color to represent each world religion.*

**TIP** Different visuals are used to represent different types of data. Line graphs are useful to compare changes over time. Bar graphs compare quantities. Pie graphs show percentages of a whole. Charts can be structured in many different ways, but they always simplify and organize information.

## APPLY THE SKILL

Turn to *Europe Geography & History,* and locate the Chapter Review. Use the "Miles of Railway Track in Selected European Countries" chart to create a line graph. To simplify the graph, include only the four countries with the highest totals. A line graph representing 1840 is shown at right.

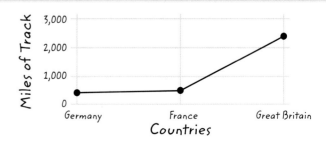

# Interpret Charts

A chart is a way to represent information or data visually. In a chart, information is organized, simplified, and summarized. To **interpret charts,** follow the steps shown at right.

**Step 1** Read the title of the chart to find out what type of information the chart presents.

**Step 2** Check the source of the data in the chart for reliability.

**Step 3** Read any headings, subtitles, or labels to understand how the chart is organized.

**Step 4** Examine the data in the chart.

**Step 5** Summarize the information.

## GUIDED MODEL

**A POPULATION OF MEDITERRANEAN CITIES (IN MILLIONS)**

| C City | 1960 | 2015 |
|---|---|---|
| C Athens, Greece | 2.2 | 3.1 |
| Barcelona, Spain | 1.9 | 5.3 |
| Istanbul, Turkey | 1.74 | 14.1 |
| Marseille, France | 0.8 | 1.6 |
| Rome, Italy | 2.33 | 3.7 |

**B Source:** UN, 2013; CIA World Factbook, 2016

**TIP** When you interpret a chart, compare the data and draw conclusions from the information. For example, in the chart above, you might conclude that Istanbul is the fastest-growing city on the Mediterranean. You might also conclude that, as a result, the city may face challenges in housing and employing its large population.

**Step 1** Read the title of the chart to find out what type of information it represents.

**TITLE A** Population of Mediterranean Cities (in millions)

**Step 2** Check the source of the data in the chart for reliability.

**SOURCE B** The UN, or United Nations, and the CIA World Factbook are well-known and reliable sources.

**Step 3** Read any headings, subtitles, or labels to understand how the chart is organized.

**C** This chart is organized in columns and rows. The rows feature major Mediterranean cities, and the columns give data for 1960 and 2015.

**Step 4** Examine the data in the chart.

The data tells how certain cities are grew over a period of 55 years. Its purpose is to report past and current data.

**Step 5** Summarize the information.

The data in this chart shows that Istanbul, Turkey, had the greatest growth and Athens, Greece, the least.

### APPLY THE SKILL

Turn to *Human & Physical Geography*, Section 1.5, "Waters of the Earth." Interpret the "Longest World Rivers" chart to answer the following questions.

1. What information does the chart present?
2. How is the chart organized?
3. Where is the Amazon River located?
4. What is the length of the Mississippi-Missouri River?
5. What is the longest river in the world?
6. Of the two Asian rivers on the chart, which is longer?

# Interpret Graphs

A graph is another way to represent information or data in picture form rather than written text. In a graph, data can be represented using numbers, symbols, or pictures. Common graphs include pie graphs, line graphs, and bar graphs. To **interpret graphs,** follow the steps shown at right.

**Step 1** Read the title of the graph and identify what type of graph it is.

**Step 2** Check the source of the data in the graph for reliability.

**Step 3** Read the labels in the graph.

**Step 4** Examine the data in the graph and look for patterns.

**Step 5** Summarize the information in the graph.

## GUIDED MODEL

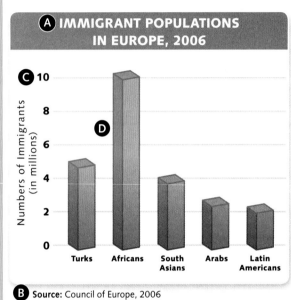

**B** Source: Council of Europe, 2006

**TIP** A pie graph is used to compare parts of a whole, and each "slice" represents a percentage. Remember that the percentages represented by the slices in a pie graph always add up to 100.

**Step 1** Read the title of the graph and identify what type of graph it is.

**TITLE A** Immigrant Populations in Europe, 2006; This is a bar graph, which is often used to compare quantities.

**Step 2** Check the source of the data in the graph for reliability.

**SOURCE B** Council of Europe, which is a reliable source on Europe.

**Step 3** Read the labels in the graph.

**C** The vertical axis is labeled "Numbers of Immigrants." Its scale goes to ten million. The horizontal axis is labeled with five immigrant groups.

**Step 4** Examine the data and look for patterns.

**D** The graph compares the numbers of immigrants to Europe from various regions in a single year.

**Step 5** Summarize the information in the graph.

The graph shows that Africans made up the largest group of immigrants in Europe in 2006. The smallest group came from Latin America.

## APPLY THE SKILL

Turn to "Compare Across Regions: World Languages" at the end of the Europe unit. Study the Number of Languages Spoken by Continent graph and use it to answer the following questions.

1. What do the colors of the bars represent?
2. What do the words written on the bars represent?
3. On which continent are the most languages spoken?
4. On which continent are the fewest languages spoken?

R32 SKILLS HANDBOOK

# Create and Interpret Time Lines

Another visual way to organize, represent, and review information is by creating a time line. When you **create a time line,** you record events and dates chronologically along an axis from left to right or from top to bottom. To create a time line, follow the steps at right.

**Step 1** Decide what your time line will show and give your time line a title.

**Step 2** Determine time line dates and place them in chronological order.

**Step 3** Plot major events on the time line in the appropriate locations.

**Step 4** Note the patterns that emerge on your time line.

## GUIDED MODEL

**Ⓐ WORLD WAR II**

**Ⓑ** 1936 — ○ 1936 Germany, Italy, and Japan form Axis Powers.

**Ⓒ** 1938

○ **Ⓓ** 1939 Germany invades Poland and defeats most of Europe.

1940

○ 1941 Japan attacks Pearl Harbor. United States enters the war.

1942

1944

○ 1945 Germany surrenders. United States bombs Japan. Japan surrenders.

1946

**TIP** When you record B.C. dates on a time line, be careful to place them in the correct order. The higher the number, the further back in time it occurred. For example, 850 B.C. took place 350 years before 500 B.C. The year A.D. 1 took place immediately after 1 B.C. No zero year exists.

**Step 1** Decide what your time line will show and give your time line a title.

Time lines often feature events that are thematically related.

*I will create a time line to record the major events that led up to and took place during World War II.*

TITLE **Ⓐ** World War II

**Step 2** Determine time line dates and place them in chronological order.

Determine your start and end dates. Make sure the interval dates in between are regular.

**Ⓑ** *The events occur from 1936 to 1945. My start date will be 1936, and my end date will be 1946.* **Ⓒ** *I will put an interval date every two years.*

**Step 3** Plot major events on the time line in the appropriate locations.

**Ⓓ** *For an event that occurs on a date that isn't labeled, I will draw a line from the time line and label the date.*

**Step 4** Note the patterns that emerge.

*On my time line, I see that Germany was powerful in the late 1930s and that the United States played an important role in the war.*

## APPLY THE SKILL

Turn to *Europe Geography & History,* Section 2.2, "Classical Greece." Study the time line and use it to answer the following questions.

1. What title might you give the time line?
2. What is the time span of the time line?
3. What are the intervals between major dates on the time line?
4. When did the Greeks defeat Persia?
5. What patterns do you see on the time line?

# Build Models

You can **build models** to show information about a place, an event, or a concept in a visual way. Models can be any visual representation of information, but the most common examples are posters, diagrams, mobiles, and dioramas. To create a poster, follow the steps shown at right.

**Step 1** Determine what your poster will show and research the topic.

**Step 2** Brainstorm ideas for your poster and sketch them.

**Step 3** Think of visual ways you can represent information on your poster.

**Step 4** Gather the supplies you need.

**Step 5** Create your poster.

## GUIDED MODEL

**Step 1** Determine what your poster will show and research the topic.

*This poster shows the structure of a manor during the Middle Ages. To create this model, the author researched the Middle Ages and the feudal system.*

**Step 2** Brainstorm ideas for your poster and sketch them.

*A rough draft or sketch is always a good idea before you begin drawing the actual poster.*

**Step 3** Think of visual ways you can represent information on your poster.

**A** *The author uses pictures and callouts to show how a manor from the Middle Ages was structured.*

**Step 4** Gather the supplies you need.

*This model required some resources on the Middle Ages, drawing paper, a pencil or pen, and paint.*

**Step 5** Create your poster.

*This model was drawn from a bird's eye view, which shows how the parts of the manor relate to each other.*

**TIP** Dioramas are models that show a three-dimensional place, event, or scene, such as a battle or extreme type of weather. A diorama is built inside a small box, such as a shoe box. The inside of the box is painted with a background, and the scene is constructed using craft supplies and common household objects.

## APPLY THE SKILL

Turn to *Human & Physical Geography*, Section 2.4, "Natural Resources." Read the "Earth's Resources" and "Categories of Resources" passages. Create a poster or other type of model to represent how Earth's natural resources are categorized and show examples of each type. A sample poster is shown at right.

### Earth's Natural Resources

 Coal    Sugarcane

 Wheat    Copper

Another type of model is a diagram, which can help observers visualize something that may not be visible in real life. Diagrams are also often used to explain how things work or to show how something is divided into parts. To build a diagram, follow these steps.

**Step 1** Determine what your diagram will show and research the topic.

**Step 2** Brainstorm ideas for your diagram and sketch them.

**Step 3** Think of visual ways you can represent information on your diagram.

**Step 4** Gather the supplies you will need.

**Step 5** Create your diagram.

## GUIDED MODEL

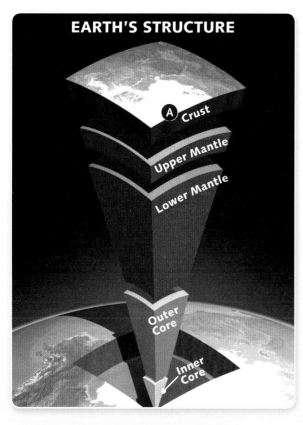

**Step 1** Determine what your diagram will show and research the topic.

*This model shows the structure of Earth's layers. It is helpful because we cannot normally see them.*

**Step 2** Brainstorm ideas for your diagram and sketch them.

*It is a good idea to determine and make a rough sketch of the approximate thicknesses of each layer ahead of time.*

**Step 3** Think of visual ways you can represent information on your diagram.

**A** *This diagram shows the layers of the earth, using a different color to represent each layer. The names of the layers are labeled.*

**Step 4** Gather the supplies you will need.

*This diagram required paints, markers, or colored pencils, as well as paper and a pen or pencil.*

**Step 5** Create your diagram.

*This model clearly shows Earth's five layers, their approximate thicknesses, and their correct order.*

**TIP** A mobile is a three-dimensional model. Mobiles are especially useful for showing how objects relate spatially to each other, such as how the planets circle the sun. Mobiles can be created by hanging objects from a simple wire hanger using yarn or thread.

### APPLY THE SKILL

Turn to *Europe Geography & History*, Section 1.2, "A Long Coastline." Read the "Exploration and Settlement" passage. Draw a diagram like the one at right to represent the structure of a European fishing village. Use visual elements rather than text to represent information.

# Create Databases

A database is a collection of information, or data, organized in a chart format so it can easily be viewed, used, and updated. You can use a computer program to **create databases** that allow you to search through the data to find only what you need. To create a database, follow these steps.

**Step 1** Determine the information your database will contain and give it a title.

**Step 2** Enter the column headings and/or row headings for your database.

**Step 3** Enter the data in the appropriate columns and rows.

## GUIDED MODEL

**Ⓐ FAMOUS EUROPEAN EXPLORERS**

| Ⓑ Explorer's Name | Explorer's Country | Date of Exploration | Place Explored |
|---|---|---|---|
| Ⓒ Bartolomeu Dias | Portugal | 1488 | Southern tip of Africa |
| Vasco de Gama | Portugal | 1498 | Southern tip of Africa; also reached India and opened up trade with Asia |
| Christopher Columbus | Italy (but he was working for Spain) | 1492 | Islands in the Caribbean; continents of North and South America |
| Jacques Cartier | France | 1530s | Northern part of North America |
| Sir Francis Drake | England | 1577 | Sailed around the world |

**TIP** You might want to use index cards to write and organize the information you want to use in your database. Once you have gathered all of your information, you can put it in a written chart format or use database software to organize it.

**Step 1** Determine the information your database will contain and give it a title.

Ⓐ This database includes data from a passage on European exploration. It records data about which explorers went to which regions. An appropriate title is "Famous European Explorers."

**Step 2** Enter the column headings and/or row headings for your database.

Remember, columns go up and down, and rows go from left to right. Database headings most commonly appear across the top or down the left side of the database, or in both places.

Ⓑ I have four categories of data, so I will use four columns, one per category. I will label each column with a heading that describes the data below it.

**Step 3** Enter the data in the appropriate columns and rows.

Ⓒ I will extract information from the text I have read and insert it in the appropriate column in the database. I will use a separate row to provide information on each explorer.

## APPLY THE SKILL

Turn to *Human & Physical Geography*, Section 2.2, "World Climate Regions." Read about the five climate regions featured in this section. Then create a database like the one at right to record the names of each climate region, a description of its weather and plant life, and a list of places that have that climate.

| WORLD CLIMATE REGIONS ||||
|---|---|---|---|
| CLIMATE REGION | WEATHER | PLANT LIFE | LOCATIONS |
|  |  |  |  |

# Create Graphic Organizers

While you read, you can **create graphic organizers** such as charts, diagrams, and time lines to take notes and explore how the information is related. To create a graphic organizer, follow the steps shown at right.

**Step 1** Determine your needs.
**Step 2** Determine what type of graphic organizer you need.
**Step 3** Draw your graphic organizer.
**Step 4** Use information from the passage, plus any relevant maps or other features, to fill in your graphic organizer.

## GUIDED MODEL

**A PROBLEM**
Earthquakes can cause severe damage and endanger people's lives.

↓

**B CONTRIBUTING FACTORS**
In some areas where earthquakes occur, buildings cannot withstand the intense shaking.

↓

**C SOLUTION**
Scientists are working hard to predict earthquakes. Engineers try to design buildings that keep people safe and minimize damage.

**TIP** You can use a cause-and-effect chain to help you understand the results of a particular event. Note that the chain can contain a single cause and multiple effects, or it can have many causes and only one effect.

**Step 1** Determine your needs.

First, think about the type of information you are recording. Then determine what you will be doing with the information. Are you gathering facts, comparing two things, or recording a series of events?

**Step 2** Determine what type of graphic organizer you need.

A chart can be used to record facts. A Venn diagram helps compare and contrast two things. A time line records a series of events over time. For this example, a problem-solution chart was created to identify a problem and understand how people are trying to solve it.

**Step 2** Draw your graphic organizer.

**Step 4** Use information from the passage, plus any relevant maps or other features to fill in your graphic organizer.

This student identified:

**A** the problem, in this case, the dangers posed by earthquakes;

**B** contributing factors, the situations that add to the problem;

**C** a possible solution to the problem.

## APPLY THE SKILL

Turn to *Europe Geography & History*, Section 3.2, "The Industrial Revolution." Read "The Revolution Begins" passage. Then create a main idea and details graphic organizer like the one at right and use it to record important information from the text.

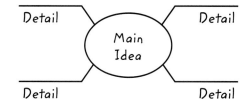

# Conduct Internet Research

When you **conduct Internet research**, you search and access source material from all over the world via the Internet. As you probably know, the World Wide Web (or "www") is part of the Internet that allows you access to online information and data. To conduct Internet research, follow the steps shown at right.

**Step 1** Choose and access a search engine.

**Step 2** Type key words or phrases into the search field.

**Step 3** Examine the search results.

**Step 4** Visit the suggested Web sites and perform more specific searches for information.

**Step 5** Use other reliable sources to verify any information you gather.

## GUIDED MODEL

**Ⓐ** | elephant habitat | Search |

**Ⓑ** **ELEPHANTS- Habitat & Distribution**
ELEPHANTS- **Habitat** & Distribution...Discover animal, environmental, and zoological career facts as you explore in-depth topic coverage via SeaWorld, ...
www.seaworld.org/.../elephants- habitat & distribution.htm -

**Elephant Facts - Defenders of Wildlife**
Get the facts on **elephant**. **Elephants** are the largest land-dwelling mammals on earth. ... of **elephants' habitat** will become significantly hotter and drier, ...
www.defenders.org › Wildlife and Habitat -

**African Elephants, African Elephant Pic ...**
African **elephants** are the largest of Earth's **land** mammals. Their enormous ears help them to keep cool in the hot African climate. ...
animals.nationalgeographic.com/animals/.../african-**elephant**/ -

**Elephants Habitat | Animal Habitats**
A quick look into the **elephant's habitat**. In studying an **elephant's habitat**, the first thing one encounters is the fact that there are several species of ...
www.animalhabitats.org/elephants_habitat/elephants_habitat.htm -

**Step 1** Choose and access a search engine.

**Step 2** Type key words into the search field.
Be as specific as possible to narrow your search and yield better results.
**Ⓐ** *I typed "elephant habitat" into the search field and clicked "search."*

**Step 3** Examine the search results.
Remember that Web addresses that end with ".edu," ".gov," and ".org" are often more reliable than ".com" sites. Review the addresses and summaries of each Web site and decide which ones to use.
**Ⓑ** *This Web site's domain is ".org", which is usually a reliable source, and the information is relevant to my topic, so I will go to this Web site.*

**Step 4** Visit the suggested Web sites and perform more specific searches for information.
*Once I'm on the site, I will search "elephant habitat preservation" to find more specific information.*

**Step 5** Use other reliable sources to verify any information you gather.
*I found the same information in an online encyclopedia and on the National Geographic Web site. This tells me the information is reliable.*

**TIP** When you use the Internet to do research for a report or project, print the source material you find online. This will provide you with a record of the Web sites you have used and allow you to review the source materials as you write your report without having to go back online.

## APPLY THE SKILL

Think of a topic from this book you would like to research. Brainstorm key words to use on a search engine. Then, with your teacher's permission, use a Web browser to conduct Internet research on your topic. If possible, print copies of the information you find. Write a brief paragraph explaining your search process and results.

| Ancient Greece | Search |

# Evaluate Internet Sources

It is important to use only the most reliable and credible information as a resource. To do so, you must **evaluate Internet sources** to be sure they are accurate and reputable. To evaluate Internet sources, follow the steps shown at right.

**Step 1** Examine the Web site's Internet address.
**Step 2** Identify the Web site's author.
**Step 3** Identify when the Web site was created or last updated.
**Step 4** Verify the information from the Web site using other reliable sources.
**Step 5** Evaluate the Internet source.

## GUIDED MODEL

**TIP** Another way to evaluate an Internet source is to determine its intended audience. Ask yourself to whom the author is writing, and study the writing style, vocabulary, and tone. Determine if the author's objective appears to be to inform, to explain, or to persuade, and whether he or she provides sufficient evidence.

**Step 1** Examine the Web site's Internet address.

**A** *This Web site's address is www.cia.gov. A government agency created the site, which makes it reliable.*

**Step 2** Identify the Web site's author.

Sites that clearly name the author are more reliable than anonymous Web sites.

**B** *The author of this Web site is the Central Intelligence Agency, or CIA. It is expert at collecting data.*

**Step 3** Identify when the Web site was created or last updated.

**C** *This Web site is updated every week, so the information is current.*

**Step 4** Verify the information from the Web site using other reliable sources.

*I found the information in an encyclopedia and on the National Geographic Web site.*

**Step 5** Evaluate the Internet source.

*I believe this is a reliable source because it is created by the government, updated frequently, and has the same information as other reliable sources.*

## APPLY THE SKILL

With your teacher's permission, visit the following Web address: www.nationalgeographic.com. Use the steps above to evaluate this Internet site as a reliable source of information. Write a brief paragraph explaining your evaluation of the site.

# Create Multimedia Presentations

You can **create multimedia presentations** on any topic using media such as photographs, video clips, and audio recordings. To create a multimedia presentation, follow the steps shown at right.

**Step 1** Determine the topic you will be presenting and the types of media you will use.

**Step 2** Research your topic.

**Step 3** Put your presentation together.

## GUIDED MODEL

**A** Unlike many species of crab, the one shown here can swim as well as walk.

**Step 1** Determine the topic you will be presenting and the types of media you will use.

Certain types of media enhance topics more effectively than others. Once you have determined your topic, decide which media work best with it. Would your topic be enhanced by images, or would audio clips or music be more relevant? Would a video clip add meaning to your presentation? What about a map or graph? You decide.

**A** This student used a photo from the **Digital Library** and wrote her own caption for it.

**Step 2** Research your topic.

Use reliable library and online sources to research your topic. Write a brief script that provides information on your topic and ties your media together.

**Step 3** Put your presentation together.

Combine your informational written script with the media you have selected. Make sure everything flows together well. You might use presentation software to display your information. Rehearse your presentation a number of times to identify and correct any problems.

**TIP** Your presentation can be based on your experiences. Conduct research to provide a context or background for the experience you intend to share. For any presentation, search the **Magazine Maker CD-ROM** for images and information, which you can import. Be sure to rehearse your presentation a few times to identify and correct any problems.

## APPLY THE SKILL

Select a topic from this text and create a multimedia presentation about it by following the steps above. Use the **Digital Library** and the **Magazine Maker CD-ROM** to create your presentation.

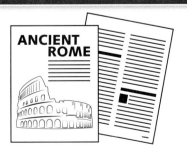

# ECONOMICS & GOVERNMENT

## PART I ECONOMICS

**agriculture** *n., the development of plants and animals to provide food.* Agricultural products include crops such as wheat, corn, and barley. Agriculture also includes animals that have been domesticated, or tamed. These animals are often called livestock, and they include cattle, sheep, pigs, and horses.

**business cycle** *n., a period during which a country's economic activities increase and then decrease in a relatively predictable pattern.* A business cycle has four phases. During expansion, businesses do well. At the peak of the cycle, economic activity begins to slow. During a contraction, economic activity continues to decrease. A contraction is also called a recession. The trough is the lowest level of economic activity. Then business starts to improve, and a new business cycle begins.

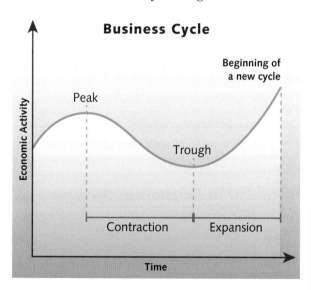

**capitalism** *n., an economic system in which private individuals or groups own the resources and produce goods for a profit.* In capitalism, private individuals or groups decide to produce goods or offer services. They offer those goods and services for sale in markets, which are places where people buy and sell. Businesses raise capital, or money, to create new products and hire workers. Capitalism is also called free enterprise because people are free to start businesses.

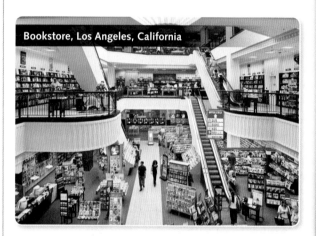

Bookstore, Los Angeles, California

**command economy** *n., an economic system in which the government controls a country's economic activities.* The government owns and controls the factories, farms, and stores in a country.

**communism** *n., an economic and political system in which the government owns and controls economic activities.* Communism is a type of command economy. In a Communist system, the government owns and operates factories, farms, and other types of economic activities. For example, in the steel industry, government officials decide what kind of steel and how much steel will be produced. North Korea, Vietnam, and Cuba have Communist systems. Communist economies have been much less efficient than free enterprise economies in producing goods and services. See **Part II Government** for the definition of communism as a political system.

R41

**corporation** *n., a company in which people own shares, or parts of the company.* A corporation sells stock, or shares of the company, to raise money to create products and services. Shareholders often receive a dividend, which is a share of the profits. Multinational corporations are corporations that operate in several countries. People purchase shares of a corporation on a stock exchange.

**depression** *n., a deep and long-lasting contraction of economic activities.* During a depression, business activity falls dramatically. Businesses hire few workers. The unemployment rate, or percent of people without work, rises. In the United States, the Great Depression lasted from 1929 until the early 1940s. A recession is also a contraction in business. It is less severe and shorter than a depression.

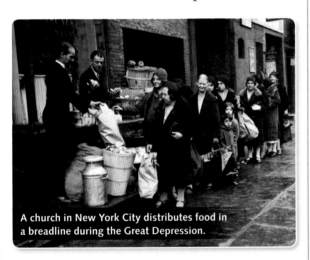
A church in New York City distributes food in a breadline during the Great Depression.

**developed nation** *n., a country with highly-developed industries, a high standard of living, and private ownership of most businesses.* Developed nations have industries that make products like automobiles and computers. Less developed countries have fewer industries and rely on agriculture. They also have lower standards of living. The United States is a developed nation. Many countries in Africa, Asia, and Latin America are less developed countries.

**economy** *n., a country's system for producing and exchanging goods and services.* A country's economic development is based on its level of economic activity. If a country has a great deal of industry, it has a high level of economic development.

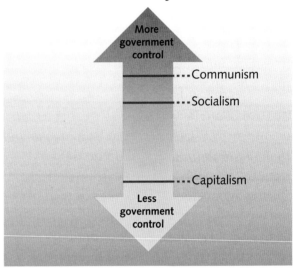

**Economic Systems**

**embargo** *n., a government ban on trade with another country.* Countries set embargoes to show their disapproval of another country's activities. For example, after North Korea tested a nuclear weapon in 2006, the United Nations placed an embargo on trade with that country.

**export** *n., a good that one country sends to another for sale or distribution.* For example, the United States exports computers to countries around the world.

**factors of production** *n., the things that go into producing a good or a service.* Economists have identified four factors of production: land, labor, capital, and entrepreneurs. Land includes all natural resources, such as oil and silver. Labor is the work that people do. Capital is the machinery and other tools that are used to create a good or service. Entrepreneurs are people who start businesses.

**free enterprise** *n., an economic system in which businesses are privately owned, people buy and sell goods in free markets, and individuals freely make decisions whether to buy or sell.* The system is also called capitalism. The United States has a free enterprise system.

In a free enterprise system, producers are motivated by self-interest. To meet consumers' demand, producers decide to manufacture a good, such as an automobile, or offer a service. By doing so, they hope to make a profit. Consumers are also motivated by self-interest. They try to buy the best goods and services they can at the lowest price offered in a market.

Government makes sure that businesses compete fairly. It also ensures that food, medicines, and other products are safe. It provides services, such as defense, that are important to a country. The government also builds infrastructure, such as transportation facilities and roads.

**Free Enterprise System**

**gross domestic product (GDP)** *n., the total value of all the goods and services that a country produces in a specific time period, such as a year.* The GDP is an important measure of an economy's strength. Economists measure the GDP by adding together four kinds of goods and services. One is the goods and services that consumers buy. Another is the machines and other items that companies buy for their businesses. A third is goods and services that government buys. Fourth is the goods and services that a country exports to other countries.

GDP per capita is a country's GDP divided by the country's population. It shows how much the country produces per person.

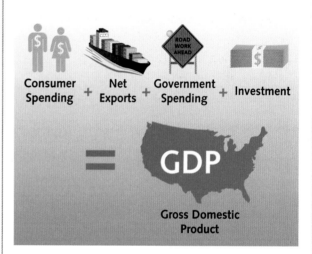

**Gross Domestic Product**

Consumer Spending + Net Exports + Government Spending + Investment = GDP

**import** *n., a good that one country receives from another for sale or distribution.* For example, the United States imports many cars and trucks from automobile manufacturers in Japan. Japan in turn, imports oil from Saudi Arabia.

**industry** *n., a group of businesses that produce a similar product or service.* For example, the film industry produces feature films. Common industries in the United States include construction, computers, pharmaceuticals, and electronics. In many industries, businesses take a raw material and turn it into a finished product. For example, in the clothing industry, businesses turn cotton, wool, and other materials into clothes.

**inflation** *n., an increase in the price of goods and services in a country.* In any given year, the average price of goods and services may go up. This is called an increase in the price level. The rate of increase is called the inflation rate. In 1980, for example, the United States had an inflation rate of about 14 percent. This means that the average prices that year were about 14 percent higher than in 1979.

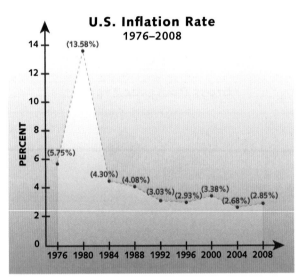

**Source:** U.S. Bureau of Labor Statistics

**manufacturing** *n., the production of physical products to be sold.* Manufacturing creates a wide variety of products, such as automobiles, steamships, airplanes, computers, and furniture. The term refers to the creation of items by machine and by hand. Light manufacturing means the creation of relatively small things, such as the circuits in computers. Heavy manufacturing means the creation of large objects, such as diesel engines for railroads. In 2010, manufacturing employed about 17 million workers in the United States and Canada.

**market economy** *n., an economic system in which people and businesses choose freely to buy and sell in markets.* A market is a place where people buy and sell goods and services. In a market economy, individuals make choices to buy and sell.

For example, an individual might decide to earn a living by making and selling T-shirts. After making them, the person would try to interest stores in selling the T-shirts. A store is a common type of market. The government establishes rules, such as the rule that stores should be safe and clean.

**mechanization** *n., the use of machines instead of humans or animals to perform tasks.* For example, in 1764, the Englishman James Hargreaves invented the spinning jenny. It was a machine that could spin yarn from wool. The machine allowed operators to create yarn much faster than they could by hand.

Spinning jenny

**monopoly** *n., the situation in which one company controls the production or selling of a product or service.* For example, suppose that a person owns the only grocery store in an isolated town. Because the store has no competition, the owner can set high prices on food. If someone else opens another grocery store in town, competition results. Because the stores are competing, they will have to lower prices and improve quality.

**national debt** *n., the total amount of money that the federal government owes.* If a government spends more money than it receives in taxes, it uses deficit spending. It must borrow money from individuals, companies, or other governments. As a government uses deficit spending, its national debt grows.

The national debt of the United States has grown rapidly since 1980. Critics say that the government should raise taxes, cut spending, or do a combination of the two to balance its budget.

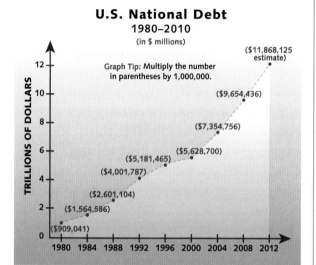

**Source:** U.S. Office of Management and Budget

**natural resources** *n., resources such as oil, coal, timber, and water that exist naturally in a place.* Having ample natural resources can help a country to develop its economy. For example, the United States has numerous natural resources, including timber and fresh water. On the other hand, Japan has become wealthy even though it has few energy resources like oil. It has developed its wealth by inventing new technologies and building quality products.

Oil pump

**opportunity cost** *n., the opportunity that a person gives up when he or she chooses to buy one item instead of another.* For example, suppose that Benita must decide between buying a new flat-screen television or taking a vacation. If she chooses the television, her opportunity cost is the vacation she did not take.

**poverty** *n., the lack of enough money to buy necessary things like food, clothing, and shelter.* Poverty is a worldwide problem. It causes disease and hunger. Poor people often do not have adequate shelter and suffer greatly during cold or hot weather. Economists say that more than one billion people in the world are poor. The U.S. government measures the percentage of its population that is poor. The graph below shows how the rate has changed over the past several years.

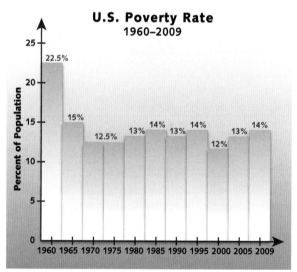

**Source:** U.S. Census Bureaus

**raw materials** *n., the materials that are used in manufacturing a final product.* Raw materials often come from natural resources, such as wood, oil, and iron ore. For example, the basic raw material in creating a table is wood. Cotton and wool are raw materials used in clothing.

**retail goods** *n., goods that are sold directly to consumers.* When you go to a store and purchase a DVD, you are buying a retail good. Today, merchants sell retail goods in a variety of ways. They sell them in stores, through vending machines, over the telephone, or on the Internet.

**service industries** *n., companies, nonprofit organizations, and government organizations that provide services rather than products.* Services include many activities, such as health care, education, financial operations, retail selling, and legal advice. In the United States, more than 75 percent of people work in service industries. Less developed countries, on the other hand, have fewer people working in service industries.

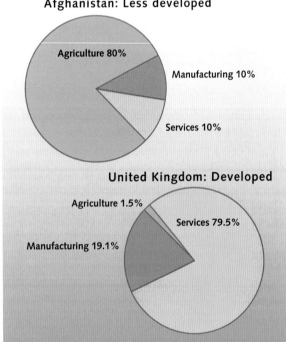

Source: www.NationMaster.com

**socialism** *n., an economic system in which the government owns and operates most businesses.* Socialism first developed in Europe in the early 1800s. Socialists wanted to eliminate poverty and improve working conditions. During the 1840s, a German socialist named Karl Marx said that the government should own all businesses. His ideas gave rise to communism.

Today, Sweden and some other European countries are often called democratic socialist countries. This means the government owns fewer businesses than in Communist countries, but it does provide many services, such as health care. In addition, taxes are high. Many economists claim that socialism is less efficient than free enterprise.

**specialization** *n., the situation in which people focus on doing tasks at which they have the most skill.* By specializing, people and countries provide goods and services efficiently. South Korea, for example, specializes in building large ships. It exchanges the revenue from selling the ships for oil produced in Saudi Arabia. Both countries benefit by doing what they do best.

**standard of living** *n., the level of economic well-being that people in a country have.* In a country with a high standard of living, people have the income to purchase such items as automobiles. In a country with a low standard of living, most of the population struggles to have enough food, clothing, and shelter.

**stock market or stock exchange** *n., a market where people buy and sell stocks and bonds.* Companies are able to raise money in one of two ways. They can sell stocks, which are shares of ownership in the company, or they can issue bonds. A bond is a written agreement to borrow money, and the company pays the money back with interest.

People may want to sell stocks or bonds to raise money or make a profit. The sale of stocks and bonds takes place on a stock exchange or a stock market. Brokers handle both the buying and selling of stocks and bonds. The largest stock market in the United States is the New York Stock Exchange on Wall Street in Manhattan.

**strike** *n., a situation in which workers stop working in order to win higher wages or benefits.* Strikes are most often called by unions. A union is an organization that represents workers in a certain industry. If a union's workers believe that wages are not high enough, they may go on strike. The strike brings the company's operations to a halt. A wildcat strike occurs when workers go on strike without the union's support.

Auto workers on strike, Naperville, Illinois, 2007

**supply and demand** *n., the economic forces that decide the price and amount of a product or service.* The supply of a good or service is the amount that a business is willing and able to offer for sale. If the company can get a higher price, it will create more of the product or service. The demand is the amount of a good or service that consumers are willing and able to buy at a given price. Usually, the demand for a good or service goes down as the price rises. The price at which supply equals demand is called the equilibrium price.

The graph shows how supply and demand work. The gray line stands for demand. The green line stands for supply. The lines meet at equilibrium, when supply equals demand. At prices above equilibrium, demand for the product goes down. At prices below equilibrium, consumer demand rises.

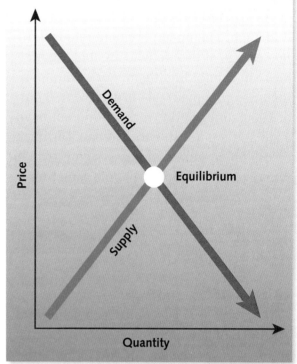

**Supply and Demand**

**tariff** *n., a tax that a country places on goods imported from another country.* Countries use tariffs to protect their own companies from competition with other countries. For example, to protect auto manufacturers, a country might place tariffs on cars from other countries. The country's own manufacturers would thrive. However, consumers might pay more for their cars because there is less price competition.

**trade** *n., the exchange of services and goods.* Trade occurs when people cannot make things themselves but can get them from other people. For example, Chile grows asparagus. It exports the vegetable to countries around the world. With the money it earns, it imports automobiles from the United States, oil from Saudi Arabia, and so forth. This is an example of international trade. The trade within a country is called domestic trade.

**unemployment rate** *n., the percentage of people in a society who cannot find work.* Unemployment occurs when a person is willing and able to work but cannot find a job. The unemployment rate is the percentage of people who are looking for jobs but cannot find them. For example, if the unemployment rate is 8 percent, 8 out of 100 people could work, but do not have a job and are not able to find one. During a recession or depression, the unemployment rate rises. Unemployed people sometimes receive government aid.

**wholesale goods** *n., goods that producers sell to other business firms, such as stores.* Wholesalers buy goods in large quantities. They then sell those goods to retailers, who sell them to customers. For example, a factory in China might produce thousands of toys for children. A wholesaler buys the toys from the factory and ships the toys to retail stores around the world. Then the retailer sells them to you, the consumer. Without wholesalers, the modern economy would not work efficiently.

## PART II GOVERNMENT

**citizenship** *n., membership in a state or a nation, with full rights and responsibilities.* A citizen of a country owes loyalty to that country. He or she is expected to perform certain duties, such as obeying the laws and paying taxes. In return, a citizen has certain rights, such as the right to vote. Civic participation is the way in which citizens participate in their government and society. Citizens let their representatives know what they think about important issues and participate in other ways.

**communism** *n., an economic and political system in which the government owns and controls economic activities.* A communist government exerts great power over its people. This type of government is called totalitarian because it totally controls its people's lives. In the Soviet Union (1917–1991), the Communist Party outlawed all other parties, and the population had no freedom of speech, religion, or other rights.

**constitution** *n., a statement that explains the basic principles and rules of an organization.* The constitution of a government explains how leaders are selected, how laws are passed, and how laws are interpreted and enforced. The U.S. Constitution was ratified, or agreed to, by the states in 1789. In 1791, the states approved of the first 10 amendments, which are known as the Bill of Rights. These amendments guarantee rights, such as the right to speak freely. The U.S. Constitution can be changed, but doing so requires significant majorities to propose amendments, and an even greater majority of states to ratify the change.

### Responsibilities of U.S. Citizens

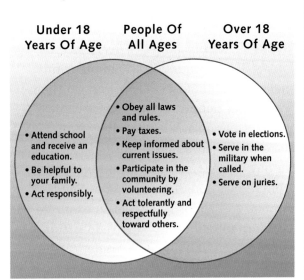

**democracy** *n., a form of government in which the citizens of a nation or state hold the power to pass laws and select leaders.* The United States is a representative democracy in which people elect fellow citizens to serve in a legislature and pass laws. They also elect their leaders, such as the President of the United States. Representative democracies are also known as democratic republics or federal republics (see definition on this page). Democracy had its origins in Greek city-states.

**dictatorship** *n., a form of government in which a ruler holds total power.* A dictator usually comes to power through force. One of the most brutal dictators was Adolf Hitler of Nazi Germany, who ruled from 1933 to 1945. He created the policy that led to the Holocaust, or the murder of 6 million Jews and other people during World War II.

Adolf Hitler

**executive branch** *n., the branch of the federal government that implements and enforces laws.* In the United States, the president is the leader of the executive branch. He or she is the commander-in-chief of the armed forces. The president signs bills from Congress into law. The executive branch then makes sure that the laws are applied. To help enforce the laws, the president has a cabinet, and each member of the cabinet is responsible for a specific area of the federal government.

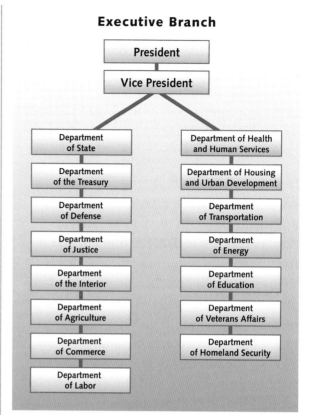

**Executive Branch**

- President
  - Vice President
    - Department of State
    - Department of the Treasury
    - Department of Defense
    - Department of Justice
    - Department of the Interior
    - Department of Agriculture
    - Department of Commerce
    - Department of Labor
    - Department of Health and Human Services
    - Department of Housing and Urban Development
    - Department of Transportation
    - Department of Energy
    - Department of Education
    - Department of Veterans Affairs
    - Department of Homeland Security

**fascism** *n., a form of government that has a dictator and controls political, economic, and other activities.* The two most well-known examples of fascism in the 20th century were Italy under Benito Mussolini (1922–1943) and Germany under Adolf Hitler (1933–1945). Mussolini came to power in Italy in 1922. He promised to solve the economic crisis that Italy faced after World War I. He soon became the country's dictator. Fascism in Italy ended in 1943 when the country surrendered during World War II.

**federal republic** *n., a form of government in which the population elects representatives to pass and carry out laws.* Many Western countries, including the United States and Canada, are federal republics. In federalism, a national, or federal, government shares power with states or provinces. The federal government has certain powers, such as national defense. The states or provinces control other activities, such as education.

**judicial branch** *n., the branch of government that interprets laws.* In the United States, the judicial branch consists of dozens of federal courts and judges. The court with the highest authority on laws is the Supreme Court of the United States. The judicial branch also includes district courts, which oversee cases at local levels.

**legislative branch** *n., the branch of government that creates, changes, or eliminates laws.* In the United States, Congress is the legislative branch, and it is divided into two Houses: the House of Representatives and the Senate. The states also have legislative branches, called state legislatures. In addition to passing laws, the U.S. Congress has other responsibilities. The Senate, for example, must approve of treaties with other countries. Democracies around the world have legislative branches. In the United Kingdom, for example, the Parliament passes laws.

Parliament, United Kingdom

**limited government; unlimited government** *n., a limited government places limits on governmental powers. An unlimited government does not place limits on governmental powers.* The United States has a limited government. The Constitution places limits on its powers. For example, the First Amendment states that the government does not have the power to limit people's free speech.

In unlimited government, the power of the government is not limited by a constitution or other document. Absolute monarchies and dictatorships are unlimited governments. From 1917 until 1991, the Soviet Union had an unlimited government.

**monarchy** *n., a form of government in which a king or a queen rules a country.* A monarch inherits the throne, or position, from his or her parent. Until the 1700s, monarchy was a common form of government. Britain, France, and Russia all had monarchs. In Russia, kings were known as czars. Then, in 1789, the French Revolution limited the powers of the French king, King Louis XVI. In 1793, the revolutionary government executed Louis and formed a republic. Through the 1800s and early 1900s, governments around the world became more democratic. They either abolished the monarchies completely or limited the powers of kings and queens.

**oligarchy** *n., a kind of government in which a small group of people holds power.* In an oligarchy, the rulers usually are either the military or the wealthy classes. Oligarchies usually rule for their own interests, not for the interests of people in lower classes. Myanmar (also known as Burma) is an oligarchy. It is ruled by a small group of military officers.

**totalitarian** *adj., a form of government in which a dictator or small group holds total control over the lives of the people in a country.* In a totalitarian country, the party in power is usually intolerant of other viewpoints and ideas. For example, in the 1930s, the Nazi Party rose to power in Germany by promising to bring back prosperity and return Germany to glory. The Nazi Party shut down all other parties, stating they were obstacles to these goals.

From the beginning of human history, people have asked questions about what life means. What will happen after I die? What is the right way to act? Religion helps people find answers to questions like those.

# World Religions Overview

A religion is an organized system of beliefs and practices. Thousands of religions exist in the world. Most teach that one or more gods, or supreme powers, exist. To help people relate to the divine power, most religions teach a set of beliefs and a moral or ethical code. These codes of conduct also teach people how to treat each other. Another way that religion affects human relationships is in the building of community. Groups of people who share beliefs often gather together to worship and celebrate important events. For a map showing the distribution of the world's religions, see Chapter 2.

**WORLD POPULATION'S RELIGIOUS AFFILIATIONS**

Christianity 33.4% | Judaism 0.2% | Islam 22.7% | Hinduism 13.8% | Buddhism 6.8% | Other 11.4% | Nonreligious 11.7%

**Source:** CIA World Factbook, 2010

# CHRISTIANITY

## Historical Origins

Christianity is based on the life and teachings of Jesus of Nazareth, also called Jesus Christ by Christians. Christians believe he was the son of God who died to save humanity from sin. He was a Jew who lived in the first century near Jerusalem, which was then part of the Roman Empire. Roman rulers put him to death fearing he might lead a revolt. The life of Jesus is recorded in the New Testament of the Christian Bible, which also contains stories of his followers and letters outlining Christian beliefs.

Detail of Jesus in Leonardo DaVinci's *The Last Supper*

## Central Beliefs

Most Christians believe there is only one God, who exists in three forms: Father, Son, and Holy Spirit. The New Testament teaches that after Jesus was executed, he rose from the dead and ascended to heaven. Christians believe they gain salvation by believing in Jesus and following his teachings. Christians gather at places of worship called churches. Their religious leaders are called either priests or ministers.

## Spread of Christianity

Jesus' followers, called disciples, carried their faith around the Mediterranean world. In the A.D. 300s, Christianity became the official religion of the Roman Empire. Later, during the period of colonization, Europeans spread Christianity around the globe. It is now the largest and most widespread religion.

Christians wave palm fronds in Managua, Nicaragua, as they celebrate Palm Sunday, the first day of Holy Week.

# ISLAM

## Historical Origins

Islam teaches that in the year 610, an Arab trader named Muhammad was visited by the angel Gabriel. The angel told Muhammad that he was God's messenger. Muslims believe that through a series of these visits, Muhammad received the words of the Qur'an, or sacred book. According to Islam, Muhammad was the last prophet that Allah, the Muslim name for God, sent to humanity. Muslims believe he is a direct descendant of Abraham, who is also the founder of Judaism.

A 17th-century Turkish ceramic tile from a mosque

## Central Beliefs

Muslims, or followers of Islam, believe there is only one God, the same God worshipped by Jews and Christians. The word *Islam* means surrender, and the goal of Islam is to surrender to the will of Allah. Muslims do this by practicing the Five Pillars of Islam. These are professing faith, praying five times a day, giving to charity, fasting, and making a journey to Mecca to reenact Abraham's dedication. Muslims worship in mosques.

## Spread of Islam

In the centuries after Muhammad's death, Muslims spread their religion by conquest. Islamic rulers took control of Southwest Asia, Central Asia, North Africa, and parts of India and Spain. Today Islam continues to spread around the world through migration and conversion. It is the world's second largest religion.

Muslims engage in prayer at a mosque in Delhi, India, during Ramadan, a holy month of fasting.

# JUDAISM

## Historical Origins

Judaism, the religion of the Jewish people, dates back more than 4,000 years. Its founder was Abraham, who lived in Mesopotamia. According to the Hebrew Bible, God told Abraham to move to Canaan in present-day Israel and Lebanon. God made an agreement with Abraham to bless his descendants. They later became known as Hebrews or Israelites. The Hebrew Bible contains books of law, history, and prophecy. Another important work is the Talmud, a collection of scholarly writings. The painting to the right, *Moses Rescued from the River Nile by Pharaoh's Daughter,* by Orazio Gentileschi, shows a famous scene from the Bible.

Moses, a descendant of Abraham, is rescued from the Nile.

## Central Beliefs

Judaism was the first major religion to teach monotheism, or the belief in one God. Jews believe God is the creator of the whole universe. Traditional Judaism emphasizes rituals and ethics, such as observing the Sabbath, daily prayer, and study, as well as giving to charity, treating others well, and pursuing justice. Judaism's holiest day is Yom Kippur, a day of fasting, prayer, and looking back on the year. Today, Jews worship in synagogues, and their leaders are called rabbis.

## Spread of Judaism

For centuries, Judaism was practiced primarily in what is present-day Israel. Several times in history, empires conquered the region and drove many Jews from the area. The last major event occurred in A.D. 135 when Rome punished Jewish rebels attempting to regain independence. As Jews spread out around the world, Judaism spread with them.

A worshipper holds up a Torah scroll during a Passover blessing at the Western Wall in Jerusalem, Israel.

# BUDDHISM

## Historical Origins

Buddhism is based on the teachings of Siddhartha Gautama, known as the Buddha or "enlightened one." He was born a prince in India in the 400s or 500s B.C. Siddhartha was protected from seeing sickness, death, poverty, or old age until he was 29. However, after he learned about suffering, he left his palace to lead a religious life. While he was meditating several years later, Buddhists believe he received enlightenment about the meaning of life.

A 14th-century painting of Buddha

## Central Beliefs

Buddhists believe that a law of cause and effect called karma controls the universe. Buddhists teach that suffering occurs because people desire what they do not have. A person who gives up desire and other negative emotions will achieve a state called nirvana, or the end of suffering. The basic beliefs of Buddhism are summarized in the Four Noble Truths. The actions that help people achieve nirvana are called the Eightfold Path.

### FOUR NOBLE TRUTHS

- Suffering is a part of life.
- Selfishness is the cause of suffering.
- It is possible to move beyond suffering.
- There is a path that leads to the end of suffering.

## Spread of Buddhism

During its first century, Buddhism spread across northern India. Over time, missionaries and travelers carried it to the Himalayas, Central Asia, and China. China spread Buddhism to Japan and Korea. In the 1800s, immigrants introduced Buddhism to the United States. In the late 20th century, the religion gained popularity in the United States and other Western countries.

Buddhist monks in Siem Reap, Cambodia, celebrate the birthday of Buddha.

# ॐ HINDUISM

## Historical Origins

Hinduism, one of the world's oldest religions, originated in India in about 1500 B.C. Scholars believe that it developed from the beliefs of a group of Indo-European people who spoke Sanskrit. The sacred writings of Hindus include the Vedas, which are poems and hymns, and the Puranas, which are sacred stories. Other Hindu texts such as the *Mahabharata*, of which the Bhagavad Gita is a part, teach Hindu beliefs in the form of epic poems.

Scene from the Bhagavad Gita, a sacred Hindu text

## Central Beliefs

In general, Hindus believe in one universal spirit called Brahman. This spirit takes the form of many deities, or gods and goddesses. The most important deities are Brahma, the creator; Vishnu, the preserver; and Shiva, the destroyer. Hindus believe that souls are constantly being reborn. Karma, the negative or positive effect of one's actions, determines if the soul moves to a higher or lower state of being. Hindu religious practice includes worship, study, and rituals such as bathing in the Ganges River.

## Spread of Hinduism

Hinduism spread from India to other parts of Asia. Today it is the main religion in Nepal, and there are significant Hindu populations in Bangladesh, Sri Lanka, and Pakistan. Nearly 80 percent of Indians are Hindus. Indian immigrants have brought Hinduism to the United States.

Women light candles in observance of Diwali, the Hindu Festival of Lights, in Jaipur, India.

# SIKHISM

## Historical Origins

Followers of Sikhism are called Sikhs, which means "student." Sikhism emerged in 1469 in Punjab, rising from the religious experience and teachings of Guru Nanak, Sikhism's first prophet. Nine gurus followed. The tenth, Guru Gobind Singh, vested his authority in the Sikh scripture—the Guru Granth Sahib—and the Khalsa—the community of initiated Sikhs. This led to the creation of a highly democratic religious community that practiced social justice.

Detail of fresco depicting Guru Nanak Sikh, Sikhism's first prophet

## Central Beliefs

Sikhs believe in one God who is formless, all-powerful, and all-loving. One can achieve unity with God through service to humanity, meditation, and honest labor. Sikhs do not use tobacco or alcohol, and they often follow a strict dress code, which includes never cutting their hair. Vaisakhi is a significant holiday during which Sikhs celebrate their history, community, and values.

## Spread of Sikhism

Sikhism is practiced by nearly 27 million people, most of whom live in the Punjab region of northwest India. However, Sikhs have immigrated all over the world, and now large numbers of Sikhs live in the United States, Canada, and the United Kingdom.

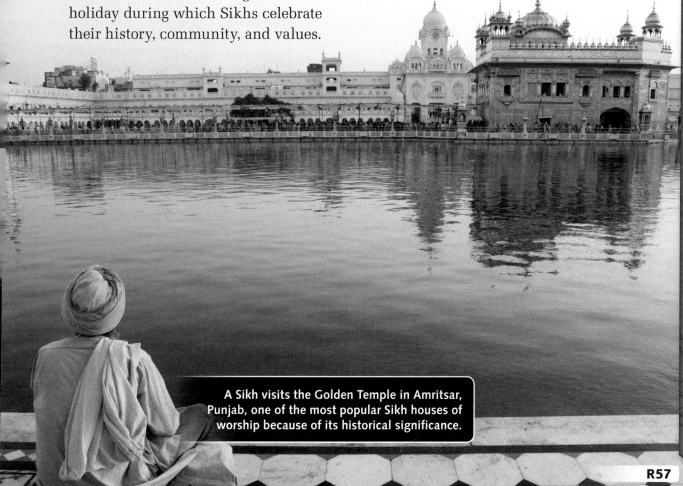

A Sikh visits the Golden Temple in Amritsar, Punjab, one of the most popular Sikh houses of worship because of its historical significance.

# CONFUCIANISM

## Historical Origins

Confucianism is an ethical system and philosophy based on the teachings of a Chinese public official and teacher named Kongfuzi. He is called Confucius in Western countries. He lived from 551 to 479 B.C. His goals were to revive traditional values and establish education as a way to improve society. The sayings and writings of Confucius were collected in a work called the *Analects*.

A 12th-century Chinese painting portrays Confucian filial piety.

## Central Beliefs

One of the most important things in Confucianism is the concept of filial piety, in which children obey and honor their parents. Confucius applied this idea to other areas. For example, subjects should obey their rulers. Confucius taught that education, right relationships, and moral behavior would create an orderly society.

## Spread of Confucianism

Confucianism became a way of life in China. Because China had a major influence on East Asia and Southeast Asia, Confucianism spread throughout these regions. It continues to be a strong cultural force in China and other Asian countries such as South Korea, Japan, and Singapore.

Celebrants participate in a festival in Qufu City, Shandong Province, China, the birthplace of Confucius.

### ONGOING ASSESSMENT

## RESEARCH LAB  GeoJournal

**Express Ideas Through Speech** Get together in a group and prepare a panel discussion in which you analyze what the major world religions have in common.

**Step 1** Appoint each person in your group to become an expert on a different religion. Each person should review the pages in the text that cover their assigned religion.

**Step 2** Each student should summarize the religion they studied. As a group, identify characteristics that all or most of the religions and ethical systems have in common.

**Step 3** Hold a panel discussion before the class and present your conclusions. After you have finished, allow time for questions.

# GLOSSARY

## A

**abolition** *n.*, the movement to end slavery, before and during the Civil War in the United States

**Aborigine** (AB uh RIHJ uh nee) *n.*, Australia's first developed culture

**absolute location** *n.*, the exact point where a place is located, identified by latitude and longitude coordinates

**abstract** *adj.*, an artistic style that stresses form and color over realism

**accommodate** *v.*, to make room for

**acknowledge** *v.*, to recognize

**Acropolis** *n.*, a rocky hill in Athens, Greece, that served as a fortress for the ancient city's most important buildings

**adapt** *v.*, to adjust or modify to fit

**adherent** *n.*, a follower of a religion, cause, or person

**adventure tourism** *n.*, a type of tourism in which travelers engage in physical activities such as mountain climbing, water sports, or hiking

**aerodynamic** *adj.*, designed to move with little wind resistance

**African National Congress (ANC)** *n.*, founded in the early 20th century, an organization of black South Africans, including Nelson Mandela, who protested discriminatory treatment

**African Union** *n.*, an organization of African countries that work toward economic progress

**aging population** *n.*, a demographic trend that occurs as the average age of a population rises

**agricultural revolution** *n.*, a period in which humans began to grow crops instead of gathering plants

**Aguinaldo, Emilio** *n.*, a leader in the movement for Filipino independence in the 1890s

**al-Qaeda** *n.*, a terrorist group based in Southwest Asia

**Alamo** *n.*, a fort in Texas where 200 Texans lost a battle during the U.S.-Mexican War

**Alexander the Great** *n.*, the conqueror who extended the Macedonian Empire and spread Greek culture throughout Eurasia from 334–323 B.C.

**alliance** *n.*, a partnership between countries

**alluvial** (a LOO vee ahl) *adj.*, describing the sediment deposited by a river

**alluvial plain** *n.*, a flat area of land next to a stream or river that floods

**alluvium** *n.*, the soil, or silt, carried by flowing water, which is ideal for farming

**Alps** *n.*, a European mountain chain

**amendment** *n.*, a formal change to a law

**ancestry** *n.*, the family one is descended from, going back in time; heritage

**Amazon River Basin** *n.*, located in South America, the largest river basin on Earth

**Andes Mountains** *n.*, mountain range extending 5,500 miles along the western side of South America

**Angel Falls** *n.*, the highest waterfall in the world, located in Venezuela

**Angkor Wat** *n.*, a large temple built in Cambodia in the 1100s dedicated to the Hindu deity Vishnu

**animal-borne** *adj.*, carried by animals

**anime** *n.*, a style of animation or cartoon developed in Japan

**animism** *n.*, a belief that everything, including objects in nature, has a soul

**annexation** *n.*, the adding of territory to a country

**Antarctic Treaty** *n.*, a 1959 agreement by 12 countries to use Antarctica for peaceful purposes and to share scientific discoveries

**apartheid** (uh PAHRT hyt) *n.*, the legal separation of the races; a system that denied black South Africans their rights

**aqueduct** *n.*, a transport system for carrying water long distances, sometimes raised on a bridge

**aquifer** *n.*, a layer of rock beneath Earth that contains water

**arable** *adj.*, fertile; suitable for farming

**Aral Sea** *n.*, a saltwater lake in Central Asia that has been greatly reduced in size due to diversion for irrigation of the rivers that flow into it

**archipelago** (ahr kuh PEH lug goh) *n.*, a chain of islands

**arid** *adj.*, very dry, having almost no rainfall

**aristocrat** *n.*, a member of the upper class

**armistice** *n.*, an agreement to stop fighting

**artifact** *n.*, an object made by humans from a past culture

**Aryans** (AIR ee uhnz) *n.*, nomads who migrated from Central Asia into the Indus Valley around 2000 B.C.

**Asia-Pacific Economic Cooperation (APEC)** *n.*, a global partnership founded in 1989 to strengthen economic ties among countries in the Pacific

**Asoka** *n.*, the leader at the height of the Mauryan Empire, around 250 B.C.

# GLOSSARY

**assimilate** *v.*, to be absorbed into a society's culture

**assimilation** *n.*, a process in which a minority group adopts the culture of the majority

**assisted migration** *n.*, a program set up in 1831 in which the British gave people money to move to Australia

**Aswan High Dam** *n.*, a dam on the Nile River completed in 1970

**Atacama Desert** *n.*, a desert located on the western side of the Andes Mountains

**Atahualpa** (AH tah WAHL pah) *n.*, the last emperor of the Incas before they were conquered by the Spanish in 1533

**atoll** *n.*, a ring-shaped reef, island, or chain of islands made of coral

**attribute** *n.*, a specific quality

**Augustus** *n.*, the first emperor of Rome in 27 B.C.

**autonomy** *n.*, a country or people's self-governance

**Aymara** (eye MAHR uh) *n.*, an indigenous culture who lives in the Andes Mountains of Peru and Bolivia

**Aztec** *n.*, a people who settled in the area of modern Mexico City, A.D. 1325–1525

# B

**ban** *v.*, to outlaw or forbid

**Bantu** *n.*, an African group who moved from west Central Africa south and east across sub-Saharan Africa from 2000–1000 B.C.

**barbarian** *n.*, a soldier or warrior considered to be culturally less developed than those being fought; the term originally comes from the German tribes who invaded the Roman Empire in A.D. 235

**Baroque period** *n.*, a period from 1600–1750 in which music had complicated patterns and themes

**barter** *v.*, to trade or exchange goods without using money

**bas-relief** *n.*, a sculpture that is slightly raised from a flat background

**basin** *n.*, a region drained by a river system

**bauxite** *n.*, the raw material used to make aluminum

**bay** *n.*, a body of water surrounded on three sides by land

**Bedouin** (BEHD u ihn) *n.*, a nomadic Arabic-speaking people of Southwest Asia

**benefit** *v.*, to be useful to

**Berlin Conference** *n.*, a meeting of European nations in 1884 to settle disputes over their colonial claims in Africa

**Berlin Wall** *n.*, wall that divided Communist East Berlin from democratic West Berlin; it was torn down in 1989

**Bhutto, Benazir** *n.*, the female prime minister of Pakistan, 1988–1999

**Biko, Stephen** *n.*, the president of a student protest organization in South Africa who was arrested and killed in 1977

**Bill of Rights** *n.*, the first ten amendments to the U.S. Constitution

**biodiversity** *n.*, the variety of species in an ecosystem

**biofuel** *n.*, an alternative fuel that is a mixture of ethanol and gasoline

**Black Sea** *n.*, an inland sea bounded by Europe, Russia, Georgia, and Turkey

**Blue Mosque** *n.*, an historic mosque built in Istanbul by Sultan Ahmed I, completed in 1617

**Bollywood** *n.*, India's film industry in Mumbai

**Bolshevik** *n.*, a political party in Russia led by Lenin that overthrew the czar in 1917

**Borobudur** *n.*, a Buddhist temple complex in central Java

**Bosporus Strait** *n.*, the waterway that passes through Istanbul and connects the Black Sea and the Sea of Marmara

**breakwater** *n.*, a barrier built to protect a harbor

**Buddhism** *n.*, a religion founded in 525 B.C. in India by Siddhartha Gautama

**bullet train** *n.*, a train that travels more than 125 miles per hour, and is shaped like a bullet

**butte** *n.*, a hill or mountain with steep sides and a flat top

**Byrd, Richard** *n.*, an American who explored Antarctica by land and air in the 1940s

**Byzantine Empire** *n.*, the eastern part of the Roman Empire after A.D. 395 that lasted about 1,000 years

# C

**Caesar, Julius** *n.*, a general who became the ruler of Rome, 46–44 B.C.

**Calypso** *n.*, a type of folk music that started in Trinidad

**canal** *n.*, a human-made waterway through land for boats and ships

**Candomblé** *n.*, a Brazilian religion that combines African spiritualism and Catholicism

**canopy** *n.*, a roof over a rain forest created by treetops

**capital** *n.*, a country's wealth and infrastructure

# GLOSSARY

**carapace** *n.*, the hard shell of an animal such as a turtle

**caravan** *n.*, a group of merchants traveling together for safety

**Cardamom Mountains** *n.*, mountain range in Cambodia

**Caribbean Sea** *n.*, a tropical sea in the Western Hemisphere bounded by Mexico, Central America, the Greater Antilles, and the Lesser Antilles

**cartographer** *n.*, a mapmaker

**cash crop** *n.*, a farm crop sold for profit

**Caspian Sea** *n.*, the largest enclosed body of water on Earth, bounded by Russia, Kazakhstan, Turkmenistan, Azerbaijan, and Iran

**caste system** *n.*, a social structure in India composed of four main levels, which was started by the Aryans

**categorize** *v.*, to group information; to classify

**Catherine the Great** *n.*, the empress of Russia from 1762 to 1796

**celadon** *adj.*, green-glazed, describing pottery made by Korean artists

**cenote** (se NO tay) *n.*, an underground pool of water

**Chang Jiang** (chahng jyahng) *n.*, the longest river in Asia, which flows through China; also known as the Yangtze

**Chang Jiang Plain** *n.*, the plains along the Chang Jiang River

**Chao Phraya River** *n.*, a river that flows through Thailand

**Chernobyl** *n.*, a city in the Ukraine in which a nuclear reactor exploded in 1986

**Choson** *n.*, a dynasty in Korea that lasted from 1392 to 1910

**Christian Bible** *n.*, the sacred book of Christianity

**Christianity** *n.*, a religion based on the life and teachings of Jesus of Nazareth

**citizen** *n.*, a person living within a territory who has rights and responsibilities granted by the government

**city-state** *n.*, an independent state made up of a city and the territories depending on it

**civil disobedience** *n.*, the nonviolent disobeying of laws

**civil war** *n.*, a war between opposing groups of citizens in the same country

**civilization** *n.*, a society with a highly developed culture, politics, and technology

**clan** *n.*, a large family-based unit with loyalty to the group

**Classical period** *n.*, a period from 1750 to 1900 in which music followed the standard rules of form and complexity, such as sonatas or symphonies

**climate** *n.*, the average condition of the atmosphere of an area over a long period of time, including temperature, precipitation, and seasonal changes

**climograph** *n.*, a graph showing a region's climate through average precipitation and temperature

**coalition** *n.*, an alliance

**coastal plains** *n.*, the lowlands next to the seacoast

**Cold War** *n.*, a long period of political tension without fighting between the United States and the Soviet Union, roughly from 1948 to 1991

**collective farm** *n.*, in the Soviet Union, a large farm where workers grew food to be distributed to the entire population.

**collide** *v.*, to crash together

**colonialism** *n.*, the practice of one country directly ruling and developing trade in a foreign territory for its own benefit

**colonize** *v.*, to build settlements and develop trade in lands that a country controls

**colony** *n.*, an area controlled by a distant country

**Columbian Exchange** *n.*, the exchange of plants, animals, and disease between the Americas, Europe, and Africa that began in the 1500s

**commerce** *n.*, the business of trading goods and services

**commercial agriculture** *n.*, the business of producing crops to sell

**commodity** *n.*, a material or good that can be bought, sold, or traded

**Common Market** *n.*, the European Economic Community formed in 1957

**commonwealth** *n.*, a nation that governs itself but is part of a larger country

**Commonwealth of Australia** *n.*, the country of Australia that became part of the British Commonwealth in 1901

**communal** *adj.*, shared

**communism** *n.*, a system of government in which a single political party controls the government and the economy

**competitive** *adj.*, able to try to win a contest or race

**complex** *n.*, a connected set of buildings

**comply** *v.*, to follow a rule or an order

**concentrated** *adj.*, gathered in one central area

**concentration camp** *n.*, an area where Jews and others were held during World War II and murdered by the Nazis

# GLOSSARY

**condensation** *n.*, the process of water vapor turning into liquid droplets due to cooling in the hydrologic cycle

**Confucianism** *n.*, an ethical system based on the teachings of Confucius

**Congo River** *n.*, a major river in Central Africa that ends in the Atlantic Ocean

**conquistador** *n.*, a Spanish soldier and leader that explored the newly discovered territories of North America, South America, Central America, and the Caribbean beginning in the 1500s, primarily in search of gold and silver

**conservation** *n.*, the protection of the environment

**Constantine** *n.*, Roman Emperor, 306–337 B.C., who made Christianity the official religion of the empire

**constitution** *n.*, a document that organizes a government and states its powers

**consumer** *n.*, a person who buys goods

**contaminated** *adj.*, infected; unfit for use because of the presence of unsafe elements

**contiguous** *adj.*, connected in one block

**continent** *n.*, a large landmass on Earth's surface; Earth has seven continents

**continental drift** *n.*, the slow movement of continents on tectonic plates

**continental shelf** *n.*, the edge of a continent that extends under water into the sea

**controversy** *n.*, a debate or quarrel

**convert** *v.*, to persuade someone to change religious beliefs

**convey** *v.*, to communicate

**convict** *n.*, a person sentenced to prison for committing a crime

**Cook Islands** *n.*, a country near New Zealand in the South Pacific made up of 15 small islands

**Cook, James** *n.*, a British navigator who explored and mapped the Pacific Islands in the 1770s

**coral island** *n.*, an island created by a gradual buildup of coral skeletons

**coral reef** *n.*, a rock-like structure built by layers of coral organisms

**cordillera** *n.*, a system of several parallel mountain ranges

**Cortés, Hernán** *n.*, the Spanish conquistador who defeated the Aztecs in 1525

**cosmopolitan** *adj.*, bringing together many different cultures and influences

**Counter-Reformation** *n.*, a movement within the Roman Catholic Church to reform its own practices

**coup** (KOO) *n.*, a sudden, illegal takeover of government by force

**Creole** *n.*, a blended language of European and non-European languages

**crevasse** *n.*, a deep, open crack in a glacier

**cricket** *n.*, a team game similar to baseball

**critical** *adj.*, extremely important, needed for survival

**Crusades** *n.*, military expeditions of the Roman Catholic Church to take back holy lands in the Middle East from Muslim control, 1096–1291

**cuisine** *n.*, the food and cooking traditions common to a certain region

**cultural hearth** *n.*, center of civilization from which ideas and technology spread

**Cultural Revolution** *n.*, social plan in China that lasted from 1966 to 1969 in which the government attempted to remove any capitalistic and anti-Communist elements from society

**culture** *n.*, a group's way of life, including types of food, shelter, clothing, language, behavior, and ideas

**culture region** *n.*, an area that is unified by common cultural traits

**cuneiform** (kyoo NEE uh form) *n.*, the earliest known form of writing, from Sumeria

**currency** *n.*, a form of money

**current** *n.*, the continuous movement of air or water in the same direction

**cyclone** *n.*, a storm with rotating winds, called a typhoon in the Eastern Hemisphere; a hurricane in other parts of the world

**czar** *n.*, term used for an emperor in Russia

# D

**dam** *n.*, a barrier that controls the flow of water

**Danube River** *n.*, a river that starts in Germany and ends at the Black Sea

**Daoism** *n.*, an ethical system that emphasizes the harmony between people and nature; also called Taoism

**Declaration of Independence** *n.*, the document declaring U.S. independence from the British Empire, adopted July 4, 1776

**deforestation** *n.*, the practice of cutting down forests for crops or urban use

**deity** *n.*, a god or goddess

**delta** *n.*, an area where a river deposits sediment as it empties into a larger body of water

# GLOSSARY

**demilitarized zone (DMZ)** *n.*, a neutral area between enemy countries, specifically between North Korea and South Korea

**democracy** *n.*, a form of government by the people, in which citizens often elect representatives to govern them

**democratization** *n.*, the process of becoming a democracy

**demographics** *n.*, the characteristics of a human population, such as age, income, and education

**deny** *v.*, to refuse to recognize

**deplete** *v.*, to drain or diminish a resource so as to make it no longer available or able to function as planned

**descendants** *n.*, future generations; relatives of a past family member

**desertification** *n.*, a gradual transition from fertile to less productive land

**developed nations** *n.*, countries with a high per capita gross domestic product

**developing nations** *n.*, countries with a low per capita gross domestic product

**dialect** *n.*, a regional variation of a main language

**Diaspora** *n.*, the movement of Jews away from Israel

**dictator** *n.*, a ruler with complete control

**diffusion** *n.*, a spreading out

**diplomacy** *n.*, discussion between groups or countries to resolve disputes or disagreements

**discrimination** *n.*, unfair treatment of an individual or group based on factors other than ability

**displace** *v.*, to force a people to leave their homes

**disregard** *v.*, to ignore

**dissolve** *v.*, to break up

**distinct** *adj.*, easily recognized

**distort** *v.*, to change the usual shape or appearance; twist out of shape

**distribute** *v.*, to spread an asset or resource amongst a larger group

**diversify** *v.*, to give variety to

**diversity** *n.*, variety

**domestic policy** *n.*, a government's plan for running affairs within its own borders

**domesticate** *v.*, to keep and use animals as a source of labor and food

**dormant** *adj.*, inactive, referring to a volcano

**drought** *n.*, a long period with little or no precipitation

**Dry Pampas** *n.*, dry grasslands in western Argentina

**Dubai** *n.*, a young and wealthy city and state in United Arab Emirates on the Persian Gulf

**due process** *n.*, in the United States, rules that authorities must follow in dealing with citizens

**Dutch East India Company** *n.*, a Netherlands company that dominated the spice trade in Southeast Asia

**dynamic** *adj.*, continuously changing

**dynastic cycle** *n.*, a pattern in the rise and fall of dynasties in Chinese history

**dynasty** *n.*, a series of rulers from the same family

# E

**earthquake** *n.*, a shaking of Earth's crust generally caused by the collision or sliding of tectonic plates

**Eastern Hemisphere** *n.*, the half of Earth east of the prime meridian

**ecologist** *n.*, a scientist who studies the relationships between living things and their environment

**economic globalization** *n.*, the practice of economic activities being conducted across national borders

**economic sector** *n.*, a subdivision or smaller part of an economy, such as industry and agriculture

**economy** *n.*, a system in which people produce, sell, and buy things

**ecosystem** *n.*, a community of living organisms and their natural environment, or habitat

**ecotourism** *n.*, a way of visiting natural areas that conserves the natural resources of the region

**El Niño** (ehl NEEN yoh) *n.*, a reversal of usual wind and ocean currents

**elevation** *n.*, the height of a physical feature above sea level

**eliminate** *v.*, to get rid of

**elite** *adj.*, superior

**Emancipation Proclamation** *n.*, an 1863 document that freed all slaves living in Confederate held territory during the U.S. Civil War

**emergence** *n.*, the development and widespread use of something new

**emigrate** *v.*, to leave one's home country to live in another country

**emirate** *n.*, a state in the country of the United Arab Emirates

**empire** *n.*, a group of peoples or states ruled by a strong single ruler

# GLOSSARY

**Enduring Voices Project** *n.*, a National Geographic project to study and preserve languages at risk of dying out

**enhance** *v.*, to improve the quality of something

**Enlightenment** *n.*, a social movement of the 1700s that worked for the education and rights of the individual

**enlist** *v.*, to volunteer for military service

**entrepreneur** *n.*, a person who starts up a new business

**entrepreneurship** *n.*, the characteristics of creativity and risk existing in a person or society

**epic poem** *n.*, a long poem that tells of a hero's adventures

**epidemic** *n.*, an outbreak of a disease affecting a great number of the population in a particular community

**equator** *n.*, an imaginary circle around Earth that is the same distance from the North and South Poles and divides Earth in half; the center or 0° line of latitude

**equinox** *n.*, when day and night are of equal length; occurs twice each year on March 21 or 22 and September 21 or 22

**erode** *v.*, to wear away

**erosion** *n.*, the process by which rocks and soil slowly break apart and are worn away

**erratic** *adj.*, inconsistent or irregular

**eruption** *n.*, a blast or explosion

**escarpment** *n.*, a steep slope

**essential** *adj.*, necessary

**establish** *v.*, to put into place

**ethanol** *n.*, a liquid alcohol removed from sugarcane or corn that can be used as a fuel alone or blended with gasoline

**ethical system** *n.*, a belief system that teaches moral behavior

**ethnic group** *n.*, a group of people who share a common culture, language, and sometimes racial heritage

**ethnobotanist** *n.*, a scientist who studies the relationship between cultures and plants

**euro** *n.*, the common currency of the European Union

**European Union (EU)** *n.*, an economic organization composed of 27 European member countries (2011)

**eurozone** *n.*, the countries that have adopted the euro as money

**evaporation** *n.*, the process of water turning into vapor and rising into the atmosphere due to the sun's heating; part of the hydrologic cycle

**evident** *adj.*, clearly present or observable

**excavate** *v.*, to carefully uncover or dig up

**exchange** *v.*, to convert money into another currency

**exile** *n.*, a state of absence from one's home country

**exile** *v.*, to force to leave a country

**exoskeleton** *n.*, a hard external covering that protects corals and other sea animals

**expand** *v.*, to make larger

**expedition** *n.*, a journey or voyage of some length, usually to other lands

**exploit** *v.*, to take advantage of; to use selfishly or unfairly for one's own profit

**export revenue** *n.*, money or income received for goods sold to another country

**export** *v.*, to send to another country for aid or profit

**extent** *n.*, the distance or degree to which something spreads

**extinct** *adj.*, completely gone, such as a species of animal or plant

**extinction** *n.*, the dying out of a species or type of living thing

**extremist** *n.*, a person with religious or political views that are outside the range of majority opinion

# F

**factory system** *n.*, a way of working in which each person works on only one part of a product

**failed state** *n.*, a country in which government, economic institutions, and order have broken down

**fallout** *n.*, radioactive particles from a nuclear explosion that fall through the atmosphere

**famine** *n.*, a long period of food shortage

**fault** *n.*, a fracture in Earth's crust

**federal republic** *n.*, a democratic form of government in which voters elect representatives and the central government shares power with the states

**federal system** *n.*, a system of government with a strong central government and local government units

**feral** *adj.*, wild, having gone wild after it was once domesticated

**fertile** *adj.*, able to produce a great deal of fruit, crops, or offspring

**Fertile Crescent** *n.*, the area around the Tigris and Euphrates rivers

**fertility rate** *n.*, the average number of children born per woman at a certain period of time

# GLOSSARY

**fertilizer** *n.*, a substance added to soil to enrich it

**feudal system** *n.*, a social structure during the Middle Ages consisting of a king, lords, vassals, and serfs

**fiber optics** *n.*, glass fibers used to send digital code quickly across great distances

**first language** *n.*, the language learned by the children of a people

**fjord** (fee ORD) *n.*, a deep, narrow bay

**floodplain** *n.*, the low-lying land next to rivers, formed by sediment left by flooding

**Foja Mountains** *n.*, a mountain range on the island of New Guinea in eastern Indonesia

**food security** *n.*, easy access to enough food

**foremost** *adj.*, first in rank or importance, leading

**fortify** *v.*, to strengthen

**fossil** *n.*, the preserved remains of ancient animals and plants

**fragmented country** *n.*, a country that is physically divided into separate parts, such as a chain of islands, and/or politically or culturally divided

**free enterprise economy** *n.*, a system in which privately-owned businesses create goods and services; also called a market economy or capitalism

**free trade agreement** *n.*, a treaty between countries that improves trade by limiting taxes on that trade

**free trade** *n.*, trade that does not impose taxes on imports

**French Indochina** *n.*, France's combined colonies of Vietnam, Cambodia, Laos, and others in Southeast Asia from 1887 through 1955

**fuel cell** *n.*, a small unit that, like a battery, combines chemicals to make energy

**fuse** *v.*, to blend

# G

**Gadsden Purchase** *n.*, a sale in 1853 from Mexico to the United States that established the current U.S. southwestern border

**Gandhi** (GAHN dee), **Mohandas** *n.*, Indian leader who pushed for India's independence from the British through civil disobedience in the 1930s; considered the father of modern India

**Ganges Delta** *n.*, fertile area where the Ganges River flows into the Bay of Bengal

**gaucho** *n.*, a cowboy of South America

**gauge** *n.*, the measurement of the width of railroad tracks

**generation** *n.*, a group of individuals who are born and live about the same time

**Genghis Khan** (JEHNG-gihs KAHN) *n.*, the Mongol ruler who established an empire in Central Asia in the early 1200s

**genre** *n.*, a form of literature such as poem, play, or novel

**geoglyph** *n.*, a large geometric design or animal shape drawn on the ground

**Geographic Information Systems (GIS)** *n.*, computer-based devices that show data about specific locations

**geographic pattern** *n.*, a similarity among places

**geothermal energy** *n.*, heat energy from within the earth that can be turned into electricity

**ger** *n.*, a portable tent made of felt

**Gettysburg Address** *n.*, the speech Abraham Lincoln made in 1863 honoring soldiers who died at the Battle of Gettysburg during the U.S. Civil War

**Giza** *n.*, a city on the Nile where ten pyramids were built in ancient Egypt

**glacier** *n.*, a large mass of ice and packed snow

**glasnost** (GLAHS-nuhst) *n.*, the Soviet Union's policy of openness that encouraged people to speak openly about government, introduced by Mikhail Gorbachev in the 1980s

**global** *adj.*, worldwide

**Global Positioning System (GPS)** *n.*, a satellite system based in space that finds absolute location and time anywhere on Earth

**global warming** *n.*, the increase in Earth's average temperature since the mid-20th century

**globalization** *n.*, the development of a world economy based on free trade and the use of foreign labor

**globe** *n.*, a three-dimensional, or spherical, model of Earth

**Gobi desert** *n.*, the largest desert in Asia, covering much of southern Mongolia and extending into China

**golden age** *n.*, a period of great wealth, culture, and democracy in Greece

**Golden Quadrilateral (GQ)** *n.*, a superhighway that connects four major cities in India

**Gorbachev, Mikhail** (mih KYL GAWR buh chawf) *n.*, a leader in the Soviet Union 1985–1991; President from 1990 until it was dissolved in 1991

**gorge** *n.*, a deep, narrow passage surrounded by steep cliffs

**government** *n.*, an organization that keeps order, sets rules, and provides services for a society

# GLOSSARY

**Grand Bazaar** *n.*, a center of shops and trading in Istanbul, Turkey

**Grand Canyon** *n.*, rock formation in southwestern United States that has been cut deeply by the Colorado River over millions of years

**grasslands** *n.*, wide open areas used for grazing and crops

**Great Barrier Reef** *n.*, a huge ecosystem off the coast of Australia made up of coral reefs

**Great Depression** *n.*, worldwide economic downturn in the 1930s, marked by poverty and high unemployment

**Great Escarpment** *n.*, the steep slope from the plateau of Southern Africa to the coastal plains

**Great Lakes** *n.*, five large freshwater lakes between Canada and the United States

**Great Leap Forward** *n.*, Mao Zedong's plan to make China's economy grow faster, 1958–1961

**Great Plains** *n.*, a flat area of land east of the Rocky Mountains

**Great Pyramid of Khufu** *n.*, the oldest and tallest pyramid in Egypt, built at Giza around 2550 B.C.

**Great Recession** *n.*, a downturn in the economies of the United States, Canada, and other countries, beginning in 2007

**Great Rift Valley** *n.*, a wide valley in East Africa, part of a chain of valleys formed when tectonic plates separated

**Great Wall of China** *n.*, a stone wall in northern China over 4,500 miles long, built to repel invaders from the north

**Great Zimbabwe** (zim BAH bwe) *n.*, a walled city of stone that the Shona people built in Southern Africa between 1200 and 1450

**greenhouse gas** *n.*, a gas that traps the sun's heat over Earth

**griot** (GREE oh) *n.*, a traditional African storyteller

**gross domestic product (GDP)** *n.*, the total value of all goods and services produced in a country in a given year

**Guaraní** (GWAH rah NEE) *n.*, an indigenous people who live in the lowlands on the Paraguay and Paraná rivers in South America

**guest worker** *n.*, a temporary laborer who migrates to work in another country

**guillotine** (GHEE uh teen) *n.*, a machine used to execute people during the French Revolution

**Gutenberg, Johannes** *n.*, a German printer who invented the printing press in 1450

# H

**habitat** *n.*, the natural environment of a living plant or animal

**Hagia Sophia** *n.*, a museum in Istanbul, Turkey, that was originally built as a church during the Roman Empire and later served as a mosque.

**half-life** *n.*, the time needed for half the atoms in a radioactive substance to decay

**Hammurabi** *n.*, the king who developed a code of law in Babylonia in the 1700s B.C.

**Han** *n.*, a Chinese dynasty that lasted from 206 B.C. to A.D. 220

**Harappan** *adj.*, related to the first urban civilization in South Asia along the Indus River around 2600 B.C.

**harbor** *n.*, a place where ships can land protected from the open sea

**Hatshepsut** (hat SHEHP soot) *n.*, a female pharaoh in Egypt around 1500 B.C.

**Hebrew Bible** *n.*, the sacred book of Judaism

**hemisphere** *n.*, one-half of Earth

**henna** *n.*, a reddish powder used to create designs on skin

**hereditary** *adj.*, passed on through the family

**heritage** *n.*, a tradition passed down from ancestors

**Hermitage Museum** *n.*, a museum of art and culture in St. Petersburg, Russia

**Hidalgo, Miguel** *n.*, a Catholic priest who led a revolt in Mexico in 1810

**hieroglyphics** (HY ruh GLIHF ihks) *n.*, an ancient system of writing that uses pictures and symbols

**highlands** *n.*, areas of high mountainous land

**Himalaya Mountains** *n.*, the highest mountain range in the world, located in South Asia

**Hinduism** *n.*, religion practiced by more than 60 percent of the population of South Asia

**Hitler, Adolf** *n.*, German head of state from 1933 to 1945

**Ho Chi Minh** *n.*, the leader of North Vietnam who wanted to unify North and South Vietnam under communism; these efforts started the Vietnam War

**Holocaust** *n.*, the mass slaughter by the Nazis of six million Jews and others during World War II

**homelands** *n.*, during Apartheid, separate areas within South Africa where black South Africans were forced to live

**hotspot** *n.*, an unusually hot part of Earth's mantle

**Huang He** (hwahng huh) *n.*, China's second largest river, also called the Yellow River

# GLOSSARY

**Human Development Index (HDI)** *n.*, a set of data used by geographers to compare quality of life in different countries including health, education, and standard of living

**human rights** *n.*, the political, economic, and cultural rights that all people should have

**hunter-gatherer** *n.*, a person who hunts animals and gathers plants and fruits for food

**hurricane** *n.*, a strong storm with swirling winds and heavy rainfall

**Hussein, Saddam** *n.*, the president of Iraq from 1979 to 2003

**hybrid** *n.*, a vehicle that can use either electricity or gas to run

**hydroelectric power** *n.*, a source of energy that uses flowing water to produce electricity

## I

**ice shelf** *n.*, a floating sheet of ice attached to a landmass

**immigrant** *n.*, a person who takes up permanent residence in another country

**immigrate** *v.*, to move to a new country or region

**immigration** *n.*, the permanent movement of a person to a different country

**impact** *n.*, an effect that produces change

**imperialism** *n.*, the practice of extending a nation's influence by controlling other territories

**Impressionism** *n.*, an artistic style in which artists used light and color in short strokes to capture a moment in time

**incentive** *n.*, the reason or motive to do something

**incorporate** *v.*, to include; to combine with something already formed

**indigenous** *adj.*, native to the area in which something is found

**indulgence** *n.*, a fee paid to the church to relax the penalty for a sin during the Middle Ages

**Indus River** *n.*, a river in the western part of South Asia

**Industrial Revolution** *n.*, a period in the 1700s and 1800s in which workers in factories began to use machines and power tools for large-scale industry

**industrialization** *n.*, the shift to large-scale production using machines

**industrialize** *v.*, to develop manufacturing

**infectious** *adj.*, capable of spreading rapidly to others

**infrastructure** *n.*, the basic systems of a society such as roads, bridges, sewers, and electricity

**inner islands** *n.*, islands in Indonesia that include Java, Madura, and Bali

**Institutional Revolutionary Party (PRI)** *n.*, a political party that controlled Mexico's government from 1929 to 2000

**interact** *v.*, to affect other people and be affected by them

**interior** *n.*, the land that is away from a seacoast

**Internet** *n.*, a communications network

**intersection** *n.*, a place where people meet or paths cross

**intifada** *n.*, an uprising or rebellion, usually meant to reference the Palestinian revolt against Israel

**invader** *n.*, an enemy who enters a country by force

**invasive species** *n.*, non-native plants or animals introduced to a new area, intentionally or unintentionally, that disturb habitats of native life forms

**Iron Curtain** *n.*, an imaginary boundary that separated Communist and non-Communist countries in Europe during the Cold War

**Irrawaddy River** *n.*, a river that flows through Myanmar

**irrigation** *n.*, the process of redirecting water to crops through channels and ditches

**Isis** *n.*, an important Egyptian goddess

**Islam** *n.*, religion founded in Saudi Arabia in the early A.D. 600s

**isolated** *adj.*, cut off from others

**isolation** *n.*, separation or being set apart from others

**isthmus** *n.*, a narrow strip connecting two large land areas

## J

**Jainism** *n.*, a religion in northwest India started in the late A.D. 500s

**Java** *n.*, one of the inner islands located in Southeast Asia

**Jerusalem** *n.*, the capital city of Israel

**Jiang Jieshi** (jee ahng jee shee) *n.*, a Chinese nationalist leader who fought Communists

**Jinnah, Mohammed Ali** *n.*, a leader of the All-India-Muslim League who helped found Pakistan and was Pakistan's first Governor-General from 1947–1948

# GLOSSARY

## K

**Kalahari** *n.*, a major desert in southern Africa

**Kanto Plain** *n.*, flat area good for agriculture and industry located east of the Japanese Alps

**karma** *n.*, in Hinduism, the negative or positive effect one receives as a result of his or her actions

**Kenyatta, Jomo** (JOH moh ken YAA taa) *n.*, an African leader who helped gain Kenya's independence and became its first elected leader in 1963

**Khmer Empire** *n.*, the largest and longest-lasting empire in Cambodia from the 800s to 1430

**Kievan Rus** *n.*, the state established by the Varangian Russes in 882 that became part of modern Russia

**Kilimanjaro** *n.*, an inactive volcano in Tanzania that is 19,340 feet tall

**kimono** *n.*, traditional Japanese women's clothing

**Kingman Reef** *n.*, an unspoiled coral reef in the Line Islands between Hawaii and American Samoa

**kinship** *n.*, blood or family relationship

**Knesset** (kuh NEH set) *n.*, the Israeli parliament

**Koguryo** (koh gur YOO) *n.*, a kingdom in northern Korea in 37 B.C.

**Kongo** *n.*, a state founded in 1390 in Central Africa

**Koryo** *n.*, a dynasty that ruled Korea from 935 to 1392

**Kremlin** *n.*, a historic complex of palaces, armories, and churches in Moscow and the seat of Russian government

**Krishna** *n.*, in Hinduism, a deity in the bodily form of Vishnu

**Kunlun Mountains** *n.*, a mountain range in East Asia

**Kurd** *n.*, a member of a non-Arab ethnic group in Southwest Asia

## L

**labor force** *n.*, the number of people available to work

**land bridge** *n.*, a strip of land connecting two landmasses

**land reform** *n.*, the breaking up of large estates to give land to the poor

**landlocked** *adj.*, surrounded by land on all sides, with no direct access to a seacoast

**landmass** *n.*, a very large area of land

**language diffusion** *n.*, the spread of languages from their original home

**language family** *n.*, a group of related languages

**latitude** *n.*, an imaginary line around Earth that runs east to west, showing location relative to the equator

**launch** *v.*, to start

**legume** *n.*, peas or beans

**Lenin, V.I.** *n.*, the Bolshevik leader who overthrew the czar in the Russian Revolution of 1917 and led the new government

**liberate** *v.*, to set someone or something free

**lingua franca** (LEEN gwa FRAWN kah) *n.*, the common language among several groups of people

**linguist** *n.*, a language scientist

**literacy rate** *n.*, the percentage of people who can read and write

**literate** *adj.*, able to read and write

**Llanos** *n.*, grasslands in northern South America

**Locke, John** *n.*, an English philosopher of the late 1600s who helped inspire the American Revolution and the Enlightenment

**locks** *n.*, devices in canals used to raise or lower ships between the waterways being connected

**loess** (less) *n.*, a yellow silt or sediment that forms thick deposits

**longitude** *n.*, an imaginary line running north to south from the North Pole to the South Pole that shows location relative to the prime meridian

**Lost Boys of Sudan** *n.*, a group of young men of Sudan who were orphaned by civil war and stuck together to escape the violence

**Lost Decade** *n.*, the 1990s in Japan, when production declined because businesses were deeply in debt

**Louisiana Purchase** *n.*, the land Thomas Jefferson bought in 1803, doubling the size of the United States

**L'Ouverture, Toussaint** (too SAN loh ver CHOOR) *n.*, a former slave who led Haiti's successful revolt for independence from France

**lowland** *n.*, a low-lying area

**lucrative** *adj.*, making money, or profitable

**Luther, Martin** *n.*, a German monk whose actions in 1517 led to the Reformation to address corruption in the Roman Catholic Church

## M

**Machu Picchu** (MAH choo PEE choo) *n.*, a complex city built on a mountain by the Incas in the 1400s

**Madura** *n.*, one of the inner islands located in Southeast Asia

# GLOSSARY

**magnetic levitation (Maglev)** *adj.*, type of train that rides on a cushion of air over tracks laid with many powerful magnets

**mainland** *n.*, the land connected to a continent, usually in reference to countries with both continental land and islands.

**maintain** *v.*, to preserve and carry on

**Malay Peninsula** *n.*, a peninsula in Southeast Asia that includes parts of Thailand, Myanmar, and the mainland section of Malaysia

**Malaysia** *n.*, a country in Southeast Asia at the bottom of the Malay Peninsula and the island of Borneo

**malnutrition** *n.*, the lack of enough food or nourishment

**Mandela, Nelson** *n.*, a leader of the African National Congress who was jailed for fighting apartheid, but who became the president of South Africa in 1994

**manga** *n.*, a type of Japanese comic book

**Manifest Destiny** *n.*, the idea that the United States had the right to expand its territory to the Pacific Ocean

**Manila** *n.*, the capital of the Philippines

**manufacturing** *n.*, the use of machines to make raw materials into usable products

**Mao Zedong** (MOW dzuh dahng) *n.*, the chairman of the Communist Party who led China from 1949–1976

**Maori** (MOW ree) *n.*, the native people of New Zealand

**map** *n.*, a two-dimensional, flat representation of Earth

**marine** *adj.*, sea-based

**marine life** *n.*, the plants and animals living in the ocean

**marine reserve** *n.*, an ocean area set aside to protect ocean life from humans

**maritime** *adj.*, related to the sea

**marketing** *n.*, advertising and promotion for a product or business

**marsupial** *n.*, a mammal whose females carry their babies in a pouch

**martial law** *n.*, government maintained by military power

**mass media** *n.*, communication from a single source with the potential to reach large audiences

**Maya** *n.*, a civilization that lived in the Yucatán and northern Central America from 100 B.C. to A.D. 900

**medicinal plant** *n.*, plant used to treat illness

**meditation** *n.*, the practice of using concentration to quiet and control thoughts

**Mediterranean climate** *n.*, a climate that has hot, dry summers and mild, rainy winters

**megacity** *n.*, a large city with more than 10 million people

**Mekong River** *n.*, the longest river in Southeast Asia; flows through Myanmar, Laos, and Thailand

**Mesa Central** *n.*, in Mexico, the southern area of the Mexican Plateau

**messiah** *n.*, a leader or savior

**mestizo** *n.*, a person who has mixed European and Native American ancestry

**methane** *n.*, a colorless, odorless natural gas released from carbon

**metropolitan area** *adj.*, populated place around a city that includes the city limits and the surrounding communities

**Mexican Cession** *n.*, the land ranging from Texas to California given by Mexico to the United States in the 1848 Treaty of Guadalupe Hidalgo

**Mexican Plateau** *n.*, a flat area of land that lies between the two ranges of the Sierra Madre Mountains in Mexico

**microcredit** *n.*, a small loan of money

**microlending** *n.*, the practice of making small loans to people starting their own businesses

**Middle Ages** *n.*, a period in Western Europe after the fall of the Roman Empire, from about 500 to 1500

**Middle Passage** *n.*, the months-long trip across the Atlantic Ocean in which enslaved Africans were brought to European colonies in the Americas

**migrate** *v.*, to move from one place to another

**migration** *n.*, the movement from one place to another

**military dictatorship** *n.*, a form of government in which the army runs the government

**mineral** *n.*, a solid natural substance found in rocks and Earth that is inorganic and has its own set of properties

**missionary** *n.*, a person sent by a religious organization to convert others to that religion

**mobile** *adj.*, movable

**modernization** *n.*, policies and actions designed to bring a country up to date in technology and other areas

**modify** *v.*, to change, or make less extreme

**Mongol Empire** *n.*, an empire established in Central Asia in the early 1200s by Genghis Khan

**monk** *n.*, a man who devotes himself to religious work

# GLOSSARY

**monopoly** *n.*, the complete control of the market for a service or product

**monotheism** (MAHN uh thee ihz uhm) *n.*, a belief in one god

**monotheistic** *adj.*, related to a religious belief in one god

**monotheistic religion** *n.*, a system of belief in one god or deity

**monsoon** *n.*, a seasonal wind that brings intense rainfall during part of the year

**Montezuma** *n.*, an Aztec ruler who was killed by the Spanish conquistador Hernán Cortés

**moral** *adj.*, right and good, describing human behavior

**mosque** *n.*, a Muslim place of worship

**mouth** *n.*, the place where a river empties into a sea

**movable type** *n.*, a printing invention in which individual characters can be moved to create different pages of text

**multinational corporation** *n.*, a large company based in one country that establishes branches in several others

**multi-party democracy** *n.*, a political system in which elections include candidates from more than one party

**multitudes** *n.*, large numbers

**Munduruku** (moon doo ROO koo) *n.*, an indigenous people of Brazil

**mural** *n.*, a large painting on a wall

# N

**Nairobi** (ny ROE bee) *n.*, Kenya's capital city

**Napoleon** *n.*, Napoleon Bonaparte, the leader of France who conquered other European countries to build an empire, 1804–1815

**National Action Party (PAN)** *n.*, a political party in Mexico that won election in 2000

**nationalism** *n.*, a strong sense of loyalty to one's country

**nationalize** *v.*, to place a private industry under government control

**natural rights** *n.*, rights such as life, liberty, and property that people possess at birth

**naturalization** *n.*, the process that someone born in another country follows to become a citizen

**navigable** *adj.*, wide or deep enough to be traveled easily by boats and ships

**navigation** *n.*, the science of finding position and planning routes

**Nazi Germany** *n.*, Germany as led by the Nazi party from 1933–1945

**N'Dour, Youssou** *n.*, a famous griot from West Africa who plays Mbalax music

**Nebuchadnezzar** (nehb buh kuhd NEHZ uhr) *n.*, king of Babylonia, 605–562 B.C.

**neutrality** *n.*, the refusal to take sides or become involved

**New Guinea** *n.*, the world's second largest island, located in the southwest Pacific

**New South Wales** *n.*, in Australia, a colony built by convicts in 1788

**Nile River** *n.*, the longest river in the world; flows through Egypt and Africa for 4,000 miles

**Nkrumah, Kwame** (KWAA may en KROO mah) *n.*, an African leader in the 1950s and 1960s who helped Ghana gain independence

**nocturnal** *adj.*, active at night rather than during the day

**nomad** *n.*, a person who moves from place to place

**nonrenewable** *adj.*, cannot reproduce quickly enough to keep up with its use

**nonrenewable fossil fuel** *n.*, a source of energy such as oil, natural gas, or coal that is in limited supply

**nonrenewable resource** *n.*, a source of energy that is limited and cannot be replaced, such as oil

**North American Free Trade Agreement (NAFTA)** *n.*, a 1994 agreement that made trade and investment easier among Canada, Mexico, and the United States

**North Anatolian Fault** *n.*, a fracture in Earth's crust that runs east and west, just south of the Black Sea

**North Atlantic Drift** *n.*, a warm ocean current that warms the waters around the northwest part of Russia

**North China Plain** *n.*, the plain along the Huang He River

**North Pole** *n.*, the northernmost point on Earth, opposite the South Pole, where all lines of longitude meet

**Northern European Plain** *n.*, a vast lowland that stretches from France to Russia

**Northern Hemisphere** *n.*, the half of Earth north of the equator

**novel** *n.*, a long work of fiction with complex characters and plot

# GLOSSARY

## O

**oasis** *n.*, a fertile place with water in a dry desert area

**occupy** *v.*, to take over

**Okavango Delta** (oh kuh VAANG oh) *n.*, an inland delta in Botswana, where the Okavango River empties into a swamp

**Olmec** *n.*, an organized society who lived along the southern coast of the Gulf of Mexico in 1000 B.C.

**one-child policy** *n.*, the Chinese law of 1979 limiting families living in urban areas to one child

**opera** *n.*, a performance that tells a story through words and music

**oppose** *v.*, to object to

**oral tradition** *n.*, the passing of stories or histories by word of mouth from one generation to the next

**Orange Revolution** *n.*, the Ukraine's peaceful removal of its prime minister in 2004

**Osman** *n.*, the first leader of the Turks in the 1300s, for whom the Ottomans were named

**Ottoman Empire** *n.*, a large, wealthy empire from 1453 to 1923 centered in what is now Turkey

**outer islands** *n.*, islands in Indonesia that include Sumatra, Borneo, New Guinea, and others

**outrigger canoe** *n.*, a boat with an attached float that helps it balance

**outsourcing** *n.*, the shifting of jobs to workers outside of a company, often to a foreign country

**overpopulation** *n.*, the condition of too many people living in one place

## P

**Paekche** (PAHK CHAY) *n.*, a kingdom in southwest Korea in 18 B.C.

**pagoda** *n.*, multistoried religious structure found in Asian countries, usually used for religious purposes

**Palestine Liberation Organization (PLO)** *n.*, an organization created by Palestinian leaders

**Pampas** *n.*, the grassy plain in Argentina

**Pan-Africanism** *n.*, a movement in the early 1900s to unite African people

**Panama Canal Zone** *n.*, the area in which the Panama Canal was built

**pandemic** *n.*, an outbreak of a disease that spreads over a wide geographic area

**Papua New Guinea** *n.*, a country in the South Pacific that is the eastern half of the island of New Guinea and includes nearby islands

**papyrus** *n.*, a paper-like material invented in ancient Egypt

**Parliament** *n.*, the legislative branch of India's government

**parliamentary democracy** *n.*, a government system in which the chief executive is the prime minister, chosen by the party with the most seats in Parliament

**Partition** *n.*, refers to the division of South Asia into the independent countries of India and Pakistan

**patrician** *n.*, a wealthy landowner in ancient Rome

**Pearl Harbor** *n.*, a U.S. naval base in Hawaii that the Japanese bombed in 1941 and that brought about U.S. entry into World War II

**peat** *n.*, the material from very old decayed plants that burns like coal

**peninsula** *n.*, a body of land surrounded on three sides by water

**perestroika** (pehr ih STORY kuh) *n.*, reform in the structure of the economy introduced by Mikhail Gorbachev in the Soviet Union in 1985

**permafrost** *n.*, permanently frozen ground

**perspective** *n.* an artistic way of showing objects as they appear to people in terms of relative distance or depth, as if in three dimensions

**pesticide** *n.*, a chemical that kills harmful insects and weeds

**Peter the Great** *n.*, Peter Romanov, who ruled Russia as czar from 1682 to 1725

**petrochemicals** *n.*, products made from petroleum, or oil

**petroleum** *n.*, raw material used to produce oil

**pharaoh** *n.*, a king in ancient Egypt

**philosopher** *n.*, a person who examines questions about the universe and searches for the truth

**pictograph** *n.*, a painted picture used to communicate

**pilgrimage** *n.*, a religious journey

**pioneer** *n.*, a settler of new land

**pipeline** *n.*, a series of connected pipes used to transport liquids or gases

**Pizarro, Francisco** *n.*, a Spanish conquistador who in 1533 overthrew the Incan emperor and founded the city of Lima, Peru

# GLOSSARY

**plain** *n.*, a level area on Earth's surface

**plantation** *n.*, a large farm that grows crops for profit

**plate** *n.*, a rigid section of Earth's crust that can move independently

**plateau** *n.*, a plain high above sea level that usually has a cliff on all sides

**plebeian** *n.*, a farmer or lower class person in ancient Rome

**poach** *v.*, to hunt or fish illegally

**poacher** *n.*, a person who hunts or fishes illegally

**poaching** *n.*, illegal hunting or fishing

**polder** *n.*, land in the Netherlands that has been reclaimed from the sea for farming

**policy** *n.*, the official guidelines and procedures of an organization or government

**pollution** *n.*, chemical or physical waste that creates an unclean or dirty environment

**Polynesian Triangle** *n.*, a large area in the South Pacific that includes many islands

**polytheism** (PAHL ee thee ihz uhm) *n.*, the belief in more than one god

**polytheistic religion** *n.*, a system of belief in many deities or gods

**popular culture** *n.*, the arts, music, and other elements of everyday life in a region

**porcelain** *n.*, a type of strong ceramic pottery

**port** *n.*, a harbor for ships where goods are exchanged

**Port-au-Prince** *n.*, the capital city of Haiti

**potential** *n.*, possibility

**precipitation** *n.*, the process of water falling to Earth as rain, snow, or hail

**predominant** *adj.*, main, most common, superior to others

**prehistoric** *adj.*, before written history

**preserve** *v.*, to protect

**pride** *n.*, a group of lions that live together

**prime meridian** *n.*, the line of 0° longitude that runs from the North to the South Pole, and passes through Greenwich, England

**privatization** *n.*, the process of government-owned businesses becoming privately owned

**profitable** *adj.*, making money, financially successful

**projection** *n.*, a way of showing Earth's curved surface on a flat map

**promote** *v.*, to encourage

**propaganda** *n.*, information made to influence people's opinions or advance an organization or party's ideas

**proportional representation** *n.*, a system in which a political party gets the same percentage of seats as its percentage of votes

**prosperous** *adj.*, economically strong

**protest** *v.*, to object to

**province** *n.*, a smaller part of a larger country, especially in Canada

**push-pull factors** *n.*, reasons why people migrate; "push" factors cause them to leave; "pull" factors make them come to a place

**pyramid** *n.*, a stone monument built as a tomb in ancient Egypt

# Q

**qanat** (kuh NOT) *n.*, a human-made underground tunnel on the Iranian Plateau, used for carrying water from the mountains

**Qin** (chihn) *n.*, the dynasty that ruled China from 221–206 B.C.

**Qur'an** *n.*, the sacred book of Islam

**Quechua** (KEHCH wah) *n.*, a people who live in the Andes Mountains of Peru, Ecuador, and Bolivia

# R

**radical** *n.*, a person who wants an extreme change or holds an extreme political position

**radioactive** *adj.*, giving off energy caused by the breakdown of atoms

**rain forest** *n.*, a forest with warm temperatures, high humidity, and thick vegetation that receives more than 100 inches of rain per year

**rain shadow** *n.*, a dry region on one side of a mountain range

**rainshadow effect** *n.*, process in which moist air rises up a mountain range and then cools and falls as precipitation, leaving the other side of the range mostly dry

**Ramses II** *n.*, Egyptian pharaoh who reigned around 1185 B.C. and expanded Egypt's empire

# GLOSSARY

**raw materials** *n.*, unfinished or natural materials such as minerals, oil, or coal used to make finished products

**Re** *n.*, the sun god of the ancient Egyptians

**rebellion** *n.*, a revolt or resistance to authority

**recession** *n.*, a slowdown in economic growth

**reclaim** *v.*, to take back

**Reconstruction** *n.*, the effort to rebuild and reunite the United States following the Civil War

**reform** *n.*, change aimed at correcting a problem

**Reformation** *n.*, the movement in the 1500s to reform Christianity

**refuge** *n.*, a safe place

**refugee** *n.*, a person who flees a place to find safety

**region** *n.*, a group of places with common traits

**regulate** *v.*, to control

**reign** *n.*, the period of rule for a king, queen, emperor, or empress

**Reign of Terror** *n.*, a movement in France led by Maximilien Robespierre in which 40,000 people were beheaded in 1793–94

**reincarnation** *n.*, the birth of the soul into another physical life

**relative location** *n.*, the position of a place in relation to other places

**reliable** *adj.*, dependable or trustworthy

**relief** *n.*, the change in elevation from one place to another

**religious tolerance** *n.*, the acceptance of different religions to be practiced at the same time, without prejudice

**relocate** *v.*, to move

**remittance** *n.*, money sent to a person in another place

**remote** *adj.*, hard to reach, isolated

**Renaissance** *n.*, meaning "rebirth," a period in the 1300–1500s in which culture and the arts flourished

**renewable energy** *n.*, energy from sources that do not run out, such as wind, sun, and water

**renewable resource** *n.*, a raw material or energy source that replaces itself over time

**reparation** *n.*, after a war, money paid as punishment, usually by the aggressors in the conflict

**republic** *n.*, a form of government in which officials are elected by the people to govern

**reserve** *n.*, land set aside for a special purpose such as farming, preserving habitats, or housing specific groups of people; a future supply (of oil)

**reservoir** *n.*, a large artificial lake that stores water

**resistance** *n.*, opposition

**restore** *v.*, to bring back

**retreat** *v.*, to go backward, not forward

**revenue** *n.*, income

**reverse** *v.*, to go in the opposite direction

**revolution** *n.*, an action to overthrow a government by citizens or colonists

**Rhine River** *n.*, a river that starts in Switzerland and ends in the North Sea

**rift valley** *n.*, a deep valley formed when Earth's crust separated, as in East Africa

**Ring of Fire** *n.*, an area along the rim of the Pacific Ocean where tectonic plates meet, causing many active volcanoes and earthquakes

**Rio de Janeiro** *n.*, a city in Brazil that will host the 2016 Olympic Games

**ritual** *n.*, a formal action that is regularly repeated

**rivalry** *n.*, competition or opposition between people

**river basin** *n.*, a low area drained by a river

**Romantic Period** *n.*, an artistic period in the early 1800s when artists painted landscapes and natural scenes to convey emotions

**roots** *n.*, cultural origins

**Rub al Khali** *n.*, a large desert in southern Saudi Arabia

**Russian Revolution** *n.*, the revolution in Russia in 1917 that overthrew the czar and put the Bolsheviks in power

**Russification** *n.*, the policy of putting Russians in charge of Soviet republics during the 1970s and 1980s

**ruthless** *adj.*, cruel

# S

**Sahara Desert** *n.*, the largest hot desert in the world, covering most of North Africa

**Sahel** (saa HEL) *n.*, in Sub-Saharan Africa, a semiarid grassland that separates the Sahara in the north from tropical grasslands in the south

**salinization** *n.*, the building up of salt in soil

**Samoa** *n.*, a country governing the western Samoan islands in the South Pacific Ocean

**samurai** (SAM uh ry) *n.*, a skilled Japanese warrior

# GLOSSARY

**sanitation** *n.*, measures such as sewers to protect public health

**Sanskrit** *n.*, the language of the Aryans, which became the basis for many languages in South Asia

**Santa Anna** *n.*, the Mexican president and general who won the battle of the Alamo but lost the U.S.-Mexican War

**São Paolo** *n.*, the largest city in Brazil

**sarcophagus** *n.*, a coffin

**sari** *n.*, a traditional Indian garment for women worn wrapped around the body

**saturate** *v.*, to soak thoroughly

**savanna** *n.*, grassland, as in the south of Sub-Saharan Africa

**scale** *n.*, the part of a map that indicates how big an area of Earth is shown

**scarcity** *n.*, a shortage of something

**scientific station** *n.*, a place to carry out research

**scorched earth policy** *n.*, in 1812, practice in which Russian troops, as they retreated from Napoleon's army, burned crops and resources that could supply the enemy

**seafarer** *n.*, a sea traveler

**secede** *v.*, to formally withdraw

**secular** *adj.*, worldly, not connected to a religion

**segregation** *n.*, separation by race

**seismic** *adj.*, having to do with earthquake activity or movement

**seize** *v.*, to take control of

**self-rule** *n.*, the government of a country by its own people

**semiarid** *adj.*, somewhat dry, with very little rainfall

**serf** *n.*, from the 1500s to the 1800s, a poor Russian or European peasant farmer who rented land from a landlord and had few rights

**shalwar-kameez** *n.*, a long shirt with loose-fitting pants worn in India and Southwest Asia

**Shang** *n.*, a family whose dynasty ruled China from 1766–1050 B.C.

**sheikh** *n.*, an Arab leader

**Shi Huangdi** (shee hwahng dee) *n.*, the ruler of the Qin dynasty, who became China's first emperor in 221 B.C.

**Shi'ite** (SHEE eyt) *adj.*, a branch of Muslims who believe that religious leaders should be descendants of Muhammad

**Shinto** *n.*, a native religion of Japan that is similar to animism

**shogun** *n.*, a military governor of Japan

**Siberia** *n.*, a huge region in central and eastern Russia

**significant** *adj.*, important

**Sikhism** *n.*, a religion in India started in the late A.D. 1400s

**Silk Roads** *n.*, ancient trade routes that connected Southwest and Central Asia with China

**Silla** (SIHL uh) *n.*, a kingdom in southeastern Korea in 57 B.C.

**silt** *n.*, fine particles of soil deposited along riverbanks

**slash-and-burn** *n.*, a farming method of clearing land by cutting down and burning forest and vegetation

**Slavs** *n.*, people who came from around the Black Sea or Poland and settled in the Ukraine and western Russia around A.D. 800

**slum** *n.*, an area in a city that is crowded, with poor housing and bad living conditions

**smartphone** *n.*, a handheld device that combines communication and software applications

**socialism** *n.*, a system of government in which the government controls economic resources

**solstice** *n.*, the point at which the sun is farthest north or farthest south of the Equator; the beginning of summer and winter

**South Pole** *n.*, the southernmost point on Earth, opposite the North Pole, where all lines of longitude meet

**Southern Alps** *n.*, a mountain range in New Zealand

**Southern Hemisphere** *n.*, the half of Earth south of the equator

**sovereignty** *n.*, a country's control over its own affairs

**Soviet Union** *n.*, the Union of Soviet Socialist Republics, a country made up of Russia and other Eurasian states, from 1922 to 1991

**soybean** *n.*, a type of bean grown for food and industrial products

**spatial thinking** *n.*, a way of thinking about space on Earth's surface, including where places are located and why they are there

# GLOSSARY

**Special Economic Zone** *n.*, an area in China that was allowed to develop a market economy with less government control of business

**sprawl** *v.*, to spread out

**standard of living** *n.*, the level of goods, services, and material comforts of people in a country

**staple** *n.*, a basic part of people's diets

**state** *n.*, a defined territory with its own government

**steel** *n.*, a strong metal made from iron combined with other metals

**steppe** *n.*, a very large plain of dry grassland

**strait** *n.*, a narrow waterway that connects two bodies of water

**strike** *n.*, a work stoppage by employees who refuse to work

**subcontinent** *n.*, separate region of a continent

**subsistence farmers** *n.*, farmers who grow food for their families to eat, not to sell

**subsistence farming** *n.*, farming to grow only enough food for families to eat, not to sell

**subsistence fishing** *n.*, fishing for enough food to live on, not for profit

**suffrage** *n.*, the right to vote

**Suleyman I** *n.*, the emperor of the Ottoman Empire in the mid-1500s

**sultan** *n.*, a leader or ruler of the Ottoman Empire

**Sumatra** *n.*, one of the outer islands in Southeast Asia

**Sunni** (SOO nee) *adj.*, a branch of Islam that believes the religious leaders should be chosen from those most qualified, as opposed to being descendants of Muhammad

**surplus** *n.*, extra

**suspension bridge** *n.*, a bridge that is hung from two or more cables

**sustainable** *adj.*, capable of being continued without damaging the environment or using up resources permanently

**Swahili** *n.*, the Bantu language that is mainly spoken in East Africa, also knows as Kiswahwali

**Sydney** *n.*, the largest city in Australia and the state capital of New South Wales

**symbol** *n.*, an object or idea that can be used to represent another object or idea

## T

**taiga** (TY guh) *n.*, the large forest area that stretches through northern Russia, Canada, and other northern countries

**Taino** (TY noh) *n.*, the native people of the Caribbean

**Taj Mahal** *n.*, a famous building built as a tomb for Shah Jahan's wife in India in the 1600s; now a UNESCO World Heritage site

**Taliban** *n.*, a group of Pashtuns in Afghanistan who began ruling Afghanistan in 1996

**Taman Negara National Park** *n.*, a national park in Malaysia within one of the world's oldest rain forests

**tariff** *n.*, a tax on imports and exports

**tax** *n.*, a fee paid to a government for public services

**tectonic plate** *n.*, a section of Earth's crust that floats on Earth's mantle

**temperate** *adj.*, mild, in terms of climate

**terra cotta** *n.*, baked clay

**terrace** *n.*, a flat surface that is built into a hillside

**terraced** *adj.*, flat fields cut into slopes or mountainsides

**terrain** *n.*, the physical features of the land

**terrorism** *n.*, a type of warfare using violence to achieve political results, typically carried out by individuals or small groups

**terrorist** *n.*, a person who uses violence to achieve political results

**textile** *adj.*, related to cloth or clothing

**theme** *n.*, topic

**thirty-eighth parallel** *n.*, the border along the 38° north latitude that separates North and South Korea, created in 1945

**Three Gorges Dam** *n.*, the largest dam in the world, located on the Chang Jiang River in China

**Tibetan Plateau** *n.*, a vast high plateau in Central Asia

**Timbuktu** *n.*, a city in West Africa that was a center of education in the 1200s

**tolerance** *n.*, acceptance of others' beliefs

**tomb** *n.*, a burial place

**topography** *n.*, physical features of land

**tornado** *n.*, a storm with powerful winds that follows an unpredictable path

# GLOSSARY

**totalitarian** *adj.*, relating to a government ruled by a dictator that requires complete obedience to the state

**tourism** *n.*, the travel business or industry

**Trail of Tears** *n.*, the route the Cherokees took during their forced migration from the southeast United States to Oklahoma in the 1830s

**trans-Atlantic slave trade** *n.*, the business of trading slaves taken from Africa across the Atlantic Ocean to the Americas starting in the 1500s

**trans-Saharan** *adj.*, across the Sahara desert

**Trans-Siberian Railroad** *n.*, the world's longest continuous railroad, linking Moscow with eastern Russia across Siberia

**transcontinental** *adj.*, across an entire continent

**transform** *v.*, to remake or change

**transition** *n.*, a change from one activity or stage to another

**transition zone** *n.*, an area between two geographic regions that has characteristics of both

**transpiration** *n.*, a process by which plants and trees release water vapor into the air

**transport** *v.*, to send from one place to another

**transportation corridor** *n.*, a land or water route to move people and goods from place to place easily

**treaty** *n.*, an agreement between two or more countries

**Treaty of Guadalupe Hidalgo** *n.*, the agreement in 1848 in which Mexico gave up the area from Texas to California to the United States

**Treaty of Tordesillas** (tor duh SEE uhs) *n.*, a treaty in 1494 that divided South American land between the Spanish and Portuguese

**Treaty of Versailles** *n.*, a peace treaty that ended World War I in 1919

**trench** *n.*, a long ditch that protects soldiers from enemy gunfire

**trend** *n.*, a change over time in a specific direction

**triangular trade** *n.*, trade among three continents: the Americas, Europe, and Africa

**tributary** *n.*, a small river that drains into a larger river

**tribute** *n.*, fees paid to another ruler or country for protection or as a token of submission

**troubadour** *n.*, a singer during the Middle Ages who performed songs about knights and love

**tsunami** (soo NAH mee) *n.*, a large, powerful ocean wave

**Tuareg** (TWAH rehg) *n.*, a semi-nomadic people who travel across the Sahara in caravans, trading salt

**tundra** *n.*, flat treeless land found in arctic and subarctic regions

**Tupinambá** (too pee NAAM baa) *n.*, a people who lived near the mouth of the Amazon River and along the Atlantic Coast around 3000 B.C.

**typhoon** *n.*, a dangerous tropical storm with heavy rains and high winds, known as a hurricane in the Western Hemisphere

**tyranny** *n.*, a harsh government

# U

**United Arab Emirates** *n.*, a country located on the Arabian Peninsula at the Persian Gulf

**United Nations (UN)** *n.*, an organization of countries formed in 1945 to keep peace among countries and protect human rights

**Universal Declaration of Human Rights** *n.*, an agreement approved by the United Nations that defines the rights that people all over the world should have

**uplands** *n.*, hills, mountains, and plateaus

**Ur** *n.*, an important Sumerian city-state, 2800–1850 B.C.

**Ural Mountains** *n.*, a mountain range that separates the Northern European Plain from the West Siberian Plain in Russia

**urban** *adj.*, describing or related to the city or suburbs

**utilize** *v.*, to make practical use of

# V

**vaccine** *n.*, a treatment to increase immunity to a particular disease

**vegetation** *n.*, plant life

**venue** *n.*, the location for an organized event

**veto** *v.*, to reject a decision made by another government body

**viceroy** *n.*, a governor of Spain's colonies in the Americas who represented the Spanish king and queen

**vocational** *adj.*, related to job skills

**volcano** *n.*, a mountain that erupts in an explosion of molten rock, gases, and ash

**vulnerable** *adj.*, open, able to be hurt by outside forces

# GLOSSARY

## W

**wallaby** *n.*, a marsupial animal that is smaller than a kangaroo

**wat** *n.*, a Buddhist temple in Southeast Asia

**waterway** *n.*, a navigable route for traveling and transport

**weapon of mass destruction (WMD)** *n.*, a weapon that causes great harm to large numbers of people

**weather** *n.*, the condition of the atmosphere at a particular time, including temperature, precipitation, and humidity for a particular day or week

**Western Hemisphere** *n.*, the half of Earth west of the prime meridian

**Wet Pampas** *n.*, humid grasslands in eastern Argentina

**Wilkes, Charles** *n.*, an American who surveyed the coastline of Antarctica in the late 1830s

**wind turbine** *n.*, an engine powered by wind to generate electricity

## Y

**Yanomami** (yaa noh MAA mee) *n.*, an indigenous people who still live in the Amazon River Basin as hunter-gatherers

**yurt** *n.*, a traditional felt tent of Central Asia

## Z

**zaibatsu** (zeye BAHT sue) *n.*, family-run organizations in Japan that owned several businesses

**Zambezi River** *n.*, a river in Southern Africa that flows through many south-central countries to the Indian Ocean

**Zheng He** (jung huh) *n.*, an admiral in the Chinese navy who led seven naval voyages exploring lands outside China; his last voyage began in A.D. 1431

**Zhou** (joh) *n.*, the dynasty that ruled China from 1050–221 B.C.

**zoologist** *n.*, a scientist who studies animals

# GLOSARIO

## A

**abolition** [abolicionismo] s., movimiento para terminar con la esclavitud, antes y después de la Guerra Civil en los Estados Unidos

**Aborigine** [aborigen] s., primera cultura desarrollada de Australia

**absolute location** [ubicación absoluta] s., punto exacto donde está ubicado un lugar, identificado por medio de las coordenadas de latitud y longitud

**abstract** [abstracto] adj., estilo artístico que enfatiza la forma y el color por sobre el realismo

**accommodate** [albergar] v., alojar

**acknowledge** [reconocer] v., admitir, agradecer

**Acropolis** [Acrópolis] s., colina rocosa ubicada en Atenas, Grecia, que servía de fortaleza para los edificios más importantes de la ciudad antigua

**adapt** [adaptar] v., ajustar o modificar para que sea apropiado

**adherent** [partidario] s., seguidor de una religión, causa o persona

**adventure tourism** [turismo aventura] s., tipo de turismo en el cual los viajeros realizan actividades físicas, como montañismo, deportes acuáticos o senderismo

**aerodynamic** [aerodinámico] adj., diseñado para moverse con poca resistencia del viento

**African National Congress (ANC)** [Congreso Nacional Africano (CNA)] s., organización de sudafricanos negros (entre ellos, Nelson Mandela) que protestaron en contra del trato discriminatorio a principios del siglo XX

**African Union** [Unión Africana] s., organización de países africanos que trabajan para lograr el progreso económico

**aging population** [envejecimiento de la población] s., tendencia demográfica que ocurre cuando aumenta la edad media de una población

**agricultural revolution** [revolución agrícola] n., período en el cual los seres humanos comenzaron a cultivar en lugar de recolectar plantas

**Aguinaldo, Emilio** [Aguinaldo, Emilio] s., líder del movimiento por la independencia filipina en la década de 1890

**Alamo** [El Álamo] s., fuerte ubicado en Texas donde 200 texanos perdieron una batalla durante la guerra entre Estados Unidos y México

**Alexander the Great** [Alejandro Magno] s., conquistador que extendió el Imperio Macedónico y difundió la cultura griega por Eurasia desde 334–323 A.C.

**alliance** [alianza] s., sociedad entre países

**alluvial** [aluvial] adj., dicho del sedimento depositado por un río

**alluvial plain** [llanura aluvial] s., área de tierra plana ubicada junto a un arroyo o río que se desborda

**alluvium** [aluvión] s., suelo o cieno arrastrado por el agua que fluye, que es ideal para los cultivos

**Alps** [Alpes] s., cadena montañosa europea

**al-Qaeda** [Al Qaeda] s., grupo terrorista con sede en el suroeste de Asia

**Amazon River Basin** [cuenca del río Amazonas] s., la cuenca fluvial más grande de la Tierra, ubicada en América del Sur

**amendment** [enmienda] s., cambio formal que se hace a una ley

**ancestry** [ascendencia] s., familia de la cual se desciende, remontándose al pasado; herencia

**Andes Mountains** [cordillera de los Andes] s., cordillera que se extiende 5,500 millas a lo largo del costado occidental de América del Sur

**Angel Falls** [Salto Ángel] s., el salto de agua más alto del mundo, ubicado en Venezuela

**Angkor Wat** [Angkor Wat] s., templo grande construido en Camboya en el siglo XII, dedicado a la deidad hindú Vishnú

**animal-borne** [de tracción animal] adj., transportado por animales

**anime** [anime] s., estilo de animación o historieta desarrollado en Japón

**animism** [animismo] s., creencia de que todas las cosas, incluidos los objetos naturales, tienen alma

**annexation** [anexión] s., acción de añadir territorio a un país

**Antarctic Treaty** [Tratado Antártico] s., acuerdo de 1959 entre 12 países para usar la Antártida con fines pacíficos y para compartir los descubrimientos científicos

**apartheid** [apartheid] s., separación legal de las razas; sistema que privaba de sus derechos a los sudafricanos negros

**aqueduct** [acueducto] s., sistema de transporte para llevar agua a grandes distancias, a veces elevado sobre un puente

**aquifer** [acuífero] s., capa de roca subterránea que contiene agua

**arable** [cultivable] adj., fértil; apropiado para la agricultura

**Aral Sea** [mar Aral] s., lago de agua salada ubicado en el centro de Asia, cuyo tamaño se ha reducido mucho como consecuencia del desvío para irrigación de los ríos que desembocan en él

**archipelago** [archipiélago] s., cadena de islas

**arid** [árido] adj., muy seco, que casi no recibe lluvia

**aristocrat** [aristócrata] s., miembro de la clase alta

**armistice** [armisticio] s., acuerdo para dejar de luchar

# GLOSARIO

**artifact** [artefacto] *s.*, objeto hecho por seres humanos de una cultura pasada

**Aryans** [arios] *s.*, nómades que migraron desde el centro de Asia al valle del Indo alrededor del año 2000 A.C.

**Asia-Pacific Economic Cooperation (APEC)** [Foro de Cooperación Económica Asia-Pacífico] *s.*, asociación global fundada en 1989 para fortalecer los vínculos económicos entre los países del Pacífico

**Asoka** [Asoka] *s.*, líder durante el apogeo del Imperio Maurya, alrededor del año 250 A.C.

**assimilate** [asimilarse] *v.*, incorporarse en la cultura de una sociedad

**assimilation** [asimilación] *s.*, proceso en el cual un grupo minoritario adopta la cultura de la mayoría

**assisted migration** [migración asistida] *s.*, programa establecido en 1831 en el cual los británicos ofrecían dinero a la gente para que se mudara a Australia

**Aswan High Dam** [Presa Alta de Asuán] *s.*, presa ubicada en el río Nilo, terminada en 1970

**Atacama Desert** [desierto de Atacama] *s.*, desierto ubicado en el lado occidental de la cordillera de los Andes

**Atahualpa** [Atahualpa] *s.*, el último emperador de los incas antes de ser conquistados por los españoles en 1533

**atoll** [atolón] *s.*, isla, arrecife o cadena de islas con forma de anillo hechas de coral

**attribute** [atributo] *s.*, cualidad específica

**Augustus** [Augusto] *s.*, primer emperador de Roma, en 27 A.C.

**autonomy** [autonomía] *s.*, auto-gobierno de un país o pueblo

**Aymara** [aimara] *s.*, cultura indígena que habita en la cordillera de los Andes de Perú y Bolivia

**Aztec** [aztecas] *s.*, pueblo que se estableció en el área de la actual ciudad de México, 1325–1525 D.C.

# B

**ban** [proscribir] *v.*, declarar fuera de la ley o prohibir

**Bantu** [bantúes] *s.*, grupo africano que se trasladó desde el oeste de África central hacia el sur y hacia el este, a través del África subsahariana, entre 2000 y 1000 A.C.

**barbarian** [bárbaro] *s.*, soldado o guerrero considerado culturalmente menos desarrollado que aquellos contra los que luchaba; el término proviene originalmente de las tribus germanas que invadieron el Imperio Romano en el año 235 D.C.

**Baroque period** [Barroco] *s.*, período entre 1600 y 1750 en el cual la música presentaba esquemas y temas complicados

**barter** [canjear] *v.*, comerciar o intercambiar bienes sin utilizar dinero

**basin** [cuenca] *s.*, región bañada por un sistema fluvial

**bas-relief** [bajorrelieve] *s.*, escultura que sobresale ligeramente de un fondo plano

**bauxite** [bauxita] *s.*, materia prima que se utiliza para fabricar aluminio

**bay** [bahía] *s.*, masa de agua rodeada por tierra en tres de sus lados

**Bedouin** [beduinos] *s.*, pueblo nómade de habla árabe del suroeste de Asia

**benefit** [beneficiar] *v.*, hacer bien

**Berlin Conference** [Conferencia de Berlín] *s.*, encuentro de las naciones europeas en 1884 para resolver los conflictos sobre sus reclamos coloniales en África

**Berlin Wall** [Muro de Berlín] *n.*, muro que dividía Berlín Este communista de Berlín Oeste democrático

**Bhutto, Benazir** [Bhutto, Benazir] *s.*, primera ministra mujer de Pakistán, 1988–1999

**Biko, Stephen** [Biko, Stephen] *s.*, presidente de una organización sudafricana de protesta estudiantil que fue arrestado y asesinado en 1977

**Bill of Rights** [Declaración de Derechos] *s.*, primeras diez enmiendas a la Constitución estadounidense

**biodiversity** [biodiversidad] *s.*, variedad de especies que viven en un ecosistema

**biofuel** [biocombustible] *s.*, combustible alternativo que es una mezcla de etanol y gasolina

**Black Sea** [mar Negro] *s.*, mar interior que limita con Europa, Rusia, Georgia y Turquía

**Blue Mosque** [Mezquita Azul] *s.*, mezquita histórica construida en Estambul por el Sultán Ahmed I, terminada en 1617

**Bollywood** [Bollywood] *s.*, industria cinematográfica de la India, ubicada en Bombay

**Bolshevik** [bolchevique] *s.*, partido político de Rusia liderado por Lenin que derrocó al zar en 1917

**Borobudur** [Borobudur] *s.*, complejo de templos budistas ubicado en el centro de Java

**Bosporus Strait** [estrecho de Bósforo] *s.*, paso que atraviesa Estambul y conecta el mar Negro y el mar de de Mármara

**breakwater** [rompeolas] *s.*, barrera construida para proteger un puerto

**Buddhism** [budismo] *s.*, religión fundada por Siddhartha Gautama en 525 A.C. en India

**bullet train** [tren bala] *s.*, tren que viaja a más de 125 millas por hora y cuya forma es similar a una bala

**butte** [cerro testigo] *s.*, colina o montaña con laderas escarpadas y cima plana

# GLOSARIO

**Byrd, Richard** [Byrd, Richard] *s.*, estadounidense que exploró la Antártida por cielo y tierra durante la década de 1940

**Byzantine Empire** [Imperio Bizantino] *s.*, parte oriental del Imperio Romano después de 395 D.C. que duró aproximadamente 1,000 años

# C

**Caesar, Julius** [César, Julio] *s.*, general que se convirtió en gobernante de Roma, 46–44 A.C.

**Calypso** [calipso] *s.*, tipo de música folclórica que se inició en Trinidad

**canal** [canal] *s.*, vía fluvial construida por el hombre a través de la tierra para botes y barcos

**Candomblé** [candomblé] *s.*, religión brasileña que combina el espiritualismo africano y el catolicismo

**canopy** [enramada] *s.*, techo creado sobre un bosque tropical por las copas de los árboles

**capital** [capital] *s.*, riqueza e infraestructura de un país

**carapace** [caparazón] *s.*, cubierta dura de ciertos animales, como la tortuga

**caravan** [caravana] *s.*, grupo de mercaderes que viajan juntos por razones de seguridad

**Cardamom Mountains** [colinas Cardamom] *s.*, cadena de montañas ubicada en Camboya

**Caribbean Sea** [mar Caribe] *s.*, mar tropical ubicado en el Hemisferio Occidental, que limita con México, América Central, las Antillas Mayores y las Antillas Menores

**cartographer** [cartógrafo] *s.*, persona que hace mapas

**cash crop** [cultivo comercial] *s.*, cultivo que se vende para obtener una ganancia

**Caspian Sea** [mar Caspio] *s.*, la masa de agua endorreica más grande de la Tierra, que limita con Rusia, Kazajistán, Turkmenistán, Azerbaiyán e Irán

**caste system** [sistema de castas] *s.*, en India, estructura social compuesta por cuatro niveles principales, que fue instaurada por los arios

**categorize** [categorizar] *v.*, agrupar información; clasificar

**Catherine the Great** [Catalina la Grande] *s.*, emperadora de Rusia desde 1762 hasta 1796

**celadon** [celadón] *adj.*, con barniz vítreo de color verde, referido a la cerámica realizada por los artistas coreanos

**cenote** [cenote] *s.*, un depósito de agua subterránea

**Chang Jiang** [Chang Jiang] *s.*, el río más largo de Asia, que fluye a través de China; también llamado Yangtsé

**Chang Jiang Plain** [llanura de Chang Jiang] *s.*, llanuras ubicadas junto al río Chang Jiang

**Chao Phraya River** [río Chao Phraya] *s.*, río que fluye a través de Tailandia

**Chernobyl** [Chernóbil] *s.*, ciudad ubicada en Ucrania, en la cual un reactor nuclear explotó en 1986

**Choson** [Choson] *s.*, dinastía coreana que duró desde 1392 hasta 1910

**Christian Bible** [Biblia cristiana] *s.*, libro sagrado del cristianismo

**Christianity** [cristianismo] *s.*, religión basada en la vida y en las enseñanzas de Jesús de Nazaret

**citizen** [ciudadano] *s.*, persona que vive dentro de un territorio y tiene derechos y responsabilidades garantizadas por el gobierno

**city-state** [ciudad estado] *s.*, estado independiente compuesto por una ciudad y los territorios que dependen de ella

**civil disobedience** [desobediencia civil] *s.*, desobediencia no violenta de las leyes

**civil war** [guerra civil] *s.*, guerra entre grupos opuestos de ciudadanos del mismo país

**civilization** [civilización] *s.*, sociedad con cultura, política y tecnología altamente desarrolladas

**clan** [clan] *s.*, unidad grande basada en la familia con lealtad hacia el grupo

**Classical period** [Clasicismo] *s.*, período comprendido entre 1750 y 1900 en el cual la música seguía las reglas establecidas para la forma y la complejidad, como en el caso de las sonatas o las sinfonías

**climate** [clima] *s.*, el promedio de las condiciones de la atmósfera de un área durante un largo período de tiempo, incluidas la temperatura, la precipitación y los cambios estacionales

**climograph** [gráfica del clima] *s.*, gráfica que muestra el clima de una región a través de la precipitación y la temperatura medias

**coalition** [coalición] *s.*, alianza

**coastal plains** [llanuras costeras] *s.*, tierras bajas ubicadas junto a la orilla del mar

**Cold War** [Guerra Fría] *s.*, largo período de tensión política sin lucha armada entre los Estados Unidos y la Unión Soviética, desde aproximadamente 1948 hasta 1991

**collective farm** [granja colectiva] *s.*, en la Unión Soviética, una granja grande donde los trabajadores cultivaban alimentos para distribuirlos a toda la población

**collide** [colisionar] *v.*, chocar

**colonialism** [colonialismo] *s.*, práctica de un país que directamente gobierna y desarrolla el comercio en un territorio extranjero para su propio beneficio

**colonize** [colonizar] *v.*, construir asentamientos y desarrollar el comercio en tierras que controla un país

**colony** [colonia] *s.*, área controlada por un país lejano

# GLOSARIO

**Columbian Exchange** [Intercambio Colombino] *s.*, intercambio de plantas, animales y enfermedades entre las Américas, Europa y África que comenzó en el siglo XVI

**commerce** [comercio] *s.*, negocio de comprar y vender bienes y servicios

**commercial agriculture** [agricultura comercial] *s.*, negocio de producir cultivos para vender

**commodity** [mercancía] *s.*, material u objeto que se puede comprar, vender o comerciar

**Common Market** [Mercado Común] *s.*, Comunidad Económica Europea formada en 1957

**commonwealth** [mancomunidad] *s.*, nación que se gobierna a sí misma pero es parte de un país mayor

**Commonwealth of Australia** [Mancomunidad de Australia] *s.*, la nación de Australia, que entró a formar parte de la Mancomunidad Británica de Naciones en 1901

**communal** [comunal] *adj.*, compartido

**communism** [comunismo] *s.*, sistema de gobierno en el cual un único partido político controla el gobierno y la economía

**competitive** [competitivo] *adj.*, capaz de intentar ganar un concurso o una carrera

**complex** [complejo] *s.*, conjunto de edificios vinculados

**comply** [acatar] *v.*, seguir una regla o una orden

**concentrated** [concentrado] *adj.*, reunido en un área central

**concentration camp** [campo de concentración] *s.*, área donde los judíos y otras personas fueron detenidos durante la Segunda Guerra Mundial y asesinados por los nazis

**condensation** [condensación] *s.*, proceso por el cual el vapor de agua se convierte en gotitas de líquido debido al enfriamiento durante el ciclo hidrológico

**Confucianism** [confucianismo] *s.*, sistema ético basado en las enseñanzas de Confucio

**Congo River** [río Congo] *s.*, río principal ubicado en África Central que desemboca en el océano Atlántico

**conquistador** [conquistador] *s.*, soldado y líder español que exploraba los territorios recientemente descubiertos de América del Norte, América del Sur, América Central y el Caribe a partir del siglo XVI, fundamentalmente en busca de oro y plata

**conservation** [conservación] *s.*, protección del medio ambiente

**Constantine** [Constantino] *s.*, emperador romano, 306–337 A.C., que hizo del cristianismo la religión oficial del imperio

**constitution** [constitución] *s.*, documento que organiza un gobierno y enuncia sus poderes

**consumer** [consumidor] *s.*, persona que compra bienes

**contaminated** [contaminado] *adj.*, infectado; inapropiado para el uso debido a la presencia de elementos peligrosos

**contiguous** [contiguo] *adj.*, conectado en un bloque

**continent** [continente] *s.*, gran masa de tierra sobre la superficie de la Tierra; la Tierra tiene siete continentes

**continental drift** [deriva continental] *s.*, movimiento lento de los continentes sobre las placas tectónicas

**continental shelf** [plataforma continental] *s.*, borde de un continente que se extiende bajo el agua y se adentra en el mar

**controversy** [controversia] *s.*, debate o disputa

**convert** [convertir] *v.*, persuadir a alguien para que cambie sus creencias religiosas

**convey** [transmitir] *v.*, comunicar

**convict** [convicto] *s.*, persona sentenciada a prisión por cometer un crimen

**Cook Islands** [islas Cook] *s.*, país ubicado cerca de Nueva Zelanda, en el Pacífico Sur, formado por 15 islas pequeñas

**Cook, James** [Cook, James] *s.*, navegante británico que exploró y cartografió las Islas del Pacífico en la década de 1770

**coral island** [isla de coral] *s.*, isla creada por la acumulación gradual de esqueletos de coral

**coral reef** [arrecife de coral] *s.*, estructura parecida a una roca construida por capas de organismos de coral

**cordillera** [cordillera] *s.*, sistema de varias cadenas de montañas paralelas

**Cortés, Hernán** [Cortés, Hernán] *s.*, conquistador español que derrotó a los aztecas en 1525

**cosmopolitan** [cosmopolita] *adj.*, que reúne muchas culturas e influencias diferentes

**Counter-Reformation** [Contrarreforma] *s.*, movimiento dentro de la Iglesia Católica Romana para reformar sus propias prácticas

**coup** [golpe de estado] *s.*, toma repentina e ilegal del gobierno por medio de la fuerza

**Creole** [criollo] *s.*, lengua que mezcla elementos de idiomas europeos y no europeos

**crevasse** [grieta] *s.*, abertura profunda en un glaciar

**cricket** [cricket] *s.*, juego de equipo similar al béisbol

**critical** [crítico] *adj.*, extremadamente importante, necesario para sobrevivir

**Crusades** [Cruzadas] *s.*, expediciones militares de la Iglesia Católica Romana para recuperar las tierras santas ubicadas en Medio Oriente que estaban bajo control musulmán, 1096–1291

**cuisine** [cocina] *s.*, alimentos y tradiciones culinarias comunes a cierta región

**cultural hearth** [centro cultural] *s.*, centro de civilización desde el cual se difunden las ideas y la tecnología

# GLOSARIO

**Cultural Revolution** [Revolución Cultural] s., en China, plan social que duró desde 1966 hasta 1969, en el cual el gobierno intentó eliminar todos los elementos capitalistas y anticomunistas de la sociedad

**culture** [cultura] s., modo de vida de un grupo, que incluye el tipo de alimentación, vivienda, vestido, idioma, comportamiento e ideas

**culture region** [región cultural] s., área que está unificada por rasgos culturales comunes

**cuneiform** [cuneiforme] s., primera forma de escritura conocida, proveniente de Sumeria

**currency** [moneda] s., forma de dinero

**current** [corriente] s., movimiento continuo de aire o agua que fluye en la misma dirección

**cyclone** [ciclón] s., tormenta con vientos giratorios, llamada tifón en el Hemisferio Oriental; denominada huracán en otras partes del mundo

**czar** [zar] s., término utilizado en Rusia para designar al emperador

# D

**dam** [presa] s., barrera que controla el flujo de agua

**Danube River** [río Danubio] s., río que nace en Alemania y desemboca en el mar Negro

**Daoism** [taoísmo] s., sistema ético que enfatiza la armonía entre las personas y la naturaleza

**Declaration of Independence** [Declaración de Independencia] s., documento que declara que los Estados Unidos son independientes del Imperio Británico, adoptado el 4 de julio de 1776

**deforestation** [deforestación] s., práctica de talar bosques para despejar la tierra y utilizarla para cultivos o uso urbano

**deity** [deidad] s., dios o diosa

**delta** [delta] s., área donde un río deposita sedimentos cuando desemboca en una masa de agua más grande

**demilitarized zone (DMZ)** [zona desmilitarizada] s., área neutral entre países enemigos, específicamente entre Corea del Norte y Corea del Sur

**democracy** [democracia] s., forma de gobierno del pueblo, en la cual los ciudadanos suelen elegir representantes para que los gobiernen

**democratization** [democratización] s., proceso de convertirse en una democracia

**demographics** [demografía] s., características de una población humana, tales como la edad, el ingreso y la educación

**deny** [negar] v., no reconocer

**deplete** [agotar] v., extraer o disminuir un recurso hasta que no esté disponible o no sea capaz de funcionar como planeado

**descendants** [descendientes] s., generaciones futuras; parientes de un miembro de la familia que vivió en el pasado

**desertification** [desertificación] s., transición gradual de la tierra fértil a tierra menos productiva

**developed nations** [naciones desarrolladas] s., países con un alto producto interno bruto per cápita

**developing nations** [naciones en vías de desarrollo] s., países con un bajo producto interno bruto per cápita

**dialect** [dialecto] s., variante regional de un idioma principal

**Diaspora** [Diáspora] s., dispersión de los judíos por todo el mundo

**dictator** [dictador] s., gobernante con control total

**diffusion** [difusión] s., propagación

**diplomacy** [diplomacia] s., discusión entre grupos o países para resolver disputas o desacuerdos

**discrimination** [discriminación] s., trato injusto de un individuo o grupo en base a factores distintos de la habilidad

**displace** [desplazar] v., obligar a un pueblo a dejar su hogar

**disregard** [desconocer] v., ignorar

**dissolve** [disolver] v., deshacer

**distinct** [definido] adj., fácilmente reconocible

**distort** [distorsionar] v., modificar la forma o apariencia usual; deformar

**distribute** [distribuir] v., repartir un bien o recurso entre un grupo mayor

**diversify** [diversificar] v., dar variedad a algo

**diversity** [diversidad] s., variedad

**domestic policy** [política interna] s., plan de un gobierno para manejar los asuntos dentro de sus propias fronteras

**domesticate** [domesticar] v., criar y usar animales como fuente de mano de obra y alimento

**dormant** [inactivo] adj., latente, referido a un volcán

**drought** [sequía] s., largo período con escasa o ninguna precipitación

**Dry Pampas** [Pampa seca] s., región árida ubicada en el oeste de Argentina

**Dubai** [Dubái] s., ciudad y estado joven y rico de los Emiratos Árabes Unidos, en el golfo Pérsico

**due process** [debido proceso] s., en los Estados Unidos, reglas que las autoridades deben seguir al tratar con los ciudadanos

**Dutch East India Company** [Compañía Neerlandesa de las Indias Orientales] s., compañía de los Países Bajos que dominaba el comercio de especias en el sureste de Asia

# GLOSARIO

**dynamic** [dinámico] *adj.*, que cambia continuamente

**dynastic cycle** [ciclo dinástico] *s.*, patrón de ascenso y caída de las dinastías en la historia china

**dynasty** [dinastía] *s.*, serie de gobernantes de la misma familia

# E

**earthquake** [terremoto] *s.*, sacudida de la corteza de la Tierra generalmente causada por el choque o el deslizamiento de las placas tectónicas

**Eastern Hemisphere** [Hemisferio Oriental] *s.*, la mitad de la Tierra ubicada al este del primer meridiano

**ecologist** [ecologista] *s.*, científico que estudia las relaciones entre los seres vivos y su medio ambiente

**economic globalization** [globalización económica] *s.*, práctica de actividades económicas que se realizan a través de las fronteras nacionales

**economic sector** [sector económico] *s.*, subdivisión o parte más pequeña de una economía, tal como la industria y la agricultura

**economy** [economía] *s.*, sistema en el cual las personas producen, venden y compran cosas

**ecosystem** [ecosistema] *s.*, comunidad de organismos vivos y su medio ambiente o hábitat natural

**ecotourism** [ecoturismo] *s.*, manera de visitar las áreas naturales que conserva los recursos naturales de la región

**El Niño** [El Niño] *s.*, inversión del viento y las corrientes oceánicas usuales

**elevation** [elevación] *s.*, altura de un accidente geográfico sobre el nivel del mar

**eliminate** [eliminar] *v.*, deshacerse de

**elite** [elite] *s.*, clase superior

**Emancipation Proclamation** [Proclama de Emancipación] *s.*, documento de 1863 que liberó a todos los esclavos que vivían en el territorio en poder de los Confederados durante la Guerra Civil estadounidense

**emergence** [aparición] *s.*, desarrollo y uso extendido de algo nuevo

**emigrate** [emigrar] *v.*, dejar el país natal para vivir en otro país

**emirate** [emirato] *s.*, estado del país de los Emiratos Árabes Unidos

**empire** [imperio] *s.*, grupo de pueblos o estados gobernados por un único gobernante fuerte

**Enduring Voices Project** [Proyecto Voces Perdurables] *s.*, proyecto de National Geographic para estudiar y conservar las lenguas que están en riesgo de desaparición

**enhance** [realzar] *v.*, mejorar la calidad de algo

**Enlightenment** [Ilustración] *s.*, movimiento social del siglo XVIII que trabajó a favor de la educación y de los derechos del individuo

**enlist** [alistarse] *v.*, presentarse como voluntario para el servicio militar

**entrepreneur** [empresario] *s.*, persona que comienza un negocio nuevo

**entrepreneurship** [espíritu empresarial] *s.*, características de creatividad y riesgo que existen en una persona o sociedad

**epic poem** [poema épico] *s.*, poema largo que relata las aventuras de un héroe

**epidemic** [epidemia] *s.*, brote de una enfermedad que afecta a una gran parte de la población de una comunidad en particular

**equator** [ecuador] *s.*, círculo imaginario alrededor de la Tierra que está a la misma distancia del Polo Norte y del Polo Sur y divide la Tierra por la mitad; la línea central o de 0° de latitud

**equinox** [equinoccio] *s.*, momento en que el día y la noche tienen la misma duración; ocurre dos veces al año, el 21 o 22 de marzo y el 21 o 22 de septiembre

**erode** [erosionar] *v.*, desgastar

**erosion** [erosión] *s.*, proceso por el cual las rocas y el suelo se rompen lentamente y se desgastan

**erratic** [errático] *adj.*, inconsistente o irregular

**eruption** [erupción] *s.*, estallido o explosión

**escarpment** [escarpe] *s.*, pendiente pronunciada

**essential** [esencial] *adj.*, necesario

**establish** [establecer] *v.*, instituir

**ethanol** [etanol] *s.*, alcohol líquido extraído de la caña de azúcar o del maíz que se puede usar como combustible, solo o mezclado con gasolina

**ethical system** [sistema ético] *s.*, sistema de creencias que enseña conductas morales

**ethnic group** [grupo étnico] *s.*, grupo de personas que tienen una cultura y una lengua en común y a veces comparten la misma herencia racial

**ethnobotanist** [etnobotánico] *s.*, científico que estudia la relación entre las culturas y las plantas

**euro** [euro] *s.*, moneda común de la Unión Europea

**European Union (EU)** [Unión Europea (UE)] *s.*, organización económica compuesta por 27 países miembros europeos (2011)

**eurozone** [eurozona] *s.*, países que han adoptado el euro como moneda

**evaporation** [evaporación] *s.*, proceso por el cual el agua se convierte en vapor y sube a la atmósfera debido al calentamiento del Sol; una parte del ciclo hidrológico

**evident** [evidente] *adj.*, claramente presente u observable

# GLOSARIO

**excavate** [excavar] *v.*, quitar la tierra cuidadosamente o hacer un hoyo

**exchange** [cambiar] *v.*, convertir el dinero en otra moneda

**exile** [exilar] *v.*, obligar a alguien a dejar un país

**exile** [exilio] *s.*, situación en la que se está ausente del propio país natal

**exoskeleton** [exoesqueleto] *s.*, cubierta externa dura que protege a los corales y a otros animales marinos

**expand** [ampliar] *v.*, hacer más grande

**expedition** [expedición] *s.*, travesía o viaje de cierta extensión, usualmente a otras tierras

**exploit** [explotar] *v.*, aprovecharse de algo; usar de manera egoísta o injusta para beneficio propio

**export** [exportar] *v.*, enviar a otro país para obtener ayuda o un beneficio

**export revenue** [ingresos por exportaciones] *s.*, dinero o ingresos recibidos por los bienes vendidos a otro país

**extent** [extensión] *s.*, distancia o grado hasta el cual se difunde algo

**extinct** [extinto] *adj.*, completamente desaparecido, tal como una especie animal o vegetal

**extinction** [extinción] *s.*, desaparición de una especie o de un tipo de ser vivo

**extremist** [extremista] *s.*, persona con opiniones religiosas o políticas que están fuera del rango de la opinión de la mayoría

# F

**factory system** [sistema fabril] *s.*, modo de trabajar en el cual cada persona trabaja en solo una parte de un producto

**failed state** [estado fallido] *s.*, país en el cual el gobierno, las instituciones económicas y el orden han colapsado

**fallout** [lluvia radiactiva] *s.*, partículas radiactivas provenientes de una explosión nuclear que caen a través de la atmósfera

**famine** [hambruna] *s.*, largo período de escasez de alimentos

**fault** [falla] *s.*, fractura en la corteza de la Tierra

**federal republic** [república federal] *s.*, forma democrática de gobierno en la cual los votantes eligen representantes y el gobierno central comparte el poder con los estados

**federal system** [sistema federal] *s.*, sistema de gobierno con un gobierno central fuerte y unidades gubernamentales locales

**feral** [asilvestrado] *adj.*, salvaje, que se ha convertido en salvaje después de haber sido domesticado

**fertile** [fértil] *adj.*, capaz de producir una gran abundancia de frutos, cultivos o crías

**Fertile Crescent** [Creciente Fértil] *s.*, área alrededor de los ríos Tigris y Éufrates

**fertility rate** [tasa de fertilidad] *s.*, promedio de niños nacidos por cada mujer en un cierto período de tiempo

**fertilizer** [fertilizante] *s.*, sustancia añadida al suelo para enriquecerlo

**feudal system** [sistema feudal] *s.*, durante la Edad Media, estructura social que consistía en un rey, señores, vasallos y siervos

**fiber optics** [fibra óptica] *s.*, fibras de vidrio utilizadas para enviar un código digital rápidamente a través de grandes distancias

**first language** [lengua materna] *s.*, idioma que aprenden los niños de un pueblo

**fjord** [fiordo] *s.*, bahía angosta y profunda ubicada

**floodplain** [planicie aluvial] *s.*, terreno bajo a la orilla de los ríos, formado por el sedimento dejado por la inundación

**Foja Mountains** [montañas Foja] *s.*, cadena de montañas ubicada en la isla de Nueva Guinea, en el este de Indonesia

**food security** [seguridad alimentaria] *s.*, fácil acceso a alimentos suficientes

**foremost** [principal] *adj.*, primero en rango o importancia, que lidera

**fortify** [fortificar] *v.*, fortalecer

**fossil** [fósil] *s.*, restos conservados de plantas y animales antiguos

**fragmented country** [país fragmentado] *s.*, país que está físicamente dividido en partes separadas, tales como una cadena de islas, y/o política o culturalmente dividido

**free enterprise economy** [economía de libre empresa] *s.*, sistema en el cual las empresas de propiedad privada producen bienes y servicios, también llamada economía de mercado o capitalismo

**free trade** [libre comercio] *s.*, comercio que no grava las importaciones con impuestos

**free trade agreement** [acuerdo de libre comercio] *s.*, tratado entre países que mejora el comercio al limitar los impuestos a dicho comercio

**French Indochina** [Indochina Francesa] *s.*, grupo de colonias pertenecientes a Francia, formado por Vietnam, Camboya, Laos y otras, ubicadas en el sureste de Asia, desde 1887 hasta 1955

**fuel cell** [celda de combustible] *s.*, unidad pequeña que, al igual que una pila o batería, combina sustancias químicas para producir energía

**fuse** [fusionar] *v.*, mezclar

# GLOSARIO

## G

**Gadsden Purchase** [Compra de Gadsden] *s.*, venta de territorio realizada en 1853 por México a los Estados Unidos que estableció la frontera suroccidental estadounidense actual

**Gandhi, Mohandas** [Gandhi, Mahatma] *s.*, líder indio que impulsó la independencia de la India respecto del Reino Unido mediante la desobediencia civil, en la década de 1930; considerado el padre de la India moderna

**Ganges Delta** [delta del Ganges] *s.*, área fértil donde el río Ganges desemboca en la bahía de Bengala

**gaucho** [gaucho] *s.*, vaquero de América del Sur

**gauge** [trocha] *s.*, medida del ancho de las vías del ferrocarril

**generation** [generación] *s.*, grupo de individuos que nacen y viven aproximadamente en la misma época

**Genghis Khan** [Gengis Kan] *s.*, gobernante mongol que estableció un imperio en el centro de Asia, a principios del siglo XIII

**genre** [género] *s.*, forma literaria, como un poema, una obra de teatro o una novela

**geoglyph** [geoglifo] *s.*, figura geométrica grande o forma de animal dibujada sobre el suelo

**Geographic Information Systems (GIS)** [Sistemas de Información Geográfica (SIG)] *s.*, aparatos computarizados que presentan datos sobre lugares específicos

**geographic pattern** [patrón geográfico] *s.*, similitud entre lugares

**geothermal energy** [energía geotérmica] *s.*, energía térmica proveniente del interior de la Tierra que se puede convertir en electricidad

**ger** [ger] *s.*, tienda portátil hecha de fieltro

**Gettysburg Address** [Discurso de Gettysburg] *s.*, discurso que Abraham Lincoln pronunció en 1863 en honor a los soldados que murieron en la batalla de Gettysburg durante la Guerra Civil estadounidense

**Giza** [Giza] *s.*, ciudad ubicada junto al Nilo, donde se construyeron diez pirámides en el antiguo Egipto

**Glacier** [glaciar] *s.*, masa grande de hielo y nieve acumulada

**glasnost** [glásnost] *s.*, política de la Unión Soviética de apertura que animó al pueblo a hablar abiertamente acerca del gobierno, introducida por Mijaíl Gorbachov en la década de 1980

**global** [global] *adj.*, mundial

**Global Positioning System (GPS)** [Sistema de Posicionamiento Global (GPS)] *s.*, sistema de satélites con base en el espacio que encuentra la ubicación absoluta y la hora de cualquier lugar de la Tierra

**global warming** [calentamiento global] *s.*, aumento de la temperatura media de la Tierra desde mediados del siglo XX

**globalization** [globalización] *s.*, desarrollo de una economía mundial basada en el libre comercio y el uso de mano de obra extranjera

**globe** [globo terráqueo] *s.*, modelo tridimensional, o esférico, de la Tierra

**Gobi desert** [desierto de Gobi] *s.*, el desierto más grande de Asia, que cubre gran parte del sur de Mongolia y se extiende hasta el interior de China

**golden age** [edad dorada] *s.*, período de gran riqueza, cultura y democracia en Grecia

**Golden Quadrilateral (GQ)** [Cuadrilátero de Oro] *s.*, superautopista que conecta cuatro ciudades importantes de la India

**Gorbachev, Mikhail** [Gorbachov, Mijaíl] *s.*, líder de la Unión Soviética, 1985–1991; presidente desde 1990 hasta que fue disuelta en 1991

**gorge** [desfiladero] *s.*, paso profundo y angosto rodeado por acantilados empinados

**government** [gobierno] *s.*, organización que mantiene el orden, establece reglas y proporciona servicios para una sociedad

**Grand Bazaar** [Gran Bazar] *s.*, centro de tiendas y comercios en Estambul, Turquía

**Grand Canyon** [Gran Cañón] *s.*, formación rocosa ubicada en el suroeste de los Estados Unidos que el río Colorado ha excavado profundamente durante millones de años

**grasslands** [praderas] *s.*, áreas abiertas y amplias apropiadas para el pastoreo y los cultivos

**Great Barrier Reef** [Gran Barrera de Coral] *s.*, ecosistema inmenso, ubicado cerca de la costa de Australia, hecho de arrecifes de coral

**Great Depression** [Gran Depresión] *s.*, descenso económico mundial ocurrido en la década de 1930, marcado por la pobreza y una alta tasa de desempleo

**Great Escarpment** [Gran Escarpe] *s.*, pendiente pronunciada que se extiende desde la meseta del sur de África hasta las llanuras costeras

**Great Lakes** [Grandes Lagos] *s.*, cinco lagos grandes de agua dulce ubicados entre Canadá y los Estados Unidos

**Great Leap Forward** [Gran Salto Adelante] *s.*, plan de Mao Zedong para hacer que la economía de China creciera más rápidamente, 1958–1961

**Great Plains** [Grandes Llanuras] *s.*, área de tierra baja y llana ubicada al este de las montañas Rocosas

**Great Pyramid of Khufu** [Gran Pirámide de Keops] *s.*, la pirámide más antigua y más alta de Egipto, construida en Guiza alrededor de 2550 A.C.

# GLOSARIO

**Great Recession** [Gran Recesión] *s.*, descenso de las economías de los Estados Unidos, Canadá y otros países, que comenzó en 2007

**Great Rift Valley** [Gran Valle del Rift] *s.*, valle ancho ubicado en África Oriental, parte de una cadena de valles formados cuando las placas tectónicas se separaron

**Great Wall of China** [Gran Muralla China] *s.*, muro de piedra ubicado en el norte de China de más de 4,500 millas de largo, construido para rechazar a los invasores provenientes del norte

**Great Zimbabwe** [Gran Zimbabue] *s.*, ciudad amurallada de piedra que el pueblo shona construyó en el sur de África entre 1200 y 1450

**greenhouse gas** [gas invernadero] *s.*, gas que atrapa el calor del Sol sobre la Tierra

**griot** [griot] *s.*, narrador africano tradicional

**gross domestic product (GDP)** [producto interno bruto (PIB)] *s.*, valor total de todos los bienes y servicios producidos en un país en un año dado

**Guaraní** [guaraníes] *s.*, pueblo indígena que vive en los valles de los ríos Paraguay y Paraná, en América del Sur

**guest worker** [trabajador huésped] *s.*, trabajador temporal que migra para trabajar en otro país

**Guillotine** [guillotina] *s.*, máquina utilizada para ejecutar a las personas durante la Revolución Francesa

**Gutenberg, Johannes** [Gutenberg, Johannes] *s.*, impresor alemán que inventó la imprenta en 1450

# H

**habitat** [hábitat] *s.*, medio ambiente natural de una planta o animal vivo

**Hagia Sophia** [Santa Sofía] *s.*, museo ubicado en Estambul, Turquía, que se construyó originalmente como iglesia, durante el Imperio Romano, y luego sirvió de mezquita

**half-life** [vida media] *s.*, tiempo necesario para que la mitad de los átomos de una sustancia radiactiva se desintegren

**Hammurabi** [Hammurabi] *s.*, rey que desarrolló un código de leyes en Babilonia, en el siglo XVIII A.C.

**Han** [Han] *s.*, dinastía china que duró desde 206 A.C. hasta 220 D.C.

**Harappan** [Harappa] *s.*, primera civilización urbana del sur de Asia, desarrollada junto al río Indo alrededor de 2600 A.C.

**harbor** [embarcadero] *s.*, lugar donde los barcos pueden atracar protegidos del mar abierto

**Hatshepsut** [Hatshepsut] *s.*, mujer que fue faraón de Egipto alrededor de 1500 A.C.

**Hebrew Bible** [Biblia hebrea] *s.*, libro sagrado del judaísmo

**hemisphere** [hemisferio] *s.*, una mitad de la Tierra

**henna** [alheña] *s.*, polvo rojizo usado para crear diseños sobre la piel

**hereditary** [hereditario] *adj.*, que se transmite de padres a hijos

**heritage** [herencia] *s.*, tradición que se transmite de los ancestros a los descendientes

**Hermitage Museum** [museo Hermitage] *s.*, museo de arte y cultura ubicado en San Petersburgo, Rusia

**Hidalgo, Miguel** [Hidalgo, Miguel] *s.*, sacerdote católico que lideró una revuelta en México en 1810

**hieroglyphics** [jeroglíficos] *s.*, sistema antiguo de escritura que emplea imágenes y símbolos

**highlands** [tierras altas] *s.*, áreas de terreno montañoso alto

**Himalaya Mountains** [Himalaya] *s.*, la cordillera más alta del mundo, ubicada en el sur de Asia

**Hinduism** [hinduismo] *s.*, religión practicada por más del 60 por ciento de la población de Asia del Sur

**Hitler, Adolf** [Hitler, Adolf] *s.*, jefe de estado alemán desde 1933 hasta 1945

**Ho Chi Minh** [Ho Chi Minh] *s.*, líder de Vietnam del Norte que quería unificar Vietnam del Norte y Vietnam del Sur bajo el comunismo; estos intentos dieron comienzo a la Guerra de Vietnam

**Holocaust** [Holocausto] *s.*, asesinato masivo cometido por los nazis contra seis millones de judíos y otras personas durante la Segunda Guerra Mundial

**homelands** [terruños] *s.*, durante el apartheid, áreas separadas dentro de Sudáfrica donde se obligaba a vivir a los sudafricanos negros

**hotspot** [punto caliente] *s.*, parte inusualmente caliente del manto de la Tierra

**Huang He** [Huang He] *s.*, el segundo río más largo de China, también llamado río Amarillo

**Human Development Index (HDI)** [Índice de Desarrollo Humano (IDH)] *s.*, conjunto de datos utilizados por los geógrafos para comparar la calidad de vida en diferentes países, incluyendo salud, educación y estándar de vida

**human rights** [derechos humanos] *s.*, derechos políticos, económicos y culturales que todas las personas deben tener

**hunter-gatherer** [cazador-recolector] *s.*, persona que caza animales y recolecta plantas y frutos para alimentarse

**hurricane** [huracán] *s.*, tormenta fuerte con vientos giratorios y lluvia intensa

**Hussein, Saddam** [Hussein, Saddam] *s.*, presidente de Iraq desde 1979 hasta 2003

**hybrid** [híbrido] *s.*, vehículo que puede funcionar a electricidad o gasolina

# GLOSARIO

**hydroelectric power** [energía hidroeléctrica] s., fuente de energía que emplea agua en movimiento para producir electricidad

## I

**ice shelf** [capa de hielo] s., lámina flotante de hielo adherida a una masa de tierra

**immigrate** [inmigrar] v., mudarse a otro país o región

**immigration** [inmigración] s., mudanza permanente de una persona a otro país

**impact** [impacto] s., efecto que produce un cambio

**imperialism** [imperialismo] s., práctica de extender la influencia de una nación controlando otros territorios

**Impressionism** [Impresionismo] s., estilo artístico en el cual los artistas usaban la luz y el color en pinceladas cortas para capturar un momento de tiempo

**incentive** [incentivo] s., razón o motivo para hacer algo

**incorporate** [incorporar] v., incluir; combinar con algo ya formado

**indigenous** [indígena] adj., originario del área donde algo se encuentra

**indulgence** [indulgencia] s., durante la Edad Media, tarifa pagada a la iglesia para disminuir el castigo por un pecado

**Indus River** [río Indo] s., río ubicado en la parte occidental de Asia del Sur

**Industrial Revolution** [Revolución Industrial] s., período de los siglos XVIII y XIX durante el cual los trabajadores de las fábricas comenzaron a usar máquinas y herramientas eléctricas para producir a gran escala

**industrialization** [industrialización] s., paso a una producción a gran escala mediante el uso de máquinas

**industrialize** [industrializar] v., desarrollar la manufactura

**infectious** [infeccioso] adj., capaz de propagarse rápidamente a otros

**infrastructure** [infraestructura] s., sistemas básicos de una sociedad, tales como las carreteras, los puentes, las cloacas y el tendido eléctrico

**inner islands** [islas interiores] s., islas de Indonesia entre las que se incluyen Java, Madura y Bali

**Institutional Revolutionary Party (PRI)** [Partido Revolucionario Institucional (PRI)] s., partido político que controló el gobierno de México desde 1929 hasta 2000

**interact** [interactuar] v., afectar a otras personas y verse afectado por ellas

**interior** [interior] s., tierra que está lejos de la costa del mar

**Internet** n., una red de comunicación

**intersection** [encrucijada] s., lugar donde se encuentran las personas o se cruzan los caminos

**intifada** [intifada] s., levantamiento o rebelión, usualmente se utiliza para referirse a la revuelta palestina en contra de Israel

**invader** [invasor] s., enemigo que entra a un país por la fuerza

**invasive species** [especies invasoras] s., plantas o animales no nativos introducidos en un área nueva, intencionalmente o no, que alteran los hábitats de los seres vivos nativos

**Iron Curtain** [Cortina de Hierro] s., en Europa, frontera imaginaria que separaba los países comunistas de los no comunistas durante la Guerra Fría

**Irrawaddy River** [río Irawadi] s., río que fluye a través de Myanmar

**irrigation** [irrigación] s., proceso de redirigir el agua hacia los cultivos a través de canales y zanjas

**Isis** [Isis] s., diosa egipcia importante

**Islam** [islamismo] s., religión fundada en Arabia Saudita a principios del siglo VII D.C.

**isolated** [aislado] adj., separado de los demás

**isolation** [aislamiento] s., separarse o ser colocado aparte de los demás

**isthmus** [istmo] s., franja estrecha de tierra que une dos áreas grandes de tierra

## J

**Jainism** [jainismo] s., religión del noroeste de India que se inició a fines del siglo VI D.C.

**Java** [Java] s., una de las islas interiores del Sureste Asiático

**Jerusalem** [Jerusalén] s., ciudad capital de Israel

**Jiang Jieshi** [Chiang Kai-shek] s., líder nacionalista chino que combatió contra los comunistas

**Jinnah, Mohammed Ali** [Jinnah, Mohammed Ali] s., líder de la Liga Musulmana Pan India que ayudó a fundar Pakistán y fue el primer gobernador general de Pakistán desde 1947–1948

## K

**Kalahari** [Kalahari] s., desierto importante del sur de África

**Kanto Plain** [llanura de Kanto] s., área llana y propicia para la agricultura e industria ubicada en el este de los Alpes japoneses

**karma** [karma] s., en el hinduismo, efecto positivo o negativo que una persona recibe como consecuencia de sus actos

**Kenyatta, Jomo** [Kenyatta, Jomo] s., líder africano que ayudó a obtener la independencia de Kenia y en 1963 se convirtió en su primer gobernante electo

# GLOSARIO

**Khmer Empire** [Imperio Jemer] *s.*, el imperio más vasto y duradero de Camboya, que se extendió desde inicios del siglo IX hasta 1430

**Kievan Rus** [Rus de Kiev] *s.*, estado fundado por los rusos Varegos en 882, que pasó a formar parte de la Rusia moderna

**Kilimanjaro** [Kilimanyaro] *s.*, volcán inactivo de Tanzania, de 19,340 pies de altura

**kimono** [kimono] *s.*, vestimenta femenina tradicional del Japón

**Kingman Reef** [arrecife Kingman] *s.*, arrecife de coral impoluto que forma parte de las Islas de la Línea, ubicadas entre Hawái y Samoa Americana

**kinship** [parentesco] *s.*, vínculo por consanguinidad o relación familiar

**Knesset** [Knéset] *s.*, parlamento del Estado de Israel

**Koguryo** [Goguryeo] *s.*, reino ubicado en el norte de Corea, en el año 37 A.C.

**Kongo** [Congo] *s.*, estado del centro de África, fundado en 1390

**Koryo** [Goryeo] *s.*, dinastía que gobernó Corea de 935 a 1392

**Kremlin** [Kremlin] *s.*, complejo histórico de palacios, arsenales e iglesias ubicado en Moscú, sede del gobierno ruso

**Krishna** [Krishna] *s.*, en el hinduismo, deidad encarnada en Visnú

**Kunlun Mountains** [montañas de Kunlun] *s.*, cordillera del este de Asia

**Kurd** [kurdo] *s.*, miembro de un grupo étnico no árabe del sudoeste de Asia

# L

**L'Ouverture, Toussaint** [L'Ouverture, Toussaint] *s.*, ex esclavo que lideró la exitosa revuelta haitiana para lograr independizarse de Francia

**labor force** [fuerza laboral] *s.*, cantidad de personas disponibles para trabajar

**land bridge** [puente de tierra] *s.*, franja de tierra que conecta dos grandes masas de tierra

**land reform** [reforma agraria] *s.*, división de las fincas de gran extensión para dar tierras a los pobres

**landlocked** [sin litoral] *adj.*, rodeado de tierras, sin acceso directo al mar

**landmass** [masa de tierra] *s.*, área muy extensa de tierra

**language diffusion** [difusión de la lengua] *s.*, la expansión de una lengua desde su lugar de origen

**language family** [familia de lenguas] *s.*, grupo de lenguas relacionadas

**latitude** [latitud] *s.*, línea imaginaria que se extiende de este a oeste alrededor de la Tierra y que indica la ubicación en relación con el ecuador

**launch** [lanzamiento] *v.*, para empezar

**legume** [legumbre] *s.*, arvejas o frijoles

**Lenin, V.I.** [Lenin, V.I.] *s.*, líder bolchevique que destituyó al zar en la Revolución Rusa de 1917 y tomó el mando del gobierno nuevo

**liberate** [liberar] *v.*, conceder la libertad a algo o a alguien

**lingua franca** [lengua franca] *s.*, lengua común a varios grupos de personas

**linguist** [lingüista] *s.*, científico del lenguaje

**literacy rate** [alfabetismo] *s.*, porcentaje de personas que saben leer y escribir

**literate** [alfabetizado] *adj.*, que puede leer y escribir

**Llanos** [llanos] *s.*, praderas del norte de Sudamérica

**Locke, John** [Locke, John] *s.*, filósofo inglés de fines del siglo XVII que contribuyó a inspirar la Revolución Norteamericana y la Ilustración

**locks** [esclusas] *s.*, compartimentos que se usan en los canales para subir o bajar los barcos entre vías fluviales conectadas

**loess** [loes] *s.*, sedimento o limo de color amarillo que forma depósitos de gran espesor

**longitude** [longitud] *s.*, línea imaginaria que corre del Polo Norte al Polo Sur y que indica la ubicación en relación con el primer meridiano

**Lost Decade** [Década Perdida] *s.*, en Japón, la década de 1990, en la que disminuyó la producción porque las empresas estaban fuertemente endeudadas

**Lost Boys of Sudan** [Niños Perdidos de Sudán] *s.*, grupo de jóvenes de Sudán que quedaron huérfanos como consecuencia de la guerra civil y se mantuvieron unidos para escapar de la violencia

**Louisiana Purchase** [Compra de la Luisiana] *s.*, tierras compradas por Thomas Jefferson en 1803 que duplicaron el tamaño de los Estados Unidos

**lowland** [tierras bajas] *s.*, área de poca altura

**lucrative** [lucrativo] *adj.*, que hace ganar dinero o produce ganancia

**Luther, Martin** [Lutero, Martín] *s.*, monje alemán cuyos actos condujeron, en 1517, a la Reforma para enfrentar la corrupción de la Iglesia Católica

# M

**Machu Picchu** [Machu Picchu] *s.*, ciudad compleja construida por los Incas sobre una montaña en el siglo XV

**Madura** [Madura] *s.*, una de las islas interiores ubicadas en el sudeste de Asia

# GLOSARIO

**magnetic levitation (Maglev)** [levitación magnética] *adj.*, tipo de tren que se desliza sobre un colchón de aire por encima de vías dotadas de muchos imanes potentes

**mainland** [masa territorial] *s.*, tierra unida a un continente; usualmente se utiliza en relación con los países que poseen tanto un sector continental como islas

**maintain** [mantener] *v.*, conservar y continuar

**Malay Peninsula** [Península Malaya] *s.*, península del Sureste Asiático que abarca partes de Tailandia, Myanmar y la parte continental de Malasia

**Malaysia** [Malasia] *s.*, país del Sureste Asiático, situado en el extremo inferior de la península malaya y en la isla de Borneo

**malnutrition** [desnutrición] *s.*, insuficiencia de alimentos o nutrientes

**Mandela, Nelson** [Mandela, Nelson] *s.*, líder del Congreso Nacional Africano, encarcelado por luchar contra el apartheid, elegido presidente de Sudáfrica en 1994

**manga** [manga] *s.*, tipo de libro de historietas japonés

**Manifest Destiny** [Destino Manifiesto] *s.*, idea de que Estados Unidos tiene el derecho de expandir su territorio hacia el océano Pacífico

**Manila** [Manila] *s.*, capital de las Filipinas

**manufacturing** [manufactura] *s.*, uso de máquinas para convertir las materias primas en productos útiles

**Mao Zedong** [Mao Zedong] *s.*, presidente del Partido Comunista que gobernó China de 1949 a 1976

**Maori** [maoríes] *s.*, pueblo nativo de Nueva Zelanda

**map** [mapa] *s.*, representación plana, bidimensional, de la Tierra

**marine** [marino] *adj.*, perteneciente al mar

**marine life** [vida marina] *s.*, plantas y animales que viven en el océano

**marine reserve** [reserva marina] *s.*, área del océano resguardada para proteger a los animales marinos de los seres humanos

**maritime** [marítimo] *adj.*, relacionado con el mar

**marketing** [marketing] *s.*, publicidad y promoción de un producto o negocio

**marsupial** [marsupial] *s.*, mamífero cuyas hembras transportan a las crías en una bolsa abdominal

**martial law** [ley marcial] *s.*, gobierno que se mantiene por el poder militar

**mass media** [medios masivos de comunicación] *s.*, comunicación proveniente de una única fuente, con el potencial de llegar a grandes audiencias

**Maya** [maya] *s.*, civilización que vivió en el Yucatán y en la parte norte de Centroamérica de 100 A.C. a 900 D.C.

**medicinal plant** [planta medicinal] *s.*, planta que se usa para tratar enfermedades

**meditation** [meditación] *s.*, práctica de usar la concentración para calmar los pensamientos y controlarlos

**Mediterranean climate** [clima mediterráneo] *s.*, clima de veranos calurosos y secos e inviernos templados y lluviosos

**megacity** [megalópolis] *s.*, ciudad grande, que tiene más de 10 millones de habitantes

**Mekong River** [río Mekong] *s.*, el río más largo del Sureste Asiático; fluye a través de Myanmar, Laos y Tailandia

**Mesa Central** [Mesa Central] *s.*, en México, la parte sur de la meseta mexicana

**messiah** [mesías] *s.*, líder o salvador

**mestizo** [mestizo] *s.*, persona que tiene una mezcla de ancestros europeos y americanos nativos

**methane** [metano] *s.*, gas natural incoloro que se libera a partir del carbono

**metropolitan area** [área metropolitana] *adj.*, sitio poblado alrededor de una ciudad que incluye los límites de la ciudad y las comunidades que la rodean

**Mexican Cession** [Cesión Mexicana] *s.*, tierras que abarcan desde Texas hasta California, entregadas por México a los Estados Unidos en 1848, de acuerdo con el tratado de Guadalupe Hidalgo

**Mexican Plateau** [Meseta Mexicana] *s.*, área de tierras llanas que se encuentran entre las dos cordilleras montañosas de Sierra Madre, en México

**microcredit** [microcrédito] *s.*, préstamo de una suma pequeña de dinero

**microlending** [micropréstamo] *s.*, práctica de otorgar préstamos de sumas pequeñas de dinero a las personas que inician sus propios negocios

**Middle Ages** [Edad Media] *s.*, período en Europa occidental posterior a la caída del Imperio Romano, desde aproximadamente 500 a 1500

**Middle Passage** [Pasaje del Medio] *s.*, viaje para cruzar el océano Atlántico, que tomaba meses, en el cual los africanos esclavizados eran llevados a las colonias europeas en América

**migrate** [emigrar] *v.*, trasladarse de un lugar a otro

**migration** [migración] *s.*, traslado de un lugar a otro

**military dictatorship** [dictadura militar] *s.*, forma de gobierno en la que el ejército ejerce el control del gobierno

**mineral** [mineral] *s.*, sustancia sólida y natural que se encuentra en las rocas y en la Tierra; es inorgánica y tiene un conjunto propio de propiedades

**missionary** [misionero] *s.*, persona enviada por una iglesia a convertir a otras personas a esa religión

**mobile** [móvil] *adj.*, que se puede mover

# GLOSARIO

**modernization** [modernización] s., políticas y acciones diseñadas para que un país se actualice tanto tecnológicamente como en otras áreas

**modify** [modificar] v., cambiar o hacer menos extremo

**Mongol Empire** [Imperio Mongol] s., imperio establecido en Asia Central por Gengis Kan a comienzos del siglo XIII

**monk** [monje] s., persona que dedica su vida a tareas religiosas

**monopoly** [monopolio] s., control total del mercado para un servicio o producto

**monotheism** [monoteísmo] s., creencia en un solo dios

**monotheistic** [monoteísta] adj., relativo a la creencia religiosa en un solo dios

**monotheistic religion** [religión monoteísta] s., sistema de creencias basadas en un solo dios o deidad

**monsoon** [monzón] s., viento estacional que trae lluvias intensas durante parte del año

**Montezuma** [Moctezuma] s., líder azteca asesinado por el conquistador español Hernán Cortés

**moral** [moral] adj., correcto y bueno, referido al comportamiento humano

**mosque** [mezquita] s., templo musulmán

**mouth** [desembocadura] s., lugar donde un río desemboca en el mar

**movable type** [tipo móvil] s., invento de aplicación en la imprenta, que permite mover los carácteres individuales para crear distintas páginas de texto

**multinational corporation** [corporación multinacional] s., empresa grande que tiene su base en un país y que abre sucursales en muchos otros países

**multi-party democracy** [democracia multipartidaria] s., sistema político en el cual las elecciones incluyen candidatos de más de un partido

**multitudes** [multitudes] s., grandes cantidades

**Mundurukú** [mundurukú] s., pueblo indígena de Brasil

**mural** [mural] s., pintura de gran tamaño realizada sobre una pared

# N

**N'Dour, Youssou** [N'Dour, Youssou] s., famoso griot de África occidental que interpreta música Mbalax

**Nairobi** [Nairobi] s., capital de Kenia

**Napoleon** [Napoleón] s., emperador de Francia que conquistó otros países europeos y formó un imperio, 1804–1815

**National Action Party (PAN)** [Partido Acción Nacional (PAN)] s., partido político mexicano que ganó las elecciones del año 2000

**nationalism** [nacionalismo] s., profundo sentimiento de lealtad al propio país

**nationalize** [nacionalizar] v., dar al gobierno el control de una empresa privada

**natural rights** [derechos naturales] s., derechos como la vida, la libertad y la propiedad, que las personas poseen desde su nacimiento

**naturalization** [naturalización] s., proceso que permite que una persona nacida en otro país se convierta en ciudadano

**navigable** [navegable] adj., suficientemente ancho o profundo para que los barcos o botes puedan navegar sin inconvenientes

**navigation** [navegación] s., ciencia de averiguar la posición y planear rutas marítimas

**Nazi Germany** [Alemania nazi] s., Alemania bajo el régimen del partido nazi, de 1933 a 1945

**Nebuchadnezzar** [Nabucodonosor] s., rey de Babilonia, 605–562 A.C.

**neutrality** [neutralidad] s., negativa a tomar partido o a involucrarse

**New Guinea** [Nueva Guinea] s., la segunda isla más grande del mundo, ubicada en el sudoeste del océano Pacífico

**New South Wales** [Nueva Gales del Sur] s., colonia construida por convictos en Australia, en 1788

**Nile River** [río Nilo] s., el río más largo del mundo; fluye 4,000 millas a través de Egipto y África

**Nkrumah, Kwame** [Nkrumah, Kwame] s., líder africano de las décadas de 1950 y 1960, que contribuyó a lograr la independencia de Ghana

**nocturnal** [nocturno] adj., activo de noche en lugar de durante el día

**nomad** [nómada] s., persona que se desplaza de un lugar a otro

**nonrenewable** [no renovable] adj., que no se puede reproducir con la misma rapidez con que se lo usa

**nonrenewable fossil fuel** [combustible fósil no renovable] s., fuente de energía, como el petróleo, el gas natural o el carbón, cuya provisión es limitada

**nonrenewable resource** [recurso no renovable] s., fuente de energía que es limitada, y no se puede reemplazar, como el petróleo

**North American Free Trade Agreement (NAFTA)** [Tratado de Libre Comercio de América del Norte (NAFTA)] s., acuerdo firmado en 1994 que facilitó el comercio y la inversión entre Canadá, México y los Estados Unidos

**North Anatolian Fault** [Falla del Norte de Anatolia] s., fractura de la corteza terrestre que se extiende al este y al oeste, justo al sur del mar Negro

# GLOSARIO

**North Atlantic Drift** [Corriente del Atlántico Norte] s., corriente marina cálida que calienta las aguas que bañan la parte noroeste de Rusia

**North China Plain** [Llanura del Norte de China] s., llanura que se extiende a lo largo del río Huang He

**North Pole** [Polo Norte] s., punto ubicado en el extremo norte de la Tierra, opuesto al Polo Sur, donde convergen todas las líneas de longitud

**Northern European Plain** [Llanura del Norte de Europa] s., vastas tierras bajas que se extienden desde Francia hasta Rusia

**Northern Hemisphere** [Hemisferio Norte] s., la mitad de la Tierra que se encuentra al norte del ecuador

**novel** [novela] s., extensa obra de ficción, con trama y personajes complejos

## O

**oasis** [oasis] s., sitio fértil con agua ubicado en un área seca y desértica

**occupy** [ocupar] v., apoderarse

**Okavango Delta** [delta del Okavango] s., delta del interior de Botsuana, donde el río Okavango desemboca en un pantano

**Olmec** [olmecas] s., sociedad organizada que vivió junto a la costa sur del golfo de México en 1000 A.C.

**one-child policy** [política de hijo único] s., ley china de 1979 que limitaba a las familias que vivían en áreas urbanas a tener solamente un hijo

**opera** [ópera] s., representación que cuenta una historia mediante música y palabras

**oppose** [oponerse] v., objetar

**oral tradition** [tradición oral] s., transmisión verbal de historias o relatos de una generación a la siguiente

**Orange Revolution** [Revolución Naranja] s., destitución pacífica del primer ministro de Ucrania en 2004

**Osman** [Osman] s., primer líder de los turcos en el siglo XIV, por cuyo nombre pasaron a ser llamados otomanos

**Ottoman Empire** [Imperio Otomano] s., vasto y rico imperio que existió desde 1453 hasta 1923, centrado en el territorio actual de Turquía

**outer islands** [islas exteriores] s., islas de Indonesia que incluyen a Sumatra, Borneo, Nueva Guinea y otras

**outrigger canoe** [canoa hawaiana] s., bote que tiene adosado un flotador que le da estabilidad

**outsourcing** [subcontratar] s., transferir empleos a trabajadores que no pertenecen a la compañía, que a menudo se encuentran en un país extranjero

**overpopulation** [superpoblación] s., situación en la que demasiadas personas viven en un mismo lugar

## P

**Paekche** [Baekje] s., reino del sudoeste de Corea, en el año 18 A.C.

**pagoda** [pagoda] s., estructura religiosa de varios pisos que se encuentra en los países asiáticos, a menudo utilizada con fines religiosos

**Palestine Liberation Organization (PLO)** [Organización para la Liberación de Palestina (OLP)] s., organización creada por líderes palestinos

**Pampas** [Pampa] s., llanuras cubiertas de pasto de la Argentina

**Pan-Africanism** [Panafricanismo] s., movimiento surgido en los inicios del siglo XX para unir a los pueblos africanos

**Panama Canal Zone** [Zona del Canal de Panamá] s., área en la que se construyó el canal de Panamá

**pandemic** [pandemia] s., brote de una enfermedad que se propaga por una vasta área geográfica

**Papua New Guinea** [Papúa Nueva Guinea] s., país del Pacífico Sur que abarca la mitad oriental de la isla de Nueva Guinea y las islas cercanas

**papyrus** [papiro] s., material similar al papel inventado en el antiguo Egipto

**Parliament** [Parlamento] s., poder legislativo del gobierno de la India

**parliamentary democracy** [democracia parlamentaria] s., sistema de gobierno en el cual el poder ejecutivo está presidido por el primer ministro, que es elegido por el partido que posee la mayoría de los escaños del Parlamento

**Partition** [Partición] s., se refiere a la división del sur de Asia en los países independientes de India y Paquistán

**patrician** [patricio] s., rico terrateniente de la antigua Roma

**Pearl Harbor** [Pearl Harbor] s., base naval de los EE.UU. ubicada en Hawái, bombardeada por los japoneses en 1941, lo cual provocó el ingreso de EE.UU. en la Segunda Guerra Mundial

**peat** [turba] s., material que se forma a partir de la descomposición de plantas muy antiguas y que arde como el carbón

**peninsula** [península] s., masa de tierra rodeada por agua en tres de sus lados

**perestroika** [perestroika] s., reformas en la estructura económica de la Unión Soviética introducidas por Mijaíl Gorbachov en 1985

**permafrost** [permafrost] s., suelo que está permanentemente congelado

**perspective** [perspectiva] s. *modo artístico de mostrar los objetos de la manera en que son vistos por las personas*, en términos de distancia o profundidad relativa, como si estuvieran en tres dimensiones

# GLOSARIO

**pesticide** [pesticida] *s.*, sustancia química que mata insectos y malezas nocivas

**Peter the Great** [Pedro el Grande] *s.*, Pedro Romanov, zar que gobernó Rusia desde 1682 hasta 1725

**petrochemicals** [petroquímicos] *s.*, productos elaborados a partir del petróleo

**petroleum** [petróleo] *s.*, materia prima que se usa para producir combustibles

**pharaoh** [faraón] *s.*, rey en el antiguo Egipto

**philosopher** [filósofo] *s.*, persona que examina las preguntas sobre el universo y busca la verdad

**pictograph** [pictograma] *s.*, imagen pintada usada para comunicar

**pilgrimage** [peregrinaje] *s.*, viaje religioso

**pioneer** [pionero] *s.*, colono de tierras nuevas

**pipeline** [tubería] *s.*, serie de tubos o caños conectados para transportar líquidos o gases

**Pizarro, Francisco** [Pizarro, Francisco] *s.*, conquistador español que en 1533 derrocó al emperador y fundó la ciudad de Lima, Perú

**plain** [llanura] *s.*, área plana de la superficie terrestre

**plantation** [plantación] *s.*, granja de gran tamaño que produce cultivos para obtener ganancias

**plate** [placa] *s.*, sección rígida de la corteza terrestre que se puede mover de manera independiente

**plateau** [meseta] *s.*, llanura ubicada a gran altura sobre el nivel del mar que a menudo tiene un precipicio en todos sus lados

**plebeian** [plebeyo] *s.*, agricultor o persona de clase baja de la antigua Roma

**poach** [caza o pesca furtiva] *v.*, cazar o pescar ilegalmente

**poacher** [cazador o pescador furtivo] *s.*, persona que caza o pesca de manera ilegal

**poaching** [cazar o pescar furtivamente] *s.*, caza o pesca ilegal

**polder** [pólder] *s.*, tierras de los Países Bajos ganadas al mar que se destinan a la agricultura

**policy** [política] *s.*, pautas y procedimientos oficiales de una organización o gobierno

**pollution** [contaminación] *s.*, desechos químicos o físicos que generan un medio ambiente sucio o poco limpio

**Polynesian Triangle** [Triángulo Polinésico] *s.*, vasta área del Pacífico Sur que abarca muchas islas

**polytheism** [politeísmo] *s.*, creencia en más de un dios

**polytheistic religion** [religión politeísta] *s.*, sistema de creencias basadas en varios dioses o deidades

**popular culture** [cultura popular] *s.*, artes, música y otros elementos de la vida cotidiana de una región

**porcelain** [porcelana] *s.*, tipo de cerámica dura

**port** [puerto] *s.*, embarcadero para barcos donde se intercambian mercancías

**Port-au-Prince** [Puerto Príncipe] *s.*, capital de Haití

**potential** [potencial] *s.*, posibilidad

**precipitation** [precipitación] *s.*, proceso que hace caer agua sobre la Tierra, en forma de lluvia, nieve o granizo

**predominant** [predominante] *adj.*, principal, más común, superior a los demás

**prehistoric** [prehistórico] *adj.*, anterior a la historia escrita

**preserve** [preservar] *v.*, proteger

**pride** [manada] *s.*, grupo de leones que viven en comunidad

**prime meridian** [primer meridiano] *s.*, línea de longitud de 0° que se extiende desde Polo Norte al Polo Sur y que pasa por Greenwich, Inglaterra

**privatization** [privatización] *s.*, proceso por el que las empresas que eran propiedad del gobierno pasan a manos privadas

**profitable** [rentable] *adj.*, que hace ganar dinero, financieramente exitoso

**projection** [proyección] *s.*, modo de mostrar la superficie curva de la Tierra sobre un mapa plano

**promote** [fomentar] *v.*, animar, estimular

**propaganda** [propaganda] *s.*, información que se difunde para influir sobre la opinión de las personas o para promover las ideas de un partido u organización

**proportional representation** [representación proporcional] *s.*, sistema en el cual un partido político consigue un porcentaje de escaños igual al porcentaje de votos que obtuvo

**prosperous** [próspero] *adj.*, económicamente fuerte

**protest** [protestar] *v.*, objetar

**province** [provincia] *s.*, parte más pequeña en que se divide un país, especialmente Canadá

**push-pull factors** [factores de atracción y repulsión] *s.*, motivos por los que las personas emigran; los factores de "repulsión" las hacen partir de un sitio, los factores de "atracción" las hacen ir hacia otro sitio

**pyramid** [pirámide] *s.*, monumento construido en roca que servía de tumba en el antiguo Egipto

# Q

**qanat** [qanat] *s.*, túnel subterráneo construido por el hombre en la meseta de Irán, usado para transportar agua desde las montañas

**Qin** [Qin] *s.*, dinastía que gobernó China durante el período 221–206 a.c.

**Quechua** [quechuas] *s.*, pueblo que vive en la cordillera de los Andes de Perú, Ecuador y Bolivia

**Qur'an** [Corán] *s.*, libro sagrado del islamismo

# GLOSARIO

## R

**radical** [radical] *s.*, persona que busca un cambio extremo o sostiene una posición política extrema

**radioactive** [radiactivo] *adj.*, que emite energía producida por la ruptura de un átomo

**rain forest** [bosque tropical] *s.*, bosque de temperatura cálida, humedad elevada y vegetación espesa que recibe más de 100 pulgadas de lluvia al año

**rain shadow** [sombra orográfica] *s.*, región seca ubicada sobre uno de los lados de una cordillera

**rainshadow effect** [efecto de la sombra orográfica] *s.*, proceso en el cual el aire húmedo asciende por una ladera de la cordillera y luego se enfría y cae en forma de precipitación, dejando el otro lado de la cordillera mayormente seco

**Ramses II** [Ramsés II] *s.*, faraón egipcio que reinó alrededor del año 1185 A.C. y expandió los límites del imperio egipcio

**raw materials** [materias primas] *s.*, materiales naturales o sin terminar, como minerales, petróleo o carbón, que se usan para elaborar productos terminados

**Re** [Ra] *s.*, dios solar de los antiguos egipcios

**rebellion** [rebelión] *s.*, revuelta o resistencia a la autoridad

**recession** [recesión] *s.*, desaceleración del crecimiento económico

**reclaim** [reclamar] *v.*, volver a tomar

**Reconstruction** [Reconstrucción] *s.*, esfuerzo realizado para reconstruir y unificar a los Estados Unidos después de la guerra civil

**reform** [reforma] *s.*, cambio que apunta a corregir un problema

**Reformation** [Reforma] *s.*, movimiento que surgió en el siglo XVI para reformar el cristianismo

**refuge** [refugio] *s.*, lugar seguro

**refugee** [refugiado] *s.*, persona que huye de un lugar para estar a salvo

**region** [región] *s.*, conjunto de sitios con características comunes

**regulate** [regular] *v.*, controlar

**reign** [reinado] *s.*, período de mando de un rey, reina, emperador o emperadora

**Reign of Terror** [El Terror] *s.*, movimiento francés liderado por Maximilien Robespierre, en el cual fueron decapitadas 40,000 personas, durante el período 1793–94

**reincarnation** [reencarnación] *n.*, el nacimiento de un alma en otra vida

**relative location** [ubicación relativa] *s.*, la posición de un lugar en relación con otros

**reliable** [confiable] *adj.*, fiable o de confianza

**relief** [relieve] *s.*, cambio en la elevación de un lugar a otro

**religious tolerance** [tolerancia religiosa] *s.*, aceptación de distintas religiones para que sean profesadas al mismo tiempo, sin prejuicios

**relocate** [relocalizar] *v.*, trasladar

**remittance** [remesa] *s.*, dinero enviado a una persona que se encuentra en otro lugar

**remote** [remoto] *adj.*, difícil de llegar, aislado

**Renaissance** [Renacimiento] *s.*, período que se desarrolló entre los siglos XIV y XVI, donde florecieron el arte y la cultura

**renewable energy** [energía renovable] *s.*, energía obtenida a partir de fuentes que no se agotan, como el viento, el Sol y el agua

**renewable resource** [recurso renovable] *s.*, materia prima o fuente de energía que se reemplaza a sí misma con el paso del tiempo

**reparation** [reparación] *s.*, dinero que, después de una guerra, pagan como castigo normalmente los agresores que iniciaron el conflicto

**republic** [república] *s.*, forma de gobierno en la cual las personas eligen funcionarios para que gobiernen

**reserve** [reserva] *s.*, tierras destinadas a propósitos especiales, como la agricultura, la preservación de los hábitats o para ser usadas como vivienda por determinados grupos de personas; futuro suministro (de petróleo)

**reservoir** [embalse] *s.*, lago artificial de gran tamaño para almacenar agua

**resistance** [resistencia] *n.*, oposición

**restore** [restaurar] *v.*, recuperar

**retreat** [retirarse] *v.*, ir hacia atrás, no hacia adelante

**revenue** [rentas] *s.*, ingresos

**reverse** [retroceder] *v.*, ir en la dirección opuesta

**Revolution** [revolución] *s.*, acción de ciudadanos o colonos cuyo objetivo es derrocar un gobierno

**Rhine River** [río Rin] *s.*, río que nace en Suiza y desemboca en el mar del Norte

**rift valley** [valle de fisura] *s.*, valle profundo que se formó al separarse la corteza terrestre, como en África Oriental

**Ring of Fire** [Anillo de Fuego] *s.*, área que se extiende a lo largo de las riberas del océano Pacífico, donde chocan las placas tectónicas, lo que genera terremotos y una gran actividad volcánica

**Rio de Janeiro** [Río de Janeiro] *s.*, ciudad de Brasil que albergará los juegos olímpicos de 2016

**ritual** [ritual] *s.*, acto formal que se repite regularmente

**rivalry** [rivalidad] *s.*, competencia u oposición entre personas

# GLOSARIO

**river basin** [cuenca de un río] *s.*, área baja por la que fluye un río

**Romantic Period** [Romanticismo] *s.*, período artístico de comienzos del siglo XIX, en el cual los artistas pintaban paisajes y escenas de la naturaleza para transmitir emociones

**roots** [raíces] *s.*, orígenes culturales

**Rub al Khali** [Rub al-Jali] *s.*, vasto desierto ubicado al sur de Arabia Saudita

**Russian Revolution** [Revolución Rusa] *s.*, revolución que tuvo lugar en Rusia en 1917, en la cual el zar fue depuesto y los bolcheviques tomaron el poder

**Russification** [rusificación] *s.*, política de designar ciudadanos rusos a cargo de las repúblicas soviéticas durante las décadas de 1970 y 1980

**ruthless** [despiadado] *adj.*, cruel

# S

**Sahara Desert** [desierto de Sahara] *s.*, el desierto más grande del mundo, que cubre la mayor parte del norte de África

**Sahel** [Sahel] *s.*, pradera semiárida del África subsahariana, limitada al norte por el Sahara y al sur por las praderas tropicales

**salinization** [salinización] *s.*, acumulación de sal en el suelo

**Samoa** [Samoa] *s.*, país que comprende las islas occidentales de Samoa, ubicadas al sur del océano Pacífico

**samurai** [samurái] *s.*, diestro guerrero japonés

**sanitation** [sanidad] *s.*, medidas tomadas para proteger la salud pública, como la red cloacal

**Sanskrit** [sánscrito] *s.*, idioma de los arios, que se convirtió en la base de numerosas lenguas del sur de Asia

**Santa Anna** [Santa Anna] *s.*, presidente y general mexicano, que venció en la batalla de El Álamo, pero perdió la guerra entre los Estados Unidos y México

**São Paolo** [San Pablo] *s.*, la ciudad más grande de Brasil

**sarcophagus** [sarcófago] *s.*, ataúd

**sari** [sari] *s.*, vestido tradicional femenino de la India, que se enrolla alrededor del cuerpo

**saturate** [saturar] *v.*, remojar completamente

**savanna** [sabana] *s.*, pradera, como la del sur del África subsahariana

**scale** [escala] *s.*, parte de un mapa que indica el tamaño en el que se muestra un área de la Tierra

**scarcity** [escasez] *s.*, falta o carencia de algo

**scientific station** [estación científica] *s.*, sitio para desarrollar una investigación

**scorched earth policy** [táctica de tierra quemada] *s.*, práctica llevada adelante por las tropas rusas en 1812, quienes, a medida que retrocedían ante el avance del ejército de Napoleón, quemaban los cultivos y todos los recursos que pudieran servir al enemigo para abastecerse

**seafarer** [navegante] *s.*, persona que viaja por el mar

**secede** [separarse] *v.*, retirarse formalmente, dejar de ser parte

**secular** [secular] *adj.*, terrenal, sin vínculo con una religión

**segregation** [segregación] *s.*, separación por raza

**seismic** [sísmico] *adj.*, relacionado con la actividad o movimiento producidos por los terremotos

**seize** [apoderarse] *v.*, tomar el control

**self-rule** [autonomía] *s.*, gobierno ejercido por los propios habitantes de un país

**semiarid** [semiárido] *adj.*, algo seco, con muy poca lluvia

**serf** [siervo] *s.*, campesino ruso o europeo, pobre y con pocos derechos, que alquilaba tierras a un terrateniente entre los siglos XVI y XIX

**shalwar-kameez** [shalwar-kameez] *s.*, camisa larga con pantalones holgados que se usa en la India y en el suroeste de Asia

**Shang** [Shang] *s.*, familia cuya dinastía gobernó China desde 1766 hasta 1050 A.C.

**sheikh** [jeque] *s.*, líder árabe

**Shi Huangdi** [Qin Shi Huang] *s.*, líder de la dinastía Qin, que se convirtió en el primer emperador de la China en 221 A.C.

**Shi'ite** [chiita] *adj.*, rama de musulmanes que considera que los líderes religiosos deben ser descendientes de Mahoma

**Shinto** [sintoísmo] *s.*, religión nativa del Japón, similar al animismo

**shogun** [sogún] *s.*, gobernador militar japonés

**Siberia** [Siberia] *s.*, enorme región del centro y este de Rusia

**significant** [significativo] *adj.*, importante

**Sikhism** [sijismo] *s.*, religión que surgió a fines del siglo XV en la India

**Silk Roads** [rutas de la seda] *s.*, antiguas rutas comerciales que unían el sudoeste y el centro de Asia con China

**Silla** [Silla] *s.*, reino ubicado en el sudeste de Corea en el año 57 A.C.

**silt** [limo] *s.*, partículas finas de suelo que se depositan a lo largo de las márgenes de los ríos

**slash-and-burn** [tala y quema] *s.*, método agrícola que consiste en despejar las tierras talando y quemando el bosque y la vegetación

# GLOSARIO

**Slavs** [eslavos] *s.*, pueblo originario de los alrededores del mar Negro o Polonia, que se instaló en Ucrania y el oeste de Rusia alrededor del año 800 D.C.

**slum** [barriada] *s.*, área densamente poblada de una ciudad, con viviendas precarias y malas condiciones de vida

**smartphone** [teléfono inteligente] *s.*, dispositivo portátil que combina la comunicación con aplicaciones de software

**socialism** [socialismo] *s.*, sistema de gobierno en el que el gobierno controla los recursos económicos

**solstice** [solsticio] *s.*, punto en que el Sol se encuentra a la distancia máxima, al sur o al norte, del ecuador; inicio del invierno y del verano

**South Pole** [Polo Sur] *s.*, punto más austral de la Tierra, opuesto al Polo Norte, donde convergen todas las líneas de longitud

**Southern Alps** [Alpes del Sur] *s.*, cordillera situada en Nueva Zelanda

**Southern Hemisphere** [Hemisferio Sur] *s.*, la mitad de la Tierra que se encuentra al sur del ecuador

**sovereignty** [soberanía] *s.*, control de un país sobre sus propios asuntos

**Soviet Union** [Unión Soviética] *s.*, Unión de las Repúblicas Socialistas Soviéticas, país formado por Rusia y otros estados euroasiáticos, que existió desde 1922 hasta 1991

**soybean** [soja] *s.*, tipo de frijol que se cultiva como alimento y para elaborar productos industriales

**spatial thinking** [pensamiento espacial] *s.*, manera de pensar en el espacio que está sobre la superficie de la Tierra, incluyendo la ubicación de los distintos lugares y por qué se encuentran allí

**Special Economic Zone** [Zona Económica Especial] *s.*, área de China en la que se permitió el desarrollo de una economía de mercado, con menor control de los negocios por parte del gobierno

**sprawl** [expandirse] *v.*, extenderse

**standard of living** [nivel de vida] *s.*, nivel de acceso de los habitantes de un país a bienes, servicios y comodidades materiales

**staple** [alimento básico] *s.*, constituyente básico de la dieta de las personas

**state** [estado] *s.*, territorio determinado que posee un gobierno propio

**steel** [acero] *s.*, metal de gran dureza elaborado a partir del hierro combinado con otros metales

**steppe** [estepa] *s.*, llanura muy extensa de praderas secas

**strait** [estrecho] *s.*, vía fluvial angosta que conecta dos masas de agua

**strike** [huelga] *s.*, interrupción del trabajo por parte de empleados que se niegan a trabajar

**subcontinent** [subcontinente] *s.*, región separada de un continente

**subsistence farmers** [agricultores de subsistencia] *s.*, agricultores que producen cultivos para alimentar a sus familias, no para vender

**subsistence farming** [agricultura de subsistencia] *s.*, agricultura que solamente produce cultivos para que las familias se alimenten, no para vender

**subsistence fishing** [pesca de subsistencia] *s.*, pescar para tener comida para poder vivir, no para obtener ganancias

**suffrage** [sufragio] *s.*, derecho al voto

**Suleyman I** [Suleyman I] *s.*, emperador del Imperio Otomano a mediados del siglo XVI

**sultan** [sultán] *s.*, líder o gobernante del Imperio Otomano

**Sumatra** [Sumatra] *s.*, una de las islas exteriores del sudeste de Asia

**Sunni** [sunita] *adj.*, rama del islamismo que sostiene que los líderes religiosos deben ser escogidos entre los candidatos más capacitados, a diferencia de hacerlo entre los descendientes de Mahoma

**surplus** [excedente] *s.*, extra

**suspension bridge** [puente colgante] *n.*, colgados de un puente que se cuelga de dos o más cables

**sustainable** [sustentable] *adj.*, capaz de ser continuado sin dañar el medio ambiente o sin agotar los recursos de manera permanente

**Swahili** [suajili] *s.*, lengua bantú hablada mayormente en África Oriental, también llamada kiswahili

**Sydney** [Sídney] *s.*, la ciudad más grande de Australia, capital del estado de Nueva Gales del Sur

**symbol** [símbolo] *s.*, objeto o idea que se puede usar para representar otro objeto o idea

# T

**taiga** [taiga] *s.*, extensa área de bosques que se extiende a través del norte de Rusia, Canadá y otros países del norte

**Taino** [taínos] *s.*, pueblo nativo del Caribe

**Taj Mahal** [Taj Mahal] *s.*, famoso edificio construido en la India en el siglo XVII, para servir como tumba de la esposa de Shah Jahan; actualmente reconocido por la UNESCO como patrimonio de la humanidad

**Taliban** [talibán] *s.*, grupo de pashtunes de Afganistán que comenzaron a gobernar el país en 1996

**Taman Negara National Park** [Parque Nacional Taman Negara] *s.*, parque nacional de Malasia situado dentro de uno de los bosques tropicales más antiguos del mundo

**tariff** [arancel] *s.*, impuesto sobre las importaciones y exportaciones

# GLOSARIO

**tax** [impuesto] s., suma que se paga al gobierno para contar con servicios públicos

**tectonic plate** [placa tectónica] s., sección de la corteza terrestre que flota sobre el manto terrestre

**temperate** [templado] adj., benigno, en términos del clima

**terra cotta** [terracota] s., arcilla endurecida al horno

**terrace** [terraza] s., superficie llana construida sobre la ladera de un monte

**terraced** [abancalado] adj., campo llano excavado en la pendiente o ladera de una montaña

**terrain** [terreno] s., características físicas de la tierra

**terrorism** [terrorismo] s., tipo de guerra que se vale de la violencia para obtener resultados políticos; es normalmente empleada por grupos reducidos o individuos

**terrorist** [terrorista] s., persona que emplea la violencia para obtener resultados políticos

**textile** [textil] adj., relativo a la tela o ropa

**theme** [tema] s., tópico

**thirty-eighth parallel** [paralelo treinta y ocho] s., frontera que separa Corea del Norte y Corea del Sur, establecida en 1945 a lo largo del paralelo de 38° de latitud norte

**Three Gorges Dam** [presa de las Tres Gargantas] s., la presa más grande del mundo, emplazada sobre el río Chang Jiang, en China

**Tibetan Plateau** [meseta tibetana] s., meseta vasta y de gran elevación situada en Asia Central

**Timbuktu** [Tombuctú] s., ciudad de África Occidental que fue un centro educativo en el siglo XIII

**tolerance** [tolerancia] s., aceptación de las creencias de los demás

**tomb** [tumba] s., lugar donde se realiza un entierro

**topography** [topografía] s., características físicas de la tierra

**tornado** [tornado] s., tormenta de vientos muy fuertes que sigue una trayectoria impredecible

**totalitarian** [totalitario] adj., relacionado con un gobierno dirigido por un dictador que exige obediencia absoluta al estado

**tourism** [turismo] s., industria o negocio de los viajes

**Trail of Tears** [Sendero de Lágrimas] s., ruta que siguieron los cheroquis durante su forzada emigración desde el sudeste de los Estados Unidos hasta Oklahoma, en la década de 1830

**trans-Atlantic slave trade** [tráfico transatlántico de esclavos] s., negocio de traficar a América, a través del océano Atlántico, esclavos originarios de África, que se inició en el siglo XVI

**transcontinental** [transcontinental] adj., que atraviesa todo un continente

**transform** [transformar] v., rehacer o cambiar

**transition** [transición] s., cambio de una actividad o etapa a otra

**transition zone** [zona de transición] s., área situada entre dos regiones geográficas y que posee características de ambas

**transpiration** [transpiración] s., proceso mediante el cual las plantas y los árboles liberan vapor de agua en el aire

**transport** [transportar] v., enviar de un lugar a otro

**transportation corridor** [corredor de transporte] s., ruta terrestre o marítima para trasladar personas o mercancías de un lugar a otro con facilidad

**trans-Saharan** [transahariano] adj., que atraviesa el desierto del Sahara

**Trans-Siberian Railroad** [ferrocarril transiberiano] n., el ferrocarril de servicio continuo más largo del mundo, que une Moscú con el este de Rusia, atravesando Siberia

**treaty** [tratado] s., acuerdo entre dos o más países

**Treaty of Guadalupe Hidalgo** [Tratado de Guadalupe Hidalgo] s., acuerdo firmado en 1848, por el cual México cedió a los Estados Unidos el área que se extiende desde Texas hasta California

**Treaty of Tordesillas** [Tratado de Tordesillas] s., tratado firmado entre españoles y portugueses en 1494, a través del cual se dividieron la posesión de las tierras de Sudamérica

**Treaty of Versailles** [Tratado de Versalles] s., tratado de paz que puso fin, en 1919, a la Primera Guerra Mundial

**trench** [trinchera] s., zanja extensa que protege a los soldados del fuego enemigo

**trend** [tendencia] s., cambio que se produce en determinada dirección con el paso del tiempo

**triangular trade** [comercio triangular] s., comercio entre tres continentes: América, Europa y África

**tributary** [tributario] s., río pequeño que fluye hacia un río más grande

**tribute** [tributo] s., sumas pagadas a otro país o gobernante a cambio de protección o como muestra de sumisión

**troubadour** [trovador] s., cantante de la Edad Media que interpretaba canciones sobre caballeros y el amor

**tsunami** [tsunami] s., ola enorme y muy potente que se forma en el océano

**Tuareg** [tuareg] s., pueblo semi nómada que se desplaza en caravanas a través del Sahara, comerciando sal

**tundra** [tundra] s., tierras llanas y sin árboles que se encuentran en las regiones árticas y subárticas

# GLOSARIO

**Tupinambá** [tupinambás] s., pueblo que vivió cerca de la desembocadura del río Amazonas y a lo largo de la costa atlántica hacia el año 3000 A.C.

**typhoon** [tifón] s., tormenta tropical peligrosa, que trae lluvias copiosas y vientos muy fuertes, que en el Hemisferio Occidental se denomina huracán

**tyranny** [tiranía] s., gobierno duro o severo

## U

**United Arab Emirates** [Emiratos Árabes Unidos] s., país situado en la península arábiga, en el golfo pérsico

**United Nations (UN)** [Naciones Unidas (ONU)] s., organización de países formada en 1945, con el objetivo de mantener la paz entre los países y proteger los derechos humanos

**Universal Declaration of Human Rights** [Declaración Universal de los Derechos Humanos] s., acuerdo aprobado por las Naciones Unidas que define los derechos que deben tener todas las personas del mundo

**uplands** [tierras altas] s., colinas, montañas y mesetas

**Ur** [Ur] s., importante ciudad estado sumeria, 2800–1850 A.C.

**Ural Mountains** [montes Urales] s., cordillera que separa la Llanura del Norte de Europa de la Llanura de Siberia Occidental, en Rusia

**urban** [urbano] adj., que describe o está relacionado con la ciudad o los suburbios

**utilize** [utilizar] v., hacer uso práctico de algo

## V

**vaccine** [vacuna] s., tratamiento para incrementar la inmunidad a una enfermedad determinada

**vegetation** [vegetación] s., formas de vida vegetal

**venue** [sede] s., ubicación donde tiene lugar un suceso programado

**veto** [veto] v., rechazar una decisión tomada por otro órgano de gobierno

**viceroy** [virrey] s., gobernador de las colonias españolas en América, en representación del rey y la reina de España

**vocational** [profesional] adj., relacionado con las destrezas laborales

**volcano** [volcán] s., montaña que, al entrar en erupción, explota y lanza roca derretida, gases y ceniza

**vulnerable** [vulnerable] adj., abierto, que puede ser lastimado por fuerzas externas

## W

**wallaby** [walabí] s., marsupial más pequeño que un canguro

**wat** [wat] s., templo budista del Sureste Asiático

**waterway** [vía fluvial] s., ruta navegable que se usa para los viajes y el transporte

**weapon of mass destruction (WMD)** [arma de destrucción masiva] s., arma que produce un daño inmenso a grandes cantidades de personas

**weather** [tiempo atmosférico] s., condiciones de la atmósfera en un momento determinado, incluyendo la temperatura, precipitación y humedad de un día o una semana determinada

**Western Hemisphere** [Hemisferio Occidental] s., la mitad de la Tierra que se encuentra al oeste del primer meridiano

**Wet Pampas** [Pampa húmeda] s., región húmeda en el este de la Argentina

**Wilkes, Charles** [Wilkes, Charles] s., estadounidense que exploró la costa de la Antártida a fines de la década de 1830

**wind turbine** [aerogenerador] s., motor propulsado por el viento para generar electricidad

## Y

**Yanomami** [yanomamis] s., pueblo indígena cuyos habitantes son cazadores recolectores y continúan viviendo en la cuenca del río Amazonas

**yurt** [yurta] s., carpa tradicional de fieltro de Asia Central

## Z

**zaibatsu** [zaibatsu] s., organizaciones dirigidas por familias japonesas, propietarias de numerosos negocios

**Zambezi River** [río Zambeze] s., río del sur de África, que fluye hacia el océano Índico atravesando varios países situados en el sur del centro de África

**Zheng He** [Zheng He] s., almirante de la marina china que lideró siete expediciones navales para explorar las tierras situadas más allá de China; su último viaje se inició en 1431 D.C.

**Zhou** [Zhou] s., dinastía que gobernó China en el período 1050–221 A.C.

**zoologist** [zoólogo] s., científico que estudia los animales

# INDEX

## A
**abolition,** 98
**Aboriginal Land Rights Act, 1976,** 678
**Aborigines,** 656–657, 678
  land taken by Britain, 661, 678
**absolute location,** 14, 19
**abstract art,** 286
**Acropolis, Greece,** 256v
**Adamawa Highlands, Nigeria,** 362, 364
**adapt,** 192
**adherents,** 426
**Adulis, Ethiopia,** 372
**adventure tourism,** 680–681, 681v
**Afghanistan,** 412–413m, 414m, 451m, 456–457, 456m
  economic indicators of, 62, 63v
  mineral resources in, 314
  schools and education in, 458–459, 459v
  U.S. invasion of, 101
**Africa**
  climate of, 47m
  Congo River Basin, 199
  cultural influence, 164, 166, 216
  independence movements in, 376–377
  religions of, 61m
  slave markets in, 208
  in triangular trade, 152–153
  as two regions, 17, 17m, 400
  wildlife in, 368–369
  *see also* Southwest Asia and North Africa; Sub-Saharan Africa
**African culture, slave trade and spread of,** 164, 166, 216, 375
**African National Congress (ANC),** 403
**African slave trade,** 372–373
**African Union (AU),** 377, 385, 385v, 401
**Afrikaners,** 402
**Age of Exploration,** 268–269
**aging population**
  in China, 577
  in Europe, 298
**agricultural regions, in North America,** 81m
**agricultural revolution,** 370, 422
**agriculture,** R41
  in Africa, 396–397
  as an economic sector, 128
  in Argentina, 224–225
  in Caribbean, 162v
  in Central America, 174–175, 174v
  in Central Asia, 318
  in Chile, 226–227
  climate and, 70, 312, 313, 604
  commercial, 80
  in the Great Plains, 80, 81m
  in Mexico, 84
  in Nile River Valley, 430, 431
  in Pacific islands, 652–653
  in Peru, 228
  in rain forests, 146
  slash-and-burn, 204–205, 205v
  in South America, 204–205
  in South Asia, 476–477, 477v
  in Southern Africa, 367
  in the steppes, 334–335, 335v
  terracing, 16v, 201
  *see also* cash crops; farming; subsistence farming
**Aguinaldo, Emilio,** 613, 613v
**ahimsa,** 486
**air masses,** 45
**Akbar the Great,** Mughal ruler, 484
**Aksum Empire,** 372
**al-Qadhafi, Colonel Mu'ammar,** 449
**al-Qaeda,** 456
**Alamo, Texas,** 108
**Albania,** 245m, 246m, 279m, 293m
**Alexander II,** 326
**Alexander the Great,** 259, 423, 433
  empire of, 259m
**Algeria,** 412–413m, 414m, 451m
**Ali, Ben,** Tunisian president, 448
**Allende, Salvador,** 223
**alliances,** 100
**Allies (in World War I),** 100, 278
**alluvial,** 372
**alluvial plain,** 417
**alluvium,** 528
**alpaca,** 228
**Alps,** 242v, 247, 250
**Altair (computer),** 121
**alternative energy,** 26–27v, 26
  in India, 505
**altitude sickness,** 472
**altitude zones,** 192v, 193
**Amazon rain forest,** 194, 198–199, 240
**Amazon River, South America,** 43v, 186v, 194–195, 195m
**Amazon River Basin,** 190m, 191, 193, 194, 195m
**amendments,** 94, 125
**American bison,** 88
**American Revolution,** 92–93
  French Revolution and, 272
**American Samoa,** 646m, 654m, 659m, 672–673v
**Amorites,** 423
**Amsterdam-Rhine Canal,** 182
**Amundsen, Roald,** 674
**Analects (of Confucius),** 542, 543
**Anatolian Plateau,** 420–421, 421m, 428
**ancestry,** 126
**ancient Ghana,** 373m
**ancient Greece,** 256–257, 420–421
**ancient Israel,** 444–445
**Anderson, Greg,** 662–663, 662v, 663v
**Andes Mountains, South America,** 190–91, 190m, 190v, 193, 194
  Aymara culture in, 217
  rain shadow of, 196
**Angel Falls, Venezuela,** 191
**Angkor Wat, Cambodia,** 594v, 608–698, 608v, 620, 621v
**Angola,** 357m, 358m, 367, 367m, 377m, 392v, 393m
**Angostura Address,** 1819, 210
**animal-borne,** 536
**anime (Japanese animation),** 570–571
**animism,** 562, 563, 620
**annexation,** 108
**Antarctic Treaty,** 1959, 675
**Antarctica,** 646m, 647, 654m, 659m, 674–675, 674–675v, 675m
  adventure tourism in, 680
  defined as region, 17, 17m
  penguins in, 536–537v
  research stations in, 674–675v, 675
**Anthony, Katey Walter,** 316–317
**Antigua and Barbuda,** 141m, 142m, 143m, 145m, 157m
**Antilles.** *See* Lesser Antilles
**apartheid,** 275, 402–403
**Apennines,** 250
**aqueducts,** 263
**aquifers,** 86–87, 481
**Arab-Israel Wars,** 452, 453
**Arabian Peninsula,** 415, 418–419, 419m, 461m
  *see also* Dubai; Kuwait; Saudi Arabia; United Arab Emirates
**Arabian Sea,** 469m, 470m, 471m, 475m, 477m, 487m, 516m
**arable,** 476v
**Aral Sea,** 318–319, 318v, 319v
**Archduke Franz Ferdinand,** 276
**archipelago,** 142v, 143, 530
**architecture**
  Greek, 258
  in Istanbul, 442–443v
  in Thailand, 622
  western, in Russia, 322–323v, 338
**Arctic Circle,** 311
**Arctic Ocean,** 40, 42
**Argentina,** 189m, 190m, 191m, 193m, 205m
  agriculture and economy of, 224–225
  in Antarctica, 675m
  cuisine in, 221
  government changes in, 222
  indigenous populations in, 217v
  land use in, 225m, 687m
  ranching in, 686–687
**arid,** 318, 415
**Armenia,** 309m, 310m, 311, 311m, 313m, 315m, 345m
**armistice,** 557
**art, European,** 286–287
**artifacts,** 126, 202
**Aruba,** 141m, 142m, 143m, 157

# INDEX

**Aryan migration,** 482–483, 484, 493m, 497
**Aryans,** 482
**Asia,** 40–41, 40m
  climate of, 47m
  immigration from, to North America, 116
  religions of, 61m
  as three regions, 17, 17m
  *see also* East Asia; South Asia; Southeast Asia; Southwest Asia and North Africa
**Asian culture, Russia and,** 322–323
**Asian-Pacific Economic Cooperation (APEC),** 676–677, 677m
**Asiatic lion,** 639v
**Askum Empire,** 373m
**Asoka, Mauryan ruler,** 484, 484v
**assimilate / assimilation,** 678
**assisted migration,** 660
**Aswan High Dam,** 431
**Atacama Desert, Chile,** 196–197v
**Atahualpa, Incan emperor,** 206
**Athena,** 258, 258v
**Athens, Greece,** 253v, 256, 257, 258
**Atlantic coast,** 21m
**Atlantic Ocean,** 42
**Atlas Mountains,** 415
**atolls,** 647
**atomic bombs,** 100, 278
**attribute,** 622
**Auckland City, New Zealand,** 682v
**Augustus, Roman emperor,** 262, 262v
**Australia,** 40, 40m, 646m, 654m, 659m
  adventure tourism in, 680–681, 681v
  in Antarctica, 675m
  climate of, 47m
  cultural diversity of, 671
  cultures of, 656–657, 668–675
  government, 678–679
  grassland use in, 687m
  history of, 656–661
  immigration and, 299, 670–671, 671v
  indigenous populations, 678–679
  languages of, 662–663
  non-native species in, 650–651
  physical geography, 646–746, 652–655
  plants and animals in, 648–651
  rainfall in, 646m, 647
  regional trade patterns, 676–677
  religions of, 61m
  vanishing languages of, 662–663
**Australia, the Pacific Realm, and Antarctica,** 642–688, 645m, 646m, 659m
  defined as region, 17, 17m
  indigenous plants and animals in, 648–649
  new trade patterns in, 676–677
  *see also* Antarctica; Australia; Pacific Realm

**Austria,** 245m, 246m, 279m, 293m
**autumn equinox,** 35, 35v
**Axis Powers (in World War II),** 100, 278
**Aymara culture, Peru,** 217
**Azerbaijan,** 309m, 310m, 311, 311m, 313m, 315m, 328, 341, 345, 345m
**azimuthal map projection,** 24, 24m
**Aztec Empire,** 103, 103m, 104–105, 126

# B

**Babur, Mughal ruler,** 484
**Babylonian Empire,** 423
**Baghdad, Iraq,** 417
**Bahamas,** 140m, 142m, 143m, 145m, 157m, 165v
**Bahrain,** 412–413m, 414, 451m
**Baja Peninsula, Mexico,** 84
**Bali,** 621
**Baltic Sea,** 338
**Bangalore, India,** 506, 506v, 508v
**Bangkok, Thailand,** 599, 622–623v
**Bangladesh,** 469m, 470m, 471m, 475m, 477m, 487, 487m, 516m
  consumer goods in, 507v
  division from Pakistan, 488
  drought in, 481
  poisoned water in, 480
  poverty in, 514–515
**banking, international, in Dubai,** 446
**Banks, Ken,** 138, 184
**Bantu migrations,** 370–371, 371m
**Barbados,** 141m, 142m, 143m, 145m, 157, 165v
**barbarians,** 262
**Barcelona, Spain,** 253v
**Baroque period (music),** 287
**barter, on Silk Roads,** 545
**bas-reliefs,** 609
**basin,** 358, 359
**Bauxite, in Malaysia,** 602
**Bay of Bengal,** 469m, 470m, 471m, 475m, 477m, 487m, 516m
**bay(s),** 249
**Bedouin,** 440
**Beethoven, Ludwig van,** 287
**Beijing, China,** 14, 238, 239v, 528, 529m
**Belarus,** 245m, 246m, 279m, 293m, 328, 347
**Belem, Brazil,** 70, 71v
**Belgium,** 245m, 246m, 279m, 284, 293m
**Belize,** 140m, 142m, 143m, 145m, 146m, 157m, 160m, 165v
**Belize Barrier Reef, Belize,** 655
**Bell, Katy Croff,** 242v
**Benin,** 357m, 358m, 363m, 372, 377m, 393m
**Berlin Conference,** 376
**Berlin Wall,** 279v, 341
**Bermuda,** 157

**Bhagavad Gita,** 490–491, 496
**Bhutan,** 469m, 470m, 471m, 473m, 475m, 477m, 487m, 500, 507v, 516m
**Bhutto, Benazir,** 513
**Bhutto, Zulfikar,** 512
**Bible, Christian,** 424–425, R52
**Bible, Hebrew,** 424, R54
**Biko, Stephen,** 403
**Bilaukataung Mountain Range, Thailand,** 602, 603m
**Bill and Melinda Gates Foundation,** 399
**Bill of Rights,** 94
**bin Laden, Osama,** 101
**biodiversity,** 194, 602, 606–607, 648
  *see also* rain forests
**biofuel,** 231
**biological resources,** 50, 69m
**Black Sea,** 313, 313m, 320, 429, 442
**blizzards,** 48, 49m
**Blue Mosque, Istanbul,** 442, 443v
**Blue Nile River,** 430
**Boers,** 402
**Bolívar, Simón,** 210–211, 211v
**Bolivia,** 189m, 190m, 191m, 193m, 205m
  Aymara culture in, 217
  indigenous populations in, 217v
  Quechua culture in, 201, 217
**Bollywood, influence of,** 502–503, 503v
**Bolsheviks,** 326, 328
**Bonaire,** 141m, 142m, 143m, 150–151v
**Bonaparte, Napoleon,** 273, 273v
**Borneo, Indonesia,** 597m, 598m, 601m, 605m, 611m
**Borobudur temple, Java, Indonesia,** 608v, 609
**Bosnia and Herzegovina,** 245m, 246m, 279m, 293m
**Bosporus Strait,** 442–443, 443m
**Boston Tea Party,** 92
**Botswana,** 357m, 358m, 367m, 368, 377m, 392v, 393, 393m, 404m
**Brahma, creator of the universe,** 496
**Brahmans,** 483, 496, 497
**Brahmaputra River, Bangladesh,** 471
**Brazil,** 189m, 190m, 191m, 193m, 205m, 236m
  climate in, 70–71, 71v
  deforestation in, 198v, 199
  economic indicators, 63v
  ethanol production, 123
  growing economy, 230–231
  indigenous populations in, 217v
  local religion in, 220
  Munduruku culture in, 217
  Portuguese claims in, 209m
  slave trade and, 208–209, 375
**Brazil Current,** 196, 197m
**bread,** 462–463
**breakwaters,** 444

# INDEX

**Britain**
  and Aborigines, 678
  colonialism in South Africa, 402
  colonization in North America, 88, 89, 89m, 90–91
  colonization of Australia and the Pacific islands, 660–661
  energy conservation, 123
  India's independence from, 488–489
  taking Maori land, 657, 682
  in World War I, 277
  *see also* Great Britain; United Kingdom
**brook,** 42
**Brunei,** 597m, 598m, 601m, 605m, 611m
**Brunelleschi's Dome,** 304v
**bubonic plague,** 264
**Buddha,** 486
**Buddhism,** 60v, 61m, 486v, R55
  in East Asia, 562
  in South Asia, 484
  in South Korea, 586–587v
  in Southeast Asia, 608–609, 620, 621
  spread of, 545, 554, 559m
  in Thailand, 622, 622v
**Bulgaria,** 245m, 246m, 279m, 293m
**bullet trains,** 572–573, 572v
**Burj Khalifa, Dubai,** 410–411v, 447v
**Burkina Faso,** 357m, 358m, 363m, 377m, 393m
**Burma,** 610, 611m
**Burundi,** 357m, 358m, 361m, 377m, 393m
**business cycle,** R41
**butte, defined,** 38–39v
**Byrd, Richard,** 675
**Byzantine Empire,** 266, 428–429
**Byzantium,** 442

## C

**Cabinda,** 357m, 358m, 365m, 367m, 377m, 393m
**Cabral, Pedro Álvares,** 207v, 208
**Caesar, Julius,** 262
**Caesarea, Israel,** 444–445, 445v
**Cairo, Egypt,** 449v
**Calderón, Felipe,** 130v, 131
**Calderón, Sila,** 159v
**Caledonian Canal,** 182
**calypso,** 166
**Cambodia,** 597m, 598m, 601m, 605m, 611m
  Asian elephants in, 626–627, 626–627v
  Hinduism in, 620, 621v
  languages in, 625v
**Cameron Highlands, Malaysia,** 602–603v
**Cameroon,** 357m, 358m, 365m, 377m, 393m
**Canada,** 77m, 78m, 79m, 81m, 89m
  biological resources, 69m
  cultural diversity in, 116
  economy of, 120–121
  languages in, 117v
  mountains of, 82–83
  natural resources of, 82–83
  plains of, 80, 81m
  Quebec, 90–91
  *see also* North America
**Canal du Midi,** 182
**canals,** 168, 250v, 251
  comparing, 182–183, 183v
**Candomblé, religion in Brazil,** 220
**canopy (forest),** 147v, 607v
**Cape Verde,** 363m
**capital,** 62, 582
**capitalism,** R41
**Caracas, Venezuela,** 210
**carapace,** 536
**caravans,** 371, 420, 441, 544
**carbon dioxide, greenhouse gas,** 198
**Cardamom Mountains, Cambodia,** 626
**Caribbean,** 140–141m, 140–183, 142m, 143m, 157m
  agriculture in, 152-153
  climate, 143m
  defined as region, 17, 17m
  fights for independence in, 156–157
  food and music, 166–167
  migration within, 176, 177v
  music and food, 166–167
  physical geography of, 142–145
  slave trade in, 374–375
  tourism in, 164–165
  in triangular trade, 152–153
**Caribbean Sea,** 44, 140m, 141m, 143
**Carpathians,** 250
**Cartier, Jacques,** 90, 268
**cartographers,** 23, 24
**cash crops,** 152–153, 397
**Caspian Sea,** 313, 313m, 344, 345
**caste system,** 483, 497, 504
**Castro, Fidel,** 158, 159v
**Catedral de Brasília,** 215v
**Catherine the Great,** 232, 324
**Catholicism**
  in Canada, 91
  in South America, 207, 217, 220
**Caucasus Mountains,** 311
**celadon (pottery),** 555v
**Celebes, Indonesia,** 597m, 598m, 601m, 605m, 611m
**cell phones,** 118
  in Southeast Asia, 635
  world-wide usage of, 590–591, 591v
**cenotes,** 86–87, 86–87v
**Central Africa,** 359, 364–365, 365m, 372
**Central African Republic,** 357m, 358m, 365m, 377m, 392v, 393, 393m
**Central America and the Caribbean,** 140–141m, 140–183, 142m, 143m, 146m, 157m
  climate, 143m
  coral reefs in, 655
  culture in, 164–169
  defined as region, 17, 17m
  earthquakes in, 144, 145v
  economics of, 158–159, 170–173, 178–179
  fights for independence in, 156–157
  food supply and, 174–175
  government in, 170–171
  history of, 152–159
  land use in, 160m
  migration from and within, 176–177
  Panama Canal, 168–169
  physical geography of, 142–147
  rain forests in, 143, 146–147, 146m
  Spanish colonization of, 105
  tourism in, 164–165, 178–179
  *see also* Caribbean
**Central Asia,** 318–319, 335v
**Central Plain, Thailand,** 599
**Central Powers (in World War I),** 100, 276
**Chad,** 357m, 358m, 363, 363m, 377m, 392v, 393m, 400
**Champlain, Samuel de,** 90, 91v
**Chang Jiang Plain, China,** 528, 529m
**Chang Jiang River, China,** 43v, 526m, 527, 528–529v, 529m, 567
  Three Gorges Dam project, 578–579, 578–579v, 579m
**Chao Phraya River, Southeast Asia,** 599, 600–601, 601m
**Charlemagne,** 264
**Charlotte, North Carolina,** 20m
**Chechnya,** 341
**checks and balances,** 95v
**Chennai, India,** 503, 509
**Chernobyl,** 346–347
**Cherokee,** 96
**Chicago, Illinois,** 13, 13v
**child labor,** 271
**Chile,** 189m, 190m, 191m, 193m, 205m, 206
  agriculture in, 226–227
  in Antarctica, 675m
  cuisine in, 221
  exported food crops, 226, 226v
  government changes in, 223
  indigenous populations in, 217v
**China,** 22m, 522v, 525m, 526m, 527m, 529m, 559m, 573m
  civil war in, 548
  communism in, 548–549
  disputed boundaries, 469m
  dynasties in history of, 538–539, 539v
  early exploration from, 546–547
  economy of, 63v, 574–575
  energy consumption in, 350, 351v
  history of, 538–549
  immigration from, 116
  influence of, 608, 609
  inventions from, 564–565
  population policy in, 576–577, 576v
  rivers in, 528–539, 529m
  Silk Roads and, 320–321
  Three Gorges Dam, 578–579,

# INDEX

578–579v, 579m
water pollution in, 481
*see also* Taiwan
**Chinatown, Chicago,** 116v
**Chinese (language), in Southeast Asia,** 624, 625v
**Chinese Communist Party,** 548, 549
**Chinese Nationalist Party,** 548
**Chobe Lion Research Institute,** 368–369
**choropleth map,** 119m
**Choson Dynasty,** 554, 555
**Christianity,** 60v, 61v
  in Europe, 284, 300m
  influence in art, 286
  in Jerusalem, 424–425
  missionaries in Africa, 376
  in Nigeria, 394
  in Russia, 322
  in South America, 207
  in Southeast Asia, 620, 621v
  spread of, 268, 426–427
**Chukchi,** 334
**Church of the Holy Sepulchre, Jerusalem,** 425v
**Cimmerians,** 320
**cities**
  coastal, 668, 669
  development along rivers, 79, 260
  European, 285
  growth of and income gap, 451
  growth of and Industrial Revolution, 271
  immigration / migration to, 669, 672, 673
  megacities, 232
  overcrowding in, 522v, 567
  secondary, 511
  urban life in Thailand, 623
  *see also* urbanization
**citizen, defined,** 64
**citizenship,** R48
  naturalization and, 134–135
  rights and responsibilities of, 124–125, 125v
**city-states,** 256, 257v, 372–373, 422
**civil disobedience,** 488
**civil rights, for indigenous people,** 679
**Civil War, U.S.,** 98–99, 99m
**civil war,** 98
  in Africa, 400–401
  in China, 548
**civilization,** 58, 102
  of ancient China, 538–539
  developed in river valleys, 482–483, 483m
**clans,** 401, 656
**Clark, William,** 96
**Classical Greece,** 258–259
**Classical period (music),** 287
**Clemenceau, Georges,** 277v
**Cleisthenes,** 257
**climate,** 44–47, 46–47v, 45m

adapting to, 335v
agriculture and, 604
and culture, 290, 291, 334–335, 500
currents and, 196–197, 311
defense of country and, 324, 325
elevation and, 44, 190m, 191, 192, 193
extreme climates, 312–313
four factors affecting, 44
landforms and, 82
latitude and, 44, 226, 227m, 312
temperate, 226
vs. weather, 44
winds and, 44, 196–197, 197v, 247, 474–475, 475m
*see also* weather
**climate, in different regions**
  Antarctica, 647
  Arabian Peninsula, 418
  Australia, 668
  Brazil compared to Russia, 70–71, 71v
  Canada, 90
  Caribbean, 143
  Central America, 143
  East Africa, 361m
  East Asia, 527, 527m
  Europe, 247m
  Korean Peninsula, 532
  Malay Peninsula, 603m
  Mediterranean, 226, 227m
  Mexico, 79m, 84
  North America, 79, 79m, 82
  Pacific islands, 652–653
  Russia, 312–313
  South America, 191, 191m
  South Asia, 471, 471m, 474–475
  Southeast Asia, 599, 604
  Southern Africa, 367
  Southwest Asia, 420–421
  Southwest Asia and North Africa, 414m, 415
**climate change**
  and endangered species, 638
  rain forests and, 198–199
  in the Sahara Desert, 415
  in South Asia, 481
  vs. global warming, 199
  *see also* global warming
**climate maps,** 70
**climate regions,** 46–47, 46–47m
**climographs,** 70, 71v
**coal,** 270
**coalition,** 454
**coast / coastline**
  cities on, 647, 668, 669
  in Southeast Asia, 599
  West African countries, 362–363
**Coastal Plain, South America,** 192
**coastal plains,** 142v, 143
  in Russia, 312
  in Southeast Asia, 604
  of Southern Africa, 366, 367
**coffee**
  from Africa, 367, 397
  from Brazil, 209, 230, 232

from Mexico, 84, 85m, 129, 133m
shade coffee, 146
**Cold War,** 100–101, 278, 328, 556
**collective farms,** 328, 548
**Colombia,** 189m, 190m, 191m, 193m, 205m, 206, 219v
  indigenous populations in, 217v
**colonialism,** 376, 488
  in Asia, 488–489, 610–611, 611m
  impact of, in Africa, 384, 394, 397
  and language use, 507
  in South Africa and apartheid, 402
  *see also* European exploration and colonization
**colonize,** 88
**colony / colonies,** 269
**Colorado River,** 83
**colors, on a map,** 18, 22m, 23
**Colosseum, Rome, Italy,** 242v
**Columbian Exchange,** 154–155, 154v, 166, 269
**Columbus, Christopher,** 152, 154–155, 155v, 206, 268v
**command economy,** 62, R41
**commerce,** 613
**commercial agriculture,** 80
**Commission on Human Rights,** 66
**commodities,** 392
**Common Market,** 292
**commonwealth,** 159
**Commonwealth of Australia,** 661
**communal,** 58v
**communism,** 65, R41, R48
  in Afghanistan, 456
  banning religion / belief, 543, 563
  in China, 548–549, 576
  in Cold War, 101
  in Cuba, 158
  in Eastern Europe, 278, 296
  in North Korea, 556, 557, 584
  in Russia, 328, 341
**Communist Eastern Europe,** 278
**Communist Party,** 277
**Comoros,** 357m, 358m, 367m, 377m, 393m
**Compare Across Regions**
  bread, 462–463
  canals, 182–183
  cell phone usage, 590–591
  citizenship, 134–135
  climate, 70–71
  deserts, 406–407
  emerging markets, 518–519
  endangered species, 638–639
  pastures and crops, 686–687
  natural resources and energy, 350–351
  the Olympics, 238–239
  world languages, 302–303
**compass, invention,** 565
**compass rose, on a map,** 18
**competitive,** 552
**complex (of buildings),** 609
**computers,** 121
  *see also* Internet access
**concentration camps,** 66, 278

INDEX **R101**

# INDEX

**condensation,** 42v
**Confederate States of America,** 98
**Confucian temple, China,** 542–543v
**Confucianism,** 542–543, 554, 562, 563, 620, R58
**Confucius,** 542
**Congo,** 357m, 358m, 365m, 377m, 392v, 393m
**Congo River,** 364–365
**Congo River Basin,** 199, 359, 364–365, 365m, 366
**Conlee, Christina,** 202–203, 203v
**conquistadores,** 104–105
**conservation,** 478
  rain forest, 178–179, 194
  see also habitat preservation
**Constantine, Roman Emperor,** 262, 426
**Constantinople,** 262, 428, 429m, 442
**constitution,** 94–95, R48
**Constitution of 1824, Mexico,** 107
**Constitution of 1917, Mexico,** 110
**consumer goods, U.S. and South Asia,** 507v
**consumers,** 295
**container cargo,** 337
**contaminated,** 346
**contiguous,** 79
**continent,** 16
**continental drift,** 37, 359
**continental margin,** 38, 39v
**continental plates,** 37v
**continental rise,** 38, 39v
**continental shelf,** 38, 39v
**convergence,** 37v
**convert / conversion (religious),** 207
**convicts, sent to colonies,** 660–661
**Cook, James,** 659, 660, 660v, 674
**Cook Islands, Polynesia,** 673
**copper, exported from Peru,** 228, 229, 229v
**coral in the Caribbean,** 150–151v
**coral islands,** 652, 653v
**coral reefs,** 647
  formation of, 654
  protecting, 654–655
**cordillera,** 82
**core, of the earth,** 36
**cormorants,** 537
**corporation,** R42
**Cortés, Hernán,** 104, 104v
**Costa Rica,** 140m, 142m, 143m, 145m, 146m, 156, 157m, 160m
  earthquakes in, 145v
  government and economics in, 170–171
**Côte d'Ivoire,** 357m, 358m, 362, 363m, 377m, 393m

**cotton,** 396v
  from Africa, 397
  from Mexico, 129
  and slave trade, 374
**Counter-Reformation,** 267
**coup,** 222, 341
**Cousteau, Jacques,** 253
**Cousteau, Pierre-Yves,** 253
**Creole,** 218
**crevasse,** 681, 681v
**cricket,** 501, 501v
**crillos,** 106, 107
**Croatia,** 245m, 246m, 279m, 293m
**crop rotation,** 476
**crops.** See agriculture; cash crops
**Crosby, Alfred,** 155
**Crusades,** 264
**crust, of the earth,** 36
**Cuba,** 140m, 141m, 142m, 143m, 145m, 157m, 158–159, 159m
  economic resources of, 158, 159m
  government in, 65
  migration from, 176
**cuisine,** 221, 290–291
  see also food
**Culhane, Thomas Taha (TH) Rassam,** 410v, 465
**cultural diversity**
  in Africa, 384, 388
  in Australia, 671
  in Dubai, 446
  in Europe, 284–285, 299
  government issues and, 628–629
  in North America, 116–117
  preserving, 391
**cultural hearths,** 422–423, 482
**cultural issues, and sea turtle preservation,** 148–149
**cultural regions,** 59v
**Cultural Revolution, China,** 549
**culture,** 8v, 58–61, 334
  caste system, 483
  climate influencing, 334–335
  conflict and, 394, 400, 454
  defining region and, 16
  development along waterways, 79, 260, 415, 528, 529
  development in isolation, 658–659
  and disease, 398
  dress and, 59v, 332–333v, 500, 501v
  early, in Mesopotamia, 422–423
  European beliefs about, 376
  media culture, 118–119
  nomadic, 204, 306v, 334, 554
  popular culture, 502–503
  railroads and spread of, 337
  slave trade and, 164, 166, 216, 375
  social class distinctions and, 106
  traditional, in Africa, 390–391
  see also architecture; food; language; literature; music; religion
**culture, arts and literature**
  European contributions to, 284–289

  influence of Greece and Rome, 258–259, 261, 286
  St. Petersburg as center of, 338
**culture, daily life**
  in Mexico, 126–127
  in South America, 220–221
  in South Asia, 500–501
**cultures, in different regions**
  Aboriginal, in Australia, 656–657
  Africa, 384–391, 390–391
  Bantu migration, 370–371
  Caribbean, 166–167
  indigenous, South America, 216–217
  Pacific islands, 658–659
  Polynesian, 672–673
  Russia, 322–323
  South America, 200–207, 216–221
**cuneiform (writing),** 423
**Curaçao,** 141m, 142m, 143m
**currency**
  in European Union, 293, 294, 295v
  paper money in China, 564
**currents,** 42, 196
  ocean, 196–197
  wind, 196–197
**Cuzco, Peru,** 200
**cyclones,** 48, 514
**Cyprus,** 412–413m, 414m, 451m
**czar,** 321
**Czech Republic,** 245m, 246m, 279m, 293m
  democracy in, 296

# D

**da Gama, Vasco,** 268
**da Vinci, Leonardo,** 266, 286v
**Dalits,** 497
**Daltry, Jenny,** 594v
**dams,** 83, 601
  Aswan, 431
  Chang Jiang, 578–579
  Hoover, 83
**Danube River,** 250
**Daoism,** 562
**Darfur, Sudan,** 400
**Darius I, King of Persia,** 257
**Darling River, Australia,** 668
**Davis, Jefferson,** 98
**Davis, Wade,** 390–391
**Day of the Dead,** 126
**de Klerk, F.W.,** 403
**Deccan Plateau,** 470m, 470v
**Declaration of Independence,** 92, 274
**Declaration of the Rights of Man and of the Citizen,** 272, 274
**Deepwater Horizon, oil spill,** 122
**deforestation,** 146, 363, 390
  in Africa, 362
  in Brazil, 198v, 199
  in Malay Peninsula, 602, 603
  weather changes and, 174
**deities,** 433, 486
**Delhi, India,** 495v

# INDEX

**deltas**
  Chao Phraya, 599, 600
  Ganges, 470m, 470v, 471
  Irrawaddy, 600
  Mekong, 600
  Niger, 394, 395
  Nile, 432
  Okavango, 368
**demilitarized zone (DMZ), Korean,** 533, 557, 557v
**democracy,** 65, 101, R49
  in Africa, 402–403
  beginnings of, 256, 257
  direct, 257
  in Europe, 256–257, 278, 293, 296–297
  in India, 504–505
  in Mexico, 130–131
  multi-party, 130, 131
  parliamentary, 448
  in Philippines, 613
  in South Korea, 556, 557
  U.S. constitution and, 94, 95v
**Democratic People's Republic of Korea,** 584
**Democratic Republic of the Congo,** 357m, 358m, 364, 365, 365m, 377m, 392v, 393m
**democratization,** 297
**demographics, changing, in Europe,** 298–299
**Denmark, coal use,** 123
**Denver, Colorado,** 13, 13v
**depression,** 100, 278, R42
**descendants,** 126, 200
**desertification,** 359, 362, 363, 415, 535
**deserts,** 82–83, 406–407, 647
  Atacama, 196–197v
  comparing deserts, 406–407
  Gobi, 406, 407v, 526m, 534, 535m, 537
  Kalahari, 359, 366, 406, 407v
  Namib, 406v
  preserving, 535
  Rub al Khali, 418, 419m
  Sahara, 358m, 359, 415
  Sonoran, 406, 407v
  Syrian, 418, 419m
  Taklimakan, 544
  Thar, 470m
**developed nations,** 506, R42
**developing nations,** 506
**Dharavi, India,** 510
**dialect,** 274, 624
**diamond exports,** 392v
**Dias, Bartolomeu,** 268
**Diaspora,** 426
**Díaz, General Porfirio,** 110
**dictators / dictatorship,** 65, 100, 158, 222, 223, R49
**dingoes,** 650–651, 650v
**diplomacy,** 338
**direct democracy,** 257

**disease**
  bubonic plague, 264
  Columbian Exchange and, 154–155
  exploration and spread of, 152–153
  improving public health and, 398–399
  native people killed by, 88, 104–105, 207, 216, 269, 657
  polluted water and, 480
  spread of, in cities, 271
**distort / distortion, on maps,** 24
**divergence,** 37v
**diverse economies,** 227
**diversity,** 116
**Dmitry, Grand Prince of Russia,** 321
**Document-Based Questions (DBQ),** 110–111, 154–155, 210–11, 274–275, 490–491, 614–615
**Dogon Village, Mali,** 354v
**Dome of the Rock, Jerusalem,** 410v, 424v
**domestic policies,** 388
  see also government policy
**domesticate / domesticated,** 422, 627v
**Dominica,** 141m, 142m, 143m, 145m, 157m
**Dominican Republic,** 140m, 141m, 142m, 143m, 145m, 157m
  dance in, 166–167v
  earthquakes in, 145v
  internal migration within, 176
  tourism in, 165v
**dormant,** 604
**Drake, Sir Francis,** 268
**drought,** 48, 80, 174, 475, 481, 584
  in Australia, 668
  ranching and, 686
**drug cartels and trade,** 131
**dry climate,** 46–47m, 46v
**dry monsoons,** 474
**Dry Pampas,** 224, 225v
**Dubai, United Arab Emirates,** 410–411v, 441, 446–447
**due process,** 125
**Dust Bowl,** 80
**Dutch.** See Netherlands
**Dutch East India Company,** 610
**dynastic cycle,** 538, 539, 539v
**dynasties**
  in China, 538–539, 539v
  in Korea, 554–555
  see also empires; kingdoms

# E

**Earle, Sylvia,** 54–55
**Earth,** 34–43
  always changing, 39
  hemispheres, 12m, 19, 19v
  landforms of, 38, 39v
  layers of, 36v
  oceans of, 42
  rotation and revolution, 34–35

  seasons in Northern Hemisphere, 35, 35v
  seasons in Southern Hemisphere, 35
  tectonic plates, 37, 37v, 40, 41, 144–145, 145m, 360, 420–421, 421m, 471, 599, 604, 605m
  view of, from space, 8v, 28– 29v
  waters of, 42–43
  world climate regions, 46, 46–47m
**Earth Day,** 72
**earthquakes,** 40–41, 144–145, 145v
  in Haiti, 144, 172, 172v, 173, 173v
  in Japan, 530, 531, 582, 583
  in Philippines, 27m
  in Ring of Fire, 40–41, 40m
  in Southeast Asia, 599
  and tectonic plate movement, 37v
**East Africa,** 359, 360–361
  empires of, 372, 373m
  modernization in Kenya, 388–389
  political map, 361m
**East African Community (EAC),** 385, 385v, 389
**East Asia,** 522–590, 525m, 526m, 527m, 529m, 559m, 573m
  cell phone usage in, 590–591, 591v
  climate of, 527, 527m
  cultures of, 562–573
  defined as region, 17, 17m
  economics of, 574–585
  government in, 574–585
  high-speed rail in, 572–573, 572v, 573m
  history of, 538–557
  physical geography of, 526–535
  population of, 527m
  wildlife in, 536–537
  see also China; Japan; Korea; Mongolia
**East India Company,** 488
**Easter Island,** 642v
**Eastern Empire (Roman),** 262
**Eastern Europe**
  communism in, 278, 296
  democracy in, 296–297
  and European Union, 291
  food in, 291
**Eastern Hemisphere,** 19, 19v
**Eastern Orthodox church,** 264, 322
**EcoFriends,** 478
**ecologist,** 600, 601
**economic activity,** 62
**Economic Community of Central African States (ECCAS),** 385
**Economic Community of West Africa (ECOWAS),** 385v
**economic development, indicators of,** 171v
**economic indicators,** 62, 63v
  see also gross domestic product (GDP)
**economic resources,** 159m
  of Cuba and Puerto Rico, 158–158, 159m
  rain forests as, 146
**economic sectors,** 62, 128–129
**economic systems,** R42

INDEX **R103**

# INDEX

**economics,** 62–63, R41–R48
African Union and, 377
of Argentina, 224–225
of Brazil, 230–231
of China, 567
of China today, 574–575
collapse of Soviet Union and, 340–341
cooperative organizations in Africa, 385, 385v
of Costa Rica, 170–171
emerging markets, 518–519
of Europe, 248, 292–295
of former communist countries, 297
growth and new trade patterns, 676–677
hosting the Olympics and, 235, 238
immigration and, 299
of Japan, 582–583
of Korea, 584
of Malaysia, 629
migration and, 176–177
natural resources and, 344–345, 393
of Nicaragua, 170–171
of North America, 120–121
of Peru, 228–229, 229v
reforms in Nigeria, 395
of Singapore, 632–633
of Southern Africa, 366–367
of St. Petersburg, Russia, 338
tourism and, 146–147, 680
unstable government and, 393
of U.S., 120–121
*see also* gross domestic product (GDP); trade
**economy / economies,** R42
capitalism, R41
command, 62, R41
communist control of, 158
defined, 62
diverse, 227
free enterprise, 159, R41, R43
free trade agreements, 171
global, 574
market, 557, 581, R44
mixed, 62
traditional, 62
western model in Japan, 552–553
**ecosystems,** 53, 146, 252–253, 478, 602
of coral reefs, 654–655
and endangered species, 638
*see also* habitats; rain forests
**ecotourism,** 179
in Africa, 369
**Ecuador,** 189m, 190m, 191m, 193m, 205m
indigenous populations in, 217v
Quechua culture in, 201, 217
**education**
in Bangladesh, 514
food security and, 175
in Kenya, 388
lack of, and poverty, 514
Peruvian classroom, 220–221v

public education in South America, 220
in South Asia, 500
*see also* literacy rate
**Egypt,** 412–413m, 414m, 451m
ancient civilization in, 432–433
Nile River Valley, 430–431
political revolt in, 449
pyramids in, 434–435
**El Niño,** 197
**El Salvador,** 140m, 142m, 143m, 145m, 146m, 156, 157m, 160m
earthquakes in, 145v
farming in, 175
tourism in, 165v
**electronics industry**
in China, 575
in Japan, 582–583v
**elevation**
altitude zones, 192v, 193
in Antarctica, 647
climate and, 44m, 190m, 191, 192, 193
farming and, 193
in Himalayas, 472, 473m
on maps, 23
vegetation and, 190m, 191
**Emancipation Proclamation,** 98, 98v
**embargo,** R42
**emergence,** 635
**emerging markets,** 518–519
**emigrate / emigrated,** 673
*see also* immigration
**emirates,** 446
**Empire of Mali,** 373m
**Empire of Songhai,** 373m
**Empire State Building, New York,** 115v
**empires,** 103, 372–373, 373m, 484
*see also* dynasties; kingdoms
**endangered species,** A6, 638–639, 638v
African elephant, 52–53v
Asian elephant, 626–627
black rhinoceros, A7
blue whale, A6
California condor, A6
giant panda, A7
hawksbill turtle, A6
polar bear, A6
sea turtle, 148–149
snow leopard, 520–521
tiger, A7
whooping crane, A7
**Endangered Species Act,** 53
**Enduring Voices Project, NG,** 662–663
**energy**
alternative sources, 26–27v, 26m, 350
consumption of, 351v
finding new sources, 122–123
natural resources and, 314, 350–351
*see also* renewable energy

**English, language of commerce,** 507
**Enlightenment,** 272, 288
**entrepreneurs,** 574
**entrepreneurship,** 62
**environment**
damage from radiation, 346–347
and society, 15
tourism and, 164–165
*see also* conservation; ecosystems; habitats; pollution
**Environmental Protection Act, India,** 481
**epic poem,** 288
**epidemic,** 105, 398
**equator,** 19, 19v
**Equatorial Guinea,** 357m, 358m, 365m, 377m, 392v, 393m
**equinox,** 35
**Erie Canal,** 97, 182
**Eritrea,** 357m, 358m, 377m, 393m
**erosion,** 39, 396–397v
**escarpment,** 366
**Estonia,** 245m, 246m, 279m, 293m
**ethanol,** 122, 123, 231
**ethical systems,** 542–543, 562
**Ethiopia,** 357m, 358m, 360, 361m, 372, 377m, 393m
economic indicators of, 63v
famine in, 396
**ethnic conflict**
in Iraq, 454
in Nigeria, 394
in Sudan, 400
**ethnic groups,** 382
in Afghanistan, 456
in Ottoman Empire, 429
**ethnobotanist,** 390
**Etruscans,** 260
**eucalyptus,** 648, 649v
**Euphrates River,** 415, 416–417, 416v, 417m, 422, 423
**Eurasia,** 322
**Eurasian plate,** 471
**Eurasian republics,** 306–352, 309m, 310m, 311m, 313m, 315m, 316m, 321m, 323m, 325m, 327m
defined as region, 17, 17m
Soviet Union and formation of, 340
*see also* Russia and the Eurasian republics
**euro,** 293, 294–295
**Europe,** 242–304, 245m, 246m, 279m, 293m
art and music in, 286–287
climate of, 44, 47m, 247, 247m
coastline of, 248–49
cuisine of, 290–291
culture of, 284–291
defined as region, 17, 17m
democracy in, 296–297
demographics of, 298–299
economics in, 292–299
French Revolution, 272–273
government, 292–298

# INDEX

history of, 256–267
industries in, 248, 271m
land use and natural resources, 251m
languages of, 284, 302, 303
literature of, 288–289
Middle Ages, 264–265
migration, immigration, 116, 176–177, 177v, 298–299
physical geography, 246–255, 249m
post World War II, 279m
regions of, 247, 250
religions of, 61m
settlement of, and water, 249
and World War I, 276–277, 277m
*see also* European exploration and colonization
**Europe Day,** 298
**European architecture,** 323
**European culture, Russia and,** 322–323
**European Economic Community (EEC),** 292
**European exploration and colonization,** 152–157, 157m, 249, 268–269, 269m, 269v
in Africa, 372, 376–377, 377m
in Australia, 657, 660–661, 676
competition between countries, 152–153, 154–155
competition for resources and, 88–89, 90–91, 152–153
diseases and, 152–153, 384
and indigenous cultures in South America, 216–217
lasting relationships from, 157
in Mexico, 104–105
and native lands taken, 657
in North America, 88–89, 89m
in Pacific islands, 659
in South America, 206–207
in South Asia, 488–489
in Southeast Asia, 610–611, 611m
and spread of religions, 426
trade routes and, 152–153
**European Union (EU),** 292–293, 293m, 294, 297
**eurozone,** 294, 295v
**evaporation,** 42v, 475
**Evenki,** 334
**excavating / excavation (archeological),** 202, 203v
**executive branch,** R49
in India, 505v
presidential power in Russia, 342
in Roman Republic, 261, 261v
in the United States, 94
**exile / exiled,** 107, 210
**exoskeleton,** 654
**expeditions,** 546
**exploration.** *See* European exploration and colonization
**export revenue,** 228, 229v
**exports,** 80, R42
dependence on, 575, 581, 582, 584
food from Peru, 228

**extinct / extinction,** 638, 650
*see also* endangered species
**extremists,** 456
**Exxon Valdez, oil spill,** 122
**Eyl, Somalia,** 400v

## F

**factors of production,** 62, R42
**factory system,** 270
**failed state,** 401
**Fairbanks, Alaska,** 44
**fallout (nuclear),** 346
**famine,** 396, 476, 584
**farming**
along rivers, 417, 422–423
collective, 328, 548
in Africa, 397
in Europe, 251
in the Great Plains, 80
in Mexico, 84
and natural flooding, 430
in New France (Canada), 90
in Nile Delta, 432
in Peru, 216v
in South America, 193, 204–205
in the steppes, 334–335, 335v
subsistence, 363, 472
terraced, 16v, 201
*see also* agriculture
**fascism,** R49
**federal government,** 504
**federal republic,** R49
**federal system,** 342–343, 343v
**federalism,** 95v
**Federated States of Micronesia,** 646m, 654m, 659m
**feral,** 651
**Fertile Crescent,** 416–417, 422–423
**fertile soil,** 146
from volcanic ash, 604
**fertility rate, in China,** 577
**fertilizer,** 224
**Festival of Pacific Arts,** 672
**feudal system,** 264–265, 265v
**fiber optics,** 121, 121v
**Fiji,** 646m, 654m, 659m
**Filipino culture,** 629
*see also* Philippines
**Finland,** 34, 245m, 246m, 279m, 293m
**fireworks,** 564, 565v
**first language,** 371
**fishing**
importance of in East Asia, 527, 531
in Pacific islands, 652–653
problems caused by overfishing, 655
subsistence, 601
**Five Pillars of Islam,** 425
**fjords,** 249
**floodplain,** 430
**floods / flooding,** 48, 474–475v
agriculture and, along rivers, 416, 417
along Nile River, 430, 432
in Amazon River Basin, 194

in Bangladesh, 514
in China, 528
from El Niños, 197
monsoon, 474, 481
natural, 417
severe erosion and, 39
**Foja Mountains, Indonesia, new species discovered in,** 606–607, 607v
**food**
bread, the universal food, 462–463
Caribbean, 166
in Mexico, 127
in regions of Europe, 290–291
in South America, 221
*see also* cuisine
**food security,** 174–175
**food shortage, in Soviet Union,** 340–341v
**food supply**
agriculture and, in Africa, 396–397
bread around the world, 462–463, 463v
land use decisions and, 686–687
in South Asia, 476, 477v, 501
**forest canopy,** 147v
**forest products,** 230
**Fort Sumter, South Carolina,** 98
**fortify,** 90v
**fossil fuels,** 198, 314, 350
**fossils,** 8–9v
dinosaur, in Mongolia, 535
in Indonesia, 612
**Four Noble Truths, of Buddhism,** 486, R55
**Fox, Vicente,** 130
**fragmented countries,** 628
**France,** 245m, 246m, 279m, 293m
and American Revolution, 92–93
in Antarctica, 675m
colonization in North America, 88, 89, 89m, 90–91
cuisine of, 290
Haiti and, 156–157, 172
immigration to, 298
nuclear power in, 123, 347
in World War I, 277
in World War II, 278
**Franklin, Benjamin,** 92
**free enterprise,** R43
**free enterprise economy, defined,** 62
**free trade agreements,** 171, 677
**freedom, in the United States,** 94
**freedom of the press, government controlled in Malaysia,** 634
**French East India Company,** 611
**French Guiana,** 189m, 190m, 191m, 193m, 205m
indigenous populations in, 217v
**French Polynesia,** 646m, 654m, 659m
**French Revolution,** 272–273
**fur trade,** 90
**Futa Jaloon Highlands, Senegal,** 362

# INDEX

## G

**Gabon,** 357m, 358m, 365m, 377m, 393m
**Gadsden Purchase,** 108, 109m
**Gambia,** 357m, 358m, 363m, 377m, 393m
**Gandhi, Mohandas,** 488, 489, 489v, 490–491, 490v
**Ganges Delta,** 470m, 470v, 471
**Ganges River,** 471, 478–479, 496
**Ganges River Basin,** 479m
**Ganges River valley,** 484
**garment industry, in Bangladesh,** 514v, 515
**gauchos,** 58v, 224, 224v
**Gaza Strip,** 453, 453m
**Gdansk, Poland,** 248v
**generation,** 662
**Genghis Khan,** 320, 321v
**genre,** 288
**geoglyphs,** 202, 203v
**Geographic Information Systems (GIS),** 13
**geographic patterns,** 13
**geographic thinking,** 12–17
  five themes, 14
  maps, 18–27
  six essential elements, 15
  spatial thinking, 12–13
  world regions, 16–17
**geography,** 12–27, 34–65
  the earth, 34–43
  economic geography, 62–63
  geographic patterns, 13
  human geography, 58–67
  physical geography, 44–55
  political geography, 64–65
  spatial thinking, 12–13
  uses of, 15
**Georgia,** 309m, 310m, 311, 311m, 313m, 315m, 328, 345, 345m
**geothermal energy,** 123, 682–683, 683m, 683v
**Germanic languages,** 284
**Germany,** 245m, 246m, 279m, 293m
  economic activity, 18m, 63v
  food in, 290
  immigration to, 298
  in World War I, 100, 276, 277
  in World War II, 100–101, 278–279
**gers,** 534–535v
**Ghana,** 357m, 358m, 363m, 372, 376, 377m, 393m
**Gilgit-Baltistan, Kashmir,** 489, 489m
**glaciers,** 79, 599, 680
**glasnost,** 340
**global climate change,** 53
**Global Issues**
  building schools, 458–459
  climate change, 198–199
  finding new energy sources, 122–123
  food supply, 174–175
  globalization, 632–633

habitat preservation, 654–655
human rights, 66–67
migration and aging populations, 298–299
pollution, 346–347
population growth, 576–577
public health, 398–399
water resources, 480–481
**global media,** 119
**global perspective,** 16
**Global Positioning System (GPS),** 14
**global warming,** 174, 350
  methane gas and, 316–317
  vs. climate change, 199
  see also climate change
**globalization,** 128
  impact of, on Mexico, 128–129
**globes,** 18–19
**Gobbo, Grace,** 390–391, 390v
**Gobi Desert, China,** 406, 407v, 526m, 534, 535m, 537
**Godin, Nesse,** 66
**gold**
  in Africa, 372
  in Australia, 660
  mining conditions, 392v
**golden age,** 258
**Golden Quadrilateral, India,** 505, 508–509, 509m
**Goodman, Beverly,** 444–445
**Gorbachev, Mikhail,** 329, 329v, 340, 341
**gorillas,** 364v
**Göta Canal,** 182
**government,** 64–65, R48–R50
  absolute monarchy, 65
  communism. See communism
  constitutional monarchy, 448
  democracy. See democracy
  dictatorship, 100, 158, 222, 223
  federal system, 342
  military rule, 222, 223
  military dictatorship, 512
  monarchy, 65, 256–257, 448
  multiparty democracy, 130, 131
  multiparty republic, 448
  oligarchy, 256–257
  parliamentary democracy, 448
  proportional representation, 342
  socialism, 328
**government policy**
  controlling industry, 128
  controlling information, 634–635
  and economics, 170–171, 222–223, 388, 389
  and mining in Peru, 228–229
  and pollution in India, 481
  and population growth, 576–77
  see also communism
**Gran Chaco floodplain,** 204
**Grand Bazaar, Istanbul,** 442, 443v
**Grand Canal, China,** 182, 528, 529, 529m

**Grand Canyon, Arizona,** 38, 74–75v, 82
**grasslands,** 191, 686–687, 687m
**Great Barrier Reef, Australia,** 654–655, 680, 681v
**Great Basin,** 82–83v
**Great Britain**
  Australia and, 676
  control over Israel, 452
  food in, 290
  immigration to, 298
  in World War II, 278
  see also Britain; United Kingdom
**Great Depression,** 100, 278, R42
**Great Escarpment,** 366
**Great Lakes, North America,** 79
**Great Leap Forward,** 549
**Great Mosque, Tunisia,** 426v
**Great Plains, U.S.,** 38, 79, 80–81
**Great Pyramid of Khufu,** 434v, 435v
**Great Rift Valley,** 360–361, 364
**Great Sphinx,** 432v
**Great Wall of China,** 14, 14v, 15v, 522–523v, 539
**Great Zimbabwe,** 373, 373m
**Greece,** 245m, 246m, 248, 256–259, 279m, 293m
  ancient, 256–257
  classical, 258–259
  debt of, 295
  golden age of, 258
  immigrants in Australia, 671
  influence of, 286
  traders from ancient Greece, 420–421
**Greek city-states,** 257v
**Green Revolution, in South Asia,** 476
**greenhouse gas,** 198, 317
**Greenwich, England,** 19
**Grenada,** 141m, 142m, 143m, 145m, 157m
**griots (storytellers),** 354–355v, 386, 387v
**gross domestic product (GDP),** 62, 238, R43
  of Bangladesh, 514
  cell phone use and, 590, 591v
  of China, 575, 575v
  defined, 62
  as economic indicator, 62, 63v
  of India, 506
  of Japan, 582
  of Mexico, 129, 129v
  Singapore, 63v
  of Southern Africa, 367, 367m
  of Taiwan, 581
**Guadeloupe,** 141m, 142m, 143m, 145m, 157m
**Guam,** 646m, 654m, 659m
**Guaraní culture,** 204, 217, 218
**Guatama, Siddhartha,** 486
**Guatemala,** 139v, 140m, 142m, 143m, 145m, 146m, 156, 157m, 160m, 164–165v

# INDEX

earthquakes in, 145v
farming in, 174–175
malnutrition in, 175
textiles, 139v
tourism in, 164–165v
**guest workers,** 441
in Dubai, 446, 447v
in oil-rich countries, 450
**Guiana Highlands,** 191
**guillotine,** 272v, 273
**Guinea,** 357m, 358m, 362, 363m, 377m, 393m
**Guinea Bissau,** 357m, 358m, 363m, 377m, 393m
**Gujranwala, Pakistan,** 511
**Gulf of Guinea,** 362
**Gulf of Mexico,** 84, 85m
oil drilling in, 81
oil spill, 55
**Gulf Stream,** 44
**gunpowder, invention of,** 564
**Gupta Empire,** 484, 485m
**Gutenberg, Johannes,** 255, 288
**Guyana,** 189m, 190m, 191m, 193m, 205m,
indigenous populations in, 217v

## H

**habitat preservation,** 52–53
Mission Blue, 55
of world's oceans, 54–55
**habitat protection,** 368–369
**habitats,** 178–179v, 368
Asian elephants, 626
loss of, and extinction, 638
oceans as, 54–55
**Haeinsa, temple of,** 555
**Hagia Sophia, Istanbul,** 442, 443v
**Haiti,** 138, 140m, 142m, 143m, 145m, 156–157, 157m
earthquakes in, 41, 144, 145v, 172, 172v, 173, 173v
economic indicators of, 63v
government in, 172–173
migration from, 176
poverty in, 172–173
**half-life,** 346
**Hamburg, Germany,** 249
**Hammurabi, King of Babylonia,** 423, 423v
**Hammurabi's Code,** 423
**Han Dynasty,** 539, 544, 554, 563
**Hanoi, Vietnam,** 619v
**Harappan civilization, South Asia,** 482, 483m, 484
**harbor,** 158
**Hargreaves, James,** 270
**Harrison, David,** 662–663, 662v, 663v
**Hatshepsut, Egyptian pharaoh,** 432–433
**Hausa-Fulani,** 394
**Helgen, Kristofer,** 606–607
**hemispheres,** 19, 19v

**Henry, Prince, of Portugal,** 268
**herders,** 534
**hereditary,** 448
**heritage,** 284
**Hermitage Museum,** 338
**Herod the Great,** 444
**Hidalgo, Miguel,** 106, 106v
**Hiebert, Fredrik,** 306, 306v
**hieroglyphics,** 102, 432, 432v
**high islands,** 652, 653v
**highlands**
Adawama, 362, 364
Cameron, 602–603v
Futa Jaloon, 362
Guiana, 191
**hill, defined,** 38
**Himalaya Mountains,** 470m, 470v, 471
exploring, 472–473
and water supply for South Asia, 481
**Hinduism,** 61m, 61v, 482, 486v, R56
conflict with Islam, 488
Ganges River in, 478
in South Asia today, 496–497, 497v
in Southeast Asia, 608–609, 620
**Hirohito,** 552v
**Hiroshige III,** 552
**Hiroshima, Japan,** 100
**Hitler, Adolf,** 100, 101, 278
invasion of Russia, 324, 325
**Ho Chi Minh,** 614
**Hogan, Zeb,** 600v, 601
**Holocaust,** 101, 278, 452
**homelands,** 403
**homolostine map projection,** 24, 25m
**Honduras,** 140m, 142m, 143m, 145m, 146m, 156, 157m, 160m
earthquakes in, 145v
farming in, 175
tourism in, 165v
**Hoover Dam,** 83
**hope spots,** 55v
**Horn of Africa,** 360, 361m
**hotspots,** 652, 663
**Houston, General Sam,** 108, 108v
**Huang He River, China,** 528, 529m
**Human Development Index (HDI),** 173
**human-environment interaction,** 14
ecosystem of the Mediterranean Sea, 252–253
habitat protection for African wildlife, 368–369
see also ecotourism; pollution; preservation; tourism
**human geography,** 58–67
**human rights,** 66–67
and EU membership, 293
organizations in Costa Rica, 170
**human systems,** 15
**humanism,** 266
**humans, earliest, in Indonesia,** 612
**humid cold climate,** 46–47m, 47v

**humid continental climate,** 247
**humid equatorial climate,** 46–47m, 46v
**humid temperate climate,** 46–47m, 46v
**Hungary,** 245m, 246m, 279m, 293m
democracy in, 296
emerging markets in, 518–519, 519v
food in, 291
**hunter-gatherers,** 204, 205
**hurricanes,** 48, 49
**Hussain, Shafqat,** 466v, 520–521
**Hussein, Saddam,** 454
**hybrid,** 122, 123
**hydroelectric power,** 314, 365, 366, 431, 578–579
**Hyksos invaders,** 432

## I

**Iberian Lynx,** 639v
**ice climate,** 46–47m, 47v
**Iceland,** 34
geothermal energy, 123
summer solstice, 34–35v
**Igbo,** 394
**illiteracy rate**
in Aborigines, 678
see also literacy rate
**immigrate / immigrant,** 116, 220
**immigration**
to Australia, 670–671, 671v
challenges of, 299
to cities, 669
to Europe, 298v, 299
influence of railroad on, 337
and naturalization, 134
to North America, 116
in Pacific islands, 673
from Philippines, 629
reasons for, 299
see also migration
**immigration laws,** 116
**imperialism,** 376
**import,** R43
**impressionism,** 286
**Inca Empire,** 187, 200–201, 200m, 206
**income gap, in oil-rich countries,** 451
**independence**
in Africa, 376–77, 388
Australian from Britain, 661
Brazilian, 209
in Central America and Caribbean, 156–157
collapse of Soviet Union and, 340–341
Haiti from France, 172
India's movement for, 488–489
Indonesia, 612
in Mexico, 106–107
Philippines, 613
in South America, 222–223
of United States, 92–93

# INDEX

**Independence Hall, Pennsylvania,** 95v
**India,** 469m, 470m, 471m, 475m, 477m, 487m, 516m
  consumer goods in, 507v
  democracy in, 504–505
  economic growth in, 506–507
  economic policy compared with Singapore, 633
  emerging markets in, 518–519
  environmental protection in, 481
  grassland use in, 687m
  highways, 508–509, 509m
  influence of, in Southeast Asia, 608
  Kashmir conflict, 489, 489m
  languages vanishing in, 663
  passenger vehicles in, 509v
  population growth in, 577
  road building in, 508–509
  see also South Asia
**India Independence Act, 1947,** 488
**Indian National Congress (INC),** 488
**Indian Ocean,** 42, 366, 367m
**Indian Removal Act, 1830,** 96
**indigenous,** 116
**indigenous cultures / peoples,** 126
  Aborigines in Australia, 656–657
  preserving, in Africa, 390–391
  rights of, 678–679
  in South America, 216–217, 217v
  see also Native Americans
**indigenous languages**
  in Australia / New Zealand, 662–663
  in South America, 218
**indigenous plants and animals,** 648
**Indonesia,** 597m, 598m, 601m, 604–605, 605m, 611m
  early empires of, 609
  government in, 628
  history of, 612
  island nations of, 604–605
  languages in, 624, 625v
  migration within, 630–631
  new species discovered in, 606–607
**indulgences,** 266
**Indus River,** 471, 482–483, 483m
**industrial pollution,** 478
**Industrial Revolution,** 120–121, 270–271
  in Russia, 326
**industrialization**
  in Japan, 552–553
  railroad and, 337
  in Singapore, 632
  in U.S., 97
**industry,** R43
  in Bangladesh, 515m
  in Brazil, 231, 231m, 232–233
  in China, 548, 549, 574v, 575
  in Europe, 248, 271m
  in Korean Peninsula, 532
  in Russia, 327m, 327v
  in South Korea, 584
  see also manufacturing
**infectious,** 399
**inflation,** R44
**information economy,** 121
**infrastructure**
  creating new in Brazil, 234
  creating new in Istanbul, 443
  improving, in India, 505, 508–509
  in Nicaragua, 170–171
  in Pakistan, 511
**inner islands, Indonesia,** 630, 631v
**Institutional Revolutionary Party (PRI),** 130
**interior,** 362
**internal migration,** 176
**International Olympic Committee (IOC),** 235
**International Polar Year 2007-2008,** 675
**International Union for Conservation of Animals (IUCN),** A6, 638
**Internet**
  and information in Malaysia, 634–635, 635v
  in Southeast Asia, 635v
  and urban life in Thailand
**Internet access,** 118, 119m
**intersection,** 164
**Inuit,** 116, 334
**invaders,** 324
**invasive species,** 650–651
**Iran,** 412–413m, 414m, 420, 451m
  war with Iraq, 454
**Iranian Plateau,** 420–421, 421m
**Iraq,** 412–413m, 414m, 417, 451m
  move toward democracy, 454
  religious / ethnic divisions, 454, 455m
**Ireland,** 245m, 246m, 279m, 290, 293m, 295
**Iron Curtain,** 278
**Irrawaddy River, Southeast Asia,** 600–601, 601m
**irrigation,** 417
**Ishtar Gate (Babylonia),** 423, 423v
**Isis,** 433, 433v
**Islam,** 60v, 61m, 285, 425, 487v, 620v, R53
  in Africa, 374
  conflict with Hinduism, 488
  in East Asia, 563
  in South Asia, 484, 486
  in Southeast Asia, 620, 621
  spread of, 427–427, 429
  see also Muslims
**Islamic Law,** 394, 448
**Island nations,** 604–605, 630–631
  see also Indonesia; Philippines
**islands**
  in Europe, 247
  formation of, 652–653, 653v
  of Japan, 530–531

**Israel,** 452–453, 453m
  creation of, 452
  democratic government in, 448
  state of, 426
**Israel, ancient,** 444–445
**Istanbul, Turkey,** 253v, 442–443, 443m
**isthmus,** 143, 256
**Italy,** 245m, 246m, 279m, 293m
  cuisine of, 290
  immigration to, 299
  language dialects in, 284
  Roman Republic, 260–261
  in World War I, 276, 277
**Iturbide, Augustín de,** 106–107, 107v
**Ivan I,** 321
**Ivan III / Ivan the Great,** 321
**Ivan IV / Ivan the Terrible,** 321, 322, 324, 324v, 326
**ivory, illegal trade in,** 626

## J

**Jacobins,** 273
**jaguars,** 639v
**Jainism,** 486v
**Jaiswal, Rakesh,** 478
**Jakarta, Indonesia,** 605
  colonization by the Netherlands, 610, 611m
**Jamaica,** 140m, 141m, 142m, 143m, 145m, 157m
  food, 166
**Jammu, Kashmir,** 489, 489m
**Japan,** 22m, 525m, 526m, 527m, 529m, 531m, 559m, 573m
  anime / manga, 570–571, 571v
  cherry blossom festivals, 592
  control of Indonesia, 612
  economic future of, 582–583
  history of, 550–553, 552–553
  industries of, 552–553
  invasion of China, 548
  physical geography of, 530
  resources and wealth in, 51
  samurai culture, 550–551
  in World War II, 100, 278
**Japanese Alps,** 531
**Java, Indonesia,** 605, 630
**Jefferson, Thomas,** 96
**Jeria, Michelle Bachelet,** 223, 223v
**Jerusalem,** 410v
**Jesus,** 424
**Jiang Jieshi,** 548
**Jinnah, Mohammed Ali,** 512
**Johnson, Lyndon,** 614
**Jolliet, Louis,** 90
**Jordan,** 412–413m, 414m, 451m
  constitutional monarchy in, 448
**Joubert, Beverly,** 368v
**Joubert, Dereck,** 368v
**Juárez, Benito,** 109, 109v
**Judaism,** 61m, 61v, 424, 426, R54

# INDEX

**judicial branch,** R50
  in India, 505v
  in Roman Republic, 261, 261v
  in Russia, 342–343
  in the United States, 94
**jute,** 514

## K

**Kabul, Afghanistan,** 456, 457v
**Kaesong, North Korea,** 585v
**Kakadu National Park, Australia,** 657v
**Kalahari,** 359
**Kalahari Basin,** 366
**Kalahari Desert, Africa,** 359, 366, 406, 407v
**Kallawaya language,** 219
**Kamarudin, Raja Petra,** 634
**Kanpur, India,** 478
**Kanto Plain, Japan,** 531
**Karachi, Pakistan,** 510, 511v
**Kariba Dam, Sambezi River,** 366
**karma,** 496
**Kashmir conflict,** 489, 489m, 505, 513
**Kazakhstan,** 309m, 310m, 311, 311m, 313m, 315m, 330m, 337m, 344, 345m
  climate in, 318
  herding culture in, 334
  iron ore in, 314
**Kennedy, John F.,** 118
**Kenya,** 357m, 358m, 360, 361, 361m, 376, 377m, 389m, 393m
  cultural diversity in, 388
  modernization in, 388–389
  wildlife reserves in, 388, 389, 389m
**Kenyatta, Jomo,** 376, 377v, 388
**Khmer Empire,** 608
**Kiev,** 320
**Kievan Rus,** 320
**Kilimanjaro,** 360
**kimonos,** 59v, 560–561v
**king penguins,** 537
**Kingdom of Benin,** 373m
**kingdoms**
  ancient in Southeast Asia, 608–609
  in Korean history, 554
  *see also* dynasties; empires
**Kingman Reef, Line Islands,** 652–653v
**kinship,** 200
**Kiribati,** 646m, 654m, 659m
**Knesset,** 448
**koala,** 645v, 648, 649v
**Koguryo kingdom,** 554
**Kolbert, Elizabeth,** 481
**Kollywood,** 503
**Kongo,** 372
**Kongo Kingdom,** 373m
**Korea,** 525m, 526m, 527m, 529m, 559m, 573m
  comparing North and South, 584–585
  early kingdoms of, 555m
  history of, 554–555, 554–557

  *see also* North Korea; South Korea
**Korean Peninsula,** 532–533, 533m
  physical geography of, 532
  resources of, 532
**Korean War,** 328, 556–557
**Koryo Dynasty,** 554
**Kovno, Russia,** 325v
**Krakatoa, Indonesia,** 41
**Kremlin, Russia,** 342–343v
**Krishna, avatar of Vishnu,** 496, 496v
**Kshatriyas,** 483, 497
**Kumar, Meira,** 504
**Kunlun Mountains, China,** 526m, 526v, 527
**Kurds, conflict in Iraq,** 454
**Kuwait,** 412–413, 414m, 451m
  Iraq invasion of, 454
  oil wealth in, 450
**Kyrgyzstan,** 309m, 310m, 311, 311m, 313m, 315m, 345m

## L

**La Paz, Bolivia,** 192v
**La Reforma,** 109
**La Tiza, Peru,** 202, 203v
**labels, on a map,** 18
**labor, as factor of production,** 62, R42
**labor force,** 670
**Lafayette, Marquis de,** 92
**Lagos, Nigeria,** 363
**Lahore, Pakistan,** 511
**lake, defined,** 42
**Lake Baikal, Russia,** 336v
**Lake Erie, United States,** 52
**Lake Tana, Ethiopia,** 430
**Lake Tanganyika,** 360
**Lake Texcoco, Mexico,** 103
**Lake Victoria, Africa,** 365m, 430
**land, as factor of production,** 62, R42
**land bridges,** 599, 656
**land grants, Mexican,** 108
**landforms,** 38–39
  of Australia, 646–647, 652–653
  of Central America and the Caribbean, 142–145
  climate and, 84, 318
  of East Asia, 526–535
  of Europe, 246–251, 249m
  isolation and, 311
  of Mexico, 84
  of North America, 78–87
  ocean, 38–39
  of Russia, 310–319
  of South America, 190–195, 224
  of South Asia, 470–473
  of Southeast Asia, 598–605
  of Southwest Asia and North Africa, 414–421, 414m, 430–431, 446–447
  of sub-Saharan Africa, 358–366
  tectonic plates and, 144, 145m
  of West Africa, 362–363

  *see also* highlands; plains, plateaus, mountains
**landlocked,** 366, 599
**landmass,** 311
**language diffusion,** 624
**language diversity**
  in Dubai, 446
  in North America, 117
  in South America, 218–219
**language families,** 218
**language migration,** 624–625
**languages,** 303v, 662
  Aboriginal, 656
  Bantu in Africa, 371
  of Canada, 125
  English, in commerce, 507
  of Europe, 284
  first, 371
  Maori, 682
  of Russia, 348m
  of South Asia, 516m, 624–625, 625v
  of sub-Saharan Africa, 385m
  vanishing, 662–663, 672
  world, 302–303
  *see also* indigenous languages; native languages
**Laos,** 597m, 598m, 601m, 605m, 611m
  languages in, 625v
**Latin,** 263
**Latin America,** 218
  as a cultural region, 59v
  immigration from, to North America, 116
  *see also* Central America and the Caribbean; Mexico; South America
**latitude**
  climate and, 44, 226, 227m, 312
  on maps and globes, 18, 19, 19v
**Latvia,** 245m, 246m, 279m, 293m
**launched,** 614
**Lebanon,** 412–413m, 414m, 451m
**Lee, General Robert E.,** 98
**legends, on a map,** 18
**legislative branch,** R50
  in India, 504, 505v
  in Roman Republic, 261, 261v
  in Russia, 342
  in the United States, 94
**legumes,** 397
**Lenin, V. I.,** 326, 328v
**Lesotho,** 357m, 358m, 367m, 377m, 393m
**Lesser Antilles,** 140m, 141m, 142m, 143m, 145m, 157m
**Lewis, Meriwether,** 96
**Lewis and Clark expedition,** 96
**liberate,** 211
**Liberia,** 357m, 358m, 363m, 377m, 393m
**Liberty Bell,** 92v
**Libya,** 412–413m, 414m, 449, 451m
**Liechtenstein,** 245m, 246m, 279m, 293m
**Lima, Peru,** 206

INDEX **R109**

# INDEX

**limited government / unlimited government,** 95v, R50
**Lin, Albert Yu-Min,** 522v
**Lincoln, Abraham,** 98, 98v
**Lindsey, Elizabeth Kapu'uwailani,** 642v, 688
**lingua franca,** 371
**literacy rate**
  in Afghanistan, 458
  as economic indicator, 63v
  in emerging markets, 518
  in Kenya, 388
  in Mexico, 131, 131v
  in South Asia, 500
**literate,** 458
**literature**
  European, 288–289
  modern, 289
**Lithuania,** 245m, 246m, 279m, 293m
**Llanos,** 191, 192
**Lloyd George, David,** 277v
**location,** 14
  challenges of remote, 315
  defense of country and, 324
  and European development, 248–249
  importance of, 13, 13v, 442–443, 608–609, 630
**locator globe,** 18, 18v
**Locke, John,** 272, 274, 288
**locks,** 169, 169v, 182
**loess,** 528
**logging**
  in rain forests, 390
  see also deforestation
**Lok Sabha,** 504
**Long March,** 539m, 548
**longitude, on maps and globes,** 18, 19, 19v
**Lost Boys of Sudan,** 400
**Lost Decade,** 582
**Louis XVI, King of France,** 272, 273
**Louisiana Purchase,** 96
**L'Ouverture, Toussaint,** 138v, 156, 156v
**Louvre, Paris, France,** 242–243v
**low islands,** 652, 653v
**Lowell, Massachusetts,** 97
**lowlands,** 204
  in Europe, 247, 248
  in South America, 204–205, 205m
**Lusitania,** 100
**Luther, Martin,** 266–267, 267v

# M

**MacArthur, General Douglas,** 556, 556v
**mace,** 612v
**Macedonia,** 245m, 246m, 279m, 293m
**Machu Picchu, Peru,** 187v, 200–201v
**Madagascar,** 357m, 358m, 367m, 377m, 393m
**magnetic compass, invention of,** 565
**magnetic levitation,** 573
**Mahabharata,** 486, 496
**Main Mountain Range, Malaysia,** 602, 603m
**mainland,** 526
**Majapahit Empire, Indonesia,** 609
**Maka language,** 219
**Makeba, Miriam,** 386
**malaria,** 398, 399, 399v
**Malaria No More,** 399
**Malawi,** 357m, 358m, 367m, 377m, 384–385v, 393m, 396
**Malay Peninsula,** 602–603
  climate of, 602, 603m
  physical geography of, 602, 603m
**Malaysia,** 597m, 598m, 599, 601m, 605m, 611m, 616m
  diversity and government issues, 629
  languages in, 624, 625v
  media and information control, 634–635
**Maldives,** 469m, 470m, 471m, 475m, 477m, 478, 487m, 507v, 516m
**Mali,** 357m, 358m, 363m, 372, 377m, 393m
**Malinche (Spanish guide and interpreter),** 104
**malnutrition,** 175, 374, 396
**Mandela, Nelson,** 67, 274, 275, 275v, 403, 403v
**manga,** 570
**Manhattan, New York,** 114-115v
**Manifest Destiny,** 96
**Manila, Philippines,** 605, 613
**mantle, of the earth,** 36
**manufacturing,** 120, R44
  automated, 121v
  in Brazil, 231, 231v
  as an economic sector, 129
  in Mexico, 128–129v, 129
  see also industry
**Mao Zedong,** 548–549m 549v
**Maori,** 659, 679, 679v, 682
**map projections,** 24–25
**maps,** 18–27
  chloropleth, 119m
  dot density, 26, 27m
  elements of, 18–19
  physical, 22–23, 22m
  point symbol, 26, 26m
  political, 22–23, 22m, 30m
  projection types, 24–25
  proportional symbol, 26, 27m
  scale on, 20–21
  thematic, 26–27
**maquiladoras,** 129
**Marie Antoinette,** 273, 273v
**marine life,** 54
**marine reserves,** 253
**maritime,** 544
**market economy,** 581, R44

**marketing,** 171
**Marquesas Islands,** 659v
**Marquette, Jacques,** 90
**Marseille, France,** 253v
**Marshall Islands,** 646m, 654m, 659m
**Marshall Plan,** 292
**marsupials,** 648
**martial law,** 580
**Martinique,** 141m, 142m, 143m, 145m, 157m, 165v
**Masai Mara National Park, Kenya,** 354v, 389v
**mass media culture,** 118–119
**Mauritania,** 357m, 358m, 363m, 377m, 393m
**Mauritius,** 357m, 358m, 377m, 393m
**Mauryan Empire,** 484, 485m
**Mayan culture / civilization,** 102, 103m
  artifacts from, 126
**Mayan temples,** 164–165v
**Mbalax (music),** 386
**Meacham, Sam,** 86–87
**Mecca,** 425
**mechanization,** R44
**media culture,** 118–119
**medicinal plants,** 648
  in African rain forests, 390–391
**meditation,** 562
**Mediterranean climate,** 226, 227m, 247
**Mediterranean Sea,** 253m
  early trade along coast of, 421
  and European trade, 248
  Ottoman Empire trade and, 429
  population of cities around, 253v
  protecting the ecosystem, 252–253
  Roman Empire, 260–263, 263m
  Silk Roads and, 544
**Medvedev, Dmitry,** 342
**megacities,** 232
**Meiji era, Japan,** 552
**Mekong Malaria Program,** 399
**Mekong River, Southeast Asia,** 600–601, 600v, 601m
**Menes, Egyptian king,** 432
**Menuhotep II, Egyptian king,** 432
**Mercator map projection,** 24, 24m
**Merv, Turkmenistan,** 320, 321m
**mesa,** 84
**Mesa Central, Mexico,** 84
**Mesa del Norte,** 84
**Mesopotamia,** 416–417, 417m, 422–423, 422m, 483
**messiah,** 424
**mestizos,** 106, 107, 126, 216
**methane gas,** 316–317, 317v
**metropolitan areas,** 623
**Mexican Cession,** 108
**Mexican Plateau,** 84

# INDEX

**Mexico,** 77m, 78m, 79m, 81m, 89m, 156
  Aztec Empire in, 103, 103m
  boundaries of, 109m
  climate of, 79, 84
  cultures in, 126–129
  democracy in, 130–131
  economy of, 63v, 128–129
  history of, 102–109
  immigration from, to U.S., 135v
  independence movement, 106–107
  Mayan culture in, 102, 103m
  Mexican Revolution, 110–111
  mountains of, 84–85
  social classes in colonial, 106
  war with U.S., 108–109
  Yucatan aquifer, 86–87
  see also North America
**Mexico City, Mexico,** 102, 103, 104, 126, 126v
**microcredit,** 397
**microlending,** 507
**Middle Ages,** 264–265
  and spread of Christianity, 426
**Middle Kingdom, Egypt,** 432
**middle passage,** 374–375, 375v
**migration**
  ancient, in South America, 202
  assisted, 660
  Bantu, in Africa, 370–371
  to Brazil, 232
  economics and, 176–177
  within Europe, 271, 299
  human, early, 8
  within Indonesia, 630–631
  from Mexico to the U.S., 129
  as movement, 14
  to Nile River Valley, 432
  in oil-rich countries, 451
  political conflict and, 176
  to Saudi Arabia, 441
  in Southern Africa, 367
  and spread of religion, 61, 426–427, 427v
  and trade in Southwest Asia and North Africa, 440–441
  see also immigration
**military, U.S., naturalization and,** 135v
**military dictatorships,** 512
**Minamoto family, Japan,** 550
**Minas Gerais, Brazil,** 232
**mineral resources,** 50, 51m, 314
  in Brazil, 209, 232
  in Malay Peninsula, 602, 603m
  in Mexico, 84, 85m, 129
  in North America, 82–83
  in Peru, 228–229
  in Russia, 314, 315m
  in Southern Africa, 367, 367m
  in sub-Saharan Africa, 392–393, 393v
  see also natural resources
**minerals,** 392
**mining,** 314, 366v
  in Brazil, 231
  open pit, 314v
  in Peru, 228–229
**Miraflor Nature Reserve,** 170–171v
**Mission Blue,** 55
**missionaries,** 61, 91, 376, 426, 427v
  in East Asia, 563
  in North America, 90, 91
  in South America, 207
**Mississippi River, U.S.,** 42, 43v, 80
  border of U.S. on, 96
  exploration of, 90
**Missouri River, U.S.,** 43v, 80
**mistral (winds),** 247
**Mitsubishi family,** 553
**Mitsui family,** 553
**mixed economy,** 62
**Miyazaki, Hayao,** 570, 571
**mobile media,** 118–119
**modernization,** 388, 508
**Mogadishu,** 372
**Mohenjo-Daro, South Asia,** 482–483
**Moldova,** 245m, 246m, 279m, 293m
**Mona Lisa (da Vinci),** 286v
**monarchy,** 65, R50
  in Saudi Arabia, 448
**Mongol Empire,** 320–321
**Mongolia,** 22m, 525m, 526m, 527m, 529m, 534–535, 559m, 573m
**Mongolian Plateau,** 534, 535m
**monks (Buddhist),** 622, 622v
**monopoly,** 208, R44
**monotheism,** 563
**monotheistic,** 424
**monotheistic religions,** 60
**monsoons,** 474–475, 475m, 527, 599
**Montenegro,** 245m, 246m, 279m, 293m
**Montezuma, Aztec ruler,** 104, 105v
**Montserrat,** 141m, 142m, 143m, 144, 144v, 145m
**moral,** 542
**Morelos, José,** 106
**Morocco,** 412–413m, 414m, 451m
**Moscow, Russia,** 321, 325, 336, 337m
  Saint Basil's Cathedral, 306v
**mosque,** 442
**Mount Etna, Italy,** 33v
**Mount Batur, Indonesia,** 604v
**Mount Everest, Asia,** 466v, 471
**Mount Fuji, Japan,** 522v
**Mount Kenya,** 360
**Mount Merapi, Java,** 594v
**Mount Rushmore, South Dakota,** 136
**Mount Sinabung, Sumatra,** 604
**Mount St. Helens, Washington,** 41v
**mountains**
  in Europe, 250
  of Japan, 530, 531
  in Mexico, 84–85
  in North America, 79, 82–83
  in South America, 190–191
  in Southeast Asia, 598m, 599
  and tectonic plate movement, 37v
**movable type,** 564
**movement,** 14
  see also immigration; migration
**Mozambique,** 357m, 358m, 367m, 377m, 393m
**Mubarak, Hosni,** 449
**Mughal Empire,** 484–485, 485m, 488
**Muhammad,** 425
**Muhammad, Askia,** 372
**mujahedeen,** 456
**multi-party democracy,** 130, 131
**multinational corporations,** 567
  in Singapore, 633
**multiparty republic, in Turkey,** 448
**multitude,** 154
**Mumbai, India,** 510
**Mundurukú,** 217
**murals,** 126
**Murmansk, Russia,** 311
**Murray River, Australia,** 668
**Musa, Mansa, King of Mali,** 372, 373v
**Muscat Sultanate, Oman,** 418–419v
**music,** 286–287
  African, 386–387
  Caribbean, 166–167
  European, 286–287
  protest music, 386
  in South Asia, 502v
  and storytelling in Africa, 354–355v
**Muslim League,** 488
**Muslim Republic of Chechnya,** 341
**Muslims**
  Crusades and, 264
  in Pakistan, 488
  Shi'ite and Sunni conflict, 454
  spread of Islam, 429
  trade and the spread of religion, 427
  see also Islam
**Myanmar,** 65, 597m, 598m, 601m, 604, 605m, 611m

# N

**Nagasaki, Japan,** 100
**Nairobi, Kenya,** 388, 388v
**Namib Desert, Namibia,** 406v
**Namibia,** 357m, 358m, 359, 367m, 377m, 392v, 393m
**Napoleon,** 273, 273v, 324, 325v
**Nasca culture,** 202–203
**National Action Party (PAN),** 130
**national debt,** R44–R45
**National Geographic**
  Anderson, Greg, NG Fellow, 662–663, 662v, 663v
  Anthony, Katey Walter, Emerging Explorer, 316–317
  Banks, Ken, Emerging Explorer, 138, 184

# INDEX

Bell, Katy Croff, Emerging Explorer, 242v
Byrd, Richard, explorer, 675
Conlee, Christina, NG Grantee, 202–203
Culhane, Thomas, Emerging Explorer, 410v, 465
Daltry, Jenny, Emerging Explorer, 594v
Davis, Wade, Explorer-in-Residence, 390–391
Earle, Sylvia, Explorer-in-Residence, 54–55
Enduring Voices Project, 219, 662–663
Genographic Project, 8v
Gobbo, Grace, Emerging Explorer, 390–391
Goodman, Beverly, Emerging Explorer, 444–445
Harrison, David, NG Fellow, 662–663, 662v, 663v
Helgen, Kristofer, Emerging Explorer, 606–607
Hiebert, Fredrik, NG Fellow, 306v
Hogan, Zeb, NG Fellow, 600v
Hussain, Shafqat, Emerging Explorer, 466v, 520–521
Joubert, Beverly, Explorer-in-Residence, 368–369
Joubert, Dereck, Explorer-in-Residence, 368–369
Lin, Albert Yu-Min, Emerging Explorer, 522v
Lindsey, Elizabeth Kapu'uwailani, NG Fellow, 642v
Meacham, Sam, NG Grantee, 86–87
Ntaiya, Kakenya, Emerging Explorer, 354v
Reinhard, Johan, Explorer-in-Residence, 203v
Sala, Enric, Explorer-in-Residence, 252–253
Salak, Kira, Emerging Explorer, 472–473
Sato, Katsufumi, Emerging Explorer, 536–537
Segura, Paola, Emerging Explorer, 186v
Shapiro, Beth, Emerging Explorer, 74v
Simoes, Cid, Emerging Explorer, 186v
Urteaga, José, Emerging Explorer, 148–149
Wallace, Scott, journalist, 199
Wells, Spencer, Explorer-in-Residence, 8v
**National Geographic magazine**
Kolbert, Elizabeth, writer, 481
Larmer, Brook, writer, 481
**National Geographic Society, preferred map projection,** 25
**National Highways Authority, India,** 508
**National Socialist German Workers' Party,** 278
**National Weather Service,** 49

**Nationalism,** 276
**nationalized (industry),** 128
**Native Americans**
colonization and, 88, 89
forced removal of, by U.S. government, 96
killed by disease, 88, 104–105
trade with, 90–91
**native animals,** 648
**native languages**
in Southeast Asia, 624–625
see also indigenous languages
**native populations**
Aborigines, 656–657
see also indigenous peoples
**natural barriers,** 311
**natural disasters,** 174
and agriculture, in Africa, 396
in Bangladesh, 514
effect on economy of Japan, 582, 583
see also earthquakes; floods/ flooding; tsunamis
**natural gas**
in North America, 80, 83
in Siberia, 344–345
**natural resources,** 50–51, 51m, R45
biological, 50, 51m
in Brazil, and trade, 208, 209
colonialism and, 488
and colonization, 376
in the Congo, 365
energy and, 350–351
of Europe, 250, 251m
as factor of production, 62
food supply and, 174, 175
improving economics in Africa, 393
of Korean Peninsula, 532, 533m
limited in Japan, 531
in Russia, 314–315, 315m, 327m, 344–345, 345m
of South Asia, 476–477, 477m
see also mineral resources; rain forests
**natural rights,** 274
**naturalization,** 125, 134–135
**Nauru,** 646m, 654m, 659m
**navigation, by Pacific islanders,** 658–659
**Nazi Germany,** 325
**N'Dour, Youssou,** 384
**Nebuchadnezzar, King of Babylonia,** 423
**Nenets,** 334, 334v
**Nepal,** 469m, 470m, 471m, 475m, 477m, 480, 487m, 507v, 516m
**Netherlands,** 245m, 246m, 251, 279m, 293m
colonization in Indonesia, 612
colonization in North America, 88, 89, 89m
colonization in South Africa, 402
colonization in Southeast Asia, 610, 611m
polders in, 248

**Neuces River, North America,** 108
**neutrality,** 100
**Neva River, Russia,** 338
**New Caledonia,** 646m, 654m, 659m
**New Delhi, India,** 480, 502v
**New France,** 88, 90–91, 91m
**New Guinea, Indonesia,** 597m, 598m, 601m, 605m, 611m
indigenous animals in, 648
**New Kingdom, Egypt,** 432–433
**New Orleans, Louisiana,** 48, 49
**New South Wales, Australia,** 660–661
**New Sweden,** 88
**New York City, U.S.,** 8v, 12v, 13
**New Zealand,** 646m, 647, 654m, 659m, 683–684
adventure tourism in, 680–681, 681v
in Antarctica, 675m
fishermen, 659v
geothermal energy, 123, 682–683, 683m, 683v
Maori in, 679, 682
non-native species in, 650–651
**newspapers,** 118
**Niagara Falls,** 74v
**Nicaragua,** 140m, 142m, 143m, 145m, 146m, 156, 157m, 160m
earthquakes in, 145v
government and economics in, 170–171
sea turtle conservation in, 146–149
tourism in, 165v
**Niger,** 357m, 358m, 362, 363m, 377m, 393m, 396, 459
**Niger River delta,** 394–395v
**Nigeria,** 357m, 358m, 362, 363m, 377m, 393m
oil in, 392v, 394–395, 395m
resources and wealth, 51
salt ponds in, 378–379v
**Nile Delta,** 432
**Nile River, Egypt,** 42, 43v, 415
**Nile River Valley,** 415, 430–431, 430v, 431m, 431v
**Nkrumah, Kwame,** 376
**Nobel Peace Prize,** 274, 275
**nocturnal,** 369
**nomadic herders,** 440
**nomads,** 204, 334
Silk Road trade and, 544
**non-governmental organizations (NGOs),** 399, 478
**nonfossil fuels,** 350
**nonrenewable fossil fuels,** 314, 350
**nonrenewable resources,** 50, 419
**Noriega, Aldo,** 203v
**Noril'sk, Russia,** 312v, 313m
**North Africa**
defined as region, 17, 17m
desertification in, 415
**North America,** 30m, 40–41, 40m, 74–136, 77m, 78m, 79m, 81m, 89m

# INDEX

changing economy of, 120–121
climate of, 46m
cultural diversity in, 116–117
defined as region, 17, 17m
history, 88–101, 102–109
immigration to, 116
media culture in, 118–119
physical geography, 78–87
religions of, 61m
*see also* Canada; Mexico; United States
**North American Free Trade Agreement (NAFTA),** 128
**North Atlantic Drift,** 247, 311
**North Atlantic Treaty Organization (NATO),** 278, 297
**North Carolina, U.S.,** 21m
**North China Plain, China,** 528, 529m
**North Island, New Zealand,** 647, 666–667v
**North Korea,** 22m, 65v, 525m, 526m, 527m, 529m, 532, 533, 557m, 559m, 573m
compared with South Korea, 584–585
government in, 65
**North Pole,** 19
**North Vietnam,** 614
**Northeast Science Station, Cherskiy, Siberia,** 316
**Northern European Plain,** 247, 251
**Northern Hemisphere,** 19, 19v, 34
seasons in, 34, 35, 35v
**Northern Mariana Islands,** 646m, 654m, 659m
**Norway,** 34, 245m, 246m, 279m, 293, 293m
in Antarctica, 675m
and European Union, 293
food in, 290
physical geography of, 249
**novel,** 288
**Novgorod, Russia,** 320
**Ntaiya, Kakenya,** 354v
**nuclear power**
Chernobyl disaster, 346–347
in France, 123, 347
Japanese disaster, 531
**nuclear weapons,** 328

# O

**oases,** 420
**Obrador, López,** 131
**occupied,** 538
**ocean currents,** 196–197, 197m, 311
and climate, 44, 247
**ocean landforms,** 38
**oceanic plates,** 37v
**oceans, as natural habitat,** 54–55
**oil**
in Arabian Peninsula, 419
in Chad, 363
and ethnic groups in Nigeria, 395m
exports from Africa, 392v
from Nigeria, 394–395
reserves in North America, 80, 81
resources in Siberia, 344–345
in Southwest Asia and North Africa, 450–451, 451m
supply and demand, 122–123
in the United Arab Emirates, 446
**oil field pump,** 50–51v
**oil spills,** 55, 81, 122
**Okavango Delta, Botswana,** 368
**Old Kingdom, Egypt,** 432
**oligarchy,** 65, R50
**Olmec culture,** 102
**Olympic National Park, Washington,** 53v
**Olympics,** 234–235
in Beijing, 565v
in Brazil, 234–235
medal count in, 238, 239v
**Oman,** 412–413m, 414m, 419m, 451m
**Omsk, Russia,** 70, 71v
**one-child policy,** 576–577
**opera,** 287
**opportunity cost,** R45
**oral tradition / storytelling,** 386, 672
**Orange Revolution,** 296–297
**Oregon Trail,** 96, 97m
**Organization for European Economic Cooperation,** 292
**Organization of African Unity (OAU),** 377
**Oriental Pearl Tower, Shanghai,** 568–569v
**Orinoco River, South America,** 204, 205m
**Osman,** 428
**Ottoman Empire,** 428–429, 429m, 442
**Outer Banks, North Carolina,** 20, 20–21v
**outer islands, Indonesia,** 630, 630v, 631v
**outrigger canoes,** 658, 658v
**outsourcing,** 506
**overnight tourism,** 164
**overpopulation,** 514
*see also* population growth

# P

**Pacific Ocean,** 42
Ring of Fire in, 40–41
**Pacific Realm,** 647
defined as region, 17, 17m
formation of islands in, 652–653, 653v
physical geography of, 647
seagoing society of, 658–659
**Paekche kingdom,** 554
**pagoda,** 580, 580v
**Pakistan,** 469m, 470m, 471m, 475m, 477m, 482, 487, 487m, 488, 516m
consumer goods in, 507v
four provinces of, 512
government changes in, 512–513, 513m
Kashmir conflict, 489, 489m
urbanization in, 510–511
**Palau,** 646m, 654m, 659m
**Palestine Liberation Organization (PLO),** 453
**Palio, traditional Italian horse race,** 284–285v, 285
**Palm Jumeira, Dubai,** 446, 446v
**palm trees,** 430v
**Pampas,** 190v, 191, 192, 686
economy of, 224–225
Wet and Dry, 224, 225v
**Pan-Africanism,** 376–377
**Pan American Games,** 234
**Panama,** 140m, 142m, 143m, 145m, 146m, 157m, 160m, 165v
**Panama Canal,** 168–169, 168–169v, 180m
Suez Canal compared to, 182–183, 183v
**Panama Canal Zone,** 168
**pandemic,** 398
**Pangaea,** 359
**paper, invention of,** 564, 565
**Papua New Guinea,** 646m, 654m, 659m, 663, 676
**papyrus,** 433
**Paraguay,** 189m, 190m, 191, 193m, 205m
Guaraní culture in, 217, 217v
**Paraguay River, South America,** 204, 205m
**Paraná River, South America,** 204
**Paris Opera, France,** 287v
**Parliament**
Canadian, 124
Indian, 504, 505, 505v
Israeli, 448
**parliamentary democracy,** 448
**Parthenon, Greece,** 258
**Partition,** 488–489
**Pashtun,** 456
**patricians,** 261, 262
**Paul Revere,** 92
**Pax Romana,** 262
**Pearl Harbor, Hawaii,** 100, 278
**peat,** 314
**Pei, I.M.,** 243v
**Pélé (soccer star),** 234v
**Peloponnesian War,** 258
**penguins,** 537, 642–643v
**peninsulares,** 106, 107
**peninsulas,** 84, 246v, 247, 256
**People's Republic of China,** 548
**perestroika,** 340
**Pericles,** 258, 259v
**permafrost,** 312, 315
methane gas and, 316–317
**Perón, Colonel Juan Domingo,** 222
**Persian Empire,** 257, 258, 423, 433
**Persian Gulf,** 418, 419m, 422
Dubai, 446–447
**perspective, in painting,** 266, 286
**Peru,** 189m, 190m, 191, 193m,

INDEX **R113**

# INDEX

205m, 212m
agriculture in, 228–229v
Aymara culture in, 216, 217
cuisine in, 221
El Niños in, 197
exported food crops, 228
indigenous populations in, 217v
Machu Picchu, 187v, 200–201v
Nasca culture in, 202–203
Quechua culture in, 201, 217
Quechua president, 222
**Peru Current,** 196, 197m
**pesticides, pollution of Aral Sea and,** 318
**Peter the Great,** 322, 324, 338
**Peterhof Palace, Russia,** 322–323v
**petrochemicals,** 450
**petroleum,** 122, 419
  see also oil
**pharaohs,** 432–433
**Philip II, King of Macedonia,** 259
**Philippines,** 597m, 598m, 601m, 605, 605m, 611m, 637m
colonization by Spain, 610, 611m
earthquakes in, 27m
ethnic diversity and government in, 629
history of, 613
independence from Spain, 613
languages in, 625v
**physical geography,** 44–55
  see also individual world regions entries
**physical systems,** 15
**Pico de Orizaba, Mexico,** 84–85v
**pictographs,** 657, 657v
**pilgrimages,** 425, 496
**Pinochet, Augusto,** 223
**pioneer trails,** 96, 96m
**pipelines, in Siberia,** 345m
**Pitcairn Islands,** 646m, 654m, 659m
**Pizarro, Francisco,** 206
**place,** 14, 15
**plains,** 38
Central Plain, Thailand, 599
of China, 528, 529, 529m
farming in, 193
Great Plains, 38, 79, 80, 81m
of Japan, 531
Northern European Plain, 247, 251
Siberian Plain, 311
  see also coastal plains; floodplain
**Plan of Ayala,** 110
**plant species, in Australia and Pacific islands,** 648
**plantations,** 88
**plate tectonics,** 359
Anatolian and Iranian plateaus, 420–421, 421m
in South Asia, 471
in Southeast Asia, 604, 605m
  see also tectonic plates
**plateaus,** 38, 360, 365
Anatolian, 420–421, 421m, 428
Deccan, 470, 470v

Iranian, 420–421, 421m
Mexican, 84
Mongolian, 534, 535m, 535v
in North America, 79, 80, 84
Tibetan, 526m, 526v, 527
**Plaza of Three Cultures, Mexico City,** 126v
**plebeians,** 261, 262
**poachers / poaching,** 53, 148–149, 368, 626
**Poland,** 245m, 246m, 279m, 293m
democracy in, 296
economy of, 297
in World War II, 100, 278
**polar climate,** 247
**polders,** 248
**political maps,** 30m
**politics**
Costa Rica and Nicaragua, 170–171
European Union benefits, 295
in India today, 504–505
  see also government
**pollution**
from agricultural chemicals, 476
and Aral Sea, 318
of Ganges River, 478–479, 478–479v
and Great Barrier Reef, 655
habitat and, 52
in Indian cities, 510
in Mediterranean Sea, 252–253
in Niger River Delta, 395
of South Asia's water supply, 480v
in Yamuna River, 480
**Polynesian culture,** 672–673
**Polynesian Triangle,** 672
**polytheism,** 563
**polytheistic religions,** 60
**popular culture,** 502
in Japan, 570–571
in South Asia, 502–503
**population**
in Haiti 1789, 161v
in Mediterranean Europe, 253m
**population density**
in Australia, 669, 669m
climate and, 312
in East Asia, 527m
in Ganges River Basin, 479m
in Indonesia, 631v
in Korean Peninsula, 532
in Russia and the Eurasian republics, 311m
in South America, 193m
in South Asia, 471, 471m
in Southeast Asia, 601m, 605m
in Thailand, 27m
in western U.S., 83v
**population growth**
China's policy on, 576–577
and food supply in Africa, 396
in India, 505, 577
in Istanbul, 442–443
in Japan, 583, 583v
in São Paulo, Brazil, 232, 233v
in Shanghai, 567

in South Asia, 511v
**porcelain,** 564, 564v
**port,** 338, 632
**Port-au-Prince, Haiti,** 172, 173
**Portugal,** 245m, 246m, 279m, 293m
exploration and colonization by, 209m, 268–269
influence in Africa, 372–373
influence in South America, 206–207, 208–209
and slave trade, 374
**pottery, celadon,** 555, 555v
**poverty,** R45
in Bangladesh, 514–515
and disease, 398
in Haiti, 172–173
in oil-rich countries, 451
in slums, India, 510, 511
  see also literacy rate; overpopulation
**Prague, Czech Republic,** 282–283v
**precipitation,** 43v
factor of climate, 70, 71v
**prehistoric,** 620
**preservation.** See habitat preservation
**Presidential Palace, Delhi, India,** 504v
**presidential power.** See executive branch
**prevailing winds, and climate,** 44
**pride (of lions),** 360–361v
**primary sector,** 62
**prime meridian,** 19, 19v
**prime minister**
in Canada, 124
in India, 504
in Singapore, 632
**printing**
early, in China, 564
Gutenberg's press, 266, 266v, 288
Korean, 555
**privatization,** 297
**profitable,** 229
**Project Snow Leopard,** 520
**projections.** See map projections
**propaganda,** 308, 329v
**prosperous,** 222
**Protestants,** 266–267
in South America, 220
**provinces,** 156
**Prussia,** 276
**public education**
in South Asia, 500
  see also education; literacy rate
**public health,** 398–399
**Pudong, Shanghai, China,** 567
**Puerto Rico,** 141m, 142m, 143m, 145m, 157m, 158–159, 159m
internal migration within, 176
**pujas,** 498v
**Punjab, India,** 486
**Puntland,** 401, 401m
**Puranas,** 486
**Putin, Vladimir,** 338, 342, 343
**Pyongyang, North Korea,** 556

# INDEX

**pyramids,** 432, 434–435
   Great Pyramid, 434–435, 434v, 435v
**Pyrenees,** 250

## Q

**qanats,** 421
**Qatar,** 412–413m, 414m, 451m
**Qin Dynasty,** 538–539
**quality of life, and Industrial Revolution,** 271
**Quebec, Canada,** 90–91
**Quebec Act, 1774,** 91
**Quebec City,** 90v
**Quechua,** 201, 217
   as official language in Peru, 218
**quetzal,** 146, 178v
**quinceañera,** 127
**Qur'an,** 425, R53

## R

**rabbits, as invasive species,** 650–651, 651m
**radiation, contamination,** 346–347
**radicals, in French Revolution,** 273
**radio,** 118
**radioactive,** 346
**railroads**
   bullet trains, 572–573, 572v
   canals and, 182
   in East Asia, 573–573, 573v
   and Industrial Revolution, 270
   Trans-Siberian, 336–337
   transcontinental, U.S., 96v, 97, 97m
**rain forests,** 53, 138, 143
   African, 199
   Amazon, 194, 198–199, 240
   in Central America, 146–147
   and climate change, 198–99
   Congo River Basin, 364–365
   conserving, 178–179
   ecotourism and, 179
   Foja Mountain, 606–607, 607v
   in Malay Peninsula, 602
   medicinal plants in, 390, 391v
   Tanzania, 390
**rain shadow,** 196, 420
**rainfall**
   in Arabian Peninsula, 415
   in Australia, 647, 647m, 668
   defining deserts, 406
   in the Great Plains, 80
   monsoon, 474, 475m
   in West Africa, 363
**rainshadow effect,** 82
**raj,** 488
**Rajya Sabha,** 504
**Ramayana,** 486
**Ramses II, Egyptian pharaoh,** 433
**Ramses temple, Egypt,** 410v
**ranching**
   in Australia, 668–669, 668v, 686–687
   in Brazil, 230
   in South America, 224, 225v, 228

**raw materials,** 50, R45
**Re,** 433
**Reagan, Ronald,** 329v
**rebellions,** 209
**recession,** 121, R41
**Reconstruction (after Civil War),** 98–99
**recycling, tourism industry and,** 164
**Red Guards, China,** 549
**Red List of Endangered Species,** 638
**Red Sea,** 42, 418, 419m
**The Reform,** 109, 130
**Reformation,** 266–267
**reforms,** 109
**refugees,** 400
**regions,** 15, 16–17, 17m
   Africa as two regions, 17
   defined, 14, 16
   in Europe, 247
   world regions, 16–17
**reign,** 323
**reincarnation,** 486
**reindeer,** 306v
**Reinhard, Johan,** 203v
**relative location,** 14
**reliable,** 635
**relief, on maps,** 23
**religions,** 60–61, 61m
   in Africa, 372
   in ancient Egypt, 433
   birthplace of major, 424–425
   conflicts of, 400, 401m, 454
   diversity in Dubai, 446
   in East Asia, 562–563, 563v
   in Europe, 274–275
   immigration and diversity of, 299
   in Istanbul, 442
   monotheistic, 60, 424, 563
   Muslim republic of Chechnya, 341
   origin and spread of, 60–61, 426–427
   polytheistic, 60, 563
   Reformation, 266–267
   in Russia, 322–323
   in South America, 207, 217, 220
   in South Asia, 486–487, 487m
   in Southeast Asia, 620–621`
   see also ethical systems
**relocation policies, in Indonesia,** 630, 631
**remittances,** 177, 629
**Renaissance,** 266–267
   music, 287
**renewable energy,** 122–123, 350
   biofuel, 231
   geothermal, 682–683, 683m, 683v
   hydroelectric, 314, 365, 366, 431, 578–579
   solar power, 26–27v, 123
   wind, 12v, 81, 122
**renewable resources,** 50
**republic,** 106, 260
**Republic of China,** 548

**Republic of China (Taiwan),** 580–581, 581m
**Republic of Korea,** 584
**Republic of South Africa, human rights in,** 66–67
**reserves,** 170v, 450
   Aborigines forced to live on, 678
   in Africa, 388, 389m
   marine, 253
**reservoir,** 578
**residency, and naturalization,** 134
**resistance,** 614
**resources**
   in Bangladesh, 514, 515m
   in Brazil, 230, 231, 231m
   in Canada, 69m
   colonization and competition for, 88–89, 90–91, 152–153, 384
   of Mexico, 84, 85m, 133m
   in North America, 80, 82–83
   in South America, 193m
   triangular trade and, 152–153
   see also economic resources; natural resources; water
**restore,** 53, 626
**retail goods,** R45–R46
**retreated,** 556
**revenue,** 344
**revolution, of the earth,** 34–35
**revolution (governmental),** 92
**Rhine River,** 250, 264
**rickshaws,** 619v
**rift valleys,** 359, 360–361
**right to vote,** 124, 125
**rights, in U.S. constitution,** 94
**rights of citizens,** 124–125, 125v
**Ring of Fire,** 40–41, 40m
   Indonesia in, 599
   Japan, 530
   Philippines in, 599
**Rio de Janeiro, Brazil,** 12v, 234–235v, 235, 238
**Rio Grande River, North America,** 108
**rituals,** 620
**river basins**
   Amazon, 190m, 191, 193, 194, 195m, 204–205
   Congo, 199, 359, 364–365, 365m, 366
   Ganges, 479m
**Rivera, Diego,** 126
**rivers,** 42
   development along, 79, 260, 415, 528, 529, 529m
   in Europe, 249m
   in North America, 79
   in Southeast Asia, 600–601, 601m
   world's longest, 43v
**Riyadh, Saudi Arabia,** 451
**Robespierre, Maximilien,** 273
**Robinson map projection,** 24, 25m
**Rocky Mountains, U.S.,** 38, 82
**Roman Catholicism,** 322
   influence of, in Middle Ages, 264–265

# INDEX

in Mexico, 126
in South America, 207
in Southeast Asia, 620, 621, 621v
see also Catholicism
**Roman Empire,** 261, 262-263, 263m
Egypt in, 433
and spread of Christianity, 426
**Roman Republic, influence and size of,** 261
**Romance languages,** 284
**Romania,** 245m, 246m, 279m, 293m
**Romanov, Nicholas,** 324v
**Romanov, Peter,** 323
**Romantic period,** 286, 287
**Rome, Italy,** 248, 253v
hills of, 260m
history of, 260-61
legacy of, 262-263
republic of, 260-261
**Roosevelt, Franklin,** 118
**roots,** 216
**rotation, of the earth,** 34-35
**Rub al Khali Desert,** 418, 419m
**Rudd, Kevin,** 678v
**runoff,** 43v
**Russia and the Eurasian republics,** 306-352, 309m, 310m, 311m, 313m, 315m, 316m, 321m, 323m, 325m, 327m
Aral Sea, 318-319
Chernobyl disaster, 346-347
climate in, 70-71, 71v, 312-313, 313m, 334-335
cultures of, 322-323, 323v, 334-339
defined as region, 17, 17m
early settlement of, 320
economics in, 341-345
energy consumption in, 350, 351v
European or Asian, 322-323, 323m
food in, 291
government of, 340-343
history of, 320-329
languages in, 348m
under Mongol rule, 320-321, 322
natural resources of, 344-345
physical geography of, 310-311, 314-319
see also Siberia; Soviet Union
**Russian Revolution,** 326, 328
**Russification,** 340
**Rwanda,** 357m, 358m, 361m, 377m, 393m

# S

**Sacramento, California,** 44
**Sahara Desert,** 358m, 359, 415
highway in, 441v
salt traders in, 440-441
trans-Saharan trade, 372
**Sahel,** 359, 362, 365, 400
**Sailendra dynasty, Indonesia,** 609
**St. Augustine, Florida,** 88
**Saint Basil's Cathedral, Moscow,** 306-307v
**St. Isaac's Cathedral, Russia,** 338, 339v
**St. Johns,** 141m, 142m, 143m
**St. Kitts and Nevis,** 141m, 142m, 143m, 145m, 157m
**St. Lawrence River, North America,** 79
**St. Lucia,** 141m, 142m, 143m, 145m, 157m, 162v
tourism in, 165v
**St. Maarten,** 141m, 142m, 143m
**St. Martin,** 141m, 142m, 143m
**St. Petersburg, Russia,** 322, 338-339
**St. Sophia cathedral, Russia,** 320v
**St. Vincent and the Grenadines,** 141m, 142m, 143m, 145m, 157m
**Sala, Enric,** 252-253
**Salak, Kira,** 472-473
**salinization,** 668
**Saliva language,** 219
**salt**
in Arabian Peninsula, 418
salt ponds in Nigeria, 378-379v
trade in Africa, 372, 440-441
**Samarkand, Uzbekistan,** 320, 321m
**Sambezi River,** 366, 367m
**Samburu National Reserve, Kenya,** 52-53v, 389v
**Samoa,** 646m, 654m, 659m, 676
**samurai,** 550v, 551v
**samurai culture,** 550-551
**San Jose, Peru,** 170, 216v
**sanitation, disease and,** 480
**Sanskrit,** 482
**Santa Anna, Antonio López de,** 107, 107v, 108
**Santa Fe Trail,** 96, 97m
**Santiago, Chile,** 221v
**São Paulo, Brazil,** 186v, 232-233
**Sao Tome & Principe,** 357m, 358m, 365m, 377m, 393m
**sarcophagus,** 435
**sari,** 499v
**Sato, Katsufumi,** 536-537
**Saudi Arabia,** 412-413m, 414m, 418-419, 419m, 451m
government in, 65
monarchy in, 448
oil wealth in, 450
**savanna,** 358, 359, 360, 361
**scale, on a map,** 18, 20-21
large, medium, small, 20-21m
**Scandinavian countries.** See Denmark, Finland, Norway, Sweden
**Scandola Natural Reserve, Mediterranean Sea,** 253
**scarcity,** 152
**Schools for Africa,** 459
**scientific stations,** 675
**Scotland, coastal rock formations,** 254-255v
**Scott, Robert Falcon,** 674
**sea, defined,** 42
**Sea of Marmara,** 442
**sea turtles,** 148-149, 252, 537
**seafarers,** 658-659
**secondary cities,** 511
**secondary sector,** 62
**secular,** 323
**segregation,** 402
**Segura, Paola,** 186v
**seismic activity,** 144
**self-rule,** 452
**semiarid,** 318, 534
**Senegal,** 362, 363m
**Seoul, South Korea,** 532, 556, 585v
**separation of powers,** 95v
**September 11, 2001,** 101, 101v, 456
**Serbia,** 245m, 246m, 279m, 293m
**Serengeti National Park, Tanzania,** 354v
**serfdom,** 326
**serfs,** 264, 326, 326v
**service industries,** R46
**service industry,** 121, 129
**shalwar-kameez,** 500, 501v
**Shang Dynasty,** 538
**Shanghai, China,** 528-529v, 567v, 568-569v
growth of, 566-567
**Shanghai World Financial Center, China,** 567
**sharia, Islamic law,** 394
**sheik,** 446
**Shi Huangdi (Qin ruler),** 538-539
tomb of, 539, 539v, 540-541v
**Shi'ite Muslims,** 454
**Shinto,** 562, 563
**shipbuilding,** 584
**Shiva, the destroyer of the universe,** 496
**shogun,** 550
**Shona,** 373
**Siberia,** 309m, 310m, 311m, 313m, 314v, 315m, 316m, 323m, 337m, 345m
climate in, 335v
lakes in, 316-317
native peoples of, 334
oil wealth in, 344
Trans-Siberian Railroad, 336-337
vegetation and climate in, 312, 313
see also Russia and the Eurasian republics
**Siberian Plain,** 311
**Sichuan Basin, China,** 528
**Sierra Leone,** 357m, 358m, 363m, 377m, 393m
**Sierra Nevada Mountains, U.S.,** 44, 82
**Sikhism,** 60v, 61v, 486v, R57
**Silk Roads,** 306, 320-321, 321m, 539, 544-545, 545m, 562
**Silla kingdom,** 554
**silt,** 417
**Simoes, Cid,** 186v
**Singapore,** 62-63, 63v, 633v
economic growth of, 632-633

# INDEX

languages in, 625v
shipping in, 63v
**Sioux,** 89m
**Siqueiros, David Alfaro,** 126
**sirocco (winds),** 247
**slash-and-burn agriculture,** 204–205v, 205
**slave trade,** 208–209, 374–375
African culture and, 164, 166, 216, 375
see also Columbian Exchange, triangular trade
**slavery**
in Brazil, 208–209
and Civil War in U.S., 98–99
in Haiti, 156, 172
in Mexico, 105
in North America, 88–89
in South America, 207
triangular trade and, 152–153
**Slavic languages,** 284
**Slavs, early settlement of Russia and,** 320
**Slovakia,** 245m, 246m, 279m, 293m
**Slovenia,** 245m, 246m, 279m, 293m
**Slumdog Millionaire,** 503v
**slums,** 233
in India, 510, 510–511v
in Pakistan, 511
**smart phones,** 119
**Snowman Trek,** 472–473, 473m
**Sobaek Mountains, South Korea,** 23v
**soccer,** 221
**socialism,** 328, R46
**Socialist Republic of Vietnam,** 614
**society.** See culture
**soil, fertile,** 251
from lava, 41
**solar power,** 26–27v, 123
**Solomon Islands,** 646m, 654m, 659m
**Solon, Athenian statesman,** 256
**solstice,** 34, 35
**Somalia,** 357m, 358m, 361m, 377m, 384, 393m, 400–401
clan-based conflict in, 401
political map, 401m
**Somaliland,** 401, 401m
**Song Dynasty,** 564
**Songhai,** 372
**Sonoran Desert, U.S.,** 46v, 406, 407v
**South Africa,** 366, 367, 367m
apartheid in, 402–403
diamond exports, 392v
human rights in, 66–67
natural resources of, 402
voting in, 64v
**South America,** 40, 40m, 186–240, 186–240, 189m, 190m, 191m, 193m, 197m, 205m, 207m, 218m
agriculture in, 224–225, 226–227, 228–229
climate and rain forests, 198–199

climate of, 46m, 196–197, 198–199
cultures in, 200–207, 216–221
daily life in, 220–221
defined as region, 17, 17m
economics in, 224–229, 230–233
government in, 222–223
history of, 200–211
Inca culture and history in, 200–201
independence movements in, 210–211
indigenous cultures in, 216–217
language diversity in, 218–219
lowlands culture in, 204–205
Nasca culture in, 202–203
Olympics in, 234–235
physical geography, 190–199
Portuguese influence in, 206–209
religions of, 61m
slavery in, 208–209
Spanish influence in, 206–207
see also names of individual countries
**South Asia,** 466–520, 466–520, 469m, 470m, 471m, 475m, 477m, 487m, 516m
cultures in, 496–503
defined as region, 17, 17m
early civilizations, 482–483, 483m
empires of, 485m
governments in, 510–515
history of, 482–489
languages of, 516m
natural resources, 476–477, 477m
physical geography of, 470–479
religion in, 486–487, 487m
traditions and modernism, 498–499
water crisis in, 480–481
see also India; Pakistan
**South Island, New Zealand,** 647
**South Korea,** 22m, 525m, 526m, 527m, 529m, 532, 533, 557m, 559m, 573m
compared with North Korea, 584–585
government of, 556, 557
**South Pole,** 19, 647, 674, 675mv
expeditions to, 674–675
**South Sudan,** 357m, 358m, 361m, 377m, 393m
**South Vietnam,** 614
**Southeast Asia,** 594–640, 597m, 598m, 601m, 605m, 611m
colonial rule in, 610–611, 611m
cultures of, 620–627
defined as region, 17, 17m
diversity and government issues, 628–629
history of, 608–615
island nations of, 604–605, 605m
Malay Peninsula, 602–603
physical geography, 598–605
plants and animals in, 639, 640
population density in, 605m
religious traditions in, 620–621
river systems, 600–601, 601m
see also Cambodia; Indonesia; Malaysia; Philippines; Thailand; Vietnam

**Southern Africa,** 359, 366–367
city-states of, 372
natural resources of, 367, 367m
**Southern African Development Community,** 385v
**Southern Hemisphere,** 19, 19v, 647
seasons in, 34
**Southwest Asia and North Africa,** 410–459, 412–413m, 414m, 451m
cultures of, 440–447
defined as region, 17, 17m
economics of, 450–451
government in, 448–459
history of, 422–429
income gap and oil wealth, 450–451
Israeli / Palestinian tensions, 452–453
Nile River Valley, 430–435
Ottoman Empire in, 428–429
religious importance of, 424–427, 444–445
see also Afghanistan; Dubai; Egypt; Iraq; North Africa; Saudi Arabia
**sovereignty,** 293
**Soviet Union,** 296, 328–329
in Cold War, 101
collapse of, 340–341
in World War II, 278
see also Russia
**soybeans,** 224, 225v
**space race,** 328
**Spain,** 245m, 246m, 279m, 293m
Caribbean and, 158–159
colonization in North America, 88, 89, 89m
colonization in Philippines, 613
colonization in Southeast Asia, 610, 611m
conquistadores, 104–105
exploration and trade, 152–153
influence of, in South America, 206–207
Mexico and, 106–107, 156
**Spanish-American War,** 158
**Sparta, Greece,** 256, 257
**spatial thinking,** 12–13, 15
**Special Economic Zone (SEZ),** 567, 574
**specialization,** R46
**species, discovering new,** 606–607
**Spice Islands.** See Indonesia
**sports**
the Olympics, 238–239
in South America, 221
in South Asia, 501
**spring equinox,** 35, 35v
**Sputnik,** 328
**Sri Lanka,** 469m, 470m, 471m, 475m, 477m, 487m, 516m
consumer goods in, 507v
**Srivijaya Empire, Indonesia,** 609
**Stalin, Josef,** 328
**Stalingrad, Russia,** 325
**standard of living,** 173, R46
in emerging markets, 518
in Soviet Union, 329
see also gross domestic product (GDP)

# INDEX

**staple (food),** 155, 290
**state,** 320
**State Duma,** 342
**steel,** 231
**Stephenson, George,** 270
**steppes,** 311
  agriculture in, 334
  farming in, 334–335, 335v
  Mongolian, 534
  of West Africa, 362–363
**stick chart, for navigation,** 658
**stock market / stock exchange,** R46
**Stolen Generation,** 678
**storytelling,** 354–355v, 386, 672
**strait,** 442
**Strait of Malacca,** 597, 598, 603, 609
**stream,** 42
**strikes,** 326, R47
**Sub-Saharan Africa,** 354–408, 357m, 358m, 377m, 393m
  agriculture in, 396–397
  apartheid in, 402–403
  Bantu migration in, 370–371
  colonization in, 376–377
  cultures of, 370–71, 384–391
  defined as region, 17, 17m
  empires and city-states, 372–373, 373m
  health issues in, 398–399
  history of, 370–377
  languages in, 385m
  natural resources of, 393m
  need for schools in, 459
  physical geography of, 358–369
  rain forest in, 364–365
  resources of, 392–397
  the Rift Valley, 360–361
  slave trade and, 374–375
  wildlife in, 368–369
  *see also* Kenya; Nigeria; Somalia; South Africa; Sudan
**subcontinent,** 471
**subduction,** 37v
**subsistence farming,** 193
  defined, 84
  in the Himalayas, 472
  in Mexico, 84
**subsistence fishing,** 601
**Sudan,** 357m, 358m, 361m, 377m, 393m, 400–401
  famine in, 396
  oil exports, 392
  political map, 401m
**Sudras,** 483, 497
**Suez Canal, Panama Canal compared to,** 182–183, 183v
**suffrage**
  in Saudi Arabia, 448
  *see also* right to vote
**sugarcane,** 196v, 230, 230v
  ethanol production, 231
  in Mexico, 129
  and slave trade, 208–209, 208v, 374
**Suleyman I, Ottoman ruler,** 428

**sultans,** 429
**Sumatra, Indonesia,** 597m, 598m, 601m, 605m, 611m
  Asian elephants in, 626–627v
  colonization by Britain, 610, 611m
  tsunami in, 599
**Sumer,** 422–423
**summer solstice,** 34, 35
**Sundiata, King of Mali,** 372
**Sunni Muslims,** 454
**supply and demand,** R47
**Suriname,** 189m, 190m, 191m, 193m, 205m
  indigenous populations in, 217v
**surplus (crops),** 175
**Suryavarman II, King in Southeast Asia,** 608
**suspension bridge,** 201, 201v
**sustainable,** 87, 476
**Swahili,** 371
**Swaziland,** 357m, 358m, 367m, 377m, 393m
**Sweden,** 34, 245m, 246m, 279m, 293m
  colonization in North America, 88, 89, 89m
  food in, 290
**Switzerland,** 245m, 246m, 279m, 293m
**Sydney, Australia,** 642v, 669, 669v, 670v
**symbols,** 491
  on a map, 18
  as part of culture, 58
**Syria,** 412–413m, 414m, 451m
**Syrian Desert,** 418, 419m

# T

**taiga,** 313
**Taino,** 166
**Taiwan,** 22m, 525m, 526m, 527m, 529m, 559m, 573m, 580–581, 581m
  physical geography of, 581m
**Taj Mahal, India,** 466v, 485, 485v
**Tajikistan,** 309m, 310m, 311, 311m, 313m, 315m, 345m
**Taklimakan Desert,** 544
**Taliban,** 456
**Tang Dynasty,** 554, 564
**Tanzania,** 357m, 358m, 361, 361m, 377m, 393m
**tariff,** 292, R47
**Tarquin, Etruscan king,** 260
**Taurus Mountains,** 415, 421
**tax,** 92
**taxation, Ottoman wealth from,** 429
**technology**
  digital, 121
  media culture, 118–119
**technology industry, in Singapore,** 632
**tectonic plates,** 37, 144–145, 145m
  movement of, 37v, 40–41, 360
  in Southeast Asia, 599

  *see also* plate tectonics
**Tehran, Iran,** 421
**television,** 118
**temperate (climate),** 79, 226
**temperatures**
  climate change and, 199
  factor of climate, 70, 71v
**Temple Emanu-el, New York,** 427v
**Tenochtitlán, Mexico,** 102v, 103, 104
**terra cotta,** 539, 540–541v
**terraced farming,** 16v, 201
**terrain,** 168, 336
**terrorism / terrorists,** 101, 456
**tertiary sector,** 62
**Texas, boundary wars with Mexico,** 108–109
**Texas, Republic of,** 108, 109m
**textiles**
  Guatemalan, 138–139v
  industry in China, 575
  industry in U.S., 97, 270
**Tezuka, Osamu (manga artist),** 570
**Thailand,** 597m, 598m, 601m, 605m, 611m
  Buddhism in, 621
  floating market, 595v
  languages in, 625v
  population density, 27m
  Thai culture, 622–623
**Thar Desert,** 470m
**theme, in maps,** 26–27
**38th parallel,** 556, 557, 557m
**Three Estates,** 272
**Three Gorges Dam, China,** 578–579, 578–579v, 579m
**Three Guarantees (Mexico),** 106–107
**three kingdoms, in Korea,** 554
**Tiber River,** 260, 260m
**Tibetan Plateau,** 526m, 526v, 527
**tidal pools,** 56–57v
**tierra caliente,** 192v, 193
**tierra fría,** 192v, 193
**tierra helada,** 192v, 193
**tierra templada,** 192v, 193
**tigers,** 639v
**Tigris River,** 415, 416–417, 422, 423, 417m
**Timbuktu,** 372
**time lines, historical**
  African history, 372–373, 376–377
  Australia and Pacific islands, 660–661
  Central American and Caribbean, 156–159
  China, 538–539, 548–549
  Europe, 258–259, 262–263, 266–267, 272–273, 276–277
  Korea, 554–557
  North America, 88–89, 92–93, 98–99, 106–109
  Russia, 320–321, 324–325, 328–329
  South America, 206–209v
  South Asia, 482–485, 488–489
  Southeast Asia, 608–609, 610–611

# INDEX

Southwest Asia and North Africa, 422–423, 432–433
**title, on a map,** 18
**tobacco, and slave trade,** 374
**Togo,** 357m, 358m, 363m, 377m, 393m
**Tokelau,** 646m, 654m, 659m
**Tokugawa Ieyasu,** 550
**Tokugawa shoguns,** 550
**Toledo, Alejandro,** 222, 222v
**tolerance,** 116, 484
**Tonga,** 646m, 654m, 659m
**Topkapi Palace, Istanbul,** 428v, 429v
**topography, affecting language groups,** 218
**tornado,** 48, 48v
**totalitarian,** 454, R50
**tourism**
  adventure, 680–681, 681v
  in Brazil, 233
  in Central America and Caribbean, 162v, 164–165
  coral reef preservation and, 654, 655
  in Costa Rica, 170
  in Dubai, 446
  ecotourism, 179
  environmental effects of, 146–147, 178–179
  in Great Barrier Reef, Australia, 654
  in Mexico, 129
  pollution from, 87
  rain forest, 146–147, 601
  in Russia, 338
**towns, growth of,** 264
**trade,** 16, R47
  African slave trade, 208–209, 372–373, 374–375
  Arab trade in Africa, 371
  and colonialism in Southeast Asia, 610–611, 611m
  exploration and, 152–153
  illegal drugs, 131
  importance in Indonesia, 612
  importance in Singapore, 62–63
  importance of location and, 248, 608–609
  and migration, 440–441
  new patterns in the Pacific, 676–677
  Renaissance and, 266
  Silk Roads, 320–321, 544–545, 545m
  in Southwest Asia and North Africa, 420–421, 440–441
  and spread of religion, 61, 426–427
  triangular, 152–153
**traditional economy,** 62
**Trail of Tears,** 96
**trains, bullet,** 572–573, 572v
**trans-Atlantic slave trade,** 374
**trans-Saharan trade,** 372
**Trans-Siberian Railroad,** 336–337, 337m
**Transantarctic Mountains,** 674
**transcontinental,** 97

**transcontinental railroad,** 96v, 97, 97m
**transform,** 37v
**transition zone,** 362–363v
**transpiration,** 198
**transportation corridors,** 384
**treaty,** 206
**Treaty of Guadalupe Hidalgo,** 109v
**Treaty of Maastricht, 1992,** 292
**Treaty of Paris, 1783,** 93, 158
**Treaty of Tordesillas, 1494,** 206, 207, 208
**Treaty of Versailles, 1919,** 277
**Treaty of Waitangi, 1840,** 679
**trenches,** 276
**trends,** 631
**triangular trade,** 152–153, 153m, 154–155
**tribe, identification, in Africa,** 384
**tribunes,** 261
**tributary,** 194v, 195v
**tribute,** 103, 320, 546
**Trinidad and Tobago,** 141m, 142m, 143m, 145m, 157m
  music and dance in, 166–167
**Tripitaka Koreana,** 555, 555v
**Triple Entente,** 276
**tropical diseases, in Africa,** 398
**troubadours,** 287
**tsunami,** 41, 445, 530, 530v, 598v, 599
  discovering ancient, 444–445
  in Japan, 582, 583
**Tuareg, salt traders,** 440–441
**tundra climate,** 46–47m, 47v
**Tunisia,** 412–413m, 414m, 451m
  political revolt in, 448–449
**Tupinambá, farming culture of,** 205
**Turkey,** 345, 412–413m, 414m, 420, 429, 443m, 451m
  democratic government in 448
  Ottoman Empire in, 428–429
**Turkmenistan,** 309m, 310m, 311, 311m, 313m, 315m, 320, 344, 345m
  Silk Road excavation, 306
**Tutankhamen,** 433v
**Tuvalu,** 646m, 654m, 659m
**Twelve Tables, of Rome,** 261
**typesetting,** 564
**typhoons,** 532, 598v, 599

## U

**Uganda,** 357m, 358m, 361m, 377m, 393m
**Ukraine,** 245m, 246m, 279m, 293m, 320, 328
  Chernobyl, 346, 347
  food in, 291
  Orange Revolution in, 296–297
**Ulaanbaatar, Mongolia,** 534, 535m
**Uluru, Australia,** 656v, 657
**unemployment rate,** R48

**UNESCO World Heritage sites,** 352
  Easter Island, 642v
  Machu Picchu, 200–201
  temple of Haeinsa, 555
**Union of South Africa,** 402
**Union of Soviet Socialist Republics (USSR),** 328
**United Arab Emirates,** 412–413m, 414m, 419m, 451m
  Dubai, 446–447
  oil wealth in, 450
**United Kingdom,** 245m, 246m, 279m, 293m
  in Antarctica, 675m
  government in, 65
  see also Britain, Great Britain
**United Nations,** 66, 401, 480
  division of British Mandate, 452
  Korea and, 556, 557
**United Nations Children's Emergency Fund (UNICEF),** 458, 459
**United States**
  annexation of Texas, 108, 109m
  in Antarctica, 675, 675m
  borders of, 93, 93m
  citizens' rights, 124–125
  Civil War in, 98–99, 99m
  climate in western, 45m
  consumer goods in, 507v
  cultures in, 116–119
  democratic government, 65, 124–125
  eastern, 21m
  economics in, 63v, 120–123
  energy consumption, 122, 350, 351v
  expansion of, 96–97, 97m
  extreme weather in, 48–49, 49m
  government compared to Russian, 343v
  grassland use in, 687m
  industrialization in, 97
  Internet access in, 119m
  languages in, 117v
  migration to, from Caribbean, 176–177, 177v
  mountains of, 82–83
  passenger vehicles in, 509v
  in Philippines, 613
  population density, 83m
  revolution, 92–93
  trade with Central America, 171
  war with Mexico, 108–109
  weather in western, 45m
  in World War I, 100, 277
  in World War II, 100–101, 278
  see also North America
**United States Capitol building,** 74v
**United States Constitution,** 94–95
**Universal Declaration of Human Rights,** 66, 67v
**Untouchables,** 497
**uplands,** 246v, 247
**Ur,** 422–423
**Ural Mountains,** 310v, 311, 322

# INDEX

**urban,** 668
**urbanization**
  in South Asia, 510–511
  *see also* Cities, growth of
**Urteaga, José,** 148–149
**Uruguay,** 189m, 190m, 191m, 193m, 205m
  cuisine in, 221
**Uzbekistan,** 309m, 310m, 311, 311m, 313m, 315m, 318, 320, 345m

## V

**vaccines,** 399
**Vaisyas,** 483, 497
**valley, defined,** 38
**Valley of Mexico,** 102
**Vanuatu,** 646m, 654m, 659m
**Vancouver, Canada,** 12v
**vaqueros,** 127
**Varanasi, India,** 496
**Varangian Russes,** 320
**Vedas,** 482, 486
**vegetation,** 191
  climate and, 46, 312–313
  elevation and, 191
**Venezuela,** 189m, 190m, 191m, 193m, 205m
  indigenous populations in, 217v
**Venice, Italy,** 250v
**venues (Olympic),** 234
**viceroys,** 156
**Vietnam,** 597m, 598m, 601m, 605m, 611m
  culture of, 608
  Dai Viet kingdom, 609
  languages in, 625v
**Vietnam War,** 118, 328, 614–615
**Vikings, early settlement of Russia and,** 320
**Villa, Pancho,** 110, 111, 111v
**Virgil,** 288
**Virgin Islands,** 141m, 142m, 143m
**Vishnu, preserver of the universe,** 491, 491v, 496, 609
**Vladivostok, Russia,** 336, 337m
**vocational (schooling),** 458
**volcanoes,** 360
  in Central America and the Caribbean, 144–145
  and creation of islands, 604, 605m
  defined, 41
  and formation of Pacific islands, 652–653, 653v
  in Ring of Fire, 40–41, 40m, 41, 41v, 530, 599
  in Southeast Asia, 599
  tectonic plate movement and, 37v, 604, 605m
**von Bismarck, Otto,** 276v
**voting,** 124v
**voyageurs,** 90, 91

## W

**wallaby,** 606
**Wallace, Scott,** 199
**Wari,** 202
**Warsaw, Poland,** 296–297v
**Warsaw Pact,** 278
**Washington, George,** 92–93
**wat,** 622, 622v
**water,** 42–43
  in aquifers, 86–87, 481
  clean, 510
  conflict over control of, 489
  crisis in South Asia, 480–481
  drinking, 510
  essential nature of, 42
  food supply and, 174
  fresh, 42, 259
  hydrologic cycle, 42–43v
  salt, 42
  supply of, in North America, 83
  *see also* oceans
**waterways,** 251
**Watt, James,** 270
**weather,** 45
  extreme, 48–49
  *see also* climate change; global warming
**weather forecaster,** 45
**weather prediction,** 49
**Wells, Spencer,** 8
**West Africa,** 362–363
  empires of, 372, 373m
  political map, 363m
**West Bank,** 453, 453m
**Western architecture, in Russia,** 322–323v
**Western Empire (Roman),** 262
**Western Hemisphere,** 12m, 19, 19v
**Western Sahara,** 357m, 358m, 363m, 377m, 393m
**Western Wall, Jerusalem,** 424v
**wet monsoons,** 474
**Wet Pampas,** 224, 225v
**wheat, in Australia and Argentina,** 686–687
**White Nile River,** 430
**wholesale goods,** R48
**wildlife**
  in Africa, 360, 361, 368–369, 369v
  in Antarctica, 680
  in Australia and the Pacific Islands, 648
  in Congo rain forest, 364
  in Korean Peninsula, 533
  in Malay Peninsula, 602
  new species in Indonesia, 606–607
  protecting Asian elephants, 626–627
  tracking in East Asia, 536–537
  *see also* endangered species; rain forests
**wildlife reserves,** 369m, 388, 389m
**Wilkes, Charles,** 674
**Wilson, Woodrow,** 277v
**wind currents,** 196–197, 197m
**wind energy,** 81, 122
**wind patterns, monsoon,** 474–475, 475m
**wind speed, in hurricanes,** 48
**wind turbine,** 122, 122v, 123v
**winds**
  and climate, 44
  mistral, 247
  sirocco, 247
**Winkel Tripel map projection,** 24, 25m
**winter solstice,** 34, 35
**wool,** 676–677v
  exported from Australia, 660
**World Bank,** 318
**World Health Organization,** 399
**World War I,** 100, 276–277
**World War II,** 100–101, 278–279
  China and Japan in, 548
  European economics after, 292
  Holocaust, 452
  Indonesia and, 612
  Philippines and, 613
**writing**
  cuneiform, 423
  hieroglyphics, 432

## Y

**Yakut,** 334
**Yamuna River,** 480
**Yanacocha, Peru,** 229
**Yangtze River.** *See* Chang Jiang River, China
**Yanomami, farming culture of,** 205
**Yanukovych, Viktor,** 296, 297
**Yellow River, China.** *See* Huang He River, China
**Yemen,** 412–413m, 414m, 419m, 451m
**Yenisey-Angara River, Asia,** 43v
**Yew, Lee Juan, Prime Minister Singapore,** 632
**Yoruba,** 394
**Yosemite National Park, California,** 44–45v
**Yucatan Peninsula, Mexico,** 84, 86–87
  Mayan culture in, 102
**Yugoslavia, civil war in,** 296
**yurts,** 306v, 335v
**Yushchenko, Viktor,** 297

## Z

**Zagros Mountains,** 415, 420v
**zaibatsu,** 552, 553
**Zambia,** 357m, 358m, 367m, 377m, 393m
**Zapata, Emiliano,** 110
**Zapatista Liberation Army,** 130, 131
**Zheng He, Chinese explorer,** 16, 546–547, 565
**Zhou Dynasty,** 538
**Zimbabwe,** 357m, 358m, 367, 377m, 393m

# INDEX

## SKILLS INDEX

### A
**analyze audios,** 287
**analyze cause and effect,** 55, 93, 109, 161, 173, 180, 249, 267, 301, 317, 321, 330, 377, 381, 395, 397, 399, 405, 436, 455, 457, 492, 513, 517, 553, 558, 567, 616, 629, R15
**analyze data,** 68, 129, 132, 147, 161, 165, 171, 177, 181, 217, 233, 236, 281, 301, 303, 349, 351, 375, 405, 407, 447, 451, 511, 517, 519, 558, 563, 575, 591, 635, 636, 639, 685, R21
**analyze models,** 35
**analyze primary and secondary sources,** 543, 571, R25
**analyze time lines,** 557
**analyze visuals,** 35, 37, 49, 53, 83, 87, 103, 105, 121, 149, 169, 177, 179, 180, 201, 203, 235, 279, 285, 317, 319, 323, 329, 335, 363, 391, 419, 431, 435, 441, 445, 479, 505, 537, 543, 551, 571, 607, 623, 649, 653, 657, 663, 673, R26
*see also* critical viewing

### B
**build models,** R33–R34

### C
**categorize,** 15, 107, 599, R7
**cause and effect.** *See* analyze cause and effect
**charts.** *See* create charts; interpret charts
**compare,** 91, 183, 461, 475, 477, 483, 501, 605, 679
**compare and contrast,** 15, 25, 30, 35, 47, 61, 105, 109, 112, 117, 119, 160, 179, 181, 212, 247, 249, 265, 295, 303, 325, 331, 335, 343, 348, 367, 393, 403, 407, 436, 441, 497, 505, 519, 527, 533, 551, 580, 589, 609, 621, 647, 665, 687, R14
**compare maps,** 547
**comparisons,** 347, 349, 351
**conduct Internet research,** 113, 237, 295, 365, 423, 449, R38
**create a magazine,** 520, 592
**create a multimedia presentation,** 133, R40
**create a slide show,** 281
**create a web page,** 589
**create brochures,** 339
**create charts,** 65, 161, 181, 183, 205, 213, 301, 315, 349, 405, 493, 607, 685, R30, R37
**create databases,** R36
**create diagrams,** 461
**create graphic organizers,** R37
**create graphs,** 147, 487, 637, R30
**create sketch maps,** 27, 184, 184, 381, R29
**create time lines,** 381, 617, R33
**create visuals,** 517, R34–R35
**create a digital presentation,** 31
**create a photo essay,** 389, 499
**create a photo gallery,** 240
**create a poster,** 161, 237, 559
**create a visual catalog,** 72
**create a visual overview,** 69
**create a visual presentation,** 665
**critical viewing,** 13, 14, 20, 23, 26, 28–29, 34, 44, 50, 52, 55, 56–57, 63, 64, 82, 84, 87, 90, 96, 100, 102, 107, 116, 118, 120, 124, 127, 128, 130, 147, 149, 164, 167, 168, 176, 192, 196, 200, 216, 219, 220, 221, 222, 224, 226, 228, 229, 230, 233, 234, 248, 252, 254–255, 260, 268, 270, 273, 283, 286, 288, 289, 290, 292, 296, 312, 314, 318, 322–323, 326, 334, 336, 339, 340, 342–343, 364, 366, 370, 374, 378–379, 384, 388, 389, 392, 394, 400, 402, 416, 418, 420, 425, 428, 430, 433, 434, 440–441, 442–443, 445, 446, 447, 449, 450, 455, 457, 459, 475, 478, 496, 497, 498, 500, 502, 503, 504, 506, 508, 510, 512, 514, 528, 530, 532, 537, 540–541, 544, 546, 551, 553, 562, 564, 565, 566, 568–569, 570, 572, 574, 578, 580, 582, 585, 586–587, 602, 604, 613, 620, 623, 624, 626, 628, 630, 631, 633, 634, 649, 650, 652–653, 655, 656, 661, 663, 668, 669, 670, 674, 676, 678, 679, 681, 682, 683

### D
**data.** *See* analyze data
**describe,** 53, 199, 201, 203, 235, 463, 549
**describe culture,** 221, 501
**describe geographic information,** 30, 51, 83, 197, 212, 359, 415, 429, R8
**design a poster,** 352
**design an anime storyboard,** 589
**details.** *See* find details
**distinguish fact and opinion,** R22
**Document-Based Questions practice,** 110–111, 154–155, 210, 211, 274–275, 490–491, 614–615
**draw a map,** 464
**draw conclusions,** 23, 49, 68, 91, 107, 110, 131, 135, 160, 173, 177, 191, 193, 213, 219, 231, 251, 259, 267, 280, 281, 285, 293, 300, 329, 345, 349, 371, 380, 389, 401, 404, 417, 421, 437, 447, 471, 483, 516, 534, 535, 537, 555, 558, 563, 575, 583, 588, 589, 591, 613, 617, 625, 637, 639, 655, 665, 685, R17

### E
**evaluate,** 207, 271, 279, 280, 349, 405, 427, 461, 507, 513, 517, 529, 539, 677, R23
**evaluate Internet sources,** R39
**explain,** 217, 239, 351, 421, 427, 435, 437, 460, 663, 671, 673, 675
**express ideas through speech,** 125, 273, 343, 423, R9
*see also* turn and talk

### F
**fact and opinion.** *See* distinguish fact and opinion
**find details,** 31, R3
**find main ideas,** 30, 68, 112, 132, 160, 180, 212, 236, 280, 300, 330, 348, 380, 404, 436, 460, 492, 516, 558, 588, 616, 636, 664, 684, R3
**find main ideas and details,** 69, 101, R3
**form and support opinions,** 67, 123, 231, 235, 287, 380, 389, 539, 577, 579, R19

### G
**gather and share information,** 331
**graphs.** *See* create graphs; interpret graphs

### H
**hold a folk-tale festival,** 352
**host a geography bee,** 520
**host a panel discussion,** 464
**human-environment interaction,** 23, 31, 41, 47, 49, 53, 68, 87, 149, 193, 231, 271, 313, 319, 335, 359, 385, 404, 419, 431, 445, 475, 477, 515, 599, 623, 627, 647, 653, 680

### I
**identify,** 89, 213, 223, 229, 239, 292, 319, 343, 425, 471, 507, 659, 664, 685, 687
**identify problems and solutions,** 131, 132, 165, 269, 297, 299, 625, 629, R20
**Internet research.** *See* conduct Internet research; evaluate Internet sources
**interpret charts,** 43, 63, 68, 112, 165, 181, 193, 253, 261, 301, 375, 393, 473, 477, 492, 589, 627, R31

INDEX **R121**

# INDEX

**interpret data,** 369, 481
**interpret graphs,** 71, 131, 135, 199, 229, 233, 236, 299, 327, 405, 459, 460, 509, 563, 577, 581, 583, 631, 685, R32
**interpret maps,** 17, 19, 21, 23, 45, 47, 49, 51, 69, 79, 81, 83, 85, 89, 91, 99, 103, 113, 133, 143, 145, 157, 159, 160, 180, 191, 195, 197, 209, 212, 219, 225, 227, 231, 237, 247, 251, 263, 269, 271, 280, 293, 300, 311, 313, 320, 323, 325, 330, 337, 348, 361, 363, 365, 367, 371, 373, 377, 381, 385, 389, 395, 401, 404, 417, 421, 431, 437, 451, 453, 455, 461, 493, 516, 527, 555, 557, 559, 588, 616, 617, 637, 647, 651, 659, 664, 669, 675, 677, 683, 684, R27–R28
**interpret maps and time lines,** 489, R27, R28, R33
**interpret models,** 43, 317
**interpret primary sources,** 31, 213
**interpret tables,** 132, 281, 331, 349, 380, 436, 558, 636, 665
**interpret time lines,** 157, 159, 207, 209, 433, 483, 549, 609, 611, 661, R33
**interpret visuals,** 265

## J

**journal entries.** *See* write journal entries

## L

**location,** 13, 19, 21, 25, 43, 51, 85, 89, 97, 99, 191, 195, 205, 219, 227, 237, 263, 330, 359, 363, 377, 381, 395, 404, 415, 417, 437, 451, 455, 471, 475, 477, 487, 489, 507, 513, 529, 531, 613, 664, 669, 675
**location, finding,** 143, 145, 157, 169

## M

**main ideas.** *See* find main ideas
**make generalizations,** 125, 236, 331, 349, 385, 431, 459, 460, 463, 485, 492, 515, 517, 558, 603, 609, 636, 665, 671, 681, R18
**make inferences,** 13, 15, 16, 19, 24, 26, 27, 31, 35, 37, 41, 45, 46, 53, 61, 65, 68, 71, 81, 83, 85, 87, 91, 93, 94, 97, 99, 103, 107, 109, 112, 113, 119, 121, 123, 125, 127, 129, 145, 146, 153, 157, 159, 161, 169, 171, 173, 175, 181, 193, 201, 202, 205, 213, 217, 233, 235, 237, 251, 253, 259, 265, 277, 278, 281, 285, 295, 297, 299, 300, 320, 323, 329, 331, 337, 340, 345, 348, 361, 369, 371, 373, 377, 381, 399, 415, 419, 429, 436, 443, 446, 453, 457, 460, 461, 463, 474, 479, 481, 493, 499, 503, 505, 511, 513, 515, 516, 531, 537, 539, 543, 545, 553, 557, 559, 565, 567, 573, 575, 577, 581, 584, 589, 599, 601, 603, 607, 613, 616, 617, 623, 629, 633, 635, 649, 655, 657, 661, 663, 673, 675, 677, 679, 684, 687, R16
**make predictions,** 117, 135, 239, 347, 391, 399, 559, 567, 583, 635, 683, R13
**make thematic maps,** 545
**maps.** *See* create sketch maps; interpret maps
**models.** *See* build models; interpret models
**monitor comprehension,** 19, 23, 35, 41, 42, 45, 51, 53, 55, 59, 61, 65, 67, 79, 84, 97, 103, 105, 131, 149, 153, 164, 166, 191, 195, 197, 199, 201, 209, 217, 219, 221, 225, 229, 247, 253, 259, 267, 269, 277, 287, 293, 299, 311, 318, 325, 329, 337, 345, 347, 359, 363, 365, 367, 391, 393, 399, 403, 415, 421, 425, 445, 451, 459, 473, 478, 480, 483, 485, 497, 507, 509, 529, 533, 535, 539, 545, 550, 563, 567, 576, 581, 599, 601, 610, 621, 623, 627, 632, 647, 653, 655, 671, 677, 680
**movement,** 81, 153, 207, 209, 280, 325, 327, 337, 345, 367, 371, 375, 403, 427, 433, 441, 453, 457, 485, 493, 509, 549, 557, 573, 611, 625, 651, 657, 661, 671
**multimedia presentation.** *See* create a multimedia presentation

## N

**notes.** *See* outline and take notes

## O

**opinions.** *See* distinguish fact and opinion; form and support opinions
**outline and take notes,** R4, R10

## P

**place,** 27, 37, 45, 59, 69, 117, 127, 171, 321, 373, 389, 393, 421, 425, 473, 605, 607, 631, 684
**play a game,** 408
**pose and answer questions,** 17, 95, 149, 313, 345, 445, 487, 537, R12
**pose questions,** 479
**predictions.** *See* make predictions
**prepare a guidebook,** 688
**prepare a multimedia presentation,** 240
**primary sources.** *See* analyze primary and secondary sources
**problems and solutions.** *See* identify problems and solutions

## R

**recognize bias,** R22
**region,** 17, 30, 61, 63, 79, 119, 129, 159, 235, 277, 300, 348, 381, 401, 429, 516, 535, 573, 588, 601, 603, 611, 617, 627, 684
**reports.** *See* write a report
**research and compare,** 639
**research and create charts,** 183, 303, 407, 463
**research and create graphs,** 71, 135, 519
**research and draw conclusions,** 239
**research and make comparisons,** 351, 591
**research and write a report,** 687
**research archaeological sites,** 437
**review,** 155

## S

**secondary sources.** *See* analyze primary and secondary sources
**sequence events,** 653, R6
**solutions.** *See* identify problems and solutions
**sources.** *See* analyze primary and secondary sources; evaluate Internet sources
**speech.** *See* express ideas through speech
**summarize,** 13, 21, 37, 39, 49, 55, 62, 65, 81, 87, 89, 93, 97, 99, 101, 103, 105, 107, 113, 116, 122, 123, 127, 133, 143, 144, 153, 159, 167, 169, 171, 173, 174, 175, 177, 179, 183, 194, 198, 199, 207, 209, 219, 223, 225, 227, 229, 231, 233, 236, 249, 251, 257, 261, 263, 264, 267, 271, 273, 289, 291, 295, 297, 298, 313, 315, 323, 326, 335, 338, 343, 347, 361, 369, 371, 373, 375, 380, 385, 386, 389, 391, 395, 397, 398, 401, 403, 417, 419, 423, 431, 433, 435, 436, 441, 449, 454, 457, 459, 471, 476, 481, 487, 489, 492, 493, 497, 501, 515, 527, 531, 543, 549, 553, 555, 570, 573, 575, 577, 579, 583, 601, 603, 605, 609, 613, 616, 621, 625, 631, 633, 635, 637, 648, 651, 654, 655, 657, 659, 661, 663, 664, 669, 673, 679, 683, R5
**synthesize,** 21, 31, 63, 69, 113, 121, 127, 129, 132, 133, 161, 181, 213, 237, 257, 281, 301, 323, 327, 331, 349, 382, 385, 405, 437, 461, 493, 517, 535, 553, 559, 581, 617, 621, 631, 637, 665, 685, R24

# INDEX

## T

**take notes.** *See* outline and take notes
**time lines.** *See* create time lines; interpret time lines
**turn and talk,** 39, 59, 143, 175, 205, 223, 253, 261, 285, 291, 340, 389, 449, 489, 503, 565, 585

## V

**visual vocabulary,** 16, 38, 58, 59, 121, 123, 147, 152, 169, 170–171, 178, 201, 203, 250, 257, 279, 310, 344, 360, 362, 396, 470, 476, 498, 526, 534, 580, 598, 600, 622, 627, 657, 658, 681
**visuals.** *See* analyze visuals

## W

**write a blog entry,** 221
**write a briefing,** 408
**write a comparison speech,** 113
**write a feature article,** 184, 240
**write a letter,** 405, 517, 665
**write a memo,** 637
**write a migration history,** 72
**write a paragraph,** 101, 515
**write a press release,** 617
**write a public service announcement,** 69
**write a response,** 497
**write a speech,** 133, 301, 493, 559
**write a travel article,** 437
**write a travel brochure,** 31
**write a travel journal,** 408
**write a travelogue,** 685
**write a World Heritage Guide,** 352
**write about endangered species,** 520
**write about geographic themes,** 15
**write an action plan,** 41
**write an email,** 331
**write analyses,** 633
**write and design a web page,** 181
**write comparisons,** 347, 397, 481
**write discussion questions,** 461
**write field notes,** 464
**write journal entries,** 213, 349, R11
**write reports,** 67, 167, 289, 375, 443, 551, 688, R10
**write tour notes,** 281

# ACKNOWLEDGMENTS

## Text Acknowledgments

210: Excerpts from *El Libertador: Writings of Simón Bolívar* by Simón Bolívar, edited by David Bushness, translated by Fred Fornoff. Copyright © 2003 by Oxford University Press. Reprinted by permission of Oxford University Press. All rights reserved.

490: Excerpts from *The Illustrated Bhagavad Gita*, translated by Ranchor Prime. Copyright © 2003 by Godsfield Press, text © by Ranchor Prime. Reprinted by permission of Godsfield Press.

542: Excerpts from *The Analects of Confucius*, translated by Simon Leys. Copyright © 1997 by Pierre Ryckmans. Used by permission of W. W. Norton & Company, Inc.

639: Data from the International Union for Conservation of Nature (IUCN) Red List of Threatened Species by IUCN. Data copyright © 2008 by the IUCN Red List of Threatened Species. Reprinted by kind permission of IUCN.

### National Geographic School Publishing

National Geographic School Publishing gratefully acknowledges the contributions of the following National Geographic Explorers to our program and to our planet:

Greg Anderson, National Geographic Fellow
Katey Walter Anthony, 2009 National Geographic Emerging Explorer
Ken Banks, 2010 National Geographic Emerging Explorer
Katy Croff Bell, 2006 National Geographic Emerging Explorer
Christina Conlee, National Geographic Grantee
Alexandra Cousteau, 2008 National Geographic Emerging Explorer
Thomas Taha Rassam (TH) Culhane, 2009 National Geographic Emerging Explorer
Jenny Daltry, 2005 National Geographic Emerging Explorer
Wade Davis, National Geographic Explorer-in-Residence
Sylvia Earle, National Geographic Explorer-in-Residence
Grace Gobbo, 2010 National Geographic Emerging Explorer
Beverly Goodman, 2009 National Geographic Emerging Explorer
David Harrison, National Geographic Fellow
Kristofer Helgen, 2009 National Geographic Emerging Explorer
Fredrik Hiebert, National Geographic Fellow
Zeb Hogan, National Geographic Fellow
Shafqat Hussain, 2009 National Geographic Emerging Explorer
Beverly and Dereck Joubert, National Geographic Explorers-in-Residence
Albert Lin, 2010 National Geographic Emerging Explorer
Elizabeth Kapu'uwailani Lindsey, National Geographic Fellow
Sam Meacham, National Geographic Grantee
Kakenya Ntaiya, 2010 National Geographic Emerging Explorer
Johan Reinhard, National Geographic Explorer-in-Residence
Enric Sala, National Geographic Explorer-in-Residence
Kira Salak, 2005 National Geographic Emerging Explorer
Katsufumi Sato, 2009 National Geographic Emerging Explorer
Cid Simoes and Paola Segura, 2008 National Geographic Emerging Explorers
Beth Shapiro, 2010 National Geographic Emerging Explorer
José Urteaga, 2010 National Geographic Emerging Explorer
Spencer Wells, National Geographic Explorer-in-Residence

## Photographic Credits

iii (l) ©Innovative Images/National Geographic School Publishing (c) ©Mary Lynne Ashley/National Geographic School Publishing (r) ©Martin Photography/National Geographic School Publishing. iv (l) ©Aesthetic Life Studio/National Geographic School Publishing (c) ©Gary Donnelly/National Geographic School Publishing (r) ©Cliento Photography/National Geographic School Publishing. vi Top to Bottom Left Side: ©Chris Ranier ©Gemma Atwal ©Ken Banks ©Rebecca Hale/National Geographic Stock ©Christina Conlee ©The Ocean Foundation/National Geographic Stock. Top to Bottom Right Side: ©Sybille Frütel Culhane ©Kevin Krug ©Mark Theissen/National Geographic Stock ©Tyrone Turner/National Geographic Stock ©Adrian Jackson ©G. Anker ©Courtesy of the Jane Goodall Institute. vii Top to Bottom Left Side: ©Chris Ranier ©Rebecca Hale, National Geographic Stock ©Brant Allen ©Rolex Awards ©Beverly Joubert/National Geographic Stock. Top to Bottom Right Side: ©Beverly Joubert/National Geographic Stock ©Calit2, Erik Jepsen ©Ka'uila Barber ©National Geographic Society, Explorer Programs and Strategic Initiatives ©Sharon Farmer ©Mark Thiessen/National Geographic Stock. viii Top to Bottom Left Side: ©Rebecca Hale/National Geographic Stock ©Lana Eklund ©Katsufumi Sato ©Victor Sanchez de Fuentes. Top to Bottom Right Side: ©Beth Shapiro ©Victor Sanchez de Fuentes ©Rachel Etherington ©David Evans/National Geographic Society. ix ©Richard Barnes/National Geographic Stock. x ©John Burcham, National Geographic Stock. xi ©Raul Touzon, National Geographic Stock. xii ©Rod Smith/National Geographic My Shot/National Geographic Stock. xiii ©Richard List, Corbis. xiv ©Photolibrary. xv ©David Alan Harvey/National Geographic Stock. xvi ©Smar Jodha/National Geographic My Shot/National Geographic Stock. xvii ©Kenji Kondo/epa/Corbis. xvii ©Gavin Hellier/Alamy. xix ©Nigel Pavitt, Corbis. xx ©R. Wallace/Stock Photos/Corbis. A1 ©Michael Dunning/Photographer's Choice/Getty Images. A6 (bl) ©Mark Hamblin/Photolibrary (br) ©Thomas Marent/Minden Pictures/National Geographic Stock (c) ©Gerard Soury/Photolibrary (t) ©John Cancalosi/Photolibrary. A7 (bl) ©Adam Jones/Getty Images (br) ©Klaus Nigge/National Geographic Stock (c) ©DLILLC/Corbis (t) ©Thomas Marent/Minden Pictures/National Geographic Stock. 1 (bkg) ©Franck Guiziou/Hemis/Corbis (b, l, r) ©Mark Thiessen/National Geographic Society. 2 (bl) ©Sharon Farmer (br) ©Ka'uila Barber (cl) ©G. Anker (cr) ©Rolex Awards (tl) ©National Geographic Stock (tr) ©Calit2, Erik Jepsen. 3 (b) ©Mark Thiessen/National Geographic Stock (c) ©Ken Banks (tl)

©Gemma Atwal (tr) ©Tyrone Turner/National Geographic Stock. 4 (bkg) ©Suman Bajpeyi/National Geographic My Shot/National Geographic Stock (l) ©Jennifer Shaffer/National Geographic School Publishing (t) ©Fresco JLinga/National Geographic My Shot/National Geographic Stock. 5 ©Scott S. Warren/National Geographic Stock. 6 (l) ©Susan Byrd/National Geographic My Shot/National Geographic Stock (r) ©Mitchell Funk/Photographer's Choice /Getty Images. 7 ©Nigel Pavitt, Corbis. 8 (bkg) ©Richard Barnes/National Geographic Stock (bl) ©Mitchell Funk/Photographer's Choice/Getty Images (cl) ©NASA Goddard Space Flight Center. (tl) ©David Evans/National Geographic Society. 10 ©Stephen Alvarez/National Geographic Stock. 11 ©Sunpix Travel/Alamy. 12 (bc) ©Mike Theiss/National Geographic Stock (cl) ©Blakeley/Alamy (cr) ©Michael S. Yamashita/National Geographic Stock 13 (tl) ©John Wark/Wark Photography, Inc. (tr) ©Mark Remaley/Precision Aerial Photo. 15 (b) ©Michael S. Yamashita/National Geographic Stock (t) ©Ma Wenxiao/Sinopictures/Photolibrary. 16 ©Susan Byrd/National Geographic My Shot/National Geographic Stock. 20 ©Stephen Alvarez/National Geographic Stock. 23 ©PictureLake/Alamy. 26 ©Brooks Kraft/Corbis. 28 (bc) ©James Forte/National Geographic Stock (bl) ©Kenneth Garrett/National Geographic Stock (br) ©Michael Poliza/National Geographic Stock (bkg) ©National Geographic Maps. 29 (bcl) ©N.C. Wyeth/National Geographic Stock (bcr) ©Justin Guarlglia/National Geographic Stock (bl) ©Kenneth Garrett/National Geographic Stock (br) ©Abraham Nowitz/National Geographic Stock 33 (b) ©David Trood/Getty Images (tr) ©Peter Carsten/National Geographic Stock. 34 ©Andrew Hasson/Alamy. 38 ©George H.H. Huey/Corbis. 41 (tc) ©Images & Volcans/Photo Researchers, Inc. (tr) ©Chris Cheadle/Getty Images. 44 ©Bill Hatcher/National Geographic Stock. 46 (bc) ©Michael Doolittle/Alamy (cl) ©Daniel Dempster Photography/Alamy (tl) ©Tom Bean/Alamy. 47 (tl) ©Frank Krahmer/Corbis (tr) ©imagebroker/Alamy. 50 ©Panoramic Images/Getty Images. 52 ©Michael Nichols/National Geographic Stock. 53 ©Russ Bishop/Alamy. 54 ©Kip Evans Photography. 55 ©NASA Goddard Space Flight Center. 56 (bc) ©Gordon Wiltsie/National Geographic Stock (bl) ©Ralph Lee Hopkins/National Geographic Stock (bkg) ©George Grall/National Geographic Stock (br) ©Paul Nicklen/National Geographic Stock 57 (bcl) ©Norbert Rosing/National Geographic Stock (bcr) ©William Albert Allard/National Geographic Stock (bl) ©Stuart Franklin/National Geographic Stock (br) ©Priit Vesilind/National Geographic Stock. 58 (b) ©David R. Frazier Photolibrary, Inc./Alamy (cl) ©Olivier Asselin/Alamy. 59 ©Greg Elms/Lonely Planet Images. 63 ©John Stanmeyer/National Geographic Stock. 64 ©Nic Bothma/epa/Corbis. 65 ©Alain Nogues/Corbis. 67 ©Michael Dunning/Photographer's Choice/Getty Images. 72 ©Romeo Gacad/AFP/Getty Images. 74 (b) ©Tetra Images/Corbis (bkg) ©John Burcham, National Geographic Stock (c) ©Walter Meayers Edwards, National Geographic Stock (t) ©Beth Shapiro. 77 ©Petra Engle/National Geographic Stock. 78 (b) ©Phil Schermeister/National Geographic Stock (bkg) ©Menno Boermans/Aurora Photos/Corbis (t) ©SeBuKi/Alamy. 80 ©Mike Grandmaison/Corbis. 82 ©Bill Hatcher/National Geographic Stock. 84 ©Daniel H. Bailey/Corbis. 86 (bkg) ©Jean-Pierre Lescourret/Corbis (cr) ©Mauricio Ramos. 88 ©Jon Arnold Images Ltd/Alamy. 89 ©American School Private Collection/Peter Newark American Pictures/The Bridgeman Art Library Nationality. 90 ©Thomas Sbampato/Photolibrary. 91 ©The Granger Collection. 92 (l) ©The Granger Collection (r) ©photostock1/Alamy. 93 ©Visions LLC/Photolibrary. 95 ©William Manning/Corbis. 96 ©Bettmann/Corbis. 98 (l) ©Corbis (r) ©American School Private Collection/Courtesy of Swann Auction Galleries/ The Bridgeman Art Library. 99 ©Corbis. 100 ©Bettmann/Corbis. 101 ©Lynn Johnson/National Geographic Stock. 102 ©The Art Archive/Museo Ciudad Mexico/Gianni Dagli Orti. 103 (l) ©Kenneth Garrett/National Geographic Stock (r) ©David R. Frazier Photolibrary, Inc./Alamy. 104 ©The Stapleton Collection/The Bridgeman Art Library. 105 ©The Stapleton Collection/The Bridgeman Art Library. 106 ©The Granger Collection. 107 (b) ©The Granger Collection (t) ©Look and Learn Magazine Ltd/The Bridgeman Art Library. 108 (l) ©North Wind Picture Archives/Alamy (r) ©Randy Faris/Corbis. 109 (l) ©Corbis (r) ©Charles & Josette Lenars/Corbis. 110 ©North Wind Picture Archives/Alamy. 111 ©Hulton Archive/Getty Images. 114 ©Joe McNally/National Geographic Stock. 116 ©Jennifer Shaffer/National Geographic School Publishing. 117 ©Mike Theiss/National Geographic Society Image Sales. 118 ©Jennifer Shaffer/National Geographic School Publishing/Art Institute of Chicago. 120 ©Car Culture/Corbis. 123 (c) ©James Forte/National Geographic Stock (tr) ©Michael Dunning/Photographer's Choice/Getty Images. 124 ©Marjorie Kamys Cotera/Daemmrich Photography/The Image Works. 125 (b) ©Robb Kottmyer (t) ©Roger Meno. 126 ©Tono Labra/Photolibrary. 127 ©Keith Dannemiller/Alamy. 128 ©STR/Reuters/Corbis. 130 ©Alfredo Guerrero/epa/Corbis. 131 ©Blaine Harrington III/Alamy. 136 ©Corbis Premium RF/Alamy. 138 (b) ©Georgios Kollidas/Alamy (bkg) ©Raul Touzon, National Geographic Stock (c) ©Stephen Alvarez, National Geographic Stock. 141 ©Konrad Wothe/Minden Pictures. 142 (b) ©Jon Arnold Images Ltd/Alamy (bkg) ©Menno Boermans/Aurora Photos/Corbis (t) ©Danny Lehman/Corbis. 144 ©Dr. Richard Roscoe/Visuals Unlimited, Inc. 147 ©Stuart Westmorland/Corbis. 148 (b) ©Paul Hoekman (t) ©Bryan Wallace. 150 (bc) ©Roy Toft/National Geographic Stock (bl) ©Michael Nichols/National Geographic Stock (bkg) ©Paul Sutherland/National Geographic Stock (br) ©Steve Winter/National Geographic Society Image Sales. 151 (bcl) ©Michael Melford/National Geographic Stock (bcr) ©Bobby Haas/National Geographic Stock (bl) ©Christian Ziegler/National Geographic Stock (br) ©Roy Toft/National Geographic Stock. 152 ©Steve Winter/National Geographic Stock. 155 ©The Bridgeman Art Library. 156 (br) ©Georgios Kollidas/Alamy (r) ©Hemis/Alamy. 157 ©Stuwdamdorp/Alamy. 158 (bl) ©Creativ Studio Heinemann/Westend61/Corbis (c) ©Interfoto/Alamy. 159 (br) ©Reuters/Corbis (tl) ©Bettmann/Corbis (tl) ©Creativ Studio Heinemann/Westend61/Corbis. 162 ©Walter Bibikow/JAI/Corbis. 164 ©Martin Gray/National Geographic Stock. 165 ©Danita Delimont/Alamy. 166 ©Nico Tondini/Photolibrary. 167 ©Rick Gerharter/Lonely Planet Images. 168 ©National Geographic Stock. 169 ©Frans Lanting/Corbis. 170 ©Photolibrary. 172 ©Yuan Man/Xinhua Press/Corbis. 173 ©Logan Abassi/UN Handout/Corbis. 174 ©JS Callahan/tropicalpix/Alamy. 175 ©Michael Dunning/Photographer's Choice/Getty Images. 176 ©Lonely Planet Images /Alamy. 178 (bkg) ©Christian Heeb/Aurora Photos (c) ©Roy Toft/National Geographic Stock (r) ©Micahel & Patricia Fogden/Minden Pictures/National Geographic Stock. 179 ©Arterra Picture Library/Alamy. 183 (l) ©Jacques Marais/Getty Images (r) ©Danny Lehman/Corbis. 184 ©Christian Zeigler/National Geographic Stock. 186 (b) ©Frans Lanting/Corbis (bkg) ©Rod Smith/National Geographic My Shot/National Geographic Stock (cl) ©David R. Frazier Photolibrary, Inc./Alamy (tl) ©Photograph by

Victor Sanchez de Fuentes. 189 ©Nick Gordon/Oxford Scientific (OSF)/Photolibrary. 190 (bkg) ©Menno Boermans/Aurora Photos/Corbis (bl) ©Colin Monteath/Minden Pictures/National Geographic Stock (br) ©John Eastcott and Yva Momatiuk/National Geographic Stock. 192 ©Ivan Kashinsk/National Geographic Stock. 196 (bkg) ©Melissa Farlow/National Geographic Stock (br) ©Aldo Sessa/Tango Stock/Getty Images. 199 (br) ©Michael Nichols/National Geographic Stock (tr) ©Michael Dunning/Photographer's Choice/Getty Images. 200 ©Ethan Welty/Aurora Photos/Alamy. 201 ©Gnter Wamser/F1online digitale Bildagentur GmbH/Alamy. 202 ©Christina Conlee. 203 (b) ©Christina Conlee (t) ©Robert Clark/National Geographic Stock. 206 (bc) ©Maria Stenzel/National Geographic Stock (c) ©Peruvian School/Museo Arqueologia, Lima, Peru/Boltin Picture Library/The Bridgeman Art Library International (tr) ©McConnell, James Edwin/Private Collection /Look and Learn/The Bridgeman Art Library International. 207 ©Cro Magnon/Alamy. 208 (bc) ©The Art Archive/Bibliothéque des Arts Décoratifs Paris/Gianni Dagli Orti (tr) ©The Art Archive/Bibliothéque des Arts Décoratifs Paris/Gianni Dagli Orti. 209 ©The Art Archive/Kharbine-Tapabor. 211 ©Luis Marden/National Geographic Stock. 215 ©Florian Kopp/imagebroker/Alamy. 216 ©Paolo Aguilar/epa/Corbis. 219 ©Corey Wise/Lonely Planet Images/Getty Images. 220 ©Mike Theiss/National Geographic Stock. 221 ©Richard Nowitz/National Geographic Stock. 223 ©Ivan Alvarado/Reuters/Corbis. 224 ©Jeremy Hoare/Alamy. 225 (b) ©James P. Blair/National Geographic Stock (c) ©Nicolas Misculin/Reuters (t) ©Kit Houghton/Corbis. 226 ©Jennifer Shaffer/National Geographic School Publishing. 228 ©Keren Su/Corbis. 229 (t) ©Imaginechina/Corbis. 231 ©Robert Clark/National Geographic Stock. 232 ©Sebastiao Moreira/epa/Corbis. 234 (b) ©Mike Theiss/National Geographic Stock (bl) ©Charles Dharapak/Pool/Reuters. 240 ©Pete McBride/National Geographic Stock. 242 (b) ©Bob Krist/National Geographic Stock (bkg) ©Richard List/Corbis (c) ©Anne Keiser/National Geographic Stock. 245 ©Atlantide Phototravel/Corbis. 246 (b) ©Yann Arthus-Bertrand/Corbis (bkg) ©Menno Boermans/Aurora Photos/Corbis (t) ©Douglas Pearson/Corbis. 248 ©Owi-Diasign/Photolibrary. 250 ©National Geographic Stock. 252 ©Octavio Aburto. 254 (bc) ©Anne Keiser/National Geographic Stock (bl) ©Agnieszka Pruszek/National Geographic My Shot/National Geographic Stock (bkg) ©Jim Richardson/National Geographic Stock (br) ©Steve Raymer/National Geographic Stock. 255 (bcl) ©Greg Dale/National Geographic Stock (bcr) ©James P. Blair/National Geographic Stock (bl) ©Richard Nowitz/National Geographic Stock (br) ©James L. Stanfield/National Geographic Stock. 256 ©Panoramic Images/National Geographic Stock. 258 (l) ©Richard Nowitz/National Geographic Stock (r) ©The Art Gallery Collection/Alamy. 259 ©PoodlesRock/Corbis. 260 ©Jean-Pierre Lescourret/Corbis. 262 (l) ©Hoberman Collection/Corbis (r) ©The Bridgeman Art Library International. 263 ©North Wind Picture Archives/Alamy. 266 (bl) ©The Bridgeman Art Library (br) ©Underwood & Underwood/Corbis. 267 (bl) ©Doug Taylor/Alamy (br) ©The Bridgeman Art Library (t) ©The Bridgeman Art Library. 268 ©Richard Schlect/National Geographic Stock. 270 ©Paul Thompson/Corbis. 272 (l) ©The Bridgeman Art Library (r) ©The Gallery Collection/Corbis. 273 (b) ©Peter Horree/Alamy (t) ©The Bridgeman Art Library. 274 ©The Bridgeman Art Library. 275 ©Scanfoto/X00729/Reuters/Corbis. 276 (l) ©Stefano Bianchetti/Corbis (r) ©Clynt Garnham/Alamy. 277 (l) ©Michael Nicholson/Corbis (r) ©Michael Nicholson/Corbis. 279 ©DC Premiumstock/Alamy. 282 ©Rudy Sulgan/Corbis. 284 ©MARKA/Alamy. 286 (l) ©Leonardo da Vinci (1452-1519) Louvre, Paris, France/ Giraudon/The Bridgeman Art Library (r) Claude Monet (1840-1926) Musee Marmottan, Paris, France/Giraudon/The Bridgeman Art Library Nationality. 287 ©Arnaud Chicurel/Hemis/Corbis. 288 ©The Gallery Collection/Corbis. 289 ©Columbia/The Kobal Collection. 290 ©Jon Arnold/JAI/Corbis. 291 ©Sergiy Koshevarov/StockPhotoPro. 292 ©Photolibrary. 294 ©Paul Seheult/Eye Ubiquitous/Corbis. 295 ©Perutskyi Petro/Shutterstock Photos. 296 ©Gregory Wrona/Alamy. 299 ©Michael Dunning/Photographer's Choice/Getty Images. 304 ©Grand Tour/Corbis. 306 (b) ©Gerd Ludwig/Corbis (bkg) ©Photolibrary (c) ©Gordon Wiltsie, National Geographic Stock (t) ©Rebecca Hale, National Geographic Stock. 309 ©Klaus Nigge/National Geographic Stock. 310 (bkg) ©Menno Boermans/Aurora Photos/Corbis (br) ©Bruno Morandi/Robert Harding World Imagery/Corbis (tl) ©Maxim Toporskiy/Alamy. 312 ©Denis Sinyakov/Reuters/Corbis. 314 ©Cary Wolinsky/National Geographic Stock. 316 ©National Geographic Stock. 318 ©Gerd Ludwig/National Geographic Stock. 319 (tl) ©U.S. Geological Survey (tr) ©NASA. 320 (l) ©Sisse Brimberg/National Geographic Society (r) ©James L. Stanfield/National Geographic Society. 321 (l) ©Massimo Pizzotti/Getty (r) ©Dallas and John Heaton/Photolibrary. 322 ©Richard Klune/Corbis. 323 ©imagebroker/Alamy. 324 ©The Bridgeman Art Library (r) ©Cary Wolinsky/National Geographic Society. 325 (l) ©The Bridgeman Art Library. 326 ©North Wind Picture Archives/Alamy. 328 (l) ©Bettmann/Corbis (r) ©The Art Archive. 329 (b) ©Bettmann/Corbis (t) ©Thomas Johnson/Sygma/Corbis. 332 ©Paul Harris/JAI/Corbis. 334 ©Arne Hodalic/Corbis. 335 (tc) ©Michael Runkel/Robert Harding World Imagery/Corbis (tr) ©Maria Stenzel/National Geographic Stock (tr) ©Sean Sprague/Photolibrary. 336 ©Olaf Meinhardt/Visum/Fotofinder. 339 ©Kristel Richard/Grand Tour/Corbis. 340 ©Shepard Sherbell/Corbis Saba. 342 ©Imagesource/Photolibrary. 344 ©Oleg Nikishin/Stringer/Getty Images. 347 (c) ©iStockphoto ©Michael Dunning/Photographer's Choice/Getty Images. 352 ©iStockphoto. 354 (b) ©Ingo Arndt/Minden Pictures/National Geographic Stock (bkg) ©David Alan Harvey/National Geographic Stock (c) ©Mitsuaki Iwago/Minden Pictures/National Geographic Stock (t) ©Kakenya Ntaiya. 357 ©Top-Pics TBK/Alamy. 358 (bkg) ©Menno Boermans/Aurora Photos/Corbis (bl) ©tbkmedia/Alamy (tr) ©Michael Poliza/National Geographic Stock. 360 ©Michael Nichols/National Geographic Stock. 362 ©Philippe Bourseiller/Getty Images. 364 ©Ian Nichols/National Geographic Stock. 366 ©Mike Hutchings/Reuters. 368 (bc) ©Beverly Joubert/National Geographic Stock (bl) ©Beverly Joubert/National Geographic Stock. 370 ©Gerald Hoberman/Hoberman Collection UK/Photolibrary. 372 (bl) ©The Trustees of the British Museum/Art Resource (br) ©ADB Travel/dbimages/Alamy. 373 ©HIP/Art Resource. 376 ©Private Collection/Look and Learn/The Bridgeman Art Library International. 377 (bl) ©Mary Evans Picture Library/The Image Works (br) ©Bruce Dale/National Geographic Stock. 378 (bc) ©James L. Stanfield/National Geographic Stock (bl) ©Tim Laman/National Geographic Stock (bkg) ©George Steinmetz/National Geographic Stock (br) ©Annie Griffiths/National Geographic Stock. 379 (bcl) ©Roy Toft/National Geographic Stock (bcr) ©Tino Soriano/National Geographic Stock (bl) ©Jodi Cobb/National Geographic Stock (br) ©Ed Kashi/National Geographic Stock. 382 ©Ralph Lee Hopkins/

National Geographic Stock. 384 ©Vanessa Burger/Images of Africa Photobank Alamy. 386 ©Paul Gilham–FIFA/FIFA via Getty Images. 387 ©David Alan Harvey/National Geographic Stock. 388 (bc) ©Nigel Pavitt/John Warburton-Lee Photography/Alamy (bl) ©Sean Sprague/Still Pictures/Photolibrary. 389 (cl) ©Michael Nichols/National Geographic Stock (tr) ©Suzi Eszterhas/Minden Pictures/National Geographic Stock. 390 ©Jane Goodall Institute. 391 (bkg) ©Gerry Ellis/ Minden Pictures/National Geographic Stock (br) ©Wade Davis/Ryan Hill. 392 ©Finbarr O'Reilly/Reuters. 394 ©George Steinmetz/Corbis. 396 (b) ©Pascal Maitre/National Geographic Stock (tr) ©Joerg Boethling/Alamy. 398 ©Louise Gubb/Corbis. 399 ©Michael Dunning/Photographer's Choice/Getty Images. 400 ©Frederic Courbet/Still Pictures/Photolibrary. 402 ©Ulrich Doering/Alamy. 403 ©Trinity Mirror/Mirrorpix/Alamy. 406 (bl) ©Chris Stenger/FN/Minden Pictures/National Geographic Stock (br) ©Walker, Lewis W./National Geographic Stock. 407 (bkg) ©Clement Philippe/Arterra Picture Library/Alamy (bkg) ©Tim Fitzharris/Minden Pictures/National Geographic Stock (cf) ©Mattias Klum /National Geographic Stock (l) ©Tom Vezo/Minden Pictures/National Geographic Stock (rbkg) ©Ted Wood/Aurora Photos (rf) ©Thomas Lehne/Alamy. 408 ©Anup Shah/Corbis. 410 (b) ©Martin Gray/National Geographic Stock (bkg) ©Smar Jodha/National Geographic My Shot/National Geographic Stock (c) ©David Boyer/National Geographic Stock (t) ©Thomas Culhane. 413 ©Vanessa Lefort/National Geographic My Shot/National Geographic Stock. 414 (bkg) ©Menno Boermans/Aurora Photos/Corbis (cl) ©Gary Cook/Alamy (cr) ©Peter Adams/Getty Images. 416 ©Ed Kashi/National Geographic Stock. 418 ©Fischer Gunter/WoodyStock/Alamy. 420 ©Chris Bradley/Axiom/photolibrary. 422 (bl) ©Victor R. Boswell, Jr/National Geographic Stock (br) ©Scala/Art Resource 423 (bl) ©Erich Lessing/Art Resource (br) ©Corbis. 424 ©Richard Nowitz/National Geographic Stock. 425 ©Oliver Weiken/epa/Corbis. 426 ©Bachmann Bachmann/F1 Online/photolibrary. 427 (bl) ©James Brunker/Alamy (tr) ©Kordcom Kordcom/age fotostock/photolibrary. 428 ©Yann Arthus-Bertrand/Corbis. 429 ©The Art Archive/Topkapi Museum Istanbul/Dagli Orti. 430 ©Gwill Owen/Sylvia Cordaiy Photo Library Ltd /Alamy. 431 ©NASA/JSC/Gateway to Astronaut Photography of Earth. 432 (bl) ©Paul Sutcliffe/Alamy. (br) ©Erich Lessing/Art Resource. 433 (bc) ©Mary Jelliffe/Ancient Art & Architecture Collection Ltd./Alamy (tr) ©Kenneth Garrett/National Geographic Stock. 434 ©Kenneth Garrett/National Geographic Stock. 438 ©Keren Su/Corbis. 440 ©Alberto Arzoz/Axiom/Aurora Photos. 441 ©Walter Bibikow/Jon Arnold Images Ltd./Alamy. 442 ©Gavin Hellier/Alamy. 443 ©David Bathgate/Corbis. 444 (b) ©Radius Images/Corbis (bl) ©G. Anker. 445 ©Hanan Isachar/Corbis. 446 ©NASA/Science Faction/Corbis. 447 ©Matthias Seifert/Reuters/Corbis. 449 (tr) ©Umit Bektas/Reuters/Corbis ©Felipe Trueba/epa european pressphoto agency. 450 ©Mohammad Berno/Document Iran/Corbis. 452 ©David Rubinger/Time & Life Pictures/Getty Images. 454 ©Sabah Arar/AFP/Getty Images. 457 ©Tim Gurney/Alamy. 459 ©Shehzad Noorani/Stillpictures/Aurora Photos. 463 ©Mark Thiessen/National Geographic Stock. 464 ©Bildarchiv Preussischer Kulturbesitz/Art Resource. 466 (b) ©Peter Adams/Corbis (bkg) ©Tibor Bognar/agefotostock (c) ©Kenji Kondo/epa/Corbis (t) ©Hussain RAE photos. 469 ©Lynn M. Stone/Nature Picture Library. 470 (bkg) ©Menno Boermans/Aurora Photos/Corbis (bl) ©Tiziana and Gianni Baldizzone/Corbis (l) ©Dinodia Images/Alamy (tc) ©Stephen Sharnoff/National Geographic Stock. 472 (bkg) ©Bobby Model/National Geographic Stock (l) ©James L. Stanfield/National Geographic Stock. 474 ©Frederic Soltan/Sygma/Corbis. 476 ©Lynsey Addario/National Geographic Image Collection. 478 ©Prakash Singh/AFP/Getty Image. 479 ©Blue Legacy International. 481 ©Michael Dunning/Photographer's Choice/Getty Images. 482 ©Luca Tettoni/Corbis. 483 (b) ©The Schoyen Collection (t) ©The Schoyen Collection. 484 (l) ©Silvio Fiore/SuperStock (r) ©The Trustees of the British Museum. 485 ©Thomas Retterath/Getty Images. 488 ©Jeremy Horner/Corbis. 489 ©Lineair/Photolibrary. 490 ©Bettmann/Corbis. 491 ©Art Directors & TRIP/Alamy. 494 ©Harish Tyagi/epa/Corbis. 496 ©Louise Batalla Duran/Alamy. 497 ©Bruce Dale/National Geographic Stock. 498 (bkg) ©Ed Kashi/National Geographic Stock (l) ©Jodi Cobb/National Geographic Stock. 500 ©Ajay Verma/Reuters/Corbis. 501 (b) ©Foodfolio–StockFood Munich (t) ©Abraham Nowitz/National Geographic Stock. 502 (bkg) ©David Cumming/Eye Ubiquitous/Corbis (bl) ©Robert Wallis/Corbis (bc) ©Stephen Romilly/Alamy. 503 (b) ©Dharma Productions/The Kobal Collection (t) ©Frazer Harrison/Getty Images. 504 ©Eric Feferberg/Pool/Reuters. 506 ©Fredrik Renander/Alamy. 507 (t1) ©Ed Kashi/National Geographic Stock (t2) ©Ed Kashi/National Geographic Stock (t2) ©Fridmar Damm/Corbis (t3) ©Akhtar Soomro/Deanpictures/The Image Works (t4) ©Akhtar Soomro/Deanpictures/The Image Works (t5) ©Andrew Holbrooke/Corbis (t7) ©National Geographic Maps (t8) ©National Geographic Maps. 508 ©Ed Kashi/National Geographic Image Collection. 510 ©Sajjad Hussain/AFP/Getty Images. 511 ©Dinodia Photo Library. 512 ©Akhtar Soomro/Deanpictures/The Image Works. 514 ©Andrew Holbrooke/Corbis. 522 (b) ©Michael Nichols/National Geographic Stock (bkg) ©George Steinmetz/National Geographic Stock (c) ©Gavin Hellier/Alamy (t) ©Albert Lin. 525 ©Mitsuaki Iwago/Minden Pictures/National Geographic Stock. 526 (b) ©Chun Ki Leung/National Geographic My Shot /National Geographic Stock (bkg) ©Menno Boermans/Aurora Photos/Corbis (t) ©Wang Jianjun/TAO Images Limited/Alamy. 528 ©Fritz Hoffmann/National Geographic Stock. 530 ©Reuters/Mainichi Shimbun. 532 ©Toby Adamson/Axiom Photographic Agency/Getty Images. 534 ©Alison Wright/National Geographic Stock. 536 (b) ©Unterthiner, Stefano/National Geographic Stock (cl) ©Katsufumi Sato/National Geographic Society. 538 (bl) ©Richard Swiecki/Royal Ontario Museum/Corbis (br) ©Ira Block/National Geographic Stock. 539 (bc) ©O. Louis Mazzatenta/National Geographic Stock (tc) ©Atlantide Phototravel/Corbis. 540 (bc) ©Michael S. Yamashita/National Geographic Stock (bl) ©Justin Guariglia/National Geographic Stock (bkg) ©O. Louis Mazzatenta/National Geographic Stock (br) ©Michael S. Yamashita/National Geographic Stock. 541 (bcl) ©Kenneth Ginn/National Geographic Stock (bcr) ©Kate Staszczak/National Geographic My Shot/National Geographic Stock (bl) ©Michael S. Yamashita/National Geographic Stock (br) ©Ira Block/National Geographic Stock. 542 ©Shiwei/Best View Stock/photolibrary. 544 ©Redlink/Corbis. 546 (bkg) ©Gregory A. Harlin/National Geographic Stock (b) ©National Geographic Maps. 548 ©Wendy Connett/Alamy. 549 ©Rob Howard/Corbis. 550 ©Ira Block/National Geographic Stock. 551 ©Ira Block/National Geographic Stock. 552 ©Asian Art & Archaeology, Inc./Corbis. 553 (b) ©Private Collection/Peter Newark Military Pictures/The Bridgeman Art Library International. 554 (bl) ©H. Edward Kim/National Geographic Stock (br) ©Korea News Service/Reuters/Corbis. 555 (bl) ©John Van Hasselt/Sygma/Corbis ©(br) ©The Trustees of The British Museum/Art Resource. 556

(bl) ©Bettmann/Corbis. 557 (l) ©Bruce Burkhardt/Corbis (r) ©Michael S. Yamashita/National Geographic Stock. 560 ©B.S.P.I./Corbis. 562 ©Buena Vista Images/Getty Images. 564 (bl) ©The Trustees of the British Museum/Art Resource. 565 ©View Stock/Alamy. 566 ©Justin Guariglia/National Geographic Stock. 568 (bc) ©Mark Leong/National Geographic Stock (bl) ©Richard Nowitz/National Geographic Stock (bkg) ©Scott S. Warren/National Geographic Stock (br) ©Jason Teale/National Geographic My Shot/National Geographic Stock. 569 (bcl) ©Pete Ryan/National Geographic Stock (bl) ©Michael S. Yamashita/National Geographic Stock (bcr) ©Thomas J. Abercrombie/National Geographic Stock (br) ©Jodi Cobb/National Geographic Stock. 570 (bl) ©Everett Kennedy Brown/epa/Corbis (cl) ©Everett Kennedy Brown/epa/Corbis. 571 ©Studio Ghibli/Tokuma Shoten/The Kobal Collection. 572 (b) ©Jose Fuste Raga/Corbis. 574 (b) ©Jianan Yu/Reuters/Corbis. 577 ©Michael Dunning/Photographer's Choice/Getty Images. 578 ©epa/Corbis. 580 (bl) ©Nicky LohReuters/Corbis (bkg) ©Justin Guariglia/National Geographic Stock. 582 ©Asia File/Alamy. 585 (bkg) ©SJ. Kim/Flickr/Getty Images (tr) ©Jane Sweeney/JAI/Corbis. 586 (bc) ©Phil Iossifidis/National Geographic My Shot/National Geographic Stock (bl) ©Paul Chesley/National Geographic Stock (bkg) ©Alicia Pudsey/National Geographic My Shot/National Geographic Stock (br) ©Arkadiusz Dudzinski/National Geographic My Shot/National Geographic Stock. 592 ©John Woodcock/iStock Vectors/Getty Images. 586 (bcl) ©Winfield Parks/National Geographic Stock (bcr) ©Paul Chesley/National Geographic Stock (bl) ©Martin Gray/National Geographic Stock (br) ©Justin Guariglia/National Geographic Stock 593 ©Natureworld/Alamy. 594 (b) ©Paul Chesley/National Geographic Stock (bkg) ©Nigel Pavitt/Corbis (c) ©John Stanmeyer LLC, National Geographic Stock (t) ©Kevin Krug. 597 ©Tui De Roy/Minden Pictures/National Geographic Society Image Sales. 598 (bkg) ©Menno Boermans/Aurora Photos/Corbis (bl) ©NASA (tr) ©Mary Plage/Photolibrary. 600 ©Brant Allen. 602 (bkg) ©Neil Rabinowitz/Corbis (bl) ©James P. Blair/National Geographic Society Image Sales. 604 ©John Stanmeyer/National Geographic Society Image Sales. 606 ©Tim Laman/National Geographic Society Image Sales. 607 (bkg) ©Tim Laman/National Geographic Society Image Sales (cl) ©Tim Laman/National Geographic Society Image Sales (cr) ©Tim Laman/National Geographic Society Image Sales. 608 (c) ©Robert Clark/National Geographic Society Image Sales (cb) ©Robert Harding Picture Library Ltd/Alamy. 609 (br) ©View Stock/Alamy. 612 (bl) ©Stringer/Indonesia/Reuters/Corbis (r) ©Lindsay Hebberd/Corbis. 613 ©FPG/Hulton Archive/Getty Images. 615 ©W.E. Garrett/National Geographic Stock. 618 ©Jose Fuste Raga/Corbis. 620 (b) ©James Nachtwey VII/National Geographic Society Image Sales. 621 (tl) ©Friedrich Stark/Alamy (tr) ©Robert Clark/National Geographic Society Image Sales. 622 (bkg) ©Danita Delimont/Gallo Images/Getty Images (bl) ©Jon Arnold Images Ltd/Alamy. 623 (cr) ©WoodyStock/Alamy. 624 ©Photolibrary. 625 ©National Geographic School Publishing. 626 (b) ©Ulet Ifansasti/Getty Images. 627 (tl) ©Buddy Mays/Alamy. 628 ©Imagemore Co., Ltd./Corbis. ©AFP/Getty Images 629 ©AFP/Getty Images. 630 ©Peter Adams/JAI/Corbis. 631 ©John Stanmeyer/National Geographic Society Image Sales. 633 (c)© Lauca Images/Alamy (t) ©Michael Dunning/Photographer's Choice/Getty Images. 634 ©Tengku Mohd Yusof/Alamy. 639 (b) ©Theo Aloffs/Minden Pictures/National Geographic Stock (cl) ©Suzi Eszterhas/Minden Pictures/National Geographic Stock (cr) ©John Cancalosi/Photolibrary (tl) ©Tom Brakefield/Photolibrary (tr) ©Top-Pics TBK/Alamy. 642 (b) ©R. Wallace/Stock Photos /Corbis (bkg) ©Yva Momatiuk & John Eastcott/Minden Pictures/National Geographic Stock (c) ©Pixtal Images/Photolibrary (t) ©Ka'uila Barber. 645 ©Mitsuaki Iwago/Minden Pictures/National Geographic Stock. 646 (bkg) ©Menno Boermans/Aurora Photos/Corbis (bl) ©Fred Bavendam/Minden Pictures/National Geographic Stock (tr) ©Michel Renaudeau/Photolibrary. 648 ©Kevin Schafer/Alamy. 649 ©Look Die Bildagentur der Fotografen GmbH/Alamy. 650 (b) ©Martin Harvey/Corbis (t) ©John Carnemolla/Corbis. 652 ©Brian Skerry/National Geographic Society Image Sales. 655 (t) ©Michael Dunning/Photographer's Choice/Getty Images (c) ©Jeff Hunter/Getty Images. 656 ©Catherine Karnow/Corbis. 657 ©Belinda Wright/National Geographic Society Image Sales. 658 (b) ©Rob Howard/Corbis (t) ©Walter Meayers Edwards/National Geographic Society Image Sales. 659 (c) ©Jean Leo Dugast/Sygma/Corbis (t) ©Image Source/Corbis. 660 ©Patrick Eden/Alamy. 661 (bl) ©Corbis (l) ©Caro/Alamy (t) ©The Bridgeman Art Library. 662 ©Chris Ranier. 663 ©Chris Ranier. 666 ©Massimo Ripani/Grand Tour/Corbis. 668 ©Amy Toensing/National Geographic Stock. 669 ©David Ball/Alamy. 670 ©Stringer/Australia/X01245/Reuters/Corbis. 672 ©Randy Olson/National Geographic Stock. 674 ©George F. Mobley, National Geographic Stock. 676 ©Bates Littlehales/National Geographic Stock. 678 ©POOL/Reuters/Corbis. 679 ©Phil Walter/Getty Images. 681 (bkg) ©David Doubilet/National Geographic Stock (tl) ©David McLain/National Geographic Stock (tr) ©David Wall/Alamy. 682 ©David L. Moore/Alamy. 683 ©Tui De Roy/Minden Pictures/National Geographic Stock. 688 ©Paul Nicklen/National Geographic Stock. R39 ©Fresco JLinga/National Geographic My Shot/National Geographic Stock. R40 ©George Grall/National Geographic Stock. R41 ©SERDAR/Alamy. R42 ©Bettmann/Corbis. R44 ©Hulton Archive/Getty Images. R45 ©Svabo/Alamy. R47 ©Scott Olson/Getty Images. R49 ©Lordprice Collection/Alamy. R50 ©Parbul TV via Reuters TV/Reuters/Corbis. R51 ©Markus Altmann/Corbis. R52 (b) ©Mario Lopez/epa/Corbis (t) ©Alinari Archives/Corbis. R53 (b) ©Anindito Mukherjee/epa/Corbis (t) ©Werner Forman/Art Resource. R54 (b) ©Ronen Zvulun/Reuters/Corbis (t) ©The Art Archive/Museo del Prado Madrid. R55 (b) ©Mak Remissa/epa/Corbis (t) ©Rubin Museum of Art/Art Resource. R56 (b) ©Joe McNally/National Geographic Stock (t) ©Biju/Alamy. R57 (b) ©Robert Harding World Imagery/Corbis (t) ©Art Directors & TRIP/Alamy. R58 (b) ©Christian Kober/Photolibrary (t) ©National Palace Museum Taiwan/The Art Archive.

## Map Credits

Mapping Specialists, LTD., Madison, WI.
National Geographic Maps, National Geographic Society

## Illustrator Credits

Precision Graphics